SELECTED MATERIAL FROM

BIOLOGY

VOLUME THREE

EVOLUTION, DIVERSITY AND ECOLOGY

SELECTED MATERIAL FROM

BIOLOGY

VOLUME THREE
EVOLUTION, DIVERSITY AND ECOLOGY

ROBERT J. BROOKER
UNIVERSITY OF MINNESOTA

ERIC P. WIDMAIER
BOSTON UNIVERSITY

LINDA GRAHAM
UNIVERSITY OF WISCONSIN

PETER STILING
UNIVERSITY OF SOUTH FLORIDA

 **Learning Solutions**

Boston   Burr Ridge, IL   Dubuque, IA   New York   San Francisco   St. Louis
Bangkok   Bogotá   Caracas   Lisbon   London   Madrid
Mexico City   Milan   New Delhi   Seoul   Singapore   Sydney   Taipei   Toronto

The **McGraw·Hill** Companies

SELECTED MATERIAL FROM BIOLOGY
VOLUME THREE: *EVOLUTION, DIVERSITY AND ECOLOGY*

This book is a McGraw-Hill Learning Solutions textbook and contains select material from *Biology* by Robert J. Brooker, Eric P. Widmaier, Linda Graham and Peter Stiling. Copyright © 2008 by The McGraw-Hill Companies, Inc. Reprinted with permission of the publisher. Many custom published texts are modified versions or adaptations of our best-selling textbooks. Some adaptations are printed in black and white to keep prices at a minimum, while others are in color.

2 3 4 5 6 7 8 9 0 DOW DOW 0 9 8 7

ISBN 13: 978-0-07-335333-3
ISBN 10: 0-07-335333-7

Editor: Shirley Grall
Production Editor: Jessica Portz
Printer/Binder: RR Donnelley

ABOUT THE AUTHORS

Robert J. Brooker

Robert J. Brooker (Ph.D., Yale University) received his B.A. in biology at Wittenberg University in 1978. At Harvard, he studied the lactose permease, the product of the *lacY* gene of the *lac* operon. He continues working on transporters at the University of Minnesota, where he is a Professor in the Department of Genetics, Cell Biology, and Development and has an active research laboratory. At the University of Minnesota, Dr. Brooker teaches undergraduate courses in biology, genetics, and cell biology. In addition to many other publications, he has written two editions of the undergraduate genetics text *Genetics: Analysis & Principles*, McGraw-Hill, copyright 2005.

Eric P. Widmaier

Eric P. Widmaier received his Ph.D. in 1984 in endocrinology from the University of California at San Francisco. His research is focused on the control of body mass and metabolism in mammals, the mechanisms of hormone action, and the postnatal development of adrenal gland function. Dr. Widmaier is currently Professor of Biology at Boston University. Among other publications, he is a co-author of *Vander's Human Physiology: The Mechanisms of Body Function*, 10th edition, published by McGraw-Hill, copyright 2006.

Linda E. Graham

Linda E. Graham received her Ph.D. in botany from the University of Michigan, Ann Arbor. Her research explores the evolutionary origin of land-adapted plants, focusing on their cell and molecular biology as well as ecological interactions. Dr. Graham is now Professor of Botany at the University of Wisconsin-Madison. She teaches undergraduate courses in biology and plant biology. She is the co-author of, among other publications, *Algae*, copyright 2000, a major's textbook on algal biology, and *Plant Biology*, copyright 2006, both published by Prentice Hall/Pearson.

Peter D. Stiling

Peter Stiling obtained his Ph.D. from University College, Cardiff, Wales in 1979. Subsequently, he became a Post-doc at Florida State University and later spent two years as a lecturer at the University of the West Indies, Trinidad. During this time, he began photographing and writing about butterflies and other insects, which led to publication of several books on local insects. Dr. Stiling is currently a Professor of Biology at the University of South Florida at Tampa. He teaches graduate and undergraduate courses in ecology and environmental science. He has published many scientific papers and is the author of *Ecology: Theories and Applications*, 4th edition by Prentice Hall/Pearson, copyright 2002. Dr. Stiling's research interests include plant-insect relationships, parasite-host relationships, biological control, restoration ecology, and the effects of elevated carbon dioxide levels on plant herbivore interactions.

The authors are grateful for the help, support,
and patience of their families, friends, and students,
Deb, Dan, Nate, and Sarah Brooker,
Maria, Carrie, and Rick Widmaier,
Jim, Michael, and Melissa Graham, and
Jacqui, Zoe, Leah, and Jenna Stiling.

Left to right: Eric Widmaier, Linda Graham, Peter Stiling, and Rob Brooker

BRIEF CONTENTS

A New Biology Book with a Modern Perspective

In addition to being active researchers and experienced writers, our author team has taught majors biology for years. We have taught with the same books that you have. Our goal in creating something new is to offer something better—a comprehensive, modern textbook featuring an evolutionary focus with an emphasis on scientific inquiry.

Through our classroom experiences and research work, we became inspired by the prospect that a new Biology text could move biology education forward. In listening to educators and students, it became clear that we needed to concentrate our efforts on seven crucial areas. These are described briefly below. We will return to each in more detail in the pages that follow.

1. **Experimentation** During the 1970s and 1980s, biological information began to expand at an exponential rate. Biology textbooks grew and to some extent the content suffered as the scientific process was squeezed out by the avalanche of new details. We are committed to striking a better balance between general concepts and experimentation by showing the connection between scientific inquiry and the principles that are learned from such experimentation, especially through the Feature Investigation sections in every chapter.

2. **Modern content** Science is a moving target. Although the content and organization of our Biology textbook is not a dramatic departure from other books, we have added modern content that will better prepare students for future careers in biology. Toward this end, we have received content reviews from over three hundred faculty members from around the world. We are convinced they have helped us produce a book with the most up-to-date content possible.

Striking examples where we feel the content demonstrates a modern approach with an emphasis on recent experimentation include the following:

- Chapter 6 of our Cell Biology Unit explores cell biology at the level of "systems biology" in which the cell is viewed as a group of interacting parts. This allows students to understand how the parts of a cell work together.

- Our Genetics Unit takes a "molecular first" approach so that students will first understand what a gene is, and then consider how genes affect the traits of organisms.

- The Evolution Unit often takes a molecular perspective, and highlights cladistic methods to generate evolutionary relationships. This approach connects evolution at the molecular level and at the level of organisms in their native environments.

- The Diversity Unit has incorporated the newest information regarding evolutionary relationships among modern species. The connection between evolutionary innovation and reproductive success allows students to appreciate why organisms have certain types of traits.

- In our Plant Biology Unit, a much more modern connection has been made between plant structure, function and genetics. Recent information from *Arabidopsis* is often discussed.

- Each chapter in our Animal Biology Unit ends with a section on the modern Impact on Public Health, including the molecular basis of many diseases. In addition, Neuroscience is covered as a mini-Unit of its own, with three complete chapters incorporating some of the most recent information in this exciting area of biology.

- The Ecology Unit also incorporates an evolutionary theme and has an expanded discussion of species interactions. This approach provides students with a deeper understanding of evolutionary adaptations that organisms have.

3. **Evolutionary Perspective** A study of the processes and outcomes of evolution serves to unify the field of biology and the units of our text. Whether describing evolutionary mechanisms at the molecular level or surveying the diversity of life through a view of modern systematics, an understanding of evolution serves to connect and integrate the disciplines of biology.

4. **A Visual Outline** We were determined to create a new art program using both graphics and photography to serve as a "visual outline." We have worked with a large team of scientific illustrators, photographers, educators and students to build an accurate, up-to-date and visually appealing new illustration program that is easy to follow, realistic and instructive.

5. **Ensuring Accuracy** We chose to work as a team of authors to create this new Biology text because the information in our discipline is increasing so rapidly. Each member of our team has experience researching and writing in our respective areas, allowing us to combine efforts to stay abreast of the field. Likewise, we have worked with a much larger team of reviewers, advisors, editors, and accuracy checkers to ensure that this text is as current and accurate as humanly possible.

6. **A Learning System** Starting with a simple outline at the beginning of each chapter, we have focused on the crucial topics in a clear and easy to follow manner. We emphasize

critical thinking and active learning by constantly returning to how science is done, and through several pedagogical devices including our Biological Inquiry Questions which appear often in figure legends throughout the text. We end each chapter with a thorough review section which returns to our outline and emphasizes higher level learning through multiple question types.

7. **Media—Teaching and Learning with Technology**
Our new book is accompanied by a vast array of electronic teaching and learning tools. We have focused on creating new student content that is built upon learning outcomes and assessing student performance. We have also created an unprecedented array of presentation and course management tools to enable instructors to enhance their lectures and manage their classrooms more effectively. Finally, we are committed to offering several electronic book and customized print options to best fit your needs.

Experimentation in Biology Reveals General Principles

Biology is the study of life. The primary way that biologists study life is through experimentation. In this textbook, we have maintained a parallel focus on the general principles of biology and experimentation.

Each chapter is divided into a few sections focusing on general principles in biology. These sections begin with an overview of why the topic is important. We then describe the features of the topic that we and our reviewers have felt are the most important and sometimes the most difficult to grasp.

In describing the principles of biology, we have woven experimentation into each chapter. To be prepared for a career in biology, students need to understand the techniques that are used in biology and bolster their critical thinking skills. Each chapter has a Feature Investigation that shows the steps in the scientific process (for example, see Figure 10.13, p. 200). These investigations include a description of the methods and end with an analysis of the data. This deeper approach allows students to appreciate how the general principles of biology were derived from experiments.

In addition, many scientists are mentioned throughout each chapter with a brief description of how their work contributed to the general principles of biology, reinforcing a sense that biology is an enterprise carried out by scientists around the world. This will prepare students for their next step into the scientific literature in future courses.

An Emphasis on Evolution Provides a Modern Perspective

Evolution is the unifying theme that connects the various areas of biology. We have chosen to explain this theme from a modern perspective by relating the information in each chapter to the genetic material, namely the genomes of organisms. Like-

wise, because most genes encode proteins, a logical extension is to also relate information to proteomes—the collection of proteins that a cell or organism can make.

We use our Genomes and Proteomes subsections as one way to integrate the various disciplines of biology. For example, let's consider our Cell Biology Unit. In this unit, we emphasize how gene regulation is responsible for the differences between a nerve and muscle cell and how descent with modification occurs at the protein level to produce families of proteins with related cellular functions. Likewise, let's consider our Ecology Unit. In this unit we explain how the modification of particular genes has enabled organisms to compete effectively in their environment. The Ecology Unit also considers how the genomes of organisms have evolved in response to environmental changes over many generations.

As a team, our authors are committed to the idea of integrating the various disciplines in Biology, not separating them. We feel it important, fun, engaging, and actually easy to highlight the evolutionary theme of biology by relating information in each chapter to Genomes and Proteomes. They are intended to provide "perspective". We keep returning to the idea that everything in biology stems from the evolution of genomes, and that genomes primarily encode proteins that ultimately provide organisms with their traits. We feel strongly that this approach provides a modern perspective that will serve our readers well in their future careers. The Genomes and Proteomes subsections are just one way that we integrate the various fields of biology. Genes, proteins, and the molecular mechanisms of life are discussed throughout the entire book.

Textbook Illustrations Are a Key to Learning

In discussions with many of our students, we have come to realize that many, probably most students, are visual learners. They read the textbook for information, but when it comes time to study, their main emphasis is on the figures. Likewise, instructors often rely heavily on good illustrations for their lectures. Therefore, a top priority in the development of our Biology textbook has been the conceptualization and rendering of the illustrations.

As you will see when you scan through this book, the illustrations are very easy to follow, particularly those that have multiple steps. We have taken the attitude that students should be able to look at the figures and understand what is going on, without having to glance back and forth between the text and art. Many figures contain text boxes that explain what the illustration is showing. In those figures with multiple steps, the boxes are numbered so that students will understand that the steps occur in a particular order. In some cases, the numbering was critical when the illustration involved features that could not be presented in a linear manner. For example, a description of hearing in Figure 45.7 is much easier to follow because we have guided the student through the process by using numbered text boxes.

Likewise, technology can help us to engage, to educate and even to inspire our students. As you will see when you skim through the pages, the drawings in this textbook are technologically advanced. They are primarily intended to educate the student and we have maintained a commitment to simplicity. Even so, the illustrations in this textbook are also aimed at being interesting and inspiring. Art elements are drawn with a strong sense of realism and three-dimensionality. The illustrations come to life, which, after all, is important to students who are interested in the study of life. We expect students to occasionally look at a figure and think, "Wow, that's cool!" We invite you to skim through the pages and see for yourself.

Accuracy Is a Top Priority

Inaccuracies in a science textbook come from two primary sources. The first is human error. Authors, editors, and illustrators occasionally make mistakes. Fortunately, such mistakes are rare. Each chapter of our Biology book has been read by dozens of people, including several accuracy checkers whose sole job was to find mistakes. Using our 360-degree developmental process (p. viii), we have worked to ensure an unprecedented level of content accuracy in our textbook.

A second and more common source of error is out-of-date material. Biology is continually changing. New information arrives on a daily basis. Some of this information makes us realize that past information was incorrect. Therefore, textbooks that fail to maintain a current perspective become progressively more inaccurate and misleading. Having an experienced author team with extensive research credentials helps to build a textbook with the most current and accurate information. In addition, over three-hundred faculty members have reviewed our Biology textbook for its content.

We are confident that our book has the most modern content that the industry has to offer. This modern perspective pushes the accuracy of our book to the highest standard possible. In future editions, we will continue to strive for cutting-edge content that maintains a high level of accuracy. In future editions, we will continue to employ the help of many reviewers and accuracy checkers to maintain our commitment to modern content and accuracy.

Our Review Process Ensures a Textbook with the Right Content

If the best writer in the world wrote a textbook single-handedly without input from others, the book would not turn out well. Extensive and open-minded reviews are essential to producing a book that is superior. As we developed our book, we took the attitude that we must always return to a previous draft, analyze it critically, and then revise it accordingly.

From an author's perspective, the review process can be pretty daunting. We turn in chapters that we think are letter perfect, and then receive back reviews that make us painfully aware that writing a textbook is harder than it looks. At the start, we created a process that would make it easier for the editors and reviewers to critically evaluate our work. The first-draft chapters were sent to many outside reviewers who are faculty that either teach General Biology, are experts in the topics found in the chapter, or both. Each first-draft chapter was reviewed by up to 15 different people.

The reviews were collected and provided to the author of a given chapter and the editorial staff. Our editors read the chapters and the reviews, and then gave each author advice on how to make the next draft better. A second important type of input also occurred at the first-draft stage. For each unit, we conducted Focus Groups in which faculty members who had reviewed the chapters of an entire unit came together for a two-day meeting to discuss the chapters with the authors and editors. While written feedback is great, face-to-face discussion often brings out bigger-picture issues that may not be found in written reviews.

At the second-draft stage, we decided upon an innovation that profoundly enhanced the quality of our Biology book. Although the illustrations in a Biology textbook are very expensive to make, we realized that a strong connection between the text and illustrations is critical to produce a superior textbook. Instead of making the illustrations at the final-draft stage, which is typical of textbook publishing, we made them very early in the process. This helped us in two ways. First, reviewers could see the art as the book developed, and make critical changes to it. Second, it allowed the authors and editors to develop a keen sense of consistency between the text and art.

Also at the second-draft stage, we assembled the text and illustrations into a format that looked like a chapter from an actual textbook. We had to keep reminding the reviewers that "These are not finalized chapters. These are early drafts that we want you to critically evaluate." This format allowed the reviewers, authors, and editors to understand how the pieces of the book would fit together. Although it was an exhaustive and rigorous process, adopting this step early in the writing process allowed us to produce a book with a sharp consistency between the text and figures.

We Are Committed to Serving Teachers and Learners

Writing a new textbook is a daunting task. To accurately and thoroughly cover a course as wide ranging as biology, we felt it was essential that our team reflect the diversity of the field. We saw an opportunity to reach students at an early stage in their education and provide their biology training with a solid and up-to-date foundation. We have worked to balance coverage of classic research with recent discoveries that extend biological concepts in surprising new directions or that forge new concepts. Some new discoveries were selected because they highlight scientific controversies, showing students that we don't have all the answers yet. There is still a lot of work for new generations of biologists. With this in mind, we've also spotlighted discoveries made by diverse people doing research in different countries to illustrate the global nature of modern biological science.

As active teachers and writers, one of the great joys of this process for us is that we have been able to meet many more educators and students during the creation of this textbook. It is humbling to see the level of dedication our peers bring to their teaching. Likewise, it is encouraging to see the energy and enthusiasm so many students bring to their studies. We hope this book and its media package will serve to aid both faculty and students in meeting the challenges of this dynamic and exciting course. For us, this remains a work in progress and we encourage you to let us know what you think of our efforts and what we can do to serve you better.

Rob Brooker   brook005@umn.edu
Eric Widmaier   widmaier@bu.edu
Linda Graham   lkgraham@wisc.edu
Peter Stiling   pstiling@cas.usf.edu

The Next Step in Textbook Development

- 10 developmental editors
- 7 developmental focus groups
- 7 art focus groups
- 25+ accuracy checkers
- more than 1,200 reviews by over 350 reviewers across the world
- 11 multiple-day symposia with over 215 majors biology educators participating
- an art development team who worked closely with the authors
- media board of consultants ·
- 3 photo consultants

The following groups of individuals have been instrumental in ensuring the highest standard of content and accuracy in this textbook. We are deeply indebted to them for their tireless efforts.

Developmental Focus Groups

Cell Unit

Russell Borski,
North Carolina State University
Peter Fajer,
Florida State University
Brad Mehrtens,
University of Illinois– Urbana/Champaign
Randall Walikonis,
University of Connecticut
Sue Simon Westendorf,
Ohio University
Mark Staves,
Grand Valley State University

Genetics Unit

Karl Aufderheide,
Texas A&M University
John Doctor,
Duquesne University
Arlene Larson,
University of Colorado– Denver
Subhash Minocha,
University of New Hampshire
John Osterman,
University of Nebraska– Lincoln
Jill Reid,
Virginia Commonwealth University

Evolution Unit

Mark Decker,
University of Minnesota– Minneapolis
Robert Dill,
Bergen Community College
Jennifer Regan,
University of Southern Mississippi
Michelle Shuster,
New Mexico State University
Fred Wasserman,
Boston University

Diversity Unit

Ernest DuBrul,
University of Toledo
Roland Dute,
Auburn University
Florence Gleason,
University of Minnesota– St. Paul
Ann Rushing,
Baylor University
Randall Yoder,
Lamar University

Plants Unit

Fred Essig,
University of South Florida
Steve Herbert,
University of Wyoming
Mike Muller,
University of Illinois– Chicago
Stuart Reichler,
University of Texas–Austin
Scott Russell,
University of Oklahoma
Rani Vajravelu,
University of Central Florida

Animals Unit

Linda Collins,
University of Tennessee– Chattanooga
William Collins,
Stony Brook University
David Kurjiaka,
Ohio University
Phil Stephens,
Villanova University
David Tam,
University of North Texas
Charles Walcott,
Cornell University

Ecology Unit

James Adams,
Dalton State College
Stanley Faeth,
Arizona State University
Barbara Frase,
Bradley University
Daniel Moon,
University of North Florida
Dan Tinker,
University of Wyoming

Media Focus Group

Russell Borski,
North Carolina State University
Mark Decker,
University of Minnesota
Jon Glase,
Cornell University
John Merrill,
Michigan State University
Melissa Michael,
University of Illinois– Urbana/Champaign
Randall Phillis,
University of Massachusetts–Amherst
Mitch Price,
Pennsylvania State University

Accuracy Checkers

David Asch,
  Youngstown State University
Karl Aufderheide,
  Texas A&M University
Deborah Brooker
Linda Collins,
  *University of Tennessee–
  Chattanooga*
Mark Decker,
  University of Minnesota
Laura DiCaprio,
  Ohio University
Marjorie Doyle,
  *University of Wisconsin–
  Madison*
Peter Fajer,
  Florida State University
Pete Franco,
  University of Minnesota
Barbara Frase,
  Bradley University

John Graham,
  Bowling Green State University
Eunsoo Kim,
  *University of Wisconsin–
  Madison*
Arlene Larson,
  *University of Colorado–
  Denver*
David Pennock,
  Miami University
Anthony M. Rossi,
  University of North Florida
Martin Silberberg,
  McGraw-Hill chemistry author
Kevin Strang,
  *University of Wisconsin–
  Madison*
Fred Wasserman,
  Boston University
Jane E. Wissinger,
  University of Minnesota

Class Testers

We would like to thank the students and faculty at Ohio University, UCLA, and Calvin College for class testing our book.

End-of-Chapter Questions

Robert Dill,
  Bergen Community College
Arlene Larson,
  *University of Colorado–
  Denver*

Jennifer Regan,
  *University of Southern
  Mississippi*

Photo Consultants

John Osterman,
  *University of Nebraska–
  Lincoln*
Sue Simon Westendorf,
  Ohio University

Kevin Strang,
  *University of Wisconsin–
  Madison*

General Biology Symposia

Every year McGraw-Hill conducts several General Biology Symposia, which are attended by instructors from across the country. These events are an opportunity for editors from McGraw-Hill to gather information about the needs and challenges of instructors teaching the major's biology course. It also offers a forum for the attendees to exchange ideas and experiences with colleagues they might not have otherwise met. The feedback we have received has been invaluable, and has contributed to the development of Biology and its supplements.

2006

Michael Bell, *Richland College*
Scott Bowling, *Auburn University*
Peter Busher, *Boston University*
Allison Cleveland,
  *University of South Florida–
  Tampa*
Sehoya Cotner,
  University of Minnesota
Kathyrn Dickson,
  *California State College–
  Fullerton*
Cathy Donald-Whitney,
  *Collin County Community
  College*
Stanley Faeth,
  Arizona State University
Karen Gerhart,
  University of California–Davis
William Glider,
  University of Nebraska– Lincoln
Stan Guffey,
  The University of Tennessee
Bernard Hauser,
  *University of Florida–
  Gainesville*
Mark Hens,
  *University of North Carolina–
  Greensboro*
James Hickey,
  *Miami University of Ohio–
  Oxford*
Sherry Krayesky,
  *University of Louisiana–
  Lafayette*
Brenda Leady,
  University of Toledo
Michael Meighan,
  *University of California–
  Berkeley*
Comer Patterson,
  Texas A&M University
Debra Pires,
  *University of California–
  Los Angeles*
Robert Simons,
  *University of California–
  Los Angeles*
Steven D. Skopik,
  University of Delaware
Ashok Upadhyaya,
  *University of South Florida–
  Tampa*
Anthony Uzwiak,
  Rutgers University
Dave Williams,
  *Valencia Community College–
  East Campus*
Jay Zimmerman,
  St. John's University

2005

Donald Buckley,
  Quinnipiac University
Arthur Buikema,
  Virginia Polytechnic Institute
Anne Bullerjahn,
  Owens Community College
Garry Davies,
  *University of Alaska–
  Anchorage*
Marilyn Hart,
  Minnesota State University
Daniel Flisser,
  Camden County College
Elizabeth Godrick,
  Boston University
Miriam Golbert,
  College of the Canyons
Sherry Harrel,
  Eastern Kentucky University
William Hoese,
  *California State University–
  Fullerton*
Margaret Horton,
  *University of North Carolina
  at Greensboro*
Carol Hurney,
  James Madison University
James Luken,
  Coastal Carolina University
Mark Lyford,
  University of Wyoming
Gail McKenzie,
  Jefferson State Junior College
Melissa Michael,
  *University of Illinois at
  Urbana-Champaign*
Subhash C. Minocha,
  University of New Hampshire
Leonore Neary,
  Joliet Junior College
K. Sata Sathasivan,
  University of Texas at Austin
David Senseman,
  *University of Texas–
  San Antonio*
Sukanya Subramanian,
  *Collin County Community
  College*
Randall Terry,
  Lamar University
Sharon Thoma,
  *University of Wisconsin–
  Madison*
William Tyler,
  *Indian River Community
  College*

2004

Jonathan Akin,
Northwestern State University of Louisiana

David Asch,
Youngstown State University

Diane Bassham,
Iowa State University

Donald Buckley,
Quinnipiac University

Ruth Buskirk,
University of Texas, Austin

Charles Creutz,
University of Toledo

Lydia Daniels,
University of Pittsburgh

Laura DiCaprio,
Ohio University

Michael Dini,
Texas Tech University

John Doctor,
Duquesne University

Ernest DuBrul,
University of Toledo

John Elam,
Florida State University

Samuel Hammer,
Boston University

Marilyn Hart,
Minnesota State University

Marc Hirrel,
University of Central Arkansas

Carol Johnson,
Texas A&M University

Dan Krane,
Wright State University

Karin Krieger,
University of Wisconsin–Green Bay

Josephine Kurdziel,
University of Michigan

Martha Lundell,
University of Texas, San Antonio

Roberta Maxwell,
University of North Carolina–Greensboro

John Merrill,
Michigan State University

Melissa Michael,
University of Illinois at Urbana-Champaign

Peter Niewarowski,
University of Akron

Ronald Patterson,
Michigan State University

Peggy Pollak,
Northern Arizona University

Uwe Pott,
University of Wisconsin, Green Bay

Mitch Price,
Pennsylvania State University

Steven Runge,
University of Central Arkansas

Thomas Shafer,
University of North Carolina, Wilmington

Richard Showman,
University of South Carolina

Michèle Shuster,
New Mexico State University

Dessie Underwood,
California State University–Long Beach

Mike Wade,
Indiana University

Elizabeth Willott,
University of Arizona

Carl Wolfe,
University of North Carolina, Charlotte

Reviewers

James K. Adams,
Dalton State College

Sylvester Allred,
Northern Arizona University

Jonathan W. Armbruster,
Auburn University

Joseph E. Armstrong,
Illinois State University

David K. Asch,
Youngstown State University

Amir M. Assadi-Rad,
Delta College

Karl J. Aufderheide,
Texas A&M University

Anita Davelos Baines,
University of Texas–Pan American

Lisa M. Baird,
University of San Diego

Diane Bassham,
Iowa State University

Donald Baud,
University of Memphis

Vernon W. Bauer,
Francis Marion University

Ruth E. Beattie,
University of Kentucky

Michael C. Bell,
Richland College

Steve Berg,
Winona State University

Arlene G. Billock,
University of Louisiana at Lafayette

Kristopher A. Blee,
California State University, Chico

Heidi B. Borgeas,
University of Tampa

Russell Borski,
North Carolina State University

Scott A. Bowling,
Auburn University

Robert Boyd,
Auburn University

Eldon J. Braun,
University of Arizona

Michael Breed,
University of Colorado, Boulder

Randy Brewton,
University of Tennessee, Knoxville

Peggy Brickman,
University of Georgia

Cheryl Briggs,
University of California, Berkeley

Peter S. Brown,
Mesa Community College

Mark Browning,
Purdue University

Cedric O. Buckley,
Jackson State University

Don Buckley,
Quinnipiac University

Arthur L. Buikema, Jr.,
Virginia Tech University

Anne Bullerjahn,
Owens Community College

Ray D. Burkett,
Southeast Tennessee Community College

Stephen P. Bush,
Coastal Carolina University

Peter E. Busher,
Boston University

Jeff Carmichael,
University of North Dakota

Clint E. Carter,
Vanderbilt University

Patrick A. Carter,
Washington State University

Merri Lynn Casem,
California State University, Fullerton

Domenic Castignetti,
Loyola University of Chicago

Maria V. Cattell

David T. Champlin,
University of Southern Maine

Jung H. Choi,
Georgia Institute of Technology

Curtis Clark,
Cal Poly Pomona

Allison Cleveland,
University of South Florida

Janice J. Clymer,
San Diego Mesa College

Linda T. Collins,
University of Tennessee at Chattanooga

Jay L. Comeaux,
Louisiana State University

Bob Connor II,
Owens Community College

Daniel Costa,
University of California at Santa Cruz

Sehoya Cotner,
University of Minnesota

Mack E. Crayton III,
Xavier University of Louisiana

Louis Crescitelli,
Bergen Community College

Charles Creutz,
University of Toledo

Karen A. Curto,
University of Pittsburgh

Mark A. Davis,
Macalester College

Mark D. Decker,
University of Minnesota

Jeffery P. Demuth,
Indiana University

Phil Denette,
Delgado Community College

Donald W. Deters,
Bowling Green State University

Hudson R. DeYoe,
University of Texas–Pan American

Laura DiCaprio,
Ohio University

Randy DiDomenico,
University of Colorado, Boulder

Robert S. Dill,
Bergen Community College

Kevin Dixon,
University of Illinois–Urbana/Champaign

John S. Doctor,
Duquesne University

Michael Meighan,
University of California, Berkeley

Douglas Meikle,
Miami University

Allen F. Mensinger,
University of Minnesota, Duluth

John Merrill,
Michigan State University

Richard Merritt,
Houston Community College

Brian T. Miller,
Middle Tennessee State University

Hugh A. Miller III,
East Tennessee State University

Thomas E. Miller,
Florida State University

Sarah L. Milton,
Florida Atlantic University

Dennis J. Minchella,
Purdue University

Subhash C. Minocha,
University of New Hampshire

Patricia Mire,
University of Louisiana at Lafayette

Daniela S. Monk,
Washington State University

Daniel C. Moon,
University of North Florida

Janice Moore,
Colorado State University

Mathew D. Moran,
Hendrix College

Jorge A. Moreno,
University of Colorado, Boulder

Roderick M. Morgan,
Grand Valley State University

James V. Moroney,
Louisiana State University

Molly R. Morris,
Ohio University

Michael Muller,
University of Illinois at Chicago

Michelle Mynlieff,
Marquette University

Allan D. Nelson,
Tarleton State University

Raymond L. Neubauer,
University of Texas at Austin

Jacalyn S. Newman,
University of Pittsburgh

Colleen J. Nolan,
St. Mary's University

Shawn E. Nordell,
St. Louis University

Margaret Nsofor,
Southern Illinois University, Carbondale

Dennis W. Nyberg,
University of Illinois at Chicago

Nicole S. Obert,
University of Illinois, Urbana-Champaign

David G. Oppenheimer,
University of Florida

John C. Osterman,
University of Nebraska–Lincoln

Brian Palestis,
Wagner College

Julie M. Palmer,
University of Texas at Austin

C. O. Patterson,
Texas A&M University

Ronald J. Patterson,
Michigan State University

Linda M. Peck,
University of Findlay

David Pennock,
Miami University

Shelley W. Penrod,
North Harris College

Beverly J. Perry,
Houston Community College System

Chris Petersen,
College of the Atlantic

Jay Phelan,
UCLA

Eric R. Pianka,
The University of Texas at Austin

Thomas Pitzer,
Florida International University

Peggy E. Pollak,
Northern Arizona University

Richard B. Primack,
Boston University

Lynda Randa,
College of Dupage

Marceau Ratard,
Delgado Community College

Robert S. Rawding,
Gannon University

Jennifer Regan,
University of Southern Mississippi

Stuart Reichler,
University of Texas at Austin

Jill D. Reid,
Virginia Commonwealth University

Anne E. Reilly,
Florida Atlantic University

Linda R. Richardson,
Blinn College

Laurel Roberts,
University of Pittsburgh

Kenneth R. Robinson,
Purdue University

Chris Ross,
Kansas State University

Anthony M. Rossi,
University of North Florida

Kenneth H. Roux,
Florida State University

Ann E. Rushing,
Baylor University

Scott Russell,
University of Oklahoma

Christina T. Russin,
Northwestern University

Charles L. Rutherford,
Virginia Tech University

Margaret Saha,
College of William and Mary

Kanagasabapathi Sathasivan,
The University of Texas at Austin

Stephen G. Saupe,
College of St. Benedict

Jon B. Scales,
Midwestern State University

Daniel C. Scheirer,
Northeastern University

H. Jochen Schenk,
California State University, Fullerton

John Schiefelbein,
University of Michigan

Deemah N. Schirf,
University of Texas at San Antonio

Mark Schlueter,
College of Saint Mary

Scott Schuette,
Southern Illinois University, Carbondale

Dean D. Schwartz,
Auburn University

Timothy E. Shannon,
Francis Marion University

Richard M. Showman,
University of South Carolina

Michele Shuster,
New Mexico State University

Robert Simons,
UCLA

J. Henry Slone,
Francis Marion University

Phillip Snider, Jr.,
Gadsden State Community College

Nancy G. Solomon,
Miami University

Lekha Sreedhar,
University of Missouri–Kansas City

Bruce Stallsmith,
University of Alabama, Huntsville

Susan J. Stamler,
College of Dupage

Mark P. Staves,
Grand Valley State University

William Stein,
Binghamton University

Philip J. Stephens,
Villanova University

Antony Stretton,
University of Wisconsin–Madison

Gregory W. Stunz,
Texas A&M University–Corpus Christi

Julie Sutherland,
College of Dupage

David Tam,
University of North Texas

Roy A. Tassava,
Ohio State University

Sharon Thoma,
University of Wisconsin–Madison

Shawn A. Thomas,
College of St. Benedict/ St. John's University

Daniel B. Tinker,
University of Wyoming

Marty Tracey,
Florida International University

Marsha Turell,
Houston Community College

J. M. Turbeville,
Virginia Commonwealth University

Rani Vajravelu,
University of Central Florida

Neal J. Voelz,
St. Cloud State University

Samuel E. Wages,
South Plains College

Jyoti R. Wagle,
Houston Community College System–Central

Charles Walcott,
Cornell University

Randall Walikonis,
University of Connecticut

Jeffrey A. Walker,
University of Southern Maine

Delon E. Washo-Krupps,
Arizona State University
Frederick Wasserman,
Boston University
Steven A. Wasserman,
*University of California,
San Diego*
R. Douglas Watson,
*University of Alabama
at Birmingham*
Cindy Martinez Wedig,
*University of Texas–
Pan American*

Arthur E. Weis,
University of California–Irvine
Sue Simon Westendorf,
Ohio University
Howard Whiteman,
Murray State University
Susan Whittemore,
Keene State College
David L. Wilson,
University of Miami
Robert Winning,
*Eastern Michigan
University*

Michelle D. Withers,
*Louisiana State
University*
Clarence C. Wolfe,
*Northern Virginia Community
College*
Gene K. Wong,
Quinnipiac University
Richard P. Wunderlin,
University of South Florida

Joanna Wysocka-Diller,
Auburn University
H. Randall Yoder,
Lamar University
Marilyn Yoder,
*University of Missouri–
Kansas City*
Scott D. Zimmerman,
*Southwest Missouri State
University*

International Reviewers

Heather Addy,
University of Calgary
Mari L. Acevedo,
*University of Puerto Rico
at Arecibo*
Heather E. Allison,
University of Liverpool, UK
David Backhouse,
University of New England
Andrew Bendall,
University of Guelph
Marinda Bloom,
*Stellenbosch University,
South Africa*
Tony Bradshaw,
Oxford-Brookes University, UK
Alison Campbell,
University of Waikato
Bruce Campbell,
Okanagan College

Clara E. Carrasco, Ph.D.,
*University of Puerto Rico–
Ponce Campus*
Keith Charnley,
University of Bath, UK
Ian Cock,
Griffith University
Margaret Cooley,
University of NSW
R. S. Currah,
University of Alberta
Logan Donaldson,
York University
Theo Elzenga,
*Rijks Universiteit Groningen,
Netherlands*
Neil C. Haave,
University of Alberta

Tom Haffie,
University of Western Ontario
Louise M. Hafner,
*Queensland University
of Technology*
Annika F. M. Haywood,
*Memorial University
of Newfoundland*
William Huddleston,
University of Calgary
Shin-Sung Kang,
KyungDuk University
Wendy J. Keenleyside,
University of Guelph
Christopher J. Kennedy,
Simon Fraser University
Bob Lauder,
Lancaster University

Richard C. Leegood,
Sheffield University, UK
Thomas H. MacRae,
Dalhousie University
R. Ian Menz,
Flinders University
Kirsten Poling,
University of Windsor
Jim Provan,
Queens University, Belfast, UK
Richard Roy,
McGill University
Han A.B. Wösten,
*Utrecht University,
Netherlands*

ACKNOWLEDGMENTS

The lives of most science-textbook authors do not revolve around an analysis of writing techniques. Instead, we are people who understand science and are inspired by it, and we want to communicate that information to other people. Simply put, we need a lot of help to get it right.

Editors are a key component that help the authors modify the content of their book so it is logical, easy to read, and inspiring. The editorial team for this Biology textbook has been a catalyst that kept this project rolling. The members played various roles in the editorial process. Lisa Bruflodt (Senior Developmental Editor) has been the master organizer. Frankly, this is a ridiculously hard job. Coordinating the efforts of dozens of people and keeping them on schedule is not always fun. Lisa's success at keeping us on schedule has been truly amazing. We are also grateful to Kris Tibbetts (Director of Development) who was involved in the early steps of the book, and kept the focus groups on track.

Our Biology book also has had 6 additional developmental editors who scrutinized each draft of their respective chapters with an emphasis on improving content, clarity, and readability. These developmental editors analyzed educational materials and reviewers' comments, and gave the authors advice on how to improve succeeding drafts. They also provided a list of the general principles that most instructors want in their Biology textbook. These general principles have been a cornerstone for the organization of our chapters.

Suzanne Olivier (Lead Freelance Developmental Editor) did an outstanding job of coordinating the staff of developmental editors. She also played an important role in editing chapters in the Genetics and Plant Biology units. Her early editing of the Genetics Unit, in particular, set the tone for many of the pedagogical features that became established throughout the entire textbook. Other developmental editors focused on particular units. Alice Fugate was involved with the Chemistry, Cell Biology, and Animal Biology Units. Her knack for getting the level of the writing appropriate for majors biology was invaluable, as was her attention to detail. Joni Fraser focused on the Cell Biology, Diversity, and Ecology Units. Somehow she successfully juggled the tasks of addressing all the reviewers' concerns while maintaining the necessary chapter length. Patricia Longoria played an important role in the early editing of the Diversity and Plant Biology Units. Patricia contributed many useful ideas for content and expression, and her unfailing enthusiasm smoothed the way over rocky parts of the process. Robin Fox edited three key chapters in the Genetics Unit. Her attention to detail and the explanation of mathematical principles were invaluable. And finally, Alan Titche was also involved with Animal Biology Unit, and played a major role in developing some of the most challenging chapters in that Unit. We would also like to thank Dr. Jim Deshler and Dr. Mary Erskine for their valuable contributions to the writing of several chapters in the Animal Biology Unit.

Deborah Brooker (Art/Text Coordinating Editor) analyzed all of the chapters in the textbook with one primary question in mind. Do the written text and figures tell a parallel story? With excruciating care, she made sure that the text and figures are consistent, and that the figures, by themselves, are accurate and easy to follow.

Imagineering Media Services Inc., of Ontario, Canada, did a fantastic job of illustrating our Biology book. They were involved early in the process by first making rough sketches based on the material in the first drafts, and then later progressed to drawings with finer detail. Their ability to make realistic, three-dimensional drawings is second to none. We're particularly grateful to Kierstan Hong, who provided a critical line of communication between the publisher, authors, and illustrators throughout most of this process, and also to Mark Mykytiuk, who also played a lead role in overseeing the art development. We would also like to gratefully acknowledge our photography researchers at Pronk & Associates of Ontario, Canada, and particularly to Fiona D'souza for keeping us on schedule. Likewise, we are grateful to John Leland, Photo Research Coordinator, at McGraw-Hill for his coordination of the photo selection process.

We would also like to thank our advisors and contributors:

Media Board of Advisors

Mark Decker,
University of Minnesota–Minneapolis

Naomi Friedman, *Developmental Editor*

Jon Glase,
Cornell University

John Merrill,
Michigan State University

Melissa Michael,
University of Illinois–Urbana/Champaign

Randall Phyllis,
University of Massachusetts

Mitch Price,
Pennsylvania State University

Tutorial Questions

Scott Bowling,
Auburn University

Don Buckley,
Quinnipiac University

Ernest DuBrul,
University of Toledo

Frederick B. Essig,
University of Florida

Jon Glase,
Cornell University

Norman A. Johnson
University of Massachusetts–Amherst

Kari Beth Krieger,
University of Wisconsin–Green Bay

Patricia Mire,
University of Louisiana–Lafayette

Allan Smits,
Quinnipiac University

Test Questions

Russell Borski,
   *North Carolina State
   University*
Robert Dunn,
   *North Carolina State
   University*
John Godwin,
   *North Carolina State
   University*
Mary Beth Hawkins,
   *North Carolina State
   University*
Harold Heatwole,
   *North Carolina State
   University*
James Mickle,
   *North Carolina State
   University*
Gerald Van Dyke,
   *North Carolina State
   University ARIS*

ARIS

Brad Mehrtens,
   *University of Illinois–
   Urbana/Champaign*

Instructor's Manual

Daniel Moon,
   University of North Florida

Student Study Guide

Michelle Shuster,
   New Mexico State University
Amy Marion,
   New Mexico State University

Lecture Outlines

Brenda Leady,
   University of Toledo

Animations

Kevin Dixon,
   *University of Illinois–
   Urbana/Champaign*

Another important aspect of the editorial process is the actual design, presentation, and layout of materials. It's confusing if the text and art aren't on the same page, or if a figure is too large or two small. We are indebted to the tireless efforts of Joyce Berendes (Lead Project Manager) and Wayne Harms (Design Manager) of McGraw-Hill. Their artistic talents, ability to size and arrange figures, and attention to the consistency of the figures have been remarkable. We also wish to thank John Joran (Designer) who cleverly crafted both the interior and exterior designs.

We would like to acknowledge the ongoing efforts of the superb marketing staff at McGraw-Hill. Kent Peterson (Vice President, Director of Marketing) oversees a talented staff of people who work tirelessly to promote our book. Special thanks to Chad Grall (Marketing Director), Wayne Vincent (Internet Marketing Manager), Debra Hash (Senior Marketing Manager) and Heather Wagner (Systems and Promotions Marketing Manager) for their ideas and enthusiasm for this book. The proposal of making a video website for the book *www.brookerbiology.com* was scary for the authors but actually turned out to be fun.

Finally, other staff members at McGraw-Hill Higher Education have ensured that the authors and editors were provided with adequate resources to achieve the goal of producing a superior textbook. These include Kurt Strand (President, Science, Engineering, and Math), Marty Lange (Vice President, Editor-in-Chief), Michael Lange (Vice President, New Product Launches), Janice Roerig-Blong (Publisher) and Patrick Reidy (Executive Editor). In particular, Michael and Patrick communicated with the authors on a regular basis regarding the progress of this project. They attended most of the focus groups and author meetings, and even provided occasional input regarding the content of the book. The bottom line is that the author team is grateful that you have believed in this project, and have provided us with the resources to make it happen.

Student Supplements

Designed to help students maximize their learning experience in biology—we offer the following options to students:

ARIS (Assessment, Review, and Instruction System) is an electronic study system that offers students a digital portal of knowledge. Students can readily access a variety of **digital learning objects** which include:

- chapter level quizzing
- pretests
- animations
- videos
- flashcards
- answers to Biological Inquiry Questions
- answers to all end-of-chapter questions
- MP3 and MP4 downloads of selected content
- learning outcomes and assessment capability woven around key content

Student Study Guide
ISBN: 0-07-299588-2

Helping students focus their time and energy on important concepts, the study guide offers students a variety of tools:

1. Practice Questions—approximately 10–12 multiple choice questions
2. Active Learning Questions—approximately 5–8 open-ended questions that ask the student to explore

something and delve into content a little deeper, reinforcing content through experiential learning.

3. Outline/Summary of Fundamental Concepts—efficient listing of key concepts.
4. Key Terms
5. Strategies for Difficult Concepts

Content Delivery Flexibility

Brooker et al., *Biology* is available in many formats in addition to the traditional textbook to give instructors and students more choices when deciding on the format of their biology text. Choices include:

Volumes

The complete text has been split into three natural segments to allow instructors more flexibility and students more purchasing options.

Volume 1—Units 1 (Chemistry), 2 (Cell), and 3 (Genetics)
ISBN 0-07-335332-9
Volume 2—Units 6 (Plants) and 7 (Animals)
ISBN 0-07-335331-0
Volume 3—Units 4 (Evolution), 5 (Diversity), and 8 (Ecology)
ISBN 0-07-335333-7

Color Custom by Chapter

For even more flexibility, we offer the Brooker: *Biology* text in a full-color, custom version that allows instructors to pick the chapters they want included. Students pay for only what the instructor chooses.

eBook

The entire text is available electronically through the ARIS website. This electronic text offers not only the text in a digital format but includes embedded links to figures, tables, animations, and videos to make full use of the digital tools available and further enhance student understanding.

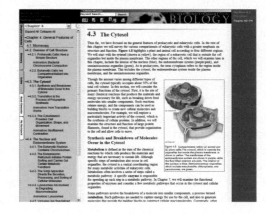

Instructor Supplements

Dedicated to providing high quality and effective supplements for instructors, the following Instructor supplements were developed for *Biology:*

ARIS with Presentation Center

Assessment, Review, and Instruction System, also known as ARIS, is an electronic homework and course management system designed for greater flexibility, power, and ease of use than any other system. Whether you are looking for a preplanned course or one you can customize to fit your course needs, ARIS is your solution.

In addition to having access to all student digital learning objects, ARIS allows instructors to:

Build Assignments

- Choose from pre-built assignments or create your own custom content by importing your own content or editing an existing assignment from the pre-built assignment.

- Assignments can include quiz questions, animations, and videos . . . anything found on the website.
- Create announcements and utilize full course or individual student communication tools
- Assign **unique multi-level tutorial questions** developed by content experts that provide intelligent feedback through a series of questions to help students truly understand a concept; not just repeat an answer.

Track Student Progress

- Assignments are automatically graded
- Gradebook functionality allows full course management including:
  - Dropping the lowest grades
  - Weighting grades / manually adjusting grades
  - Exporting your gradebook to Excel, WebCT or BlackBoard

- Manipulating data allowing you to track student progress through multiple reports

Offer More Flexibility

- **Sharing Course Materials with Colleagues** — Instructors can create and share course materials and assignments with colleagues with a few clicks of the mouse allowing for multiple section courses with many instructors (and TAs) to continually be in synch if desired.
- **Integration with BlackBoard or WebCT**—once a student is registered in the course, all student activity within McGraw-Hill's ARIS is automatically recorded and available to the instructor through a fully integrated grade book that can be downloaded to Excel, WebCT, or Blackboard.

Presentation Center

Build instructional materials wherever, whenever, and however you want!

ARIS Presentation Center is an online digital library containing assets such as photos, artwork, animations, PowerPoints, and other media types that can be used to create customized lectures, visually enhanced tests and quizzes, compelling course websites, or attractive printed support materials.

Access to your book, access to all books!

The Presentation Center library includes thousands of assets from many McGraw-Hill titles. This ever-growing resource gives instructors the power to utilize assets specific to an adopted textbook as well as content from all other books in the library.

Nothing could be easier!

Accessed from the instructor side of your textbook's ARIS website, Presentation Center's dynamic search engine allows you to explore by discipline, course, textbook chapter, asset type, or keyword. Simply browse, select, and download the files you need to build engaging course materials. All assets are copyright McGraw-Hill Higher Education but can be used by instructors for classroom purposes.

Instructor's Testing and Resource CD-ROM

ISBN: 0-07-295658-5

This cross-platform CD-ROM provides these resources for instructors:

- **Instructor's Manual**—This manual contains instructional strategies and activities, student misconceptions,

etymology of key terms, "Beyond the Book" interesting facts, and sources for additional web resources.

- **Test Bank**—The test bank offers multiple-choice and true/false questions that can be used for homework assignments or the preparation of exams.
- **Computerized Test Bank**—This software can be utilized to quickly create customized exams. The user-friendly program allows instructors to sort questions by format or level of difficulty; edit existing questions or add new ones; and scramble questions and answer keys for multiple versions of the same test.

Student Response System

Wireless technology brings interactivity into the classroom or lecture hall. Instructors and students receive immediate feedback through wireless response pads that are easy to use and engage students. This system can be used by instructors to:

- Take attendance
- Administer quizzes and tests
- Create a lecture with intermittent questions
- Manage lectures and student comprehension through the use of the gradebook
- Integrate interactivity into their PowerPoint presentations

Transparencies

ISBN: 0-07-295657-7

This boxed set of overhead transparencies includes every piece of line art in the textbook. The images have been modified to ensure maximum readability in both small and large classroom settings.

BIOLOGY LABORATORY MANUAL

Darrell S. Vodopich, *Baylor University*
Randy Moore, *University of Minnesota*
ISBN: 0-07-332398-5

This laboratory manual is designed to accompany Brooker et al: *Biology*. The experiments and procedures are simple, safe, easy to perform, and especially appropriate for large classes. Few experiments require a second class-meeting to complete the procedure. Each exercise includes many photographs, traditional topics, and experiments that help students learn about life. Procedures within each exercise are numerous and discrete so that an exercise can be tailored to the needs of the students, the style of the instructor, and the facilities available.

BIOLOGICAL INVESTIGATIONS LAB MANUAL

Warren D. Dolphin, *Iowa State University*
ISBN: 0-07-332399-3

Developed to accompany Brooker et al: *Biology*, this lab manual focuses on labs that are investigative and ask students to use more critical thinking and hands-on learning. The author emphasizes investigative, quantitative, and comparative approaches to studying the life sciences.

A VISUAL JOURNEY

Our art program was painstakingly designed in conjunction with the text development to ensure 1) each concept is accurately portrayed, 2) consistency is maintained between the text and art, and 3) it's appropriately placed on the page. The art serves as a visual outline for students, often offering textboxes that explain difficult concepts. For multistep processes, these textboxes are numbered so that the student can easily follow the process from beginning to end.

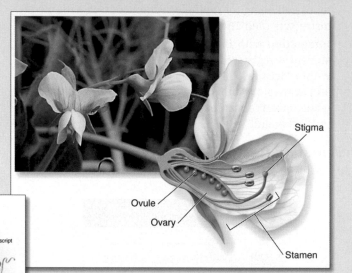

1 **Initiation:**
The promoter functions as a recognition site for sigma factor. RNA polymerase is bound to sigma factor, which causes it to recognize the promoter. Following binding, the DNA is unwound into a bubble known as the open complex.

2 **Elongation/synthesis of the RNA transcript:**
Sigma factor is released and RNA polymerase slides along the DNA in an open complex to synthesize RNA.

3 **Termination:**
When RNA polymerase reaches the terminator, it and the RNA transcript dissociate from the DNA.

(a) Stages of transcription

Figure 12.6 **Stages of transcription.** (a) Transcription can be divided into initiation, elongation, and termination. The inset emphasizes the direction of RNA synthesis and base pairing between the DNA template strand and RNA. (b) Three-dimensional structure of a bacterial RNA polymerase.

(b) Structure of a bacterial RNA polymerase

and birds maintain a relatively constant body temperature in spite of changing environmental temperatures (Figure 1.2d), while reptiles and amphibians do not. By comparison, all organisms continually regulate their cellular metabolism so that nutrient molecules are used at an appropriate rate, and new cellular components are synthesized when they are needed.

Growth and Development All living things grow and develop; **growth** produces more or larger cells, while **development** produces organisms with a defined set of characteristics. Among unicellular organisms such as bacteria, new cells are relatively small, and they increase in volume by the synthesis of additional cellular components. Multicellular organisms, such as

1 Pathogens produce distinctive elicitor compounds, which are the products of *Avr* genes.

2 Plant membrane or cytosolic receptors (*R* gene products) bind elicitors.

3 The binding of elicitors causes the production of H_2O_2 and NO. H_2O_2 kills pathogens and stimulates cell-wall strengthening.

4 Together, H_2O_2 and NO stimulate production of defense compounds and alarm signals, and induce cell death. Visible necrotic areas of dead cells appear where pathogen growth has been stopped.

AN INTRODUCTION TO BIOLOGY   5

The members of the same species are closely related genetically. In Units VI and VII, we will examine plants and animals at the level of cells, tissues, organs, and complete organisms.

7. **Population:** A group of organisms of the same species that occupy the same environment is called a **population**.
8. **Community:** A biological **community** is an assemblage of populations of different species. The types of species that are found in a community are determined by the environment and by the interactions of species with each other.
9. **Ecosystem:** Researchers may extend their work beyond living organisms and also study the environment. Ecologists analyze **ecosystems**, which are formed by

interactions of a community of organisms with their physical environment. Unit VIII considers biology from populations to ecosystems.
10. **Biosphere:** The **biosphere** includes all of the places on the Earth where living organisms exist, encompassing the air, water, and land.

Modern Forms of Life Are Connected by an Evolutionary History

Life began on Earth as primitive cells about 3.5 to 4 billion years ago. Since that time, those primitive cells underwent evolutionary changes that ultimately gave rise to the species we see today.

living organisms can be analyzed in a hierarchical manner, starting with the tiniest level of organization, and progressing to levels that are physically much larger and more complex. Figure 1.3 depicts a scientist's view of biological organization at different levels.

1. **Atoms:** An **atom** is the smallest component of an element that has the chemical properties of the element. All matter is composed of atoms.
2. **Molecules and macromolecules:** As discussed in Unit I, atoms bond with each other to form **molecules**. When many molecules bond together to form a polymer, this is called a **macromolecule**. Carbohydrates, proteins, and

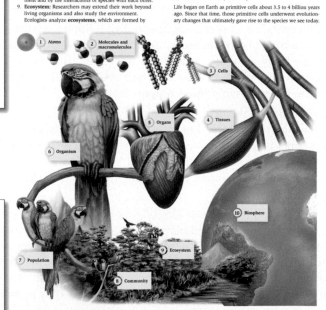

Figure 1.3 The levels of biological organization.

BIOLOGICAL INQUIRY QUESTIONS

These questions are designed to help students delve more deeply into a concept or experimental approach described in the art. These questions challenge a student to analyze the content of the figure they are looking at.

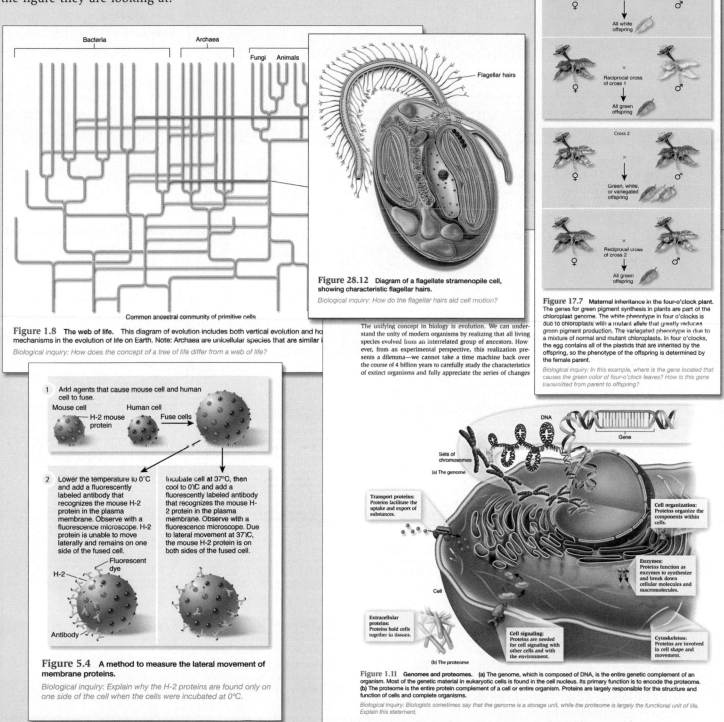

Figure 1.8 The web of life. This diagram of evolution includes both vertical evolution and horizontal mechanisms in the evolution of life on Earth. Note: Archaea are unicellular species that are similar in...

Biological inquiry: How does the concept of a tree of life differ from a web of life?

Figure 28.12 Diagram of a flagellate stramenopile cell, showing characteristic flagellar hairs.

Biological inquiry: How do the flagellar hairs aid cell motion?

The unifying concept in biology is evolution. We can understand the unity of modern organisms by realizing that all living species evolved from an interrelated group of ancestors. However, from an experimental perspective, this realization presents a dilemma—we cannot take a time machine back over the course of 4 billion years to carefully study the characteristics of extinct organisms and fully appreciate the series of changes

Figure 17.7 Maternal inheritance in the four-o'clock plant. The genes for green pigment synthesis in plants are part of the chloroplast genome. The white phenotype in four o'clocks is due to chloroplasts with a mutant allele that greatly reduces green pigment production. The variegated phenotype is due to a mixture of normal and mutant chloroplasts. In four o'clocks, the egg contains all of the plastids that are inherited by the offspring, so the phenotype of the offspring is determined by the female parent.

Biological inquiry: In this example, where is the gene located that causes the green color of four-o'clock leaves? How is this gene transmitted from parent to offspring?

Figure 5.4 A method to measure the lateral movement of membrane proteins.

Biological inquiry: Explain why the H-2 proteins are found only on one side of the cell when the cells were incubated at 0°C.

Figure 1.11 Genomes and proteomes. (a) The genome, which is composed of DNA, is the entire genetic complement of an organism. Most of the genetic material in eukaryotic cells is found in the cell nucleus. Its primary function is to encode the proteome. (b) The proteome is the entire protein complement of a cell or entire organism. Proteins are largely responsible for the structure and function of cells and complete organisms.

Biological inquiry: Biologists sometimes say that the genome is a storage unit, while the proteome is largely the functional unit of life. Explain this statement.

Experimental & Modern Content

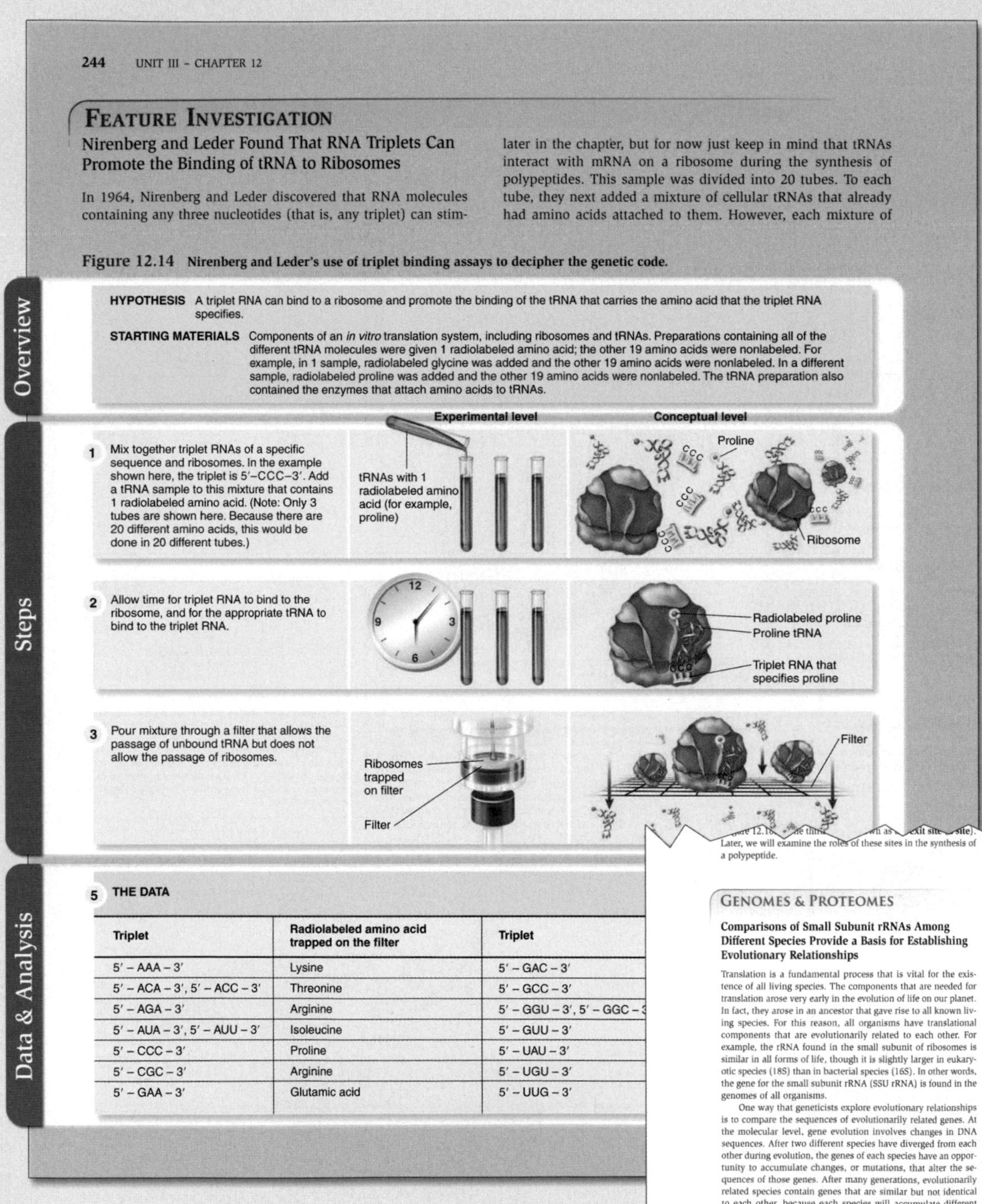

Feature Investigation

Focusing on hypothesis testing and discovery-based science, the Feature Investigations describe a key experiment, including 1) an overview of the hypothesis or goal of the experiment, 2) the steps of the experiment, and 3) ending with an analysis of data. This encourages an appreciation of the scientific process.

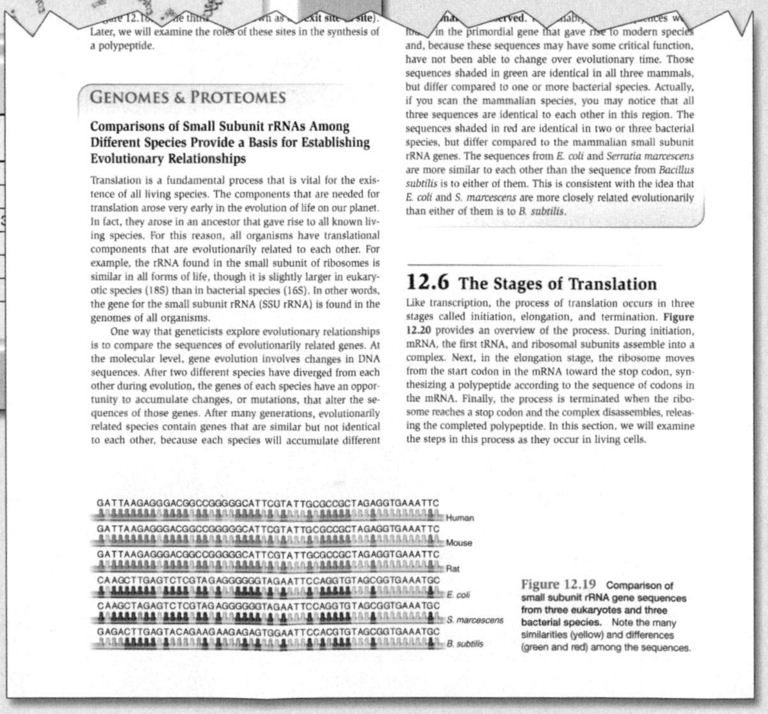

FEATURE INVESTIGATION

Nirenberg and Leder Found That RNA Triplets Can Promote the Binding of tRNA to Ribosomes

In 1964, Nirenberg and Leder discovered that RNA molecules containing any three nucleotides (that is, any triplet) can stimulate later in the chapter, but for now just keep in mind that tRNAs interact with mRNA on a ribosome during the synthesis of polypeptides. This sample was divided into 20 tubes. To each tube, they next added a mixture of cellular tRNAs that already had amino acids attached to them. However, each mixture of

Figure 12.14 Nirenberg and Leder's use of triplet binding assays to decipher the genetic code.

Overview

HYPOTHESIS A triplet RNA can bind to a ribosome and promote the binding of the tRNA that carries the amino acid that the triplet RNA specifies.

STARTING MATERIALS Components of an *in vitro* translation system, including ribosomes and tRNAs. Preparations containing all of the different tRNA molecules were given 1 radiolabeled amino acid; the other 19 amino acids were nonlabeled. For example, in 1 sample, radiolabeled glycine was added and the other 19 amino acids were nonlabeled. In a different sample, radiolabeled proline was added and the other 19 amino acids were nonlabeled. The tRNA preparation also contained the enzymes that attach amino acids to tRNAs.

Steps

Experimental level   Conceptual level

1 Mix together triplet RNAs of a specific sequence and ribosomes. In the example shown here, the triplet is 5′–CCC–3′. Add a tRNA sample to this mixture that contains 1 radiolabeled amino acid. (Note: Only 3 tubes are shown here. Because there are 20 different amino acids, this would be done in 20 different tubes.)

tRNAs with 1 radiolabeled amino acid (for example, proline)

Proline

Ribosome

2 Allow time for triplet RNA to bind to the ribosome, and for the appropriate tRNA to bind to the triplet RNA.

Radiolabeled proline Proline tRNA

Triplet RNA that specifies proline

3 Pour mixture through a filter that allows the passage of unbound tRNA but does not allow the passage of ribosomes.

Ribosomes trapped on filter

Filter

Filter

Data & Analysis

5 **THE DATA**

Triplet	Radiolabeled amino acid trapped on the filter	Triplet
5′ – AAA – 3′	Lysine	5′ – GAC – 3′
5′ – ACA – 3′, 5′ – ACC – 3′	Threonine	5′ – GCC – 3′
5′ – AGA – 3′	Arginine	5′ – GGU – 3′, 5′ – GGC – 3′
5′ – AUA – 3′, 5′ – AUU – 3′	Isoleucine	5′ – GUU – 3′
5′ – CCC – 3′	Proline	5′ – UAU – 3′
5′ – CGC – 3′	Arginine	5′ – UGU – 3′
5′ – GAA – 3′	Glutamic acid	5′ – UUG – 3′

Later, we will examine the roles of these sites in the synthesis of a polypeptide.

GENOMES & PROTEOMES

Comparisons of Small Subunit rRNAs Among Different Species Provide a Basis for Establishing Evolutionary Relationships

Translation is a fundamental process that is vital for the existence of all living species. The components that are needed for translation arose very early in the evolution of life on our planet. In fact, they arose in an ancestor that gave rise to all known living species. For this reason, all organisms have translational components that are evolutionarily related to each other. For example, the rRNA found in the small subunit of ribosomes is similar in all forms of life, though it is slightly larger in eukaryotic species (18S) than in bacterial species (16S). In other words, the gene for the small subunit rRNA (SSU rRNA) is found in the genomes of all organisms.

One way that geneticists explore evolutionary relationships is to compare the sequences of evolutionarily related genes. At the molecular level, gene evolution involves changes in DNA sequences. After two different species have diverged from each other during evolution, the genes of each species have an opportunity to accumulate changes, or mutations, that alter the sequences of those genes. After many generations, evolutionarily related species contain genes that are similar but not identical to each other, because each species will accumulate different

...served. ...ably ...quences w... fo...in the primordial gene that gave rise to modern species and, because these sequences may have some critical function, have not been able to change over evolutionary time. Those sequences shaded in green are identical in all three mammals, but differ compared to one or more bacterial species. Actually, if you scan the mammalian species, you may notice that all three sequences are identical to each other in this region. The sequences shaded in red are identical in two or three bacterial species, but differ compared to the mammalian small subunit rRNA genes. The sequences from *E. coli* and *Serratia marcescens* are more similar to each other than the sequence from *Bacillus subtilis* is to either of them. This is consistent with the idea that *E. coli* and *S. marcescens* are more closely related evolutionarily than either of them is to *B. subtilis*.

12.6 The Stages of Translation

Like transcription, the process of translation occurs in three stages called initiation, elongation, and termination. **Figure 12.20** provides an overview of the process. During initiation, mRNA, the first tRNA, and ribosomal subunits assemble into a complex. Next, in the elongation stage, the ribosome moves from the start codon in the mRNA toward the stop codon, synthesizing a polypeptide according to the sequence of codons in the mRNA. Finally, the process is terminated when the ribosome reaches a stop codon and the complex disassembles, releasing the completed polypeptide. In this section, we will examine the steps in this process as they occur in living cells.

GATTAAGAGGGACGGCCGGGGGCATTCGTATTGCGCCGCTAGAGGTGAAATTC Human
GATTAAGAGGGACGGCCGGGGGCATTCGTATTGCGCCGCTAGAGGTGAAATTC Mouse
GATTAAGAGGGACGGCCGGGGGCATTCGTATTGCGCCGCTAGAGGTGAAATTC Rat
CAAGCTTGAGTCTCGTAGAGGGGGGTAGAATTCCAGGTGTAGCGGTGAAATGC E. coli
CAAGCTTGAGTCTCGTAGAGGGGGGTAGAATTCCAGGTGTAGCGGTGAAATGC S. marcescens
GAGAGCTTGAGTACAGAAGAAGAGAGTGGAATTCCACGTGTAGCGGTGAAATGC B. subtilis

Figure 12.19 Comparison of small subunit rRNA gene sequences from three eukaryotes and three bacterial species. Note the many similarities (yellow) and differences (green and red) among the sequences.

GENOMES AND PROTEOMES

Providing an evolutionary foundation for our understanding of biology, each Genomes and Proteomes subsection describes modern information regarding the genomic composition of organisms and how this relates to proteomes (their protein composition) and evolution.

END-OF-CHAPTER MATERIALS

The end-of-chapter materials offer students many different opportunities to focus in on key concepts and help them work at improving their knowledge:

Chapter Summary
The Chapter Summary provides the student with an overview of the biological principles and experimental approaches that have been described in the chapter. The summary is organized according to the sections of each chapter, and presents a bulleted list of key concepts.

Test Yourself
These multiple-choice questions are designed to provide students with the sense of how well they understand the material in the chapter. Answers are provided on the ARIS website.

Conceptual Questions
The aim of conceptual questions is to test a student's knowledge of biological principles, such as how a biological mechanism works or the features of a biological process or structure.

Experimental Questions
These questions challenge the student to consider the experiments found in a chapter and to critically evaluate technical procedures and analyze biological data.

Collaborative Questions
Broad in nature, students may benefit by discussing these questions with their peers.

CHAPTER SUMMARY

3.1 The Carbon Atom and the Study of Organic Molecules

- Organic chemistry is the science of studying carbon-containing molecules, which are found in living organisms. Wöhler's work with urea marked the birth of organic chemistry. (Figure 3.1)
- One property of the carbon atom that makes life possible is its ability to form four covalent bonds with other atoms. Carbon can form both polar and nonpolar bonds. The combination of different elements and different types of bonds allows a vast number of organic compounds to be formed from only a few chemical elements. (Figures 3.2, 3.3)
- Organic molecules may occur in various shapes. The structures of molecules determine their functions.

3.2 Classes of Organic Molecules and Macromolecules

- The four major classes of organic molecules are carbohydrates, lipids, proteins, and nucleic acids. Macromolecules are large organic molecules that are composed of many thousands of atoms. Some macromolecules are polymers because they are formed by linking together many smaller molecules called monomers.
- Carbohydrates are composed of carbon, hydrogen, and oxygen atoms. Most cells can break down carbohydrates, releasing energy and storing it in newly created bonds in ATP.
- Carbohydrates include monosaccharides (the simplest sugars), disaccharides, and polysaccharides. The polysaccharides starch (in plant cells) and glycogen (in animal cells) provide an efficient means of storing energy. The plant polysaccharide cellulose serves a support or structural function. (Figures 3.6, 3.7, 3.8)

TEST YOURSELF

1. Molecules that contain the element _____ are considered organic molecules.
   - a. hydrogen
   - b. carbon
   - c. oxygen
   - d. nitrogen
   - e. calcium

2. _____ was the first scientist to synthesize an organic molecule. The organic molecule synthesized was _____.
   - a. Kolbe, urea
   - b. Wöhler, urea
   - c. Wöhler, acetic acid
   - d. Kolbe, acetic acid
   - e. Wöhler, glucose

3. The versatility of carbon to serve as the backbone for a variety of different molecules is due to
   - a. the ability of carbon atoms to form four covalent bonds.
   - b. the fact that carbon usually forms ionic bonds with many different atoms.
   - c. the abundance of carbon in the environment.
   - d. the ability of carbon to form covalent bonds with many different types of atoms.
   - e. both a and d.

CONCEPTUAL QUESTIONS

1. Define isomers.
2. List the four classes of organic molecules and give a function of each.
3. Explain the difference between saturated and unsaturated fatty acids.
4. List the seven characteristics of life and explain a little about each.
5. Give the levels of organization from the simplest to most complex.
6. Discuss the difference between discovery-based science and hypothesis testing.
7. What are the steps in the scientific method, also called hypothesis testing?
8. When conducting an experiment, explain how a control sample and an experimental sample differ from each other.

EXPERIMENTAL QUESTIONS

1. Before the experiments conducted by Anfinsen, what were the common beliefs among scientists about protein folding?
2. Explain the hypothesis tested by Anfinsen.
3. Why did Anfinsen use urea and β-mercaptoethanol in his experiments? Explain the result that was crucial to the discovery that the tertiary structure of a protein is dependent on the primary structure.
4. List the seven characteristics of life and explain a little about each.
5. Give the levels of organization from the simplest to most complex.
6. List the taxonomic groups from most inclusive to least inclusive.
7. Explain how actin filaments are involved in movement.
8. Explain the function of the Golgi apparatus.

COLLABORATIVE QUESTIONS

1. Discuss several types of carbohydrates.
2. Discuss some of the roles that proteins play in organisms. Discuss several differences between plant and animal cells.
3. Discuss the relationship between the nucleus, the rough endoplasmic reticulum, and the Golgi apparatus.
4. Discuss the two categories of transport proteins found in plasma membranes.

www.brookerbiology.com
This website includes answers to the Biological Inquiry questions found in the figure legends and all end-of-chapter questions.

Chapter Summary

Test Yourself

Conceptual Questions

Experimental Questions

Collaborative Questions

CONTENTS

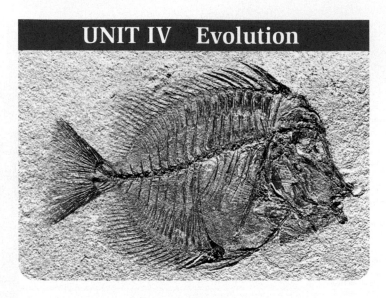

UNIT V Diversity

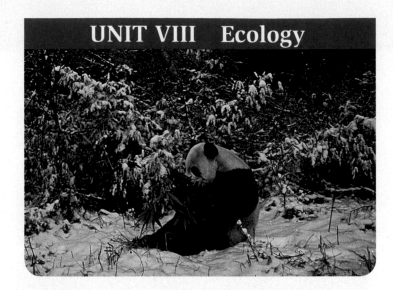

UNIT VIII   Ecology

22

ORIGIN AND HISTORY OF LIFE

CHAPTER OUTLINE

A fossil fish. This approximately 50-million-year-old fossil of a unicorn fish (*Naso rectifrons*) is an example of the many different kinds of organisms that have existed since the origin of life on Earth.

Astronomers now think that the universe began with an explosion called the Big Bang about 13.7 billion years ago, when the first clouds of the elements hydrogen and helium were formed. Gravitational forces collapsed these clouds to create stars that converted hydrogen and helium into heavier elements, including carbon, nitrogen, and oxygen, which are the building blocks of life on Earth. These elements were returned to interstellar space by exploding stars called supernovas, forming clouds in which simple molecules such as water, carbon monoxide, and hydrocarbons were formed. The clouds then collapsed to make a new generation of stars and solar systems.

Our solar system began about 4.6 billion years ago after one or more local supernova explosions. According to one widely accepted scenario, about 500 planetesimals (asteroids and comets) occupied the region where Venus, Earth, and Mars are now found. The Earth, which is estimated to be 4.55 billion years old, grew from the accumulation of planetesimals over a period of 100–200 million years. For the first half billion years or so after its formation, the Earth was too hot to allow water to accumulate on its surface. By 4 billion years ago, the Earth had cooled enough for the outer layers of the planet to solidify and for oceans to form.

The period between 4.0 and 3.5 billion years ago marked the emergence of life on our planet. Though scientists can never be certain how the first primitive life forms arose, plausible hypotheses have emerged from our understanding of modern life. The first forms of life that we know about produced well-preserved microscopic fossils that were found in western Australia. These fossils, which are 3.5 billion years old, resemble photosynthetic bacteria called cyanobacteria that live today (**Figure 22.1**).

The first section of this chapter will survey a variety of hypotheses regarding the potential origins of biological molecules and living cells. Keep in mind that many of these hypotheses are speculative, and are being changed as new information comes to light. Starting 3.5 billion years ago, the formation of fossils (such as the one shown in the chapter opening photo) has provided biologists with a history of life on Earth from its earliest beginnings to the present day. The last section of this chapter surveys a time line for the history of life. This chapter emphasizes when particular forms of life arose. Later chapters in this unit examine the mechanisms by which populations of organisms change over the course of many generations. This process, termed **biological evolution**, involves genetic changes that occur over the course of many generations. Such genetic modifications often lead to dramatic changes in traits and even the formation of new species.

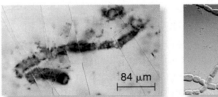

(a) Fossil prokaryote **(b) Modern cyanobacteria**

Figure 22.1 Earliest fossils and living cyanobacteria. **(a)** A fossilized prokaryote about 3.5 billion years old that is thought to be an early cyanobacterium. **(b)** A modern cyanobacterium, which has a similar morphology. Cyanobacteria are unicellular, but they aggregate into chains as shown here.

22.1 Origin of Life on Earth

As we learned in Units I, II, and III, living cells are complex collections of molecules and macromolecules. DNA stores the information for the amino acid sequence of proteins, RNA acts as an intermediary in the process of protein synthesis, and proteins form the foundation for the structure and activities of living cells. Life as we now know it requires this interplay between DNA, RNA, and proteins for its existence and perpetuation. On modern Earth, all living cells are made from pre-existing cells.

But how did life get started? As described in Chapter 1, living organisms have several characteristics that distinguish them from nonliving materials. Because DNA, RNA, and proteins are the central players in the enterprise of life, scientists who are interested in the origin of life have focused much of their attention on the formation of these macromolecules and their building blocks, namely nucleotides and amino acids. To understand the origin of life, we can view the process as occurring in four overlapping stages:

Stage 1: Nucleotides and amino acids were produced prior to the existence of cells.

Stage 2: Nucleotides and amino acids became polymerized to form DNA, RNA, and proteins.

Stage 3: Polymers became enclosed in membranes.

Stage 4: Polymers enclosed in membranes evolved cellular properties.

Obviously we cannot take a time machine back 4 billion years and determine with certainty how these events occurred. Instead, scientists study the existence of modern life, as well as geological processes and fossils, and speculate about the conditions that existed on primitive Earth. This approach has led researchers to a variety of hypotheses regarding the origin of life, none of which can be firmly verified. Nevertheless, certain possibilities are becoming more plausible, and perhaps even compelling. In this section, we will consider a few scientific viewpoints that wrestle with the question, How did life begin?

Stage 1: Several Scientific Hypotheses Have Been Proposed to Explain the Origin of Organic Molecules

Let's begin our inquiry into the first stage of the origin of life by considering how nucleotides and amino acids may have been generated prior to the existence of living cells. In the 1920s, the Russian biochemist Alexander Oparin and the Scottish biologist John Haldane independently proposed that organic molecules such as nucleotides and amino acids arose spontaneously under the conditions that occurred on primitive Earth. According to this hypothesis, the spontaneous appearance of organic molecules produced what they called a "primordial soup," which eventually gave rise to living cells.

The conditions on primitive Earth, which were much different than they are today, may have been more conducive than modern conditions to the spontaneous formation of organic molecules. In particular, scientists think that little oxygen was present, and the atmosphere instead contained inorganic carbon dioxide, nitrogen gas, and water vapor. Current hypotheses suggest that organic molecules, and eventually macromolecules, formed spontaneously; this is termed prebiotic (before life) or abiotic (without life) synthesis. These slowly forming organic molecules accumulated because there was little free oxygen, so they were not spontaneously oxidized, and there were as yet no living organisms, so they were also not metabolized. The slow accumulation of these molecules in the early oceans over a long period of time formed what is now called the **prebiotic soup**. The formation of this medium was a key event that preceded the origin of life.

Though most scientists agree that life originated from the assemblage of nonliving matter on primitive Earth, the mechanism of how and where these molecules originated is widely debated. Many intriguing hypotheses have been proposed. Keep in mind that these hypotheses are not mutually exclusive. Indeed, more than one mechanism may have contributed to the formation of a prebiotic soup. A few of the more widely debated ideas are described next.

Reducing Atmosphere Hypothesis Based largely on geological data, many scientists in the 1950s thought that the atmosphere on primitive Earth was rich in water vapor (H_2O), hydrogen gas (H_2), methane (CH_4), and ammonia (NH_3). These components, along with a lack of atmospheric oxygen, produce a reducing atmosphere because methane and ammonia readily give up electrons and thereby reduce other molecules. As described in Chapters 7 and 8, such oxidation-reduction reactions are required for the formation of complex organic molecules from simple inorganic molecules.

In 1953, Stanley Miller, a student in the laboratory of Harold Urey, was the first scientist to use experimentation to test whether the prebiotic synthesis of organic molecules is possible. His experimental apparatus was intended to simulate the conditions on primitive Earth that were postulated in the 1950s (**Figure 22.2**). Water vapor from a flask of boiling water rose into another chamber containing H_2, CH_4, and NH_3. Miller inserted two electrodes that sent electrical discharges into the chamber to simulate lightning bolts. A condenser jacket cooled some of the gases from the chamber, causing droplets to form that dropped into a trap. He then took samples from this trap for chemical analysis. In his first experiments, he observed the formation of hydrogen cyanide (HCN) and formaldehyde (CH_2O). Such molecules are precursors of more complex organic molecules. These precursors also combined to make larger molecules such as the amino acid glycine. Later experiments by Miller and others demonstrated the formation of sugars, many types of amino acids, and nitrogenous bases found in nucleic acids (for example, adenine).

The studies of Miller and Urey were particularly important because they were the first attempt to apply scientific experimentation to our quest to understand the origin of life. Their pioneering strategy was to determine if a scientific hypothesis is plausible, although it cannot prove that an event in the past really happened that way. In spite of the importance of these studies, critics of the reducing atmosphere hypothesis have argued that Miller and Urey were wrong about the composition of primitive Earth's environment.

Since the 1950s, ideas about the atmosphere on early Earth have changed. More recently, many scientists have suggested that the atmosphere on primitive Earth was not reducing, but instead was a neutral environment composed mostly of carbon monoxide (CO), carbon dioxide (CO_2), nitrogen gas (N_2), and H_2O. These newer ideas are derived from studies of volcanic gas, which has much more CO_2 and N_2 than CH_4 and NH_3, and from the observation that UV radiation destroys CH_4 and NH_3, so that these molecules would have been short-lived on primitive Earth. Nevertheless, since the experiments of Urey, many newer investigations have shown that organic molecules can be made under a variety of conditions. Using different combinations of gases based on these corrected assumptions of atmospheric conditions on primitive Earth, researchers have still achieved similar results. Namely, organic molecules can be made prebiotically from a neutral environment composed primarily of CO, CO_2, N_2, and H_2O.

At this time, the idea that organic molecules were made in the atmosphere of primitive Earth is plausible. Even so, other scientists have proposed alternative mechanisms for the production of organic molecules before the emergence of life.

Extraterrestrial Hypothesis Many scientists have argued that sufficient organic carbon would have been present in the asteroids and comets that reached the surface of early Earth to make them instrumental in stocking the prebiotic soup. Modern evidence in support of this idea comes from the study of these fallen bodies, called meteorites. A significant proportion of meteorites belong to a class known as carbonaceous chondrites. Such meteorites contain a substantial amount of organic carbon, including amino acids and nucleic acid bases. Based on this observation, scientists have postulated that such meteorites could have transported a significant amount of organic molecules to primitive Earth.

However, opponents of this hypothesis argue that most of this material would have been destroyed by the intense heating that accompanies the passage of large bodies through the atmosphere and their subsequent collision with the surface of the Earth. The degree to which heat would have destroyed the organic molecules remains a matter of controversy.

Deep-Sea Vent Hypothesis In 1988, the German organic chemist Günter Wächtershäuser proposed that key organic molecules may have originated in deep-sea vents, which are cracks in the Earth's surface where superheated water rich in metal ions and hydrogen sulfide (H_2S) mixes abruptly with cold sea water. These vents release hot gaseous substances from the interior of the Earth at temperatures in excess of 300°C (572°F). Supporters of this hypothesis propose that biologically important molecules may have been formed in the temperature gradient between the extremely hot vent water and the cold water that surrounds the vent at the bottom of the ocean (**Figure 22.3a**).

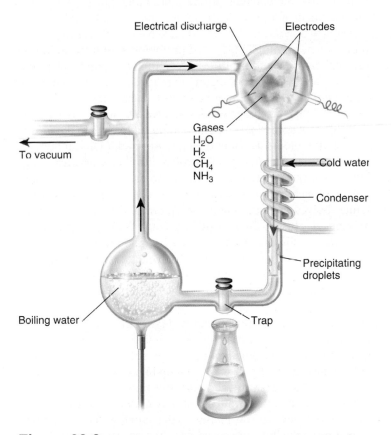

Figure 22.2 Testing the reducing atmosphere hypothesis for the origin of life—the Miller and Urey experiment.

Biological inquiry: With regard to the origin of life, why are biologists interested in the abiotic synthesis of organic molecules?

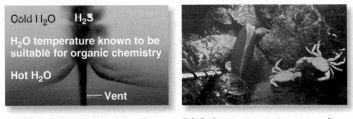

(a) Deep-sea vent hypothesis   **(b) A deep-sea vent community**

Figure 22.3 The deep-sea vent hypothesis for the origin of life. (a) Deep-sea vents are cracks in the Earth's surface that release hot gases such as hydrogen sulfide (H_2S). This heats the water near the vent and creates a gradient between the very hot water adjacent to the vent and the cold water that is farther away from the vent. The synthesis of organic molecules can occur in this gradient. (b) Photograph of a biological community near a deep-sea vent consisting of giant tube worms, clams, and crabs.

Experimentally, the temperatures within this gradient are known to be suitable for the synthesis of molecules that form components of biological molecules. For example, the reaction between iron and H_2S yields pyrites and H_2, and has been shown to provide the energy necessary for the reduction of N_2 to NH_3. Nitrogen is an essential ingredient of the molecular building blocks of life, amino acids and nucleic acids. But N_2, which is found abundantly on Earth, is chemically inert, so it is unlikely to have given rise to life. Most scientists believe instead that NH_3 was required to help life get started.

Interestingly, complex biological communities are found in the vicinity of modern deep sea vents. Various types of fish, worms, crabs, clams, shrimps, and bacteria are found in significant abundance in those areas (**Figure 22.3b**). Unlike most other forms of life on our planet, these organisms receive their energy from chemicals in the vent and not from the sun.

Stage 2: Organic Polymers May Have Formed on the Surface of Clay

The preceding three hypotheses provide reasonable mechanisms whereby small organic molecules could have accumulated on primitive Earth. Scientists hypothesize that the second stage in the origin of life was a period in which simple organic molecules polymerized to form more complex organic polymers such as DNA, RNA, or proteins. Most ideas regarding the origin of life assume that polymers with lengths of at least 30–60 monomers are needed to store enough information to make a viable genetic system. Experimentally, the prebiotic synthesis of such polymers is not possible in aqueous solutions because hydrolysis competes with polymerization. For this reason, many scientists have speculated that the synthesis of polymers did not occur in a prebiotic soup, but instead took place on a solid surface or in evaporating tidal pools.

In 1951, John Bernal first suggested that the prebiotic synthesis of polymers took place on clay. In his book *The Physical Basis of Life* he wrote that "clays, muds and inorganic crystals are powerful means to concentrate and polymerize organic molecules." Many clay minerals are known to bind organic molecules such as nucleotides and amino acids. In addition, negative charges within the clay itself attract metal divalent cations, such as Mg^{2+}, that can catalyze the chemical reactions that produce polymers.

Experimentally, many research groups have demonstrated the formation of nucleic acid polymers and polypeptides on the surface of clay, given the presence of monomer building blocks. During the prebiotic synthesis of RNA, the purine bases of the nucleotides interact with the silicate surfaces of the clay. Divalent cations, such as Mg^{2+}, bind the nucleotides to the negative surfaces of the clay, thereby positioning the nucleotides in a way that promotes bond formation between the phosphate of one nucleotide and the ribose sugar of an adjacent nucleotide. In this way, polymers such as RNA may have been formed.

Stage 3: Cell-Like Structures May Have Originated When Polymers Were Enclosed by a Boundary

The third stage in the origin of living cells is believed to be the formation of a boundary that separated the environment from internal polymers, such as RNA. The term **protobiont** (or prebiont) is used to describe the first nonliving structures that evolved into living cells. Protobionts had four characteristics that put them on the path to living cells:

1. A boundary, such as a membrane, separated the external environment from the internal contents of the protobiont.
2. Polymers inside the protobiont contained information.
3. Polymers inside the protobiont had enzymatic functions.
4. The protobionts were capable of self-replication.

Scientists envision protobionts as aggregates of prebiotically produced molecules and macromolecules that acquired a boundary, such as lipid bilayer, that allowed them to maintain an internal chemical environment distinct from that of their surroundings. Protobionts were not capable of precise self-reproduction like living cells, but probably could divide to increase in number. Such protobionts are thought to have exhibited basic metabolic pathways in which the structures of organic molecules were changed. In particular, the polymers inside protobionts must have gained the enzymatic ability to link together organic building blocks to create new polymers. This would have been a critical step in the process that eventually provided protobionts with the ability to self-replicate. According to this scenario, the ability of protobionts to self-replicate became more refined over time, and metabolic pathways became more complex. Eventually, these structures exhibited the characteristics that we attribute to living cells.

Different scenarios have been proposed to explain the formation of protobionts. Russian biologist Aleksandr Oparin hypothesized in 1924 that living cells evolved from **coacervates**, droplets that form spontaneously from the association of charged polymers such as proteins, carbohydrates, or nucleic acids. Their name derives from the Latin *coacervare*, meaning to assemble together or cluster. Coacervates measure 1–100 μm (micrometers) across, possess osmotic properties, and are surrounded by a tight skin of water molecules (**Figure 22.4a**). This boundary allows the selective absorption of simple molecules from the surrounding medium.

If enzymes are trapped within coacervates, they can perform primitive metabolic functions (**Figure 22.4b**). For example, researchers have made coacervates containing the enzyme glycogen phosphorylase. When glucose-1-phosphate was made available to the coacervates, it was taken up into them and starch was produced. The starch merged with the wall of the coacervates, which increased in size and eventually divided into two. When the enzyme amylase was included, the starch was broken down to maltose, which was released from the coacervates.

An alternative pathway to protobionts is the formation of **microspheres**, which are small water-filled vesicles surrounded by a macromolecular boundary (**Figure 22.4c**). If hot solutions

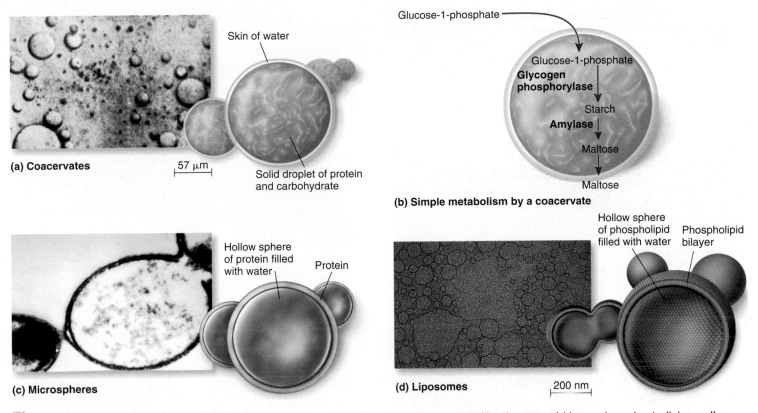

Figure 22.4 **Protobionts and their lifelike functions.**   Primitive cell-like structures like these could have given rise to living cells. **(a)** This micrograph shows coacervates made by Oparin from a mixture of gelatine (composed primarily of protein) and gum arabic (composed of protein and carbohydrate). **(b)** The illustration shows simple metabolism that can be performed by coacervates. **(c)** Micrograph and illustration of microspheres, which are water-filled spheres of macromolecules such as protein. **(d)** An electron micrograph and illustration of liposomes. Each liposome is made of a phospholipid bilayer surrounding an aqueous compartment.

of proteins are cooled under the correct conditions, they produce microspheres about 2 μm in diameter that are hollow and have an outer layer of protein.

Similar experiments have been done with lipids. When certain types of lipids are dissolved in water, they spontaneously form **liposomes**, which are vesicles surrounded by a lipid bilayer (**Figure 22.4d**). Interestingly, in 2003, Martin Hanczyc, Shelly Fujikawa, and Jack Szostak showed that clay can catalyze the formation of liposomes that grow and divide, a primitive form of self-replication. Furthermore, if RNA was on the surface of the clay, the researchers discovered that liposomes were formed that enclosed RNA. These experiments are exciting because they showed that the formation of membrane vesicles containing RNA molecules is a plausible route to the first living cells based on simple physical and chemical forces.

Stage 4: Cellular Characteristics May Have Evolved via Chemical Selection, Beginning with an RNA World

The majority of scientists favor RNA as the first macromolecule that was found in protobionts. Unlike other polymers, RNA exhibits three key functions. First, RNA has the ability to store

information in its nucleotide sequence. Second, due to base pairing, its nucleotide sequence has the capacity for replication. And third, RNA can perform a variety of enzymatic functions. The results of many experiments have shown that RNA molecules can function as **ribozymes**, acting as enzymes to synthesize the macromolecules found in living cells. By comparison, DNA and proteins are not known to have all three attributes. DNA is not known to have enzymatic activity, and proteins are not known to undergo self-replication. Thus, RNA appears to be the most self-sufficient substance of living matter. RNA can perform functions that are characteristic of proteins, and at the same time can serve as genetic material with replicative and informational functions.

But how did the RNA molecules that were first made prebiotically evolve into more complex molecules that produced cell-like characteristics? Researchers propose that a process called chemical selection was responsible for the increased complexity. **Chemical selection** occurs when a chemical within a mixture has special properties or advantages that cause it to increase in number compared to other chemicals in the mixture. Initially, scientists speculate that the special properties that enabled certain RNA molecules to undergo chemical selection were its ability to self-replicate and to perform other enzymatic functions.

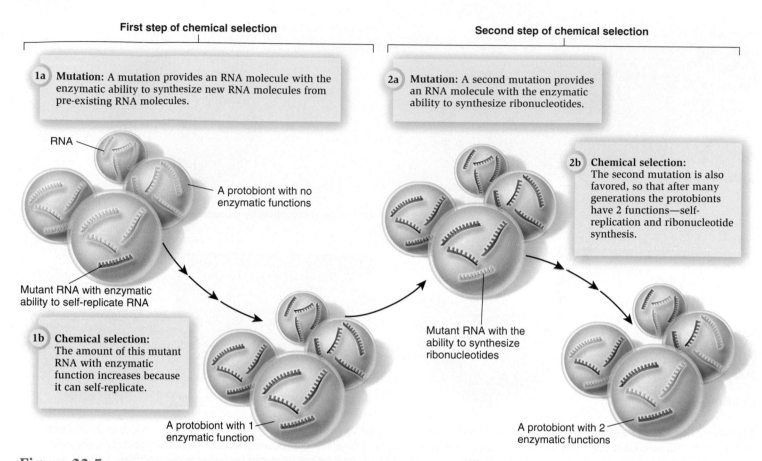

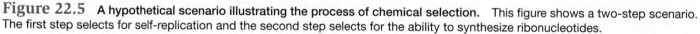

Figure 22.5 **A hypothetical scenario illustrating the process of chemical selection.** This figure shows a two-step scenario. The first step selects for self-replication and the second step selects for the ability to synthesize ribonucleotides.

Biological inquiry: What is meant by the term chemical selection?

Because cells are a complex mixture of different macromolecules that carry out many enzymatic functions, researchers cannot establish with certainty the specific sequence of steps that produced modern cells. However, as a way to understand the concept of chemical selection, let's consider a hypothetical scenario showing two steps of chemical selection. **Figure 22.5** shows a group of protobionts that contain RNA molecules that were made prebiotically. RNA molecules inside these protobionts can be used as templates for the prebiotic synthesis of complementary RNA molecules. Such a process of self-replication, however, would be very slow because it would not be catalyzed by enzymes in the protobiont. In a first step of chemical selection, the sequence of one of the RNA molecules has undergone a mutation that gives it the enzymatic ability to attach nucleotides together, using RNA molecules as a template. This protobiont would have an advantage over the others because it would be capable of faster self-replication of its RNA molecules. Over time, due to its enhanced rate of replication, this type of protobiont would increase in number compared to the others. Eventually, the group of protobionts shown in the figure contains only this type of enzymatically functional RNA.

A second step of chemical selection is also shown in Figure 22.5 (right side). A second mutation in an RNA molecule could produce the enzymatic function that would promote the synthesis of ribonucleotides, the building blocks of RNA. This protobiont would have the advantage of not having to rely on the prebiotic synthesis of ribonucleotides, which also is a very slow process. Therefore, the protobiont having the ability to both self-replicate its RNA molecules and synthesize ribonucleotides would have an advantage over a protobiont that could only self-replicate. Over time, the faster rate of ribonucleotide synthesis and self-replication would cause an increase in the numbers of the protobionts with both functions.

The **RNA world** is a hypothetical period on primitive Earth when both the information needed for life and the enzymatic activity of living cells were contained solely in RNA molecules. In this scenario, lipid membranes enclosing RNA exhibited the properties of life due to RNA genomes that were copied and maintained through the catalytic function of RNA molecules. Over time, scientists envision that mutations occurred in these RNA molecules, occasionally presenting new functional possibilities. Chemical selection for these new functions would have eventually produced an increase in complexity in these cells, with RNA molecules accruing activities such as peptide bond formation and other enzymatic functions.

But is an RNA world a plausible scenario? As described next in the Feature Investigation, chemical selection of RNA molecules can occur experimentally.

FEATURE INVESTIGATION

Bartel and Szostak Demonstrated Chemical Selection in the Laboratory

Remarkably, scientists have been able to perform experiments in the laboratory that can select for RNA molecules with a particular function. The first such study by David Bartel and Jack Szostak was conducted in 1993 (**Figure 22.6**). Using molecular techniques, they synthesized a mixture of 10^{15} RNA molecules that we will call the long RNA molecules. Each long RNA in this mixture contained two regions. The first region at the 5′ end was a constant region that formed a stem-loop structure. Its sequence was identical among all 10^{15} molecules. The constant region was next to a second region that was 220 nucleotides in length. A key feature of the second region is that its sequence was variable among the 10^{15} molecules. The researchers hypothesized that this variation could occasionally result in a long RNA molecule with the enzymatic ability to catalyze a phosphoester bond, which is a covalent bond between two adjacent nucleotides.

They also made another type of RNA molecule, which we will call the short RNA, with two important properties. First, the short RNA had a region that was complementary to a site near the constant region of the long RNA molecules. Second, the short RNA had a tag sequence that caused it to bind tightly to column packing material referred to as beads.

To begin this experiment, the researchers incubated a large number of the short and long RNA molecules together. During this incubation period, long and short RNA molecules would hydrogen bond to each other due to their complementary regions. Although hydrogen bonding is not permanent, this step allowed the long and short RNAs to recognize each other for a short period of time. The researchers reasoned that a long RNA with the enzymatic ability to form covalent phosphoester bonds may make this interaction more permanent by catalyzing a phosphoester bond between the long and short RNA molecules. Following this incubation, the mixture of RNAs was passed through a column with beads that specifically bound the short RNA. The aim of this approach was to select for longer RNA molecules that had covalently bonded to the short RNA molecule (see Conceptual Level of Figure 22.6).

The vast majority of RNAs would not have the enzymatic ability to catalyze a phosphoester bond. These would pass out of the column at step 2, because hydrogen bonding between the long and short RNAs is not sufficient to hold them together for very long. Such long RNAs would be discarded. Long RNAs with the ability to catalyze a phosphoester bond to the short RNA would remain bound to the column beads at step 2. These enzymatic RNAs were then flushed out at step 3 to generate a mixture of RNAs termed pool #1. The researchers expected this pool to contain several different long RNA molecules with varying abilities to catalyze a phosphoester bond.

To further the chemical selection process, the scientists used the first pool of long RNA molecules flushed out at step 3 to make more long RNA molecules. This was accomplished via PCR. This next batch also had the constant and variable regions.

Figure 22.6 Bartel and Szostak demonstrated chemical selection for RNA molecules that can catalyze phosphoester bond formation.

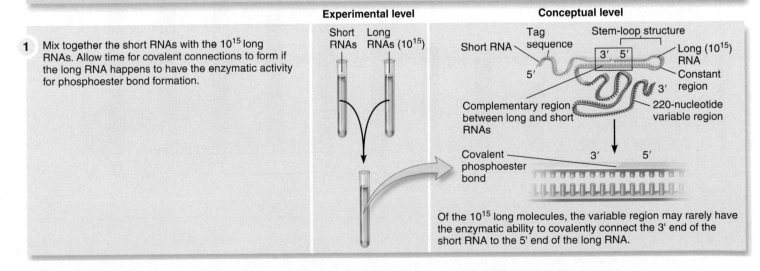

HYPOTHESIS Among a large pool of RNA molecules, some of them may contain the enzymatic ability to catalyze a phosphoester bond; these can be selected for in the laboratory.

STARTING MATERIALS Many copies of short RNA were synthesized that had a tag sequence that binds tightly to column packing material called beads. Also, a population of 10^{15} long RNA molecules was made that contained a constant region with a stem-loop structure and a 220-nucleotide variable region. Note: The variable regions of the long RNAs were made using a PCR step that caused mutations in this region.

Experimental level   **Conceptual level**

1   Mix together the short RNAs with the 10^{15} long RNAs. Allow time for covalent connections to form if the long RNA happens to have the enzymatic activity for phosphoester bond formation.

Short RNAs   Long RNAs (10^{15})

Tag sequence   Stem-loop structure
Short RNA   3′ 5′   Long (10^{15}) RNA
5′   Constant region
Complementary region between long and short RNAs   3′   220-nucleotide variable region
Covalent phosphoester bond   3′   5′

Of the 10^{15} long molecules, the variable region may rarely have the enzymatic ability to covalently connect the 3′ end of the short RNA to the 5′ end of the long RNA.

2 Pass the mixture through a column of beads that binds the tag sequence found on the short RNA. Add additional liquid to flush out long RNAs that are not covalently attached to short RNAs.

Column

Column bead

Tag sequence

Tag sequences promote the binding of the short RNA to the column beads. Long RNAs that are covalently attached to a short RNA will also be bound.

Discard these long RNAs.

This long RNA does not bind to the beads because the variable region does not possess the enzymatic ability to covalently attach to short RNA.

3 Add a low pH solution to prevent the tag sequence from binding to the beads. This causes the tightly bound HNAs to be flushed out of the column.

Low pH wash

4 The flushed-out RNAs are termed pool #1. Use pool #1 to make a second batch of long RNA molecules. This involved a PCR step using reverse transcriptase to make cDNA. The PCR primers recognized the beginning and end of the long RNA sequence and copied only this region. The cDNA was then used as a template to make long RNA via RNA polymerase.

Refer back to Figure 20.08 for a description of PCR.

Pool #1

5 Repeat procedure to generate 10 consecutive pools of RNA molecules.

Pool #1 Pool #2 Pool #3 Pool #4 Pool #5 Pool #6 Pool #7 Pool #8 Pool #9 Pool #10

6 Test a sample of the original population and each of the 10 pools for the enzymatic ability to catalyze a phosphoester bond.

Gap

Covalent bond

7 THE DATA

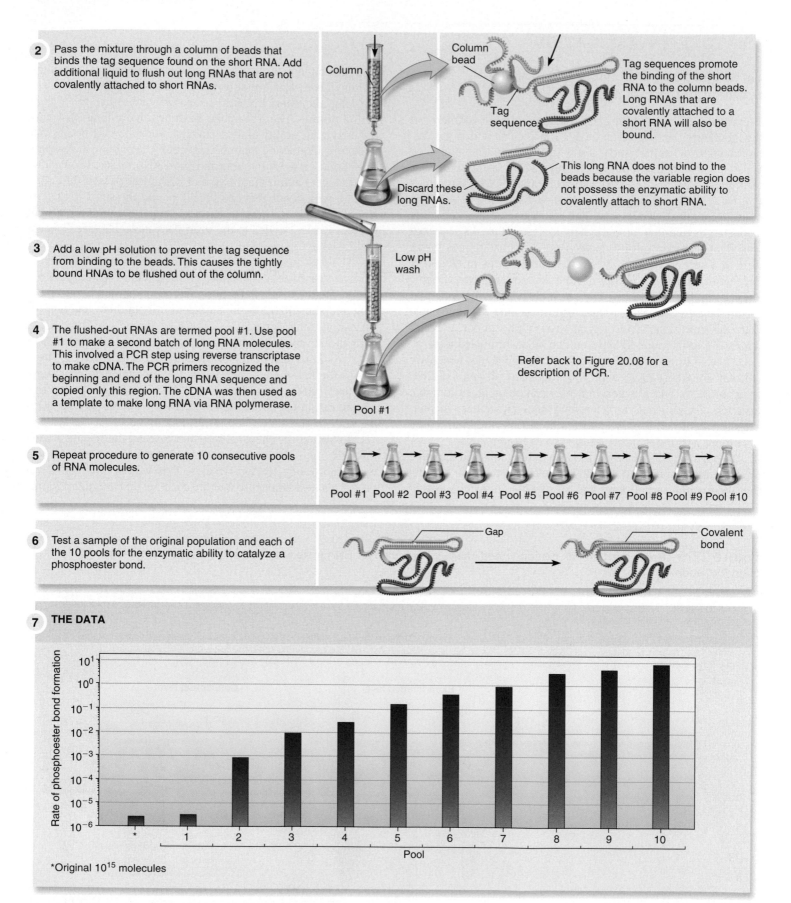

Rate of phosphoester bond formation

10^1
10^0
10^{-1}
10^{-2}
10^{-3}
10^{-4}
10^{-5}
10^{-6}

*   1   2   3   4   5   6   7   8   9   10

Pool

*Original 10^{15} molecules

However, the variable regions were derived from the variable regions of pool #1 RNA molecules that had enzymatic activity. The researchers reasoned that additional variation might occasionally produce an RNA molecule with improved enzymatic activity. This second batch of long RNA molecules was subjected to the same steps as was the first batch of 10^{15} molecules. In this case, the group of long molecules flushed out at step 3 was termed pool #2. This protocol was followed eight more times to generate 10 consecutive pools of RNA molecules. The researchers then analyzed the original random collection of 10^{15} RNA molecules and each of the 10 pools for the enzymatic ability to catalyze a phosphoester bond. As seen in the data, each successive pool became enriched for molecules with higher enzymatic activity. Pool #10 showed enzymatic activity that was approximately 3 million times higher than the original random pool of molecules!

Much like the work of Miller and Urey, these experiments demonstrated the feasibility of another phase of the prebiotic evolutionary process, in this case evolution via chemical selection. The results showed that chemical selection improves the functional characteristics of a group of RNA molecules over time by increasing the proportions of those molecules with enhanced function.

The RNA World Was Superseded by the Modern DNA/RNA/Protein World

Assuming that an RNA world was the origin of life, researchers have asked the question, Why and how did the RNA world evolve into the DNA/RNA/protein world we see today? Many potential pathways to a DNA/RNA/protein world are possible. Some researchers have argued that proteins evolved in parallel with RNA molecules. However, many proponents of the RNA world hypothesis suggest that DNA and proteins evolved later. If so, the RNA world may have been superseded by a DNA/RNA world or an RNA/protein world before the emergence of the modern DNA/RNA/protein world. Let's now consider the advantages of a DNA/RNA/protein world as opposed to the simpler RNA world, and how this modern biological world came into being.

Information Storage As we discussed previously, RNA can store information in its base sequence. So why did DNA take over that function, as is the case in modern cells? During the RNA world, RNA had to perform two roles, informational and catalytic. Scientists have speculated that the incorporation of DNA into cells would have relieved RNA of its informational role and thereby allowed RNA to perform a greater variety of other functions. For example, if DNA stored the information for the synthesis of RNA molecules, such RNA molecules could bind cofactors, have modified bases, or bind peptides that might enhance their catalytic function. Cells with both DNA and RNA would have had an advantage over those with just RNA, and so they would have been selected. Another advantage of DNA is stability. Compared to RNA, DNA is less likely to suffer mutations.

A second issue is, How did the DNA world come into being? Scientists have proposed that an ancestral RNA molecule had the ability to make DNA using RNA as a template. This function, known as reverse transcriptase activity, is described in Chapter 18. Interestingly, modern eukaryotic cells can use RNA as a template to make DNA. For example, telomerase, which is described in Chapter 11, copies the ends of chromosomes using an RNA template.

Metabolism and Other Cellular Functions Now let's consider the origin of proteins. The emergence of proteins as catalytic entities may have been a great advantage to early cells. Due to the many different chemistries of the 20 amino acids, proteins have vastly greater catalytic potential and efficiency than do RNA molecules, again providing a major advantage to cells that had both proteins and RNA. In modern cells, proteins have taken over most but not all catalytic functions. In addition, proteins can perform other important tasks. For example, cytoskeletal proteins carry out structural roles, and certain membrane proteins are responsible for the uptake of substances into living cells.

The answer to the question of how proteins came into being after an RNA world seems to be rooted in RNA function. Chemical selection experiments have shown that RNA molecules can catalyze the formation of peptide bonds and even attach amino acids to primitive tRNA molecules. Interestingly, modern protein synthesis still involves a central role for RNA. First, mRNA provides the information for a polypeptide sequence. Second, tRNA molecules act as adaptors for the formation of a polypeptide chain. And finally, ribosomes containing rRNA provide an arena for polypeptide synthesis. Furthermore, RNA within the ribosome acts as a ribozyme to catalyze peptide bond formation. Taken together, the analysis of translation in modern cells is consistent with an evolutionary history in which RNA molecules were instrumental in the emergence and formation of proteins.

22.2 History of Life on Earth

Thus far, we have considered how the first primitive cells may have come into existence. Recall that the first fossils of single-celled organisms were preserved approximately 3.5 billion years ago (see Figure 22.1). In this section, we will examine some of the major changes in life that have occurred since that time. As you will learn, the period from 3.5 billion years ago to the present has seen dramatic changes in the composition of life on Earth.

We will begin with a brief description of the geological changes that occurred on Earth that have impacted the emergence of new forms of life. Then we will examine how fossils are formed, and how they provide a fascinating yet incomplete journey through the history of life.

Many Environmental and Biological Changes Have Occurred Since the Origin of the Earth

The **geological timescale** is a time line of the Earth's history from its origin about 4.55 billion years ago to the present. This time line is subdivided into four eons, and then further subdivided into many eras. The first three eons are collectively known as the Precambrian. **Figure 22.7** provides the geological timescale and describes some of the major events that occurred during the history of life. The names of several eons and eras end in *-zoic* (meaning animal life) because we often recognize these time intervals on the basis of animal life. We will examine these time periods later in this chapter.

The changes that have occurred in living organisms over the past 4 billion years are the result of two interactive processes. First, as discussed in the next several chapters, genetic changes in organisms can affect their characteristics. Such changes often have an important impact on the ability of organisms to survive in their native environment. Second, the environment on Earth has undergone dramatic changes. Such environmental changes have profoundly influenced the types of organisms that have existed during different periods of time. As we will examine later, environmental influences can be both positive and negative. In some cases, a change can allow new types of organisms to come into being. Alternatively, environmental changes can cause the **extinction** of a species or group of species. A recurring pattern seen in the history of life is the emergence of new species and the extinction of other species. In some cases, biological changes and extinctions are correlated with major environmental changes, which include the following:

- *Climate/Temperature:* During the first 2.5 billion years of its existence, the surface of the Earth gradually cooled. However, during the last 2 billion years, the Earth has undergone major fluctuations in temperature, producing Ice Ages that alternate with warmer periods. Furthermore, the temperature on Earth is not uniform, which produces many different environments where the temperatures are quite different, such as a tropical rain forest and the North Pole.

- *Atmosphere:* The chemical composition of the gases surrounding the Earth has changed substantially over the past 4 billion years. One notable change involves oxygen. The emergence of organisms that were capable of photosynthesis added oxygen to the atmosphere. Prior to 2.5 billion years ago, relatively little oxygen was in the atmosphere. Levels of oxygen in the form of O_2 began to rise significantly at about this time. Our current atmosphere contains about 21% O_2.

- *Landmasses:* As the Earth cooled, landmasses formed that were surrounded by bodies of water. This created two different environments, terrestrial and aquatic. Furthermore, over the course of billions of years, the major landmasses, known as the continents, have shifted their positions, changed their shapes, and in some cases have become separated from each other. This phenomenon, called **continental drift**, is shown in **Figure 22.8**.

- *Floods:* Catastrophic floods have periodically had major impacts on the organisms in the flooded regions, sometimes causing extinction.

- *Glaciation:* On a periodic basis, glaciers have moved across continents and altered the composition of species on those landmasses. Glaciation also affects the water level of oceans.

- *Volcanic eruptions:* The eruptions of volcanoes can negatively impact the species in the vicinity of the eruption, sometimes causing extinctions. In addition, volcanic eruptions in the ocean can lead to the formation of new islands. Massive eruptions may also spew so much debris into the atmosphere that they can affect global temperatures and limit solar radiation, which limits photosynthetic production.

- *Meteorite impacts:* During its long history, the Earth has been struck by many meteorites. Large meteorites have had substantial impacts on the Earth's environment.

The effects of one or more of these changes have sometimes caused many species to go extinct at the same time. Such events are called **mass extinctions**. Five large mass extinctions occurred near the end of the Ordovician, Devonian, Permian, Triassic, and Cretaceous periods. Geological time periods are often based on the occurrences of these mass extinctions. The rapid extinction of many modern species due to human activities is sometimes referred to as the sixth mass extinction.

The study of the history of life involves an analysis of both environmental (geological) and biological changes that have occurred over the past 4 billion years. A key observation that provides us with a window into our biological past is the identification and examination of fossils, which are described next.

Fossils Provide a Glimpse into the History of Life

Fossils are the recognizable remains of past life on Earth. They can take many forms, including bones, shells, and leaves, the impression of cells, or other evidence, such as tracks or burrows. Scientists who study fossils are called **paleontologists** (*paleo* means ancient). The fossil record provides a record of the life forms that existed during particular geological periods. For example, rocks formed during the Proterozoic eon may have fossils of relatively simple organisms, such as bacteria, algae, and wormlike animals. By comparison, fossils formed during the Phanerozoic eon include complex animals and plants such as mammals and trees. Though the remains of dinosaurs often spark a childhood interest in fossils, the great bulk of the fossil record is dominated by the remnants of animals with shells and by the microscopic remains of plants and animals. These are the fossils studied by most paleontologists.

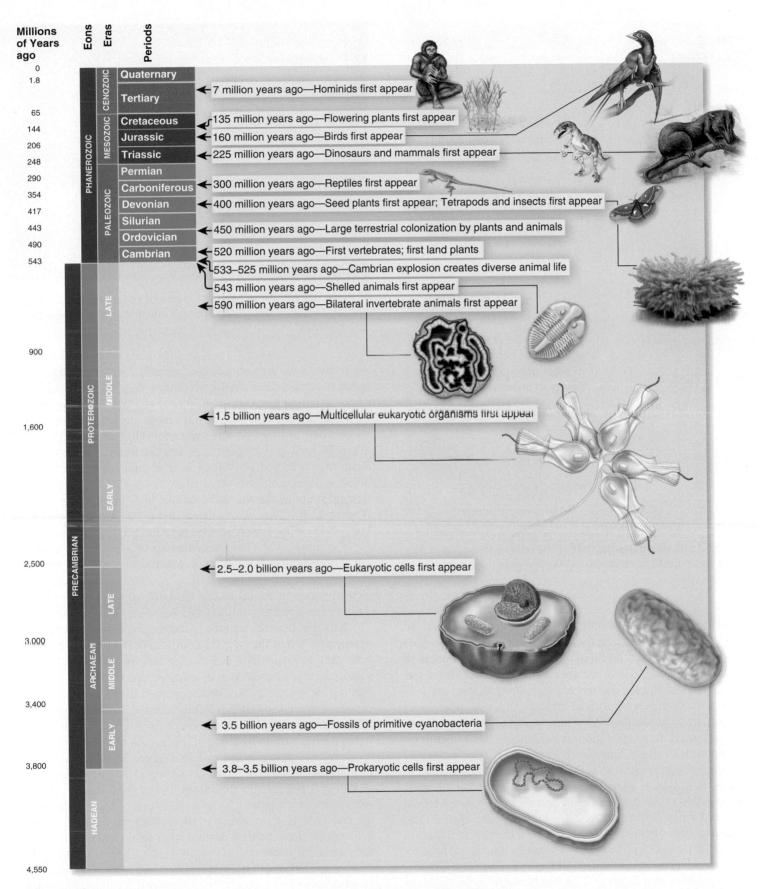

Figure 22.7 The geological timescale and an overview of the history of life on Earth.

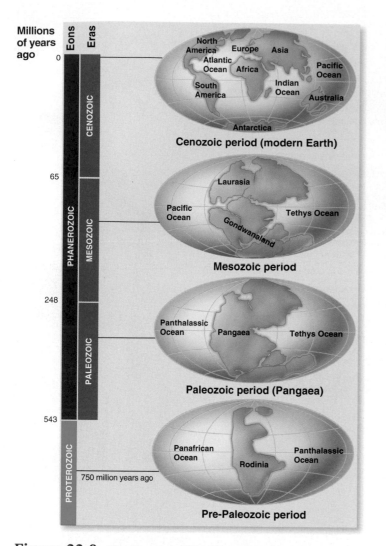

Figure 22.8 Continental drift. The relative locations of the continents on Earth have changed over the past hundreds of millions of years.

Figure 22.9 An example of layers of sedimentary rock that contain fossils. This photo shows a site that contains several layers of rock.

Because our understanding of the history of life is derived primarily from the fossil record, it is important to appreciate how fossils are formed and dated, and to understand some of the inherent biases in fossil analyses. Let's begin with fossil formation. Many of the rocks observed by paleontologists are sedimentary rocks that were formed from particles of older rocks broken apart by water or wind. These particles, such as gravel, sand, and mud, may settle and bury living and dead organisms at the bottoms of rivers, lakes, and oceans. Over time, more particles pile up, and sediments at the bottom of the pile eventually become rock. Gravel particles form rock called conglomerate, sand becomes sandstone, and mud becomes shale. Most fossils are formed when organisms are buried quickly, and then during the process of sedimentary rock formation, their hard parts are gradually replaced over millions of years by minerals, producing a recognizable representation of the original organism.

Information that helps to date fossils began to be collected in the mid-1600s, when Danish scientist Nicholas Steno studied the relative positions of sedimentary rocks. He noted that solid particles settle from a fluid according to their relative weight or size. The largest or heaviest settle first, and the smallest or lightest settle last. Because sedimentary rocks are formed particle by particle and bed by bed, the layers are piled one on top of the other. Thus, in any sequence of layered rocks, a given bed is older than any bed on top of it, and younger than any bed below it. Paleontologists often study changes in life-forms over time by studying the fossils in various beds from bottom to top (**Figure 22.9**). The more ancient life-forms are found in the lower beds, while newer species are found in the upper beds.

A common way to estimate the age of a fossil is by analyzing the elemental isotopes within the accompanying rock, a process called **radioisotope dating**. As discussed in Chapter 2, elements may be found in two or more forms called isotopes. Radioactive isotopes are unstable and decay at a specific rate. The **half-life** of a radioisotope is the length of time required for exactly one-half of the original isotope to decay (**Figure 22.10a**). Each radioactive isotope has its own unique half-life. Within a sample of rock, scientists can measure the amount of a given radioactive isotope as well as the amount of the isotope that is produced when the original isotope decays. For dating geological materials, several types of isotope decay patterns are particularly useful: carbon to nitrogen, potassium to argon, rubidium to strontium, and uranium to lead (**Figure 22.10b**). Except for recent fossils in which carbon-14 dating can be employed, fossil dating is not usually conducted on the fossil itself or on the sedimentary rock in which the fossil is found. Most commonly, igneous rock in the vicinity of the sedimentary rock is dated. For example, igneous rock derived from an ancient lava flow initially contains uranium-235 but no lead-207. By comparing the relative proportions of uranium-235 and lead-207, the age of the igneous rock can be accurately determined. Even so, paleontologists expect the fossil record to underestimate the actual date

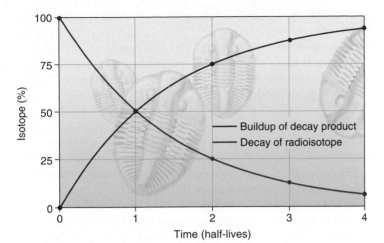

(a) Decay of a radioisotope

Radioisotope	Decay product	Half-life (years)	Useful dating range (years)
Carbon-14	Nitrogen-14	5,730	100–30,000
Potassium-40	Argon-40	1.3 billion	100,000–4.5 billion
Rubidium-87	Strontium-87	47 billion	10 million–4.5 billion
Uranium-235	Lead-207	710 million	10 million–4.5 billion
Uranium-238	Lead-206	4.5 billion	10 million–4.5 billion

(b) Radioisotopes that are useful for geological dating

Figure 22.10 Radioisotope dating of fossils. (a) Rocks can be dated by measuring the relative amounts of a radioisotope and its decay product that they contain. (b) These five isotopes are particularly useful for the dating of fossils.

Table 22.1	Biases That Occur in the Fossil Record
Factor	**Description**
Anatomy	Organisms with hard body parts, such as animals with a thick shell, are more likely to be preserved than are organisms composed of soft tissues.
Size	The remains of larger organisms are more likely than those of smaller organisms to be found as fossils.
Number	Species that existed in greater numbers are more likely to be preserved within the fossil record than are those that existed in smaller numbers.
Environment	Inland species are less likely to become fossilized than are those that lived in a marine environment or near the edge of water.
Geology	Due to the chemistry of fossilization, certain organisms are more likely to be preserved than are other organisms. In addition, parts of two or more different species may be subjected to mixing prior to fossilization. Such an event may distort the fossil record.
Time	Organisms that lived relatively recently are more likely to be found as fossils than are organisms that lived very long ago.
Paleontology	Certain types of fossils may be more interesting to paleontologists. In addition, a significant bias exists with regard to the locations where paleontologists search for fossils. For example, if a paleontologist were interested in dinosaurs, he or she would search in regions where other dinosaur fossils have already been found. Often, these are places where organized and academic paleontology has been going on most intensely, particularly in Europe and the U.S.

that a species came into existence because they are unlikely to find the first member of a particular species.

Before ending our discussion of fossils, we should consider some factors that impart a bias to the fossil record (**Table 22.1**). First, certain organisms are more likely to become fossilized than are others. Factors such as anatomy, size, number, and the environment and time in which they lived play important roles in determining the likelihood that an organism will be preserved in the fossil record. In addition, geological processes may create biases because they may favor the fossilization of certain types of organisms. Such processes can also create confusion because they may cause parts of different species to be mixed after death. Finally, unintentional biases arise that are related to the efforts of paleontologists. For example, scientific interests may favor searching for and analyzing certain species over others.

Though the fossil record should not be viewed as a comprehensive, balanced story of the history of life, it has provided a wealth of information regarding the types of life that existed in the distant past. The rest of this chapter will survey the emer-

gence of life-forms from 3.5 billion years ago to the present. In addition, the fossil record has provided compelling evidence for the theory of evolution. We will begin to examine this topic in Chapter 23.

Prokaryotic Cells Arose During the Archaean Eon

The Archaean (meaning "ancient") eon was a time when diverse microbial life flourished in the primordial oceans. As mentioned earlier in this chapter, the first known fossils of living cells were preserved in rocks that are dated 3.5 billion years old, though scientists postulate that cells arose many millions of years prior to this time. Based on the morphology of fossilized remains, these first cells were prokaryotic, lacking a true nucleus. During the more than 1 billion years of the Archaean eon, all life-forms were prokaryotic. At the time of their emergence, hardly any free oxygen (O_2) was in the Earth's atmosphere. Therefore, the single-celled microorganisms of this eon almost certainly used only **anaerobic** (without oxygen) metabolism.

As discussed in Chapter 4, prokaryotes are divided into two groups: bacteria and archaea. Bacteria (also called eubacteria) are the most prevalent prokaryotic organisms on modern Earth.

Many species of archaea (also called archaebacteria) have also been identified, though they are less common than bacteria, and tend to occupy extreme environments such as hot springs. Both bacteria and archaea share fundamental similarities, indicating that they are derived from a common ancestor. Even so, certain differences indicate that these two types of prokaryotes diverged from each other quite early in the history of life. In particular, bacteria and archaea show some interesting differences with regard to metabolism, lipid composition, and genetic pathways.

An important factor that greatly influenced the emergence of prokaryotic, and eventually eukaryotic species, is energy. As we learned in Unit II, living cells require energy to survive and reproduce. Organisms may follow two different strategies to obtain energy. Some are **heterotrophs**, which means that their energy is derived from eating other organisms or materials from other organisms. In this case, the energy of chemical bonds within the biological molecules of the food provides a source of energy. Alternatively, many organisms are **autotrophs** and have metabolic pathways that directly harness energy from either inorganic molecules or light. Among modern species, plants are an important example of autotrophs. As discussed in Chapter 8, plants can directly absorb light energy and use it (via metabolic pathways) to synthesize organic molecules such as glucose. On modern Earth, heterotrophs ultimately rely on autotrophs for the production of food.

Even though the first identified fossils are those of organisms that resembled photosynthetic cyanobacteria, scientists have hypothesized that the first living cells were actually heterotrophs. To be an autotroph, a cell must have the metabolic pathways that are needed to directly use energy to make organic molecules. Instead, evolutionary biologists have speculated that it would have been simpler for the first primitive cells to use the organic molecules in the prebiotic soup as a source of food. However, because these organic molecules were made by prebiotic processes that were very slow, the proliferation of heterotrophs is postulated to have gradually exhausted this supply of available organic matter. As the organic supply dwindled, cells that evolved the ability to synthesize organic molecules from inorganic sources would have had a growth advantage. For example, cells arose that were capable of photosynthesis. These early photosynthetic cells are proposed to be similar to modern cyanobacteria.

Then why were cyanobacteria preserved in fossils, while fossils of their heterotrophic ancestors were not? The most likely answer is related to their manner of growth. Interestingly, certain cyanobacteria promote the formation of a layered structure called a **stromatolite** (**Figure 22.11**). The aquatic environment where these cyanobacteria survive is rich in minerals such as calcium. The cyanobacteria, which grow in large mats, deplete the carbon dioxide in the surrounding water. This causes calcium carbonate to precipitate over the growing mat of bacterial cells, preserving those cells in the lower layers. The bacteria continue to grow upward to produce a new layer. Over time, many layers of sediment may be formed. As shown in Figure 22.11, this process still occurs today in places such as Shark Bay in western Australia, which is renowned for the stromatolite turfs rising along its beaches.

The emergence and proliferation of ancient cyanobacteria had two critical consequences. First, the autotrophic nature of these bacteria enabled them to produce organic molecules from carbon dioxide. This prevented the depletion of organic food-

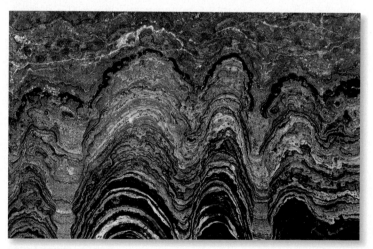

(a) Fossil stromatolite

(b) Modern stromatolites

Figure 22.11 Fossil and modern stromatolites: evidence of autotrophic cyanobacteria. Each stromatolite is a rocklike structure, typically 1 meter in diameter. **(a)** Section of a fossilized stromatolite. These layers are mats of mineralized cyanobacteria, one layer on top of the other. The existence of fossil stromatolites provides evidence of early autotrophic organisms, which produced organic molecules and oxygen near the beginning of the history of life on Earth. **(b)** Modern stromatolites that have formed in western Australia.

stuffs that would have been exhausted if only heterotrophs existed. Second, cyanobacteria produce oxygen as a waste product of photosynthesis. During the Archaean and Proterozoic eons, the activity of cyanobacteria led to the gradual rise in atmospheric oxygen that we discussed earlier. The increase in atmospheric oxygen spelled doom for many prokaryotic groups. Anaerobic species became restricted to a few anoxic (without oxygen) environments such as deep within the soil. However, oxygen enabled the formation of new **aerobic** (with oxygen) prokaryotic species as well as the emergence and eventual explosion of eukaryotic life-forms, which are described next.

GENOMES & PROTEOMES

The Origin of Eukaryotic Cells During the Proterozoic Eon Involved a Union Between Bacterial and Archaeal Cells

Eukaryotic cells arose during the Proterozoic eon, which began 2.5 billion years ago and ended 543 million years ago (see Figure 22.7). The origin of the first eukaryotic cell is a matter of debate. In modern eukaryotic cells, genetic material is found in three distinct organelles. All eukaryotic cells contain DNA in the nucleus and mitochondria, and plant and algal cells also have DNA in their chloroplasts. To address the issue of the origin of eukaryotic species, scientists have paid great attention to the properties of the DNA found in these three organelles, and to how that DNA compares with DNA found in modern prokaryotic species. Many eukaryotic and prokaryotic genomes have been completely sequenced, as discussed in greater detail in Chapter 21. These DNA sequences are then compared to each other to determine evolutionary relationships. From such studies, the nuclear, mitochondrial, and chloroplast genomes appear to be derived from once-separate cells that came together.

Let's begin with the nuclear genome. From a genome perspective, both bacteria and archaea have contributed substantially to modern eukaryotic genomes. Eukaryotic genes encoding proteins involved in metabolic pathways and lipid biosynthesis appear to be derived from ancient bacteria, while genes involved with transcription and translation are derived from an archaeal ancestor. To account for this observation, several hypotheses have been proposed, the most widely accepted of which involve an association between ancient bacteria and archaea. Such relationships could have been symbiotic or endosymbiotic. A **symbiotic** relationship is one in which two different species live in direct contact with each other. For example, some scientists have postulated an ancient symbiotic relationship in which a bacterium and an archaeon formed a close association (**Figure 22.12a**). This eventually led to a fusion event that combined the genetic material of the two organisms. Over time, selection favored the retention of bacterial genes involved in metabolism and lipid biosynthesis and archaeal genes concerned with transcription and translation. A second possible scenario is an **endosymbiotic** relationship, in which one organism lives inside the other. According to this idea, one prokaryotic cell engulfed another, which became an endosymbiont (**Figure 22.12b**). For example, an ancient archaeon may have engulfed a bacterium, maintaining the bacterium in its cytoplasm as an endosymbiont. This process may have occurred via endocytosis, which is described in Chapter 5. Over time, genes were transferred to the archaeal host cell.

Let's now consider the origin of mitochondrial and chloroplast genomes. In 1905, a Russian botanist, Konstantin Mereschkowsky, was the first to suggest that such organelles may have an endosymbiotic origin. However, the question of endosymbiosis was largely ignored until researchers in the 1950s discovered that chloroplasts and mitochondria contain their own genetic material. The issue of endosymbiosis was revived and hotly debated when in 1981 Lynn Margulis published a book presenting evidence in support of this hypothesis entitled *Origin of Eukaryotic Cells*. During the 1970s and 1980s, the advent of molecular genetic techniques allowed researchers to analyze genes from mitochondria, chloroplasts, and prokaryotic species. The data are consistent with the endosymbiotic origin of these organelles.

Mitochondria found in eukaryotic cells are derived from a bacterial species that resembles modern α-proteobacteria, a diverse group of bacteria that can carry out oxidative phosphorylation that is used to make ATP. One possibility is that an endosymbiotic event involving this bacterial species created the first eukaryotic cell, and that the mitochondrion is a remnant of that event. Alternatively, symbiosis or endosymbiosis may have produced the first eukaryotic cell, and then a subsequent endosymbiosis resulted in mitochondria (see Figure 22.12). Over the next few years, the sequencing of many prokaryotic and mitochondrial genomes may help to resolve this controversy. DNA-sequencing data indicate that chloroplasts were derived from an endosymbiotic relationship between a primitive eukaryotic cell and a cyanobacterium.

Experimental support for an endosymbiotic origin of eukaryotic cells has also come from the study of proteobacteria. Curiously, an endosymbiotic relationship involving two different proteobacteria was recently reported, which demonstrates the feasibility of an endosymbiotic relationship among prokaryotic species. In mealy bugs, bacteria survive within the cytoplasm of large host cells of a specialized organ called a bacteriome. Recent analyses of these bacteria have shown that the bacteria inside the host cells share their own endosymbiotic relationship. In particular, γ-proteobacteria live inside β-proteobacteria. Such an observation is consistent with the idea that an endosymbiotic relationship could have given rise to the first eukaryotic species over 2 billion years ago.

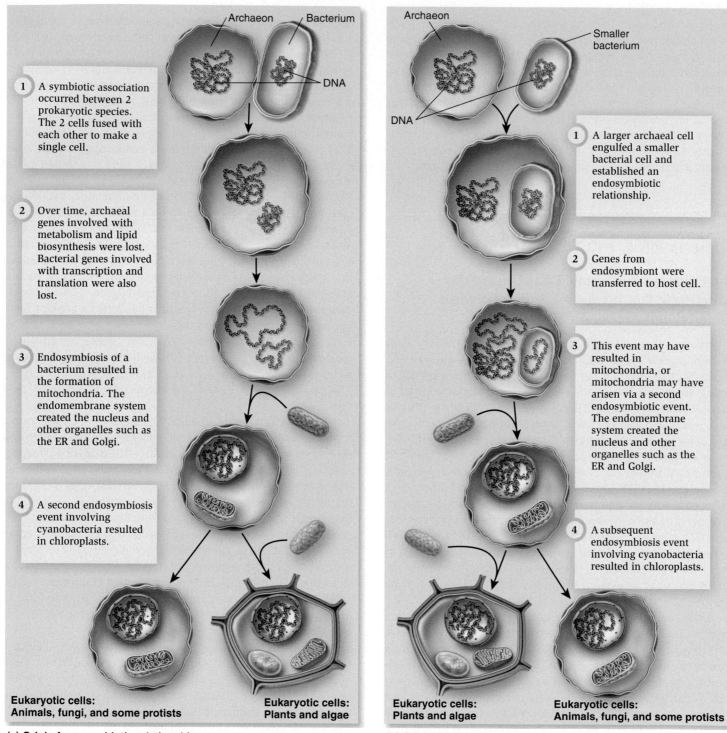

① A symbiotic association occurred between 2 prokaryotic species. The 2 cells fused with each other to make a single cell.

② Over time, archaeal genes involved with metabolism and lipid biosynthesis were lost. Bacterial genes involved with transcription and translation were also lost.

③ Endosymbiosis of a bacterium resulted in the formation of mitochondria. The endomembrane system created the nucleus and other organelles such as the ER and Golgi.

④ A second endosymbiosis event involving cyanobacteria resulted in chloroplasts.

Archaeon   Bacterium

DNA

Eukaryotic cells: Animals, fungi, and some protists

Eukaryotic cells: Plants and algae

(a) Origin from symbiotic relationship

Archaeon

Smaller bacterium

DNA

① A larger archaeal cell engulfed a smaller bacterial cell and established an endosymbiotic relationship.

② Genes from endosymbiont were transferred to host cell.

③ This event may have resulted in mitochondria, or mitochondria may have arisen via a second endosymbiotic event. The endomembrane system created the nucleus and other organelles such as the ER and Golgi.

④ A subsequent endosymbiosis event involving cyanobacteria resulted in chloroplasts.

Eukaryotic cells: Plants and algae

Eukaryotic cells: Animals, fungi, and some protists

(b) Origin from endosymbiotic relationship

Figure 22.12   Possible symbiotic or endosymbiotic relationships that gave rise to the first eukaryotic cells.

Biological inquiry: What is the fundamental difference between the scenarios described in this figure?

Multicellular Eukaryotes and the Earliest Animals Arose During the Proterozoic Eon

Multicellular eukaryotes are thought to have first emerged about 1.5 billion years ago, in the middle of the Proterozoic eon, although the oldest fossils are dated at approximately 1.2 billion years old. Simple multicellular organisms could have originated in two different ways. One possibility is that several individual cells found each other and aggregated to form a colony. Cellular slime molds, discussed in Chapter 28, are examples of modern organisms in which groups of single-celled organisms can come together to form a small multicellular organism. According to the fossil record, such organisms have remained very simple for hundreds of millions of years. Alternatively, another way that multicellularity can occur is when a single cell divides, and the resulting cells stick together. This pattern occurs in many simple multicellular organisms, such as algae and fungi, as well as in species with more complex body plans, such as plants and animals. Biologists cannot be certain whether the first multicellular organisms arose by an aggregation process or by cell division and adhesion. However, the development of complex, multicellular organisms now occurs by cell division and adhesion.

An interesting example that compares unicellular organisms to more complex multicellular organisms is found among species of volvocine green algae that are related evolutionarily. These algae exist as unicellular species, as small clumps of cells of the same cell type, or as larger groups of cells with two distinct cell types. **Figure 22.13** compares four species of volvocine algae. *Chlamydomonas reinhardtii* is a unicellular alga (Figure 22.13a). It is called a biflagellate because the cells possess two flagella. *Gonium pectorale* is a multicellular organism composed of eight cells (Figure 22.13b). This simple multicellular organism is formed from a single cell by cell division and adhesion. All of the cells in this species are of the same cell type—biflagellate. Other volvocine algae have evolved into larger and

more complex organisms. *Pleaodorina californica* has 64 to 128 cells (Figure 22.13c), while *Volvox aureus* has about 1,000 to 2,000 cells (Figure 22.13d). A new feature of these more complex organisms is that they have two cell types—somatic and reproductive cells. The somatic cells are biflagellate cells, while the reproductive cells are not. When comparing *P. californica* and *V. aureus*, *V. aureus* has a higher percentage of somatic cells than does *P. californica*.

Overall, an analysis of these four species of algae illustrates three important principles that are found among complex multicellular species. First, such organisms arise from a single cell that divides to produce daughter cells that adhere to one another. Second, the daughter cells can follow different fates, thereby producing multicellular organisms with different cell types. Third, as organisms get larger, more cells tend to be somatic cells. The somatic cells carry out the activities that are required for the survival of the multicellular organism. The reproductive cells are specialized for the sole purpose of producing offspring.

Toward the end of the Proterozoic eon, multicellular animals emerged. The first animals were **invertebrates**, soft-bodied animals without backbones. Among modern animals, most species, except for organisms such as sponges and corals, exhibit bilateral symmetry. This is a two-sided body plan with a left and right side that are mirror images of each other. Because each side of the body has appendages such as legs, one advantage of bilateral symmetry is that it facilitates locomotion. Bilateral animals also have anterior and posterior ends, with the mouth at the anterior end, and dorsal and ventral sides as described in Chapter 19. In 2004, Jun-Yuan Chen, David Bottjer, and colleagues discovered the earliest known ancestor of animals with bilateral symmetry. It was a minute creature barely visible to the naked eye that was shaped like a flattened helmet (**Figure 22.14**). This fossil was found in south China and is approximately 580–600 million years old.

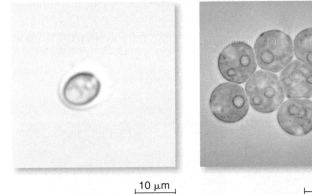

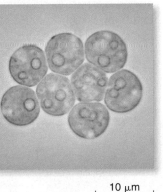

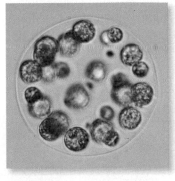

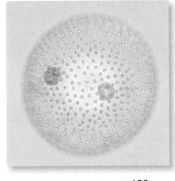

(a) *Chlamydomonas reinhardtii*, a unicellular alga

(b) *Gonium pectorale*, composed of 8 identical cells

(c) *Pleaodorina californica*, composed of 64 to 128 cells, has 2 cell types, somatic and reproductive

(d) *Volvox aureus*, composed of about 1,000 to 2,000 cells, has 2 cell types, somatic and reproductive

Figure 22.13   Variation in the level of multicellularity among volvocine algae.

Biological inquiry: Describe three changes that occur in these four types of algae.

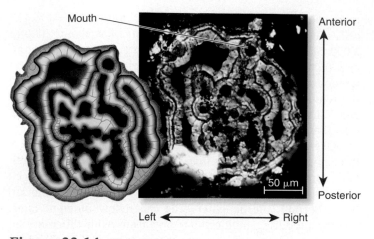

Figure 22.14 **Fossil of an early invertebrate animal showing bilateral symmetry.**   This fossil of an early animal, *Vernanimalcula guizouena*, is dated from 580 to 600 million years ago.

Phanerozoic Eon: The Paleozoic Era Saw the Diversification of Invertebrates and the Colonization of Land by Plants and Animals

The proliferation of multicellular eukaryotic life has been extensive during the Phanerozoic eon, which started 543 million years ago (mya) and extends to the present day. Phanerozoic means "well-displayed life," referring to the abundance of fossils of plants and animals that have been identified from this eon. As described earlier in Figure 22.7, the Phanerozoic eon is subdivided into three eras, the Paleozoic, Mesozoic, and Cenozoic. Because they are relatively recent and we have many fossils from these eras, each of them is further subdivided into periods. We will consider each era with its associated conditions and prevalent forms of life separately.

The term Paleozoic means "ancient animal life." The Paleozoic era covers the time from 543 to 248 mya, and is subdivided into six periods, known as the Cambrian, Ordovician, Silurian, Devonian, Carboniferous, and Permian. Periods are usually named after regions where rocks and fossils of that age were first discovered.

Cambrian Period (543 to 490 mya) The Cambrian climate was generally warm and wet, with no evidence of ice at the poles. During the Cambrian period, an event called the **Cambrian explosion** occurred in which there was an abrupt increase (on a geological scale) in the diversity of animal species. Many fossils telling the story of the Cambrian explosion were found in a rock bed in the Canadian Rockies called the Burgess Shale, which was discovered by Charles Walcott in 1909. At this site, both soft- and hard-bodied (shelled) invertebrates were buried in an underwater mudslide and preserved in water that was so deep and oxygen-free that decomposition was minimal (**Figure 22.15a**). The excellent preservation of these softer tissues is what makes this deposit unique (**Figure 22.15b**).

In the middle of the Cambrian period, all of the existing major types of marine invertebrates arose, plus many others that no longer exist. The Cambrian explosion generated over 100 major animal groups with significantly different body plans, but only about 30 of these occur in modern species. Examples of those that still exist include echinoderms (sea urchins and starfish), arthropods (insects, spiders, crustaceans), mollusks (clams and snails), and **chordates** (organisms with a spinal chord). Interestingly, although many new species of animals have arisen since this time, these later species have not shown a major reorganization of body plan, but instead exhibit variations on themes that were established during the Cambrian explosion. Approximately 520 million years ago, the first **vertebrates** (animals with backbones) appeared.

The cause of the Cambrian explosion is not understood. Because it occurred shortly after marine animals evolved shells, some scientists have speculated that the changes observed in animal species may have allowed them to exploit new environments and thereby evolve adaptations that would be beneficial in those environments. Alternatively, others have suggested that the increase in diversity may be related to atmospheric oxygen levels. During this period, oxygen levels were increasing, and perhaps more complex body plans became possible only after the atmospheric oxygen surpassed a certain threshold. In addition, as atmospheric oxygen reached its present levels, an ozone (O_3) layer was created that screens out ultraviolet radiation, thereby allowing complex life to live in shallow water and eventually on land.

(a) The Burgess Shale

(b) A fossilized arthropod, *Marella*

Figure 22.15 **The Cambrian explosion and the Burgess Shale.**   (a) This photograph shows the original site in the Canadian Rockies discovered by Charles Walcott. Since its discovery, this site has been made into a quarry for the identification of fossils. (b) A fossil of an extinct arthropod, *Marella*, that was collected at this site.

Ordovician Period (490 to 443 mya) Like the Cambrian period, the climate of the early and middle parts of the Ordovician period was warm, and the atmosphere was very moist. This period had a diverse group of marine invertebrates, including trilobites and brachiopods (**Figure 22.16**). Marine communities consisted of invertebrates, algae, primitive jawless fish (a type of early vertebrate), mollusks, and corals. Fossil evidence also suggests that primitive land plants and arthropods may have first invaded the land during this period.

Toward the end of the Ordovician period, the climate changed rather dramatically. Large glaciers formed, which drained the relatively shallow oceans, causing the water levels to drop. This resulted in a mass extinction—estimates suggest that over 60% of the existing marine invertebrates became extinct during this period.

Silurian Period (443 to 417 mya) In contrast to the dramatic climate changes observed during the Ordovician period, the climate during the Silurian was relatively stable. The glaciers largely melted, which caused the ocean levels to rise.

No new major types of invertebrate animals appeared during this period, but significant changes were observed among existing vertebrate and plant species. Many new types of fish have been observed in the fossil record. In addition, the coral reefs made their first appearance during this period.

The Silurian marked a large colonization by terrestrial plants and animals. For this to occur, organisms evolved adaptations that prevented them from drying out, such as external cuticles. Ancestral relatives of spiders and centipedes became prevalent. Also, the earliest fossils of **vascular plants**, which can transport water, sugar, and salts throughout the plant body, were observed in this period.

Devonian Period (417 to 354 mya) In the Devonian, generally dry conditions occurred across much of the northern landmasses. However, the southern hemisphere was mostly covered by cool, temperate oceans.

The Devonian saw a major increase in the number of terrestrial species. At first, the vegetation consisted primarily of small plants, only a meter tall or less. Later, ferns, horsetails and **seed plants**, such as gymnosperms, also emerged. By the end of the Devonian, the first trees and forests were formed.

A major expansion of terrestrial animals also occurred. Insects emerged and other invertebrates became plentiful. In addition, **tetrapods**, which are vertebrates with four legs, came into existence. These first tetrapods were amphibians, living on land but requiring water to reproduce because of their jelly-coated eggs.

In the oceans, many types of invertebrates flourished, including brachiopods, echinoderms, and corals. This period is sometimes called the Age of Fishes, as many new types of fish emerged. During a period of approximately 20 million years near the end of the Devonian period, a prolonged series of extinctions eliminated many marine species. The cause of this mass extinction is not well understood.

Carboniferous Period (354 to 290 mya) The term Carboniferous refers to the rich coal deposits that were the result of the vegetation and climate of this period. The Carboniferous had the ideal conditions for the beginnings of coal. It was a cooler period and much of the land was covered by forest swamps. Coal was formed over many millions of years from compressed layers of rotting vegetation.

Plants and animals further diversified during the Carboniferous period. Very large plants and trees became prevalent. For example, tree ferns such as *Psaronius* grew to a height of 15 meters or more (**Figure 22.17**). The first flying insects emerged. For example, giant dragonflies with a wingspan of over two feet inhabited the forest swamps. Terrestrial vertebrates also became more diverse. Amphibians were very prevalent. One innovation that seemed particularly beneficial was the amniotic egg, which is covered with a leathery or hard shell. This prevented the desiccation of the embryo inside. This innovation was probably critical for the emergence of reptiles, which occurred during this period.

Permian Period (290 to 248 mya) At the beginning of the Permian, continental drift had brought much of the total land together, fused into a supercontinent known as Pangaea (see Figure 22.8). The interior regions of Pangaea were probably dry, with great seasonal fluctuations. The forests of fernlike plants shifted to gymnosperms. Species resembling modern conifers first appeared in the fossil record. During this period, amphibians were prevalent, but reptiles became the dominant vertebrate species. The first mammal-like reptiles also appeared.

(a) Trilobite

(b) Brachiopod

Figure 22.16 Shelled, invertebrate fossils of the Ordovician period.

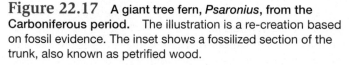

Figure 22.17   A giant tree fern, *Psaronius*, from the **Carboniferous period.**   The illustration is a re-creation based on fossil evidence. The inset shows a fossilized section of the trunk, also known as petrified wood.

Figure 22.18   *Megazostrodon*, the first known mammal of the Triassic period.   The illustration is a re-creation based on fossilized skeletons.

At the end of the Permian period, the largest known mass extinction in the history of the Earth occurred. As a result of this event, 90–95% of marine species were eliminated. Similarly, a large proportion of terrestrial species became extinct. The cause of this extinction is the subject of much research and controversy. One possibility is that glaciation destroyed the habitats of terrestrial species and lowered ocean levels, which would have created greater competition among marine species. Another hypothesis is that enormous volcanic eruptions in Siberia produced large ash clouds that abruptly changed the climate on Earth.

Phanerozoic Eon: The Mesozoic Era Saw the Rise and Fall of the Dinosaurs

The Permian extinction marks the division between the Paleozoic and Mesozoic eras. Mesozoic means "middle animals." It was a time period that saw great changes in animal and plant species. This era is sometimes called the Age of Dinosaurs, as dinosaurs flourished during this time. The climate during the Mesozoic era was consistently hot, and terrestrial environments were relatively dry. Little if any ice was found at either pole. The Mesozoic is divided into three periods, the Triassic, Jurassic, and Cretaceous, which we will consider separately.

Triassic Period (248 to 206 mya) Reptiles were plentiful in this period, including new groups such as crocodiles and turtles. Some mammal-like reptiles continued to survive from the Permian period. The first dinosaurs emerged during the middle of the Triassic, as did the first true **mammals** such as the small Megazostrodon (**Figure 22.18**). Gymnosperms were the dominant land plant. Volcanic eruptions near the end of the Triassic are thought to have led to global warming, resulting in mass extinctions that eliminated many marine and terrestrial species.

Jurassic Period (206 to 144 mya) Gymnosperms, such as conifers, continued to be the dominant vegetation in this period. Dinosaurs became the dominant land vertebrate. Some dinosaurs attained enormous sizes, such as the massive *Brachiosaurus* that reached a length of 25 m (80 ft) and weighed up to 100 tons! The first known bird, *Archaeopteryx* (**Figure 22.19**), emerged in the Jurassic period, and mammals continued to exist although they were not prevalent.

Cretaceous Period (144 to 65 mya) On land, dinosaurs continued to be the dominant animals in this period. The earliest **flowering plants**, called angiosperms, which form seeds within a protective chamber, emerged and began to diversify.

The end of the Cretaceous also witnessed another mass extinction, which brought an end to many previously successful groups of organisms, such as dinosaurs. The dinosaurs and many other species abruptly died out. As with the Permian extinction, the cause of this mass extinction is also still debated. One plausible hypothesis suggests that a large meteorite or asteroid hit the Yucatán Peninsula of Mexico, lifting massive amounts of debris into the air and thereby blocking the sunlight from reaching the Earth's surface. Such a dense haze could have cooled the Earth's surface by 11–15°C (20–30°F). Evidence is also relatively strong that volcanic eruptions were the primary culprit for this mass extinction. Currently, both the volcanic and meteorite impact scenarios are reasonable hypotheses to explain the Cretaceous mass extinction, and both may have been contributing factors.

Phanerozoic Eon: Mammals and Flowering Plants Diversified During the Cenozoic Era

The Cenozoic era spans the most recent 65 million years. The Cenozoic is divided into two periods, the Tertiary and Quaternary.

Figure 22.19 A fossil of the first known bird, *Archaeopteryx*, which emerged in the Jurassic period.

In many parts of the world, tropical conditions were replaced by a colder, drier climate. The Cenozoic is sometimes called the Age of Mammals, because during this time mammals became the largest terrestrial animals. However, this phrase is perhaps misleading, because the Cenozoic era has also seen an amazing diversification of many types of organisms, including birds, fish, insects, and flowering plants.

Tertiary Period (65 to 1.8 mya) On land, the mammals, which survived from the Cretaceous period, began to diversify rapidly.

The diversification of mammals occurred during the early part of the Tertiary period. Whales emerged during this period. Likewise, birds and terrestrial insects also diversified. Angiosperms became the dominant land plant, and insects became important for their pollination. In the seas, fish also diversified, and sharks became abundant.

Toward the end of Tertiary period, about 7 million years ago, hominids came into existence. **Hominids** include modern humans, chimpanzees, gorillas, and orangutans plus all of their recent ancestors. By comparison, the term **hominin** refers to a subset of hominids, namely modern humans, extinct human species (for example, of the *Homo* genus), and our immediate ancestors. In 2002, a fossil of the earliest known hominid, *Sahelanthropus tchadensis*, was discovered in Central Africa. This fossil was dated at between 6 and 7 million years old. Another early hominid genus, called *Australopithecus*, first emerged in Africa about 4 million years ago. They walked upright, and had a protruding jaw, prominent eyebrow ridges, and a small braincase.

Quaternary Period (1.8 mya to Present) Periodic Ice Ages have been prevalent during the last 1.8 million years, covering much of Europe and North America. This period has witnessed the widespread extinction of many species of mammals, particularly larger species. Certain species of hominids became increasingly more human-like. Near the beginning of the Quaternary period, fossils were discovered of *Homo habilis*, who was called the "handy man" because tools have been found with the fossil remains. Fossils that are classified as *Homo sapiens* first appeared about 130,000 years ago. The evolution of hominids is discussed in more detail in Chapter 34.

Chapter Summary

- Life began on Earth from nonliving material between 3.5 and 4.0 billion years ago. (Figure 22.1)

22.1 Origin of Life on Earth

- The first stage in the formation of life involved the synthesis of organic molecules to form a prebiotic soup. Possible scenarios of how this occurred are the reducing atmosphere, extraterrestrial, and deep-sea vent hypotheses. (Figures 22.2, 22.3)

- The second stage was the bonding of organic molecules to form polymers. This is thought to have occurred on the surface of clay.

- The third stage in the evolution of the first living cells occurred when polymers became enclosed in a structure that separated them from the external environment. Such structures, called protobionts, may have initially been coacervates, microspheres, or liposomes. (Figure 22.4)

- The fourth and final stage that led to the first living cells was chemical selection in which molecules with useful functional properties such as self-replication and other enzymatic functions increased in number. (Figure 22.5)

- The precursors of living cells as well as the first living cells themselves are thought to have used RNA for information storage and for carrying out enzymatic functions. This earliest phase of life is termed the RNA world.

- Bartel and Szostak demonstrated that chemical selection for RNA molecules that can catalyze phosphoester bond formation is possible experimentally. (Figure 22.6)

- The RNA world was later superseded by a DNA/RNA/protein world.

22.2 History of Life on Earth

- The geological timescale, which is divided into four eons and many eras and periods, charts the major events that occurred during the history of life. (Figure 22.7)

- The formation and extinction of new species, as well as mass extinctions, are often correlated with changes in temperature, atmosphere, and landmass locations, as well as floods, glaciation, volcanic eruptions, and meteorite impacts. (Figure 22.8)

- Fossils, which are preserved remnants of past life-forms, are formed in sedimentary rock. Radioisotope dating is one way to estimate the age of a fossil. The fossil record is incomplete and has several biases. (Figures 22.9, 22.10, Table 22.1)

- During the Archaean eon, the two groups of prokaryotes, bacteria and archaea, arose. The first prokaryotes were anaerobic heterotrophs. Later organisms, such as cyanobacteria, became phototrophs and produced oxygen. Cyanobacteria become preserved in structures called stromatolites. (Figure 22.11)

- Eukaryotic cells arose during the Proterozoic eon. This origin involved a union between bacterial and archaeal cells. The origin of mitochondria and chloroplasts was an endosymbiotic relationship. (Figure 22.12)

- Multicellular eukaryotes evolved during the Proterozoic eon, and first emerged about 1.5 billion years ago. Multicellularity now occurs via cell division and the adherence of the resulting cells to each other. A multicellular organism can produce multiple cell types. (Figure 22.13)

- The first bilateral animal emerged toward the end of the Proterozoic eon, approximately 580–600 million years ago. (Figure 22.14)

- The Phanerozoic eon is subdivided into the Paleozoic, Mesozoic, and Cenozoic eras. During the Paleozoic era, invertebrates greatly diversified, particularly during the Cambrian explosion, and the land became colonized by plants. Terrestrial vertebrates, including tetrapods, also became more diverse. (Figures 22.15, 22.16, 22.17)

- Dinosaurs were prevalent during the Mesozoic Era, particularly during the Jurassic period. Mammals and birds also emerged. (Figures 22.18, 22.19)

- During the Cenozoic Era, mammals diversified and flowering plants became the dominant species. The first hominids emerged approximately 7 million years ago.

Test Yourself

1. The prebiotic soup was
   a. the assemblage of unicellular prokaryotes and eukaryotes that existed in the oceans of primitive Earth.
   b. the accumulation of organic molecules in the oceans of primitive Earth.
   c. the mixture of organic molecules that was found in the cytoplasm of the earliest cells on Earth.
   d. a pool of nucleic acids that contained the genetic information for the earliest organisms.
   e. none of the above.

2. Which of the following is not a characteristic of protobionts that was necessary for the evolution of living cells?
   a. a membrane-like boundary separating the external environment from an internal environment
   b. polymers capable of functioning in information storage
   c. polymers capable of enzymatic activity
   d. self-replication
   e. compartmentalization of metabolic activity

3. RNA is believed to be the first functional macromolecule in protobionts because it
   a. is easier to synthesize compared to other macromolecules.
   b. has the ability to store information, self-replicate, and perform enzymatic activity.
   c. is the simplest of the macromolecules commonly found in living cells.
   d. all of the above
   e. a and c only

4. The movement of landmasses that have changed their positions, shapes, and association with other landmasses is called
   a. glaciation.   d. biogeography.
   b. Pangaea.   e. geological scale.
   c. continental drift.

5. Paleontologists estimate the dates of fossils by
   a. the layer of rock in which the fossils are found.
   b. analysis of radioactive isotopes found in nearby rock.
   c. the complexity of the body plan of the organism.
   d. all of the above
   e. a and b only

6. The fossil record does not give us a complete picture of the history of life because
   a. not all past organisms have become fossilized.
   b. only organisms with hard skeletons can become fossilized.
   c. fossils of very small organisms have not been found.
   d. fossils of early organisms are located too deep in the crust of the Earth to be found.
   e. all of the above.

7. The endosymbiosis hypothesis explaining the evolution of eukaryotic cells is supported by
   a. DNA-sequencing analysis comparing bacterial genomes, mitochondrial genomes, and eukaryotic nuclear genomes.
   b. naturally occurring examples of endosymbiotic relationships between bacterial cells and eukaryotic cells.
   c. the presence of DNA in mitochondria and chloroplasts.
   d. all of the above
   e. a and b only

8. Which of the following explanations of the evolution of multicellularity in eukaryotes is seen in the development of complex, multicellular organisms today?
   a. endosymbiosis
   b. aggregation of cells to form a colony
   c. division of cells with the resulting cells sticking together
   d. multiple cell types aggregating to form a complex organism
   e. none of the above

9. The earliest fossils of vascular plants occurred during the _____ period.
   a. Ordovician   d. Triassic
   b. Silurian   e. Jurassic
   c. Devonian

10. The appearance of the first hominids date to the _____ period.
    a. Triassic   d. Tertiary
    b. Jurassic   e. Quaternary
    c. Cretaceous

Conceptual Questions

1. What are the four stages in which the origin of life occurred?

2. Define radioisotope dating and half-life.

3. Briefly discuss the Cambrian explosion.

Experimental Questions

1. What is chemical selection? What was the hypothesis tested by Bartel and Szostak?

2. In conducting the selection experiment among pools of long RNA molecules with various catalytic abilities, what was the purpose of using the short RNA molecules?

3. What were the results of the experiment conducted by Bartel and Szostak? What impact did this study have on our understanding of the evolution of life on Earth?

Collaborative Questions

1. Discuss three possible hypotheses of how organic molecules were first formed.

2. Discuss three periods in geological time.

---

www.brookerbiology.com

This website includes answers to the Biological Inquiry questions found in the figure legends and all end-of-chapter questions.

23

AN INTRODUCTION TO EVOLUTION

CHAPTER OUTLINE

Selective breeding. Many crop plants, which are produced after many generations of selective breeding, illustrate how the traits of organisms can dramatically change over time.

Organic life beneath the shoreless waves
Was born and nurs'd in Ocean's pearly caves
First forms minute, unseen by spheric glass,
Move on the mud, or pierce the watery mass;
These, as successive generations bloom,
New powers acquire, and larger limbs assume;
Whence countless groups of vegetation spring,
And breathing realms of fin, and feet, and wing.

From *The Temple of Nature* by Erasmus Darwin (1731–1802), grandfather of Charles Darwin. Published posthumously in 1803.

The word evolution is often associated with a process that involves change. **Biological evolution** is a heritable change in one or more characteristics of a population or species across many generations. Evolution can be viewed on a small scale as it relates to changes in a single gene in a population over time, or it can be viewed on a larger scale as it relates to the formation of new species or groups of related species. It is helpful to begin our discussion of evolution with a definition of a species. Unfortunately, as we will examine in Chapter 25, a precise definition of species is not always possible. As a working definition, biologists define **species** as a group of related organisms that share a distinctive form. Among species that reproduce sexually, such as plants and animals, members of the same species are capable of interbreeding to produce viable and fertile offspring. The term **population** refers to members of the same species that are likely to encounter each other, and so have the opportunity to interbreed. Some of the emphasis in the study of evolution is on understanding how populations change over the course of many generations to produce new species.

In the first part of this chapter, we will examine the history of evolutionary thought and some of the basic tenets of evolution, particularly those that were proposed by Charles Darwin in the mid-1800s. Though the theory of evolution has been refined over the past 150 years or so, the fundamental principle of evolution has remained unchanged, and has provided a cornerstone for our understanding of biology. Theodosius Dobzhansky, an influential evolutionary scientist of the 1900s, once said, "Nothing in biology makes sense except in the light of evolution." The extraordinarily diverse and seemingly bizarre array of species on our planet can be explained within the context of evolution. As is the case with all scientific theories, biological evolution is called a theory because it is supported by a substantial body of evidence and because it explains a wide range of observations. The theory of evolution provides answers to many questions related to the diversity of life.

In the second part of this chapter, we will survey the extensive data that show the results of evolution. These data not only support the theory of evolution but also allow us to understand the interrelatedness of different species, whose similarities are often related to descent from a common ancestor. Much of the early evidence supporting evolution came from visual observations and comparisons of modern and extinct species.

More recently, advances in molecular genetics, particularly those related to DNA sequencing and genomics, have revolutionized the study of evolution. Scientists now have information that allows us to understand how evolution involves changes in the DNA of a given species. These changes affect both a species' genes and the proteins they encode. The term **molecular evolution** refers to the molecular changes in genetic material that underlie the phenotypic changes associated with evolution.

A theme of this textbook, namely Genomes & Proteomes, is rooted in an understanding of these changes. In the last section of this chapter, we consider some of the exciting new information that helps us to appreciate evolutionary change at the molecular level. In the following chapters of this unit, we will examine how such changes are acted upon by evolutionary forces in ways that alter the traits of a given species and may eventually lead to the formation of new species.

23.1 The Theory of Evolution

Undoubtedly, the question "Where did we come from?" has been asked and debated by people for thousands, perhaps tens of thousands of years. Many of the earliest ideas regarding the existence of living organisms were strongly influenced by religion and philosophy, and these ideas suggested that all forms of life have remained the same since their creation. In the 1600s, however, scholars in Europe began a revolution that created the basis of empirical and scientific thought. **Empirical thought** relies on observation to form an idea or hypothesis, rather than trying to understand life from a nonphysical or spiritual point of view. The shift toward empirical thought encouraged scholars to look for the basic rationale behind a given object or phenomenon.

In the mid- to late-1600s, the first scientist to carry out a thorough study of the natural world was an Englishman named John Ray, who developed an early classification system for plants and animals based on anatomy and physiology. He established the modern concept of a species, noting that organisms of one species do not interbreed with members of another, and used it as the basic unit of his classification system. Ray's ideas on classification were later extended by the Swedish naturalist Carolus Linnaeus. Neither Ray nor Linnaeus proposed that evolutionary change promotes the formation of new species. However, their systematic classification of plants and animals helped scholars of this period perceive the similarities and differences among living organisms.

Late in the 1700s, a small number of European scientists began to quietly suggest that life-forms are not fixed. A French zoologist, George Buffon, actually said that living things do change through time. However, Buffon was careful to hide his views in a 44-volume natural history book series. While he was a quiet pioneer in asserting that species can change over generations, he publicly rejected the idea that one species could evolve into another species.

Around the same time, a French naturalist named Jean-Baptiste Lamarck suggested an intimate relationship between variation and evolution. By examining fossils, he realized that some species had remained the same over the millennia and others had changed. Lamarck hypothesized that species change over the course of many generations by adapting to new environments. He believed that living things evolved in a continuously upward direction, from dead matter, through simple to more complex forms, toward human "perfection." According to

Lamarck, organisms altered their behavior in response to environmental change. He thought behavioral changes modified traits, and he hypothesized that such modified traits were inherited by offspring. He called this idea the **inheritance of acquired characteristics**. For example, according to Lamarck's hypothesis, giraffes developed their elongated necks and front legs by feeding on high tree leaves. The exercise of stretching up to the leaves altered the neck and legs, and Lamarck presumed that these acquired characteristics were transmitted to offspring. However, further research has rejected Lamarck's idea that acquired traits can be inherited.

Interestingly, Erasmus Darwin, the grandfather of Charles Darwin, who was a contemporary of Buffon and Lamarck, was an early advocate of evolutionary change. He was a physician, a plant biologist, and also a poet. He was aware that modern species were different from many fossil types and also saw how plant and animal breeders used breeding practices to change the traits of domesticated species (see chapter opener photo). He knew that offspring inherited features from their parents, and went so far as to say that life on Earth could be descended from a common ancestor.

Overall, Charles Darwin's many scientific predecessors set the stage for the theory of evolution. With this historical introduction, we will now consider Darwin's observations, and the tenets that provide the foundation for this theory.

Darwin Suggested That Species Are Derived from Pre-Existing Species

Charles Darwin, a British naturalist born in 1809, played a key role in developing the theory that existing species have evolved from pre-existing species. Darwin's unique perspective and his ability to formulate evolutionary theory were shaped by several different fields of study, including ideas of his time about physical and biological processes.

Two main geological hypotheses predominated in the early 19th century. Catastrophism was first proposed by French zoologist and paleontologist Baron Georges Cuvier to explain the age of the Earth. Cuvier suggested that the Earth was just 6,000 years old, and that only catastrophic events had changed its geological structure. This idea fit well with religious teachings. Alternatively, uniformitarianism, proposed by James Hutton and promoted by Sir Charles Lyell, suggested that changes in the Earth are directly caused by recurring events. For example, they suggested that geological processes such as erosion existed in the past, and that they happened then at the same gradual rate as they do now. For such slow geological processes to eventually lead to substantial changes in the Earth's characteristics, a great deal of time was required, so Hutton and Lyell were the first to propose that the age of the Earth was well beyond 6,000 years. The ideas of Hutton and Lyell helped to shape Darwin's view of the world.

Darwin's thinking was also influenced by a paper published in 1798 called *Essay on the Principle of Population* by Thomas Malthus, an English economist. Malthus asserted that the pop-

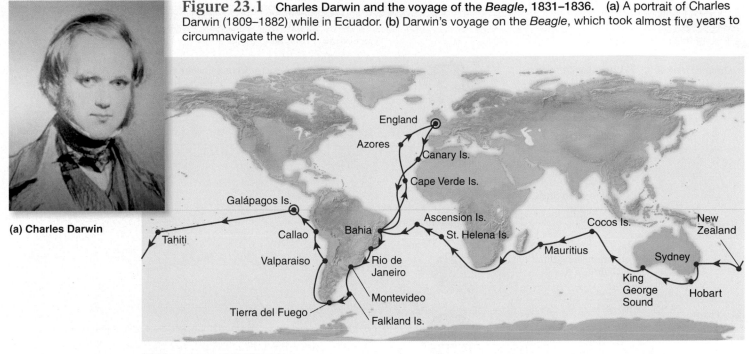

Figure 23.1 Charles Darwin and the voyage of the *Beagle*, 1831–1836.   (a) A portrait of Charles Darwin (1809–1882) while in Ecuador. (b) Darwin's voyage on the *Beagle*, which took almost five years to circumnavigate the world.

(a) Charles Darwin

(b) The voyage of the *Beagle*

ulation size of humans can, at best, increase linearly due to increased land usage and improvements in agriculture, while our reproductive potential is exponential (for example, doubling with each generation). He argued that famine, war, and disease will limit population growth, especially among the poor. An important message from Malthus's work was that only a fraction of any population will survive and reproduce.

Darwin's evolutionary ideas were most influenced by his own experiences and observations. His famous voyage on the *Beagle*, which lasted from 1831 to 1836 (**Figure 23.1**), involved a careful examination of many different species. The main mission of the *Beagle* was to map the coastline of southern South America and take oceanographic measurements. Darwin's job was to record the weather, geological features, plants, animals, fossils, rocks, minerals, and indigenous people. He also collected many specimens of plants and animals, which had to be carefully packed and labeled.

Though Darwin made many interesting observations on his journey, he was particularly struck by the distinctive traits of island species that provided them with ways to better exploit their native environment. For example, Darwin observed several species of finches found on the Galápagos Islands. We now know that these finches all evolved from a single species similar to the blue-black grassquit finch (*Volatina jacarina*), commonly found along the Pacific Coast of South America. Once on the Galápagos Islands, the finches' ability to survive in their new habitat depended, in part, on changes in the size and shape of their bills over many generations. These specializations enabled succeeding generations to better obtain food (**Table 23.1**). For example, the ground and vegetarian finches have sturdy, crush-

ing bills that they use to crush various sizes of seeds or buds. The tree finches have grasping bills that they use to pick up insects from trees. The mangrove, woodpecker, warbler, and cactus finches have pointed, probing bills. The first three of these use their probing bills to search for insects in crevices; the cactus finches use their probing bills to open cactus fruits and eat the seeds. One species, the woodpecker finch, even uses twigs or cactus spines to extract insect larvae from holes in dead tree branches. Darwin clearly saw the similarities among these species, yet he noted the differences that provided them with specialized feeding strategies.

With an understanding of geology and population growth, and his observations from his voyage on the *Beagle*, Darwin had formulated his theory of evolution by the mid-1840s. However, he had cataloged and described all of the species he had collected on his *Beagle* voyage except for one type of barnacle. Some have speculated that Darwin may have felt that he should establish himself as an expert on one species before making generalizations about all of them. Therefore, he spent several additional years studying barnacles. During this time, the geologist Charles Lyell, who had greatly influenced Darwin's thinking, strongly encouraged Darwin to publish his theory of evolution. In 1856, Darwin began to write a long book to explain his ideas. In 1858, however, Alfred Wallace, a naturalist working in the East Indies, sent Darwin an unpublished manuscript to read prior to its publication. In it, Wallace proposed the same ideas concerning evolution. Darwin therefore quickly excerpted some of his own writings on this subject, and two papers, one by Darwin and one by Wallace, were published in the Proceedings of the Linnaean Society of London.

Table 23.1	A Comparison of Beak Type and Diet Among the Galápagos Finches Darwin Studied			
Type of finch	**Species**		**Type of beak**	**Diet**
Ground finches	Large ground finch (*Geospiza magnirostris*)		Crushing	Seeds—Ground finches have crushing beaks to crush various sizes of seeds; large beaks can crush large seeds, while smaller beaks are better for crushing small seeds.
	Medium ground finch (*G. fortis*)			
	Small ground finch (*G. fuliginosa*)			
	Sharp-billed ground finch (*G. difficilis*)			
Vegetarian finch	Vegetarian finch (*Platyspiza crassirostris*)		Crushing	Buds—Vegetarian finches have crushing beaks to pull buds from branches.
Tree finches	Large tree finch (*Camarhynchus psittacula*)		Grasping	Insects—Tree finches have grasping beaks to pick insects from trees. Those with heavier beaks can also break apart wood in search of insects.
	Medium tree finch (*Camarhynchus pauper*)			
	Small tree finch (*Camarhynchus parvulus*)			
	Mangrove finch (*Cactospiza heliobates*)		Probing	Insects—These finches have probing beaks to search for insects in crevices and then to pick them up. The woodpecker finch can also use a cactus spine for probing.
	Woodpecker finch (*Cactospiza pallidus*)			
Warbler finch	Warbler finch (*Certhidea olivacea*)		Probing	
Cactus finches	Large cactus finch (*G. conirostris*)		Probing	Seeds—Cactus finches have probing beaks to open cactus fruits and take out seeds.
	Cactus finch (*G. scandens*)			

These papers were not widely recognized. A short time later, however, Darwin finished his book, *The Origin of Species*, which described his ideas in greater detail and included observational support. This book, which received high praise from many scientists and scorn from others, started a great debate concerning evolution. Although some of his ideas were incomplete due to the fact that the genetic basis of traits was not understood at that time, Darwin's work remains one of the most important contributions to our understanding of biology.

The Theory of Evolution Explains How Natural Selection Acting on Genetic Variation Can Change Populations over Many Generations

The fundamental principle that underlies evolution is that biological species do not have a fixed, static existence but instead exhibit changing characteristics over the course of many generations. Darwin hypothesized that existing life-forms on our planet are the product of the modification of pre-existing life-forms. He expressed this concept of biological evolution as "the theory of descent with modification through variation and natural selection." As its name suggests, evolution is based on two factors: (1) variation within a given species and (2) forces of nature, which are termed natural selection. During the process of **natural selection,** certain individuals are less likely to survive and reproduce in a particular environment, while other individuals with traits that make them better suited to their native environment tend to flourish and reproduce. According to this idea, nature "selects" those individuals possessing certain traits that favor reproductive success. Over long periods of time, this process of natural selection eventually leads to **adaptation**, which is a form of evolutionary change in which a population's characteristics change to make its members better suited to their native environment. Note that natural selection affects the survival and/or reproduction of individuals, which over time affects the evolution of populations and ultimately species.

The genetic basis for variation within a species was not understood at the time Darwin proposed his theory of evolution by natural selection. In fact, Darwin's theory preceded, by a few years, Mendel's pioneering work in genetics. Even so, Darwin and many other people before him observed that offspring resemble their parents more than they do unrelated individuals. Therefore, he assumed that some traits are passed from parent to offspring.

Since the time of Darwin, the study of genetics has allowed scientists to understand the relationship between traits and inheritance. Genetic variation is a consistent feature of natural populations. Such variation may involve differences in genes, changes in chromosome structure, and alterations in chromosome number. In contrast to Lamarck's ideas, we now know that these genetic differences are not produced by an individual's behavior or other response to its environment. Rather, genetic variation is caused by random mutations that alter the genetic composition of particular individuals. In Chapter 24, we will consider how these random mutations are acted upon by evolutionary forces to change the genetic composition of populations over the course of many generations.

Based on Darwin's ideas regarding natural selection and a more modern understanding of genetics, biologists in the 1920s to 1940s were able to frame the theory of evolution in a modern perspective termed neo-Darwinism or the **modern synthesis of evolution**. Within a given population of interbreeding organisms, natural variation exists that is caused by random changes in the genetic material. Such genetic changes may affect the phenotype of an individual in a positive, negative, or neutral way. If a genetic change promotes an individual's survival and/or ability to successfully reproduce, natural selection may increase the prevalence of that trait in future generations.

For example, let's consider a population of finches that migrates from the South American mainland to a Galápagos island. The seeds produced on this island are larger than those produced on the mainland. Birds with larger beaks would be better able to feed on these seeds and so would be more likely to survive and pass on that trait to their offspring. Therefore, in succeeding generations, the population will tend to have a greater proportion of finches with larger beaks (**Figure 23.2**). Alternatively, if a genetic change happens to be detrimental to an individual's survival and/or reproduction, natural selection is likely to eliminate this type of variation. For example, if a mutation occurred that causes a finch in the same environment to have a slightly smaller beak, this bird would be less likely to survive and pass on this change to its offspring. Natural selection, which acts upon variation involving many different genes, may ultimately produce a new species with a combination of traits that are quite different from those of the original species, such as finches with larger beaks. In other words, the newer species has evolved from a pre-existing species. Now let's look at some other examples of change in organisms over time.

23.2 Observations of Evolutionary Change

Over the past 150 years, the research community has learned that no known concept other than descent with modification from a common ancestor can scientifically account for the diversity and unity of life on our planet. Observations regarding the theory of evolution can be gleaned from many sources (**Table 23.2**). Historically, the first descriptions of biological evolution came from studies of the fossil record, the distribution of living organisms on our planet, selective breeding experiments, and the comparison of similar anatomical features in different species. More recently, additional examples of the theory of evolution have been revealed at the molecular level. By comparing DNA sequences from many different species, evolutionary biologists have gained great insight into the relationship between the evolution of species and the associated changes in the genetic material. In this section, we will survey a variety of observations that show the process of evolutionary change.

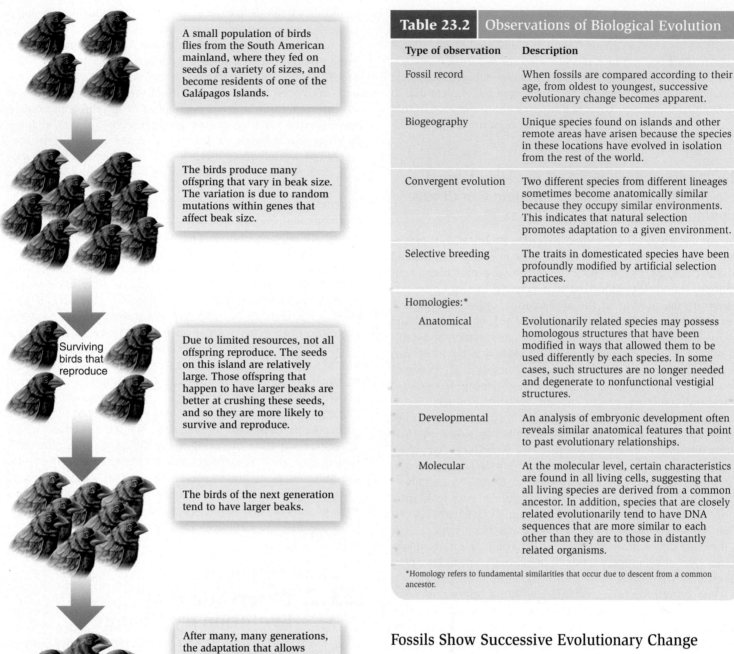

A small population of birds flies from the South American mainland, where they fed on seeds of a variety of sizes, and become residents of one of the Galápagos Islands.

The birds produce many offspring that vary in beak size. The variation is due to random mutations within genes that affect beak size.

Surviving birds that reproduce

Due to limited resources, not all offspring reproduce. The seeds on this island are relatively large. Those offspring that happen to have larger beaks are better at crushing these seeds, and so they are more likely to survive and reproduce.

The birds of the next generation tend to have larger beaks.

After many, many generations, the adaptation that allows success in feeding on larger seeds has created a new species with larger beaks, as well as other modified traits, such as changes in color, that are suited to the new environment.

Figure 23.2 Evolutionary adaptation to a new environment via natural selection.   The example shown here involves a species of finch adapting to a new environment on one of the Galápagos Islands. The plants on this island produce larger seeds than do the plants on the mainland from which the birds had originated. According to Darwin's theory of evolution by natural selection, the process of adaptation eventually led to the formation of a new species with larger beaks that were better suited to crushing the large seeds in its new environment. .

Table 23.2	Observations of Biological Evolution
Type of observation	**Description**
Fossil record	When fossils are compared according to their age, from oldest to youngest, successive evolutionary change becomes apparent.
Biogeography	Unique species found on islands and other remote areas have arisen because the species in these locations have evolved in isolation from the rest of the world.
Convergent evolution	Two different species from different lineages sometimes become anatomically similar because they occupy similar environments. This indicates that natural selection promotes adaptation to a given environment.
Selective breeding	The traits in domesticated species have been profoundly modified by artificial selection practices.
Homologies:*	
Anatomical	Evolutionarily related species may possess homologous structures that have been modified in ways that allowed them to be used differently by each species. In some cases, such structures are no longer needed and degenerate to nonfunctional vestigial structures.
Developmental	An analysis of embryonic development often reveals similar anatomical features that point to past evolutionary relationships.
Molecular	At the molecular level, certain characteristics are found in all living cells, suggesting that all living species are derived from a common ancestor. In addition, species that are closely related evolutionarily tend to have DNA sequences that are more similar to each other than they are to those in distantly related organisms.

*Homology refers to fundamental similarities that occur due to descent from a common ancestor.

Fossils Show Successive Evolutionary Change

As discussed in Chapter 22, the fossil record reveals a history of life from its earliest beginnings some 3.5–4.0 billion years ago. Today, scientists have access to a far more extensive fossil record than was available to Darwin or other scientists of his time. Even though the fossil record is still incomplete, the many fossils that have been discovered often provide detailed information regarding evolutionary change in a series of related organisms. When fossils are compared according to their age, from oldest to youngest, successive evolutionary change becomes apparent.

Let's consider a few examples in which paleontologists have observed evolutionary change. In 2005, fossils of *Tiktaalik roseae*, nicknamed fishapod, were discovered by Edward Daeschler, Neil Shubin, and Farish Jenkins that illuminate the steps that led to the evolution of tetrapods, which are animals with four legs.

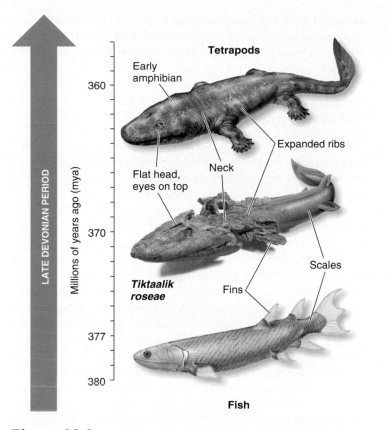

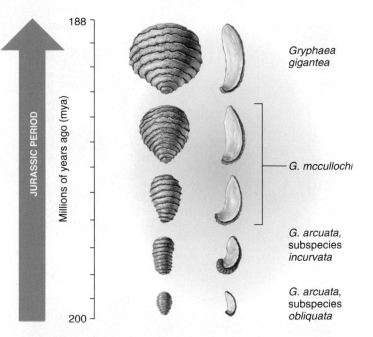

Figure 23.4 Evolutionary changes in the shape of oyster shells. During a 12-million-year period in the early Jurassic, the shells of oysters of the extinct genus *Gryphaea* became larger and flatter. Scientists have hypothesized that such a change was an adaptation to disruptive water currents.

Biological inquiry: If a population of oysters was transferred to a shallow bay with calm water currents, how might the population change over the course of many generations?

Figure 23.3 Evolutionary change in the tetrapod lineage, showing a transitional form. This figure shows two early tetrapod ancestors, a Devonian fish and the transitional form *Tiktaalik roseae*, as well as one of their descendents, an early amphibian. An analysis of the fossils shows that *T. roseae*, also known as a fishapod, had both fish and amphibian characteristics, so it probably was able to survive brief periods out of the water.

T. roseae is called a **transitional form** because it provides a link between earlier species and many later species (**Figure 23.3**). In this case, the fishapod is a transitional form between fish, which are aquatic animals, and tetrapods, which are usually terrestrial animals. Unlike a true fish, *T. roseae* had a broad skull, a flexible neck, and eyes mounted on the top of its head like a crocodile. Its interlocking rib cage suggests it had lungs. Perhaps the most surprising discovery was that its pectoral fins (those on the side of the body) revealed the beginnings of a primitive wrist and five finger-like bones. These appendages would have been adequate for *T. roseae* to peek its head above the water from shallow river bottoms and look for prey. During the Devonian period, 417–354 million years ago, this could have been an important advantage in the marshy floodplains of large rivers.

As a second example, certain oysters began to undergo a change in shell structure about 200 million years ago. Oysters with smaller, curved shells were superseded by oysters with larger, flatter shells (**Figure 23.4**). This change was observed over the course of 12 million years during the early Jurassic period, when water currents became stronger. Scientists have hypothesized that larger, flatter shells are more stable in disrup-

tive water currents, so these shells were better adapted to the environmental change.

One of the best-studied examples of evolutionary change is our third example, that of the horse family. Modern members include horses, zebras, and donkeys. These species, which are large, long-legged animals adapted to living in open grasslands, are the remaining descendants of a long lineage that produced many extinct species since its origin approximately 55 million years ago. Examination of the horse lineage provides a particularly interesting case of how evolution involves adaptation to changing environments.

The earliest known fossils of the horse family (termed *Hyracotherium*) revealed that the animals were small with short legs and broad feet (**Figure 23.5**). Such early horses lived in wooded habitats and are thought to have eaten leaves and herbs. Between the time of these first members of the horse family and modern horses, the fossil record has revealed adaptive changes in size, foot anatomy, and tooth morphology. The first horses were the size of dogs, while modern horses typically weigh more than a half ton. *Hyracotherium* had four toes on its front feet and three on its hind feet. Instead of hooves, these toes were encased in fleshy pads. By comparison, the feet of modern horses have a single toe, enclosed in a tough, bony hoof. The fossil record shows an increase in the length of the central toe, the development of a bony hoof, and the loss of the other toes.

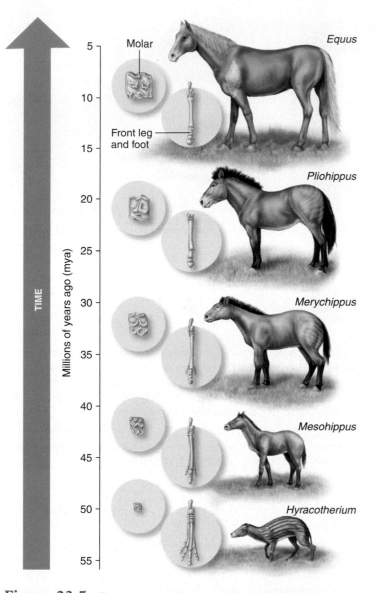

Figure 23.5 **Evolutionary changes that led to the modern horse.** The major changes that occurred in horses' body size, foot anatomy, and tooth morphology were adaptations to the changing environment in which the horses and their ancestors lived over the last 55 million years of their evolution. Note: This figure is meant to emphasize changes that led to modern horses. The evolutionary pathway that produced modern horses involves several branches, and is described in Chapter 26.

Finally, the teeth of *Hyracotherium* were relatively small compared to those of modern horses. Over the course of millions of years, horse teeth have increased in size and developed a complex pattern of ridges on their molars.

These changes in horse characteristics can be attributed to natural selection producing adaptations to changing global climates. Over North America, where much of horse evolution occurred, large areas changed from dense forests to grasslands. Their increase in size and changes in their foot structure allowed horses to escape predators and travel great distances in search

of food. The changes seen in horse's teeth are consistent with a dietary shift from eating more tender leaves to eating grasses and other vegetation that are more abrasive and require more chewing.

Biogeography Indicates That Species in a Given Area Have Evolved from Pre-Existing Species

Biogeography is the study of the geographical distribution of extinct and modern species. Patterns of past evolution are often found in the natural geographical distribution of related species. From such studies, scientists have discovered that isolated continents and island groups have evolved their own distinct plant and animal communities. As mentioned earlier in this chapter, Darwin himself observed several species of finches found on the Galápagos Islands that had unique characteristics, such as beak shapes, when compared with similar finches found on the mainland. As we discussed, scientists now think that these island species evolved from mainland birds that had migrated to the islands and then became adapted to a variety of feeding habits. One example of this was shown in Figure 23.2.

Around the world, islands, which are often isolated from other landmasses, provide numerous examples in which geography has played a key role in the evolution of new species. Islands often have many species of plants and animals that are **endemic**, which means they are naturally found only in a particular location. Most endemic island species have closely related relatives on nearby islands or the mainland. For example, consider the island fox (*Urocyon littoralis*), which lives on the Channel Islands located off the coast of southern California between Los Angeles and Santa Barbara (**Figure 23.6**). This type of fox is found nowhere else in the world. It weighs about 3–6 pounds and feeds largely on insects, mice, and fruits. The island fox evolved from the mainland gray fox (*Urocyon cinereoargenteus*), which is much larger, usually 7–11 pounds. During the last Ice Age, about 16,000–18,000 years ago, the Santa Barbara channel was frozen and narrow enough for ancestors of the mainland gray fox to cross over to the Channel Islands. When the Ice Age ended, the ice melted and sea levels rose, causing the foxes to be cut off from the mainland. Over the last 16,000–20,000 years, these foxes evolved into the smaller island fox. The smaller size of the island fox is an example of island dwarfing, the phenomenon in which the size of large animals isolated on an island shrinks dramatically over many generations. It is a form of natural selection in which smaller size provides a survival advantage probably because of limited food.

The evolution of major animal groups is also correlated with known changes in the distribution of landmasses on the Earth. The first mammals arose approximately 225 million years ago. The fossils of the first placental mammals, which are those that have long internal gestation and give birth to offspring that are well developed, are only 80 million years old. The first mammals arose when the area that is now Australia was still connected to the other continents. However, the first placental mammals evolved after continental drift had separated Australia from the

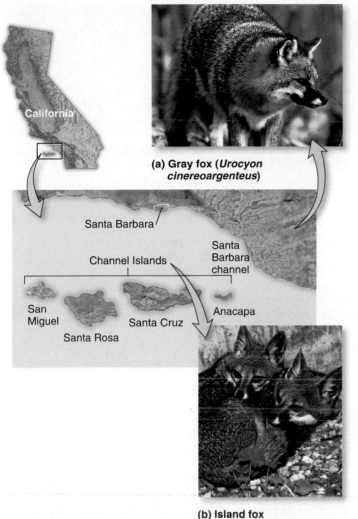

Figure 23.6 **The evolution of an endemic island species from a mainland species.** (a) The smaller island fox found on the Channel Islands evolved from (b) the gray fox found on the California mainland.

other continents (refer back to Figure 22.8). Except for a few species of bats and rodents that have migrated to Australia more recently, Australia lacks any of the larger, terrestrial placental mammals. This observation is consistent with the idea that placental mammals first arose somewhere other than Australia, and that the barrier of a large ocean prevented most terrestrial placental mammals from migrating there. Instead, Australia has more than 100 species of kangaroos, koalas, and other marsupials. Marsupials are a group of mammal species in which young are born in a very immature condition and then develop further in the mother's abdominal pouch, which covers the mammary glands. Most species of Australian marsupials are not found on other continents. Evolutionary theory is consistent with the idea that the existence of these unique Australian species is due to their having evolved in isolation from the rest of the world for millions of years.

Convergent Evolution Suggests Adaptation to the Environment

The process of natural selection is also evident in the study of plants and animals that have similar characteristics, even though they are not closely related evolutionarily. This similarity is due to **convergent evolution**, in which two different species from different lineages show similar characteristics because they occupy similar environments. For example, the long snout and tongue of both the giant anteater (*Myrmecophaga tridactyla*), found in South America, and the echidna (*Tachyglossus aculeatus*), found in Australia, are similar yet independently evolved adaptations that enable these animals to feed on ants (**Figure 23.7a**). The giant anteater is a placental mammal, while the echidna is an egg-laying mammal, so they are not closely related evolutionarily. Another example involves aerial rootlets found in vines such as English ivy (*Hedera helix*) and wintercreeper (*Euonymus fortunei*) (**Figure 23.7b**). Based on differences in their structures, these aerial rootlets appear to have developed independently as an effective means to cling to the support on which a vine attaches itself.

A third example of convergent evolution is revealed by the molecular analysis of fish that live in very cold water. Antifreeze proteins allow certain species of fish to survive the subfreezing temperatures of Arctic and Antarctic waters by preventing the formation of ice crystals in their blood. By studying these fish, researchers have determined that they are an interesting case of convergent evolution (**Figure 23.7c**). Among different species of fish, one of five different genes has independently evolved to produce antifreeze proteins. For example, in the sea raven (*Hemitripterus americanus*), the antifreeze protein is rich in the amino acid cysteine, and the secondary structure of the protein is in a β sheet conformation. In contrast, the antifreeze protein in the longhorn sculpin (*Trematomus nicolai*) evolved from an entirely different gene. The antifreeze protein in this species is rich in the amino acid glutamine, and the secondary structure of the protein is largely composed of α helices.

The similar characteristics in the examples shown in Figure 23.7, which are the result of convergent evolution, are called **analogous structures** or **convergent traits**. They represent cases in which structures have arisen independently, two or more times, because species have occupied similar types of environments on the Earth. By comparison, homologous structures have a single evolutionary origin.

Selective Breeding Is a Human-Driven Form of Natural Selection

The term **selective breeding** refers to programs and procedures designed to modify traits in domesticated species. This practice, also called **artificial selection**, is related to natural selection. In forming his theory of natural selection, Charles Darwin was influenced by his observations of selective breeding by pigeon breeders. The primary difference between natural and artificial selection is how the parents are chosen. Natural selection is due to natural variation in reproductive success.

(a) Their long snouts and tongues enable these species, the giant anteater (left) and the echidna (right), to feed on ants.

(b) Their aerial rootlets allow these vines, English ivy (left) and wintercreeper (right), to climb up supports.

(c) Antifreeze proteins enable these fish, the sea raven (left) and the longhorn sculpin (right), to survive in frigid waters. The antifreeze proteins in these 2 species evolved from entirely different genes.

Figure 23.7 **Examples of convergent evolution.** All three pairs of organisms shown in this figure are not closely related evolutionarily, but occupy similar environments, suggesting that natural selection promotes the formation of similar adaptations that are well suited to a particular environment.

(a) Bulldog   **(b) Greyhound**   **(c) Dachshund**

Figure 23.8 **Common breeds of dogs that have been obtained by selective breeding.** By selecting parents carrying the alleles that influence traits desirable to humans, dog breeders have produced breeds with distinctive features. For example, the bulldog has alleles that give it short legs and a flat face. All the dogs shown in this figure carry the same kinds of genes (for example, genes that affect their size, shape, and fur color). However, the alleles for many of these genes are different among these dogs, thereby allowing humans to select for or against them and produce breeds with strikingly different phenotypes.

Organisms that are able to survive and reproduce are more likely to pass their genes on to future generations. Nature determines or "chooses" which individuals will be successful parents. In artificial selection, the breeder chooses as parents those individuals that possess traits desirable to humans.

The underlying phenomenon that makes selective breeding possible is genetic variation. Within a group of individuals of the same species, variation may exist in a trait of interest. For selective breeding to be successful, the underlying cause of the phenotypic variation must be related to differences in the **alleles**,

different forms of a particular gene, that determine the trait. The breeder will choose parents with desirable phenotypic characteristics. For centuries, humans have been practicing selective breeding to obtain domesticated species with interesting or agriculturally useful characteristics. For example, many common breeds of dog are the result of selective breeding strategies (**Figure 23.8**). All dogs are members of the same species, *Canis familiaris*, so they can be interbred to produce offspring. Selective breeding can dramatically modify the traits in a species. When you compare certain breeds of dogs (for example, a

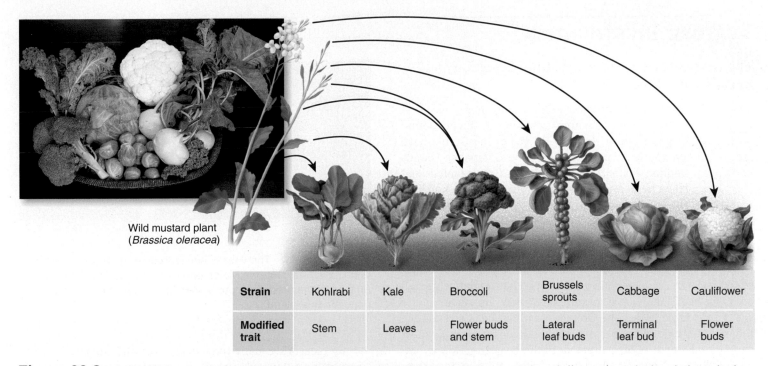

Strain	Kohlrabi	Kale	Broccoli	Brussels sprouts	Cabbage	Cauliflower
Modified trait	Stem	Leaves	Flower buds and stem	Lateral leaf buds	Terminal leaf bud	Flower buds

Wild mustard plant
(*Brassica oleracea*)

Figure 23.9   Crop plants developed by selective breeding of the wild mustard plant.   Although these six agricultural plants look quite different from each other, they carry many of the same alleles as the wild mustard plant. However, they differ among each other in alleles that affect the formation of flowers, buds, stems, and leaves.

greyhound and a dachshund), they hardly look like members of the same species!

Likewise, most of the food we eat is obtained from species that have been profoundly modified by selective breeding strategies. This includes products such as grains, fruits, vegetables, meat, milk, and juices. For example, **Figure 23.9** illustrates how certain characteristics in the wild mustard plant (*Brassica oleracea*) have been modified by selective breeding to create several varieties of domesticated crops, including broccoli, Brussels sprouts, and cauliflower. This plant is native to Europe and Asia, and plant breeders began to modify its traits approximately 4,000 years ago. As seen here, certain traits in the domestic strains differ considerably from those of the original wild species. These varieties are all members of the same species. They can interbreed to produce viable offspring. For example, in the grocery store you may have seen brocciflower, which is produced from a cross between broccoli and cauliflower.

As a final example, **Figure 23.10** shows the results of an artificial selection experiment on corn begun at the Illinois Experiment Station in 1896, even before the rediscovery of Mendel's laws. This study began with 163 ears of corn with an oil content ranging from 4 to 6%. In each of 80 succeeding generations, corn plants were divided into two separate groups. In one group, members with the highest oil content in the kernels were chosen as parents of the next generation. In the other group, members with the lowest oil content were chosen. After many generations, the oil content in the first group rose to over 18%. In the other group, it dropped to less than 1%. These results show that selective breeding can modify a trait in a very directed manner.

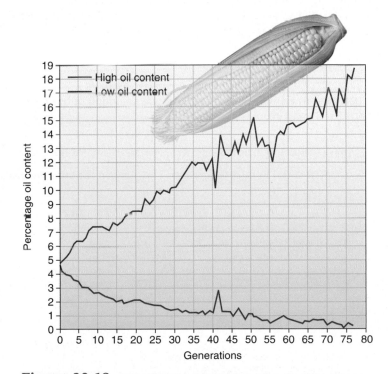

Figure 23.10   Results of selective breeding for oil content in corn plants.   In this example, corn plants were selected for breeding based on high or low oil content of the kernels. Over the course of many generations, this had a major impact on the amount of corn oil—an agriculturally important product—that is made by the two groups of plants.

Biological inquiry: When comparing Figures 23.8, 23.9, and 23.10, what general effects of artificial selection do you observe?

FEATURE INVESTIGATION

The Grants Have Observed Natural Selection in Galápagos Finches

Since 1973, Peter Grant, Rosemary Grant, and their colleagues have studied the process of natural selection in finches found on the Galápagos Islands. For over 30 years, the Grants have focused much of their work on one of the Galápagos Islands known as Daphne Major (**Figure 23.11a**). This small island (0.34 km²) has a moderate degree of isolation (it is 8 km from the nearest island), an undisturbed habitat, and a resident population of the finch *Geospiza fortis*, the medium ground finch (**Figure 23.11b**).

To study natural selection, the Grants have observed various traits in finches over the course of many years. One example is beak size. The medium ground finch has a relatively small crushing beak, allowing it to more easily feed on small, tender seeds (see Table 23.1). The Grants quantified beak size among the medium ground finches of Daphne Major by carefully measuring beak depth (a measurement of the beak from top to bottom, at its base) on individual birds. During the course of their studies, they compared the beak sizes of parents and offspring by examining many broods over several years. The depth of the beak was transmitted from parents to offspring, regardless of environmental conditions, indicating that differences in beak sizes are due to genetic differences in the population. In other words, they found that beak depth was a heritable trait.

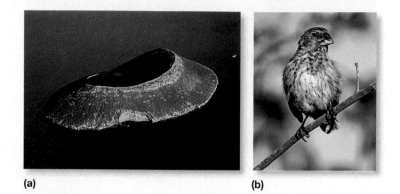

(a)   (b)

Figure 23.11 The Grants' investigation of natural selection in finches. (a) Daphne Major, one of the Galápagos Islands. (b) One of the medium ground finches (*Geospiza fortis*) that populate this island.

By measuring many birds every year, the Grants were able to assemble a detailed portrait of natural selection in action. In the study shown in **Figure 23.12**, they measured beak depth from 1976 to 1978. In the wet year of 1976, the plants of Daphne Major produced an abundance of the small seeds that these finches could easily eat. However, a drought occurred in 1977. During this year, the plants on Daphne Major tended to produce few of the smaller seeds, which the finches rapidly consumed.

Figure 23.12 The Grants and natural selection of beak size among the medium ground finch. The results are those recorded from 1976 and 1978.

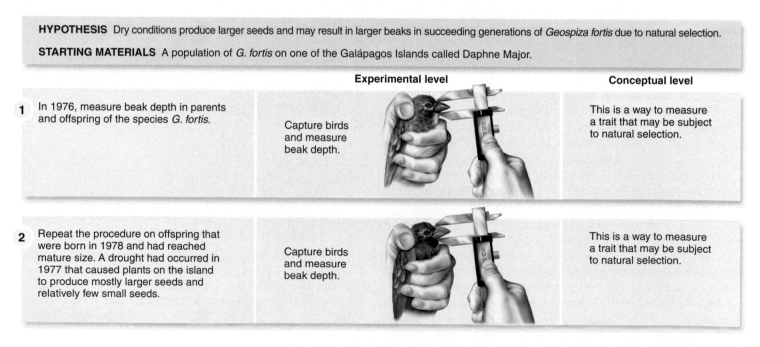

HYPOTHESIS Dry conditions produce larger seeds and may result in larger beaks in succeeding generations of *Geospiza fortis* due to natural selection.

STARTING MATERIALS A population of *G. fortis* on one of the Galápagos Islands called Daphne Major.

Experimental level   Conceptual level

1 In 1976, measure beak depth in parents and offspring of the species *G. fortis*.

Capture birds and measure beak depth.

This is a way to measure a trait that may be subject to natural selection.

2 Repeat the procedure on offspring that were born in 1978 and had reached mature size. A drought had occurred in 1977 that caused plants on the island to produce mostly larger seeds and relatively few small seeds.

Capture birds and measure beak depth.

This is a way to measure a trait that may be subject to natural selection.

3 **THE DATA**

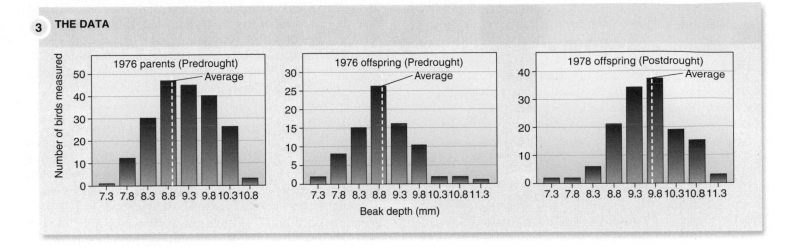

To survive, the finches resorted to eating larger, drier seeds, which are harder to crush. As a result, birds with larger beaks were more likely to survive because they were better at breaking open these large seeds. In the year after the drought, the average beak depth of birds in the population increased, because the surviving birds with larger beaks passed this trait on to their offspring. Overall, these results illustrate the power of natural selection to alter the nature of a trait, in this case beak depth, in a given population.

A Comparison of Anatomical, Developmental, and Molecular Homologies Shows Evolution of Related Species from a Common Ancestor

Let's now consider other widespread observations of the process of evolution among living organisms. In biology, the term **homology** refers to a fundamental similarity that occurs due to descent from a common ancestor. As a result of evolution, homology is often observed between different species. Two species may have a similar trait because the trait was originally found in a common ancestor. As described next, such homologies may involve anatomical, developmental, or molecular features.

Anatomical Homologies As noted by Theodosius Dobzhansky, the theory of evolution provides a sensible framework for understanding the diversity of life. Many observations regarding anatomical features of plants and animals simply cannot be understood in any meaningful scientific way except as a result of evolution. A comparison of vertebrate anatomy is a case in point. An examination of the limbs of modern vertebrate species reveals similarities that indicate that the same set of bones has undergone evolutionary changes to become the bones used today for many different purposes. As seen in **Figure 23.13**, the forelimbs of vertebrates have a strikingly similar pattern of bone arrangements. These are termed **homologous structures** because they are considered to be derived from a common ancestor. The forearm has developed different uses among various vertebrates, including grasping, walking, flying, swimming, and climbing.

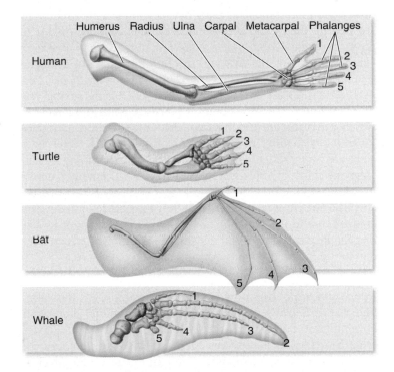

Figure 23.13   **An example of anatomical homology: homologous structures found in vertebrates.**   The same set of bones is found in the human arm, turtle arm, bat wing, and whale flipper, although these bones have been modified in ways that allow them to perform different functions. This homology suggests that all of these animals evolved from a common ancestor.

Table 23.3 | Examples of Vestigial Structures

Organism	Vestigial structure(s)
Humans	Muscles to wiggle ears and tail bone in adult, rudimentary gill ridges in embryo
Boa constrictors	Skeletal remnants of hip bones and hind legs
Whales	Skeletal remnants of a pelvis
Manatees	Fingernails on the flippers
Dandelion	An asexually reproducing plant (*Taraxicum officinale*) that still makes flowers, even though it does not produce fertile seeds.
Hornbills and cuckoos	In certain families of birds, both of the common carotid arteries are nonfunctional. They are present as fibrous cords. Their vascular function has been assumed by other vessels.

The theory of evolution explains how these animals have descended from a common ancestor and how natural selection has modified the same initial pattern of bones in ways that ultimately allowed them to be used for several different purposes.

Another observation of evolution is the phenomenon of **vestigial structures**, which are anatomical features that have no apparent function but resemble structures of their presumed ancestors. Table 23.3 describes several examples. An interesting case is found in humans. People have a complete set of muscles for moving their ears, even though most people are unable to do so. By comparison, many modern mammals can move their ears, and presumably this was an important trait in a distant human ancestor. Within the context of evolutionary theory, vestigial structures are evolutionary relics. Organisms having vestigial structures share a common ancestry with organisms in which the structure is functional. Natural selection maintains functional structures in a population of individuals. However, if a species changes its lifestyle so that the structure loses its purpose, the selection forces that would normally keep the structure in a functional condition are no longer present. When this occurs, the structure may degenerate over the course of many generations due to the accumulation of mutations that limit its size and shape. Natural selection may eventually eliminate such traits due to the inefficiency of producing unused structures.

Developmental Homologies Another example of homology due to evolution is the way that animals undergo embryonic development. Species that differ substantially at the adult stage often bear striking similarities during early stages of embryonic development. These temporary similarities are called developmental homologies. In addition, evolutionary history is revealed during development in certain organisms such as vertebrates. For example, if we consider human development, several features are seen in the embryo that are not present at birth. Human embryos possess rudimentary gill ridges like a fish embryo, even though human embryos receive oxygen via the umbilical cord. The presence of gill ridges indicates that humans evolved from

an aquatic animal with gill slits. A second observation is that every human embryo has a long bony tail. It is difficult to see the advantage of such a structure *in utero*, but easy to understand its presence assuming that an ancestor to the human lineage possessed a long tail. These observations, and many others, illustrate that development has evolved over time.

Molecular Homologies Our last example of homology due to evolution involves molecular studies. When scientists examine the features of cells at the molecular level, similarities called **molecular homologies** are found which indicate that living species evolved from a common ancestor or interrelated group of common ancestors. For example, all living species use DNA to store information. RNA molecules such as mRNA, tRNA, and rRNA are used to access that information, and proteins are the functional products of most genes. Furthermore, certain biochemical pathways are found in all or nearly all species, although minor changes in the structure and function of proteins involved in these pathways have occurred. For example, all species that use oxygen, which constitutes the great majority of species on our planet, have similar proteins that together make up an electron transport chain and an ATP synthase. In addition, nearly all living organisms can metabolize glucose via a glycolytic pathway that is described in Chapter 7. Taken together, these types of observations indicate that such molecular phenomena arose very early in the origin of life, and have been passed to all or nearly all modern forms.

The most compelling observation at the molecular level indicating that modern life-forms are derived from a common ancestor is revealed by analyzing genetic sequences and finding genetic homologies, or similar genes. The same type of gene is often found in diverse organisms. Furthermore, the degree to which a genetic sequence from different species is similar reflects the evolutionary relatedness of those species. As an example, let's consider a gene that encodes the p53 protein that plays a role in preventing cancer (see Chapter 14). **Figure 23.14** shows a short amino acid sequence that makes up part of the p53 protein from a variety of species, including five mammals, one bird, and three fish. The top sequence is the human p53 sequence, and the right column describes the percentages of amino acids within the entire sequence that are identical to the entire human sequence. Amino acids in the other species that are identical to humans are highlighted in orange. The sequences from the two monkeys are closest to humans, followed by the other two mammalian species (rabbit and dog). The three fish sequences are the least similar to the human sequence, but you may notice that the fish sequences tend to be similar to each other. Taken together, the data shown in Figure 23.14 illustrate two critical points regarding gene evolution. First, certain genes are found in a diverse array of species such as mammals, birds, and fish. Second, the sequences of closely related species tend to be more similar to each other than they are to distantly related species. The mechanism for this second observation is described in the next section.

	Short amino acid sequence within the p53 protein	Percentages of amino acids in the whole p53 protein that are identical to human p53
Human (*Homo sapiens*)	Val Pro Ser Gln Lys Thr Tyr Gln Gly Ser Tyr Gly Phe Arg Leu Gly Phe Leu His Ser Gly Thr	100
Rhesus monkey (*Macaca mulatta*)	Val Pro Ser Gln Lys Thr Tyr His Gly Ser Tyr Gly Phe Arg Leu Gly Phe Leu His Ser Gly Thr	95
Green monkey (*Cercopithecus aethiops*)	Val Pro Ser Gln Lys Thr Tyr His Gly Ser Tyr Gly Phe Arg Leu Gly Phe Leu His Ser Gly Thr	95
Rabbit (*Oryctolagus cuniculus*)	Val Pro Ser Gln Lys Thr Tyr His Gly Asn Tyr Gly Phe Arg Leu Gly Phe Leu His Ser Gly Thr	86
Dog (*Canis familiaris*)	Val Pro Ser Pro Lys Thr Tyr Pro Gly Thr Tyr Gly Phe Arg Leu Gly Phe Leu His Ser Gly Thr	80
Chicken (*Gallus gallus*)	Val Pro Ser Thr Glu Asp Tyr Gly Gly Asp Phe Asp Phe Arg Val Gly Phe Val Glu Ala Gly Thr	53
Channel catfish (*Ictalurus punctatus*)	Val Pro Val Thr Ser Asp Tyr Pro Gly Leu Leu Asn Phe Thr Leu His Phe Gln Glu Ser Ser Gly	48
European flounder (*Platichthys flesus*)	Val Pro Val Val Thr Asp Tyr Pro Gly Glu Tyr Gly Phe Gln Leu Arg Phe Gln Lys Ser Gly Thr	46
Congo puffer fish (*Tetraodon miurus*)	Val Pro Val Thr Thr Asp Tyr Pro Gly Glu Tyr Gly Phe Lys Leu Arg Phe Gln Lys Ser Gly Thr	41

Figure 23.14 **An example of genetic homology: a comparison of a short amino acid sequence within the p53 protein from nine different animals.** This figure compares a short region of the p53 protein, which plays a role in preventing cancer. Amino acids are represented by three-letter abbreviations. The orange-colored amino acids in the sequences are identical to those in the human sequence. The numbers in the right column indicate the percentage of amino acids within the whole p53 protein that is identical with the human p53 protein, which is 393 amino acids in length. For example, 95% of the amino acids, or 373 of 393, are identical between the p53 sequence found in humans and that in Rhesus monkeys.

Biological inquiry: In the sequence shown in this figure, how many amino acid differences are there between the following pairs: Rhesus and green monkeys, Congo puffer fish and European flounder, and Rhesus monkey and Congo puffer fish? What do these differences tell you about the evolutionary relationships among these four species?

23.3 The Molecular Processes That Underlie Evolution

Historically, the study of evolution was based on comparing the anatomies of extinct and modern species to identify similarities between related species. However, the advent of molecular approaches for analyzing DNA sequences has revolutionized the field of evolutionary biology. Now we can analyze how changes in the genetic material are associated with changes in phenotype, and how those changes have led to the formation of new species. In this section, we will examine some of the molecular changes in the genetic material that are associated with evolution.

Homologous Genes Are Derived from a Common Ancestral Gene

When two genes are derived from the same ancestral gene, they are called **homologous genes**. The analysis of homologous genes reveals the molecular details of evolutionary change. As an example, let's consider a gene in two different species of bacteria that encodes a transport protein involved in the uptake of metal ions into bacterial cells. Such genes, which are homologous yet from different species, are said to be **orthologs** of each other. Millions of years ago, these two species had a common ancestor (**Figure 23.15**). Over time, the common ancestor diverged

into additional species, eventually evolving into *Escherichia coli*, *Clostridium acetylbutylicum*, and many other species. Since this divergence, the metal transporter gene has accumulated mutations that alter its sequence, though the similarity between the *E. coli* and the *C. acetylbutylicum* genes remains striking. In this case, the two sequences are similar because they were derived from the same ancestral gene, but they are not identical due to the independent accumulation of different random mutations.

Gene Duplications Create Gene Families

Orthologs are examples of evolutionary change occurring in separate species. Demonstrations of evolutionary change can also be found within a single species. Two or more homologous genes found within a single species are termed **paralogs** of each other. Rare gene duplication events can produce multiple copies of a gene and ultimately lead to the formation of a gene family. A **gene family** consists of two or more copies of paralogous genes within the genome of a single organism. A well-studied example of a gene family is the globin gene family found in humans and many other animal species. The globin genes encode polypeptides that are subunits of proteins that function in oxygen binding. One such protein is hemoglobin that is found in red blood cells and carries oxygen throughout the body.

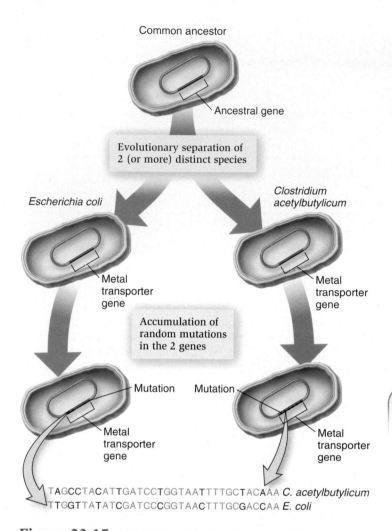

Common ancestor

Ancestral gene

Evolutionary separation of
2 (or more) distinct species

Escherichia coli

*Clostridium
acetylbutylicum*

Metal
transporter
gene

Metal
transporter
gene

Accumulation of
random mutations
in the 2 genes

Mutation   Mutation

Metal
transporter
gene

Metal
transporter
gene

TAGCCTACATTGATCCTGGTAATTTTGCTACAAA *C. acetylbutylicum*
TTGGTTATATCGATCCCGGTAACTTTGCGACCAA *E. coli*

Figure 23.15   **The evolution of orthologs, homologous
genes from different species.**   After two species diverged from
each other, the genes accumulated random mutations that
created similar but not identical gene sequences called orthologs.
These orthologs in *E. coli* and *C. acetylbutylicum* encode metal
transporters. Only one of the two DNA strands is shown from
each of the genes. Bases that are identical between the two
genes are shown in orange.

The globin gene family is composed of 14 genes that were originally derived from a single ancestral globin gene. According to
an evolutionary analysis, the ancestral globin gene first duplicated between 500 to 600 million years ago. Since that time,
additional duplication events and chromosomal rearrangements
have produced the current number of 14 genes on three different human chromosomes (refer back to Figure 21.8).

Gene families have been important in the evolution of
traits. Even though all of the globin polypeptides are subunits
of proteins that play a role in oxygen binding, the accumulation
of changes in the various family members has created globins
that are more specialized in their function. For example, myoglobin is better at binding and storing oxygen in muscle cells,
whereas the hemoglobins are better at binding and transporting
oxygen via the red blood cells. Also, different globin genes are

expressed during different stages of human development. The
epsilon (ε)- and zeta (ζ)-globin genes are expressed very early in
embryonic life, while the gamma (γ)-globin genes exhibit maximal expression during the second and third trimesters of gestation. Following birth, the γ-globin gene is turned off and the
β-globin gene is turned on. These differences in the expression
of the globin genes reflect the differences in the oxygen transport needs of humans during the embryonic, fetal, and postpar-
tum stages of life.

What is the evolutionary significance of the globin gene
family regarding adaptation? Internal gestation is one way that
animals have adapted to a terrestrial environment. On land, egg
cells and small embryos are very susceptible to drying out if
they are not protected in some way. Species such as birds and
reptiles lay eggs with a protective shell around them. Most mammals, however, have adjusted to a terrestrial environment by
evolving the adaptation of internal gestation. The ability to develop young internally has been an important factor in the survival and proliferation of mammals. The embryonic and fetal
forms of hemoglobin allow the embryo and fetus to capture
oxygen from the bloodstream of the mother.

GENOMES & PROTEOMES

New Genes in Eukaryotes Have Evolved via Exon Shuffling

Thus far we have considered how evolutionary change results
in the formation of related genes, which are described as orthologs and paralogs. Evolutionary mechanisms are also revealed
when the parts of genes that encode protein domains are compared within a single species. Many proteins, particularly those
found in eukaryotic species, have a modular structure composed of two or more domains with different functions. For example, certain transcription factors have discrete domains involved with hormone binding, dimerization, and DNA binding.
As described in Chapter 13, the glucocorticoid receptor has a
domain that binds the hormone, a second domain that facilitates protein dimerization, and a third domain that allows the
glucocorticoid receptor to bind to glucocorticoid response elements (GREs) next to genes. By comparing the modular structure of eukaryotic proteins with the genes that encode them,
geneticists have discovered that each domain tends to be encoded by one exon, or by a series of two or more adjacent exons.
As we learned in Chapter 12, exons contain the coding sequences
of a gene, which are separated by noncoding introns.

During the evolution of eukaryotic species, many new genes
have been created by a type of mutation known as **exon shuffling**. During this process, an exon and the flanking introns are
inserted into a gene, thereby producing a new gene that encodes
a protein with an additional domain (**Figure 23.16**). This process
may also involve the duplication and rearrangement of exons.
Exon shuffling results in novel genes that express proteins with
diverse functional modules. Such proteins can then alter traits
in the organism that can be acted upon by natural selection.

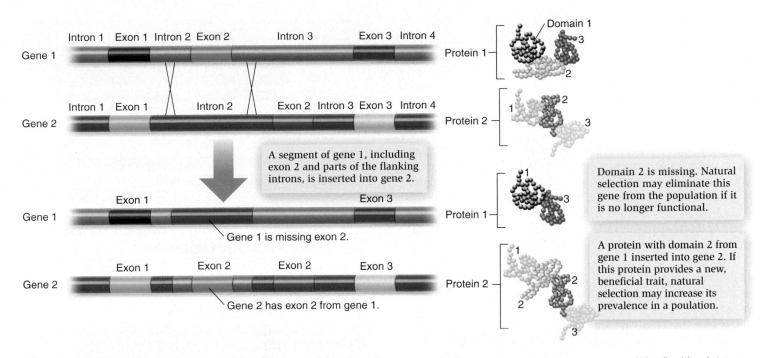

Figure 23.16 **The process of exon shuffling.** In this example, a segment of one gene containing an exon and the flanking introns has been inserted into another gene. A rare, abnormal crossing-over event called nonhomologous recombination may cause this to happen. This results in proteins that have new combinations of domains and possibly new combinations of functions.

Biological inquiry. What is the evolutionary advantage of exon shuffling?

Exon shuffling may occur by more than one mechanism. One possibility is that a double crossover could promote the insertion of an exon into another gene (see Figure 23.16). This is called nonhomologous or illegitimate recombination because the two regions involved in the crossover are not homologous to each other. Alternatively, transposable elements that are described in Chapter 21 may promote the movement of exons into other genes.

Horizontal Gene Transfer Also Contributes to the Evolution of Species

At the molecular level, the type of evolutionary change depicted in Figures 23.14 through 23.16 is called **vertical evolution**. In these cases, species evolve from pre-existing species by the accumulation of gene mutations, gene duplications, and exon shuffling. Vertical evolution involves genetic changes in a series of ancestors that form a lineage. In addition to vertical evolution, species accumulate genetic changes by another process called **horizontal gene transfer**, which involves the exchange of genetic material among different species.

Figure 23.17 illustrates one possible mechanism for horizontal gene transfer. In this example, a eukaryotic cell has engulfed a bacterial cell by endocytosis. During the degradation of the bacterium in an endocytotic vesicle, a bacterial gene happens to escape to the nucleus of the cell, where it is inserted into

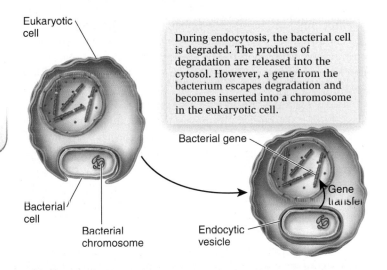

Figure 23.17 Horizontal gene transfer from a bacterium to a eukaryote. In this example, a bacterium is engulfed by a eukaryotic cell, and a bacterial gene is transferred to one of the eukaryotic chromosomes.

one of the chromosomes. In this way, a gene has been transferred from a bacterial species to a eukaryotic species. By analyzing gene sequences among many different species, researchers have discovered that horizontal gene transfer is a common phenomenon. This process can occur from prokaryotes to eukaryotes, from eukaryotes to prokaryotes, between different species of prokaryotes, and between different species of eukaryotes.

Therefore, when we view evolution, it is not simply a matter of one species evolving into one or more new species via the accumulation of random mutations. It also involves the horizontal transfer of genes among different species, enabling those species to acquire new traits that foster the evolutionary process.

Gene transfer among bacterial species is relatively widespread. As discussed in Chapter 18, bacterial species may carry out three natural mechanisms of gene transfer known as conjugation, transformation, and transduction. By analyzing the genomes of bacterial species, scientists have determined that many genes within a given bacterial genome are derived from horizontal gene transfer. Genome studies have suggested that as much as 20–30% of the variation in the genetic composition of modern prokaryotic species can be attributed to this process. For example, in *E. coli* and *Salmonella typhimurium*, roughly 17% of their genes have been acquired via horizontal gene transfer during the past 100 million years. The roles of these acquired genes are quite varied, though they commonly involve functions that are readily acted upon by natural selection. These include genes that confer antibiotic resistance, the ability to degrade toxic compounds, and pathogenicity (the ability to cause disease).

Evolution at the Genomic Level Involves Changes in Chromosome Structure and Number

Thus far, we have considered several ways that a species might acquire new genes. These include random mutations within preexisting genes, gene duplications to create gene families, exon shuffling, and horizontal gene transfer. Evolution also occurs at the genomic level, involving changes in chromosome structure and number. When comparing the chromosomes of closely related species, changes in chromosome structure and/or number are common.

As an example, **Figure 23.18** compares the banding patterns of the three largest chromosomes in humans and the corresponding chromosomes in chimpanzees, gorillas, and orangutans. (See Chapter 15 for a description of chromosome banding.) The banding patterns are strikingly similar because these species are closely related evolutionarily. Even so, you can see some interesting differences. Humans have one large chromosome 2, but this chromosome is divided into two separate chromosomes in the other three species. This explains why human cells have 23 pairs of chromosomes while ape cells have 24. The fusion of the two smaller chromosomes during the development of the

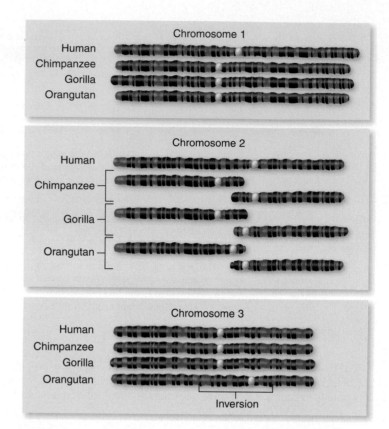

Figure 23.18   **An example of genomic evolution.**   This figure is a comparison of banding patterns among the three largest human chromosomes and the corresponding chromosomes in apes. It is a schematic drawing of Giemsa-stained chromosomes. The differences between these chromosomes illustrate the changes that have occurred during the evolution of these related species.

human lineage may have caused this difference in chromosome pair numbers. Another interesting change in chromosome structure is seen in chromosome 3. The banding patterns among humans, chimpanzees, and gorillas are very similar, but the orangutan has a large inversion that flips the arrangement of bands in the centromeric region. As discussed in Chapter 25, changes in chromosome structure and number may affect the ability of two organisms to breed with one another. In this way, such changes have been important in the establishment of new species.

CHAPTER SUMMARY

- Biological evolution is a heritable change in one or more characteristics of a population or species across many generations.

23.1 The Theory of Evolution

- Charles Darwin proposed the theory of evolution based on his understanding of geology and population growth and his observations of species in their natural settings. His voyage on the *Beagle*, during which he studied many species including finches on the Galápagos Islands, was particularly influential. (Figure 23.1, Table 23.1)

- Darwin's theory of evolution involves descent with modification that leads to better adaptation to environmental conditions. According to the modern synthesis, natural selection acts on existing genetic variation over the course of many generations to produce populations of organisms with traits that promote greater reproductive success (adaptations). (Figure 23.2)

23.2 Observations of Evolutionary Change

- Observations of evolution include the fossil record, biogeography, convergent evolution, selective breeding, and homologies. (Table 23.2)

- Fossils show successive evolutionary change over long periods of time. The fossil record often reveals transitional forms that link past ancestors to modern species. (Figures 23.3, 23.4, 23.5)

- The geographical distribution of species, biogeography, provides information on how certain species are evolutionarily related to each other. Often, when populations become isolated, they evolve into a new species. (Figure 23.6)

- Convergent evolution involves independent adaptations resulting in analogous structures that are similar because organisms have evolved in similar environments. (Figure 23.7)

- Selective breeding, also known as artificial selection, illustrates how changes in genetic variation over the course of many generations can dramatically change the traits of organisms. (Figures 23.8, 23.9, 23.10)

- The Grants showed that natural selection can promote changes in beak size in the medium ground finch. (Figures 23.11, 23.12)

- Homologous structures are similar because they are derived from the same ancestral structure. The set of bones in the arms of vertebrates is one example. (Figure 23.13)

- Vestigial structures are found in species because they are derived from structures that were once functional, but have degenerated because they no longer have use in a modern species. (Table 23.3)

- Homology also occurs during development and at the molecular level of gene and protein sequences. (Figure 23.14)

23.3 The Molecular Processes That Underlie Evolution

- Molecular evolution refers to the molecular changes in genetic material that underlie the phenotypic changes associated with evolution.

- Homologous genes are derived from the same ancestral gene. They accumulate random mutations that make their sequences similar but somewhat different. Orthologs are homologous genes in different species. (Figure 23.15)

- Paralogs are homologous genes in the same species, produced by gene duplication events. Paralogs constitute a gene family. An example is the globin gene family that promoted the evolutionary adaptation of internal gestation.

- Exon shuffling is a form of mutation in which exons are inserted into genes and thereby create proteins with additional functional domains. (Figure 23.16)

- Another mechanism that creates genetic variation is horizontal gene transfer, in which genetic material is transferred between different species. Such genetic changes are acted upon by natural selection. (Figure 23.17)

- Evolution is also associated with changes in chromosome structure and chromosome number. (Figure 23.18)

TEST YOURSELF

1. The process involving changes in one or more characteristics of a population that are heritable and occur across many generations is called
   a. natural selection.
   b. sexual selection.
   c. population genetics.
   d. biological evolution.
   e. inheritance of acquired characteristics.

2. Lamarck's vision of evolution differed from Darwin's in that Lamarck believed
   a. living things evolved in an upward direction.
   b. behavioral changes modified heritable traits.
   c. genetic differences among individuals in the population allowed for evolution.
   d. a and b only
   e. none of the above.

3. Which of the following scientists influenced Darwin's views on the nature of population growth?
   a. Cuvier
   b. Malthus
   c. Lyell
   d. Hutton
   e. Wallace

4. An evolutionary change in which an organism's characteristics change in ways that make it better suited to its environment is
   a. natural selection.
   b. an adaptation.
   c. an acquired characteristic.
   d. evolution.
   e. both a and c.

5. Vestigial structures are anatomical structures
   a. that have more than one function.
   b. that have no function.
   c. that look similar in different species but have different functions.
   d. that have the same function in different species but have very different appearances.
   e. of the body wall.

6. Which of the following is an example of developmental homologies seen in human embryonic development?
   a. gill ridges
   b. umbilical cord
   c. tail
   d. both a and c
   e. all of the above

7. Two or more homologous genes found within a particular species are called
   a. homozygous.
   b. orthologs.
   c. paralogs.
   d. heterologs.
   e. duplicates.

8. The phenomenon of exon shuffling
   a. creates new gene products by changing the pattern of intron removal in a particular gene.
   b. creates new genes by inserting exons and flanking introns into a different gene sequence, thereby introducing a new domain in the gene product.
   c. rearranges the sequence of exons in a single gene.
   d. rearranges the introns in a particular gene, creating new gene products.
   e. both a and d.

9. Horizontal gene transfer is
   a. the transmission of genetic information from parent to offspring.
   b. the exchange of genetic material among individuals of the same species.
   c. the exchange of genetic material between mates.
   d. the exchange of genetic material among individuals of different species.
   e. none of the above.

10. Genetic variation can increase as a result of
    a. random mutations in genes.
    b. exon shuffling.
    c. gene duplication.
    d. horizontal gene transfer.
    e. all of the above.

CONCEPTUAL QUESTIONS

1. Briefly describe the various observations that support the theory of biological evolution.

2. Define convergent evolution and give an example.

3. Explain how homologous forelimbs of vertebrates support the idea of biological evolution.

EXPERIMENTAL QUESTIONS

1. What features of Daphne Major made it a suitable field site for studying the effects of natural selection?

2. Why is beak depth in finches a good trait for a study of natural selection? What environmental conditions were important to allow the Grants to collect information concerning natural selection?

3. What were the results of the Grants' study following the drought in 1977? What impact did these results have on the theory of evolution?

COLLABORATIVE QUESTIONS

1. Discuss evolution and how it occurs.

2. Discuss horizontal gene transfer.

---

www.brookerbiology.com

This website includes answers to the Biological Inquiry questions found in the figure legends and all end-of-chapter questions.

24

POPULATION GENETICS

CHAPTER OUTLINE

Colorful African cichlids. The choice of mates among populations of cichlids may depend on color.

Population genetics is the study of genes and genotypes in a population. The central issue in population genetics is genetic variation. Population geneticists want to know the extent of genetic variation within populations, why it exists, and how it changes over the course of many generations. Population genetics helps us to understand how underlying genetic variation is related to phenotypic variation, and other issues such as mate preference (see chapter opening photo).

Population genetics emerged as a branch of genetics in the 1920s and 1930s. Its mathematical foundations were developed by theoreticians who extended the principles of Mendel and Darwin by deriving equations to explain the occurrence of genotypes within populations. These foundations can be largely attributed to British evolutionary biologists J. B. S. Haldane and Ronald Fisher, and American geneticist Sewall Wright. As we will see, several researchers who analyzed the genetic composition of natural and experimental populations provided support for their mathematical theories. More recently, population geneticists have used techniques to probe genetic variation at the molecular level. In addition, the staggering improvement in computer technology has aided population geneticists in the analysis of their genetic theories and data.

In this chapter, we will explore the extent of genetic variation that occurs in populations and how such variation is subject to change. In many cases, such changes are associated with adaptations, which are characteristics of a species that have evolved over a long period of time by the process of natural selection. The concept of adaptation was discussed in Chapter 23.

This chapter will examine the various ways that natural selection leads to adaptation.

24.1 Genes in Populations

Population genetics is an extension of our understanding of Darwin's theory of natural selection, Mendel's laws of inheritance, and newer studies in molecular genetics. All of the genes in a population make up its **gene pool.** Each member of the population receives its genes from its parents, which, in turn, are members of the gene pool. Individuals that reproduce contribute to the gene pool of the next generation. Population geneticists study the genetic variation within the gene pool, and how such variation changes from one generation to the next. The emphasis is often focused on an understanding of variation in alleles between members of a population. As discussed in Chapter 16, alleles are different forms of the same gene. In this section, we will examine some of the general features of populations and gene pools.

A Population Is a Group of Interbreeding Individuals

A **population** is a group of individuals of the same species that can interbreed with one another. Certain species occupy a wide geographic range and are divided into discrete populations. For example, distinct populations of a given species may be located on different continents. (A more detailed description of populations and their native environments is given in Chapter 56.)

Another common situation is that a large mountain or some other type of geographic barrier may separate two or more populations on the same continent.

Populations are dynamic units that change from one generation to the next. They may change in number, geographic location, and genetic composition. Natural populations may go through cycles of "feast or famine," during which the population gains or loses individuals. In addition, natural predators or disease may periodically decrease the size of a population significantly, and then later the population may rebound to its original size. Populations or individuals within populations may migrate to a new site and establish a distinct population at a new location that may differ in environment from the original site.

As population sizes and locations change, their genetic composition generally changes as well. Some of the genetic changes involve adaptive evolution, which means that a species is better adapted to its environment, making it more likely to survive and reproduce. For example, a population of mammals may move from a warmer to a colder geographic location. Over the course of many generations, adaptive evolution may change the population such that the fur of the animals is thicker and provides better insulation against the colder temperatures.

Figure 24.1   An example of polymorphism: the two color variations found in the orchid *Dactylorhiza sambucina*.

GENOMES & PROTEOMES

Genes in Natural Populations Are Usually Polymorphic

The term **polymorphism** (meaning many forms) refers to the phenomenon that many traits display variation within a population. Historically, polymorphism first referred to variation in phenotypes. Polymorphisms in color and pattern have long attracted the attention of population geneticists. **Figure 24.1** illustrates a striking example of polymorphism in the elder-flowered orchid (*Dactylorhiza sambucina*). Throughout the range of this species in Europe, both yellow- and red-flowered individuals are prevalent.

From a genetic perspective, polymorphism is due to two or more alleles that influence the phenotype of the individual that inherits them. In other words, it is due to genetic variation. Geneticists also use the term polymorphism to describe the variation in genes; this is sometimes called genetic polymorphism. A gene that commonly exists as two or more alleles in a population is described as a **polymorphic gene**. By comparison, a **monomorphic gene** exists predominantly as a single allele in a population. By convention, when 99% or more of the alleles of a given gene are identical, the gene is considered to be monomorphic. Said another way, a polymorphic gene must have one or more additional alleles that make up more than 1% of the alleles in the population.

At the level of a particular gene, a polymorphism may involve various types of changes such as a deletion of a significant region of the gene, a duplication of a region, or a change in a single nucleotide. This last phenomenon is called a single-nucleotide polymorphism (SNP). SNPs ("snips") are the smallest type of genetic change that can occur within a given gene, and they are also the most common. In human populations, for example, SNPs represent 90% of all the variation in human DNA sequences that occurs among different people. Current estimates indicate that SNPs with a frequency of 1% or more are found very frequently in genes. In humans, a gene that is 2,000–3,000 bp in length will, on average, contain 10 different SNPs in the human population. The high frequency of SNPs indicates that polymorphism is the norm for most human genes. Likewise, relatively large, healthy populations of nearly all species exhibit a high level of genetic variation, as evidenced by the occurrence of SNPs within most genes. As discussed later in this chapter, genetic variation provides the raw material for populations to evolve over the course of many generations.

Population Genetics Is Concerned with Allele and Genotype Frequencies

To analyze genetic variation in populations, one approach is to consider the frequency of alleles in a quantitative way. Two fundamental calculations are central to population genetics: **allele frequencies** and **genotype frequencies**. Allele and genotype frequencies are defined as:

$$\text{Allele frequency} = \frac{\text{Number of copies of a specific allele in a population}}{\text{Total number of all alleles for that gene in a population}}$$

$$\text{Genotype frequency} = \frac{\text{Number of individuals with a particular genotype in a population}}{\text{Total number of individuals in a population}}$$

Though these two frequencies are related, make sure you keep in mind a clear distinction between them. As an example, let's consider a population of 100 four o'clock plants with the following genotypes:

49 red-flowered plants with the genotype *RR*   49/100

42 pink-flowered plants with the genotype *Rr*

9 white-flowered plants with the genotype *rr*

When calculating an allele frequency for diploid species, remember that homozygous individuals have two copies of an allele, whereas heterozygotes have only one. For example, in tallying the *r* allele, each of the 42 heterozygotes has one copy of the *r* allele, and each white-flowered plant has two copies. Therefore, the allele frequency for *r* equals

$$\text{Frequency of } r = \frac{(Rr) + 2(rr)}{2(RR) + 2(Rr) + 2(rr)}$$

$$\text{Frequency of } r = \frac{42 + (2)(9)}{(2)(49) + (2)(42) + (2)(9)}$$

$$= \frac{60}{200} = 0.3, \text{ or } 30\%$$

This result tells us that the allele frequency of *r* is 0.3. In other words, 30% of the alleles for this gene in the population are the *r* allele.

Let's now calculate the genotype frequency of *rr* (white-flowered) plants.

$$\text{Frequency of } rr = \frac{9}{49 + 42 + 9}$$

$$= \frac{9}{100} = 0.09, \text{ or } 9\%$$

We see that 9% of the individuals in this population have white flowers.

Allele and genotype frequencies are always less than or equal to one (that is, less than or equal to 100%). If a gene is monomorphic, the allele frequency for the single allele will equal or be close to a value of 1.0. For polymorphic genes, if we add up the frequencies for all the alleles in the population, we should obtain a value of 1.0. In our four o'clock example, the allele frequency of *r* equals 0.3. Therefore, we can calculate the frequency of the other allele, *R*, as equal to 1.0 − 0.3 = 0.7, because they must add up to 1.0.

The Hardy-Weinberg Equation Relates Allele and Genotype Frequencies in a Population

In 1908, Godfrey Harold Hardy, an English mathematician, and Wilhelm Weinberg, a German physician, independently derived a simple mathematical expression called the Hardy-Weinberg equation that relates allele and genotype frequencies when they are not changing. Let's examine the Hardy-Weinberg equation using the population of four o'clock plants that we have just considered. If the allele frequency of *R* is denoted by the variable *p*, and the allele frequency of *r* by *q*, then

$$p + q = 1$$

For example, if *p* = 0.7, then *q* must be 0.3. In other words, if the allele frequency of *R* equals 70%, the remaining 30% of alleles must be *r*, because together they equal 100%.

For a gene that exists in two alleles, the Hardy-Weinberg equation states that

$(p + q)^2 = 1$ (Note: the number 2 in this equation reflects the fact that the genotype is due to the inheritance of two alleles, one from each parent.)

Therefore

$$p^2 + 2pq + q^2 = 1 \text{ (the \textbf{Hardy-Weinberg equation})}$$

If we apply this equation to our flower color gene, then

p^2 equals the genotype frequency of *RR*

$2pq$ equals the genotype frequency of *Rr*

q^2 equals the genotype frequency of *rr*

If *p* = 0.7 and *q* = 0.3, then

Frequency of $RR = p^2 = (0.7)^2 = 0.49$

Frequency of $Rr = 2pq = 2(0.7)(0.3) = 0.42$

Frequency of $rr = q^2 = (0.3)^2 = 0.09$

In other words, if the allele frequency of *R* is 70% and the allele frequency of *r* is 30%, the expected genotype frequency of *RR* is 49%, *Rr* is 42%, and *rr* is 9%.

To see the relationship between allele frequencies and genotypes in a population, **Figure 24.2** considers the relationship between allele frequencies and the way that gametes combine to produce genotypes. The Hardy-Weinberg equation reflects the way gametes combine randomly with each other to produce offspring. In a population, the frequency of a gamete carrying a particular allele is equal to the allele frequency in that population. For example, if the allele frequency of *R* equals 0.7, the frequency of a gamete carrying the *R* allele also equals 0.7. The frequency of producing an *RR* homozygote, which produces red flowers, is 0.7 × 0.7 = 0.49, or 49%. The probability of inheriting both *r* alleles, which produces white flowers, is 0.3 × 0.3 = 0.09, or 9%. In our Punnett square, two different gamete combinations can produce heterozygotes with pink flowers (Figure 24.2). An offspring could inherit the *R* allele from pollen and *r* from the egg, or *R* from the egg and *r* from pollen. Therefore, the frequency of heterozygotes is *pq* + *pq*, which equals 2*pq*. In our example, this is 2(0.7)(0.3) = 0.42, or 42%.

The Hardy-Weinberg equation predicts an **equilibrium** of unchanging allele and genotype frequencies in a population. If a population is in equilibrium, it is not adapting and evolution is not occurring. However, this prediction is valid only if certain conditions are met in a population. These conditions require that evolutionary mechanisms, those forces that can change allele and

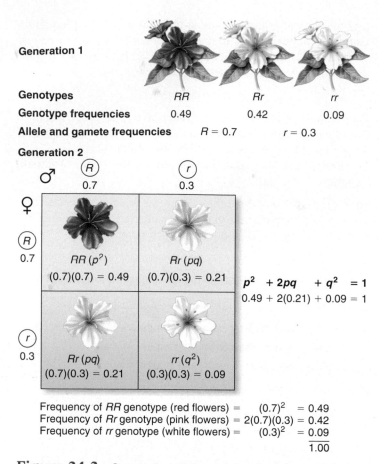

Figure 24.2 Comparing allele and genotype frequencies in a population with the Hardy-Weinberg equation and a Punnett square. A population of four o'clock plants has allele and gamete frequencies of 0.7 for the *R* allele and 0.3 for the *r* allele. Knowing the allele frequencies allows us to calculate the genotype frequencies in the population.

Biological inquiry: What would be the frequency of pink flowers in a population where the allele frequency of R is 0.4 and the population is in Hardy-Weinberg equilibrium? Assume that R and r are the only two alleles.

genotype frequencies, are not acting on a population. With regard to a particular gene of interest, these conditions are:

- The population is so large that allele frequencies do not change due to random sampling error.
- The members of the population mate with each other without regard to their phenotypes and genotypes.
- No migration occurs between different populations.
- No survival or reproductive advantage exists for any of the genotypes—in other words, no natural selection occurs.
- No new mutations occur.

In reality, no population satisfies the Hardy-Weinberg equilibrium completely. Nevertheless, in large natural populations with little migration and negligible natural selection, the Hardy-Weinberg equilibrium may be nearly approximated for certain

genes. However, researchers often discover instead that allele and genotype frequencies for one or more genes in a given species are not in Hardy-Weinberg equilibrium. In such cases, we would say that the population is in disequilibrium—in other words, evolutionary mechanisms are affecting the population. When this occurs, population geneticists may wish to identify the reason(s) why disequilibrium has occurred because this may impact the future survival of the species.

24.2 Evolutionary Mechanisms and Their Effects on Populations

The genetic variation in all natural populations changes over the course of many generations. The term **microevolution** is used to describe changes in a population's gene pool from generation to generation. Such change is rooted in two related phenomena (**Table 24.1**). First, the introduction of new genetic variation into a population is one essential aspect of microevolution. As discussed in Chapter 23, genetic variation can originate by a variety of molecular mechanisms. New alleles of pre-existing genes can arise by random mutation, and new genes can be introduced into a population by gene duplication, exon shuffling, and horizontal gene transfer. Such mutations, albeit rare, provide a continuous source of new variation to populations. In 1926, the Russian geneticist Sergei Tshetverikov was the first to suggest that random mutations are the raw material for evolution, but they do not constitute evolution itself. Mutations clearly supply new genetic variation to a population. However, due to their low rate of occurrence, mutations do not act as a major force in promoting widespread changes in a population. If mutations were the only type of change occurring in a population, that population would not evolve because mutations are so rare.

In this section we will discuss the second phenomenon that is required for microevolution, the action of evolutionary mechanisms that alter the prevalence of a given allele or genotype in a population. These mechanisms are natural selection, random genetic drift, migration, and nonrandom mating (Table 24.1). The collective contributions of these evolutionary mechanisms over the course of many generations have the potential to promote widespread genetic changes in a population.

To consider the effects of these evolutionary mechanisms, we will examine how they may affect the type of genetic variation that occurs when a gene exists in two alleles in a population. As you will learn, these mechanisms may cause one allele or the other allele to be favored, or they may create a balance where both alleles are maintained in a population. Although we will discuss the effects of these mechanisms on genetic variation involving alleles of a single gene caused by mutation, keep in mind that these same evolutionary mechanisms can also affect the frequencies of new genes that arise in a population by gene duplication, exon shuffling, and horizontal gene transfer.

Table 24.1	Factors That Govern Microevolution
Sources of new genetic variation*	
New alleles	Random mutations within pre-existing genes introduce new alleles into populations, but at a very low rate. New mutations may be beneficial, neutral, or deleterious. For alleles to rise to a significant percentage in a population, evolutionary mechanisms, such as natural selection, random genetic drift, and migration, must operate on them.
Gene duplication	Abnormal crossover events and transposable elements may increase the number of copies of a gene. Over time, the additional copies can accumulate random mutations and create a gene family.
Exon shuffling	Abnormal crossover events and transposable elements may promote gene rearrangements in which one or more exons from one gene are inserted into another gene. The protein encoded by such a gene may display a novel function and can then be acted upon by evolutionary mechanisms.
Horizontal gene transfer	Genes from one species may be introduced into another species. Events such as endocytosis and interspecies mating may promote this phenomenon.
Evolutionary mechanisms that alter existing genetic variation	
Natural selection	The phenomenon in which the environment selects for individuals that possess certain traits. Natural selection can favor the survival of members with beneficial traits or disfavor the survival of individuals with unfavorable traits. As a type of natural selection, sexual selection favors traits that increase the reproductive success of individuals.
Random genetic drift	This is a change in genetic variation from generation to generation due to random sampling error. Allele frequencies may change as a matter of chance from one generation to the next. This is much more likely to occur in a small population.
Migration	Migration can occur between two different populations that have different allele frequencies. The introduction of migrants into a recipient population may change the allele frequencies of that population.
Nonrandom mating	The phenomenon in which individuals select mates based on their phenotypes or genetic lineage. This can alter the relative proportion of homozygotes and heterozygotes that is predicted by the Hardy-Weinberg equation, but it will not change allele frequencies.

*Described in Chapter 23.

Natural Selection Favors Individuals with the Greatest Reproductive Success

As we discussed in Chapter 23, Charles Darwin and Alfred Wallace independently proposed the theory of evolution by natural selection. According to this theory, only a certain percentage of the offspring that a species produces will survive. This "struggle for existence" results in the selective survival of individuals that have inherited certain genotypes. Such genotypes confer greater **reproductive success**. In this regard, natural selection usually acts upon two aspects of reproductive success. First, certain characteristics make organisms better adapted to their environment and more likely to survive to reproductive age; such organisms have a greater chance to reproduce and contribute offspring to the next generation. Therefore, natural selection favors individuals with adaptations that provide a survival advantage. Second, natural selection favors individuals that produce viable offspring. As discussed later in this chapter, traits that enhance the ability of individuals to reproduce are often subject to natural selection.

Let's consider how natural selection can operate when a gene exists as two alleles in a population. Keep in mind that natural selection acts on individuals, while evolution occurs at the population level. A modern description of natural selection can relate our knowledge of molecular genetics to the phenotypes of individuals.

1. Within a population, allelic variation arises from random mutations that cause differences in DNA sequences. A mutation that creates a new allele may alter the amino acid sequence of the encoded protein. This in turn may alter the function of the protein.
2. Some alleles may encode proteins that enhance an individual's survival or reproductive capability compared to that of other members of the population. For example, an allele may produce a protein that is more efficient at a higher temperature, conferring on the individual a greater probability of survival in a hot climate.
3. Individuals with beneficial alleles are more likely to survive and contribute their alleles to the gene pool of the next generation.
4. Over the course of many generations, allele frequencies of many different genes may change through natural selection, thereby significantly altering the characteristics of a population. The net result of natural selection is a population that is better adapted to its environment and more successful at reproduction.

As mentioned earlier, Haldane, Fisher, and Wright developed mathematical relationships to explain the phenomenon of natural selection. To begin our quantitative discussion of natural selection, we need to consider the concept of **Darwinian fitness**, which is the relative likelihood that a genotype will contribute to the gene pool of the next generation as compared with other genotypes. Although this property often correlates with physical fitness, the two ideas should not be confused. Darwinian fitness is a measure of reproductive success. An extremely fertile individual may have a higher Darwinian fitness than a less fertile individual that appears more physically fit.

To examine Darwinian fitness, let's consider an example of a hypothetical gene existing in A and a alleles. We can assign fitness values to each of the three possible genotypes according to their relative reproductive success. For example, let's suppose that the average reproductive successes of the three genotypes are:

> AA produces five offspring
>
> Aa produces four offspring
>
> aa produces one offspring

By convention, the genotype with the highest reproductive success is given a fitness value of 1.0. Fitness values are denoted by the variable W. The fitness values of the other genotypes are assigned values relative to this 1.0 value.

> Fitness of AA: $W_{AA} = 1.0$
>
> Fitness of Aa: $W_{Aa} = 4/5 = 0.8$
>
> Fitness of aa: $W_{aa} = 1/5 = 0.2$

Variation in fitness occurs because individuals with certain genotypes have greater reproductive success. Such genotypes exhibit a higher fitness compared to others. Natural selection acts on phenotypes that are derived from an individual's genotype.

Likewise, the effects of natural selection can be viewed at the level of a population. The average reproductive success of members of a population is called the **mean fitness of the population**. Over many generations, as individuals with higher fitness values become more prevalent, natural selection also in-

creases the mean fitness of the population. In this way, the process of natural selection results in a population of organisms that is well adapted to its native environment and likely to be successful at reproduction.

Natural Selection Can Follow Different Patterns

By studying species in their native environments, population geneticists have discovered that natural selection can occur in several ways. In most of the examples described next, natural selection leads to adaptation so that a species is better able to survive to reproductive age.

Directional Selection **Directional selection** favors individuals at one extreme of a phenotypic distribution that have greater reproductive success in a particular environment. Different phenomena may initiate the process of directional selection. One way that directional selection may arise is that a new allele may be introduced into a population by mutation, and the new allele may confer a higher fitness in individuals that carry it (**Figure 24.3**). If the homozygote carrying the favored allele has the highest fitness value, directional selection may cause this favored allele to eventually become predominant in the population, perhaps even becoming a monomorphic allele.

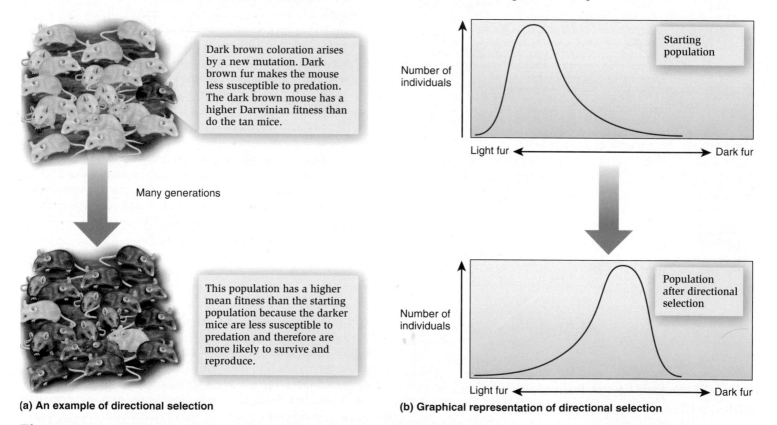

(a) An example of directional selection

(b) Graphical representation of directional selection

Figure 24.3 **Directional selection.** This pattern of natural selection selects for one extreme of a phenotype that confers the highest fitness in the population's environment. **(a)** In this example, a mutation for darker fur arises in a population of mice. This new genotype confers higher Darwinian fitness, because mice with dark fur can evade predators and are more likely to survive and reproduce. Over many generations, directional selection will favor the prevalence of darker individuals. **(b)** These graphs show the change in fur color phenotypes in this mouse population before and after directional selection.

Biological inquiry: Over the short and long run, does directional selection favor the preservation of genetic diversity?

Another possibility is that a population may be exposed to a prolonged change in its living environment. Under the new environmental conditions, the relative fitness values may change to favor one genotype, and this will promote the elimination of other genotypes. As an example, let's suppose a population of finches on a mainland already has genetic variation that affects beak size (refer back to Figure 23.2). A small number of birds migrate to an island where the seeds are generally larger than they are on the mainland. In this new environment, birds with larger beaks would have a higher fitness because they would be better able to crack open the larger seeds, and thereby survive to reproductive age. Over the course of many generations, directional selection would produce a population of birds carrying alleles that promote larger beak size.

Stabilizing Selection **Stabilizing selection** favors the survival of individuals with intermediate phenotypes. The extreme values of a trait are selected against. Stabilizing selection tends to decrease genetic diversity. An example of stabilizing selection involves clutch size (number of eggs laid) in birds, which was first proposed by British biologist David Lack in 1947. Under stabilizing selection, birds that lay too many or too few eggs per nest have lower fitness values than do those that lay an intermediate number of eggs (**Figure 24.4**). Laying too many eggs has the disadvantage that many offspring will die due to inadequate

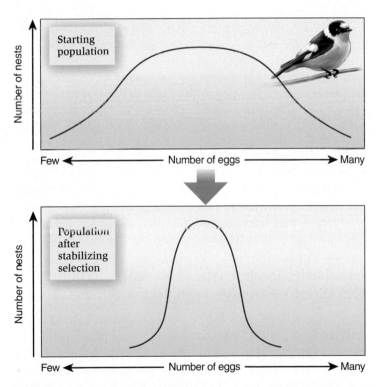

Figure 24.4 **Stabilizing selection.** In this pattern of natural selection, the extremes of a phenotypic distribution are selected against. Those individuals with intermediate traits have the highest fitness. These graphs show the results of stabilizing selection on clutch size in a population of collared flycatchers (*Ficedula albicollis*). This process results in a population with less diversity and more uniform traits.

parental care and food. In addition, the strain on the parents themselves may decrease their likelihood of survival and therefore their ability to produce more offspring. Having too few offspring, however, does not contribute many individuals to the next generation. Therefore, the most successful parents are those that produce an intermediate clutch size. In the 1980s, Swedish evolutionary biologist Lars Gustafsson and his colleagues examined the phenomenon of stabilizing selection in the collared flycatcher (*Ficedula albicollis*) on the island of Gotland south of Sweden. They discovered that Lack's hypothesis concerning an optimal clutch size appears to be true for this species.

Disruptive Selection **Disruptive selection** (also known as diversifying selection) favors the survival of two or more different genotypes that produce different phenotypes. In disruptive selection, the fitness values of a particular genotype are higher in one environment and lower in a different environment, while the fitness values of the other genotype vary in an opposite manner. Disruptive selection is likely to occur in populations that occupy diverse environments, so that some members of the species will survive in each type of environmental condition. An example involves colonial bentgrass (*Agrostis tenuis*). In certain locations where this grass is found, such as South Wales, isolated places occur where the soil is contaminated with high levels of heavy metals such as copper due to human activities such as mining. The relatively recent metal contamination has selected for the proliferation of mutant strains that show tolerance to the heavy metals. Such genetic changes enable the plants to grow on contaminated soil but tend to inhibit growth on normal, noncontaminated soil. These metal-resistant plants often grow on contaminated sites that are close to plants that grow on uncontaminated land and do not show metal tolerance (**Figure 24.5**).

In the case of metal-resistant and metal-sensitive grasses, the members of a population occupy heterogeneous environments that are geographically continuous; members of the populations can freely interbreed. In other cases, members of a single species may occupy two or more different environments that are geographically isolated from each other. Given enough time, disruptive selection due to heterogeneous environments can eventually lead to the evolution of two or more different species, a process that will be described in Chapter 25.

Balancing Selection Contrary to a popular misconception, natural selection does not always cause the elimination of "weaker" or less fit alleles. **Balancing selection** is a type of natural selection that maintains genetic diversity in a population. Over many generations, balancing selection can create a situation known as a **balanced polymorphism**, or a stable polymorphism, in which two or more alleles are kept in balance, and therefore are maintained in a population over the course of many generations.

Balancing selection does not favor one particular allele in the population. Population geneticists have identified two common ways that this pattern of selection can occur. First, for genetic variation involving a single gene, balancing selection favors the heterozygote rather than either corresponding homozygote.

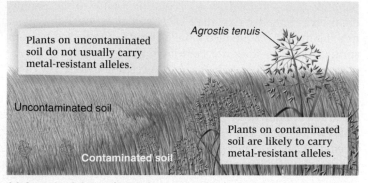

(a) Growth of *Agrostis tenuis* on uncontaminated and contaminated soil

Figure 24.5   **Disruptive selection.**   This pattern of natural selection selects for two different phenotypes, each of which is most fit in a particular environment. **(a)** In this example, mutations have created metal-resistant alleles in colonial bentgrass (*Agrostis tenuis*), that allow it to grow on soil contaminated with high levels of heavy metals such as copper. These alleles provide high fitness where the soil is contaminated, but they confer low fitness where it is not contaminated. Because both metal-resistant and metal-sensitive alleles are maintained in the population, this situation is an example of disruptive selection due to heterogeneous environments. **(b)** These graphs show the change in phenotypes in this bentgrass population before and after disruptive selection.

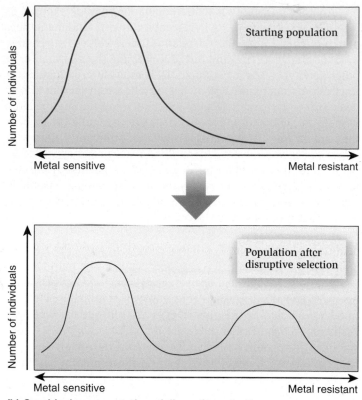

(b) Graphical representation of disruptive selection

This situation is called **heterozygote advantage**. Balanced polymorphisms can sometimes explain the high frequency of alleles that are deleterious in a homozygous condition. A classic example is the H^S allele of the human β-globin gene, which is described in Chapter 14. A homozygous $H^S H^S$ individual has sickle-cell anemia, a disease that leads to the sickling of the red blood cells. The $H^S H^S$ homozygote has a lower fitness than a homozygote with two copies of the more common β-globin allele, $H^A H^A$. However, the heterozygote, $H^A H^S$, has the highest level of fitness in areas where malaria is endemic. Compared with $H^A H^A$ homozygotes, heterozygotes have a 10–15% better chance of survival if infected by the malarial parasite, *Plasmodium falciparum*. Therefore, the H^S allele is maintained in populations where malaria is prevalent, even though the allele is detrimental in the homozygous state (**Figure 24.6**). The balanced polymorphism results in a higher mean fitness of the population. In areas where malaria is endemic, a population composed of all $H^A H^A$ individuals would have a lower mean fitness.

Negative frequency-dependent selection is a second way that natural selection can produce a balanced polymorphism. In this pattern of natural selection, the fitness of a genotype decreases when its frequency becomes higher. In other words, rare individuals have a higher fitness and common individuals have a lower fitness. Therefore, rare individuals are more likely to reproduce while common individuals are less likely, thereby producing a balanced polymorphism in which no genotype becomes too rare or too common.

An interesting example of negative frequency-dependent selection involves the elder-flowered orchid (*D. sambucina*), which was shown earlier in Figure 24.1. Throughout its range

both yellow- and red-flowered individuals are prevalent. The explanation for this polymorphism is related to its pollinators, which are mainly bumblebees such as *Bombus lapidarius* and *Bombus terrestris*. The pollinators increase their preference for the flower color of *D. sambucina* as it becomes less common in a given area. One reason why this may occur is because *D. sambucina* is a rewardless flower—it does not provide its pollinators with any reward such as sweet nectar. Pollinators are more likely to learn that the more common color of *D. sambucina* in a given area does not offer a reward, and this may explain their preference for the flower color that is less common. For example, in an area where the yellow-colored flowers are common, bumblebees may have learned that this color does not offer a reward, so they are more likely to visit red-flowered plants.

Sexual Selection Is a Type of Natural Selection That Directly Promotes Reproductive Success

Thus far we have mainly focused on examples of natural selection that favor traits that promote the survival of individuals to reproductive age. This form of natural selection often produces adaptations for survival in particular environments. Now let's turn our attention to a form of natural selection, called **sexual selection**, that is directed at certain traits of sexually reproducing species that make it more likely for individuals to find or choose a mate and/or engage in successful mating. Darwin originally described sexual selection as "the advantage that certain individuals have over others of the same sex and species solely with respect to reproduction." Within a species, members of the same sex (typically males) compete with each other for

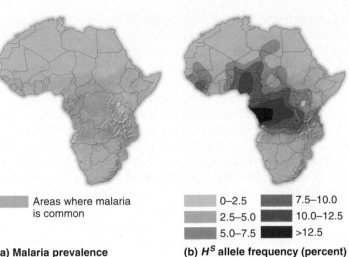

(a) Malaria prevalence   **(b) H^S allele frequency (percent)**

Areas where malaria is common

0–2.5	7.5–10.0
2.5–5.0	10.0–12.5
5.0–7.5	>12.5

Figure 24.6 Balancing selection and heterozygote advantage. In this pattern of natural selection, genetic diversity is maintained in a population. This example shows balancing selection maintaining two alleles of the β-globin gene, designated H^A and H^S, in human populations in Africa. This situation occurs due to heterozygote advantage, because the heterozygous condition confers resistance to malaria. **(a)** The geographic prevalence of malaria in Africa. **(b)** The frequency of the H^S allele of the β-globin gene in the same area. In the homozygous condition, the H^S allele causes the disease sickle-cell anemia. However, this sickle-cell allele is maintained in human populations as a balanced polymorphism, because in areas where malaria is prevalent, the heterozygote carrying one copy of the H^S allele has a higher fitness than either of the corresponding homozygotes ($H^A H^A$ and $H^S H^S$).

(a) Intrasexual selection   **(b) Intersexual selection**

Figure 24.7 Examples of the results of sexual selection, a type of natural selection. **(a)** An example of intrasexual selection. The enlarged claw of the male fiddler crab is used in direct male-to-male competition. In this photograph, a male inside a burrow is extending its claw out of the burrow to prevent another male from entering. **(b)** An example of intersexual selection. Female peacocks choose males based on their colorful and long tail feathers, and the robustness of their display.

the opportunity to mate with members of the opposite sex. Such competition results in sexual selection.

In many species of animals, sexual selection affects male characteristics more intensely than it does female. Unlike females, which tend to be fairly uniform in their reproductive success, male success tends to be more variable, with some males mating with many females and others not mating at all. (See Chapter 55 for a discussion of different mating strategies between the sexes.) Sexual selection results in the evolution of traits, called secondary sexual characteristics, that favor reproductive success. The result of this process is sometimes a significant difference between the appearances of the sexes in the same species, a situation called sexual dimorphism.

Sexual selection can be categorized as either **intrasexual selection**, between members of the same sex, or **intersexual selection**, between members of the opposite sex. Let's begin with intrasexual selection. Examples of traits that result from intrasexual selection in animals include horns in male sheep, antlers in male moose, and the enlarged claw of male fiddler crabs (**Figure 24.7a**). In many animal species, males directly compete with each other for the opportunity to mate with females, or they may battle for a particular territory. In fiddler crabs (*Uca paradussumieri*), males enter the burrows of females that are ready to mate. If another male attempts to enter the burrow, the male already inside the burrow stands in the burrow shaft and blocks the entrance with his enlarged claw.

Now let's consider an example of intersexual selection, namely female choice. This type of sexual selection often results in showy characteristics in males. **Figure 24.7b** shows a classic example that involves the Indian peacock (*Pavo cristatus*), the national bird of India. Male peacocks have long and brightly colored tail feathers, which they fan out as a mating behavior. Females select among males based on feather color and pattern and physical prowess of the display.

A less obvious type of intersexual selection is cryptic female choice, in which the female reproductive system can influence the relative success of sperm. As an example of cryptic female choice, the female genital tract of certain animals selects for sperm that tend to be genetically unrelated to the female. Sperm from males closely related to the female, such as brothers or cousins, are less successful than are sperm from genetically unrelated males. The selection for sperm may occur over the journey through the reproductive tract. The egg itself may even have mechanisms to prevent fertilization by genetically related sperm. Cryptic female choice occurs in species in which females may mate with more than one male, such as many species of reptiles and ducks. A similar mechanism is found in many plant species in which pollen from genetically related plants, perhaps from the same flower, is unsuccessful at fertilization, while pollen from unrelated plants is successful. One possible advantage of cryptic female choice is that it inhibits inbreeding, which is described later in this chapter. At the population level, cryptic female choice may promote genetic diversity by favoring interbreeding among genetically unrelated individuals.

Sexual selection is sometimes a combination of both intrasexual and intersexual selection. During breeding season, male elk (*Cervus elaphus*) become aggressive and bugle loudly to challenge other male elk. Males spar with their antlers, which usually turns into a pushing match to determine which elk is stronger. Female elk then choose the strongest bulls as their mates.

Sexual selection can explain traits that may decrease an individual's chances of survival but increase their chances of reproducing. For example, the male guppy (*Poecilia reticulata*) is brightly colored compared to the female. In nature, females prefer brightly colored males. Therefore, in places with few predators, the males tend to be brightly colored. However, in places where predators are abundant, brightly colored males are less plentiful because they are subject to predation. In this case, the relative abundance of brightly and dully colored males depends on the balance between sexual selection, which favors bright coloring, and escape from predation, which favors dull coloring.

Many animals have secondary sexual characteristics, and evolutionary biologists generally agree that sexual selection is responsible for such traits. But why should males compete, and why should females be choosy? Researchers have proposed various hypotheses to explain the underlying mechanisms. One possible reason is related to the different roles that males and females play in the nurturing of offspring. In some species, the female is the primary caregiver, while the male plays a minor role. In such species, the Darwinian fitness of males and females may be influenced by their mating behavior. Males increase their fitness by mating with multiple females. This increases their likelihood of passing their genes on to the next generation.

By comparison, females may produce relatively few offspring and their reproductive success may not be limited by the number of available males. Females will have higher fitness if they choose males that are good defenders of their territory, and have alleles that confer a survival advantage to their offspring. One measure of alleles that confer higher fitness is age. Males that live to an older age are more likely to carry beneficial alleles. Many research studies involving female choice have shown that females tend to select traits that are more likely to be well developed in older males than they are in immature males. In certain species of birds, for example, females tend to choose males with a larger repertoire of songs, which is more likely to occur in older males.

Overall, sexual selection is a form of natural selection in which the evolution of certain traits occurs differently between the two sexes. Sexual selection is not some extra force in opposition to natural selection. It is governed by the same processes involved in the evolution of traits that are not directly related to sex. Sexual selection can be directional, stabilizing, disruptive, or balancing. For example, directional selection probably played an important role in the evolution of the large and brightly colored tail of the male peacock. As described next, sexual selection can be diversifying if females select for males with different traits.

FEATURE INVESTIGATION

Seehausen and van Alphen Found That Male Coloration in African Cichlids Is Subject to Female Choice

Cichlids are tropical freshwater fish that are popular among aquarium enthusiasts. This family of fish (*Cichlidae*) is made up of more different species than is any other vertebrate family. The more than 3,000 species vary with regard to body shape, coloration, behavior, and feeding habits. By far the greatest diversity of these fish is found in Lake Victoria, Lake Malawi, and Lake Tanganyika in East Africa, where collectively more than 1,800 species are found. Lake Victoria, for example, has 500 species.

Cichlids have complex mating behavior and brood care. Females play an important role in choosing males with particular characteristics. To study the importance of female choice, population geneticists Ole Seehausen and Jacques van Alphen investigated the effect of male coloration of *Pundamilia pundamilia* and *Pundamilia nyererei*. In some locations, *P. pundamilia* and *P. nyererei* do not readily interbreed and behave like two distinct biological species, while in other places they behave like a single interbreeding species with two color morphs. They can interbreed to produce viable offspring, and both inhabit Lake Victoria. Males of both species have blackish underparts and blackish vertical bars on their sides (**Figure 24.8a**). *P. pundamilia* males are grayish white on top and on the sides, and they

Figure 24.8 Male coloration in African cichlids. (a) Two males (*Pundamilia pundamilia*, top, and *Pundamilia nyererei*, bottom) under normal illumination. (b) The same species under orange monochromatic light, which obscures their color differences.

have a metallic blue and red dorsal fin, which is the uppermost fin. By comparison, *P. nyererei* males are orange on top and yellow on their sides.

Seehausen and van Alphen hypothesized that females choose males for mates based on the males' coloration. The researchers took advantage of the observation that colors are obscured under orange monochromatic light. As seen in **Figure 24.8b**, males of

Figure 24.9 A study by Seehausen and van Alphen involving the effects of male coloration on female choice in African cichlids.

HYPOTHESIS Female African cichlids choose mates based on the males' coloration.

STARTING MATERIALS Two species of cichlid, *Pundamilia pundamilia* and *Pundamilia nyererei*, were chosen. The males differ with regard to their coloration. A total of 8 males and 8 females (4 males and 4 females from each species) were tested.

Experimental level | Conceptual level

1. Place 1 female and 2 males in an aquarium. Each male is within a separate glass enclosure. The enclosures contain 1 male from each species.

This is a method to evaluate sexual selection via female choice in 2 species of cichlid.

2. Observe potential courtship behavior for 1 hour. If a male exhibited lateral display (a courtship invitation) and then the female approached the enclosure that contained the male, this was scored as a positive encounter. This protocol was performed under normal illumination and under monochromatic illumination.

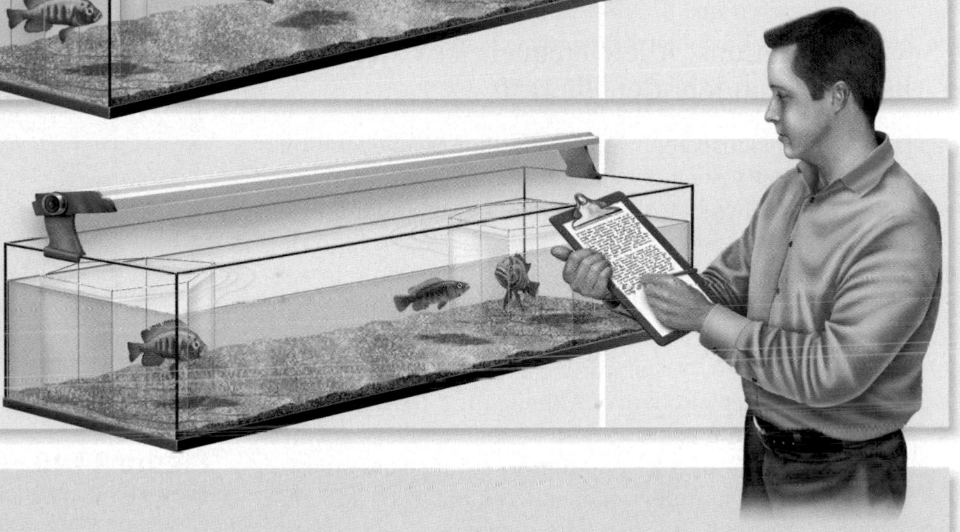

3. **THE DATA**

Female	Male	Light condition	Percentage of positive encounters*
P. pundamilia	*P. pundamilia*	Normal	16
P. pundamilia	*P. nyererei*	Normal	2
P. nyererei	*P. nyererei*	Normal	16
P. nyererei	*P. pundamilia*	Normal	5
P. pundamilia	*P. pundamilia*	Monochromatic	20
P. pundamilia	*P. nyererei*	Monochromatic	18
P. nyererei	*P. nyererei*	Monochromatic	13
P. nyererei	*P. pundamilia*	Monochromatic	18

*A positive encounter occurred when a male's lateral display was followed by the female approaching the male.

both species look similar under these conditions. As shown in **Figure 24.9**, a female of one species was placed in an aquarium that contained one male of each species within an enclosure. The males were within glass enclosures to avoid direct competition with each other, which would have likely affected female choice. The goal of the experiment was to determine which of the two males a female would prefer. Courtship between a male and female begins when a male swims toward a female and exhibits a lateral display (that is, shows the side of his body to the female). If the female is interested, she will approach the male, and then the male will display a quivering motion. Such courtship behavior was examined under normal light and under orange monochromatic light.

As seen in the data, Seehausen and van Alphen found that the females' preference for males was dramatically different depending on the illumination conditions. Under normal light, *P. pundamilia* females preferred *P. pundamilia* males, and *P. nyererei* females preferred *P. nyererei* males. However, such mating preference was lost under orange monochromatic light. If the light conditions in their native habitats are similar to the normal light used in this experiment, female choice would be expected to separate cichlids into two populations—*P. pundamilia* females mating with *P. pundamilia* males and *P. nyererei* females

mating with *P. nyererei* males. In this case, sexual selection appears to have followed a diversifying mechanism in which certain females prefer males with one color pattern while other females prefer males with a different color pattern. When this occurs, a possible outcome of such sexual selection is that it can separate one large population into smaller populations that selectively breed with each other, and eventually become distinct species. The topic of species formation is discussed in greater depth in Chapter 25.

In Small Populations, Allele Frequencies Can Be Altered by Random Genetic Drift

Thus far, we have focused on natural selection as an evolutionary mechanism that fosters genetic change. Let's now turn our attention to other ways that the gene pool of a population can change. In the 1930s, Sewall Wright played a large role in developing the concept of **random genetic drift**, which refers to changes in allele frequencies due to random sampling error. The term genetic drift is derived from the observation that allele frequencies may "drift" randomly from generation to generation as a matter of chance. Although the Darwinian fitness values of particular genotypes allow researchers to predict the allele frequencies of a population in future generations, random sampling error, or deviation between observed and predicted values, can arise due to random events that are unrelated to fitness. For example, an individual with a high fitness value may, as a matter of bad luck, not encounter a member of the opposite sex. Changes in allele frequencies due to genetic drift happen regardless of the fitness of individuals that carry those alleles. Likewise, random sampling error can influence which alleles happen to be found in the gametes that fuse with each other in a successful fertilization.

What are the effects of genetic drift? Over the long run, genetic drift favors either the loss or the fixation of an allele—when its frequency reaches 0% or 100% in a population, respectively. The rate at which an allele is either lost or fixed depends on the population size. **Figure 24.10** illustrates the potential consequences of genetic drift in one large (*N* = 1,000) and two small (*N* = 10) populations. This simulation involves the frequency of hypothetical *B* and *b* alleles of a gene for fur color in a population of mice—*B* is the black allele and *b* is the white allele. At the beginning of this hypothetical simulation, which runs for 50 generations, all of these populations had identical allele frequencies: *B* = 0.5 and *b* = 0.5. In the small populations, the allele frequencies fluctuated substantially from generation to generation. Eventually, in one simulation, the *B* allele was eliminated, while in the other, it was fixed at 100%. These small populations would then consist of only white mice or black mice, respectively. At this point, the allele has become monomorphic and cannot fluctuate any further. By comparison, the frequencies of *B* and *b* in the large population fluctu-

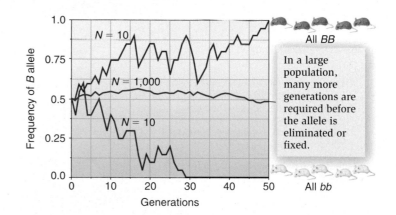

Figure 24.10 Genetic drift and population size. This graph shows a hypothetical simulation of random genetic drift and its effects on small and large populations of black (*B* allele) and white (*b* allele) mice. In all cases, the starting allele frequencies are *B* = 0.5 and *b* = 0.5. The red lines illustrate two populations of mice in which *N* = 10. The blue line shows a population in which *N* = 1,000. Genetic drift leads to random changes in allele frequencies, eventually causing either the elimination or fixation of alleles. This happens much more quickly in small populations than it does in large ones. In this simulation, genetic drift has led to small populations of all-black (*BB*) or all-white (*bb*) mice in 50 generations or less.

ated much less. As discussed in Chapter 16, the relative effects of random sampling error are much less when the sample size is large. Nevertheless, genetic drift will eventually lead to allele loss or fixation even in large populations, but this will take many more generations to occur than it does in small populations.

In nature, genetic drift may rapidly alter allele frequencies when population sizes are small. One example is called the **bottleneck effect**. A population can be reduced dramatically in size by events such as earthquakes, floods, drought, and human destruction of habitat. Such occurrences may randomly eliminate most of the members of the population without regard to genetic composition. The period of the bottleneck, when the population size is very small, may be influenced by genetic drift. This may happen primarily for two reasons. First, the surviving members may have allele frequencies that differ from those of the original population. Second, allele frequencies are expected to drift substantially during the generations when the population size is small. In extreme cases, alleles may even be eliminated. Eventually, the bottlenecked population may regain its original size. However, the new population is likely to have less genetic variation than the original one. A hypothetical example of this process is shown with a population of frogs in **Figure 24.11**. In this example, a starting population of frogs is found in three phenotypes: yellow, dark green, and striped. Due to a bottleneck caused by a drought, the dark green variety is lost from the population.

As another example, the African cheetah has lost nearly all of its genetic variation. This was likely due to a bottleneck effect. An analysis by population geneticists has suggested that a severe bottleneck occurred in this species approximately 10,000–12,000 years ago, reducing it to near extinction. The species eventually rebounded in numbers, but the bottleneck apparently reduced its genetic variation to very low levels. The modern species is monomorphic for nearly all of its genes.

Another common phenomenon in which genetic drift may have a rapid impact is the **founder effect**. This occurs when a small group of individuals separates from a larger population and establishes a colony in a new location. For example, a few individuals may migrate from a large continental population and become the founders of an island population. The founder effect differs from a bottleneck in that it occurs in a new location, although both effects reduce the size of a population. The founder effect has two important consequences. First, the founding population, which is relatively small, is expected to have less genetic variation than the larger original population from which it was derived. Second, as a matter of chance, the allele frequencies in the founding population may differ markedly from those of the original population.

Population geneticists have studied many examples in which isolated populations were founded via colonization by members of another population. For example, in the 1960s, American geneticist Victor McKusick studied allele frequencies in the Old Order Amish of Lancaster County, Pennsylvania. At that time, this was a group of about 8,000 people, descended from just three couples that immigrated to the U.S. in 1770. Among this population of 8,000, a genetic disease known as the Ellis–van Creveld syndrome (a recessive form of dwarfism) was found at a frequency of 0.07, or 7%. By comparison, this disorder is extremely rare in other human populations, even the population from which the founding members had originated. The high frequency in the Lancaster County population is a chance occurrence due to the founder effect.

The Neutral Theory of Evolution Proposes That Genetic Drift Plays an Important Role in Promoting Genetic Change

In 1968, Japanese evolutionary biologist Motoo Kimura proposed that much of the variation seen in natural populations is caused by genetic drift. Because it is a random process, genetic drift does not preferentially select for any particular allele—it can eliminate both beneficial and deleterious alleles. Much of

the time, genetic drift promotes **neutral variation**, which does not favor any particular genotype. According to Kimura's **neutral theory of evolution**, most genetic variation is due to the accumulation of neutral mutations that have attained high frequencies in a population via genetic drift.

Neutral mutations involve changes in genotypes that do not affect the phenotype of the organism, so they are not acted upon by natural selection. For example, a mutation within a structural gene that changes a glycine codon from GGG to GGC would not affect the amino acid sequence of the encoded protein.

Starting population includes 3 phenotypes of frogs: yellow, dark green, and striped.

A drought causes a bottleneck in which the population size is decreased and the dark green phenotype is lost.

Population size recovers but genetic variation is decreased, as only 2 phenotypes are left.

Figure 24.11   **A hypothetical example of the bottleneck effect.**   This example involves a population of frogs in which a drought dramatically reduces population size, resulting in a bottleneck. The bottleneck reduces the genetic diversity in the population.

Biological inquiry: How does the bottleneck effect undermine the efforts of conservation biologists who are trying to save species nearing extinction?

The resulting genotypes are equal in fitness. Because neutral mutations do not affect phenotype, they can spread throughout a population due to genetic drift (**Figure 24.12**). This theory has been called **non-Darwinian evolution** and also "survival of the luckiest" to contrast it with Darwin's "survival of the fittest" theory. Kimura agreed with Darwin that natural selection is responsible for adaptive changes in a species during evolution. His main idea is that much of the modern variation in gene sequences is explained by neutral variation rather than adaptive variation.

The sequencing of genomes from many species supports the neutral theory of evolution. When we examine changes of the coding sequence within structural genes, we find that nucleotide substitutions are more prevalent in the third base of a codon than they are in the first or second base. Mutations in the third base are often neutral because they may not change the amino acid sequence of the protein (refer back to Table 12.1). In contrast, random mutations at the first or second base are more likely to be harmful than beneficial and tend to be eliminated from a population. In addition, when mutations do change the coding sequence, they are more likely to involve conservative substitutions. For example, the difference between two alleles of a given gene may be the replacement of a nonpolar amino acid with another nonpolar amino acid. This change is conservative in the sense that it is less likely to affect protein function.

In general, the DNA sequencing of hundreds of thousands of different genes from hundreds of species has provided compelling support for the neutral theory of evolution. However, the argument is by no means resolved. Certain geneticists, called selectionists, oppose the neutralist theory. They often present persuasive theoretical arguments in favor of natural selection as the primary factor promoting genetic variation. In any case, the argument is largely a quantitative rather than a qualitative one. Each school of thought accepts that genetic drift and natural selection both play key roles in evolution. The neutralists argue that most genetic variation arises from neutral genetic mutations and genetic drift, whereas the selectionists argue that beneficial mutations and natural selection are primarily responsible.

Migration Between Two Populations Tends to Increase Genetic Variation

Earlier in this chapter, we considered how migration to a new location by a relatively small group can result in a founding population with an altered genetic composition due to genetic drift. In addition, migration between two different established populations can alter genetic variation. As a hypothetical example, let's consider two populations of a particular species of deer that are separated by a mountain range running north and south (**Figure 24.13**). On rare occasions, a few deer from the western population may travel through a narrow pass between the mountains and become members of the eastern population. If the two populations are different with regard to genetic variation, this migration will alter the frequencies of certain alleles in the eastern population. Of course, this migration could occur in the

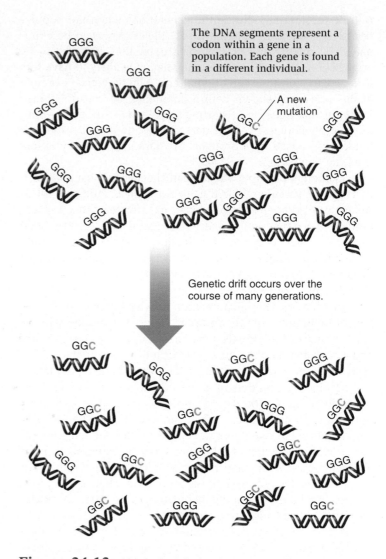

The DNA segments represent a codon within a gene in a population. Each gene is found in a different individual.

A new mutation

Genetic drift occurs over the course of many generations.

Figure 24.12 **Neutral evolution in a population.** In this example, a mutation within a gene (each gene shown represents a member of the population) changes a glycine codon from GGG to GGC, which does not affect the amino acid sequence of the encoded protein. Over the course of many generations, genetic drift may cause this neutral allele to become prevalent in the population, perhaps even monomorphic.

opposite direction as well and would then affect the western population. This phenomenon, called **gene flow**, occurs whenever individuals migrate between populations having different allele frequencies.

In nature, individuals commonly migrate in both directions. Such bidirectional migration has two important consequences. First, migration tends to reduce differences in allele frequencies between neighboring populations. In fact, population geneticists can evaluate the extent of migration between two populations by analyzing the similarities and differences between their allele frequencies. Populations that frequently mix their gene pools via migration tend to have similar allele frequencies,

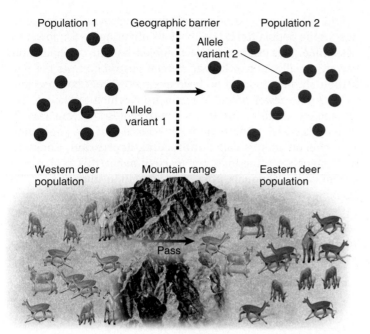

Figure 24.13 **Migration and gene flow.** In this example, two populations of a deer species are separated by a mountain range. On rare occasions, a few deer from the western population travel through a narrow pass and become members of the eastern population, thereby changing some of the allele frequencies in the population and promoting gene flow.

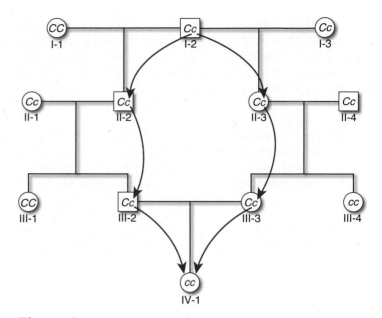

Figure 24.14 **A human pedigree containing inbreeding.** The parents of individual IV-1 are genetically related. Therefore, individual IV-1 has a higher probability of being homozygous for genes than do members of a noninbred population.

Biological inquiry: Although inbreeding by itself does not affect allele frequencies, how might inbreeding indirectly affect allele frequencies over the course of many generations if natural selection was also occurring?

whereas isolated populations are more disparate, due to the effects of natural selection and genetic drift. Second, migration tends to enhance genetic diversity within a population. As discussed earlier in this chapter, new mutations are relatively rare events. Therefore, a new mutation may arise in only one population. Migration may then introduce this new allele into a neighboring population.

Nonrandom Mating Affects the Relative Proportion of Homozygotes and Heterozygotes in a Population

As mentioned earlier in this chapter, one of the conditions required to establish the Hardy-Weinberg equilibrium is random mating. This means that individuals choose their mates irrespective of their genotypes. In many cases, particularly in human populations, this condition is frequently violated. Such **nonrandom mating** takes different forms. Assortative mating occurs when individuals with similar phenotypes are more likely to mate. If the similar phenotypes are due to similar genotypes, assortative mating tends to increase the proportion of homozygotes and decrease the proportion of heterozygotes in the population. The opposite situation, where dissimilar phenotypes mate preferentially, is called disassortative mating. This type of mating favors heterozygosity.

Another form of nonrandom mating involves the choice of mates based on their genetic history rather than their phenotypes. Individuals may choose a mate who is part of the same genetic lineage. The mating of two genetically related individuals, such as cousins, is called **inbreeding**. This sometimes occurs in human societies and is more likely to take place in nature when population size becomes very small.

In the absence of other evolutionary forces, nonrandom mating does not affect allele frequencies in a population. However, it will disrupt the balance of genotypes that is predicted by the Hardy-Weinberg equilibrium. As an example, let's consider inbreeding in a family pedigree. Figure 24.14 illustrates a human pedigree involving a mating between cousins. Individuals III-2 and III-3 are cousins and have produced the daughter labeled IV-1. She is said to be inbred, because her parents are genetically related. The parents of an inbred individual have one or more common ancestors. In the pedigree of Figure 24.14, I-2 is the grandfather of both III-2 and III-3.

Inbreeding increases the relative proportions of homozygotes and decreases the likelihood of heterozygotes in a population. This is because an inbred individual has a higher chance of being homozygous for any given gene than does a noninbred individual, because the same allele for that gene could be inherited twice from a common ancestor. For example, let's suppose that individual I-2 is a heterozygote, *Cc* (see red lines in Figure 24.14). The *c* allele could pass from I-2 to II-2 to III-2, and finally to IV-1.

Likewise, the *c* allele could pass from I-2 to II-3 to III-3, and then to IV-1. Therefore, IV-1 has a chance of being homozygous because she inherited both copies of the *c* allele from a common ancestor of both of her parents. Inbreeding does not favor any particular allele—it does not favor *c* over *C*—but it does increase the likelihood that an individual will be homozygous for any given gene.

Although inbreeding by itself does not affect allele frequencies, it may have negative consequences with regard to recessive alleles. Rare recessive alleles that are harmful in the homozygous condition are found in all populations. Such alleles do not usually pose a problem because heterozygotes carrying a rare recessive allele are also rare, making it very unlikely that two heterozygotes will mate with each other. However, when inbreeding is practiced, homozygous offspring are more likely to be produced. For example, rare recessive diseases in humans are more frequent when inbreeding occurs.

In natural populations, inbreeding will lower the mean fitness of the population if homozygous offspring have a lower fitness value. This can be a serious problem as natural populations become smaller due to human habitat destruction. As the population shrinks, inbreeding becomes more likely because individuals have fewer potential mates from which to choose. The inbreeding, in turn, produces homozygotes that are less fit, thereby decreasing the reproductive success of the population. This phenomenon is called **inbreeding depression**. Conservation biologists sometimes try to circumvent this problem by introducing individuals from one population into another. For example, the endangered Florida panther (*Felis concolor coryi*) suffers from inbreeding-related defects, which include poor sperm quality and quantity, and morphological abnormalities. To alleviate these effects, individuals from Texas have been introduced into the Florida population of panthers.

Chapter Summary

- Population genetics is the study of genes and genotypes in a population. The focus is on understanding genetic variation.

24.1 Genes in Populations

- All of the genes in a population constitute a gene pool.
- With regard to population genetics, a population is defined as a group of individuals of the same species that can interbreed with one another.
- Polymorphism refers to a genotype or phenotype that is found in two or more forms in a population. A monomorphic gene exists predominantly (>99%) as a single allele in a population. (Figure 24.1)
- An allele frequency is the number of copies of an allele divided by the total number of alleles in a population, while a genotype frequency is the number of individuals with a given genotype divided by the total number of individuals.
- The Hardy-Weinberg equation ($p^2 + 2pq + q^2 = 1$) relates allele and genotype frequencies. (Figure 24.2)
- The Hardy-Weinberg equation predicts an equilibrium if the population size is very large, mating is random, the populations do not migrate, no natural selection occurs, and no new mutations are formed.

24.2 Evolutionary Mechanisms and Their Effects on Populations

- Microevolution involves changes in a population's gene pool from one generation to the next.
- The sources of new genetic variation are random gene mutations, gene duplications, exon shuffling, and horizontal gene transfer.

- This variation is acted upon by natural selection, genetic drift, migration, and nonrandom mating to alter allele and genotype frequencies, and ultimately cause a population to evolve over many generations. (Table 24.1)
- Natural selection favors individuals with the greatest reproductive success. Darwinian fitness is a measure of reproductive success. The mean fitness of a population is its average reproductive success.
- Directional selection is a form of natural selection that favors one extreme of a phenotypic distribution. (Figure 24.3)
- Stabilizing selection is a second form of natural selection that favors an intermediate phenotype. (Figure 24.4)
- Disruptive selection is a third pattern of natural selection that favors two or more genotypes. An example is when a population occupies a diverse environment. (Figure 24.5)
- Balancing selection maintains a balanced polymorphism in a population. Examples include heterozygote advantage and negative frequency-dependent selection. (Figure 24.6)
- Sexual selection is directed at traits that make it more likely for individuals to find or choose a mate and/or engage in successful mating. This can lead to traits described as secondary sexual characteristics. (Figure 24.7)
- Seehausen and van Alphen discovered that female choice of mates in cichlids is influenced by male coloration. This is an example of sexual selection. (Figures 24.8, 24.9)
- Genetic drift involves changes in allele frequencies due to random sampling error. It occurs more rapidly in small populations and leads to either the elimination or the fixation of alleles. (Figure 24.10)
- The bottleneck effect is a form of genetic drift in which a population size is dramatically reduced and then rebounds. During the bottleneck, genetic variation may be lost from a population. (Figure 24.11)

- The neutral theory of evolution by Kimura indicates that much of the genetic variation observed in populations is due to the accumulation of neutral genetic changes. (Figure 24.12)
- Gene flow occurs when individuals migrate between different populations and cause changes in the genetic composition of the resulting populations. (Figure 24.13)
- Inbreeding is a type of nonrandom mating in which genetically related individuals mate with each other. This tends to increase the proportion of homozygotes relative to heterozygotes. When homozygotes have lower fitness, this phenomenon is called inbreeding depression. (Figure 24.14)

TEST YOURSELF

1. Population geneticists are interested in the genetic variation in populations. The most common type of genetic change that can cause polymorphism in a population is
   a. a deletion of a gene sequence.
   b. a duplication of a region of a gene.
   c. a rearrangement of a gene sequence.
   d. a single-nucleotide substitution.
   e. an inversion of a segment of a chromosome.

2. The Hardy-Weinberg equation characterizes the genotype frequencies and allele frequencies
   a. of a population that is experiencing selection for mating success.
   b. of a population that is extremely small.
   c. of a population that is very large and not evolving.
   d. of a community of species that is not evolving.
   e. of a community of species that is experiencing selection.

3. Considering the Hardy-Weinberg equation, what portion of the equation would be used to calculate the frequency of individuals that do not exhibit a disease but are carriers of a recessive genetic disorder?
   a. q
   b. p^2
   c. $2pq$
   d. q^2
   e. both b and d

4. Which of the following does not alter allele frequencies?
   a. selection
   b. immigration
   c. mutation
   d. inbreeding
   e. emigration

5. Which of the following statements is correct regarding mutations?
   a. Mutations are not important in evolution.
   b. Mutations provide the source for genetic variation that other evolutionary forces may act upon.
   c. Mutations occur at such a high rate that they promote major changes in the gene pool from one generation to the next.
   d. Mutations are insignificant when considering evolution of a large population.
   e. Mutations are of greater importance in larger populations than in smaller populations.

6. In a population of fish, body coloration varies from a light shade, almost white, to a very dark shade of green. If changes in the environment resulted in decreased predation of individuals with the lightest coloration, this would be an example of _____ selection.
   a. disruptive
   b. stabilizing
   c. directional
   d. sexual
   e. artificial

7. Considering the same population of fish described in question 6, if the stream environment included several areas of sandy, light-colored bottom areas and lots of dark-colored vegetation, both the light- and dark-colored fish would have selective advantage and increased survival. This type of scenario could explain the occurrence of
   a. genetic drift.
   b. disruptive selection.
   c. mutation.
   d. stabilizing selection.
   e. sexual selection.

8. The microevolutionary force most sensitive to population size is
   a. mutation.
   b. migration.
   c. selection.
   d. genetic drift.
   e. all of the above.

9. The neutral theory of evolution differs primarily from Darwinian evolution in that
   a. neutral theory states natural selection does not exist.
   b. neutral theory states that most of the genetic variation in a population is due to neutral mutations, which do not alter phenotypes.
   c. neutral variation alters survival and reproductive success.
   d. neutral mutations are not affected by population size.
   e. both b and c.

10. Populations that experience inbreeding may also experience
    a. a decrease in fitness due to an increased frequency of recessive genetic diseases.
    b. an increase in fitness due to increases in heterozygosity.
    c. very little genetic drift.
    d. no apparent change.
    e. increased mutation rates.

CONCEPTUAL QUESTIONS

1. Explain the five conditions that are required for Hardy-Weinberg equilibrium.
2. List and define the four types of selection.
3. Define the founder effect.

EXPERIMENTAL QUESTIONS

1. What hypothesis is tested in the Seehausen and van Alphen experiment?

2. Describe the experimental design for this study, illustrated in Figure 24.9. What was the purpose of conducting the experiment under the two different light conditions?

3. What were the results of the experiment in Figure 24.9?

COLLABORATIVE QUESTIONS

1. Discuss four sources of new genetic variation in a population.

2. Discuss various patterns of natural selection that lead to environmental adaptation and also discuss sexual selection.

www.brookerbiology.com

This website includes answers to the Biological Inquiry questions found in the figure legends and all end-of-chapter questions.

25

ORIGIN OF SPECIES

CHAPTER OUTLINE

Two different species of zebras. Grey zebra (*Equus grevyi*) is shown here on the *left* and Grant's zebra (*E. quagga*), which has fewer and thicker stripes, is shown on the *right*. This chapter will examine how different species come into existence.

The origin of living organisms was described by several ancient philosophers as the great "mystery of mysteries." Perhaps that is why so many different views have been put forth to explain the existence of living species. At the time of Aristotle (4th century B.C.E.), most people believed that some living organisms could come into being by spontaneous generation, which is the idea that nonliving materials could give rise to living organisms. For example, it was commonly believed that worms and frogs could arise from mud. By comparison, many religious teachings contend that species were divinely made and have remained the same since their creation. In contrast to these ideas, the work of Charles Darwin provided the scientific theory of evolution by descent with modification. This theory helps us to understand the diversity of life, and in particular, it presents a logical explanation for how pre-existing species can evolve into new species.

This chapter provides an exciting way to build on the information that we have considered in previous chapters. In Chapter 22, we examined how the first primitive cells in an RNA world evolved into prokaryotic cells and eventually eukaryotes. Chapter 23 surveyed the tenets on which the theory of evolution is built, and in Chapter 24, we viewed evolution on a small scale as it relates to a single gene. In this chapter, we will consider evolution on a larger scale, as it relates to the formation of new species.

To biologists, the concept of a **species** has come to mean a group of organisms that maintains a distinctive set of attributes in nature. As a student, you may already have an intuitive sense of this concept. It is obvious that giraffes and mice are different species. However, as we will learn in the first section of this chapter, the distinction between different, closely related species is often blurred in natural environments, so that it may not be easy to definitively distinguish two species (see chapter-opening photo). Among other uses, species identification is important because it allows biologists to plan for the preservation and conservation of those species.

In this chapter we will also focus on the mechanisms that promote the formation of new species, a phenomenon called **speciation**. The term **macroevolution** refers to evolutionary changes that create new species and groups of species. It concerns the diversity of organisms established over long periods of time through the evolution and extinction of many species. Macroevolution occurs by the accumulation of microevolutionary changes, those that occur in a single gene (see Chapter 24). Natural selection results in the evolution of traits that promote environmental adaptation and reproductive success. In this chapter, we will learn how the same evolutionary mechanisms that account for microevolution also play a role in the formation of new species.

25.1 Species Concepts

The number of species on Earth is astounding. A study done by biologist E. O. Wilson and colleagues in 1990 estimated the known number of species at approximately 1.4 million. However, many species have yet to be identified. This is particularly true among prokaryotic organisms. Estimates of the number of unidentified species range from 2 to 100 million!

The difficulty of identifying whether certain groups constitute unique species is often rooted in the phenomenon that a single species may exist in two distinct populations that are in the slow process of evolving into two or more different species.

The amount of time that two populations are separated will have an important impact. If the time is short, the two populations are likely to be very similar, so they would be considered the same species. If the time is long, sufficient changes may have occurred so that the two populations would show unequivocal differences that allow them to maintain their distinctive set of features in nature. When studying natural populations, evolutionary biologists are often confronted with situations where some differences between two populations are apparent, but it is difficult to decide whether the two populations truly represent separate species. When two or more groups within the same species display one or more traits that are somewhat different but not enough to warrant their placement into different species, biologists sometimes classify such groups as subspecies.

In this section, we will begin by considering different attributes that biologists examine when deciding if two groups of organisms constitute different species. In later sections, we will consider mechanisms that explain how new species arise in nature.

The Members of a Species Have a Common Set of Characteristics That Distinguish Them from Other Species

Biologists adopt methods of species identification that are based on their own experience with the organisms they study. The characteristics that a biologist uses to identify a species depend, in large part, on the species in question. For example, the traits used to distinguish insect species would be quite different from those used to identify different bacterial species. The most commonly used characteristics are physical or morphological traits, the ability to interbreed, common evolutionary lineages, and ecological factors. These different ways of considering species have led to the use of different approaches for distinguishing species called **species concepts** (**Table 25.1**). A comparison of these concepts will help you to appreciate the various factors that are involved in categorizing the species on our planet.

Phylogenetic Species Concept According to the **phylogenetic species concept***, as advocated by Quentin Wheeler and Norman Platnick, the members of a single species are identified by having a unique combination of characteristics. Historically, the first way to categorize species was based on their physical characteristics. Organisms are classified as the same species if their anatomical traits appear to be very similar. Likewise, microorganisms can be classified according to morphological traits at the cellular level. In addition, molecular features such as DNA sequences can now be used to compare organisms. An advantage of the phylogenetic species concept is that it can be applied to all types of organisms.

---

*Although it is named the phylogenetic species concept, it is not based on phylogenetic trees, which are described in Chapter 26. The phylogenetic species concept is related to other species concepts not covered here called the morphological species concept and typological species concept.

Table 25.1	Species Concepts
Concept	**Description**
Phylogenetic species concept	Various physical characteristics can be analyzed to evaluate the identity of species. These often include morphological (anatomical) traits. In the case of unicellular organisms, characteristics such as cell-wall structure and other cellular traits may be examined. Molecular characteristics can also be compared.
Biological species concept	Two species are often judged to be separate species if they are unable to interbreed in nature to produce viable, fertile offspring.
Evolutionary species concept	An analysis of ancestry may help biologists to determine if two groups are members of the same species or represent evolutionarily distinct species.
Ecological species concept	The ability of organisms to successfully occupy their own ecological niche or habitat, including their use of resources and impact on the environment, may be used to distinguish species.

Although this concept is a common way for taxonomists to categorize species, it has a few drawbacks. First, it may be difficult to decide how many traits to consider when characterizing individuals. In addition, it is difficult to analyze quantitative traits that vary in a continuous way among members of the same species. Another drawback is that the degree of dissimilarity that separates different species may not be easy to decide upon. Researchers often disagree about how much morphological difference is necessary to separate different species. Another shortcoming of the phylogenetic species concept is that members of the same species sometimes look very different, and conversely, members of different species sometimes look remarkably similar to each other. For example, **Figure 25.1a** shows two different frogs of the species *Dendrobates tinctorius*, commonly called the dyeing poison frog. This species exists in many different-colored morphs, which are individuals of the same species that have noticeably dissimilar appearances. Alternatively, **Figure 25.1b** shows two different species of frog, the Northern leopard frog (*Rana pipiens*) and the Southern leopard frog (*Rana utricularia*), which look similar.

Biological Species Concept So why would biologists describe two species, such as the Northern leopard frog and Southern leopard frog, as being different if they are morphologically similar? One reason is that they are unable to breed with each other in nature. Therefore, a second way to define a species is by the ability to interbreed. In the late 1920s, geneticist Theodosius Dobzhansky proposed that each species is reproductively isolated from other species. Such **reproductive isolation** prevents one species from successfully interbreeding with other species.

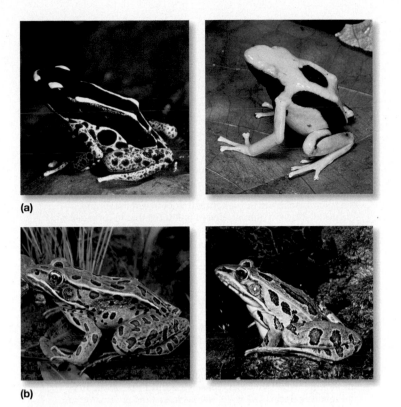

(a)

(b)

Figure 25.1 Difficulties in applying the phylogenetic species concept. In some cases, members of the same species can appear quite different. (a) Two color morphs of one species, the dyeing poison frog (*Dendrobates tinctorius*). (b) Two different species of frog, the Northern leopard frog (*Rana pipiens*, left) and the Southern leopard frog (*Rana utricularia*, right), which look similar.

In 1942, evolutionary biologist Ernst Mayr expanded on the ideas of Dobzhansky to provide a biological definition of a species. According to Mayr's **biological species concept**, a species is a group of individuals whose members have the potential to interbreed with one another in nature to produce viable, fertile offspring but cannot successfully interbreed with members of other species. As discussed later in this section, reproductive isolation among species of plants and animals can occur by an amazing variety of different mechanisms.

The biological species concept has been used to distinguish many plant and animal species, especially those that look alike but are reproductively isolated. Even so, it suffers from three main problems. First, in nature, it may be difficult to determine if two populations are reproductively isolated, particularly if they are large populations with overlapping geographical ranges. Second, taxonomists have noted many cases in which two different species can interbreed in nature yet consistently maintain themselves as separate species. For example, different species of Yucca plants, such as *Yucca pallida* and *Yucca constricta*, do interbreed in nature yet typically maintain populations with distinct characteristics. For this reason, they are viewed as distinct species based on the phylogenetic species concept. A third drawback of

the biological species concept is that it cannot be applied to asexual species such as bacteria, nor can it be applied to extinct species. Likewise, some species of plants and fungi are only known to reproduce asexually. Therefore, the biological species concept has been primarily used to distinguish closely related species of animals and plants that reproduce sexually.

Evolutionary Species Concept In 1961, American paleontologist George Gaylord Simpson proposed a species concept based on ancestry. According to the **evolutionary species concept**, a species is derived from a single lineage that is distinct from other lineages and has its own evolutionary tendencies and historical fate. A **lineage** is the genetic relationship between an individual or group of individuals and its ancestors. The evolutionary species concept is a theoretical viewpoint that is focused on the pathway that has led to the formation of each distinct species. In this regard, it can be applied to the formation of all species. We will consider the topic of evolutionary lineages in Chapter 26.

One drawback of the evolutionary species concept is that it does not provide an easy way to identify a unique species. In most cases, lineages are difficult to examine and evaluate quantitatively. The interpretation of lineages involving fossils may be particularly difficult to evaluate due to incomplete fossil remains as well as transitional forms that may be missing from the fossil record. Currently, the most common way for researchers to analyze lineages of modern species is by comparing the DNA sequences of particular genes. We can obtain samples of cells from different individuals and compare the genes within those cells to see how similar or different they are. Even so, it is difficult to decide where to draw the line when separating groups into different species. Is a 0.1% difference in their genome sequences sufficient to warrant placement into two different species, or do we need a 1% or 5% difference? A very tiny genetic change can have a dramatic effect on phenotype. Conversely, as discussed in Chapter 24, many genetic changes are neutral and have no discernable effect on phenotype. For these and other reasons, researchers cannot simply compare DNA sequences among different groups and always reach concrete conclusions as to whether or not the groups should be considered different species. Nevertheless, as we will examine in Chapter 26, sequence comparisons are very useful in evaluating whether multiple species—which are already considered distinct—are closely or distantly related evolutionarily.

Ecological Species Concept The **ecological species concept**, described by American evolutionary biologist Leigh Van Valen in 1976, is a viewpoint that considers a species within its native environment. Each species occupies an **ecological niche**, which is the unique set of habitat resources that a species requires, as well as its influence on the environment and other species. Within their own niche, members of a given species compete with each other for survival. If two organisms are very similar, their needs will overlap, which results in competition. Such competing individuals are likely to be of the same species.

According to this concept, species are formed because evolutionary mechanisms control how each type of species uses resources. This species concept is particularly useful in distinguishing bacterial species that do not reproduce sexually. Bacterial cells of the same species are likely to use the same types of resources (sugars, vitamins, etc.) and grow under the same types of conditions (temperature, pH, etc.).

Reproductive Isolating Mechanisms Help to Maintain the Distinctiveness of Each Species

Thus far we have considered four species concepts that provide biologists with approaches that are aimed at the identification of different species. In our discussion of these concepts, you may have realized that the identification of a species is not always a simple matter. With regard to plants and animals, the biological species concept proposed by Mayr has played a major role in the way that biologists study plant and animal species, partly because it identifies a possible mechanism for the process of forming new species—reproductive isolation. For this reason, much research has been done to try to understand the mechanisms that prevent interbreeding between different species, which are called **reproductive isolating mechanisms**. Keep in mind that populations do not intentionally erect these reproductive barriers. Rather, reproductive isolating mechanisms are merely a consequence of genetic changes that occur usually because a species becomes adapted to its own particular environment. The view of evolutionary biologists is that reproductive isolation typically evolves as a by-product of genetic divergence. Over time, as a species evolves its own unique characteristics, some of those traits are likely to prevent breeding with other species.

Biologists have discovered many mechanisms that prevent closely related species from interbreeding. These mechanisms fall into two categories, prezygotic and postzygotic. **Prezygotic mechanisms** prevent the formation of a zygote, while **postzygotic mechanisms** block the development of a viable and fertile individual after fertilization has taken place. **Figure 25.2** summarizes some of the more common ways that reproductive isolating mechanisms prevent reproduction between different species. When two species do produce offspring, such an offspring is called an **interspecies hybrid**.

Prezygotic Reproductive Isolating Mechanisms We will consider five types of prezygotic mechanisms. The first is an obvious way to prevent interbreeding, which is for members of different species to avoid contact with each other. This phenomenon, called habitat isolation, refers to a geographic barrier to interbreeding. For example, a large body of water may separate two different plant species that live on nearby islands. The second prezygotic isolating mechanism is temporal isolation, in which species reproduce at different times of the day or year. In the northeastern U.S., for example, the two most abundant field crickets, *Gryllus veletis* and *Gryllus pennsylvanicus* (spring and fall field crickets, respectively), do not differ in song or habitat and are morphologically very similar (**Figure 25.3**).

Prezygotic Mechanisms

Species 1   Species 2

Habitat isolation: Species may occupy different habitats, so they never come in contact with each other.

Temporal isolation: Species have different mating or flowering seasons or times of day, or become sexually mature at different times of the year.

Behavioral isolation: Sexual attraction between males and females of different animal species is limited due to differences in behavior or physiology.

Attempted mating

Mechanical isolation: Morphological features such as size, incompatible genitalia, etc., may prevent 2 members of different species from interbreeding.

Gametic isolation: Gametic transfer takes place, but the gametes fail to unite with each other. This can occur because the male and female gametes fail to attract, because they are unable to fuse, or because the male gametes are inviable in the female reproductive tract of another species. In plants, the pollen of one species usually cannot generate a pollen tube to fertilize the egg cells of another species.

Fertilization

Postzygotic Mechanisms

Hybrid inviability: The egg of one species is fertilized by the sperm from another species, but the fertilized egg fails to develop past the early embryonic stages.

Hybrid sterility: The interspecies hybrid survives, but it is sterile. For example, the mule, which is sterile, is produced from a cross between a male donkey (*Equus asinus*) and a female horse (*Equus caballus*).

Hybrid breakdown: The F_1 interspecies hybrid is viable and fertile, but succeeding generations (F_2, etc.) become increasingly inviable. This is usually due to the formation of less-fit genotypes by genetic recombination.

Interspecies hybrid

Figure 25.2 **Reproductive isolating mechanisms.** These mechanisms prevent successful breeding between different species. They can occur prior to fertilization (prezygotic) or after fertilization (postzygotic).

(a) Spring field cricket (*Gryllus veletis*)

(b) Fall field cricket (*Gryllus pennsylvanicus*)

Figure 25.3   **An example of temporal isolation.**   Interbreeding between these two species of crickets does not usually occur because *Gryllus veletis* matures in the spring, but *Gryllus pennsylvanicus* matures in the fall.

However, *G. veletis* matures in the spring, whereas *G. pennsylvanicus* matures in the fall. This minimizes interbreeding between the two species.

In the case of animals, mating behavior and anatomy often play key roles in promoting reproductive isolation. An example of the third type of isolation, behavioral isolation, is found between the eastern meadowlark (*Sturnella magna*) and western meadowlark (*Sturnella neglecta*). Both species are nearly identical in shape, coloration, and habitat, and their ranges overlap in the central U.S. (**Figure 25.4**). For many years, they were thought to be the same species. When biologists discovered that the western meadowlark is a separate species, it was given the species name *neglecta* to reflect the long delay in its recognition. In the zone of overlap, very little interspecies mating takes place between eastern and western meadowlarks, largely due to differences in their songs. The eastern meadowlark's song is a simple series of whistles, typically about four or five notes. By comparison, the song of the western meadowlark is a longer series of flutelike gurgling notes that go down the scale. These differences in songs enable meadowlarks to recognize potential mates as members of their own species.

A fourth type of isolation, called mechanical isolation, occurs when morphological features such as size or incompatible genitalia prevent two species from interbreeding. For example, male dragonflies use a pair of special appendages to grasp females during copulation. When a male tries to mate with a female of a different species, his grasping appendages do not fit her body shape.

A fifth type of prezygotic isolating mechanism can occur when two species attempt to interbreed, but the gametes fail to unite in a successful fertilization event. This phenomenon, called gametic isolation, is widespread among plant and animal species. In aquatic animals that release sperm and egg cells into the water, gametic isolation is important to prevent interspecies hybrids. For example, closely related species of sea urchins may release sperm and eggs into the water at the same time. Researchers have discovered that sea urchin sperm have a protein on their surface called bindin, which mediates sperm–egg attachment and membrane fusion. The structure of bindin is significantly different among different sea urchin species, and thereby

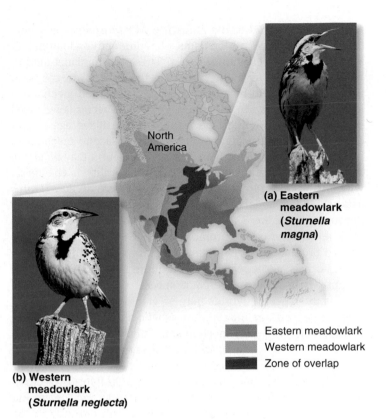

(a) Eastern meadowlark (*Sturnella magna*)

(b) Western meadowlark (*Sturnella neglecta*)

■ Eastern meadowlark
■ Western meadowlark
■ Zone of overlap

Figure 25.4   **An example of behavioral isolation.**   (a) The eastern meadowlark (*Sturnella magna*) and (b) western meadowlark (*Sturnella neglecta*) are very similar in appearance. The red region in this map shows where the two species' ranges overlap. However, very little interspecies mating takes place due to differences in their songs. Meadowlarks use their songs to recognize potential mates of their own species.

ensures that fertilization occurs between sperm and egg cells of the same species.

In flowering plants, gametic isolation is a particularly vital mechanism to achieve reproductive isolation. As discussed in Chapter 39, plant fertilization is initiated when a pollen grain lands on the stigma of a flower and sprouts a pollen tube that ultimately reaches an egg cell (look ahead to Figure 39.13). When pollen is released from a plant, it could be transferred to the stigma of many different plant species. In most cases, when a pollen grain lands on the stigma of a different species, it either fails to generate a pollen tube or the tube does not grow properly and reach the egg cell. The mechanism that controls pollen tube growth in plants has not been widely studied. The process is best understood in members of the mustard genus (*Brassica*), which includes turnips and cabbage. In these species, pollen tube growth is known to involve molecules released by pollen grains that activate receptors found in the cell wall of the stigma. If a similar mechanism is used to achieve gametic isolation between different species, the receptors that recognize molecules released by the pollen grain must be very specific for the molecules of their own species, and fail to recognize those molecules released by pollen grains of other species.

Postzygotic Reproductive Isolating Mechanisms Let's now turn to postzygotic mechanisms of reproductive isolation, of which we will discuss three types. These mechanisms tend to be less common in nature, because they are more costly in terms of energy and resources used. The first such mechanism is hybrid inviability, in which an egg of one species is fertilized by a sperm from another species, but the fertilized egg cannot develop past the early embryonic stages. A second mechanism is hybrid sterility, in which an interspecies hybrid may be viable but sterile. A classic example of hybrid sterility is the mule, which is produced by a mating between a male donkey (*Equus asinus*) and a female horse (*Equus caballus*) (**Figure 25.5**). Because a horse has 32 chromosomes per set and a donkey has 31, a mule inherits 63 chromosomes (32 + 31). When a mule tries to produce offspring, all of its chromosomes do not have pairs, so it is sterile. Note that the mule has no species name because it is not considered a species due to this sterility. Finally, interspecies hybrids may be viable and fertile, but subsequent generations may harbor genetic abnormalities that are detrimental. This third mechanism, called hybrid breakdown, can be caused by changes in chromosome structure. The chromosomes of closely related species may have structural differences from each other, such as inversions. In hybrids, a crossover may occur in the region that is inverted in one species but not the other. This will produce gametes with too little or too much genetic material. For this reason, such hybrids would often have offspring with developmental abnormalities.

Male donkey (*Equus asinus*)

×

Female horse (*Equus caballus*)

Mule

Figure 25.5 **An example of hybrid sterility.** When a male donkey (*Equus asinus*) mates with a female horse (*Equus caballus*), their offspring is a mule, which is sterile.

25.2 Mechanisms of Speciation

The formation of a new species, or speciation, is caused by genetic changes in a particular group that make it different from the species from which it was derived. As discussed in Chapter 24, random mutations in genes can be acted upon by natural selection and other evolutionary mechanisms to alter the genetic composition of a population. New species commonly evolve in this manner. In addition, interspecies matings, changes in chromosome number, and horizontal gene transfer may also cause new species to arise. In all of these cases, the underlying cause of speciation is the accumulation of genetic changes that ultimately promote enough differences so that we judge a population to constitute a unique species.

Even though genetic changes account for the phenotypic differences we observe among living organisms, such changes do not fully explain the existence of many distinct species on our planet. Why does life diversify into the more or less discrete populations that we recognize as species? Two main explanations have been proposed:

1. Species are a consequence of adaptation to different ecological niches.
2. Species of sexually reproducing organisms arise via reproductive isolation.

Depending on the species involved, one or both factors may play a dominant role in the formation of new species. In this section, we will begin by considering two different patterns of speciation that occur over the course of many generations. We will then consider several examples of speciation occurring in nature. As you will learn, adaptation and/or reproductive isolating mechanisms are critical aspects of the speciation process.

Speciation Can Occur as a Linear or a Branching Process

Before we consider mechanisms of speciation, let's first examine the patterns of speciation that are commonly observed. By studying the fossil record and the diversity of modern species, evolutionary biologists have found two different patterns of speciation. During **anagenesis** (from the Greek, *ana*, up, and *genesis*, origin), a single species is transformed into a different species over the course of many generations (**Figure 25.6a**). In this process, evolutionary mechanisms cause the characteristics of the species to change. For example, a single species of fish in an isolated lake may evolve into a different species due to changes in available food sources.

By comparison, **cladogenesis** (from the Greek *clados*, branch) involves the division of a species into two or more species. In the case of the birds shown in **Figure 25.6b**, the original mainland species has remained relatively unchanged, while populations on neighboring islands have evolved substantially different traits by natural selection. Such cladogenesis is a more common form of speciation. One reason why is related to geographic isolation. If one or more individuals move to a new location

(a) Anagenesis   **(b) Cladogenesis**

Figure 25.6   A comparison between anagenesis and cladogenesis, two patterns of speciation.   (a) Anagenesis is the change of one species into another, whereas (b) cladogenesis occurs when one population diverges into two or more different species.

Biological inquiry: Explain why cladogenesis requires reproductive isolation and anagenesis does not.

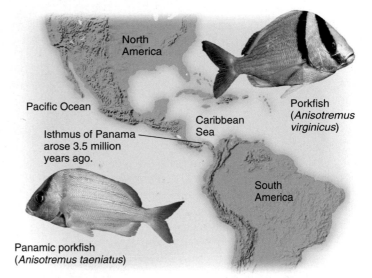

Panamic porkfish
(*Anisotremus taeniatus*)

Figure 25.7   An example of allopatric speciation.   An ancestral fish population was split into two by the formation of the Isthmus of Panama about 3.5 million years ago. Since that time, different genetic changes occurred in the two populations because of their geographic isolation; these changes eventually led to the formation of different species. The porkfish (*Anisotremus virginicus*) is found in the Caribbean Sea, whereas the Panamic porkfish (*Anisotremus taeniatus*) is found in the Pacific Ocean.

that is geographically isolated from the original population, evolutionary mechanisms such as natural selection will operate independently on the two populations. This mechanism of speciation is described next.

Geographic Isolation Can Promote Allopatric Speciation

Although sexual reproduction is not a barrier to anagenesis, it is a barrier to cladogenesis. A population will evolve as a single unit if the members can successfully breed with one another, thereby preventing its divergence into two or more discrete species. The process of cladogenesis begins only when gene flow becomes limited between two or more populations. **Allopatric speciation** (from the Greek *allos*, other, and the Latin *patria*, homeland) is thought to be the most prevalent way for cladogenesis to occur. This form of speciation occurs when some members of a species become geographically separated from the other members. In some cases, the separation may be caused by slow, geological events that eventually produce quite large geographic barriers. For example, a mountain range may emerge and split a species that occupies the lowland regions. Or a creeping glacier may divide a population. **Figure 25.7** shows an interesting example in which geological separation promoted speciation. A fish called the porkfish (*Anisotremus virginicus*) is found in the Caribbean Sea, whereas the Panamic porkfish (*Anisotremus taeniatus*) is found in the Pacific Ocean. These two species were derived from an ancestral species that was split by the formation of the Isthmus of Panama about 3.5 million years ago. Since this event, the two populations have been geographically isolated and have evolved into distinct species via natural selection and other evolutionary mechanisms.

Allopatric speciation can also occur via a second mechanism related to the founder effect that was described in Chapter 24. This occurs when a small population moves to a new location that is geographically separated from the main population. For example, a storm may force a small group of birds from a mainland to a distant island. In this case, migration between the island and the mainland population is an infrequent event. In a relatively short period of time, the founding population on the island may evolve into a new species. Different evolutionary forces may contribute to this rapid evolution. First, as discussed in Chapter 24, genetic drift may quickly lead to the random fixation of certain alleles and the elimination of other alleles from the small population. Another factor is natural selection, because the environment on the island may differ significantly from the mainland environment.

The Hawaiian Islands are a showcase of the founder effect. The islands' extreme isolation coupled with their phenomenal array of ecological niches has enabled a small number of founding species to evolve into a vast assortment of different species. Biologists have investigated several examples of **adaptive radiation**, in which a single ancestral species has evolved into a wide array of descendant species that differ greatly in their habitat, form, or behavior. For example, approximately 1,000 species of *Drosophila* are found dispersed throughout the islands. Evolutionary studies suggest that these were derived from a single colonization by one species of fruit fly! Another striking example is seen with a family of birds called honeycreepers (*Drepanidinae*) (**Figure 25.8**). Researchers estimate that the honeycreepers' ancestor arrived in Hawaii 3–7 million years ago.

Figure 25.8 **An example of the founder effect with subsequent adaptive radiation.** The honeycreepers' ancestor is believed to be related to a Eurasian rosefinch that arrived on the Hawaiian Islands approximately 3–7 million years ago. Since that time, at least 54 different species of honeycreepers (*Drepanidinae*) have evolved on the islands. A few selected examples are shown here. Adaptations to feeding have produced species with notable differences in beak morphology.

This ancestor was a single species of finch, possibly a Eurasian rosefinch (*Carpodacus* sp.) or, less likely, the North American house finch (*Carpodacus mexicanus*). At least 54 different species of honeycreepers, many of which are now extinct, have evolved from this founding event to fill available niches in the islands' habitats. Seed eaters developed stouter, stronger bills for

cracking tough husks. Insect-eating honeycreepers developed thin, warbler-like bills for picking insects from foliage or strong, hooked bills to root out wood-boring insects. And nectar-feeding honeycreepers evolved curved bills for extracting nectar from the flowers of Hawaii's endemic plants.

FEATURE INVESTIGATION

Podos Found That an Adaptation to Feeding Also May Have Promoted Reproductive Isolation in Finches

To investigate how environmental adaptation may contribute to reproductive isolation, in 2001 American evolutionary biologist Jeffrey Podos analyzed the songs of Darwin's finches on the Galápagos Islands. Like the honeycreepers, the differences in beak sizes and shapes among the various species of finches are adaptations to different feeding strategies. Podos hypothe-

sized that changes in beak morphology could also impact the songs that the birds produce, thereby having the potential to affect mate choice. The components of the vocal tract of birds, including the trachea, syrinx, and beak, work collectively to produce a bird's song. Birds actively modify the shape of their vocal tracts during singing. Beak movements are normally very rapid and precise.

Podos focused on two aspects of a bird's song. The first feature is the frequency range, which is a measure of the minimum and maximum frequencies in a bird's song. The second feature is the trill rate. A trill is a series of notes or group of notes repeated in succession. **Figure 25.9** shows a graphic depiction of the songs of Darwin's finches. As you can see, the song patterns of these finches are quite different from each other.

To quantitatively study the relationship between beak size and song, Podos first captured male finches on one of the Galápagos Islands (Santa Cruz) and measured their beak sizes (**Figure 25.10**). The birds were banded and then released into the wild; the banding provided a way to identify the birds whose beaks had already been measured. After release, the songs of the banded birds were recorded on a tape recorder, and then their range of frequencies and trill rate were analyzed. Podos then compared the data for the Galápagos finches to a large body of data that had been collected on many other bird species. This comparison was used to evaluate whether beak size, in this case, beak depth—the measurement of the beak from top to bottom, at its base—constrained either the frequency range and/or the trill rate of the finches.

The results of this comparison are shown in the data of Figure 25.10. As seen here, the relative constraint on vocal performance became higher as the beak depth became larger. This means that birds with larger beaks had a more narrow frequency range and/or a slower trill rate. Podos proposed that as jaws and beaks become adapted for strength to crack open larger seeds, they will be less able to perform the rapid movements associated with certain types of songs. In contrast, the finches whose beaks were adapted to probe for insects or eat smaller seeds had less constraint on their vocal performance. From the perspective of evolution, the changes observed in song patterns for the Galápagos finches could have played an important role in promoting reproductive isolation, because song pattern is one factor involved in mate selection in birds. Therefore, a by-product of beak adaptation for feeding is that it also appears to have an effect on song pattern, possibly promoting reproductive isolation. In this way, populations of finches would have evolved into their own distinct species.

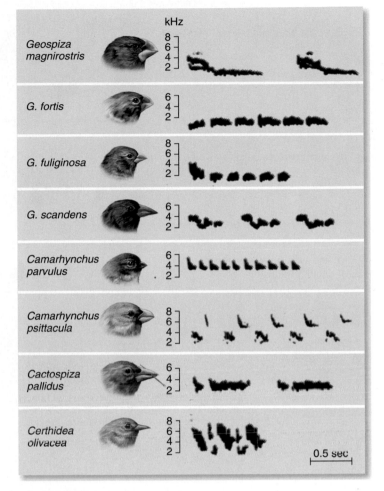

Figure 25.9 Differences in the songs of Galápagos finches. These spectrograms depict the frequency of each bird's song over time, measured in kilohertz (kHz). The songs are produced in a series of trills that have a particular pattern and occur at regular intervals. Notice the differences in frequency and trill rate between different species of birds.

Figure 25.10 The effects of beak depth on song among different species of Galápagos finches.

HYPOTHESIS Changes in beak morphology that are an adaptation to feeding may also affect the songs of Galápagos finches and thereby lead to reproductive isolation between species.

STARTING MATERIALS This study was conducted on finch populations of the Galápagos Island of Santa Cruz.

	Experimental level	Conceptual level
1 Capture male finches and measure their beak depth. Beak depth is measured at the base of beak, from top to bottom.		This is a measurement of phenotypic variation in beak size.

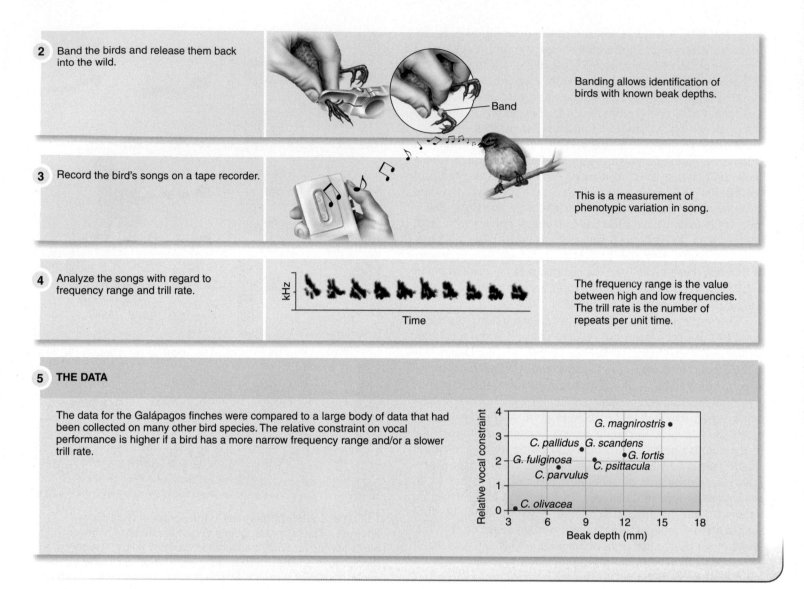

2 Band the birds and release them back into the wild.

Banding allows identification of birds with known beak depths.

3 Record the bird's songs on a tape recorder.

This is a measurement of phenotypic variation in song.

4 Analyze the songs with regard to frequency range and trill rate.

kHz

Time

The frequency range is the value between high and low frequencies. The trill rate is the number of repeats per unit time.

5 THE DATA

The data for the Galápagos finches were compared to a large body of data that had been collected on many other bird species. The relative constraint on vocal performance is higher if a bird has a more narrow frequency range and/or a slower trill rate.

Hybridization Can Occur When Reproductive Isolation Between Two Populations Is Incomplete

Speciation sometimes occurs when members of a species are only partially separated or when a species lives in one small area or moves about very little. In these cases, the geographic separation is not complete. For example, a mountain range may divide a species into two populations, but there may be breaks in the range where the two groups are connected physically. In these zones of contact, the members of two populations can interbreed, although this tends to occur infrequently. Likewise, speciation may occur among species that live in one small area even though no large geographic isolation exists. Certain organisms move about so little that 100–1,000 meters may be sufficient to limit interbreeding between neighboring groups. Plants, terrestrial snails, rodents, grasshoppers, lizards, and many flightless insects may speciate in this manner.

Prior to complete reproductive isolation, the zones where two populations can interbreed are known as **hybrid zones**. **Figure 25.11** shows a hybrid zone along a mountain pass that connects two deer populations. For speciation to occur, the amount of gene flow within hybrid zones must become very limited (refer back to Figure 24.13). But how does this happen? As the two populations accumulate different genetic changes, this may decrease the ability of individuals from different populations to mate with each other in the hybrid zone. For example, natural selection in the western deer population may favor an increase in body size that is not favored in the eastern population. Over time, as this size difference between members of the two populations becomes greater, breeding in the hybrid zone may decrease. Larger individuals may not interbreed easily with smaller individuals due to mechanical isolation. Alternatively, larger individuals may prefer larger individuals as mates, while smaller individuals may also prefer each other. Once gene

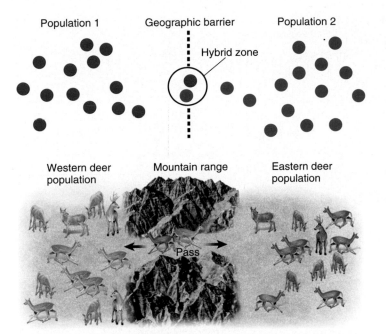

Figure 25.11 **A hypothetical hybrid zone.** A narrow mountain pass allows members of two deer populations to occasionally interbreed with each other, creating hybrids between the two populations.

flow through the hybrid zone is greatly diminished, the two populations are reproductively isolated. Over the course of many generations, such populations may evolve into distinct species.

Sympatric Speciation Occurs When Populations in Direct Contact Become Reproductively Isolated from Each Other

Thus far, we have considered how geographic isolation can result in allopatric speciation. In the early stages of this process, reproductive isolation occurs because geographic separation prevents interbreeding. After long separation periods, reproductive isolation may also involve the accumulation of genetic changes that directly affect reproduction, such as changes in breeding season or mating behavior. By comparison, **sympatric speciation** (from the Greek *sym*, together) occurs when members of a species that initially occupy the same habitat within the same range diverge into two or more different species. Sympatric speciation tends to involve abrupt genetic changes that quickly lead to the reproductive isolation of a group of individuals. Let's consider a few examples in which such sudden changes can quickly lead to the formation of a new species within the same geographic area as the original species.

One abrupt genetic change that may occur is a change in chromosome number. Plants tend to be more tolerant of changes in chromosome number than are animals, as discussed in Chapter 15. In particular, alterations in the number of sets of chromosomes, such as polyploidy, often produce healthy plants.

Such changes, which are relatively common in plants on an evolutionary timescale, can result in sympatric speciation. Polyploidy is so frequent in plants that it is a major mechanism of their speciation. In ferns and flowering plants, about 30–50% of the species are polyploid. By comparison, polyploidy is much less common in animals, but it can occur. For example, roughly 30 species of reptiles and amphibians have been identified that are polyploids derived from diploid ancestors.

As described in Chapter 15, complete nondisjunction of chromosomes during gamete formation can increase the number of chromosome sets in the same species (autopolyploidy) or in hybrids of different species (allopolyploidy). The formation of a polyploid can abruptly lead to reproductive isolation. As an example, let's consider the origin of a natural species of a plant called the common hemp nettle, *Galeopsis tetrahit*. This species is thought to be an allotetraploid derived from two diploid species, *Galeopsis pubescens* and *Galeopsis speciosa* (**Figure 25.12**). These two diploid species contain 16 chromosomes each ($2n = 16$), while *G. tetrahit* contains 32. Its origin is not known, but one possibility is that an interspecies mating between *G. pubescens* and *G. speciosa* produced an allodiploid with 16 chromosomes (one set from each species) and then the allodiploid underwent complete nondisjunction to become an allotetraploid carrying four sets of chromosomes—two from each species.

The allotetraploid is fertile, because all of its chromosomes occur in homologous pairs that can segregate evenly during meiosis. However, a cross between an allotetraploid and a diploid produces an offspring that is monoploid for one chromosome set and diploid for the other set. The chromosomes of the monoploid set cannot be evenly segregated during meiosis. These offspring are expected to be sterile, because they will produce gametes that have incomplete sets of chromosomes. This hybrid sterility, which is an example of a postzygotic isolating mechanism, causes the allotetraploid to be reproductively isolated from both diploid species. Therefore, this process could have led to the formation of a new species, *G. tetrahit*, by sympatric speciation.

Sympatric speciation may also occur when genetic changes enable members of a species to occupy a new niche within the same geographic range as the original species. An example in which this process may be occurring involves colonial bentgrass (*Agrostis tenuis*), which was discussed in Chapter 24. In certain locations such as South Wales, isolated places are found where the soil is contaminated with high levels of heavy metals such as copper. On these sites, natural selection has promoted the proliferation of plants carrying alleles that confer resistance to the heavy metals. These alleles enable *A. tenuis* to grow on contaminated soil but tend to inhibit growth on normal, non-contaminated soil. Metal-resistant populations are continuous with populations that grow on uncontaminated land and do not show metal tolerance (refer back to Figure 24.5). In recent years, the metal-tolerant plants are starting to show a change in their flowering season. Over time, if this process continues, the metal-tolerant population may evolve into a new species that cannot interbreed with the original (metal-sensitive) species.

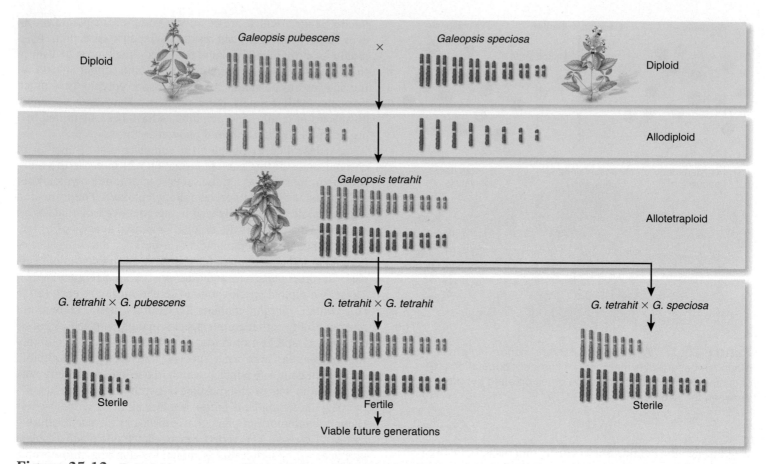

Figure 25.12   **Polyploidy and sympatric speciation.**   This example shows how polyploidy may have caused reproductive isolation between three natural species of hemp nettle. Research suggests that the diploid species *Galeopsis pubescens* and *Galeopsis speciosa* produced the allotetraploid *Galeopsis tetrahit*. If *G. tetrahit* is mated with either of the other two species, the resulting offspring would be monoploid for one chromosome set and diploid for the other set, making them sterile. Therefore, *G. tetrahit* is reproductively isolated from the diploid species, making it a new species.

Biological inquiry: Suppose that G. pubescens *was crossed to* G. tetrahit *to produce an interspecies hybrid as shown at the bottom of this figure. If this interspecies hybrid was crossed to* G. tetrahit, *how many chromosomes do you think an offspring would have? The answer you give should be a range, not a single number.*

25.3 The Pace of Speciation

Throughout the history of life on Earth, the rate of evolutionary change and speciation has not been constant, although the degree of inconstancy has been debated since the time of Darwin. Even Darwin himself suggested that evolution can be fast or slow. **Figure 25.13** illustrates contrasting views concerning the rate of evolutionary change. These ideas are not mutually exclusive but represent two different ways to consider the tempo of evolution. The concept of **gradualism** suggests that each new species evolves continuously over long spans of time (Figure 25.13a). The principal idea is that large phenotypic differences that produce new species are due to the accumulation of many small genetic changes. By comparison, the concept of **punctuated equilibrium**, advocated in the 1970s by American paleontologists and evolutionary biologists Niles Eldredge and Stephen Jay Gould, suggests that the tempo of evolution is more sporadic (Figure 25.13b). According to this idea, species exist

relatively unchanged for many generations. During this period, the species is in equilibrium with its environment. These long periods of equilibrium are punctuated by relatively short periods (that is, on a geological timescale) during which evolution occurs at a far more rapid rate, perhaps due to environmental changes via the founder effect or new mutations that are quickly acted upon by natural selection.

In reality, neither of the views presented in Figure 25.13 fully accounts for evolutionary change. The occurrence of punctuated equilibrium is often supported by the fossil record. Paleontologists rarely find a gradual transition of fossil forms. Instead, new species seem to arise rather suddenly in a layer of rocks, persist relatively unchanged for a very long period of time, and then become extinct. Scientists think that the transition period during which a previous species evolved into a new species was so short that few, if any, of the transitional members were preserved as fossils. Even so, these rapid periods of change were probably followed by long periods of equilibrium that likely

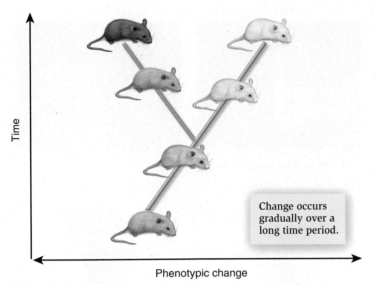

(a) Graualism

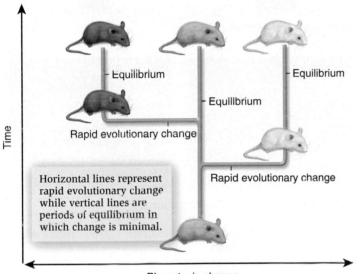

(b) Punctuated equilibrium

Figure 25.13 **A comparison of gradualism and punctuated equilibrium.** (a) During gradualism, the phenotypic characteristics of a species gradually change due to the accumulation of small genetic changes. (b) During punctuated equilibrium, species exist essentially unchanged for long periods of time, during which they are in equilibrium with their environment. These equilibrium periods are punctuated by relatively short periods of evolutionary change during which phenotypic characteristics may change rapidly.

involved the additional accumulation of many small genetic changes, consistent with gradualism.

As discussed earlier, rapid evolutionary change can be explained by genetic phenomena. Single gene mutations can have dramatic effects on phenotypic characteristics. Therefore, only a small number of new mutations may be required to alter phe-

notypic characteristics, eventually producing a group of individuals that makes up a new species. Likewise, events such as changes in chromosome number and alloploidy may abruptly create individuals with new phenotypic traits. On an evolutionary timescale, these types of events can be rather rapid, because one or only a few genetic changes can have a major impact on the phenotype of the organism.

In conjunction with genetic changes, species may also be subjected to sudden environmental shifts that quickly drive the gene pool in a particular direction via natural selection. For example, a small group may migrate to a new environment in which different alleles provide better adaptation to the surroundings. Alternatively, a species may be subjected to a relatively sudden environmental event that has a major impact on survival. For example, the climate may change or a new predator may infiltrate the geographic range of the species. Natural selection may lead to a rapid evolution of the gene pool by favoring those genetic changes that allow members of the population to survive the climatic change or to have phenotypic characteristics that allow them to avoid the predator.

Finally, another issue associated with the speed of evolution is generation time. Species of large animals with long generation times tend to evolve much more slowly than do microbes with short generations. Many new species of bacteria will come into existence during our lifetime, while new species of large animals tend to arise on a much longer timescale. This is an important consideration for people because bacteria have great environmental impact; they are decomposers of organic materials and pollutants in the environment, and they also play a role in many diseases of plants, animals, and humans.

25.4 Evo-Devo: Evolutionary Developmental Biology and the Form of New Species

As we have learned in the preceding sections of this chapter, the origin of new species involves genetic changes that lead to adaptations to environmental niches and/or to reproductive isolating mechanisms that prevent closely related species from interbreeding. These genetic changes result in morphological and physiological differences that distinguish one species from another. In recent years, many evolutionary biologists have begun to investigate how genetic variation produces species and groups of species with novel shapes and forms. The underlying reasons for such changes are often rooted in the developmental pathways that control an organism's morphology. **Evolutionary developmental biology** (referred to as **evo-devo**) is an exciting and relatively new field of biology that compares the development of different organisms in an attempt to understand ancestral relationships between organisms and the developmental mechanisms that bring about evolutionary change. During the past few decades, developmental geneticists have gained a better understanding of biological development at the molecular level.

Much of this work has involved the discovery of genes that control development in experimental organisms. As more and more organisms have been analyzed, researchers have become interested in the similarities and differences that occur between closely related and distantly related species. Evolutionary developmental biology has arisen in response to this trend.

A central question in evo-devo studies is, How do new morphological forms come into being? For example, how does a nonwebbed foot evolve into a webbed foot? Or how does a new organ, such as an eye, come into existence? As we will learn, such novelty arises through genetic changes, also called genetic innovations. Certain types of genetic innovation have been so advantageous that they have resulted in groups of many new species. For example, the innovation of feathered wings resulted in the evolution of many different species of birds. In this section, we will learn that proteins that control developmental changes, such as cell-signaling proteins and transcription factors, often play a key role in promoting the morphological changes that occur during evolution.

The Spatial Expression of Genes That Affect Development Can Have a Dramatic Effect on Phenotype

In Chapter 19, we considered the role of genetics in the development of multicellular organisms. As we learned, genes that play a role in development may influence cell division and growth, cell differentiation, cell migration, and cell death. The interplay among these four processes creates an organism with a specific body form and function. As you might imagine, developmental genes are very important to the phenotypes of individuals. They affect traits such as the shape of a bird's beak, the length of a giraffe's neck, and the size of a plant's flower. In recent years, the study of development has indicated that developmental genes are key players in the evolution of many types of traits. Changes in such genes affect traits that can be acted on by natural selection. Furthermore, variation in the expression of such genes may be commonly involved in the acquisition of new traits that promote speciation.

As an example, let's compare the formation of a chicken's foot with that of a duck. Developmental biologists have discovered that the morphological differences between a webbed and a nonwebbed foot are due to the differential expression of two different cell-signaling proteins called bone morphogenetic protein 4 (BMP4) and gremlin. The *BMP4* gene is expressed throughout the developing limb; this is shown in **Figure 25.14a**, in which the BMP4 protein is stained blue. The BMP4 protein causes cells to undergo apoptosis and die. The gremlin protein, which is stained brown in **Figure 25.14b**, inhibits the function of BMP4 and thereby allows cells to survive. In the developing chicken limb, the *gremlin* gene is expressed throughout the limb, except in the regions between each digit. Therefore, these cells die, and a chicken develops a nonwebbed foot (**Figure 25.14c**). By comparison, in the duck, *gremlin* is expressed throughout the entire

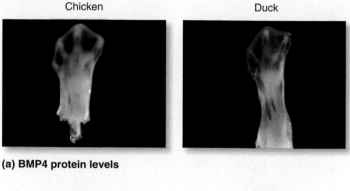

(a) BMP4 protein levels

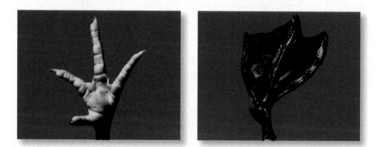

(b) Gremlin protein levels

(c) Comparison of a chicken and a duck foot

Figure 25.14   The role of cell-signaling proteins in the morphology of birds' feet.   This figure shows how changes in developmental gene expression can produce new traits. (a) Expression of the *BMP4* gene in the developing limbs. BMP4 protein is stained blue here. (b) Expression of the *gremlin* gene in the developing limbs. Gremlin protein is stained brown here. Note that *gremlin* is expressed in the interdigit region only in the duck. Gremlin inhibits BMP4, which causes programmed cell death. (c) Because BMP4 is not inhibited in the interdigit regions in the chicken, the cells in this region die, and the foot is not webbed. By comparison, inhibition of BMP4 in the interdigit regions in the duck results in a webbed foot.

limb, including the interdigit regions, which results in a webbed foot. Interestingly, researchers have been able to introduce gremlin protein into the interdigit regions of developing chicken limbs. This produces a chicken with webbed feet!

During the evolution of birds, variation in the expression of these genes determined whether or not their feet were webbed. Mutations occurred that provided variation in the expression of the *BMP4* and *gremlin* genes. In terrestrial settings, having nonwebbed feet is an advantage because these are more effective at holding onto perches, running along the ground, and snatching prey. Therefore, natural selection would maintain nonwebbed feet in terrestrial environments. This process explains the occurrence of nonwebbed feet in chickens, hawks, crows, and many other terrestrial birds. In aquatic environments, webbed feet are an advantage because they act as paddles for swimming, so genetic variation that produced webbed feet would have been promoted by natural selection. Over the course of many generations, this gave rise to the webbed feet that are now found in ducks, geese, penguins, and other aquatic birds.

But how does having webbed or nonwebbed feet influence speciation? Perhaps this trait would not directly affect the ability of two individuals to mate. However, due to natural selection, birds with webbed feet would become more prevalent in aquatic environments, while birds with nonwebbed feet would be found in terrestrial locations. Therefore, reproductive isolation would occur because the populations would occupy different environments.

The *Hox* Genes Have Been Important in the Evolution of a Variety of Body Plans Found in Different Species of Animals

The study of developmental genes has revealed interesting trends among large groups of species. *Hox* genes, which are discussed in Chapter 19, are found in all animals. Developmental biologists have speculated that genetic variation in the *Hox* genes may have been a critical event that spawned the formation of many new body types, yielding many different species. As shown in **Figure 25.15**, the number and arrangement of *Hox* genes varies considerably among different types of animals. Sponges, the simplest of animals, have at least one *Hox* gene, whereas insects typically have nine or more. In most cases, multiple *Hox* genes occur in a cluster in which the genes are close to each other along a chromosome. In mammals, such *Hox* gene clusters have been duplicated twice during the course of evolution to form four clusters, all slightly different, with a total of 38 genes.

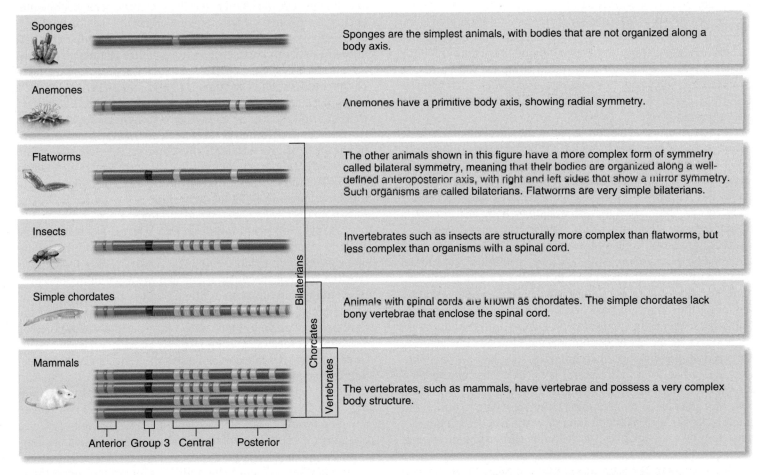

Figure 25.15   **Relationship between *Hox* gene number and body complexity in different types of animals.**   Researchers speculate that the duplication of *Hox* genes and *Hox* gene clusters played a key role in the evolution of more complex body plans in animals. A correlation seems to exist between increasing numbers of *Hox* genes and increasing complexity of body structure. The *Hox* genes are divided into four groups, called anterior, group 3, central, and posterior, based on their relative similarities. Each group is represented by a different color in this figure.

Researchers propose that increases in the number of *Hox* genes have been instrumental in the evolution of many animal species with greater complexity in body structure. To understand how, let's first consider *Hox* gene function. All *Hox* genes encode transcription factors that act as master control proteins to direct the formation of particular regions of the body. Each *Hox* gene controls a hierarchy of many genes that includes other control genes, which regulate the expression of target genes, as well as structural genes, which encode proteins that ultimately affect the morphology of the organism. The evolution of complex body plans is associated with an increase not only in the number of regulatory genes—as evidenced by the increase in *Hox* gene complexity during evolution—but also in structural genes, which ultimately affect an organism's form and function.

But how would an increase in *Hox* genes enable more complex body forms to evolve? Part of the answer lies in the spatial expression of the *Hox* genes. In fruit flies, for example, different *Hox* genes are expressed in different segments of the body along the anteroposterior axis (refer back to Figure 19.17). Therefore, an increase in the number of *Hox* genes allows each of these master control genes to become more specialized in the region that it controls. One segment in the middle of the fruit fly body can be controlled by a particular *Hox* gene and form wings and legs, while a segment in the head region can be controlled by a different *Hox* gene and develop antennae. Therefore, research suggests that one way for new, more complex body forms to evolve is by increasing the number of *Hox* genes, thereby making it possible to form many specialized parts of the body that are organized along a body axis.

Three lines of evidence support the idea that *Hox* gene complexity has been instrumental in the evolution and speciation of animals with different body patterns. First, as discussed in Chapter 19, *Hox* genes are known to control body development. Second, as described in Figure 25.15, a general trend is observed in which animals with a simpler body structure tend to have fewer *Hox* genes and fewer *Hox* clusters in their genomes than do the genomes of more complex animals. Third and finally, a comparison of *Hox* gene evolution and animal evolution bear striking parallels. Researchers can analyze *Hox* gene sequences among modern species and make estimates regarding the timing of past events. If the DNA sequences of homologous genes in two different species are fairly different, the modern genes evolved from an ancestral gene in the very distant past. In contrast, if two homologous genes are very similar, they arose from an ancestral gene that was more recent. Using this type of approach, geneticists can estimate when the first *Hox* gene arose by gene innovation; the date is difficult to pinpoint but is well over 600 million years ago. The single *Hox* gene found in the sponge has descended from this primordial *Hox* gene. In addition, gene duplications of this primordial gene produced clusters of *Hox* genes in other species. Clusters, such as those found in modern insects, were likely to be present approximately 600 million years ago. A duplication of that cluster is estimated to have occurred around 520 million years ago. Remarkably, these estimates of *Hox* gene origins correlate with major speciation events in the history of animals. As described in Chapter 22, the Cambrian explosion, which occurred from

533 to 525 million years ago, saw a phenomenal diversification in the body plan of invertebrate species. This diversification occurred after the *Hox* cluster was formed and was possibly undergoing its first duplication to create two *Hox* clusters. Also, approximately 420 million years ago, a second duplication produced species with four *Hox* clusters. This event precedes the proliferation of tetrapods—vertebrates with four limbs—that occurred during the Devonian period, approximately 417–354 million years ago. Modern tetrapods, such as mammals, have four *Hox* clusters. This second duplication may have been a critical event that led to the evolution of complex terrestrial vertebrates with four limbs, such as lizards, bears, and humans.

Developmental Genes That Affect Growth Rates Can Also Have a Dramatic Effect on Phenotype and Can Lead to the Formation of New Species

In our previous example in Figure 25.14, a difference in the spatial expression of the *gremlin* gene affected whether a bird has nonwebbed or webbed feet. In this case, the expression determined if apoptosis would occur in a defined region. Another way that genetic variation can influence morphology is by influencing the relative growth rates of different parts of the body. The pattern whereby different parts of the body grow at different rates with respect to each other is called **allometric growth**. As an example, **Figure 25.16a** compares the progressive growth of the head between a human and chimpanzee. At the fetal stage, the size and shape of the heads look fairly similar. However, after this stage, the relative growth rates of certain regions become markedly different, thereby affecting the shape and size of the head in the adult. In the chimpanzee, the jaw region grows faster, giving the chimpanzee a much larger and longer jaw. In the human, the jaw grows more slowly and the region of the skull that surrounds the brain grows slightly faster. Therefore, humans have smaller jaws but larger skulls.

Changes in growth rates can also affect the developmental stage at which one species reproduces compared to that of another species. This can occur in two ways. One possibility is that the parts of the body associated with reproduction could grow and mature faster than the rest of the body. Alternatively, reproduction could happen at the same absolute age, but the development of nonreproductive body parts could be slowed down in one species, causing it to reproduce at an earlier stage. In either case, the morphological result is the same—reproduction in the adult is observed at an earlier stage in one species than it is in another. The retention of juvenile traits in an adult organism is called **paedomorphosis** (from the Greek *paedo*, meaning "young" or "juvenile," and *morph*, relating to the form of an organism). It is particularly common among salamanders. Even though paedomorphic species have certain juvenile traits as adults, they still have the ability to reproduce successfully. For example, Cope's giant salamander (*Dicamptodon copei*) becomes mature and reproduces in the aquatic form, without changing into a terrestrial adult as do other salamander species (**Figure 25.16b**). The adult form of Cope's giant salamander has gills and a large paddle-shaped tail, features that resemble those of the tadpole stage of other salamander species.

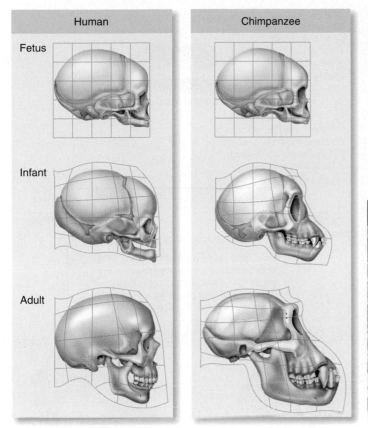

(a) **The effects of allometric growth on head morphology in the human and chimpanzee**

Figure 25.16 **Effects of growth rate on development.** **(a)** Allometric growth refers to the phenomenon in which one region of the body grows faster than another. Allometric growth explains why the heads of adult chimpanzees and humans have different shapes even though their fetal shapes are quite similar. **(b)** Paedomorphosis occurs when an adult species retains characteristics that are juvenile traits in another related species. Cope's giant salamander reproduces at the tadpole stage.

(b) **Paedomorphosis in Cope's giant salamander**

GENOMES & PROTEOMES

The Study of the *Pax6* Gene Indicates That Different Types of Eyes Evolved from a Simpler Form

Thus far in this section, we have focused on the roles of particular genes as they influence the development of the body. Explaining how a complex organ comes into existence is another major challenge for evolutionary biologists. While it is relatively easy to understand how a limb could undergo evolutionary modifications to become a wing, flipper, or arm, it is more difficult to understand how a body structure comes into being in the first place. In his book *The Origin of Species*, Charles Darwin addressed this question and admitted that the existence of complex organs was difficult to understand. His book had an entire chapter devoted to the question of eye development and evolution. As noted by Darwin, the eyes of vertebrate species are exceedingly complex, being able to adjust focus, let in different amounts of light, and detect a spectrum of colors. Darwin speculated that such complex eyes must have evolved from a simpler structure through the process of descent with modification. With amazing insight, he suggested that a very simple eye would be composed of two cell types, a photoreceptor cell and an adjacent pigment cell. The photoreceptor cell, which is a type

of nerve cell, is able to absorb light and respond to it. The function of the pigment cell is to stop the light from reaching one side of the photoreceptor cell. This primitive, two-cell arrangement would allow an organism to sense both light and the direction from which the light comes.

A primitive eye would provide an additional way for an organism to sense its environment, possibly allowing it to avoid predators or locate food. Vision is nearly universal among animals, which indicates that there must be a strong selective advantage to better eyesight. Over time, eyes could become more complex by enhancing the ability to absorb different amounts and wavelengths of light, and also by refinements in structures such as the addition of lenses that focus the incoming light.

Since the time of Darwin, many evolutionary biologists have wrestled with the question of eye evolution. From an anatomical point of view, researchers have discovered many different types of eyes. For example, the eyes of fruit flies, squid, and humans are quite different from each other. Furthermore, species that are closely related evolutionarily sometimes have different types of eyes. This observation led evolutionary biologists such as Luitfried von Salvini-Plawen and Ernst Mayr to propose that eyes may have independently arisen many different times during evolution. Based solely on morphology, such a hypothesis seemed reasonable and for many years was accepted by the scientific community.

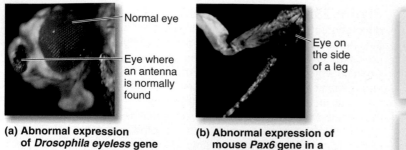

(a) Abnormal expression of *Drosophila eyeless* gene

(b) Abnormal expression of mouse *Pax6* gene in a fruit fly leg

Figure 25.17 Formation of additional eyes in *Drosophila* due to the abnormal expression of a master control gene for eye morphogenesis. **(a)** When the *Drosophila eyeless* gene is expressed in the antenna region, eyes are formed where antennae should be located. **(b)** When the mouse *Pax6* gene is expressed in the leg region of *Drosophila*, a small eye is formed there.

Biological inquiry: What do you think would happen if a researcher used genetic engineering techniques to express the Drosophila eyeless *gene at the tip of a mouse's tail?*

The situation took a dramatic turn when geneticists began to study eye development. Researchers identified a master control gene, *Pax6*,* that controls the expression of many other genes and thereby influences eye development in both rodents and humans. In mice and rats, a mutation in this gene results in small eyes. A mutation in the human *Pax6* gene causes an eye disorder called aniridia, in which most of the eye is underdeveloped. Similarly, *Drosophila* has a homologous gene named *eyeless* that also causes a defect in eye development when mutant. The *eyeless* and *Pax6* genes are homologous; they were derived from the same ancestral gene.

In 1995, Swiss geneticist Walter Gehring and his colleagues were able to show experimentally that the abnormal expression of the *eyeless* gene in other parts of the fruit fly body could promote the formation of additional eyes. For example, using genetic engineering techniques, they were able to express the *eyeless* gene in the region where antennae should form. As seen in **Figure 25.17a**, this resulted in the formation of fruit fly eyes where antennae are normally found! Remarkably, the expression of the mouse *Pax6* gene in *Drosophila* can also cause the formation of eyes in unusual places. For example, **Figure 25.17b** shows the formation of an eye on the leg of a fruit fly.

Note that the mouse *Pax6* master control gene switches on eye formation in *Drosophila*, but the eye produced is a *Drosophila* eye, not a mouse eye. This happens because the genes that are activated by the *Pax6* master control gene are all from the *Drosophila* genome. In fact, in *Drosophila*, the *Pax6* homolog called *eyeless* switches on a cascade involving 2,500 different genes required for eye morphogenesis. In more primitive eyes, the *Pax6* gene would be expected to control a cascade of fewer genes.

Taken together, these results are consistent with two ideas. First, the *Pax6* gene and its *Drosophila* homolog are master control

*Pax is an acronym for paired box. The protein encoded by this gene contains a domain called a paired box.

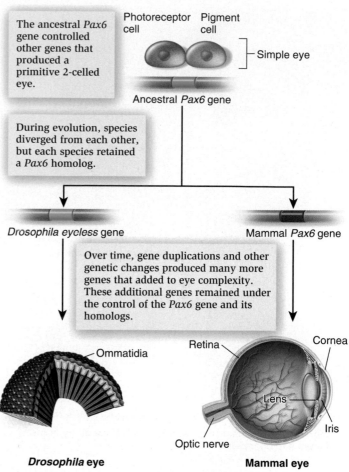

Figure 25.18 Genetic control of eye evolution. In this diagram, genetic changes, under control of the ancestral *Pax6* gene, led to the evolution of different types of eyes.

genes that promote the formation of an eye. The second idea is that the eyes of *Drosophila* and mammals are evolutionarily derived from the modification of an eye that arose once during evolution. If *Drosophila* and mammalian eyes had arisen independently, the *Pax6* gene from mice would not be expected to induce the formation of eyes in *Drosophila*.

Since the discovery of the *Pax6* and *eyeless* genes, homologs of this gene have been discovered in many different species. In all cases where it has been tested, this gene directs eye development. The *Pax6* gene and its homologs encode a transcription factor protein that controls the expression of many different genes. Gehring and colleagues have hypothesized that the eyes from many different species all evolved from a common ancestral form consisting of, as proposed by Darwin, one photoreceptor cell and one pigment cell (**Figure 25.18**). As mentioned, such a very simple eye can accomplish some rudimentary form of vision by detecting light and its direction. Eyes such as these are still found in modern species such as the larvae of certain types of mollusks. Over the course of evolution, simple eyes were transformed into more complex types of eyes by modifications that resulted in the addition of more types of cells such as lens cells and muscle cells.

<hr>

Chapter Summary

- A species is a group of organisms that maintains a distinctive set of attributes in nature. Speciation is the process whereby new species are formed. Macroevolution refers to the evolutionary changes that create new species and groups of species.

25.1 Species Concepts

- Four species concepts that are used to identify a species are phylogenetic, biological, evolutionary, and ecological. (Table 25.1)
- The phylogenetic species concept identifies species based on similarities in form, which sometimes can be misleading. (Figure 25.1)
- The biological species concept evaluates species based on their ability to interbreed.
- The evolutionary species concept considers the lineages that gave rise to species.
- The ecological species concept identifies species by their ability to occupy an ecological niche.
- By studying the biological species concept, researchers have identified many reproductive isolating mechanisms, both prezygotic and postzygotic, that prevent two different species from breeding with each other. (Figures 25.2, 25.3, 25.4, 25.5)

25.2 Mechanisms of Speciation

- Anagenesis is the evolution of one species into another species, while cladogenesis is the divergence of one species into two or more different species. (Figure 25.6)
- Allopatric speciation occurs when a population becomes geographically isolated from other populations and evolves into one or more new species. When speciation occurs multiple times due to the migration of populations into new environments, the evolutionary process is called adaptive radiation. (Figures 25.7, 25.8)
- Podos discovered that changes in beak depth, associated with evolutionary changes involved with feeding, may also promote reproductive isolation by altering the song pattern of finches. (Figures 25.9, 25.10)
- As allopatric speciation is occurring, limited reproduction between adjacent populations may take place in a hybrid zone, so that few interspecies hybrids are produced. (Figure 25.11)
- Sympatric speciation involves the formation of different species that are not geographically isolated from one another. (Figure 25.12)

25.3 The Pace of Speciation

- Gradualism refers to a pace of speciation that involves many small genetic changes, while punctuated equilibrium is a pattern of evolution in which new species arise rapidly, on an evolutionary timescale, and then remain unchanged for long periods of time. (Figure 25.13)

25.4 Evo-Devo: Evolutionary Developmental Biology and the Form of New Species

- Evolutionary developmental biology is concerned with changes in developmental patterns that have given rise to new species and groups of species. These changes often involve variation in the expression of transcription factors and cell-signaling proteins.

- The spatial expression of genes that affect development can change phenotypes dramatically, as shown by the expression of the *gremlin* and *BMP4* genes to form birds with webbed or nonwebbed feet. (Figure 25.14)
- The evolution and duplication of *Hox* genes played an important role in producing groups of animals with specific body plans that are organized along an anteroposterior axis. (Figure 25.15)
- Patterns that affect the relative growth rates of body parts are called allometric growth. Paedomorphosis occurs when an adult species retains characteristics that are juvenile traits in another related species. (Figure 25.16)
- The *Pax6* gene and its homolog in other species are master control genes that control eye development in animals. This result suggests that eyes arose only once during evolution. (Figures 25.17, 25.18)

<hr>

Test Yourself

1. Macroevolution refers to the evolutionary changes that
   a. occur in multicellular organisms.
   b. create new species and groups of species.
   c. occur over long periods of time.
   d. cause changes in allele frequencies.
   e. occur in large mammals.

2. The biological species concept classifies a species based on
   a. morphological characteristics.
   b. reproductive isolation.
   c. the niche the organism occupies in the environment.
   d. genetic relationships between an organism and its ancestor.
   e. both a and b.

3. Which of the following would be considered an example of a postzygotic isolating mechanism?
   a. incompatible genitalia
   b. different mating seasons
   c. incompatible gametes
   d. mountain range separating two populations
   e. fertilized egg fails to develop normally

4. Hybrid breakdown occurs when species hybrids
   a. do not develop past the early embryonic stages.
   b. have a reduced life span.
   c. are infertile.
   d. are fertile but produce offspring with reduced viability and fertility.
   e. produce offspring that only express the traits of one of the original species.

5. The evolution of one species into a single new species is
   a. hybridization.
   b. anagenesis.
   c. cladogenesis.
   d. adaptive radiation.
   e. microevolution.

6. Founder events may lead to rapid speciation because of
   a. differences in natural selection on the new population versus the original population.
   b. genetic differences due to genetic drift.
   c. enhanced gene flow between the new population and the original population.
   d. all of the above.
   e. a and b only.

7. A major mechanism of speciation in plants but not animals is
   a. anagenesis.
   b. polyploidy.
   c. hybrid breakdown.
   d. genetic changes that alter the organism's niche.
   e. both a and d.

8. The concept of punctuated equilibrium suggests that
   a. the rate of evolution is constant, with short time periods of no evolutionary change.
   b. evolution occurs gradually over time.
   c. small genetic changes accumulate over time to allow for phenotypic change and speciation.
   d. long periods of little evolutionary change are interrupted by short periods of major evolutionary change.
   e. both b and c.

9. Researchers suggest that an increase in the number of *Hox* genes
   a. would lead to reproductive isolation in all cases.
   b. could explain the evolution of color vision.
   c. allows for the evolution of more complex body forms.
   d. results in the decrease in the number of body segments in insects.
   e. all of the above.

10. The observation that the mammalian *Pax6* gene and the *Drosophila eyeless* gene are homologous genes that promote the formation of different types of eyes suggests that
    a. *Drosophila* eyes are more complex.
    b. mammalian eyes are more complex.
    c. eyes arose once during evolution.
    d. eyes arose at least twice during evolution.
    e. eye development is a simple process.

CONCEPTUAL QUESTIONS

1. Define species.

2. Distinguish between prezygotic isolating mechanisms and postzygotic isolating mechanisms and give an example of each.

3. Define punctuated equilibrium.

EXPERIMENTAL QUESTIONS

1. What did Podos hypothesize regarding the effects of beak size on a bird's song? How could changes in beak size and shape lead to reproductive isolation among the finches?

2. How did Podos test the hypothesis that beak morphology caused changes in the birds' songs?

3. Did the results of Podos's study support his original hypothesis? Explain. What is meant by the phrase "by-product of adaptation" and how does it apply to this particular study?

COLLABORATIVE QUESTIONS

1. Discuss how geographic isolation can lead to speciation.

2. Evolution can proceed at different rates. Discuss the two models that address the speed at which evolution occurs.

www.brookerbiology.com

This website includes answers to the Biological Inquiry questions found in the figure legends and all end-of-chapter questions.

26

TAXONOMY AND SYSTEMATICS

CHAPTER OUTLINE

26.1 Taxonomy

26.2 Systematics

A new species, the African forest elephant. In 2001, biologists discovered that this is a unique species of elephant, now named *Loxodonta cyclotis*. This is one example of how scientists have changed taxonomic groupings in response to recent discoveries in molecular systematics.

In Chapter 25, we considered the formation of new species and learned that species diversity is enormous. Biologists estimate that between 10 and 100 million species currently exist on our planet. To make some sense out of such amazing variation, and to be able to communicate with each other about it, biologists have categorized living organisms into groups of related species. **Taxonomy** (from the Greek *taxis*, order, and *nomos*, law) is the field of biology that is concerned with the theory, practice, and rules of classifying living and extinct organisms and viruses. Taxonomy results in the ordered division of species into groups based on similarities and dissimilarities in their characteristics. This task has been ongoing for over 300 years. As we discussed in Chapter 23, John Ray, who was the first scientist to carry out a thorough study of the natural world in the 1600s, developed an early classification system for plants and animals based on anatomy and physiology. Ray made the first attempt to broadly classify all known forms of life and therefore was a pioneer in the field of modern taxonomy. Ray's ideas regarding taxonomy were later extended by Carolus Linnaeus in the mid-1700s.

Systematics is the study of biological diversity and the evolutionary relationships among organisms, both extinct and modern. In 1950, the German entomologist Willi Hennig began classifying organisms in a new way. Hennig proposed that evolutionary relationships should be inferred from new features shared by descendants of a common ancestor. Since that time, biologists have applied systematics to the field of taxonomy. In other words, taxonomic groups are now based on hypotheses regarding evolutionary relationships that are derived from systematics.

As scientists obtain additional information regarding genetic variation and evolutionary trends, taxonomic groupings have changed to accommodate the new and more accurate data. For example, until recently, biologists thought there were two species of elephant in the world—the African savanna elephant (*Loxodonta africana*) and the Asian elephant (*Elephas maximus*). However, by analyzing the DNA of African elephants, a third species, now called the African forest elephant (*Loxodonta cyclotis*), was discovered (see chapter-opening photo). This surprising finding was made somewhat by accident in 2001. A DNA identification system was set up to trace ivory poachers. By studying the DNA, researchers discovered that Africa has two different *Loxodonta* elephant species. The African forest elephant is found in the forests of central and western Africa. The African savanna elephant, which is larger and has longer tusks, lives on large, dry grasslands.

In this chapter, we will begin with a discussion of taxonomy from a traditional point of view. However, because recent information has dramatically changed our view of taxonomy, we will also consider newer ways to group organisms that are more logically based on evolutionary relationships. We will then examine how the analyses of morphological and molecular genetic data are used to place species into such groups.

26.1 Taxonomy

A hierarchy is a system of organization that involves successive levels. In biology, every species is placed into several different groups. Some groups are very large and encompass many species that have fundamental similarities and, at the same time, many differences. For example, a leopard and a fruit fly are both classified as animals, though they differ in many traits. Leopards and lions are placed into a smaller group called felines (more formally named Felidae), which are predatory cats. The felines are a subset of the animal group, which has species that share many similar traits. Different species that are found in small taxonomic groups are likely to share many of the same characteristics.

In this section, we will consider how biologists use a hierarchy to group similar species. We also survey the major groups of organisms and some of their common features. As you will learn later in this chapter, a few of our traditional ideas about taxonomy have been challenged as biologists have gathered new information regarding the genetic composition of modern species. For example, taxonomists now believe that fungi and animals are much more closely related than they once did.

Taxonomy Provides a Hierarchy of Groups of Organisms

Modern taxonomy places species into progressively smaller hierarchical groups. Each group is called a **taxon** (plural, *taxa*). The taxonomic group called the **kingdom** was originally the highest and most inclusive. Linnaeus classified all life into two kingdoms, plants and animals. In 1969, American ecologist Robert Whittaker proposed a five-kingdom system in which all life was classified into the kingdoms Monera, Protista, Fungi, Plantae, and Animalia. (Monera included all prokaryotes.) However, as biologists began to learn more about the evolutionary relationships among these groups, a subject we will discuss later in this chapter, they found that these groupings did not correctly reflect the relationships among them.

In the late 1970s, based on information in the sequences of genes, American biologist Carl Woese proposed the idea of creating a category called a **domain**. Under this system, all forms of life are grouped within three domains called **Bacteria**, **Archaea**, and **Eukarya** (also spelled Eucarya) (**Figure 26.1**). The original kingdom Monera was split into two domains, Bacteria and Archaea, due to major differences between these two types of prokaryotes. The terms Bacteria and Archaea are capitalized when referring to the domains. Alternatively, these terms can also refer to prokaryotic cells and species, in which case they are not capitalized. A single bacterial cell is called a bacterium, while a single archaeal cell is an archaeon. The domain Eukarya includes the traditional four eukaryotic kingdoms, **Protista**, **Fungi**, **Plantae**, and **Animalia**. However, as described later in this chapter, this method of grouping eukaryotic species has been revised.

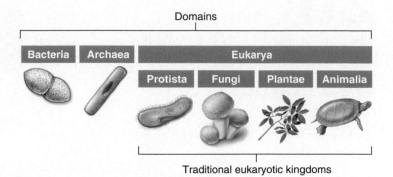

Figure 26.1   **A classification system for living and extinct organisms.**   All organisms can be grouped into three large domains, Bacteria, Archaea, or Eukarya. Eukaryotes have been traditionally divided into the four kingdoms Protista (protists), Fungi (fungi), Plantae (plants), and Animalia (animals). As discussed later, this traditional system has been revised.

The three domains of life contain millions of different species. Subdividing them into progressively smaller groups makes it easier for biologists to appreciate the relationships among living species. Proceeding down the hierarchical ladder, kingdoms are in turn subdivided into **phyla** (singular, *phylum*), each of which is divided into **classes**, then **orders**, **families**, and **genera** (singular, *genus*). Each of these taxa contains fewer species that are more similar to each other than they are to the members of the taxon above it in the hierarchy. For example, the taxon Animalia, which is at the kingdom level, has a larger number of fairly diverse species than does the class Mammalia, which contains species that are relatively similar to each other.

To further understand taxonomy, let's consider the classification of a species such as the gray wolf (*Canis lupus*) (**Figure 26.2**). The original range of *C. lupus* covered the majority of the Northern Hemisphere. Due primarily to habitat destruction, gray wolf populations are much less prevalent now, but they are still found in a few areas in North America and Eurasia. The gray wolf is placed in the domain Eukarya, and then within the kingdom Animalia, which includes the approximately 2 million species of all animals. Next, the gray wolf is classified in the phylum Chordata. The 50,000 species of animals in this group all have four common features at some stage of their development. These are a notochord (a cartilaginous rod that runs along the back of all chordates at some point in their life cycle), a tubular nerve or spinal cord located above the notochord, gill slits or arches, and a postanal tail. Examples of animals in the Chordata phylum include fishes, reptiles, and mammals.

The gray wolf is in the class Mammalia, which includes all 5,000 species of mammals. Two common features of animals in this group are hair and mammary glands. The mammary glands produce milk, which nourishes the young, and hair helps insulate the body to maintain a warm, constant body temperature. There are 26 orders of mammals; the order that includes the gray wolf is called Carnivora and has about 270 species. The gray wolf is next categorized in the family known as Canidae. This is

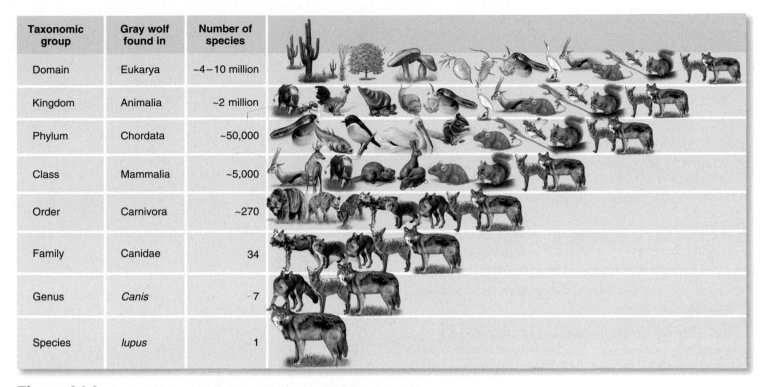

Taxonomic group	Gray wolf found in	Number of species
Domain	Eukarya	~4–10 million
Kingdom	Animalia	~2 million
Phylum	Chordata	~50,000
Class	Mammalia	~5,000
Order	Carnivora	~270
Family	Canidae	34
Genus	*Canis*	7
Species	*lupus*	1

Figure 26.2   A taxonomic classification of the gray wolf (*Canis lupus*).

a relatively small family of 34 species including different types of wolves, jackals, foxes, wild dogs, and the coyote and domestic dog. All species in the family Canidae are doglike animals. The smallest grouping that contains the gray wolf is its genus, *Canis*, which includes only two types of wolves, four types of jackals, and the coyote.

As originally advocated by Linnaeus, **binomial nomenclature** is the standard method for naming species. The scientific name of every species has two Latinized names, which are its genus name and its unique species epithet. The genus name is always capitalized, while the species epithet is not. Both names are italicized or underlined. After the first mention, the genus name is abbreviated to a single letter. For example, we would write that *Canis lupus* is the gray wolf, and in subsequent sentences the species would be referred to as *C. lupus.*

The rules for naming animal species, such as *Canis lupus*, have been established by the International Commission on Zoological Nomenclature (ICZN). The ICZN provides and regulates a uniform system of nomenclature to ensure that every animal has a unique and universally accepted scientific name. ICZN publishes the International Code of Zoological Nomenclature containing the rules accepted as governing the application of scientific names to all animals. The ICZN also provides rulings on individual nomenclatural problems brought to its attention. As long as ICZN rules are followed, new species can be named by anyone, not only by scientists. When someone believes he or she has identified a new species, that person generally does research through various publications (journals, books, etc.) to determine if it is a new species. In addition, the person needs to provide specimens to a museum where they will be preserved so someone else can verify the description or do research using the original specimens. When naming a new species, genus names are always nouns or treated as nouns, while species epithets may be either nouns or adjectives. The names often have a Latin or Greek origin and refer to characteristics of the species or to features of its habitat. For example, the genus name of the newly discovered African forest elephant, *Loxodonta*, is from Greek *loxo*, meaning slanting, and *odonta*, meaning tooth. The species epithet *cyclotis* refers to the observation that the ears of this species are rounder compared to those of *L. africana*. However, sometimes the naming of a species is not taken very seriously. For example, there is a beetle named *Agra vation* and a spider named *Draculoides bramstokeri*. International organizations are also involved in the nomenclature of species other than animals. The rules for naming plants and fungi are governed by the International Association of Plant Taxonomy (IAPT), and the naming of prokaryotes is overseen by the International Committee on Systematics of Prokaryotes (ICSP).

Organisms of the Three Domains Have Basic Similarities Yet Distinctive Differences

As discussed in Chapter 22, scientists think that all life originated from primordial prokaryotic cells, sometime between 4.0 and 3.5 billion years ago. Soon thereafter, the two prokaryotic domains, Bacteria and Archaea, diverged from each other.

Between 2.5 and 2.0 billion years ago, the first unicellular eukaryotic species came into being via an endosymbiotic or symbiotic relationship between archaea and bacteria. Eventually, multicellular eukaryotic species arose approximately 1.5 billion years ago.

Due to the evolutionary relatedness of Bacteria, Archaea, and Eukarya, the three domains of life share striking similarities:

- DNA is used as the genetic material.
- All species use the same genetic code (with only a few rare codon exceptions).
- Messenger RNA encodes the information to produce proteins.
- Transfer RNA and ribosomes are needed to synthesize proteins, using mRNA as a source of genetic information.
- All living cells are surrounded by a plasma membrane.
- Certain metabolic pathways, such as glycolysis, are found in all three domains.

Biologists believe that these traits are universal because all three domains evolved from a common prokaryotic ancestor. However, when we compare the domains of life in detail, we find that some characteristics are not shared by all three. For example, the cytoplasm of eukaryotic cells is compartmentalized into various types of organelles, while those of bacterial and archaeal cells are not. Such dissimilarities exist because major evolutionary changes have occurred since the time that the three domains diverged from each other. **Table 26.1** compares a variety of molecular and cellular characteristics among Bacteria, Archaea, and Eukarya.

On modern Earth, the domain Bacteria is a diverse collection of many species that can live almost anywhere (**Figure 26.3a**). Bacteria are so widespread that we can make only general statements about their ecology. They are found on the tops of mountains, at the bottoms of oceans, in the guts of animals, and even in the frozen rocks of Antarctica. More bacteria are found in a person's digestive tract than there are people in the world! The human mouth is home to more than 500 different species of bacteria. A key reason why bacteria have been so successful in occupying various habitats is their metabolic diversity. Different bacterial species have evolved an amazing variety of ways to obtain energy and organic molecules from their environment. Most species of bacteria decompose existing organic materials, though some can carry out photosynthesis or break down inorganic molecules to create their own organic molecules. Bacterial species can survive on many different sources of energy. Many bacteria metabolize common organic molecules such as sugars and amino acids. However, some species can use organic molecules that are less common, and are not usually broken down by archaea or eukaryotes. These include compounds found in gasoline, crude oil, pesticides, and industrial solvents!

Bacterial species come in a myriad of shapes and sizes, but individual bacterial cells are usually quite small, in the range of 1 to 5 microns long. Under the microscope, bacterial cells and archaeal cells do not possess morphological differences that

Table 26.1	Distinguishing Cellular and Molecular Features of Domains Bacteria, Archaea, and Eukarya*		
Characteristic	**Bacteria**	**Archaea**	**Eukarya**
Chromosomes	Usually circular	Circular	Usually linear
Nucleosome structure	No	No	Yes
Chromosome segregation	Fission	Fission	Mitosis/meiosis
Introns in genes	Rarely	Rarely	Commonly
Ribosomes	70S	70S	80S
Initiator tRNA	Formylmethionine	Methionine	Methionine
Operons	Yes	Yes	No
Capping of mRNA	No	No	Yes
RNA polymerases	One	Several	Three
Promoters of structural genes	–35 and –10 sequences	TATA box	TATA box
Cell compartmentalization	No	No	Yes
Membrane lipids	Ester-linked	Ether-linked	Ester-linked

* The descriptions in this table are meant to represent the general features of most species in each domain. Some exceptions are observed. For example, certain bacterial species have linear chromosomes.

would allow them to be distinguished. This is one reason why it took so long for researchers to realize that bacteria and archaea should be placed in different domains. Nevertheless, as described in Table 26.1, certain of their molecular features are different.

The domain Archaea is less diverse than Bacteria. However, this domain was discovered relatively recently, in the 1970s, and the list of archaeal species continues to expand. Thus far, many archaeal species have been found to live in extreme environments (**Figure 26.3b**). The name archaea, meaning ancient, refers to the observation that archaea tend to occupy environments that are thought to be similar to those found on the ancient Earth. Thus far, most archaea can be placed into one of three categories: extreme halophiles, methanogens, or hyperthermophiles. Extreme halophiles live in salty environments. Some species can live in water with salt concentrations above 15% (seawater averages 3.5% salinity). Methanogens release methane as a waste product of cellular metabolism. These species live at the bottom of lakes and swamps and in the intestinal tracts of animals. Methanogens living in your digestive tract cause intestinal gas. Hyperthermophiles live in extremely hot water, even temperatures over 100°C (212°F). Examples include archaea that live in hot springs and around deep-sea thermal vents. More recently, biologists have come to realize that archaea are not entirely restricted to extreme environments. For example, newer research has shown they are abundant in the open sea.

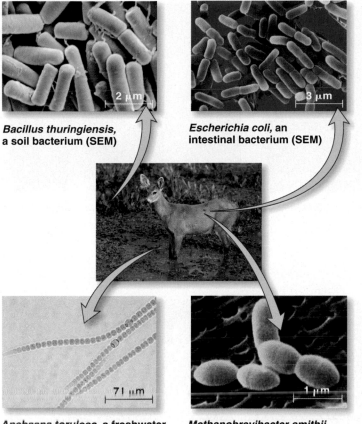

Bacillus thuringiensis, a soil bacterium (SEM)

Escherichia coli, an intestinal bacterium (SEM)

Anabaena torulosa, a freshwater bacterium (LM)

Methanobrevibacter smithii, an intestinal archaeon

(a)

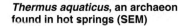

Hot springs

(b)

Thermus aquaticus, an archaeon found in hot springs (SEM)

Figure 26.3 **A comparison of selected species of bacteria and archaea.** (a) Bacteria are found in nearly all environments, such as lakes (*Anabaena torulosa*), soil (*Bacillus thuringiensis*), and even within the digestive tracts of animals (*Escherichia coli*). Archaea are also found within the digestive tracts of animals (*Methanobrevibacter smithii*) and **(b)** often occupy extreme environments such as hot springs (*Thermus aquaticus*).

Though less diverse than Bacteria, Archaea is nevertheless a diverse domain of successful species.

The third domain, Eukarya, is so diverse that it has been traditionally divided into four kingdoms, based on a set of characteristics that is unique to each kingdom. We will consider these kingdoms next.

Table 26.2	A Comparison of the Four Traditional Kingdoms of Domain Eukarya			
Characteristic	**Protista**	**Fungi**	**Plantae**	**Animalia**
Multicellular	A few species	Most species	All species	All species
Cell wall	Some species, various types	Yes, chitin	Yes, cellulose	No
Multiple cell and tissue types	Some species	Some species	Yes	Yes
Capable of photosynthesis	Some species	No	Nearly all species	No
Use of organic molecules in the environment	Some species	All species, digestion begins externally	Rarely	All species, digestion is usually within a digestive tract

The Organisms of Each Traditional Eukaryotic Kingdom Have a Distinctive Set of Characteristics

The four traditional kingdoms of the Eukarya domain, Protista, Fungi, Plantae, and Animalia, provide a broad basis for biologists to appreciate the diversity of eukaryotic species. The members of each kingdom possess certain characteristics that tend to set them apart from the other kingdoms. **Table 26.2** compares some of the common features that are found among species of these four kingdoms.

Protists are species within the traditional kingdom Protista, which contains members that are considered to be the simplest eukaryotes (**Figure 26.4**). Most species are unicellular, but some are colonial, and others are simple multicellular organisms closely related to unicellular protist species. Certain protists, such as algae, are capable of photosynthesis, while the remaining species eat bacterial or other protistan cells, or organic material suspended in water. Protists live in aquatic habitats, and they represent a critical step in the early evolution of eukaryotes. They evolved from prokaryotes and eventually gave rise to the other three kingdoms of eukaryotes.

From the perspective of taxonomy, protists are an unusually diverse group of organisms that were put together because they did not seem to belong to any other group. In some ways, the kingdom Protista was a place for leftover eukaryotic species that could not be classified into any of the other three eukaryotic kingdoms. Over 250,000 species of protists are estimated to currently exist. Many biologists are studying the evolutionary relationships among the protists, and the categorization of protists into a single kingdom has been revised. Current models are discussed later in this chapter and in Chapter 28.

The kingdom Fungi contains mushroom-forming fungi, molds, and yeasts (**Figure 26.5**). Species of fungi are present all over the world, in aquatic as well as terrestrial environments. Scientists estimate their diversity at over 100,000 species. Many fungi have symbiotic relationships with plants. These organisms

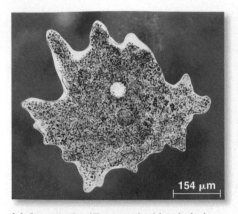

(a) An amoeba (*Entamoeba histolytica*)

(b) A paramecium (*Paramecium caudatum*) (SEM)

(c) A green alga (*Acrosiphonia coalita*)

Figure 26.4 Examples of protists.

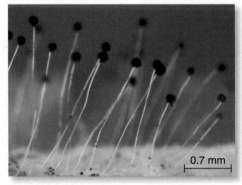

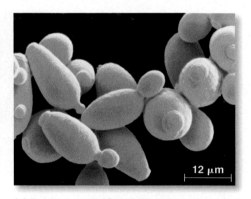

(a) A species that forms mushrooms, golden chanterelle (*Cantharellus cabaruis*)

(b) A common bread mold (*Neurospora crassa*)

(c) Baker's yeast (*Saccharomyces cerevisiae*) (SEM)

Figure 26.5 Examples of fungi.

typically secrete digestive enzymes into their environment and then transport the smaller breakdown products into their cytoplasm. Along with bacteria, fungi play an important role in the decomposition and recycling of organic materials on our planet. A distinguishing cellular feature is that fungal cells are surrounded by a cell wall that is made of the polysaccharide chitin. Though a few species such as yeast are unicellular, most fungi are multicellular, consisting of a mass of threadlike filaments called hyphae that combine to make up the fungal body called the mycelium. The mycelium usually grows underground. For many fungal species, the familiar mushrooms that we can see aboveground are actually the reproductive structures of the fungus. Fungi are discussed in more detail in Chapter 29.

The kingdom Plantae is composed of multicellular species that are nearly all capable of photosynthesis. These include mosses, ferns, conifers, and flowering plants (**Figure 26.6**). Approximately 300,000 species are thought to currently exist. Though many species are found in aquatic environments, plants are the foundation of all terrestrial habitats. Rather than using

organic molecules from their environment, nearly all plants produce their own organic molecules via photosynthesis. They are therefore the Earth's primary producers of the organic molecules that sustain both themselves and most other living species. (Certain species of bacteria and algae are also capable of photosynthesis.) At the cellular level, a distinguishing feature of plant cells is a cell wall that is made primarily of the carbohydrate cellulose. Chapters 30 and 31 provide an introduction to plants, which are covered in more detail in Unit VI.

All species of the kingdom Animalia are multicellular and rely on other organisms for their nourishment. It is a particularly diverse kingdom composed of more than 1 million species. These include sponges, worms, insects, mollusks, fish, amphibians, reptiles, birds, and mammals (**Figure 26.7**). Most ingest food and digest it in an internal cavity. Except for sponges, the bodies of animals are composed of cells organized into several types of tissues; each tissue is composed of cells that are specialized to perform a particular function. Most animals are capable of complex and relatively rapid movement compared to other organisms.

(a) Juniper moss (*Polytrichum juniperinum*)

(b) Royal fern (*Osmunda regalis*)

(c) A conifer, the Colorado blue spruce (*Picea pungens*)

(d) A flowering plant, the common sunflower (*Helianthus annuus*)

Figure 26.6 Examples of plants.

(a) Yellow sponge (*Cleona celata*)

(b) Firefly (*Photinus granulatus*)

(c) Zebra mussel (*Dreissena polymorpha*)

(d) Rainbow trout (*Oncorhynchus mykiss*)

(e) The lesser flamingo (*Phoeniconaias minor*)

Figure 26.7 Examples of animals.

(f) Red squirrel (*Tamiasciurus hudsonicus*)

A nervous system enables animals to receive environmental stimuli and respond with specialized movements. At the cellular level, a distinguishing feature is that animal cells lack a rigid cell wall such as that surrounding plant, fungal, and bacterial cells. Animal diversity is discussed in Chapters 32–34, and animal form and function are described in more detail in Unit VII.

Our traditional view of four eukaryotic kingdoms has recently changed based on a clearer understanding of the evolutionary relationships between the organisms involved. For example, protists are no longer considered to make up a single kingdom. Later in this chapter, we will examine the newer groupings of eukaryotic phyla. In the next section, we will see how the information generated by the study of systematics has been instrumental in reshaping the classification of organisms.

26.2 Systematics

Systematics is the study of biological diversity and evolutionary relationships. Biologists use systematics to develop the methods to construct taxonomic groups. For example, the classification of the gray wolf described earlier in Figure 26.2 is based on systematics. Therefore, taxonomy is one part of systematics. By studying the similarities and differences among species, biologists can gain information about **phylogeny**, which is the evolutionary history of a species or group of species. As you will learn, phylogeny helps us understand the relationships between ancestors and their descendants.

In this section, we will first consider how evolutionary biologists construct diagrams or trees that describe the evolutionary relationships among various species, both living and extinct. Biologists typically gather morphological or molecular genetic data, and then they use different mathematical strategies to analyze those data and construct evolutionary trees. Newer approaches using molecular genetic data have revolutionized our understanding of evolutionary relationships, causing biologists to revise the broad taxonomic groupings that have existed for the past few decades.

A Phylogenetic Tree Depicts the Evolutionary Relationships Between Past and Present Species

A systematic approach is followed to produce a **phylogenetic tree**—a diagram that describes a phylogeny. Such a tree is a hypothesis of the evolutionary relationships among various species, based on the information available to and gathered by systematists. Let's look at what information a phylogenetic tree contains and the form in which it is presented.

Figure 26.8 shows a hypothetical phylogenetic tree of the relationships between various bird species, in which the species are labeled A through K. The vertical axis represents time, with the oldest species at the bottom. As discussed in Chapter 25 (refer back to Figure 25.6), new species can be formed by anagenesis, in which a single species evolves into a different species, or more commonly by cladogenesis, in which a species diverges into two or more species. The nodes or branch points in a phylogenetic tree illustrate times when cladogenesis has occurred. For example, approximately 12 million years ago, species A diverged into species A and species B. Figure 26.8 also shows anagenesis. After species B split into species B and D, species D then evolved into species G by anagenesis. The tips of branches represent species that became extinct in the past, such as species B, C, and E, or modern species, such as F, I, G, J, H, and K, which are at the top of the tree. Species A and D are also extinct but gave rise to modern species.

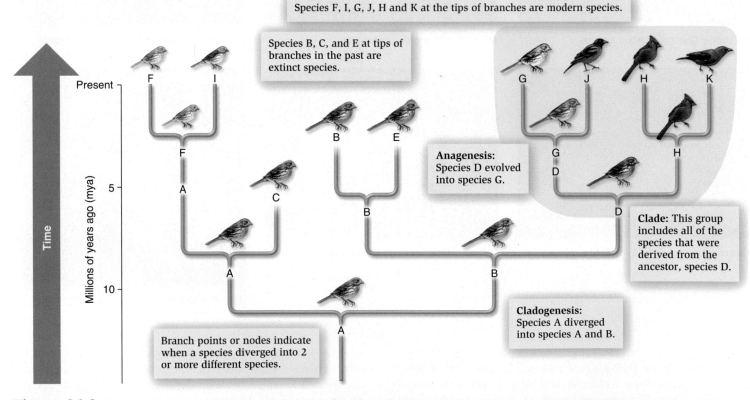

Figure 26.8 How to read a phylogenetic tree.   This hypothetical tree shows the proposed relationships between various bird species. Species are grouped into clades derived from a common ancestor.

Biological inquiry: Can two different species have more than one common ancestor?

By studying the branch points of a phylogenetic tree, researchers can group species according to common ancestry. A **monophyletic group**, also known as a **clade**, is a group of species, a taxon, consisting of the most recent common ancestor and all of its descendants. For example, the group highlighted in light green in Figure 26.8 is a clade derived from the common ancestor labeled D. The present-day descendants of a common ancestor can also be called a clade. In this case, species G, J, H, and K would form a modern clade. Likewise, the entire tree forms a clade, with species A as a common ancestor. Thus, smaller and more recent clades are subsets of larger clades.

The relationship between a clade and taxonomy depends on how far back we go to choose a common ancestor. For broader taxa, such as a kingdom, the common ancestor existed a very long time ago, on the order of hundreds of millions or even billions of years ago. For smaller taxa, such as a family or genus, the common ancestor occurred much more recently, on the order of millions or tens of millions of years ago. This idea is shown schematically in **Figure 26.9**. This small, hypothetical kingdom is a monophyletic group that contains 64 species. (Actual kingdoms are obviously larger and more complex.) The diagram emphasizes the taxa that contain the species designated number 43. The common ancestor that gave rise to this kingdom of organisms existed approximately 1 billion years ago. Over time, more recent species arose that became the common ancestors to the phylum, class, order, family, and genus that contain species number 43.

A goal of modern systematics is to create taxa according to evolutionary relationships. Systematics attempts to organize species into monophyletic groups, which means that each group includes an ancestral species and all of its descendants. Ideally, every taxon, whether it is a kingdom, phylum, class, order, family, or genus, should be a monophyletic group. However, as biologists gather more information about evolutionary relationships among extinct and modern species, we sometimes discover that previously established taxonomic groups are not monophyletic. When this occurs, efforts are made to revise taxonomy so it is consistent with evolutionary relationships.

The Study of Systematics Is Usually Based on Morphological or Genetic Homology

Systematics assumes that species that have diverged from each other in the very distant past have had a large amount of time to accumulate many genetic differences. Ultimately, such genetic changes are likely to cause differences in morphology and other attributes. By comparison, systematics also assumes that species that have diverged from each other more recently tend to be much more similar both genetically and morphologically.

As we discussed in Chapter 23, the term **homology** refers to similarities among various species that occur because the species are derived from a common ancestor. Attributes that are the result of homology are said to be homologous. For example, the wing of a bat, the arm of a human, and the front leg of a cat are homologous structures (refer back to Figure 23.13). Similarly, genes found in different species are homologous if they have been derived from the same ancestral gene (refer back to Figure 23.14).

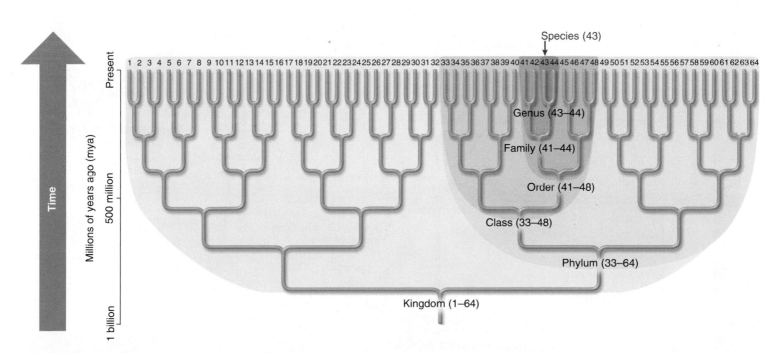

Figure 26.9 Schematic relationship between a phylogenetic tree and taxonomy.   The shaded areas highlight the kingdom, phylum, class, order, family, and genus for species number 43. All of the taxa are monophyletic groups or clades. Broader taxa, such as phyla and classes, are derived from more ancient common ancestors. Smaller taxa, such as families and genera, are derived from more recent common ancestors. These smaller taxa are subsets of the broader taxa.

In systematics, researchers identify homologous features that are shared by some species but not by others. This allows them to group species based on their similarities. Researchers usually study homology at the level of morphological traits or at the level of genes. In addition, the data they gather are viewed in light of geographic data. Because most organisms do not migrate extremely long distances, species that are closely related evolutionarily are likely to inhabit neighboring or overlapping geographic regions. Let's now consider how researchers would analyze morphological and genetic homology to construct phylogenetic trees.

Morphological Analysis Traditionally, the first studies in systematics, which occurred prior to the advent of molecular genetic techniques, focused on morphological features of extinct and modern species. To establish evolutionary relationships based on morphological homology, many traits have to be analyzed to identify similarities and differences as a way to obtain a comprehensive picture of species' relatedness. Researchers use a variety of methods to combine information about different traits.

At times, the data can be complex because traits can change more than once during evolution, and the same trait can arise independently in different species. As described in Chapter 23, convergent evolution sometimes leads to convergent or analogous traits, those that arise independently in different species due to adaptation to similar environments. For example, the giant anteater (*Myrmecophaga tridactyla*), in South America, and the echidna (*Tachyglossus aculeatus*), found in Australia, have similar adaptations, such as long snouts and tongues, that enable these animals to feed on ants (refer back to Figure 23.7). These are convergent traits because they were not derived from a common ancestor. Rather, they arose twice during evolution due to adaptation to similar environments. In systematics, convergent evolution can cause errors if a researcher assumes that a particular trait arose only once, and that all species having the trait are derived from a common ancestor.

By studying morphological features of extinct species in the fossil record, paleontologists have constructed many phylogenetic trees that chart a series of organisms that led to the existence of modern species. In this approach, the tree is based on

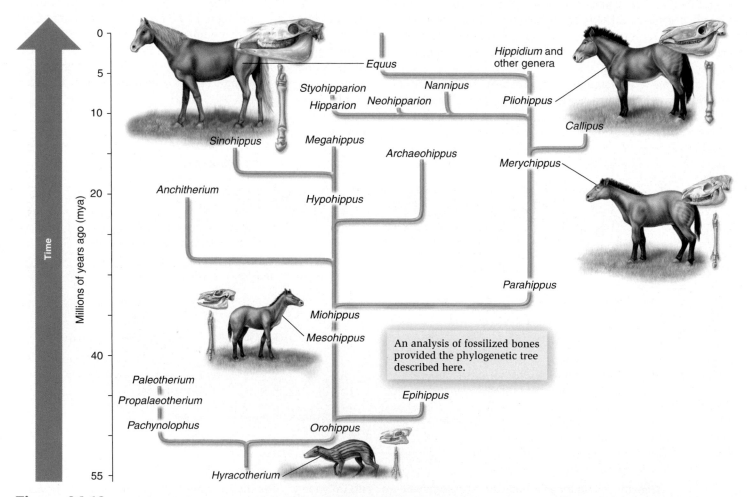

Figure 26.10 **Horse evolution.** An analysis of morphological traits was used to produce this phylogenetic tree showing the evolutionary changes that led to the modern horse. Three important morphological changes that occurred in these genera were larger size, fewer toes, and a shift toward a jaw structure suited for grazing.

morphological features that change over the course of many generations. As an example, **Figure 26.10** depicts a current hypothesis of the evolutionary changes that led to the development of the modern horse. This figure shows genera, not individual species. Many morphological features were used to construct this tree. Because hard parts of the body are preserved in the fossil record, this tree is largely based on the analysis of changes in hoof structure, lengths and shapes of various leg bones, skull shape and size, and jaw and tooth morphology. Over an evolutionary time scale, the accumulation of many genetic changes has had a dramatic impact on species' characteristics. In the genera that are depicted in this figure, a variety of morphological changes occurred such as an increase in size, fewer toes, and a modified jaw structure suitable for grazing on fibrous grasses.

Molecular Systematics and Molecular Clocks The field of **molecular systematics** involves the analysis of genetic data, such as DNA sequences, to identify and study genetic homology and construct phylogenetic trees. In 1963, Austrian biologist Emile Zuckerkandl and American chemist and Nobel laureate Linus Pauling were the first to suggest that molecular data should be used to establish evolutionary relationships. Nucleotide base sequences in DNA and amino acid sequences of proteins are particularly well suited to studying relationships, because genetic sequences change over the course of many generations due to the accumulation of mutations. Therefore, when comparing homologous genes in different organisms, DNA sequences from closely related organisms are more similar to each other than they are to sequences from distantly related species.

As discussed in Chapter 24, researchers have speculated that most genetic variation that exists in populations is neutral, meaning that it is not acted upon by natural selection. The reasoning behind this concept is that favorable mutations are likely to be very rare, and detrimental mutations are likely to be eliminated from a population by natural selection. A large body of evidence supports the idea that much of the genetic variation observed in modern species is due to the accumulation of neutral mutations. From an evolutionary point of view, if neutral mutations occur at a relatively constant rate, they can act as a **molecular clock** on which to measure evolutionary time.

Figure 26.11 illustrates the concept of a molecular clock. The graph's y-axis is a measure of the number of nucleotide sequence differences between pairs of species. The x-axis plots the amount of time that has elapsed since a pair of species shared a common ancestor. As seen in this diagram, the number of sequence differences is higher when two species shared a common ancestor in the very distant past than it is in pairs that shared a more recent common ancestor. The explanation for this phenomenon is that the gene sequences of the species accumulate independent mutations after they have diverged from each other; a longer period of time since their divergence allows for a greater accumulation of mutations, which makes their sequences more different.

Figure 26.11 suggests a linear relationship between the number of sequence changes and the time of divergence. Such a

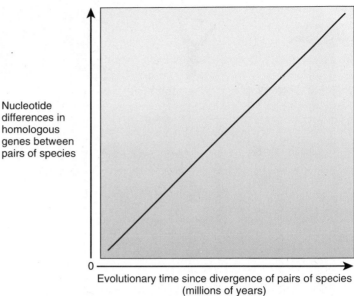

Figure 26.11 **A molecular clock.** According to the concept of a molecular clock, neutral mutations accumulate over evolutionary time. When comparing homologous genes between species, those species that diverged more recently tend to have fewer differences than do those whose common ancestor occurred in the very distant past.

relationship indicates that the observed rate of neutral mutations remains constant over millions of years. For example, a linear relationship predicts that a pair of species that has 20 nucleotide differences in a given gene sequence would have a common ancestor that is roughly twice as old as that of a pair showing 10 nucleotide differences. While actual data sometimes show a relatively linear relationship over a defined time period, evolutionary biologists do not believe that molecular clocks are perfectly linear over very long periods of time. Several factors can contribute to nonlinearity of molecular clocks. These include differences in the generation times of the species being analyzed, the presence of mutations that are acted upon by natural selection, and variation in mutation rates between different species.

To obtain reliable data, researchers must calibrate their molecular clocks. How much time does it take to accumulate a certain percentage of nucleotide changes? To perform such a calibration, researchers must have information regarding the date when two species shared a common ancestor. Such information could come from the fossil record, for instance. The genetic differences between those species are then divided by the amount of time since their last common ancestor to calculate a rate of change. For example, research suggests that humans and chimpanzees shared a common ancestor approximately 6 million years ago. The percentage of nucleotide differences between mitochondrial DNA of humans and chimpanzees is 12%. From these data, the molecular clock for changes in mitochondrial DNA sequences of primates is calibrated at roughly 2% nucleotide changes per million years.

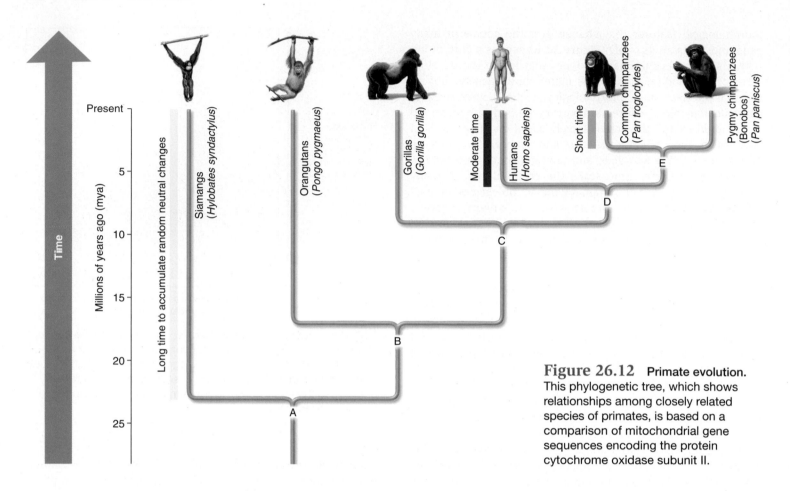

Figure 26.12 Primate evolution. This phylogenetic tree, which shows relationships among closely related species of primates, is based on a comparison of mitochondrial gene sequences encoding the protein cytochrome oxidase subunit II.

To understand the concept of a molecular clock, let's consider the evolution of some species of primates. **Figure 26.12** illustrates a simplified diagram of evolutionary relationships among several species that was derived by comparing DNA sequences in a mitochondrial gene. This gene encodes a protein called cytochrome oxidase subunit II, which is involved with cellular respiration. The vertical scale represents time, and the branch points that are labeled with letters represent common ancestors. Let's take a look at three branch points (labeled A, D, and E) and relate them to the concept of a molecular clock. The common ancestor labeled A diverged into two species that ultimately gave rise to siamangs and the other five species. Since this divergence, there has been a long time (approximately 23 million years) for the siamang genome to accumulate random neutral changes that would be different from the random changes that have occurred in the genomes of the other five species (see yellow bar in Figure 26.12). Therefore, the gene in the siamangs is more different than are the genes in the other seven species. Now let's compare humans and chimpanzees. The common ancestor that gave rise to these species is labeled D. This species diverged into two species that eventually gave rise to humans and chimpanzees. This divergence occurred a moderate time ago, approximately 6 million years ago, as illustrated by the red bar. Compared to humans and chimpanzees, humans and siamangs have more differences in their gene sequences because there has been more time for them to accumulate random neutral mutations. Finally, let's consider the two species of chimpanzees,

whose common ancestor is labeled E. Since the divergence of species E into two species, approximately 3 million years ago, the time for the molecular clock to tick (that is, accumulate random mutations) is relatively short, as depicted by the green bar in Figure 26.12. Therefore, the two modern species of chimpanzee have very similar gene sequences.

Genetic sequence information is primarily used for studying relationships between modern organisms. However, as discussed later in our Feature Investigation, sometimes DNA can be obtained from extinct organisms as well. For evolutionary comparisons, the DNA sequences of many genes have been obtained from a wide range of sources. Several genes have been used to construct phylogenetic trees. For example, the gene that encodes an RNA of the small subunit of the ribosome (SSU rRNA) is commonly analyzed to compare higher taxa such as phyla. Because rRNA is universal in all living organisms, its function must have been established at an early stage in the evolution of life on this planet, and its sequence has changed fairly slowly. The molecular clock for SSU rRNA has been calibrated at approximately 1% sequence change per 50 million years. Furthermore, SSU rRNA is a rather large molecule, so it contains a large amount of sequence information. This gene has been sequenced from thousands of different species.

Slowly changing genes such as the gene that encodes SSU rRNA are useful for evaluating distant evolutionary relationships, such as comparing plants and animals. Such genes may also be useful in assessing evolutionary relationships among bacterial species that evolve fairly rapidly due to their short generation times. Other genes have changed more rapidly because of a greater tolerance of neutral mutations. For example, the mitochondrial genome and DNA sequences within introns can more easily incur neutral mutations (compared to the coding sequences of genes), and so their sequences change frequently during evolution. More rapidly changing DNA sequences have been used to study recent evolutionary relationships, particularly among eukaryotic species such as large animals that have long generation times and tend to evolve more slowly. In these cases, slowly evolving genes may not be very useful for establishing evolutionary relationships because two closely related species are likely to have identical or nearly identical DNA sequences for such genes. Instead, sequence differences are more easily found among closely related species when the DNA sequences are more rapidly changing. Our example in Figure 26.12 of closely related species of primates uses sequence changes in the mitochondrial gene for cytochrome oxidase subunit II because this gene tends to change fairly rapidly on an evolutionary timescale.

A Cladistic Approach Is the Most Common Way to Make a Phylogenetic Tree

A **cladistic approach** reconstructs a phylogenetic tree by considering the various possible pathways of evolution and then choosing the most plausible tree. Cladistics is generally accepted as the best method available for the construction of phylogenetic trees, which are known as **cladograms.** A cladistic approach compares traits that are either shared or not shared by different species. A trait shared with a distant ancestor is called a **shared primitive character** or **symplesiomorphy.** Such traits are viewed as being older traits—ones that occurred earlier in evolution. In contrast, a **shared derived character**, or **synapomorphy**, is a trait that is shared by a group of organisms but not by a distant common ancestor. Compared to primitive characters, derived characters are more recent traits on an evolutionary timescale. For example, among mammals, only some species have flippers, such as whales and dolphins. In this case, flippers were derived from the two front limbs of an ancestral species. The word derived refers to the phenomenon that evolution involves the modification of traits in pre-existing species. In other words, newer populations of organisms are derived from changes in pre-existing populations. The basis of the cladistic approach is to analyze many shared derived characters among groups of species to deduce the pathway that gave rise to the species.

To understand the concept of primitive and shared derived characters, **Figure 26.13** shows a cladogram that compares several traits among five species of animals. Shared derived characters are used to establish a cladogram. The various colors in this tree emphasize when particular traits arose during the evolution of these species. A branch point, or node, is where two species differ in shared derived characters. The oldest common ancestor, which would now be extinct, had a notochord and gave rise to all five species. Vertebrae are a shared derived character of the lamprey, salmon, lizard, and rabbit, but not amphioxus. By com-

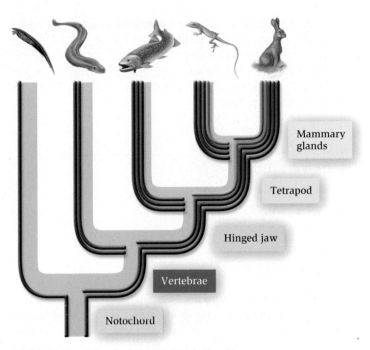

(a) Cladogram based on morphological traits

	Amphioxus	Lamprey	Salmon	Lizard	Rabbit
Notochord	Yes	Yes	Yes	Yes	Yes
Vertebrae	No	Yes	Yes	Yes	Yes
Hinged jaw	No	No	Yes	Yes	Yes
Tetrapod	No	No	No	Yes	Yes
Mammary glands	No	No	No	No	Yes

(b) Characteristics among species

Figure 26.13   Primitive versus shared derived characters involving morphological traits.   (a) This phylogenetic tree illustrates both primitive and shared derived characters in a cladogram of relationships among five animal species. (b) A comparison of characteristics among these species.

Biological inquiry: What shared derived character is common to the salmon, lizard, and rabbit, but not the lamprey?

parison, a hinged jaw is a shared derived character of the salmon, lizard, and rabbit, but not of the lamprey or amphioxus. The table shown in Figure 26.13b compares the characters that are found in these five species.

In a cladogram, an **ingroup** is a monophyletic group in which we are interested. By comparison, an **outgroup** is a species or group of species that is most closely related to an ingroup. For example, let's suppose that the salmon, lizard, and rabbit are an ingroup. The outgroup would be the lamprey, because this species is the most closely related to the ingroup but lacks one of the shared derived characters, namely a hinged jaw. All traits that are shared by the outgroup and the ingroup must have arisen in a common ancestor that predates the divergence of the two groups.

Likewise, the concept of shared derived characters can apply to molecular data such as a sequence of a gene. Let's consider an example to illustrate this idea. Our example involves molecular data obtained from seven different hypothetical species called A–G. In these species, a homologous region of DNA was sequenced as shown here:

    12345678910
A: GATAGTACCC
B: GATAGTTCCC
C: GATAGTTCCG
D: GGTATTACCC
E: GGTATAACCC
F: GGTAGTACCA
G: GGTAGTACCC

The cladogram of **Figure 26.14** is a hypothesis of how these DNA sequences arose. In this case, a mutation that changes the DNA sequence is analogous to a modification of a characteristic. Species that share such genetic changes possess shared derived characters because the new genetic sequence was derived from a more primitive sequence.

Now that we have an understanding of primitive and shared derived characters, let's consider the steps a researcher would follow to construct a cladogram using a cladistics approach.

1. **Choose the species in whose evolutionary relationships you are interested**. In a simple cladogram, such as those described in this chapter, individual species are compared to each other. In more complex cladograms, species may be grouped into larger taxa (for example, families) and compared with each other. If such grouping is done, the groups must be clades for the results to be reliable.

2. **Choose characters for comparing different species**. A character is a general feature of an organism. Characters may come in different versions called character states. For example, hair color is a character, while brown hair and red hair are character states.

3. **Determine the order of character states.** In other words, determine if a character state is primitive or derived. This information may be available by examining the fossil record, for example, but is usually done by comparing the ingroup with the outgroup. An assumption is made that the outgroup has the primitive state(s).

4. **Group species (or higher taxa) based on shared derived characters**.

5. **Build a cladogram based on the following principles**:
   • All species (or higher taxa) are placed on tips in the phylogenetic tree, not at branch points.
   • Each cladogram branch point should have a list of one or more shared derived characters that are common to all species above the branch point unless the character is later modified.
   • All shared derived characters appear together only once in a cladogram unless they arose independently during evolution more than once.

6. **Choose the most likely cladogram among possible options**. When grouping species or higher taxa, more than one cladogram may be possible. Therefore, analyzing the data and producing the most likely cladogram is a key aspect of this process. As described next, different theoretical approaches can be followed to achieve this goal.

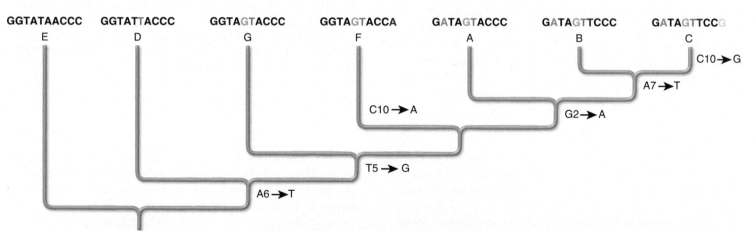

GGTATAACCC
12345678910
Proposed primitive
sequence

Figure 26.14   **Primitive versus shared derived characters involving a molecular trait.** This phylogenetic tree illustrates a cladogram of relationships involving homologous gene sequences found in seven species. Mutations that alter a primitive DNA sequence are shared among certain species and thereby allow the construction of a cladogram.

Biological inquiry: What nucleotide change is a shared derived character for species A, B, and C, but not for species G?

Different Strategies Can Be Followed to Produce the Most Likely Cladogram

The challenge in a cladistic approach is to determine the correct order of events. It may not always be obvious which traits are ancestral and came earlier, and which are derived and came later in evolution. Different approaches can be used to deduce the correct order. First, for morphological traits, a common way to deduce the order of events is to analyze fossils and determine the relative dates that certain traits arose. A second strategy that can be used to deduce the correct order is to assume that the best hypothesis is the one that requires the fewest number of evolutionary changes. This concept, called the **principle of parsimony**, states that the preferred hypothesis is the one that is the simplest. For example, if two species possess a tail, we would initially assume that a tail arose once during evolution, and that both species have descended from a common ancestor with a tail. Such a hypothesis is simpler, and more likely to be correct, than assuming that tails arose twice during evolution, and that the tails in the two species are not due to descent from a common ancestor.

The principle of parsimony can also be applied to gene sequence data, as can other methods called maximum likelihood and Bayesian analysis. In these methods, a model for the most likely changes during gene evolution is applied to the data. Maximum likelihood, for example, will produce the tree that is most likely based on the data and the application of a particular evolutionary model. These models take into account various types of factors, such as whether mutations would affect the first, second, or third nucleotide in a codon (mutations affecting the third codon are often neutral) and whether the data conform to a molecular clock. When constructing their trees, researchers often have preferences about the kinds of strategies they employ. More than one strategy may be used to construct a cladogram.

Now that we have a better understanding of synapomorphies and the steps used to build a cladogram, let's look at a hypothetical example. We will use molecular data for four taxa (A–D), where A is presumed to be the outgroup and has all of the primitive states.

```
    12345
A: GTACA (outgroup)
B: GACAG
C: GTCAA
D: GACCG
```

Given this information, three potential trees are shown in **Figure 26.15**, although more are possible. In these examples, tree 1 requires seven mutations, and tree 2 requires six, while tree 3 requires only five. Therefore, tree 3 requires the fewest number of mutations and is considered the most parsimonious. Based on the principle of parsimony, it would be the more likely choice.

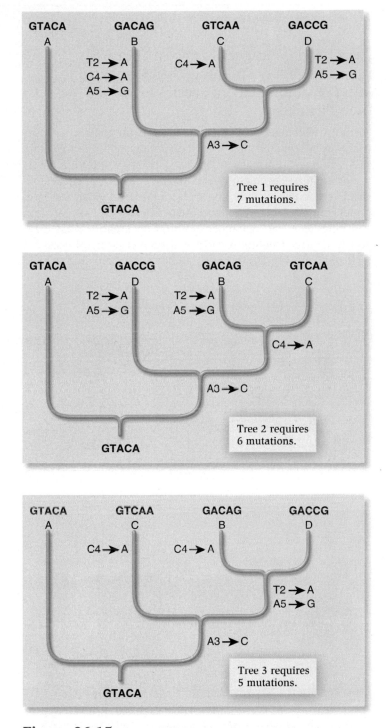

Figure 26.15 **The cladistic approach from molecular genetic data.** These are three possible evolutionary trees for the evolution of a short DNA sequence, but many more are possible. Changes in nucleotide sequence are shown along each tree. According to the principle of parsimony, tree number 3 is the more likely choice because it requires only five mutations. When constructing cladograms based on long genetic sequences, researchers use computers to generate trees with the fewest possible genetic changes.

FEATURE INVESTIGATION

Cooper and Colleagues Extracted DNA from Extinct Flightless Birds and Then Compared It with DNA from Modern Species to Reconstruct Their Phylogeny

Starting with small tissue samples from extinct species, scientists have discovered that it is occasionally possible to obtain DNA sequence information. This is called ancient DNA analysis or molecular paleontology. Since the mid-1980s, some researchers have become excited about the information derived from sequencing DNA of extinct specimens. Debate has centered on how long DNA can remain intact after an organism has died. Over time, the structure of DNA is degraded by hydrolysis and the loss of purines. Nevertheless, under certain conditions (cold temperature, low oxygen, etc.), DNA samples may be stable as long as 50,000–100,000 years. In most studies involving extinct specimens, the ancient DNA is extracted from bone, dried muscle, or preserved skin. These samples are often from museum specimens that have been gathered by paleontologists. In recent years, this approach has been used to study evolutionary relationships between modern and extinct species.

In the Feature Investigation shown in **Figure 26.16**, Alan Cooper, Cécile Mourer-Chauviré, Geoffrey Chambers, Arndt von Haeseler, Allan Wilson, and Svante Pääbo investigated the evolutionary relationships among some extinct and modern species of flightless birds. This is an example of discovery-based science. The researchers wanted to gather data to propose a hypothesis regarding the evolutionary relationships among certain species.

Figure 26.16   DNA analysis of phylogenetic relationships among modern and extinct flightless birds by Cooper and colleagues.

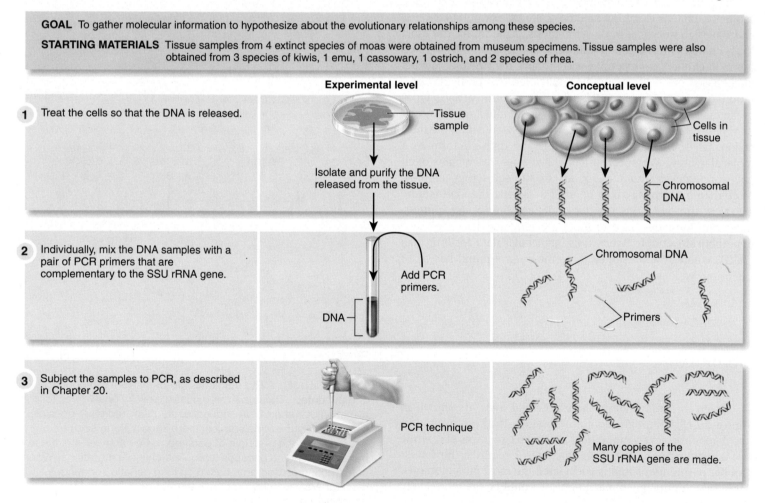

GOAL To gather molecular information to hypothesize about the evolutionary relationships among these species.

STARTING MATERIALS Tissue samples from 4 extinct species of moas were obtained from museum specimens. Tissue samples were also obtained from 3 species of kiwis, 1 emu, 1 cassowary, 1 ostrich, and 2 species of rhea.

Experimental level   **Conceptual level**

1 Treat the cells so that the DNA is released.
    Tissue sample
    Cells in tissue

    Isolate and purify the DNA released from the tissue.
    Chromosomal DNA

2 Individually, mix the DNA samples with a pair of PCR primers that are complementary to the SSU rRNA gene.
    Add PCR primers.
    DNA
    Chromosomal DNA
    Primers

3 Subject the samples to PCR, as described in Chapter 20.
    PCR technique
    Many copies of the SSU rRNA gene are made.

4 Subject the amplified DNA fragments to DNA sequencing, as described in Chapter 20.

Sequence the amplified DNA.

The amplification of the SSU rRNA gene allows it to be subjected to DNA sequencing.

5 Align the DNA sequences to each other, using computer techniques described in Chapter 21.

Align sequences using computer programs.

Align sequences to compare the degree of similarity.

6 **THE DATA**

```
Moa 1      GCTTAGCCCTAAATCCAGATACTTACCCTACACAAGTATCCGCCCGAGAACTACGAGCACAAACGCTTAAAACTCTAAGGACTTGGCGGTGCCCCAAACCCA
Kiwi 1     · · · · · · · · · · · · · · ·T·G· · · · · ·GT· · ·CT· · · ·C· · · · · · · · · · · · · · · · · · · · · · · · · · · · · · · · · · · · · · · · · · · · · ·T· · · · · ·
Emu        · · · · · · · · · · · · · · ·TT· · · · · ·C· ·T· · ·CAG· ·C· · · · · ·T· · · · · · · · · · · · · · · · · · · · · · · · · · · · · · · · · · · ·T· · · · · ·
Cassowary  · · · · · · · · · · · · · · ·TT· · · · · ·CG·TA· ·CTG· · · · · · · · · · · · · · · · · · · · · · · · · · · · · · · · · · · · · · · · · · · ·T· · · · · ·
Ostrich    · · · · · · ·T· · · · ·AT· · · · ·C· ·CT· · · · · · · · · · · · · · · · · · · · · · · · · · · · · · · · · · · · · · · · · · · · · · · · · · · · ·T· · · · · ·
Rhea 1     · · · · · · · · · · ·T· · · · · · · ·C· ·CT· · · · · · · · · · · · · · · · · · · · · · · · · · · · · · · · · · · · · · · · · · · · · · · · · · · · ·T· · · · · ·

Moa 1      CCTAGAGGAGCCTGTTCTATAATCGATAATCCACGATACACCCGACCATCCCTCGCCCGT–GCAGCCTACATACCGCCGTCCCCAGCCCGCCT--AATGAAA
Kiwi 1     · · · · · · · · · · · · · · · · ·C· · · · · · · · · ·A· · · ·T· ·T· · ·AAC–A· · · · · ·T· · · · · · · · · · · ·G· · · ·T· · · ·AA· · · · ·G·
Emu        · · · · · · · · · · · · · · · · ·C· · · · · · · · · ·A· · · ·T· ·AA–A· · · · · · · · · · · · · · · · · · · · · ·G· · · · · · · · · · · · –· · · · ·
Cassowary  · · · · · · · · · · · · · · · · ·C· · · · · · · · · ·AG· · · ·I· ·I· ·AA·IA· · · · · · · · · · · · · · · · ·G· · · · · · · · · · ·––·G· · ·G
Ostrich    · · · · · · · · · · · · · · · · · · · · · · · · ·T· · ·A· · ·C· · ·T· ·A––T· · · · · · · · · · · · · ·G· · · · · · · · · · · ·C––· · ·G
Rhea 1     · · · · · · · · · · · · · · · · ·C· · · · · · · · · ·T· ·T· · ·A·–· · · · · · · · · · · · · · · · · · · · · · · · · · · · · · · ·TA·G· · · · ·

Moa 1      G–AACAATAGCGAGCACAACAGCCCTCCCCCGCTAACAAGACAGGTCAAGGTATAGCATATGAGATGGAAGAAATGGGCTACATTTTCTAACATAGAACACC
Kiwi 1     ·–· · · ·C· · · ·A· · · · · · ·TA· ·–· ·A· · · · · · · · · · · · · · · ·C· · · · · · · · · · · · · · · · · · · · · · · · · ·A· · · · ·T· ·T
Emu        ·–· · · · · · · · · · · ·T· · · ·AC––TT· · · · · · · · · · · · · ·G· · · · · · · · · · · · · · · · · · · · · · · · · · · · · · · · · ·T·T
Cassowary  ·–· · · · · · · · · · · ·T· · · · · ·AC–·T· · · · · · · · · · · · · ·G· · · · · · · · · · · · · · · · · · · · · · · · · · · · · · · ·T· · ·
Ostrich    ·–· · · · · · · · · · · · ·T· · · ·A––· · · · · · · · · · · · · · · · · ·GAG· · · · · · · · · · · · · · · · · · · · · · · · · ·T· ·A
Rhea 1     ·–· · · ·C· ·AG· ·T· ·T· ·TA===· · · · · · · · · · · · · · · ·G· · · · · · · · · · · · · · · · · · ·TC· · · · · ·A·

Moa 1      C-------------ACGAAAGAGAAGGTGAAACCCTCCTCAAAAGGCGGATTTAGCAGTAAAAATAGAACAAGAATGCCTATTTTAAGCCCGGCCCTGGGGC
Kiwi 1     –· · · · · · · · · · · ·A· ·GGT· · · · ·T·–C· ·T·G· · · · · · · · · · · · · ·C· · · ·T· · · ·GA·T· · · · · · · · –·T· · · ·A· · · ·
Emu        –· · · · · · · · · · · · ·AG·T· · · · ·T·AC·T· ·G· · · · · · · · · · · · · · ·C· · · ·T· · · ·GA·T· · · · ·A––·T· · ·T·A· · ·
Cassowary  –· · · · · · · · · · · ·A· ·G·T· · · · ·T·A· · ·T·G· · · · · · · · · · · · · ·C· · · · · · · · ·GA·T· · · · ·A–· · · · · ·A· · · ·
Ostrich    –· · · · · · · · · · · · ·G·TA· · · · ·T·A· · · ·G· · · · · · · · · · · · · · · · · · ·T· · · ·GA·T· · · · ·–T· · · ·T· ·A· · · ·
Rhea 1     –· · · · · · · · · ·G· · · · ·GGCA· · · · · ·–AC· · · ·CG· · · · · · · · · · · · · ·G· · ·G·TC· · ·A· · ·C·C· · · · · · · · · · ·A· · · ·
```

The kiwis and moas existed in New Zealand during the Pleistocene. Species of modern kiwis still exist, but the moas are now extinct. Eleven known species of moas formerly existed. In this study, the researchers investigated the phylogenetic relationships of four extinct species of moas, which were available as museum samples, kiwis of New Zealand, and several other living species of flightless birds. These included the emu and the cassowary (both found in Australia and New Guinea), the ostrich (found in Africa and formerly Asia), and two rheas (found in South America).

Samples from the various species were subjected to PCR to amplify a region of the SSU rRNA gene. This provided enough DNA for DNA sequencing. The data in Figure 26.16 illustrate a comparison of the sequences of a continuous region of the SSU rRNA gene from these species. The first line shows the DNA sequence for one of the four extinct moa species. Below it are the sequences of several of the other species they analyzed. When the other sequences are identical to the first sequence, a dot is placed in the corresponding position. When the sequences are different, the changed nucleotide base (A, T, G, or C) is placed there.

In a few regions, the genes are different lengths. In these cases, a dash is placed at the corresponding position.

As you can see from the large number of dots, the gene sequences among these flightless birds are very similar, though some differences occur. If you look carefully at the data, you will notice that the sequence from the kiwi (a New Zealand species) is actually more similar to the sequence from the ostrich (an African species) than it is to that of the moa, which was once found in New Zealand. Likewise, the kiwi is more similar to the emu and cassowary (found in Australia and New Guinea) than to the moa. Contrary to their original expectations, the researchers concluded that the kiwis are more closely related to Australian and African flightless birds than they are to the moas. From these results, they proposed that New Zealand was colonized twice by ancestors of flightless birds. First, New Zealand was colonized by the ancestor of moas, and then at a later date by an ancestor of the kiwis, which evolved independently of the moas in Australia and New Guinea. As shown in **Figure 26.17**, the researchers constructed a new evolutionary tree that illustrates the revised relationships among these modern and extinct species.

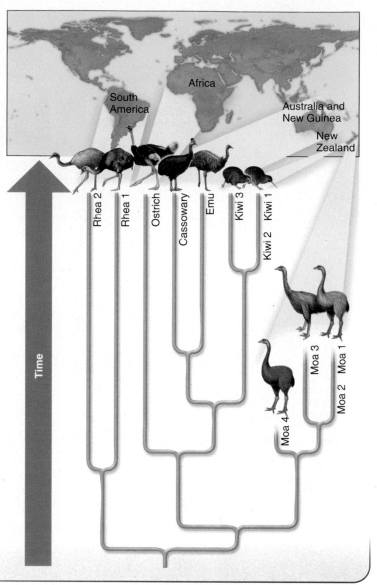

Figure 26.17 A revised phylogenetic tree of flightless birds. This phylogenetic tree was based on a comparison of DNA sequences from extinct and modern birds. The species included are moas, kiwis, emus, cassowaries, ostriches, and rheas. Moas are now extinct, but all other species still exist.

Biological inquiry: With regard to geography, why are the results in this figure surprising?

Molecular Systematics Is Changing Our View of Taxonomy

Taxonomy is a work in progress. As researchers gather new information, they sometimes discover that some of the current taxonomic groups are not monophyletic. **Figure 26.18** compares a monophyletic taxon with those that are not. As mentioned earlier, a monophyletic taxon contains a common ancestor and all of the species that are derived from that ancestor (Figure 26.18a). The ideal goal of taxonomy is to place organisms into monophyletic groups. A **paraphyletic taxon** is a group that contains a common ancestor and some, but not all, of its descendants (Figure 26.18b). In contrast, a **polyphyletic taxon** consists of members of several evolutionary lines and does not include the most recent common ancestor of the included lineages (Figure 26.18c). Over time, as we learn more about evolutionary relationships, taxonomic groups are being reorganized in an attempt to recognize monophyletic groups in biological classifications.

For higher eukaryotic taxa, such as kingdoms, molecular genetic data has shed new light regarding the placement of species into monophyletic groups. Many recent models propose several major groups, sometimes called **supergroups**, as a way to organize eukaryotes into monophyletic groups. **Figure 26.19** shows a diagram that hypothesizes eight supergroups. Of the four traditional kingdoms, both Fungi and Animalia are within the Opisthokonta supergroup, while Plantae is found within Archaeplastida. The remaining branches and supergroups used to be classified within the single kingdom Protista. Thus, molecular data and newer ways of building trees reveal that protists played a key role in the evolution of many diverse groups of eukaryotic species, producing several large monophyletic groups. A survey of the characteristics of these different types of protists will be described in Chapter 28.

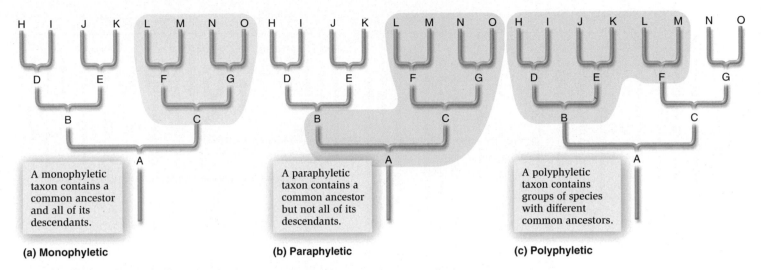

Figure 26.18 A comparison of monophyletic, paraphyletic, and polyphyletic taxonomic groups.

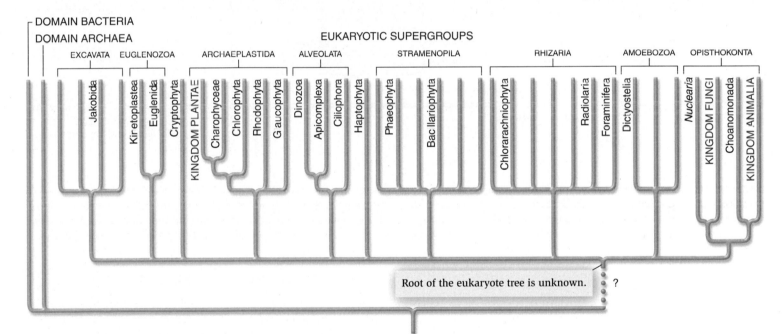

Figure 26.19 A modern cladogram for eukaryotes. This drawing should be considered a working hypothesis. The arrangement of these supergroups relative to each other is not entirely certain.

GENOMES & PROTEOMES

Due to Horizontal Gene Transfer, the Tree of Life Is Really a "Web of Life"

Thus far, we have considered various ways to construct phylogenetic trees, which describe the relationships between ancestors and their descendents. The type of evolution that is depicted in previous figures is vertical evolution, which involves changes in groups of species due to descent from a common ancestor. Since the time of Darwin, vertical evolution has been the tradi-tional way that biologists view the evolutionary process. However, over the past couple of decades researchers have come to realize that evolution is not so simple. In addition to vertical evolution, horizontal gene transfer has also played a significant role in the phylogeny of all living species.

As discussed in Chapter 23, horizontal gene transfer is the transfer of genes between different species. An analysis of many bacterial genomes has shown that horizontal gene transfer has been a major force in the evolution of prokaryotes. Horizontal gene transfer continued after the divergence of the three major domains of life, including transfer between prokaryotic and

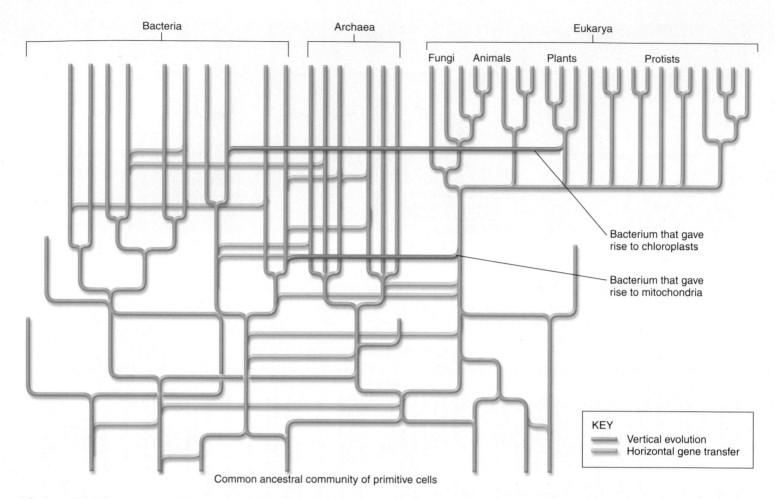

Bacterium that gave
rise to chloroplasts

Bacterium that gave
rise to mitochondria

KEY
— Vertical evolution
═ Horizontal gene transfer

Common ancestral community of primitive cells

Figure 26.20   **A web of life.**   This phylogenetic tree shows not only the modern classification of all life on Earth but includes the contribution of horizontal gene transfer in the evolution of species on our planet. This phenomenon was prevalent during the early stages of evolution when all organisms were unicellular. Horizontal gene transfer continues to be a prominent factor in the speciation of Bacteria and Archaea. Note: This tree is meant to be schematic. For example, Figure 26.19 is a more realistic representation of the evolutionary relationships among modern eukaryotic species.

eukaryotic species and among eukaryotic species. With regard to modern organisms, horizontal gene transfer is still prevalent among prokaryotic species. By comparison, this process is less common among modern eukaryotes, though it does occur. Researchers have speculated that multicellularity and sexual reproduction have presented barriers to horizontal gene transfer in most eukaryotes. For a gene to be transmitted to eukaryotic offspring, it would have to be transferred into a eukaryotic cell that is a gamete or a cell that gives rise to gametes.

Recently, scientists have debated the role of horizontal gene transfer in the earliest stages of evolution, prior to the emergence of the two prokaryotic domains. The traditional viewpoint was that the three domains of life arose from a single type of prokaryotic (or preprokaryotic) cell called the universal ancestor. However, genomic research has suggested that horizontal gene transfer may have been particularly common during the early stages of evolution on Earth, when all species were unicellular. Rather than proposing that all life arose from a single type of prokaryotic cell, horizontal gene transfer may have

been so prevalent that the universal ancestor may have actually been an ancestral community of cell lineages that evolved as a whole. If that were case, the tree of life cannot be traced back to a single universal ancestor.

Figure 26.20 illustrates a schematic scenario for the evolution of life on Earth that includes the roles of both vertical evolution and horizontal gene transfer. This has been described as a "web of life" rather than a "tree of life." Instead of a universal ancestor, a web of life began with a community of primitive cells that transferred genetic material in a horizontal fashion. Horizontal gene transfer was also prevalent during the early evolution of bacteria and archaea, and when eukaryotes first emerged as unicellular species. In modern bacteria and archaea, it remains a prominent way to foster evolutionary change. By comparison, the region of the diagram that contains most eukaryotic species has a more treelike structure, because horizontal gene transfer has become less common in these species, though it does occur occasionally.

CHAPTER SUMMARY

26.1 Taxonomy

- Taxonomy is the field of biology that is concerned with the theory, practice, and rules of classifying living and extinct organisms and viruses.

- Taxonomy places all living organisms into groups called taxa. The broadest groups are the three domains called Bacteria, Archaea, and Eukarya. (Figure 26.1, Table 26.1)

- Domains are divided into a hierarchy composed of kingdoms, phyla, classes, orders, families, genera, and species. (Figure 26.2)

- Binomial nomenclature provides each species with a scientific name that refers to its genus and species epithet.

- The two prokaryotic domains, Bacteria and Archaea, possess distinctive characteristics at the molecular level and occupy particular environments. (Figure 26.3)

- Eukaryotes were traditionally divided into four kingdoms called Protista, Fungi, Plantae, and Animalia. (Figures 26.4, 26.5, 26.6, 26.7, Table 26.2)

26.2 Systematics

- Systematics is the study of biological diversity and evolutionary relationships. The evolutionary history of a species is its phylogeny.

- A phylogenetic tree is a hypothesis that describes the phylogeny of particular species. A monophyletic group or clade includes all of the species that are derived from a common ancestor. (Figure 26.8)

- The hierarchy of taxonomy is related to the timing of common ancestors. Smaller taxa, such as families and genera, are derived from more recent common ancestors than are broader taxa such as kingdoms and phyla. (Figure 26.9)

- Morphological and genetic data are used to construct phylogenetic trees. (Figure 26.10)

- Assuming that the rate of neutral mutation is relatively constant, genetic data provide a molecular clock on which to measure evolutionary time. (Figures 26.11, 26.12)

- A cladistic approach constructs a phylogenetic tree, also called a cladogram, by considering the various possible pathways of evolution. Species are grouped together according to shared derived characters that come from primitive characters. An ingroup is a clade of interest, while an outgroup is closely related but lacks one or more shared derived characters. (Figures 26.13, 26.14)

- The cladistic approach can produce many possible cladograms. The most likely cladogram is chosen by a variety of methods such as the principle of parsimony or maximum likelihood. (Figure 26.15)

- Cooper and colleagues analyzed DNA sequences from extinct and modern flightless birds and hypothesized a phylogeny in which New Zealand was colonized twice, once by moas and later by kiwis. (Figures 26.16, 26.17)

- Molecular systematics, which involves the analysis of genetic sequences, has led to major revisions in taxonomy. Ideally, all taxa should be monophyletic, though previously established taxa sometimes turn out to be paraphyletic or polyphyletic. (Figure 26.18)

- Modern eukaryotes are now divided into eight supergroups instead of the four traditional kingdoms. (Figure 26.19)

- Due to horizontal gene transfer, the tree of life should really be described as a web of life. (Figure 26.20)

TEST YOURSELF

1. The study of biological diversity based on evolutionary relationships is
   a. paleontology.
   b. evolution.
   c. systematics.
   d. ontogeny.
   e. both a and b.

2. Which of the following is the correct order of the taxa used to classify organisms?
   a. kingdom, domain, phylum, class, order, family, genus, species
   b. domain, kingdom, class, phylum, order, family, genus, species
   c. domain, kingdom, phylum, class, family, order, genus, species
   d. domain, kingdom, phylum, class, order, family, genus, species
   e. kingdom, domain, phylum, order, class, family, species, genus

3. When considering organisms within the same taxon, which level includes organisms with the greatest similarity?
   a. kingdom
   b. class
   c. order
   d. family
   e. genus

4. Which of the following characteristics is not shared by prokaryotes and eukaryotes?
   a. DNA is the genetic material.
   b. Messenger RNA encodes the information to produce proteins.
   c. All cells are surrounded by a plasma membrane.
   d. The cytoplasm is compartmentalized into organelles.
   e. Both a and d.

5. The characteristics that define organisms in the kingdom Plantae include
   a. unicellular.
   b. photosynthetic.
   c. multicellular.
   d. a and b only.
   e. b and c only.

6. The evolutionary history of a species is
   a. ontogeny.
   b. taxonomy.
   c. evolution.
   d. phylogeny.
   e. embryology.

7. A group composed of all species derived from a common ancestor is referred to as
   a. a phylum.
   b. a monophyletic group or clade.
   c. a phenogram.
   d. an outgroup.
   e. a taxon.

8. The goal of modern taxonomy is to
   a. classify all organisms based on morphological similarities.
   b. classify all organisms in monophyletic groups.
   c. classify all organisms based solely on genetic similarities.
   d. determine the evolutionary relationships between similar species.
   e. none of the above.

9. The concept that the preferred hypothesis is the one that is the simplest is
   a. phenetics.
   b. cladistics.
   c. the principle of parsimony.
   d. maximum likelihood.
   e. both b and d.

10. Researchers believe that horizontal gene transfer is less prevalent in eukaryotes because of
    a. the presence of organelles.
    b. multicellularity.
    c. sexual reproduction.
    d. all of the above.
    e. b and c only.

CONCEPTUAL QUESTIONS

1. Explain binomial nomenclature and give an example.
2. Explain both the usefulness and potential pitfalls of morphological analysis.
3. Explain the value of a molecular clock.

EXPERIMENTAL QUESTIONS

1. What is molecular paleontology? What was the purpose of the study conducted by Cooper and colleagues?
2. What birds were examined in the Cooper study and what are their geographic distributions? Why were the different species selected for this study?

3. What results did Cooper and colleagues obtain by comparing these DNA sequences? How did the results of this study impact the proposed phylogeny of flightless birds?

COLLABORATIVE QUESTIONS

1. Discuss taxonomy and its hierarchical organization.
2. Discuss systematics and the concept of a phylogenetic tree.

---

www.brookerbiology.com

This website includes answers to the Biological Inquiry questions found in the figure legends and all end-of-chapter questions.

27

THE BACTERIA
AND ARCHAEA

A visible cyanobacterial bloom gives a pea-soup appearance to lake water.

Among Earth's life-forms, bacteria and archaea are unique in several ways. These organisms have the simplest cell structure and include the smallest known cells. Bacteria and archaea are also the most abundant organisms on Earth. About half of Earth's total biomass consists of an estimated 10^{30} individuals; just a pinch of garden soil can contain 2 billion, and about a million occur in 1 ml of seawater. Bacteria and archaea live in nearly every conceivable habitat, including extremely hot or salty waters that support no other life, and they are also Earth's most ancient organisms, having originated more than 3.5 billion years ago. Their great age and varied habitats have fostered very high diversity. Today, many millions of species of bacteria and archaea collectively display more diverse metabolic processes than occur in any other group of organisms. Many of these metabolic processes are important on a global scale, influencing Earth's climate, atmosphere, soils, water quality, and human health and technology. In this chapter we will survey the diversity, structure, reproduction, metabolism, and ecology of bacteria and archaea. This survey will illustrate major principles of diversity, including descent with modification and horizontal gene transfer.

27.1 Diversity and Evolution

As we have noted, one of the prominent features of bacteria and archaea is their astounding diversity. In the past, microbiologists studied diversity by isolating these organisms from nature and growing cultures in the laboratory. Such cultures allowed microbiologists to observe variation in structure and metabolism, major features used to classify bacteria and archaea. Today, microbiologists also use molecular techniques to detect diverse bacteria and archaea in nature. By using these new techniques,

they have discovered that bacteria and archaea are vastly more diverse than previously realized, and many new species have been discovered. For example, in 2004, gene sequencing expert Craig Venter and associates found 148 new bacteria and archaea by sequencing DNA extracted from microbial organisms from the Sargasso Sea. Similar studies of other habitats have also revealed much new diversity, though only 1% of the newly discovered species have been cultured in the laboratory. Many species of bacteria and archaea are known only as a distinctive molecular sequence.

Though much remains to be learned about the diversity of Earth's microorganisms, extensive molecular analysis has supported the concept that prokaryotic microbes can be classified into two major domains of life: the **Archaea** and **Bacteria** (also called Eubacteria) (**Figure 27.1**). The terms prokaryote and prokaryotic are often used to refer to archaeal and bacterial cells because these organisms lack nuclei and other cellular features typical of eukaryotes. As we noted in Chapter 26, in the 1970s microbiologist Carl Woese and associates proposed splitting the kingdom Monera, which had included all prokaryotes, into these two domains, based on comparisons of ribosomal RNA sequences from diverse microorganisms. In this section we will first survey the major kingdoms and phyla of the domains Archaea and Bacteria and then explore how horizontal gene transfer—the transfer of genes between different species—has influenced their evolution.

Domain Archaea Includes Inhabitants of Extremely Harsh Environments

Organisms classified in the domain Archaea, commonly known as archaea, display some unique characteristics. First, archaea possess a number of features in common with the eukaryotic nucleus

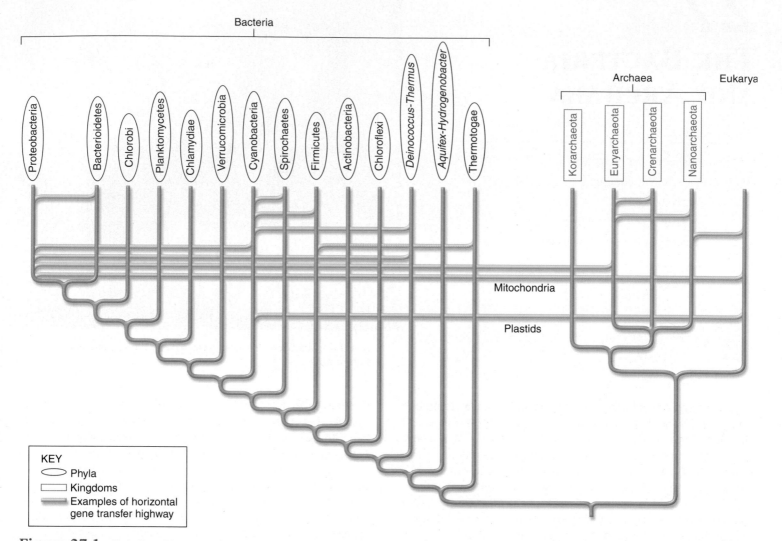

Figure 27.1   Relationships and diversification of the Bacteria and Archaea.   Bacteria and Archaea are the two domains featuring prokaryotic cells. Eukarya is the domain of organisms composed of eukaryotic cells. Each domain has diversified into multiple kingdoms and phyla. Many cases of horizontal gene transfer among phyla, kingdoms, and domains are known. These cases include the acquisition of mitochondria and plastids by eukaryotes.

and cytoplasm, suggesting common ancestry. For example, histone proteins are typically associated with the DNA of both archaea and eukaryotes, but they are absent from most bacteria. Another distinctive feature of archaea is their membrane lipids, which are formed with ether linkages (in contrast, ester linkages characterize the membrane lipids of bacteria and eukaryotes). Ether-linked membranes are resistant to damage by heat and other extreme conditions, which helps explain why many archaea are able to grow in extremely harsh environments.

Though many archaea occur in soils and surface ocean waters of moderate conditions, diverse archaea occupy habitats of very high salt content, acidity, methane levels, or temperatures that would kill most bacteria and eukaryotes. Organisms that occur primarily in extreme habitats are known as **extremophiles**. One example is the methane producer *Methanopyrus*, which grows best at deep-sea thermal vent sites where the temperature is 98°C. In fact, *Methanopyrus* is so closely adapted to

its extremely hot environment that it will not grow when the temperature is less than 84°C. Such archaea are known as hyperthermophiles. Some archaea prefer habitats having both high temperatures and extremely low pH. For example, microbiologist Thomas Brock discovered the archaeal genus *Sulfolobus* in samples taken from sulfur hot springs having a pH of 3 or lower.

Extreme halophiles ("salt-lovers") occupy evaporation ponds used to produce salt from seawater, often growing so abundantly that they color the ponds red (**Figure 27.2**). Halophiles are often red because their plasma membranes contain large amounts of rhodopsins, proteins combined with the red light-sensitive pigment known as retinal. (Similar rhodopsins play important roles in eukaryote light sensing, and they are essential for animal vision.) In bacteria and archaea, rhodopsin functions as a proton pump, a protein that can move protons and other ions across the plasma membrane.

The domain Archaea includes the kingdoms Crenarchaeota, Euryarchaeota, Korarchaeota, and Nanoarchaeota. Crenarchaeota includes *Sulfolobus* and other organisms that grow in extremely hot or cold habitats. Euryarchaeota includes methane producers and extreme halophiles. Korarchaeota is primarily known from DNA sequences found in samples from hot springs. Nanoarchaeota includes the hyperthermophile *Nanoarchaeum equitans*, which appears to be a parasite of the thermal vent crenarchaeote *Ignicoccus*. Molecular biologist Elizabeth Waters and associates sequenced the exceptionally small genome of *N. equitans* and determined that it represents an early branching archaeal lineage.

Domain Bacteria Includes Proteobacteria, Cyanobacteria, and Many Other Phyla

Molecular studies suggest the existence of 50 or so bacterial phyla (considered kingdoms by some). However, the structural and metabolic features of about half of these are unknown. Though some members of domain Bacteria live in extreme environments, many more favor moderate conditions. Many bacteria form symbiotic associations with eukaryotes and are thus of concern in medicine and agriculture. The characteristics of 14 prominent bacterial phyla are briefly summarized in **Table 27.1**.

Table 27.1	Representative Bacterial Phyla
Phyla	**Characteristics**
Thermotogae	Hyperthermophiles; an ancient and slowly evolving lineage.
Aquifex-Hydrogenobacter	Hyperthermophiles; thought to be a very ancient lineage.
Deinococcus-Thermus	Extremophiles. The genus *Deinococcus* is known for high resistance to ionizing radiation, and the genus *Thermus* inhabits hot springs habitats. *Thermus aquaticus* has been used in commercial production of Taq polymerase enzyme used in the polymerase chain reaction (PCR), an important procedure in molecular biology laboratories.
Chloroflexi	Known as the green nonsulfur bacteria; conduct photosynthesis without releasing oxygen (anoxygenic photosynthesis).
Actinobacteria	Gram-positive bacteria producing branched filaments; many form spores. *Mycobacterium tuberculosis*, the agent of tuberculosis in humans, is an example. Actinobacteria are notable antibiotic producers; over 500 different antibiotics are known from this group. The pharmaceutical industry produces antibiotics from large-scale cultures of the actinobacterium *Streptomyces*. Some fix nitrogen in association with plants.
Firmicutes	Diverse Gram-positive bacteria, some of which produce endospores.
Spirochaetes	Motile bacteria having distinctive corkscrew shapes, with flagella held close to the body. They include the pathogens *Treponema pallidum*, the agent of syphilis, and *Borrelia burgdorferi*, which causes Lyme disease.
Cyanobacteria	The oxygen-producing photosynthetic bacteria (some are also capable of anoxygenic photosynthesis). Photosynthetic pigments include chlorophyll *a* and phycobilins, which often give cells a blue-green pigmentation. Occur as unicells, colonies, unbranched filaments, and branched filaments. Many of the filamentous species produce specialized cells: dormant akinetes and heterocysts in which nitrogen fixation occurs. In waters having excess nutrients, cyanobacteria produce blooms and may release toxins harmful to the health of humans and wild and domesticated animals.
Verrucomicrobia	Notable for producing tubulin-like proteins.
Chlamydiae	Notably tiny, obligate intracellular parasites. Some cause eye disease in newborns or sexually transmitted diseases.
Planktomycetes	Reproduce by budding; cell walls lack peptidoglycan; and cytoplasm contains nucleus-like bodies.
Chlorobi	Known as the green sulfur bacteria; conduct anoxygenic photosynthesis. Use H_2S or organic compounds as photosynthetic electron donors. Store elemental sulfur, which can be further oxidized to sulfate.
Bacteroidetes	Includes representatives of diverse metabolism types; some are common in the human intestinal tract, and others are primarily aquatic.
Proteobacteria	A very large group of Gram-negative bacteria, collectively having high metabolic diversity. Includes many species important in medicine, agriculture, and industry.

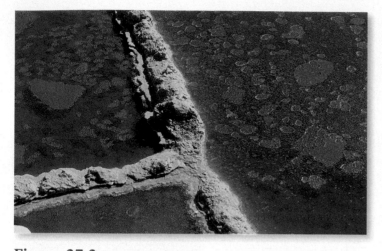

Figure 27.2 Hypersaline waters colored red by numerous halophilic archaea.

Biological inquiry: What material explains the red color of the water in this figure? ·

Figure 27.3 *Agrobacterium tumifaciens* infection causes cancer-like tumors to grow on plants.

Among these, the Proteobacteria and the Cyanobacteria are particularly diverse and relevant to eukaryotic cell evolution, global ecology, and human affairs.

Proteobacteria Though Proteobacteria share molecular and cell-wall features, this phylum displays amazing diversity of form and metabolism. Genera of this phylum are classified into five major subgroups: alpha (α), beta (β), gamma (γ), delta (δ), and epsilon (ε). As we saw in Chapter 22, the ancestry of mitochondria can be traced to the α-proteobacteria, which also include several genera noted for symbiotic relationships with animals and plants. For example, *Rhizobium* and related genera of α-proteobacteria form nutritionally beneficial associations with the roots of legume plants and are thus agriculturally important (see Chapter 37). Another α-proteobacterium, *Agrobacterium tumifaciens*, causes destructive cancer-like galls to develop on susceptible plants, including grapes and ornamental crops (**Figure 27.3**). *A. tumifaciens* induces gall formation by injecting DNA into plant cells, a property that has led to the use of the bacterium in the production of transgenic plants (see Chapter 20).

The genus *Nitrosomonas*, a soil inhabitant important in the global nitrogen cycle, represents the β-proteobacteria. *Neisseria gonorrhoeae*, the agent of the sexually transmitted disease gonorrhea, is another member of the β-proteobacteria. *Vibrio cholerae*, a γ-proteobacterium, causes cholera epidemics when drinking water becomes contaminated with animal waste during floods and other natural disasters. The γ-proteobacteria *Salmonella enterica* and *Escherichia coli* strain O157:H7 also cause human disease, and food and water are widely tested for their presence. The δ-proteobacteria include the colony-forming myxobacteria and predatory bdellovibrios, which drill their way through the cell walls of other bacteria in order to consume them. *Helicobacter pylori*, which causes stomach ulcers, belongs to the ε-proteobacteria.

Cyanobacteria The phylum Cyanobacteria contains photosynthetic bacteria that are abundant in fresh waters, oceans, and wetlands and on the surfaces of arid soils. Cyanobacteria are named for the typical blue-green (cyan) coloration of their cells (**Figure 27.4**). Blue-green pigmentation results from the presence of accessory phycobilin pigments that help chlorophyll absorb light energy. Cyanobacteria are the only prokaryotes that generate oxygen as a product of photosynthesis. Ancient cyanobacteria produced Earth's first oxygen-rich atmosphere, which allowed the rise of eukaryotes. The plastids of eukaryotic algae and plants arose from cyanobacteria.

Cyanobacteria display the greatest structural diversity found among bacterial phyla. Some occur as single cells, while others form colonies of cells held together by a thick gluey substance called mucilage (Figure 27.4a,b). Many cyanobacteria form filaments of cells that are attached end to end (Figure 27.4c). Some of the filamentous cyanobacteria produce specialized cells and display intercellular chemical communication, the hallmarks of multicellular organisms. Many cyanobacteria that grow in conditions of high light intensity produce protective brown sunscreen compounds at their surfaces (Figure 27.4d).

Cyanobacteria play essential ecological roles by producing organic carbon and fixed nitrogen (see Section 27.4). However, several kinds of cyanobacteria, notably the genera *Microcystis*, *Anabaena*, and *Cylindrospermopsis*, form nuisance growths in freshwater lakes during the warm season. Such growths, known as blooms, give the water a pea-soup appearance (see chapter-opening photo). Blooms develop when natural waters receive excess fertilizer from sewage discharges or agricultural runoff. Such blooms are becoming more common every year and are of serious concern because they may produce toxins in amounts sufficient to harm the health of humans and other animals. Consequently, it is inadvisable for people or pets to swim in or consume water that has a visible cyanobacterial bloom.

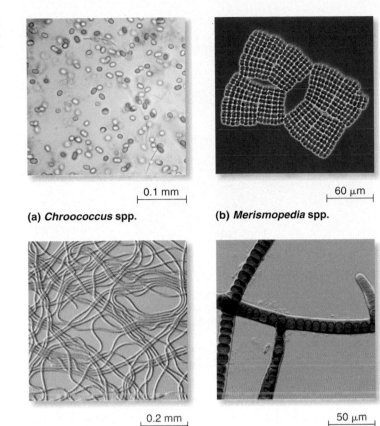

0.1 mm

(a) *Chroococcus* spp.

60 μm

(b) *Merismopedia* spp.

0.2 mm

(c) *Oscillatoria* spp.

50 μm

(d) *Stigonema* spp.

Figure 27.4   The phylum Cyanobacteria illustrates major types of microbial cell aggregations.   **(a)** The genus *Chroococcus* occurs as unicells. **(b)** The genus *Merismopedia* is a flat colony of cells held together by mucilage. **(c)** The genus *Oscillatoria* is an unbranched filament. **(d)** The genus *Stigonema* is a branched filament having a mucilage sheath, which has sunscreen compounds that are colored brown.

Now that we have learned something about the diversity of Archaea and Bacteria, let's consider the effects of gene exchanges within and between these domains.

Horizontal Gene Transfer Influences Diversity of Bacteria and Archaea

Horizontal gene transfer, also known as lateral gene transfer, is the movement of one or more genes from one species to another. This process contrasts with vertical gene transfer from parent to progeny. Horizontal gene transfer increases genetic diversity and influences the methods used to infer the phylogeny of bacteria and archaea.

Horizontal gene transfer is common among bacteria and archaea, and it can result in large genetic changes. For example, at least 17% of the genes present in the common human gut inhabitant *E. coli* came from other bacteria. In addition, genes move among the bacterial, archaeal, and eukaryotic domains. For example, about a third of the genes present in the archaeon *Methanosarcina mazei* originally came from bacteria, and there is genetic evidence for transfer of genes from Nanoarchaeota to protists. Viruses are probably important gene transfer vectors. This is illustrated by the fact that photosynthetic genes typical of oceanic cyanobacteria widely occur in marine viruses that infect these prokaryotes. These genes may evolve in the viral host before being transferred to another cyanobacterial cell.

To explore the extent of horizontal gene transfer, evolutionary microbiologists Robert Beiko, Timothy Harlow, and Mark Ragan examined the pattern of occurrence of more than 220,000 proteins encoded in 144 microbial genomes. These investigators discovered that major gene-sharing highways occur, particularly among close relatives or between microorganisms living in similar habitats (see Figure 27.1). As an example of the latter, Niels-Ulrik Frigaard, Edward DeLong, and their associates discovered a case of probable horizontal transfer between proteobacteria and euryarchaeota living in the same well-lit surface ocean waters. Proteobacteria occurring in such waters produce distinctive rhodopsins known as proteorhodopsins. As noted earlier, rhodopsins are pigment-protein complexes that enable cells to use light energy to drive proton pumps, a very useful trait for microbes occupying light-rich habitats. Because proteorhodopsins had not previously been found in euryarchaeota, the investigators were surprised to find that about 10% of euryarchaeota collected from the same area also possessed proteorhodopsin genes. This observation suggests that proteorhodopsin genes were horizontally transferred to the euryarchaeota from their proteobacterial neighbors. Acquisition of proteorhodopsin genes allowed some of the euryarchaeota to also benefit from light-driven proton pumping. This example and others have shown that horizontal gene transfer is an effective evolutionary process in bacteria and archaea, allowing them to acquire new metabolic processes despite lacking the sexual processes typical of eukaryotes. Horizontal transfer thereby increases the metabolic diversity of bacteria and archaea.

Horizontal gene transfer has the potential to interfere with human efforts to deduce evolutionary relationships. For example, if proteorhodopsin genes had been used in phylogenetic studies of ocean microorganisms, systematists might have falsely concluded that proteobacteria and euryarchaeota were closely related, because both groups possess these distinctive genes. Such a conclusion would be erroneous, because euryarchaeota did not inherit proteorhodopsin genes vertically. To understand prokaryote relationships, molecular systematists analyze ribosomal RNA (rRNA) genes and other sequences thought to move infrequently in a horizontal manner. Such analyses may more accurately reflect patterns of vertical inheritance. However, even ribosomal genes can be vulnerable to horizontal transfer. Scott Miller, Michelle Wood, and their associates discovered part of an rRNA gene from a proteobacterium within the genome of a cyanobacterium obtained from high-salinity water.

Phylogenetic trees displaying current understanding of relationships (see Figure 27.1) reveal several important concepts.

Bacteria and Archaea probably evolved from a common ancestor, and the eukaryotic nucleus and cytoplasm likely arose in an ancient archaeal organism. In addition, mitochondria and plastids originated from proteobacteria and cyanobacteria by endosymbiosis (see Chapter 22). In these cases, endosymbiosis resulted in the transfer of many genes from bacteria to eukaryotes. Finally, bacteria and archaea are amazingly diverse, but many phyla and species lack scientific names because microbiologists know so little about them. The next section surveys ways in which the better-known bacteria and archaea vary in structure and locomotion.

27.2 Structure and Motility

Bacteria and archaea have several features in common, including small size, rapid growth, and simple cellular structure. With few exceptions, bacteria and archaea are 1–5 μm in diameter and are thus known as microorganisms or microbes. (By contrast, most plant and animal cells are between 10 and 100 μm in diameter.) Small cell size limits the amount of materials that can be stored within cells but allows faster cell division. When nutrients are sufficient, many microorganisms can divide several times within a single day. This explains how bacteria can spoil food rapidly, and why infections can spread quickly within the human body. Despite these common features, prokaryotes differ in many ways. In this section we explore variation in cell structure and shape, in surface and cell wall features, and in movement. We will also learn how these characteristics influence the roles of bacteria and archaea in nature and human affairs.

Bacteria and Archaea Vary in Cellular Structure

Prokaryotic cells are much simpler than eukaryotic cells. Even so, many prokaryotes display surprising complexity of cellular structure resulting from adaptive evolution. For example, cyanobacteria and other photosynthetic bacteria are able to use light energy to produce organic compounds and typically contain large numbers of intracellular tubules known as thylakoids (**Figure 27.5**). The extensive membrane surface of the thylakoids has large amounts of chlorophyll and other components of the photosynthetic apparatus. Thylakoids thus enable photosynthetic bacteria to take maximum advantage of light energy in their environments.

Thylakoids develop by ingrowth of the plasma membrane, and in some bacteria, plasma membrane ingrowth has generated additional intriguing adaptations—magnetosomes and nucleus-like bodies—that are sometimes described as bacterial organelles. Magnetosomes are tiny crystals of an iron mineral known as magnetite, each surrounded by a membrane. These structures occur in the bacterium *Magnetospirillum* and related genera (**Figure 27.6**). In each cell, about 15 to 20 magnetosomes occur in a row, together acting as a compass needle to orient the bacteria within the Earth's magnetic field. This helps the bacteria to locate the low-oxygen habitats they prefer. Microbiologists Arash Komeili, Grant Jensen, and their colleagues used

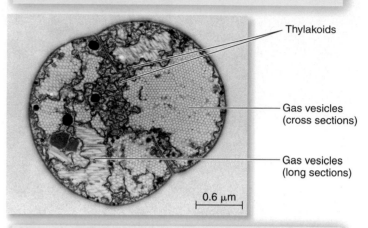

Thylakoids provide a greater surface area for chlorophyll and other molecules involved in photosynthesis.

Thylakoids

Gas vesicles (cross sections)

Gas vesicles (long sections)

0.6 μm

The gas vesicles buoy this photosynthetic organism to the lighted water surface, where it often forms conspicuous scums.

Figure 27.5   Cells of the cyanobacterial genus *Microcystis* contain photosynthetic thylakoid membranes and numerous gas vesicles.

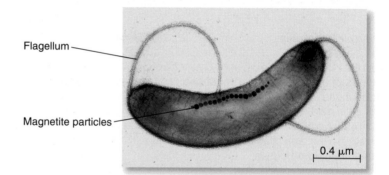

Flagellum

Magnetite particles

0.4 μm

Figure 27.6   The spirillum *Magnetospirillum magnetotacticum*.   An internal row of iron-rich magnetite crystals, each enclosed by a membrane, functions like a compass needle, allowing this bacterium to detect the Earth's magnetic field. This feature allows *M. magnetotacticum* to locate its preferred habitat, low-oxygen subsurface waters. Cells use their flagellum to move from less favorable to more attractive locations.

rapid freezing techniques and a special type of transmission electron microscope to observe the development of magnetosomes. They found that the process begins with ingrowth of the plasma membrane to form a row of spherical vesicles. If *Magnetospirillum* cells are grown in media having low iron levels, the vesicles remain empty. But if iron is available, a magnetite crystal forms within each vesicle. The investigators also observed that fibrils of an actin-like protein keep the magnetosomes aligned in a row. (Recall from Chapter 4 that actin is a major cytoskeletal protein of eukaryotes.) Mutant bacteria lacking a functional form of this protein produce magnetosomes, but they do not remain aligned in a row. Instead, magnetosomes scatter around mutant cells, disrupting their ability to detect a magnetic field.

Plasma membrane invaginations also produce nucleus-like bodies in *Gemmata obscuriglobus* and other members of the bacterial phylum Planktomycetes. In *G. obscuriglobus*, an envelope composed of a double membrane encloses all cellular DNA and some ribosomes. Although this bacterial envelope lacks nuclear pores characteristic of the eukaryotic nuclear envelope, it likely plays a similar adaptive role in isolating DNA from other cellular influences.

A final example of surprising cell structure complexity occurs in a bacterial phylum known as the Verrucomicrobia. These bacteria produce cellular proteins that are remarkably similar to eukaryotic tubulins, proteins that self-assemble into microtubules (see Chapter 4). The existence of actin- and tubulin-like proteins in prokaryotic cells suggests that eukaryotic microfilaments and microtubules originated from ancestral proteins by descent with modification.

Archaea and Bacteria Vary in Cell Shape and Arrangement

Microbial cells occur in five major shapes (**Figure 27.7**): spheres (**cocci**), rods (**bacilli**), comma-shaped cells (**vibrios**), and spiral-shaped cells that are either flexible (**spirochaetes**) or rigid (**spirilli**; see Figure 27.6). In addition, such cells can occur in several types of arrangements. Some microorganisms occur only as single cells or pairs of cells resulting from recent division (Figure 27.7a). Others occur as cell aggregates or are attached end to end to form filaments.

The shapes of bacterial cells, and the degree of their arrangement in groups, can be important diagnostic features in water and medical testing. For example, the presence of abundant rod-shaped cells in samples suggests possible contamination by *E. coli* or other proteobacteria that normally occur in the human gut. The proteobacterial genus *Neisseria* occurs as paired cocci, a feature that aids rapid detection of this organism when disease is suspected. Microorganisms may also produce surface mucilage, which plays important roles in disease and ecology.

Slimy Mucilage Often Coats Microbial Surfaces

Many microorganisms produce a coat of slimy mucilage that varies in consistency and extent. Mucilage is composed of polysaccharides, protein, or both, which are secreted from cells and serves a variety of functions. One example is the mucilage coat (known as a capsule) that helps some disease bacteria to evade the defense system of their host. You may recall that Frederick Griffith discovered the transfer of genetic material while experimenting with capsule-producing pathogenic forms and capsule-less nonpathogenic strains of the bacterium *Streptococcus pneumoniae* (refer back to Figure 11.1). The immune system cells of mice are able to destroy this bacterium only if it lacks a capsule.

Mucilage also holds together the cells of colonial microorganisms (see Figure 27.4b), helps aquatic species to float in water, binds mineral nutrients, and repels predators. Slime sheaths (see Figure 27.4d) often coat bacterial filaments, where they may help to prevent drying. Mucilage is also critical to the formation of biofilms, which are environmentally and medically important. **Biofilms** are aggregations of microorganisms that secrete adhesive mucilage, thereby gluing themselves to surfaces (**Figure 27.8**). They help microbes to remain in favorable locations for growth; otherwise body or environmental fluids would wash them away. A process known as **quorum sensing** fosters biofilm formation. During quorum sensing, individual microbes secrete small molecules having the potential to influence the behavior of nearby microbes. If enough individuals are present (a quorum), the concentration of signaling molecules builds to a level that causes collective behavior. In the case of biofilms, populations of microbes respond to chemical signals by moving to a common location and producing mucilage.

From a human standpoint, biofilms have both beneficial and harmful consequences. In aquatic and terrestrial environments, they help to stabilize and enrich sand and soil surfaces. In contrast, biofilms that form on the surfaces of animal tissues can be harmful. Dental plaque is an example of a harmful biofilm (Figure 27.8); if allowed to remain, the bacterial community secretes acids that can damage tooth enamel.

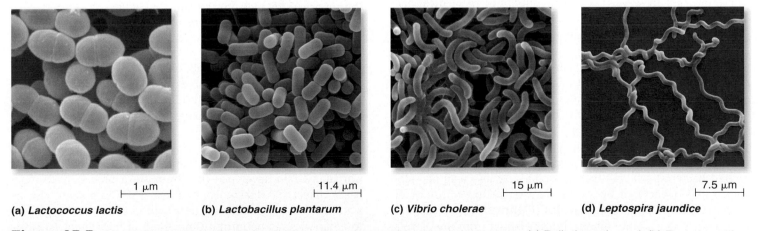

(a) *Lactococcus lactis*   1 μm

(b) *Lactobacillus plantarum*   11.4 μm

(c) *Vibrio cholerae*   15 μm

(d) *Leptospira jaundice*   7.5 μm

Figure 27.7   **Major types of microbial cell shapes.**   Scanning electron microscopic views. **(a)** Ball-shaped cocci. **(b)** Rod-shaped bacilli. **(c)** Comma-shaped vibrios. **(d)** Spiral-shaped spirochaetes.

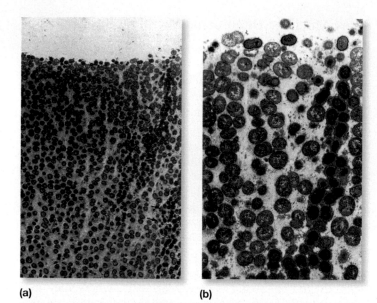

(a)   **(b)**

Figure 27.8   **A biofilm composed of a community of micro-organisms glued by mucilage to a surface.**   **(a)** This transmission electron micrograph shows a thin slice of dental plaque, with the tooth surface at the lower edge. **(b)** Close-up view showing chains of *Streptococcus sobrinus* bacterial cells, which are dark because they have been labeled with an identifying antibody. The plaque layer is about 50 μm deep.

Biofilms may also develop in industrial pipelines, where the attached microbes can contribute to corrosion by secreting enzymes that chemically degrade metal surfaces.

Archaea and Bacteria Differ in Cell-Wall Structure

Whether coated with mucilage or not, most prokaryotes possess a rigid cell wall outside the plasma membrane. Cell walls maintain cell shape and help protect against attack by viruses or predatory bacteria. Cell walls also help microbes avoid lysing in hypotonic conditions, when the solute concentration is higher inside the cell than outside. The structure and composition of cell walls vary in ways that can be important in medical and other contexts.

The cell walls of most archaea and certain bacteria are composed of protein or glycoprotein. In contrast, the polymer known as **peptidoglycan** is lacking from archaea but is an important component of most bacterial cell walls. Peptidoglycan is composed of carbohydrates that are cross-linked by peptides. Bacterial cell walls occur in two major forms that differ in their amount of peptidoglycan, staining properties, and response to antibiotics. The first type of bacterial cell wall features a relatively thick peptidoglycan layer (**Figure 27.9a**). The second bacterial cell-wall type has relatively less wall peptidoglycan and is enclosed by a thin, outer envelope whose outer leaflet is rich in **lipopolysaccharides** (**Figure 27.9b**). This outer layer envelope is a lipid bilayer, but is distinct from the plasma membrane. Lipopolysaccharides and peptidoglycan affect bacterial responses to antibiotics and sometimes disease symptoms.

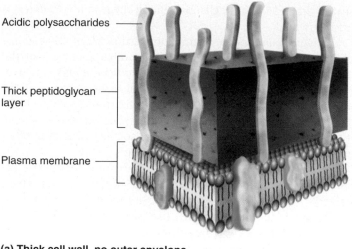

Acidic polysaccharides

Thick peptidoglycan layer

Plasma membrane

(a) Thick cell wall, no outer envelope

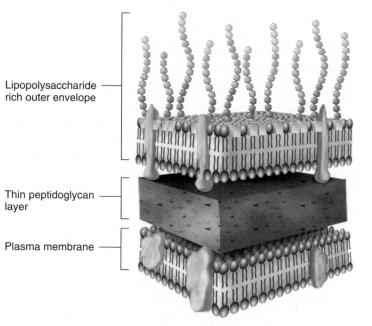

Lipopolysaccharide rich outer envelope

Thin peptidoglycan layer

Plasma membrane

(b) Thinner cell wall, with outer envelope

Figure 27.9   Cell-wall structure of Gram-positive and Gram-negative bacteria.   **(a)** The structure of the cell wall of Gram-positive bacteria. **(b)** The structure of the cell wall and lipopolysaccharide envelope typical of Gram-negative bacteria.

For example, part of the peptidoglycan covering of *Bordetella pertussis* is responsible for the extensive tissue damage associated with whooping cough. To identify the two major types of bacterial cell walls, microbiologists use a staining procedure known as the Gram stain.

The Gram Stain In the late 1800s, the Danish physician Hans Christian Gram developed the **Gram stain** procedure to more easily detect and distinguish bacteria. The Gram stain remains a useful tool to identify bacteria and predict their responses to antibiotics. To perform a Gram stain, a microbiologist starts by smearing bacteria onto a glass slide and heating it briefly to aid cell attachment. The microbiologist then floods the slide

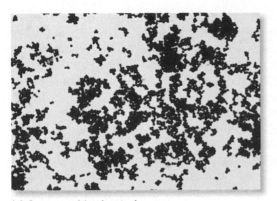
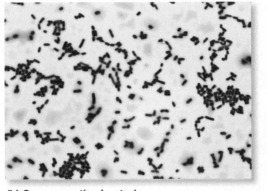

(a) Gram-positive bacteria

(b) Gram-negative bacteria

Figure 27.10 Gram-positive and Gram-negative bacteria. (a) *Streptococcus pneumoniae* stains positive (purple) with the Gram stain. (b) *Escherichia coli* stains negative (pink) when the Gram stain procedure is applied.

with crystal violet, a purple dye, followed by an iodine solution. The iodine binds the purple dye, forming an insoluble complex. Next, the microbiologist adds alcohol. The alcohol dehydrates peptidoglycan, thereby trapping the purple crystal violet-iodine complex. The alcohol is able to remove purple dye from thin peptidoglycan walls, but not dye bound in cell walls with thick peptidoglycan layers. Finally, safranin, a pink stain, is applied. At the end of the procedure, some bacteria will remain purple; these are known as Gram-positive bacteria (**Figure 27.10a**). Other types of bacteria will lose the purple stain at the alcohol step but retain the final pink stain; these are known as Gram-negative bacteria (**Figure 27.10b**). If a fluorescence microscope is available, a single-step fluorescent stain can be used to distinguish Gram-positive from Gram-negative bacteria (**Figure 27.11**). A closer look at Gram-positive and Gram-negative bacteria will reveal how these staining differences are useful.

Gram-Positive Bacteria Gram-positive bacteria occur in the phyla Firmicutes and Actinobacteria (see Figure 27.1, Table 27.1). Gram-positive bacteria typically have thick peptidoglycan cell walls lacking a lipopolysaccharide envelope. When treated by the Gram stain process, Gram-positive bacteria appear purple because their more abundant peptidoglycan traps a large amount of the crystal violet-iodine dye complex. Such bacteria are typically vulnerable to penicillin and related antibiotics, because these antibiotics interfere with peptidoglycan synthesis. An example of a Gram-positive bacterium is *S. pneumoniae* (see Figure 27.10a). Strains of this organism cause strep throat, streptococcal pneumonia, streptococcal meningitis (an infection of the spinal fluid), and eye infections. Streptococci also include the infamous "flesh-eating" bacteria that cause necrotizing fasciitis, a disease whose progression is notoriously difficult to control. The entire DNA sequence of *S. pneumoniae* has been determined, revealing the presence of many genes conferring the ability to break down several types of molecules in human tissues for use as food. The presence of these genes explains why *S. pneumoniae* can cause such varying disease symptoms. *Staphylococcus aureus* is another medically important Gram-positive bacterium, which is commonly found on human skin.

Gram-Negative Bacteria Gram-negative bacteria have thin peptidoglycan cell walls enclosed by a lipopolysaccharide envelope.

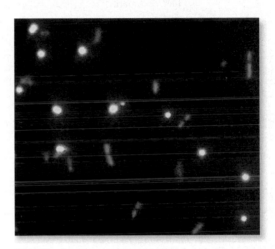

Figure 27.11 A commercial fluorescent stain can be used to distinguish Gram-positive and Gram-negative bacteria. This preparation shows green-fluorescent cells of the Gram-negative bacterium *E. coli* mixed with orange-yellow fluorescent cells of the Gram-positive bacterium *Staphylococcus aureus*.

Biological inquiry: What are the advantages and disadvantages of this fluorescence staining procedure as compared to the classical Gram stain process?

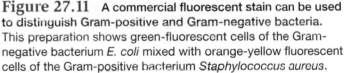

Diverse phyla of bacteria display Gram-negative staining. The outer lipopolysaccharide envelope and lower amount of cell-wall peptidoglycan enable Gram-negative bacteria to resist the effects of penicillin and chemically similar antibiotics. (Gram-negative bacteria would be treated with different antibiotics.) *E. coli* is an example of a Gram-negative bacterium that lives in the human lower intestine (see Figure 27.10b). A particularly virulent strain, *E. coli* O157:H7, is a significant cause of foodborne illness.

Microbes Display Diverse Types of Motility Structures

Many microorganisms have structures at the cell surface or within cells that enable them to change position in their environment. Motility allows microbes to move to favorable conditions within gradients of light, gases, or nutrients. In addition, motility structures allow them to respond to chemical signals emitted from other microbes during quorum sensing and mating.

Microbes move by swimming, twitching or gliding, adjusting their floatation in water, or pirating cytoskeletal proteins of invaded eukaryotic cells.

Structures known as **flagella** enable swimming behavior. Prokaryotic flagella differ from eukaryotic flagella in several ways. Prokaryotic flagella lack an internal cytoskeleton of microtubules, the motor protein dynein, and a plasma membrane covering, all features that characterize eukaryotic flagella (see Chapter 4). Unlike eukaryotic flagella, prokaryotic flagella do not repeatedly bend and straighten. Instead, prokaryotic flagella are propelled by molecular machines composed of a filament,

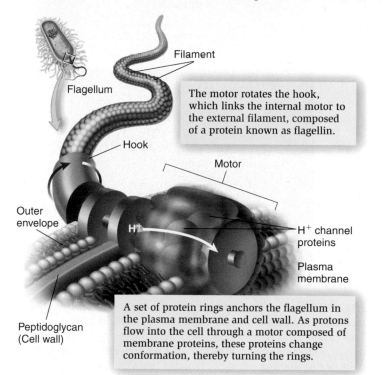

The motor rotates the hook, which links the internal motor to the external filament, composed of a protein known as flagellin.

A set of protein rings anchors the flagellum in the plasma membrane and cell wall. As protons flow into the cell through a motor composed of membrane proteins, these proteins change conformation, thereby turning the rings.

Figure 27.12   Diagram of a prokaryotic flagellum, showing motor, hook, and filament.

hook and motor that work together somewhat like a boat's outboard motor (**Figure 27.12**). Lying outside the cell, the long, stiff, curved filament acts as a propeller. The hook links the filament with the motor, a set of protein rings at the cell surface. Hydrogen ions (protons), which have been pumped out of the cytoplasm, usually via the electron transport system, diffuse back into the cell through channel proteins. This proton movement powers the turning of the hook and filament.

Prokaryotic species differ in the number and location of flagella, which may occur singly, in clumps at one pole of a bacterial cell, or emerging from around the cell (**Figure 27.13**). Differences in flagellar number and location cause microorganisms to exhibit different modes of swimming. For example, spirochaete flagella are located outside the peptidoglycan cell wall but within the confines of an outer membrane that holds them close to the cell. Rotation of these flagella causes spirochaetes to display characteristic bending, flexing, and twirling motions.

Some prokaryotes twitch or glide across surfaces, using threadlike cell surface structures known as **pili** (**Figure 27.14**). *Myxococcus xanthus* cells, for example, move by alternately extending and retracting pili from one pole or the other. This process allows directional movement toward food materials. If nutrients are low, cells of these bacteria glide together to form tiny treelike colonies, which are part of a reproductive process. As we will see, pili can also play important roles in bacterial reproduction and disease processes.

Cyanobacteria and some other bacteria that live in aquatic habitats use cytoplasmic structures known as **gas vesicles** to adjust their buoyancy. This process allows them to move up or down in the water column, an advantage in finding nutrients and avoiding harmful conditions. Gas vesicles are hollow cylinders whose water-tight walls are made of protein (see Figure 27.5). These vesicles do not actually contain gas concentrations different from the rest of the cell, but their density is lower than that of surrounding cytoplasm, helping bacteria to float. Gas vesicles can repeatedly collapse and reassemble, depending on

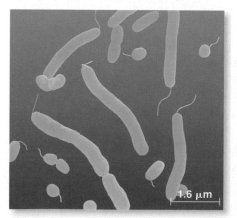

(a) *Vibrio parahaemoliticus*

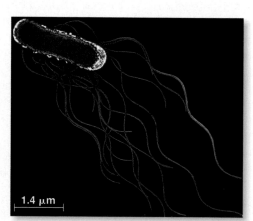

(b) *Salmonella enterica*

Figure 27.13   Microbial cells can produce one or more flagella at the poles, or numerous flagella around the periphery. **(a)** *Vibrio parahaemoliticus*, a bacterium that causes seafood poisoning, has a single short flagellum. **(b)** *Salmonella enterica*, another bacterium that causes food poisoning, has many flagella distributed around the cell periphery.

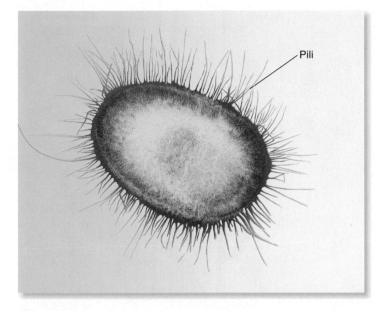

Figure 27.14 Pili extending from the surface of *Proteus mirabilis*.

27.3 Reproduction

Bacteria and archaea enlarge their populations by means of a process known as binary fission. In addition, some microbes produce tough cells that can withstand deleterious conditions for long periods in a dormant condition. Finally, although bacteria and archaea lack meiosis and other features typical of eukaryotic sexual reproduction, they are able to obtain additional genes by a variety of methods. Each of these aspects of reproduction helps to explain how microbes function in nature and human affairs.

Populations of Bacteria and Archaea Increase by Binary Fission

The cells of bacteria and archaea divide by splitting in two, a process known as **binary fission** (**Figure 27.15a**; also refer back to Figure 18.6). If sufficient nutrients are available, an entire population of identical cells can be produced from a single parental cell by repeated binary fission. This growth process allows microbes to become very numerous in water, food, or animal tissues, potentially causing harm.

Binary fission is the basis for a widely used method for detecting and counting bacteria in patient fluids, food, or water samples. Microbiologists who study the spread of disease need to quantify bacterial cells in samples taken from the environment. Medical technicians often need to count bacteria in body fluid samples to assess the likelihood of infection. But because bacterial cells are small and often unpigmented, they are difficult to count directly. One way that microbiologists count bacteria is to place a measured volume of sample into plastic dishes filled with a semisolid nutrient medium. Bacteria in the sample undergo repeated binary fission to form colonies of cells that are visible to the unaided eye (**Figure 27.15b**).

the cell's solute content. When sugars and other solutes are abundant in the cytoplasm, the high osmotic pressure collapses the vesicles. Cells in which the gas vesicles have collapsed experience increased density, so that they tend to sink. Sinking removes bacteria from damaging radiation at the water's surface and allows them to harvest inorganic minerals from deeper water. But when organic solutes made via photosynthesis are used up during cell metabolism, the gas vesicles reassemble, providing floatation to the surface allowing photosynthesis to resume. The presence of gas vesicles explains why cyanobacteria often form buoyant scums at the surfaces of ponds and lakes (see chapter-opening photo).

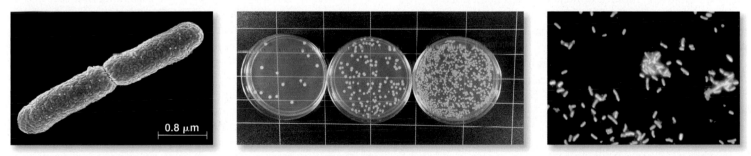

(a) Bacterium undergoing binary fission **(b) Colonies developed from single cells** **(c) Bacteria stained with fluorescent DNA-binding dye**

Figure 27.15 **Binary fission and counting microbes.** (a) Division of a bacterial cell as viewed by scanning electron microscopy. (b) When samples are spread onto the surfaces of laboratory dishes containing nutrients, single cells of bacteria or archaea may divide repeatedly to form visible colonies, which can be easily counted. The number of colonies is an estimate of the number of culturable cells in the original sample. (c) If a fluorescence microscope is available, cells can be counted directly by applying a fluorescent stain that binds to cell DNA. Each cell glows brightly when illuminated with ultraviolet light.

Biological inquiry: Which procedure would you choose to count bacteria in a sample that is known to include many species that have not as yet been cultured?

Since each colony represents a single cell that was present in the original sample, the number of colonies in the dish reflects the number of living bacteria in the original sample.

Another way to detect and count bacteria is to treat samples with a stain that binds bacterial DNA, causing cells to glow brightly when illuminated with ultraviolet light. The glowing bacteria can be viewed and counted by the use of a fluorescence microscope (**Figure 27.15c**). The fluorescence method must be used when the microbes of interest cannot be cultured in the laboratory. Microorganisms can also be stained with fluorescent reagents that bind to specific DNA sequences, allowing microbiologists to detect and count bacteria and archaea of precise genetic types. This procedure, known as fluorescence _in situ_ hybridization (FISH), is a powerful technique for finding particular types of microorganisms in mixed populations.

Some Bacteria Survive Harsh Conditions as Akinetes or Endospores

Some bacteria produce thick-walled cells known as akinetes or endospores that are able to survive unfavorable conditions in a dormant state. Akinetes and endospores develop when bacteria have experienced stress, such as low nutrients or unfavorable temperatures. Such cells are able to germinate into metabolically active cells when conditions improve again. For example, aquatic filamentous cyanobacteria often produce large, food-filled **akinetes** when winter approaches (**Figure 27.16a**). Akinetes are able to survive winter at the bottoms of lakes, and they produce new filaments in spring when they are carried by water currents to the brightly lit surface.

Endospores (**Figure 27.16b**) are cells having tough protein coats that are produced inside bacterial cells and then released when the enclosing cell dies and breaks down. Endospores can remain alive, though in a dormant state, for hundreds of years at least. Some studies suggest even longer lifetimes. For example, Raúl Cano and Monica Borucki cultured bacterial cells from _Bacillus sphaericus_ endospores that were carefully isolated from the gut of a bee that had been trapped in amber (hardened tree resin) for more than 25 million years. These and other researchers determined that DNA obtained from the cultures was identical to DNA taken directly from fossil bees. They also found that similar bacteria and endospores commonly occur in modern bees of the same species.

The ability to produce endospores allows some Gram-positive bacteria to cause serious diseases. For example, _Bacillus anthracis_ causes the disease anthrax, a potential agent in bioterrorism and germ warfare. Most cases of human anthrax result when endospores of _B. anthracis_ enter wounds, causing skin infections that are relatively easily cured by antibiotic treatment. But sometimes the endospores are inhaled, or consumed in undercooked, contaminated meat, potentially causing more serious illness or death. _Clostridium botulinum_ can contaminate the improperly canned food that has not been heated to temperatures high enough to destroy its tough endospores. When

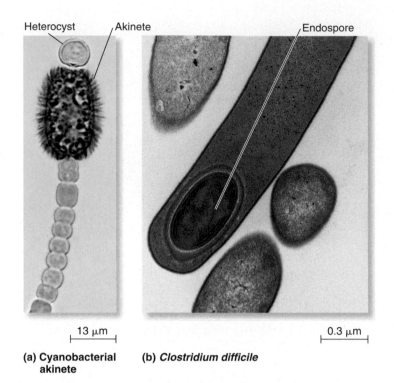

(a) Cyanobacterial akinete   **(b)** _Clostridium difficile_

Figure 27.16   Specialized cells capable of dormancy.
(a) Akinetes are thick-walled, food-filled cells produced by some cyanobacteria. They are able to resist stressful conditions and generate new populations when conditions improve. As discussed later, the heterocyst is a specialized cell in which nitrogen fixation occurs. (b) An endospore with a resistant wall develops within the cytoplasm of the bacterium _Clostridium difficile_.

the endospores germinate and bacterial cells grow in the food, they produce a deadly toxin, as well as NH_3 and CO_2 gas, which causes can lids to bulge. If humans consume the food, the toxin causes botulism, a severe type of food poisoning that can lead to respiratory and muscular paralysis. The botulism toxin has been recently marketed commercially as Botox, which is injected into the skin, where it paralyzes facial muscles, thereby reducing the appearance of wrinkles. _Clostridium tetani_ produces a nerve toxin that causes lockjaw, also known as tetanus, when bacterial cells or endospores enter wounds from soil. The ability of the genera _Bacillus_ and _Clostridium_ to produce resistant endospores helps to explain their widespread presence in nature and their potential danger to humans.

Bacteria and Archaea Obtain Genetic Material by Transduction, Transformation, and Conjugation

In Chapter 18, you were introduced to the varied ways in which microorganisms can acquire genetic material from other cells. In this chapter we have noted the ecological and evolutionary impacts of horizontal gene exchange. DNA may enter cells by means of viral vectors in the process known as **transduction**.

Microbes are able to take up DNA directly from their environments, in the process known as **transformation**. Some bacteria transmit DNA during a mating process known as **conjugation**. Pili (see Figure 27.14) help mating cells attach to each other, allowing conjugation to proceed in Gram-negative bacteria. Such gene exchange processes have contributed to the diversity of microbial nutrition and metabolism, our next topic.

27.4 Nutrition and Metabolism

Bacteria and archaea use a wide variety of materials and chemical transformations to obtain carbon and energy for growth (**Table 27.2**). These materials and transformations explain the nutritional diversity of microorganisms. Bacteria and archaea together display more diverse types of metabolism than other groups of organisms. Microbes can be classified according to type of nutrition, response to oxygen, and presence of specialized metabolic processes.

Bacteria and Archaea Display Diverse Types of Nutrition and Responses to Oxygen

Cyanobacteria and some other bacteria are **autotrophs** (meaning "self-feeders"), organisms that are able to produce all or most of their own organic compounds. Autotrophic microorganisms fall into two categories: photoautotrophs and chemoautotrophs. **Photoautotrophs** are able to use light as a source of energy for synthesis of organic compounds from CO_2 and H_2O, or H_2S. **Chemoautotrophs** are able to use energy obtained by chemical modifications of inorganic compounds to synthesize organic compounds. Such chemical modifications include nitrification (the conversion of ammonia to nitrate) and the oxidation of sulfur, iron, or hydrogen.

Heterotrophs (meaning "other feeders") are organisms that require at least one organic compound, and often more. Some microorganisms are **photoheterotrophs**, meaning that they are able to use light energy to generate ATP, but they must take in organic compounds from their environment. **Chemoheterotrophs** must obtain organic molecules for both energy and as a carbon source. Among the many types of bacterial chemoheterotrophs is *Propionibacterium acnes*, which causes acne, affecting up to 80% of adolescents in the U.S. The genome sequence of *P. acnes* has revealed numerous genes that allow it to break down skin cells and consume the products.

Microorganisms differ in their need for oxygen. **Obligate aerobes** require O_2. **Facultative aerobes** are more adaptable than obligate aerobes. They can use O_2 in aerobic respiration, obtain energy via anaerobic fermentation, or use inorganic chemical reactions to obtain energy. One fascinating example of a facultative aerobe is *Thiomargarita namibiensis*, a giant bacterium that lives in marine waters off the Namibian coast of Africa. This heterotroph obtains its energy in two ways: by oxidizing sulfide with oxygen when this is available or, when oxygen is low or unavailable, by oxidizing sulfide with nitrate. In either case, the cells convert sulfide to elemental sulfur, which is stored within the cells as large globules.

In contrast to obligate aerobes, **obligate anaerobes** such as the bacterial genus *Clostridium* are poisoned by O_2. People suffering from gas gangrene (caused by *Clostridium perfringens* and related species) are usually treated by placement in a chamber having a high oxygen content (called a hyperbaric chamber), which kills the organisms and deactivates the toxins. **Aerotolerant anaerobes** do not use O_2, but they are not poisoned by it either. These organisms obtain their energy by anaerobic respiration, which uses electron acceptors other than oxygen in electron transport processes. Anaerobic metabolic processes include denitrification (the conversion of nitrate into N_2 gas) and the reduction of manganese, iron, and sulfate, which are all important in the Earth's mineral cycles.

Bacteria Play Important Roles as Nitrogen Fixers

Many cyanobacteria and some other microbes are known as **diazotrophs** ("dinitrogen consumers") because they conduct a specialized metabolic process called **nitrogen fixation**. The removal of nitrogen from the gaseous phase is called fixation. During nitrogen fixation, the enzyme nitrogenase converts inert atmospheric gas (N_2) into ammonia (NH_3). Plants and eukaryotic algae can use ammonia (though not N_2) to produce proteins and other essential nitrogen-containing molecules. As a result, many plants have developed symbiotic relationships with diazotrophs, which provide ammonia to the plant partner. For example, nitrogen-fixing cyanobacteria typically live in leaf cavities of water ferns that float in rice paddies. The fern partner and the rice crop both take up ammonia produced by the diazotrophs. This natural fertilizer allows the rice plants to produce more protein. Many types of heterotrophic soil bacteria also fix nitrogen. Examples include *Rhizobium* and its relatives, which form close associations with the roots of legumes (see Chapter 37).

Table 27.2	Major Nutritional Types of Bacteria and Archaea		
Nutrition type	**Energy source**	**Carbon source**	**Example**
Autotroph			
Photoautotroph	Light	CO_2	Cyanobacteria
Chemoautotroph	Inorganic compounds	CO_2	*Sulfolobus*
Heterotroph			
Photoheterotroph	Light	Organic compounds	Chloroflexi
Chemoheterotroph	Organic compounds	Organic compounds	Many

Oxygen binds to nitrogenase, irreversibly disabling it. Thus, most diazotrophs are able to conduct nitrogen fixation only in low-oxygen habitats. Many cyanobacteria accomplish nitrogen fixation in specialized cells known as **heterocysts** (see Figure 27.16a). Heterocysts display many adaptations that reduce nitrogenase exposure to oxygen. These adaptations include thick walls, which reduce inward oxygen diffusion; absence of oxygen-producing photosystems; and increased respiration, which consumes oxygen.

GENOMES & PROTEOMES

Gene Expression Studies Revealed How Cyanobacteria Fix Nitrogen in Hot Springs

The microbial communities of hot springs in Yellowstone National Park and other thermal areas around the world have long fascinated microbiologists interested in the occurrence of life at high temperatures. Thermal pools characteristically display beautiful, multicolored microbial mats (**Figure 27.17**). Such mats are composed of diverse photoheterotrophs, chemoautotrophs, chemoheterotrophs, and photoautotrophs, including many types of cyanobacteria. However, in Yellowstone thermal pools where temperatures range from 50 to 70°C single-celled cyanobacteria of the genus *Synechococcus* are the only photoautotrophic organisms present. Heterocyst-producing cyanobacteria are absent from such hot waters, and few heterotrophic diazotrophs tolerate such temperatures, yet nitrogen fixation occurs in these pools. It was not clear which organisms could fix nitrogen until genomic information provided an essential clue.

Anne-Soisig Steunou, Arthur Grossman, and associates sequenced the genomes of *Synechococcus* cultures obtained from

Figure 27.17   **A thermal pool in Yellowstone National Park.** Brightly colored mats at the pool edge are microbial communities that include thermophilic cyanobacteria.

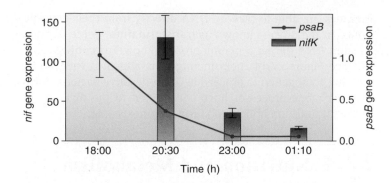

Figure 27.18   Daily changes in photosynthesis and nitrogen fixation gene expression in *Synechococcus* cultures isolated from thermal pools.

Biological inquiry: At what time of the day does the photosynthesis gene psaB reach its lowest level of expression, and when does the nitrogen fixation gene nifK reach its highest expression level?

these hot springs and discovered that *nif* (<u>ni</u>trogen <u>fi</u>xation) genes were present. They also determined that *Synechococcus nif* genes were expressed in parts of the microbial mat having temperatures near 60°C. These results indicated that *Synechococcus* conducts nitrogen fixation in the mats, producing fixed nitrogen and other compounds used by other community members. But the investigators wondered how this single-celled bacterium managed to fix nitrogen without the oxygen arising from photosynthesis disabling the functioning of nitrogenase. They tracked the expression of *Synechococcus* genes over a 24-hour period and found that after nightfall the expression of a photosynthesis gene (*psaB*) fell dramatically, while expression of nitrogen fixation genes (*nifH*, *nifD*, and *nifK*) increased (**Figure 27.18**). They concluded that *Synechococcus* turns on nitrogen fixation at dusk, when oxygen production from photosynthesis drops. But nitrogen fixation is an energy-intensive process, so how does *Synechococcus* fuel it? Steunou and colleagues used differential gene expression analyses to establish that *Synechococcus* uses fermentation, an anaerobic metabolic pathway, to supply the ATP needed for nitrogen fixation. This study not only illustrates important features of microbial metabolism but also the essential ecological roles played by bacteria and archaea, a topic that we will explore in more detail next.

27.5 Ecological Roles and Biotechnology Applications

Bacteria and archaea play several key ecological roles, including the production and cycling of carbon. Earth's carbon cycle depends on microorganisms that produce and degrade organic compounds, including methane. Bacteria also play fascinating roles as symbionts living in close associations with eukaryotes.

In this section, we focus on these diverse ecological roles and also provide examples of ways that humans use the metabolic capabilities of bacteria and archaea in biotechnology.

Bacteria and Archaea Play Important Roles in Earth's Carbon Cycle

Earth's carbon cycle is the sum of all the transformations that occur among compounds that contain carbon. (See Chapter 59 for a detailed discussion of the carbon cycle.) Bacteria and archaea are important in producing and degrading organic compounds. For example, cyanobacteria and other autotrophic bacteria are important **producers**. These bacteria, together with algae and plants, synthesize the organic compounds used by other organisms for food. **Decomposers**, also known as saprobes, include heterotrophic microorganisms (as well as fungi and animals). These organisms break down dead organisms and organic matter, releasing minerals for uptake by living things. Microbial decomposers can consume diverse materials, including oil, explosives, and pesticides, making them useful in cleaning up toxic spills and dumps.

Archaea and bacteria also influence Earth's carbon cycle by producing and consuming methane (CH_4). Methane—the major component of natural gas—is a powerful greenhouse gas, as are CO_2 and H_2O vapor. Atmospheric methane thus has the potential to alter the Earth's climate. Several groups of anaerobic archaea known as the **methanogens** convert CO_2, methyl groups, or acetate to methane and release it from their cells. Methanogens live in swampy wetlands, in deep-sea habitats, or in the digestive systems of ungulate animals, such as cattle. Marsh gas produced in wetlands is largely composed of methane, and large quantities of methane produced long ago are trapped in deep-sea and subsurface Arctic deposits.

Microbes also play a role in limiting atmospheric methane. The balance of methane in Earth's atmosphere is maintained by the activities of aerobic bacteria known as **methanotrophs**, which consume methane. Some methanotrophs live in symbiotic association with marine invertebrates in deep-sea thermal vents or hydrocarbon seep communities. Methane-consuming bacteria also live in close association with wetland plants, which release the oxygen needed by these bacteria to metabolize methane. In the absence of methanotrophs, Earth's atmosphere would be much richer in the greenhouse gas methane, which would substantially increase global temperatures.

Many Bacteria Live in Symbiotic Associations with Eukaryotes

As we have seen, many bacteria live in symbiotic associations with eukaryotic organisms, a relationship called **symbiosis**. If symbiotic associations are beneficial to both partners, the association is known as a **mutualism**. If one partner benefits at the expense of the other, the association is known as a **parasitism**, and the partner that benefits is termed a parasite or a pathogen. There are numerous examples of mutualistic and parasitic bacteria.

Mutualistic Microbes Bacteria are involved in many mutually beneficial partnerships in aquatic and terrestrial habitats. Many aquatic protists depend on bacterial partners for vitamins or other essential compounds. For example, the common green seaweed *Ulva* does not display its typical lettuce-leaf-like structure unless bacterial partners belonging to the phylum Bacteroidetes are present. The bacteria produce a compound that induces normal seaweed development. Bioluminescent bacteria form symbiotic relationships with squid and other marine animals. In deep-sea thermal vent communities, sulfur-oxidizing bacteria live within the tissues of tube worms and mussels, supplying these animals with carbon compounds used as food. Bacteriologist Cameron Curry and his associates have documented a complex land association involving four partners: ants, fungi that the ants cultivate for food, parasitic fungi that attack the food fungi, and mutualistic Actinobacteria, which produce antibiotics. These antibiotics control the growth of the fungal parasite, preventing it from destroying the ants' fungal food supply. The ants rear the useful bacteria in cavities on their body surfaces; glands near these cavities supply the bacteria with nutrients. Aphids also harbor mutualistic bacteria that help to defend against pathogens or parasites. Nitrogen-fixing cyanobacteria partner with a variety of plants (see Chapter 31) and, together with certain fungi, form many types of lichens (see Chapter 29).

Parasitic and Pathogenic Microbes **Parasites** are organisms that obtain organic compounds from living hosts. If parasitic microbes cause disease symptoms in their hosts, the microorganisms are known as **pathogens**. Cholera, leprosy, tetanus, pneumonia, whooping cough, diphtheria, Lyme disease, scarlet fever, rheumatic fever, typhoid fever, bacterial dysentery, and tooth decay are among the many examples of human diseases caused by bacterial pathogens. Bacteria also cause many plant diseases of importance in agriculture, including blights, soft rots, and wilts. How do microbiologists determine which bacteria cause these diseases? The pioneering research of the Nobel Prize–winning German physician Robert Koch provides the answer.

In the mid- to late 1800s, Koch established a series of steps to determine whether a particular organism causes a specific disease. First, the presence of the suspected pathogen must correlate with occurrence of symptoms. Next, the pathogen must be isolated from an infected host and grown in pure culture if possible, and cells from the pure culture should cause disease when inoculated into a healthy host. Finally, one should be able to isolate the same pathogen from the second-infected host. Using these steps, known as **Koch's postulates**, Koch discovered the bacterial causes of anthrax, cholera, and tuberculosis. Subsequent investigators have used Koch's postulates to establish the causes of many other infectious diseases. As recently as the 1980s, bacteriologists used Koch's postulates to establish that Legionnaires' disease is caused by a bacterial pathogen, *Legionella pneumophila*.

How Pathogenic Bacteria Attack Cells Modern research is providing new information about how bacteria attack cells. Such knowledge aids in developing strategies for disease prevention and treatment. Many pathogenic bacteria attack cells by binding to the target cell surfaces and injecting substances that help them utilize cell components. During their evolution, some pathogenic bacteria have transformed flagella into needle-like systems for injecting infection proteins into animal or plant cells as part of the infection process. Such modified flagella are known as type III secretion systems called injectisomes (**Figure 27.19a**). *Yersinia pestis* (the agent of bubonic plague) and *Salmonella enterica* (which causes salmonellosis) are examples of bacteria whose type III secretion systems allow them to attack human cells. These bacteria also induce the host cell to form a plasma membrane pocket that encloses the bacterial cell, bringing it into the target cell. Once within a host's cell, pathogenic bacteria use the cell's resources to reproduce and spread to nearby tissues.

Some other bacterial pathogens use a type IV secretion system to deliver toxins or transforming DNA into cells (**Figure 27.19b**). Examples of such bacteria that cause human disease include *Helicobacter pylori*, *Legionella pneumophila*, and *Bordetella pertussis*. The plant pathogen *Agrobacterium tumifaciens* uses a type IV secretion system to transfer DNA (T DNA) into plant cells. The bacterial T DNA encodes an enzyme that affects normal plant growth, with the result that cancer-like tumors develop (see Figure 27.3). Type IV systems evolved from pili and other components of bacterial mating. Thus, types III and IV attack systems are examples of descent with modification, the evolutionary process by which organisms acquire new features. Some experts propose that type III and IV systems may be useful in human gene therapy, to deliver DNA to target cells.

Antibiotic-Resistant Pathogens Antibiotic compounds are widely used to treat diseases caused by bacterial pathogens, but overuse of antibiotics can lead to production of resistant bacteria, a serious health problem. For example, some strains of *Staphylococcus aureus*, which causes "staph" infections, have developed resistance to penicillin and similar antibiotics. Livestock are often fed antibiotics to reduce the effects of minor bacterial infections or to prevent infections from spreading, thereby increasing the production profit margin. Unfortunately, this activity promotes the evolution of antibiotic-resistant bacterial strains. When humans become infected by such strains, antibiotic treatments are not effective, and cases have occurred in which patients' infections were resistant to all known antibiotics. The World Health Organization has advised eliminating the routine use of antibiotics in livestock.

Some Bacteria Are Useful in Industrial and Other Applications

Several industries have harnessed the metabolic capabilities of microbes obtained from nature. The food industry uses bacteria to produce chemical changes in food that improve consistency

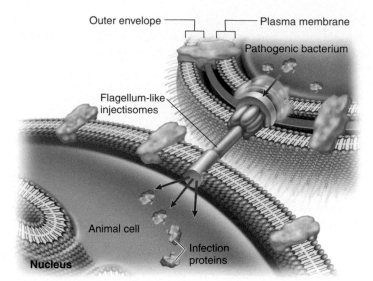

(a) Type III secretion system

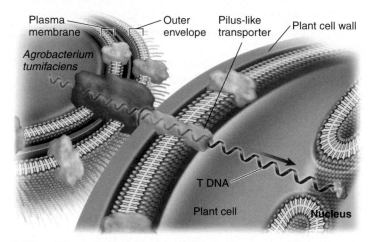

(b) Type IV secretion system

Figure 27.19 Attack systems of pathogenic bacteria. (a) The Type III secretion system functions like a syringe to inject proteins into host cells, thereby starting a disease process. (b) The Type IV secretion system forms a channel through which DNA can be transmitted from a pathogen to a host cell, in this case from the bacterium *Agrobacterium tumifaciens* into a plant cell.

or flavor—to make dairy products, including cheese and yogurt, for example. Cheese makers add pure cultures of certain bacteria to milk. The bacteria consume milk sugar (lactose) and produce lactic acid, which aids in curdling the milk.

The chemical industry produces materials such as vinegar, amino acids, enzymes, vitamins, insulin, vaccines, antibiotics, and other useful pharmaceuticals by growing particular bacteria in giant vats. For example, bacteria produce the antibiotics streptomycin, tetracycline, kanamycin, gentamycin, bacitracin, polymyxin-B, and neomycin.

The ability of microorganisms to live in harsh environments and break down organic compounds makes them very useful in treating wastewater, industrial discharges, and harmful substances such as explosives, pesticides, and oil spills. This process is known as bioremediation. *Geobacter sulfurreducens*, for example, is used to precipitate metals such as uranium from contaminated water, thereby purifying it. This bacterium uses metal ions and elemental sulfur in an electron transport process to oxidize acetate to CO_2, in the process producing ATP needed for growth and reproduction. Taking advantage of this fact, engineers drip acetate into the groundwater, thereby enriching the population of *G. sulfurreducens* to 85% of the microbial community at contaminated sites, which more efficiently precipitates uranium.

In addition to food production and waste treatment, bacteria are also used in agriculture. Several species of *Bacillus*, particularly *B. thuringiensis*, produce crystalline proteins known as Bt-toxins, which kill insects that ingest them. Tent caterpillars, potato beetles, gypsy moths, mosquitoes, and black flies are among the pests that can be controlled by Bt-toxin. For this reason, genes involved in Bt-toxin production have been engineered into some crop plants such as corn to limit pests, thereby increasing crop yields. *Deinococcus radiodurans* is another example of a bacterial species whose unusual ecological properties may prove useful to humans, as described next.

FEATURE INVESTIGATION

The Daly Experiments Revealed How Mn(II) Helps *Deinococcus radiodurans* to Avoid Radiation Damage

The bacterial phylum known as *Deinococcus-Thermus* includes *Deinococcus radiodurans*, which is unusually resistant to chemical mutagens and nuclear radiation (see Table 27.1). This bacterium can survive brief radiation doses greater than 10,000 Gray (Gy) and continuous radiation levels as high as 50 Gy/hour. (By contrast, a radiation dose of 5 Gy is lethal to humans. This trait evolved as an adaptation that aids *Deinococcus* survival in its natural arid desert habitats. Like nuclear radiation, drying and solar radiation cause chromosome breakage, and *Deinococcus* has acquired very effective methods for repairing damaged DNA.) M. J. Daly and associates were interested in learning more about how *D. radiodurans* survives treatment with high radiation, with the hope that such information might lead to better ways of treating victims of radiation sickness.

As the result of a series of experiments, Daly and colleagues learned that radiation-resistant bacteria tended to have higher levels of manganese—Mn(II)—ions than do radiation-sensitive bacteria. These facts suggested the hypothesis that *D. radiodurans* might protect itself by accumulating Mn(II) ions to high levels. In experiments reported in 2004 and described in Figure 27.20, Daly and colleagues first grew *D. radiodurans* in basic culture media containing three different levels of manganese ion (50, 100, or 250 nM Mn(II)). As they grew, the bacteria took up Mn(II) ions from the media. Researchers then inoculated bacteria from each type of medium into separate sectors of laboratory dishes containing media to which low or high levels of Mn(II) ions had been added. Next, they exposed some of the dishes to 50 Gy/hour of radiation, leaving one as a nonirradiated control. After a period of growth, irradiated bacteria grown in dishes containing high Mn(II) ion levels grew as well as the control, but irradiated bacteria grown in dishes containing low Mn(II) ion levels did not grow as well. Irradiated bacteria that had been pregrown with high Mn(II) ion levels fared particularly well.

This experiment revealed that *D. radiodurans* cells that had accumulated high levels of Mn(II) ions were better able to grow in the presence of ionizing radiation.

Figure 27.20 Manganese helps *Deinococcus radiodurans* to avoid radiation damage.

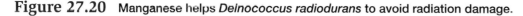

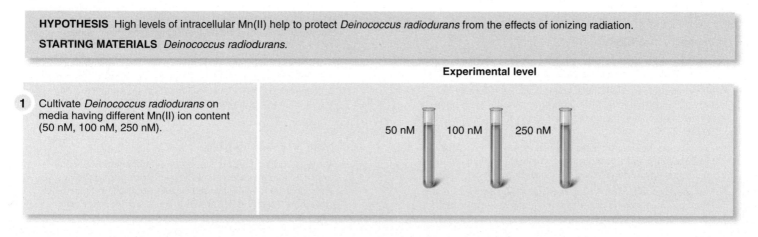

HYPOTHESIS High levels of intracellular Mn(II) help to protect *Deinococcus radiodurans* from the effects of ionizing radiation.

STARTING MATERIALS *Deinococcus radiodurans.*

Experimental level

1 Cultivate *Deinococcus radiodurans* on media having different Mn(II) ion content (50 nM, 100 nM, 250 nM).

50 nM   100 nM   250 nM

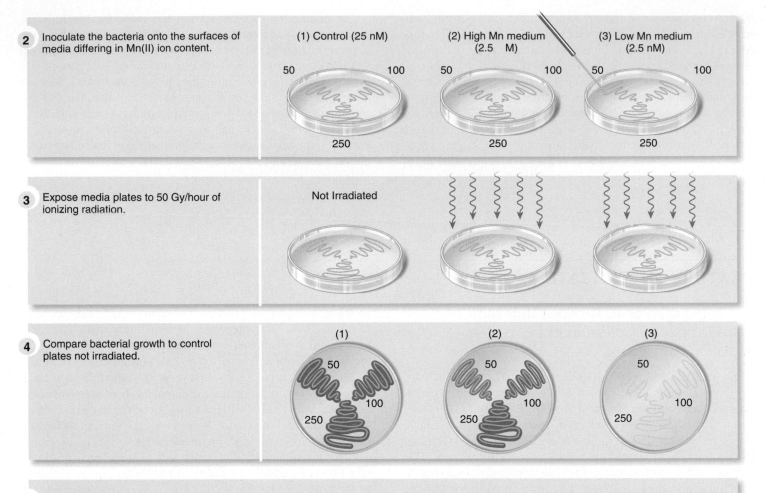

2 Inoculate the bacteria onto the surfaces of media differing in Mn(II) ion content.

(1) Control (25 nM)

(2) High Mn medium (2.5 M)

(3) Low Mn medium (2.5 nM)

3 Expose media plates to 50 Gy/hour of ionizing radiation.

Not Irradiated

4 Compare bacterial growth to control plates not irradiated.

5 THE DATA

Results from step 4: Even when exposed to ionizing radiation, *D. radiodurans* grew well when high levels of Mn(II) ions are present in its environment (compare plates 2 and 1). The higher the level of Mn accumulated by cells, the better the growth after ionizing radiation (plate 2). When environmental levels of Mn are low, cells are more readily damaged by ionizing radiation (compare plates 2 and 3).

CHAPTER SUMMARY

27.1 Diversity and Evolution

- Domains of life include Eukarya, Bacteria, and Archaea (known informally as eukaryotes, bacteria, and archaea). (Figure 27.1)

- Many representatives of the Archaea occur in extremely hot, salty, or acidic habitats. Ether-linked membranes are among the features of archaea that enable their survival in extreme habitats. (Figure 27.2)

- The domain Bacteria includes 50 or more phyla, including Proteobacteria and Cyanobacteria, which are particularly diverse and of great evolutionary and ecological importance. (Table 27.1)

- Mitochondria arose from a proteobacterial ancestor, and many modern proteobacteria occur in symbiotic relationships with animals or plants. (Figure 27.3)

- Algal and plant plastids arose from a cyanobacterial ancestor. Modern cyanobacteria play important ecological roles, including the production of harmful blooms in overfertilized waters. (Figure 27.4)

- Widespread horizontal DNA transfer has occurred among bacteria and archaea. Horizontal DNA transfer is particularly common among close relatives and inhabitants of the same environment, and it allows microorganisms to evolve rapidly.

27.2 Structure and Motility

- Bacteria and archaea are composed of prokaryotic cells that are smaller and simpler than those of eukaryotes. Even so, structures such as thylakoids, magnetosomes, and nucleus-like bodies are examples of surprising cell structure complexity. (Figures 27.5, 27.6)

- Cells of bacteria and archaea vary in shape. Major cell shape types are spherical cocci, rod-shaped bacilli, comma-shaped vibrios, and coiled spirilli and spirochaetes. Some cyanobacteria display specialized cells and intercellular communication, the hallmarks of multicellular organisms. (Figure 27.7)

- Many microbes occur within a coating of slimy mucilage, which may play a role in diseases or aid in the development of biofilms. Biofilm development is influenced by quorum sensing, a process in which group activity is coordinated by chemical communication. (Figure 27.8)

- Most prokaryotic cells possess a protective cell wall. Archaea and some bacteria have walls composed of proteins, but most bacterial cell walls contain peptidoglycan, which is composed of carbohydrates cross-linked by peptides. (Figure 27.9)

- Gram-positive bacterial cells have walls rich in peptidoglycan, while Gram-negative cells have less peptidoglycan in their walls and are enclosed by a lipopolysaccharide envelope. Gram-positive bacteria can be distinguished from Gram-negative bacteria by use of the Gram stain and other staining procedures. (Figures 27.10, 27.11)

- Motility enables microbes to change positions within their environment, which aids in locating favorable conditions for growth. Some microorganisms swim by means of flagella; others twitch or glide by the use of threadlike pili, or adjust their buoyancy in water by means of intracellular vesicles. (Figures 27.12, 27.13, 27.14)

27.3 Reproduction

- Populations of bacteria and archaea enlarge by binary fission, a simple type of cell division that provides a means by which culturable microbes can be counted. (Figure 27.15)

- Some bacteria are able to survive harsh conditions as dormant akinetes or endospores. (Figure 27.16)

- Many microorganisms obtain new DNA sequences directly from their environment (transformation), by means of viral vectors (transduction), or by a mating process (conjugation).

27.4 Nutrition and Metabolism

- Bacteria and archaea can be grouped according to nutritional type, response to oxygen, or presence of distinctive metabolic features. Major nutritional types are photoautotrophs, chemoautotrophs, photoheterotrophs, and chemoheterotrophs. (Table 27.2)

- Obligate aerobes require oxygen, while facultative aerobes are able to live with oxygen or without it by using different processes for obtaining energy. Obligate anaerobes are poisoned by oxygen, while aerotolerant anaerobes do not use oxygen but are not poisoned by it; both obtain their energy by anaerobic respiration.

- Nitrogen fixation is an example of distinctive metabolism displayed only by certain microorganisms. A number of plants display symbiotic associations with bacteria that fix nitrogen, called diazotrophs; many of these associations are ecologically or agriculturally important. (Figures 27.17, 27.18)

27.5 Ecological Roles and Biotechnology Applications

- Bacteria and archaea play key roles in Earth's carbon cycle as producers, decomposers, symbionts, or pathogens.

- Methane-producing methanogens and methane-using methanotrophs are important in the carbon cycle, and they influence the Earth's climate.

- Parasitic bacteria obtain organic compounds from living hosts, and if disease symptoms result, such bacteria are known as pathogens.

- Bacteria attack eukaryotic cells by means of modified flagella, known as Type III secretion systems, or Type IV secretion systems, which evolved from pili and other mating components. (Figure 27.19)

- Many bacteria and archaea are useful in industrial and other applications; others are used to make food products or antibiotics or to clean up polluted environments. The cellular adaptations of bacteria that are extremely resistant to ionizing radiation may suggest new ways to treat radiation sickness in humans. (Figure 27.20)

Test Yourself

1. Which of the following features is common to prokaryotic cells?
   a. a nucleus, featuring a nuclear envelope with pores
   b. mitochondria
   c. plasma membranes
   d. mitotic spindle
   e. none of the above

2. The bacterial phylum that produces oxygen gas as the result of photosynthesis is
   a. the proteobacteria.
   b. the cyanobacteria.
   c. the Gram positive bacteria.
   d. all of the above.
   e. none of the above.

3. The Gram stain is a procedure that microbiologists use to
   a. determine if a bacterial strain is a pathogen.
   b. determine if a bacterial sample can break down oil.
   c. infer the structure of a bacterial cell wall and bacterial response to antibiotics.
   d. count bacteria in medical or environmental samples.
   e. all of the above.

4. Place the following steps in the correct order, according to Koch's postulates:
   I. Determine if pure cultures of bacteria cause disease symptoms when introduced to a healthy host.
   II. Determine if disease symptoms correlate with presence of a suspected pathogen.
   III. Isolate the suspected pathogen and grow it in pure culture, free of other possible pathogens.
   IV. Attempt to isolate pathogen from second-infected hosts.
   a. II, III, IV, I
   b. II, IV, III, I
   c. III, II, I, IV
   d. II, III, I, IV
   e. I, II, III, IV

5. Cyanobacteria play what ecological role?
   a. producers
   b. consumers
   c. decomposers
   d. parasites
   e. none of the above

6. Bacterial structures that are produced by pathogenic bacteria for use in attacking host cells include
   a. Type III and IV secretion systems.
   b. magnetosomes.
   c. gas vesicles.
   d. thylakoids.
   e. none of the above.

7. The structures that enable some Gram-positive bacteria to remain dormant for extremely long periods of time are known as
   a. akinetes.
   b. endospores.
   c. biofilms.
   d. lipopolysaccharide envelopes.
   e. pili.

8. By means of what process do populations of bacteria or archaea increase their size?
   a. mitosis
   b. meiosis
   c. conjugation
   d. transduction
   e. none of the above

9. By what means do bacterial cells acquire new DNA?
   a. by conjugation, the mating of two cells of the same bacterial species
   b. by transduction, the injection of viral DNA into bacterial cells
   c. by transformation, the uptake of DNA from the environment
   d. all of the above
   e. none of the above

10. How do various types of bacteria move?
    a. by the use of flagella composed of motor, hook, and filament
    b. by means of pili, which help cells twitch or glide along a surface
    c. by using gas vesicles to regulate buoyancy in water bodies
    d. all of the above
    e. none of the above

CONCEPTUAL QUESTIONS

1. Explain why many microbial populations grow more rapidly than do eukaryotes, and how bacterial population growth influences the rate of food spoilage or infection.

2. Why does the overuse of antibiotics in medicine and agriculture result in widespread antibiotic resistance?

3. What organisms are responsible for the blue-green blooms that often occur in warm weather on lake surfaces?

EXPERIMENTAL QUESTIONS

1. What feature of *Deinococcus radiodurans* attracted the attention of researchers?

2. What hypothesis did Daly and associates develop to explain radiation resistance in *D. radiodurans*?

3. As shown in the Feature Investigation in Figure 27.20, bacterial cells were grown in media having various levels of manganese ion, which they absorbed. Later, some of these bacteria were exposed to high levels of ionizing radiation, while control bacteria were not exposed. What results of this experiment support the hypothesis that cellular manganese plays a role in radiation resistance?

COLLABORATIVE QUESTIONS

1. How would you go about cataloging the phyla of bacteria and archaea that occur in a particular place?

2. How would you go about developing a bacterial product that could be sold for remediation of a site contaminated with materials that are harmful to humans?

www.brookerbiology.com

This website includes answers to the Biological Inquiry questions found in the figure legends and all end-of-chapter questions.

28

PROTISTS

A population of green algal cells, each surrounded by a halo of protective mucilage.

Protists are eukaryotes that live in moist habitats and are mostly microscopic in size. Despite their small size, protists have a greater impact on global ecology and human affairs than most people realize. For example, the photosynthetic protists known as algae generate at least half of the oxygen in the Earth's atmosphere and produce organic compounds that feed marine and freshwater animals. The oil that fuels our cars and industry derives from pressure-cooked algae that accumulated on the ocean floor over millions of years. Today, algae are being engineered into systems for producing biofuels and cleaning pollutants from water.

Protists also include some parasites that cause serious human illnesses. For example, in 1993 the waterborne protist *Cryptosporidium parvum* sickened 400,000 people in Milwaukee, Wisconsin, costing $96 million in medical expenses and lost work time. The related protist *Plasmodium falciparum*, which is carried by mosquitoes in many warm regions of the world, causes the disease malaria. Every year, nearly 500 million people become ill with malaria and more than 2 million die of this disease. As we will see in this chapter, sequencing the genomes of these and other protist species has suggested new ways of battling such deadly pathogens.

In this chapter, we will survey protist diversity, including structural, nutritional, and ecological variations. We begin by exploring ways of informally classifying protists, by ecological roles, habitat, and motility. We then will focus on the defining features and evolutionary importance of the major protist phyla. Next, the nutritional modes and defensive adaptations of protists are discussed, and we conclude by looking at the reproductive adaptations that allow protists to exploit and thrive in a variety of environments. In the course of the discussion, you will also see the enormous role that protists play in ecosystem functioning and in human health. Our study of protists also illustrates important principles of diversity, particularly ways in which organisms modify their environments and provide essential biological services.

28.1 An Introduction to Protists

Protists are eukaryotes that are not classified in the plant, animal, or fungal kingdoms, though some protists are closely related to plants or animals or fungi (**Figure 28.1**). The term protist comes from the Greek word *protos*, meaning "first," reflecting the fact that protists were Earth's first eukaryotes. Protists display two common characteristics: They are most abundant in moist habitats, and most of them are microscopic in size. Despite these shared features, modern phylogenetic analyses based on comparative analysis of DNA sequences and cellular features reveal that protists do not form a monophyletic group. Instead, protist phyla are classified into several eukaryotic **supergroups** that each display distinctive features (**Table 28.1**). One of these supergroups contains several phyla of protists in addition to the plant kingdom, while another supergroup links certain protists to the animal and fungal kingdoms (see Figure 28.1). Such classifications reveal the great importance of protists in our understanding of eukaryotic evolution.

In addition to classification based on evolutionary relationships, protists are often informally classified according to diverse ecological roles, habitats, or type of motility. In this section, we will examine these informal classifications as a way of introducing protist diversity.

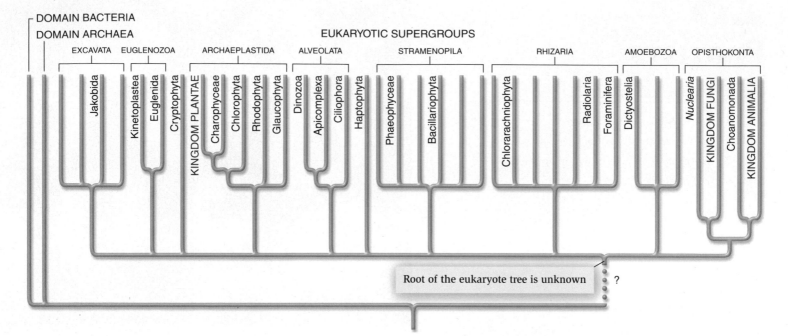

Figure 28.1 **A phylogenetic tree showing major eukaryotic supergroups.** Eight supergroup names are indicated above brackets, the names of eukaryotic kingdoms are shown in red, and major protist phyla and classes discussed in this chapter are also shown. The position of the root of the eukaryotic tree is uncertain, as are the branching patterns of supergroups and many phyla.

Protists Can Be Informally Classified According to Diverse Ecological Roles

Protists are often classified according to ecological roles into three major groups: algae, protozoa, and fungus-like protists. The term **algae** (Latin for "seaweeds") applies to about 10 phyla of protists that include both photosynthetic and nonphotosynthetic species (see Table 28.1). Photosynthetic algae produce organic compounds and oxygen, thereby providing substances used by heterotrophic organisms. For example, the single-celled algae known as diatoms often serve as food for heterotrophic protists (**Figure 28.2**). Despite the common feature of photosynthesis, algae do not form a monophyletic group descended from a single common ancestor (Figure 28.1).

Similarly, the term **protozoa** (Greek for "first life") is commonly used to describe diverse heterotrophic protists. Many protozoa are mobile by means of flagella or other specialized motility structures. Protozoa feed by absorbing small organic molecules or ingesting prey. For example, the protozoa known as ciliates move by means of many short motility structures known as cilia, and they munch on smaller cells such as diatoms (Figure 28.2). Like the algae, the protozoa do not form a monophyletic group. Instead, they are widely distributed among the major protist supergroups (Figure 28.1).

The heterotrophic **fungus-like protists** often resemble true fungi in having threadlike, filamentous bodies and absorbing nutrients from their environments (**Figure 28.3a**; see Chapter 29).

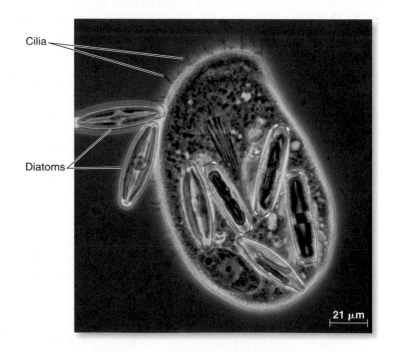

Figure 28.2 **Algal and protozoan cells.** A large, heterotrophic ciliate protozoan, whose surface is covered with motility cilia, has consumed several golden-pigmented, glass-walled algal cells known as diatoms. The algal cells were ingested by the process of phagotrophy and will be digested as food. Diatom cells that have avoided capture glide nearby.

Table 28.1	Eukaryotic Supergroups and Examples of Constituent Kingdoms, Phyla, Classes, or Species	
Supergroup	KINGDOMS, **Phyla**, classes, or *species*	**Distinguishing features**
EXCAVATA	**Jakobida** *Giardia lamblia* *Trichomonas vaginalis*	Unicellular flagellates, often with feeding groove; mitochondria highly modified in specialized parasites
EUGLENOZOA	**Kinetoplastea** (kinetoplastids) *Trypanosoma brucei* **Euglenida** (euglenoids)	Unicellular flagellates; disk-shaped mitochondrial cristae; secondary plastids (when present) derived from endosymbiotic green algae
ARCHAEPLASTIDA	**Glaucophyta** (glaucophytes) **Rhodophyta** (red algae) **Chlorophyta** (green algae) Charophyceae (charophyceans) KINGDOM PLANTAE (land plants)	Primary plastids having only two envelope membranes
ALVEOLATA	**Dinozoa** (dinoflagellates) *Pfiesteria shumwayae* **Ciliophora** (ciliates) **Apicomplexa** (apicomplexans) *Cryptosporidium parvum* *Plasmodium falciparum*	Peripheral membrane sacs (alveoli); some Dinozoa have secondary plastids derived from red algae, some have secondary plastids derived from green algae, and some have tertiary plastids derived from diatoms or cryptomonads; Apicomplexa sometimes have secondary plastids derived from red or green algae
STRAMENOPILA	**Bacillariophyta** (diatoms) Phaeophyceae (brown algae) *Phytophthora infestans* (fungus-like)	Strawlike flagellar hairs; fucoxanthin accessory pigment common in autotrophic forms
RHIZARIA	**Chlorarachniophyta** **Radiolaria** **Foraminifera**	Thin, cytoplasmic projections; secondary plastids (when present) derived from endosymbiotic green algae
AMOEBOZOA	*Entamoeba histolytica* **Dictyostelia** (a slime mold phylum) *Dictyostelium discoideum*	Amoeboid movement by pseudopodia
OPISTHOKONTA	**Choanomonada** (choanoflagellates) KINGDOM ANIMALIA KINGDOM FUNGI *Nuclearia* spp.	Swimming cells possess a single posterior flagellum
Protist phyla listed in this chapter whose supergroup affiliations are controversial or unknown:		
	Haptophyta (haptophytes)	Haptonema, derived from a flagellum that aids food-gathering and attachment; many produce calcium carbonate cell coverings; secondary plastids derived from red algae
	Cryptophyta (cryptomonads)	Flagellates, most have secondary plastids derived from red algae

While the structure and reproduction of fungus-like protists resemble those of a true fungus (**Figure 28.3b**), they are actually more closely related to diatoms than they are to fungi. These examples illustrate that the terms algae, protozoa, and fungus-like protists, while useful in describing ecological roles, lack taxonomic or evolutionary meaning. Next we will survey ways in which protists have been informally classified according to diverse habitats.

Protists Can Be Informally Classified According to Diverse Habitats

Although protists occupy nearly every type of moist habitat, they are particularly common and diverse in oceans, lakes, wetlands, and rivers. Even extreme aquatic environments such as Antarctic ice and acidic hot springs serve as habitats for some protists.

In such places protists may swim or float in open water, or live attached to surfaces such as rocks or beach sand. These different habitats influence protist structure and size.

Swimming or floating protists are members of an informal group of organisms known as plankton, a group that also includes bacteria, viruses, and small animals. The photosynthetic protists in plankton are called **phytoplankton** ("plantlike" plankton), while heterotrophic protists in plankton are described as protozoan plankton. Planktonic protists are necessarily quite small in size; otherwise they would readily sink to the bottom. Staying afloat is particularly important for phytoplankton, which need light for photosynthesis. For this reason, planktonic protists occur primarily as single cells, colonies of cells held together with mucilage, or short filaments of cells linked end to end (**Figure 28.4a–c**).

Many protists live within **periphyton**, communities of microorganisms that are attached by mucilage to underwater surfaces such as rocks, sand, and plants. Because sinking is not a problem for attached protists, these often produce multicellular bodies, such as branched filaments (**Figure 28.4d**). Although some **seaweeds** are very large single cells, most are multicellular algae, which often produce large and complex bodies (**Figure 28.4e**). Seaweeds usually grow attached to underwater surfaces such as rocks, sand, or offshore oil platforms. Seaweeds require sunlight and carbon dioxide for photosynthesis and growth, so most grow along coastal shorelines, fairly near the water surface. Even so, some seaweeds occur at amazing depths. By using a submersible vehicle, algal ecologists Mark Littler, Diane Littler, and associates found seaweeds growing 210 m beneath the surface of ocean waters near San Salvador in the Bahama Islands, the deepest known occurrence of photosynthetic eukaryotes.

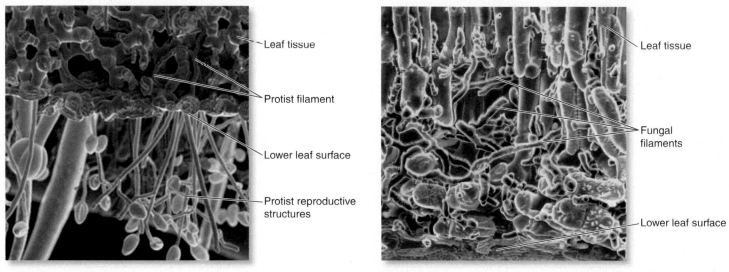

(a) Protist attacking a leaf

(b) True fungus

Figure 28.3　A fungus-like protist and a true fungus.　(a) A fungus-like protist, *Phytophthora infestans*, has absorbed enough nutrients from the plant to produce reproductive structures that emerge from pores on the lower leaf surface. The reproductive structures will eventually break off and be dispersed to other host leaves. (b) A true fungus (*Phragmidium* spp.) produces threadlike filaments that likewise grow through leaf tissues, absorbing nutrients. These nutrients will be used to produce reproductive structures that perpetuate the parasite.

Biological inquiry: How could a microbiologist determine whether unknown filaments growing in diseased plant leaves were fungus-like protists or true fungi?

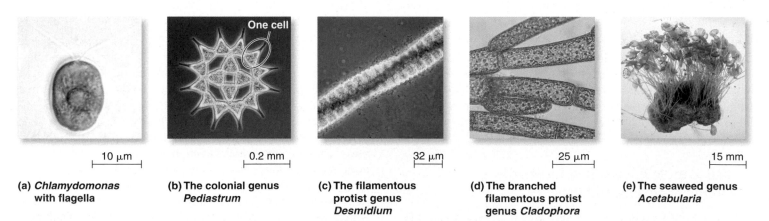

(a) *Chlamydomonas* **with flagella**

(b) The colonial genus *Pediastrum*

(c) The filamentous protist genus *Desmidium*

(d) The branched filamentous protist genus *Cladophora*

(e) The seaweed genus *Acetabularia*

Figure 28.4　Diversity of algal body types.　(a) The single-celled flagellate genus *Chlamydomonas* occurs in the phytoplankton of lakes. (b) The colonial genus *Pediastrum* is composed of several cells arranged in a lacy star shape. This arrangement aids in keeping this alga afloat in water. (c) The filamentous genus *Desmidium* occurs as a twisted row of cells. (d) The branched filamentous genus *Cladophora*, which occurs in the attached periphyton, is large enough to see with the unaided eye. (e) The relatively large seaweed genus *Acetabularia* grows on rocks and coral rubble in shallow tropical oceans.

Protists Can Be Informally Classified According to Type of Motility

Microscopic protists have evolved diverse ways to propel themselves in moist environments. Swimming by means of eukaryotic flagella, cilia, amoeboid movement, and gliding are major types of protist movements.

Many types of photosynthetic and heterotrophic protists are able to swim because they produce one or more eukaryotic flagella, cellular extensions whose movement is based on interactions between microtubules and the motor protein dynein (eukaryotic flagella are described in Chapter 4). Eukaryotic flagella rapidly bend and straighten, thereby pulling or pushing cells through water. Protists that use flagella to move in water are commonly known as **flagellates** (see Figure 28.4a). Flagellates are typically composed of one or only a few cells and are small—usually within 2–20 μm long—because flagellar motion is not powerful enough to keep larger bodies from sinking. Some flagellate protists are sedentary, living attached to underwater surfaces. These protists use flagella to collect bacteria and other small particles for consumption as food. Seaweeds and other immobile protists often produce small, flagellate reproductive cells that have a greater capacity for mobility. Flagellate reproductive cells allow these protists to mate and disperse to new habitats.

An alternate type of protist motility relies on cilia, tiny hairlike extensions on the outsides of cells. Cilia are structurally similar to eukaryotic flagella but are shorter and more abundant on cells (see Figure 28.2). Protists that move by means of cilia are known as **ciliates** and are classified in the phylum Ciliophora (see Table 28.1). The presence of many cilia allows ciliates to achieve larger sizes than flagellates yet still remain buoyant in water. The coordinated movement of cilia enables some ciliates to scamper across underwater surfaces rich in food materials, and other ciliates to generate hurricane-like vortices of water that concentrate food particles.

A third type of motility is amoeboid movement. This kind of motion involves extending protist cytoplasm into lobes, known as pseudopodia ("false feet"). Once these pseudopodia move toward a food source or other stimulus, the rest of the cytoplasm may flow after them, thereby changing the shape of the entire organism as it creeps along. Protist cells that move by pseudopodia are described as **amoebae** (**Figure 28.5**). Finally, many diatoms, the malarial parasite *Plasmodium falciparum*, and some other protists glide along surfaces in a snail-like fashion by secreting protein or carbohydrate slime. With the exception of ciliates, motility classification does not correspond with the phylogenetic classification of protists, our next topic.

28.2 Evolution and Relationships

Modern biologists study protist diversity and infer their phylogenetic relationships by comparing cellular structure and gene sequences. Because the relationships of some protists are uncertain or disputed and new protist species are continuously

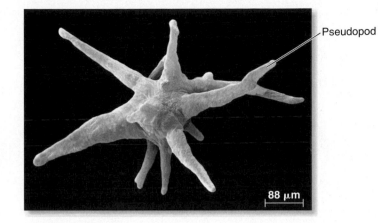

Figure 28.5 The amoebozoan genus *Pelomyxa*, showing pseudopodia.

being discovered, concepts of protist evolution and relationships are constantly changing. Even so, molecular and cellular data reveal that many protists can be classified into diverse eukaryotic supergroups (see Figure 28.1). In this section, we survey these supergroups, focusing on the defining features and evolutionary importance of the major protist phyla. We will also examine ways in which protists are important ecologically or in human affairs.

A Feeding Groove Characterizes Many Protists Classified in the Excavata

The protist supergroup known as the Excavata is related to some of Earth's earliest eukaryotes and is therefore important in understanding the early evolution of eukaryotes. The Excavata is named for a feeding groove "excavated" into the cells of many representatives, such as the genus *Jakoba* (phylum Jakobida) (**Figure 28.6**). The feeding groove is an important adaptation that allows these organisms to ingest small particles of food in their aquatic habitats. Once collected within the feeding groove, food particles are taken into cells by a type of endocytosis known as **phagotrophy** or phagocytosis. During phagotrophy, a vesicle of plasma membrane surrounds each food particle and pinches off within the cytoplasm. Enzymes within these food vesicles break the food particles down into small molecules that, upon their release into the cytoplasm, can be respired for energy. Phagotrophy is the evolutionary basis for the process of endosymbiosis, because particle ingestion provides a way for host cells to take in endosymbionts. Early in protist history, endosymbiotic proteobacterial cells gave rise to mitochondria, the organelle that is the major site of ATP synthesis. Consequently, most protists possess mitochondria, though these may be highly modified in parasitic species.

Some modern protists that are members of Excavata have become parasitic within animals, including humans. Instead of eating particles via feeding grooves, parasitic species attack host cells and absorb food molecules released from them.

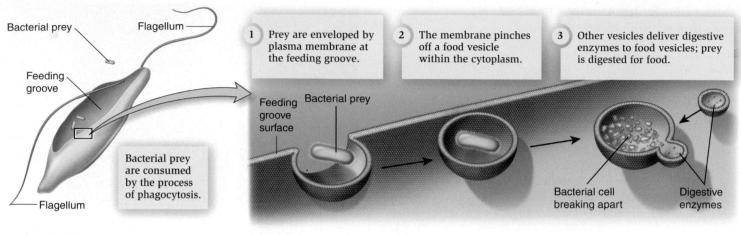

(a) Feeding groove

(b) Phagocytosis

1. Prey are enveloped by plasma membrane at the feeding groove.
2. The membrane pinches off a food vesicle within the cytoplasm.
3. Other vesicles deliver digestive enzymes to food vesicles; prey is digested for food.

Figure 28.6 Feeding groove and phagotrophy displayed by many species of supergroup Excavata. (a) Diagram of *Jakoba libera*, phylum Jakobida, showing flagella emerging from the feeding groove. (b) Diagram of phagocytosis, the process by which food particles are consumed at a feeding groove.

Biological inquiry: What happens to ingested particles after they enter feeding cells?

Trichomonas vaginalis, for example, causes a sexually transmitted infection of the human genitourinary tract. It has an undulating membrane and flagella that allow it to move over mucus-coated skin. *Giardia lamblia* causes giardiasis, an intestinal infection that can result from drinking untreated water or from unsanitary conditions in day-care centers (**Figure 28.7**). *T. vaginalis* and *G. lamblia* were once thought to lack mitochondria but are now known to possess structures that are highly modified mitochondria. In *T. vaginalis*, the modified mitochondria are called hydrogenosomes because they produce hydrogen gas. In *G. lamblia*, reduced mitochondria known as mitosomes do not themselves produce ATP, but they do generate iron-sulfur cofactors essential for ATP production in the cytosol by anaerobic processes. Microbiologist Pavel Dolezal and colleagues recently demonstrated that mitosomes and hydrogenosomes possess inner membrane protein import mechanisms otherwise found only in mitochondria. This evidence demonstrates the evolutionary origin of mitosomes and hydrogenosomes from mitochondria.

Disk-Shaped Mitochondrial Cristae Occur in Euglenozoa

The Euglenozoa is a supergroup of flagellates whose name originates from that of a common representative genus, *Euglena*. Members of the Euglenozoa feature mitochondrial cristae that are disk-shaped (**Figure 28.8**). (In contrast, the mitochondrial cristae of most eukaryotes are typically flattened or tube-shaped; refer back to Figure 4.24.) How or if these different types of cristae function differently is unknown. Euglenozoa includes the phyla informally known as kinetoplastids (Kinetoplastea) and euglenoids (Euglenida) (see Table 28.1).

Euglenoids possess unique, interlocking ribbon-like protein strips just beneath their plasma membranes (**Figure 28.9a**).

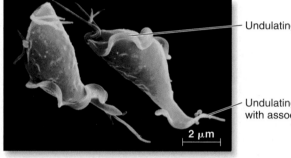

(a) *Trichomonas vaginalis*

2 μm

Undulating membrane

Undulating membrane with associated flagella

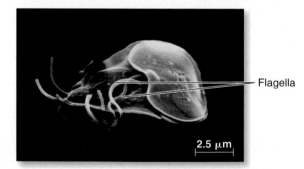

(b) *Giardia lamblia*

2.5 μm

Flagella

Figure 28.7 Parasitic members of the supergroup Excavata. (a) *Trichomonas vaginalis*. (b) *Giardia lamblia*. These specialized, parasitic flagellates absorb nutrients from living hosts.

These strips make the surface of some euglenoids so flexible that they can crawl through mud. This unique type of motility is known as euglenoid movement or metaboly. Many euglenoids are heterotrophic, but some possess green plastids and

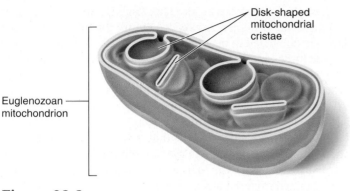

Figure 28.8 **A euglenozoan trait.** The disk-shaped mitochondrial cristae characteristic of the supergroup Euglenozoa.

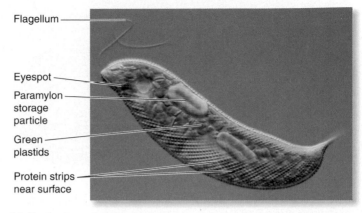

(a) *Euglena*

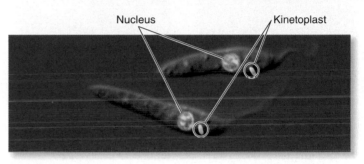

(b) *Leishmania*

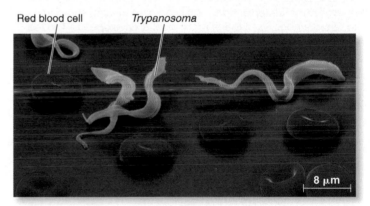

(c) *Trypanosoma*

Figure 28.9 *Euglenozoa.* **(a)** *Euglena* has helical protein ribbons near the surface, internal green plastids, a red eyespot, and white storage carbohydrate granules. **(b)** LM showing the kinetoplast DNA mass typical of kinetoplastid mitochondria. **(c)** In this SEM, several undulating kinetoplastids appear near disk-shaped red blood cells.

thus are photosynthetic. Plastids are organelles found in plant and algal cells that are distinguished by their synthetic abilities. Many euglenoids possess a light-sensing system that includes a conspicuous red structure known as an eyespot or stigma, and light-sensing molecules located in a swollen region at the base of a flagellum. Most euglenoids produce conspicuous storage carbohydrate particles known as paramylon. Euglenoids are particularly abundant and ecologically significant as photosynthesizers and phagotrophs in wetlands.

Kinetoplastids are named for an unusually large mass of DNA (known as a kinetoplast) that occurs in their single large mitochondrion (**Figure 28.9b**). These protists also feature an unusual modified peroxisome that contains glycolytic enzymes; in most eukaryotes, glycolysis occurs in the cytosol. Some kinetoplastids, such as *Trypanosoma brucei*, the causative agent of sleeping sickness, are serious pathogens of humans and other animals (**Figure 28.9c**).

Archaeplastida Are Distinguished by Plastids Arising from Primary Endosymbiosis

The ancestors of the supergroup Archaeplastida ("ancient plastids") most likely obtained plastids by the process of **primary endosymbiosis**. Such plastids are known as **primary plastids**; these can be identified by the presence of an envelope composed of two membranes (**Figure 28.10a**). Primary plastids arose when heterotrophic host cells captured cyanobacterial cells, thereby acquiring photosynthetic ability. These cyanobacterial endosymbionts evolved into plastids, in the process losing their cell walls and ability to live independently. Most species classified in Archaeplastida possess plastids, though some are heterotrophic because photosynthetic pigments have been lost.

The Archaeplastida encompasses some very diverse and ecologically important groups of organisms, including (1) the phylum Rhodophyta, informally known as red algae; (2) the phylum Chlorophyta, informally known as the green algae; (3) the kingdom Plantae, the land plants, which evolved from green algal ancestors; and (4) a small phylum of algae having blue-green plastids, the Glaucophyta, informally called glaucophytes (see Figure 28.1). Grouping the red algae, green algae, land plants,

and glaucophytes together within the supergroup Archaeplastida is based on the assumption that all primary plastids originated with a single endosymbiotic event, a notion that remains somewhat controversial. Evolutionary biologist John Stiller and several other experts have used molecular evidence to frame an alternate hypothesis that green and red primary plastids might have evolved independently as the result of separate endosymbiotic events.

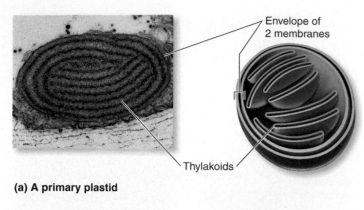

Envelope of
2 membranes

Thylakoids

(a) A primary plastid

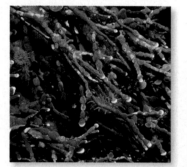

(b) *Calliarthron*

(c) *Chondrus crispus*

Figure 28.10 Red algae and primary plastid of the supergroup Archaeplastida. **(a)** A primary plastid, showing the two membranes of the primary plastid envelope. **(b)** The red algal genus *Calliarthron* has cell walls that are impregnated with calcium carbonate. This stony, white material makes the red alga appear pink. **(c)** *Chondrus crispus* is an edible red seaweed.

Red algae characteristically lack flagella, and most species are multicellular marine seaweeds (**Figure 28.10b,c**). Some red algae are cultivated in ocean waters for production of industrial and scientific materials or food. Sushi wrappers are composed of the sheetlike red algal genus *Porphyra*. Carrageenan is a complex red algal cell-wall polysaccharide that has numerous applications in the food industry, for example, keeping chocolate particles suspended evenly in ice cream or milk. Agar is a material extracted from red seaweeds that is widely used in research laboratories to solidify growth media used for cultivating microorganisms. Molecular biologists commonly use agarose, likewise obtained from red algae, to separate DNA molecules by gel electrophoresis.

Green algae of diverse structural types occur in fresh waters, the ocean, and on land (see Figure 28.4). A class of modern green algae informally known as charophyceans is closely related to the ancestor of land plants. Charophycean green algae resemble land plants in many aspects of cell structure, reproduction, and molecular biology (see Chapter 30). Thus, charophycean green algae are useful in deciphering the early evolutionary history of land plants.

Membrane Sacs Lie at the Cell Periphery of Protists Classified in Alveolata

The supergroup Alveolata includes three important phyla: (1) the Ciliophora, informally known as ciliates (see Figure 28.2); (2) the Dinozoa, informally known as dinoflagellates; and (3) the Apicomplexa, a medically important group of parasites. This supergroup is named for saclike membranous vesicles known as alveoli that are present at the cell periphery in all of these phyla (**Figure 28.11a**). The alveoli in some dinoflagellates seem empty,

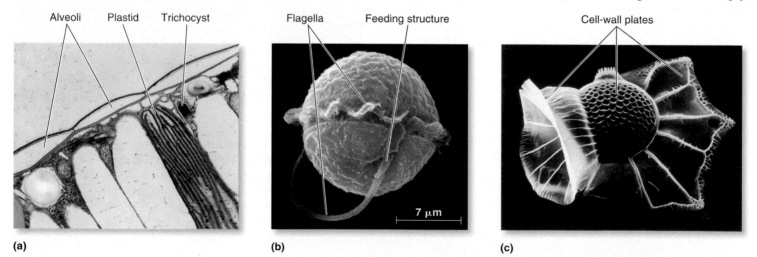

Alveoli   Plastid   Trichocyst

Flagella   Feeding structure

Cell-wall plates

7 μm

(a)   **(b)**   **(c)**

Figure 28.11 Dinoflagellates of the supergroup Alveolata and characteristic alveoli. **(a)** Sac-shaped membranous vesicles known as alveoli lie beneath the plasma membrane of a dinoflagellate, along with trichocysts ready for discharge. **(b)** The surface of *Peridiniopsis berolinensis* appears smooth because the alveoli seem empty. Two types of flagella are seen on this freshwater dinoflagellate. One flagellum coils around a cellular groove; as it moves, this flagellum causes the cell to spin. By contrast, a straight flagellum extends from the cell, acting as a rudder to determine the direction of backward or forward movement. **(c)** The alveoli of the marine dinoflagellate genus *Ornithocercus* contain cellulose cell-wall plates.

Biological inquiry: Why do parts of the cell wall of Ornithocercus *resemble sails?*

so that the cell surface appears smooth (**Figure 28.11b**). By contrast, the alveoli of many dinoflagellates contain plates of cellulose, which taken together form an armor-like enclosure (**Figure 28.11c**). These plates are often modified in ways that provide adaptive advantage, such as protection from predators or increased ability to float.

About half of dinoflagellate species possess photosynthetic plastids of diverse types (see Table 28.1), and half lack plastids and are thus heterotrophic. Dinoflagellates play particularly important roles in coastal oceans as the result of their nutrition and defenses (see Section 28.3). Ciliates are notable for a complex mating behavior known as conjugation (see Section 28.4). Apicomplexans include the malarial agent *Plasmodium falciparum* (see Section 28.4), the related protist *Cryptosporidium parvum*, and other serious pathogens of humans and other animals.

Flagellar Hairs Distinguish Stramenopila

The supergroup Stramenopila (informally known as the stramenopiles) encompasses a wide range of algae, protozoa, and fungus-like protists that usually produce flagellate cells at some point in their lives. The Stramenopila (from the Greek *stramen*, straw + *pila*, hair) is named for distinctive strawlike hairs composed of glycoprotein that occur on the surfaces of flagella (**Figure 28.12**). These flagellar hairs function something like oars to greatly increase swimming efficiency. Biologists discovered this by using antibodies to clump the flagellar hairs together, which caused hairs to detach without otherwise harming the cells. As a result of this treatment, the flagellate cells could no longer swim as fast. Stramenopiles are also informally known as heterokonts (a term meaning "different flagella") because the two flagella often present on swimming cells have slightly different structures.

Fungus-like stramenopiles cause serious diseases of seaweeds, mollusks, fish, and terrestrial crop plants. For example, the fungus-like protist *Phytophthora infestans* (see Figure 28.3) caused the historic Irish potato crop failure that dramatically influenced immigration to the U.S. in the mid-19th century and remains a serious crop pest today. A related species, *Phytophthora ramorum*, causes sudden oak death, and *Phytophthora sojae* is a serious pest of soybeans. These fungus-like protists produce flagellate reproductive cells that enable them to spread on and between plants in films of water. The genomes of these crop parasites are currently being analyzed with the goal of finding better ways to diagnose the presence of the organisms and control their spread in the environment.

Stramenopiles also include many phyla or classes of algae having golden- or brown-colored plastids. Examples include diverse glass-enclosed diatoms (Bacillariophyta) (see Figure 28.2) and giant brown seaweeds (Phaeophyceae) that form extensive kelp forests in cold and temperate coastal oceans. Kelp forests are essential nurseries for fish and shellfish.

Molecular and biochemical evidence indicates that the plastids of stramenopile algae arose by the process of **secondary endosymbiosis** from ingested red algal cells (**Figure 28.13**).

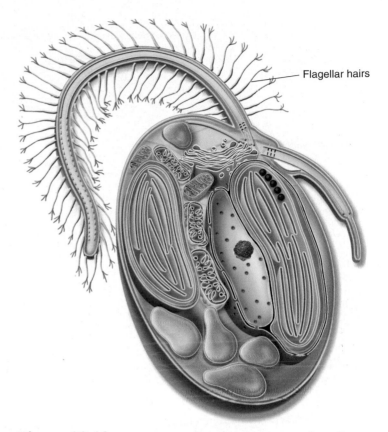

Figure 28.12   Diagram of a flagellate stramenopile cell, showing characteristic flagellar hairs.

Biological inquiry: How do the flagellar hairs aid cell motion?

Secondary endosymbiosis occurs when the endosymbiont is a eukaryote having a primary plastid. Recall that red algal cells possess primary plastids. **Secondary plastids** originate with the endosymbiotic incorporation of a eukaryotic cell having a primary plastid. The host cell digests most of the endosymbiont but retains its plastid. Secondary plastids possess envelopes composed of more than two membranes.

Molecular evidence indicates that the plastids of most photosynthetic dinoflagellates also arose by secondary endosymbiosis from red algal cells, as did plastids of the algal phyla Cryptophyta (flagellates commonly known as cryptomonads) and Haptophyta (flagellate haptophytes) (see Table 28.1). Evolutionary protistologist Patrick Keeling and some other biologists hypothesize that alveolates, cryptomonads, haptophytes, and stramenopiles originated from a single common ancestor that had acquired a secondary plastid from a red algal endosymbiont. These scientists explain the occurrence of heterotrophic species in these phyla as the result of multiple plastid loss events, and they have proposed that these phyla should be classified into a supergroup named Chromalveolata.

By contrast, biogeochemist Paul Falkowski and colleagues argue that red algal cells possess molecular features that favored their incorporation into diverse types of host cells on multiple

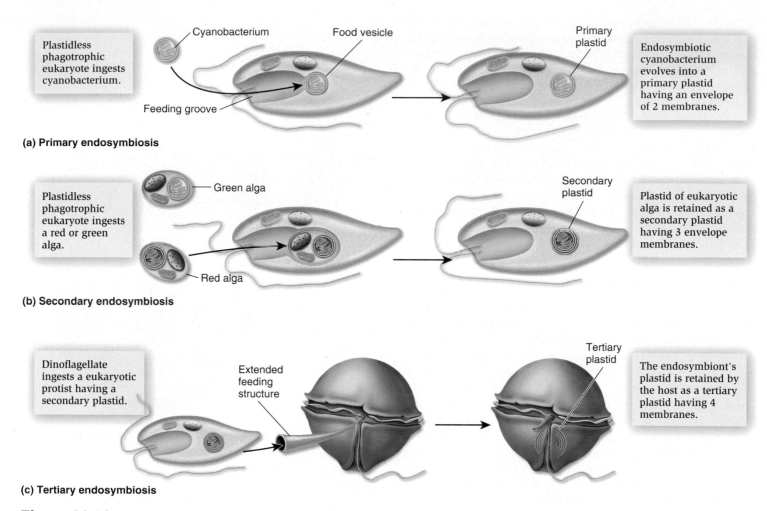

Plastidless phagotrophic eukaryote ingests cyanobacterium.

Cyanobacterium

Food vesicle

Feeding groove

Primary plastid

Endosymbiotic cyanobacterium evolves into a primary plastid having an envelope of 2 membranes.

(a) Primary endosymbiosis

Plastidless phagotrophic eukaryote ingests a red or green alga.

Green alga

Red alga

Secondary plastid

Plastid of eukaryotic alga is retained as a secondary plastid having 3 envelope membranes.

(b) Secondary endosymbiosis

Dinoflagellate ingests a eukaryotic protist having a secondary plastid.

Extended feeding structure

Tertiary plastid

The endosymbiont's plastid is retained by the host as a tertiary plastid having 4 membranes.

(c) Tertiary endosymbiosis

Figure 28.13   Primary, secondary, and tertiary endosymbiosis.

occasions. In other words, the similar plastids of photosynthetic stramenopiles, alveolates, cryptomonads, and haptophytes could have arisen by parallel evolution. If this explanation is correct, grouping these distinctive organisms into a single supergroup would not appropriately reflect phylogenetic relationships.

Relevant to this controversy is the fact that some dinoflagellates possess **tertiary plastids** acquired by engulfing diatoms or cryptomonads. Such plastids were obtained by independent occurrences of **tertiary endosymbiosis**—the acquisition by hosts of plastids from cells that already possessed secondary plastids (see Figure 28.13). In addition, many heterotrophic protists and aquatic animals have independently acquired diverse types of algal or plastid endosymbionts, because this adaptation is nutritionally valuable. The acquisition of plastids by endosymbiosis is an example of horizontal gene transfer. It is clear that endosymbiosis has been a powerful force in the evolution of protists.

Spiky Cytoplasmic Extensions Are Present on the Cells of Many Protists Classified in Rhizaria

Several groups of flagellates and amoebae that have thin, hairlike extensions of their cytoplasm, known as filose pseudopodia,

are classified into the supergroup Rhizaria (from the Greek *rhiza*, root). Rhizaria includes the phylum Chlorarachniophyta, whose spider-shaped cells possess secondary plastids obtained from endosymbiotic green algae (see Table 28.1). The Radiolaria (**Figure 28.14a**) and Foraminifera (**Figure 28.14b**) are two phyla of ocean plankton that produce exquisite mineral shells and frequently have symbiotic algal partners. The accumulated calcium carbonate shells of ancient foraminifera (and haptophyte algae) produced huge chalk deposits about 100 million years ago. Examples include notable geological formations such as the white cliffs of Dover, England (**Figure 28.14c**). Together, abundant ancient populations of foraminifera and haptophytes also produced the famous North Sea oil deposits.

Fossil foraminiferan shells are extensively used to infer past climatic conditions. For example, David Field, Mark Ohman, and fellow scientists used the abundances of different species of fossil foraminifera deposited in Southern California coastal ocean sediments over the past 1,400 years to demonstrate recent climate warming. These scientists observed that over the course of the 20th century, the number of foraminifera typical of warm climates rose dramatically, while the number of species that prefer cold climates fell.

(a) Radiolarian **(b) Foraminiferan** **(c) Fossil deposit containing rhizarians**

Figure 28.14 Living representatives of supergroup Rhizaria and fossil deposits. (a) A radiolarian, *Acanthoplegma* spp., showing long filose pseudopodia. (b) A foraminiferan, showing calcium carbonate shell with long filose pseudopodia extending from pores in the shell. (c) Fossil carbonate remains of foraminifera and haptophyte algae deposited over millions of years form the White Cliffs of Dover in the United Kingdom.

Amoebozoa Includes Many Types of Amoebae with Pseudopodia

The supergroup Amoebozoa includes many types of amoebae that move by extension of pseudopodia (see Figure 28.5). One example is the human parasite *Entamoeba histolytica*, which causes a severe form of intestinal illness. Several types of protists known as slime molds are also classified in this supergroup (see Table 28.1). For example, the phylum Dictyostelia includes the slime mold *Dictyostelium discoideum*, widely used as a model laboratory system for understanding communication among cells. These slime molds produce amoebae that feed on small particles. Under starvation conditions, these amoebae aggregate and form reproductive structures known as "fruiting bodies" because they resemble the spore-producing structures of fungi.

A Single Flagellum Occurs on Swimming Cells of Opisthokonta

The supergroup Opisthokonta is named for the presence of a single posterior flagellum on swimming cells. Because their flagellate cells exhibit this feature, the animal and fungal kingdoms are included in Opisthokonta, as are the protists known as choanoflagellates (formally, the Choanomonada; see Table 28.1). The choanoflagellates are single-celled or colonial organisms that live in fresh water or marine periphyton, which are communities of attached microorganisms (**Figure 28.15**). Choanoflagellates feature a distinctive collar surrounding the single flagellum. The collar is made of cytoplasmic extensions known as tentacles, which filter bacterial food from water currents generated by flagellar motion. Molecular evidence suggests that choanoflagellates are the modern protists most closely related to the common ancestor of animals. Thus, evolutionary biologists interested in the origin of animals study choanoflagellates for molecular clues to this important event in our evolutionary history.

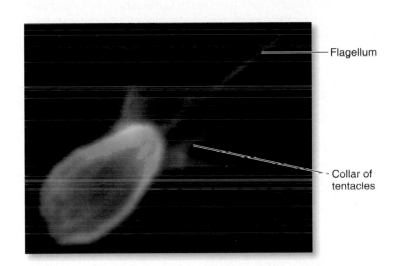

Figure 28.15 A choanoflagellate, of the supergroup Opisthokonta.

Biological inquiry: How do choanoflagellates obtain food?

The preceding surveys of protist diversity and evolutionary relationships provide the foundation for a closer look at protist functions in nature. A focus on protist nutritional, defensive, and reproductive adaptations will help explain these roles, which often influence human affairs.

28.3 Nutritional and Defensive Adaptations

Wherever you look in moist places, you will find protists playing diverse and important ecological roles. In this section we will survey protist nutritional and defensive adaptations, which help to explain their ecological functions.

Protists Display Four Basic Types of Nutrition

Protist nutrition occurs in four basic types: phagotrophy, osmotrophy, autotrophy, and mixotrophy. All of these are important in ecological food webs that connect organisms of many types. Heterotrophic protists that specialize in phagotrophy (particle feeding) are known as **phagotrophs**, whereas those relying on osmotrophy (uptake of small organic molecules) are **osmotrophs**. Photosynthetic protists are **autotrophs**, organisms that can make their own organic nutrients. **Mixotrophs** are protists that are able to use autotrophy as well as phagotrophy or osmotrophy to obtain organic nutrients. The genus *Dinobryon* (**Figure 28.16**), a photosynthetic stramenopile that lives in phytoplankton lakes and consumes enormous numbers of bacteria, is an example of a mixotroph. Mixotrophic protists may switch back and forth between autotrophy and heterotrophy, depending on conditions in their environment. If sufficient light, carbon dioxide, and other minerals are available, mixotrophs produce their own organic food. If any of these resources limits photosynthesis, or organic food is especially abundant, mixotrophs can function as heterotrophs. Mixotrophs thus have remarkable nutritional flexibility.

Heterotrophic protists that feed on nonliving organic material function as decomposers, also known as saprobes. Decomposers, which also include many prokaryotes and fungi, are essential in breaking down wastes and releasing minerals for use by other organisms. Heterotrophic protists that feed on the living cells of other organisms are parasites. Some parasites are pathogens, which cause disease in other organisms. *Trichomonas vaginalis*, *Giardia lamblia*, *Entamoeba histolytica*, and *Phytophthora infestans* are examples of pathogenic protists that have previously been described in this chapter. Humans view such protists as pests when they harm us or our agricultural animals and crops, but such protists also play important roles in nature by controlling the population growth of other organisms. Such roles illustrate the principle that organisms modify their environments and provide essential biological services. The diversity of photosynthetic protists and their storage molecules further illustrate these principles.

Algal Protists Vary in Photosynthetic Pigments and Food Storage Molecules

As we have earlier noted, algae display a surprising diversity of coloration, from gold to brown, red, and green. Why do so many pigmentation types occur? The answer is related to light availability in these protists' watery environment.

If you dive more than a few feet into lakes or the ocean, your aquatic world will appear intensely blue-green. This occurs because water absorbs the longer red to yellow wavelengths of light to a greater degree than shorter blue and green wavelengths. Little red light penetrates far into natural waters, depriving chlorophyll *a*, the photosynthetic pigment universally present in photosynthetic eukaryotes, of much of the light it would ordinarily absorb. Photosynthetic protists living in water have coped with this filtering effect by adapting their photosynthetic systems so that they capture more of the available blue-green light. Seaweeds, for example, may be colored red, gold, or brown, in addition to green, because they produce diverse accessory pigments that are able to absorb more of the light that is available underwater and transfer the energy to chlorophyll *a* for photosynthesis. The red accessory pigment phycoerythrin is abundant in plastids of red algae, and golden fucoxanthin enriches the color of golden and brown algae. Carotene and lutein have similar accessory pigment functions in green algae and were inherited by their land plant descendants, thus playing important roles in human nutrition. These and other accessory pigment adaptations explain why algae occur in so many different colors.

Photosynthetic protists also vary in the types of molecules that serve as food storages. Starch stored in the plastids of green algae and oil droplets in diatom cells are two common examples. Algae use such food storage cells for energy when light or minerals are too low to allow photosynthesis; such cells are also desirable food sources for heterotrophs. As a consequence of photosynthesis and food storage, autotrophic and mixotrophic protists play essential roles in aquatic food webs as primary producers. Such protists generate the organic food and atmospheric oxygen consumed by many other organisms.

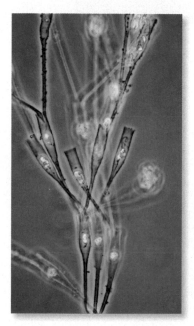

Figure 28.16

A mixotrophic protist.   The genus *Dinobryon* is a colonial flagellate that occurs in the phytoplankton of freshwater lakes. The photosynthetic cells have golden plastids and also capture and consume bacterial cells.

Protists Defend Themselves in Diverse Ways

Protists use a wide variety of defensive adaptations to ward off attack. Major types of defenses are slimy, tough, or spiny cell coverings; sharp projectiles that can be explosively shot from cells; light flashes; and toxic compounds.

Many protists have cell coverings such as slimy mucilages (see chapter-opening photo) or cell walls that provide protection from attack by herbivores or pathogens. Cell walls may also aid in preventing osmotic damage or enhance floatation in water. Rigid cellulose walls are common in brown and green algae, while slimy polysaccharide polymers form a protective matrix around red algal cells. Calcium carbonate forms a stony coat for many protist cells, including foraminifera, haptophytes, and some marine seaweeds. Ornate glassy coatings of silica protect diatoms, while metallic iron and manganese crystals armor other protists.

Several types of protist cells contain compressed protein structures known as trichocysts (see Figure 28.11a). Upon attack, trichocysts rapidly elongate into spear-shaped projectiles that are shot from protist cells, thereby discouraging herbivores from feeding. Some species of ocean dinoflagellates emit flashes of blue light when disturbed, explaining why ocean waters teeming with these protists display bioluminescence. The light flashes may deter herbivores by startling them, but when ingested the dinoflagellates make the herbivores also glow, revealing them to hungry fish. Light flashes benefit dinoflagellates by helping to reduce populations of herbivores that consume the algae.

Various protist species produce **toxins**, compounds that inhibit animal physiology. Toxins probably originated as a means of protection from relatively small herbivores that might otherwise eat protists. Dinoflagellates are probably the most important protist toxin producers; they synthesize several types of toxins that affect humans and other animals. Why does this happen? Under natural conditions, small populations of dinoflagellates produce low amounts of toxin that do not harm large organisms. Dinoflagellate toxins become dangerous to humans when people contaminate natural waters with excess mineral nutrients such as nitrogen and potassium from untreated sewage, industrial discharges, or fertilizer that washes from agricultural fields. The excess nutrients foster explosive growth of algal populations called blooms, which then produce sufficient toxin to affect birds, aquatic mammals, fish, and humans. Toxins can concentrate in organisms, and humans who ingest shellfish that have accumulated dinoflagellate toxins can suffer poisoning.

Some dinoflagellates use toxins to attack fish, which they consume for food, and excessive growth of these dinoflagellates can also be harmful to people. In 1992, ecologist JoAnn Burkholder and colleagues reported that a new type of dinoflagellate had caused major fish kills in the overly fertile waters of the Chesapeake Bay. These investigators gave the new organism the generic name *Pfiesteria* to honor the late dinoflagellate expert Lois Pfiester. The scientists described the genus *Pfiesteria* as a "phantom predator" because populations of swimming cells seemed to appear rapidly in response to the presence of fish or their excretions. *Pfiesteria piscida* and *Pfiesteria shumwayae* tend to lie dormant on the ocean bottom until chemical signals emitted by their fish prey stimulate production of swimming cells. Burkholder and colleagues reported that the dinoflagellate toxin damages the fish skins, allowing the voracious dinoflagellate predators easier access to flesh. These biologists also discovered that *Pfiesteria*-associated toxin caused amnesia and other nervous system conditions in fishers and scientists who were exposed to it. The discovery of *Pfiesteria* excited the media, which dubbed it a "killer alga" and focused attention on nutrient pollution of Chesapeake Bay, the fundamental cause of *Pfiesteria*'s excessive growth and fish kills. Because of this organism's importance to the fishing industry and human health, teams of aquatic ecologists have continued to study the genus *Pfiesteria* and its toxin.

FEATURE INVESTIGATION

Burkholder and Colleagues Demonstrated That Strains of the Dinoflagellate Genus *Pfiesteria* Are Toxic to Mammalian Cells

A team of investigators led by JoAnn Burkholder performed an experiment to determine whether or not two strains of *P. shumwayae* were toxic to mammalian cells (**Figure 28.17**). One of these strains (CCMP 2089) had earlier been reported to be nontoxic, and neither strain had been tested for its impact on mammalian cells. The experiment was conducted in a biohazard containment facility, for the safety of the investigators. In a first step, the team grew the two *Pfiesteria* strains on different food sources, because the impact of food on dinoflagellate toxicity was unclear. Both strains were grown with algal cells as food and with juvenile fish as a food source.

In a second step, the dinoflagellates grown in step 1 were transferred to tanks with fish to elicit maximal toxin production, before treating mammalian cell cultures with the dinoflagellates. Dinoflagellates were not added to a control tank of fish. Toxin was detected, and fish deaths occurred, in all of the tanks except the control. In a third step, samples from the step 2 treatments were added to mammalian cell cultures. The investigators then determined the relative levels of toxicity to the mammalian cells. They found that both strains of *P. shumwayae* subjected to both feeding treatments were toxic to mammalian cells.

Figure 28.17 Burkholder and colleagues demonstrated that some strains of *Pfiesteria shumwayae* are toxic to fish and mammalian cells.

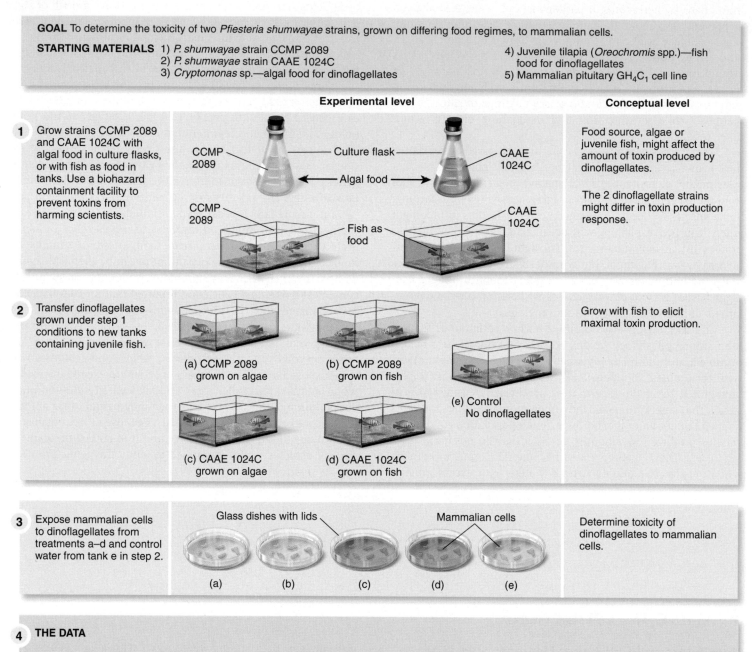

GOAL To determine the toxicity of two *Pfiesteria shumwayae* strains, grown on differing food regimes, to mammalian cells.

STARTING MATERIALS 1) *P. shumwayae* strain CCMP 2089
2) *P. shumwayae* strain CAAE 1024C
3) *Cryptomonas* sp.—algal food for dinoflagellates

4) Juvenile tilapia (*Oreochromis* spp.)—fish food for dinoflagellates
5) Mammalian pituitary GH$_4$C$_1$ cell line

Experimental level   **Conceptual level**

1 Grow strains CCMP 2089 and CAAE 1024C with algal food in culture flasks, or with fish as food in tanks. Use a biohazard containment facility to prevent toxins from harming scientists.

CCMP 2089 —— Culture flask —— CAAE 1024C
◄—— Algal food ——►
CCMP 2089   CAAE 1024C
Fish as food

Food source, algae or juvenile fish, might affect the amount of toxin produced by dinoflagellates.

The 2 dinoflagellate strains might differ in toxin production response.

2 Transfer dinoflagellates grown under step 1 conditions to new tanks containing juvenile fish.

(a) CCMP 2089 grown on algae
(b) CCMP 2089 grown on fish
(e) Control No dinoflagellates
(c) CAAE 1024C grown on algae
(d) CAAE 1024C grown on fish

Grow with fish to elicit maximal toxin production.

3 Expose mammalian cells to dinoflagellates from treatments a–d and control water from tank e in step 2.

Glass dishes with lids   Mammalian cells
(a)   (b)   (c)   (d)   (e)

Determine toxicity of dinoflagellates to mammalian cells.

4 **THE DATA**

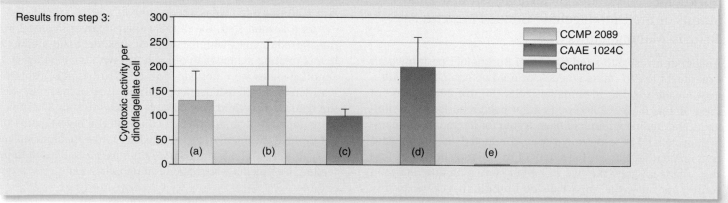

Results from step 3:

28.4 Reproductive Adaptations

Diverse reproductive adaptations allow protists to thrive in an amazing variety of environments. These include specialized asexual reproductive cells, tough-walled dormant cells that allow protists to survive periods of environmental stress, and several types of sexual life cycles.

Protist Populations Increase by Means of Asexual Reproduction

All protists are able to reproduce themselves by asexual means that involve mitotic cell divisions of parental cells to produce progeny. When resources are plentiful, repeated mitotic divisions of single-celled protists will generate large protist populations, as in the case of Chesapeake Bay *Pfiesteria* species. By contrast, multicellular protists often generate specialized asexual cells that help disperse the organisms in their environment. For example, red seaweeds disperse single nonflagellate cells that drift with the currents until they encounter a suitable substrate for attachment and growth of new seaweeds. Many other multicellular or nonmotile protists produce flagellate cells called zoospores that swim through water or water films, thereby dispersing in the environment.

Many protists produce unicellular **cysts** as the result of asexual (and in some cases, sexual) reproduction (**Figure 28.18**). Cysts often have thick, protective walls and can remain dormant through periods of unfavorable climate or low food availability. Dinoflagellates commonly produce cysts. *Pfiesteria* species lie in wait on the ocean bottom as cysts until fish prey are present. Dinoflagellate cysts in ship ballast water can be transported from one port to another, a problem that has caused harmful

dinoflagellate blooms to appear in harbors around the world. Ship captains can help to prevent such ecological disasters by heating ballast water before it is discharged from ships.

Many protozoan pathogens spread from one host to another via cysts. For example, waterborne cysts are the form in which the apicomplexan pathogen *Cryptosporidium parvum* infects humans. Water supplies that are contaminated with large numbers of cysts can sicken many people. In the case of the outbreak in Milwaukee, Wisconsin, noted at the beginning of the chapter, *C. parvum* cysts probably entered drinking water supplies in water that has drained from cattle feedlots. *Entamoeba histolytica* is a pathogen that infects people who consume food or water that is contaminated with cysts. Once inside the human digestive system, *E. histolytica* attacks intestinal cells, causing amoebic dysentery, a worldwide problem.

Sexual Reproduction Provides Multiple Benefits to Protists

Eukaryotic sexual reproduction, featuring gametes, zygotes, and meiosis first arose among protists. Sexual reproduction has not been observed in some protist phyla such as kinetoplastids and euglenoids. However, it occurs widely among green algae, red algae, apicomplexans, ciliates, and dinoflagellates. Sexual reproduction is generally adaptive because it produces diverse genotypes, thus increasing the potential for faster evolutionary response to environmental change. Many protists reap additional benefits from sexual reproduction, illustrated by several types of sexual life cycles.

Zygotic Life Cycles Most unicellular protists that reproduce sexually display what is known as a **zygotic life cycle** (**Figure 28.19**). In this type of life cycle, haploid cells transform into gametes. Some protists produce nonmotile eggs and smaller flagellate sperm. However, many other protists have gametes that look similar to each other structurally but have distinctive biochemical features and hence are known as + and − mating strains. Gametes fuse to produce thick-walled diploid zygotes, which give this type of life cycle its name. Such zygotes often have tough cell walls and can survive stressful conditions, much like cysts. Some of the earliest fossils regarded as ancient protists are probably tough-walled zygotes.

Sporic Life Cycles Many multicellular green and brown seaweeds display a **sporic life cycle**, which is also known as alternation of generations (**Figure 28.20a**). Organisms having sporic life cycles produce two types of multicellular organisms: a haploid gametophyte generation that produces gametes (sperm or eggs), and a diploid sporophyte generation that produces spores by the process of meiosis. This type of life cycle takes its name from the characteristic production of spores as the result of meiosis. Each of the two types of multicellular organisms can adapt to distinct habitats or seasonal conditions, thus allowing protists to occupy more types of environments and for longer periods.

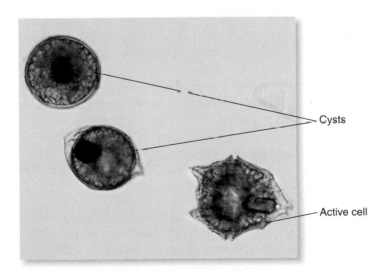

— Cysts

— Active cell

Figure 28.18   Cysts.   The round cells are dormant, tough-walled cysts of the dinoflagellate *Peridinium limbatum*. The pointed cell is an actively growing cell of the same species. As cysts develop, the outer cellulose plates present on actively growing cells are cast off.

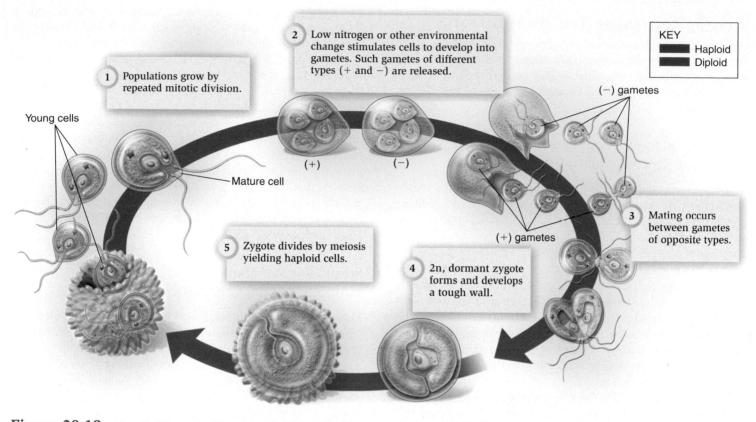

Figure 28.19 Zygotic life cycle, illustrated by the unicellular flagellate genus *Chlamydomonas*. In *Chlamydomonas*, mature cells are haploid.

Many red seaweeds display a variation of the sporic life cycle that involves alternation of three distinct multicellular generations (**Figure 28.20b**). Their unique type of sexual life cycle is an adaptation that allows red algae to cope with the lack of flagella on sperm. Because red algal sperm are unable to swim to eggs, fertilization occurs only when sperm carried by ocean currents happen to drift close to eggs. Fertilization can thus be rare. Many red algae compensate by making millions of spores that are produced by two distinct sporophyte generations.

Gametic Life Cycle and the Problem of Diatom Size Diatoms are examples of a relatively few protists known to display a **gametic life cycle** (**Figure 28.21**). In gametic life cycles, all cells except the gametes are diploid, and gametes are produced by meiosis. Diatoms use sexual reproduction not only to increase genetic variability but also to regenerate maximal species size. Many diatoms must do this as the result of a unique problem that develops during asexual reproduction by mitosis.

In many diatoms, one daughter cell arising from mitosis is smaller than the other, and it is also smaller than the parent cell. This happens because diatom cell walls are composed of two overlapping halves, much like two-part round glass laboratory dishes having lids that overlap bottoms. After each mitotic division, each daughter cell receives one-half of the parent cell wall. The daughter cell that inherits a larger, overlapping parental

"lid" then produces a new "bottom" that fits inside. This daughter cell will be the same size as its parent. However, the daughter cell that inherits the parental "bottom" uses this wall half as its lid and produces a new, even smaller "bottom." This cell will be smaller than its sibling or parent. Consequently, after many such mitotic divisions, the average cell size of diatom populations often declines over time. If diatom cells become too small, they lose the ability to survive. Sexual reproduction helps to solve this problem.

Sexual reproduction allows diatom species to recover maximal cell size, thereby preventing species extinction. Diatom cells mate within a blanket of mucilage, each partner undergoing meiotic divisions to produce gametes. The large, spherical diatom zygotes that result from fertilization (Figure 28.21b) later undergo a series of mitotic divisions to produce new diatom cells having the maximal size for the species.

Ciliate Sexual Reproduction Among protists, ciliates have one of the most complex sexual processes known. Ciliates are unusual in having two types of nuclei, a single large macronucleus and one or more smaller micronuclei. Macronuclei, which develop from micronuclei, serve as the source of information for cell function, and they divide when ciliates reproduce asexually by mitosis (**Figure 28.22a**). Sexual reproduction in ciliates begins when two cells pair and fuse longitudinally—a process known

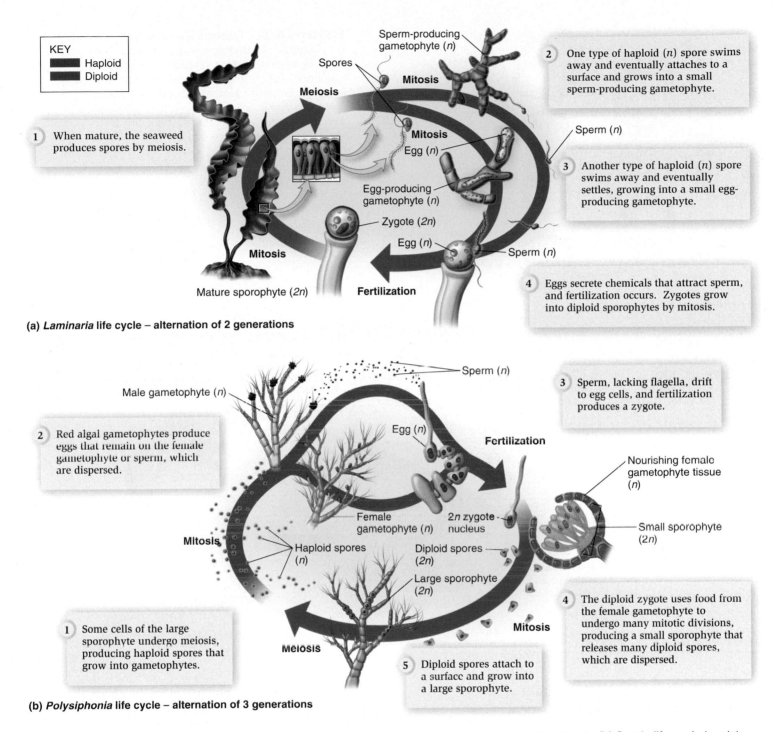

(a) *Laminaria* **life cycle – alternation of 2 generations**

KEY
Haploid
Diploid

1 When mature, the seaweed produces spores by meiosis.

Meiosis

Spores

Sperm-producing gametophyte (*n*)

Mitosis

2 One type of haploid (*n*) spore swims away and eventually attaches to a surface and grows into a small sperm-producing gametophyte.

Mitosis

Egg (*n*)

Sperm (*n*)

Egg-producing gametophyte (*n*)

3 Another type of haploid (*n*) spore swims away and eventually settles, growing into a small egg-producing gametophyte.

Zygote (*2n*)

Egg (*n*)   Sperm (*n*)

Mitosis

Mature sporophyte (*2n*)   Fertilization

4 Eggs secrete chemicals that attract sperm, and fertilization occurs. Zygotes grow into diploid sporophytes by mitosis.

(b) *Polysiphonia* **life cycle – alternation of 3 generations**

Male gametophyte (*n*)

Sperm (*n*)

3 Sperm, lacking flagella, drift to egg cells, and fertilization produces a zygote.

2 Red algal gametophytes produce eggs that remain on the female gametophyte or sperm, which are dispersed.

Egg (*n*)

Fertilization

Nourishing female gametophyte tissue (*n*)

Female gametophyte (*n*)   *2n* zygote nucleus

Small sporophyte (*2n*)

Mitosis

Haploid spores (*n*)

Diploid spores (*2n*)

Large sporophyte (*2n*)

4 The diploid zygote uses food from the female gametophyte to undergo many mitotic divisions, producing a small sporophyte that releases many diploid spores, which are dispersed.

Mitosis

1 Some cells of the large sporophyte undergo meiosis, producing haploid spores that grow into gametophytes.

Meiosis

5 Diploid spores attach to a surface and grow into a large sporophyte.

Figure 28.20 Sporic life cycles. (a) Sporic life cycle as illustrated by the brown seaweed *Laminaria*. (b) Sporic life cycle involving three alternating generations, illustrated by the red seaweed *Polysiphonia*.

as **conjugation** (**Figure 28.22b**). In *Paramecium caudatum*, the single micronucleus undergoes meiosis early in conjugation. Of the four haploid micronuclei produced, all disintegrate except one, which divides by mitosis. Then the paired ciliates exchange one of each pair of micronuclei, after which the paired cells separate and the macronuclei disintegrate. The two genetically different micronuclei in each cell then fuse, and the new diploid nuclei divide mitotically to produce four macronuclei and multiple micronuclei. Subsequent mitotic divisions occur, distributing one macronucleus plus one or more micronuclei (depending on the species) to each cell. Conjugation seems to be essential for continued existence of ciliate species.

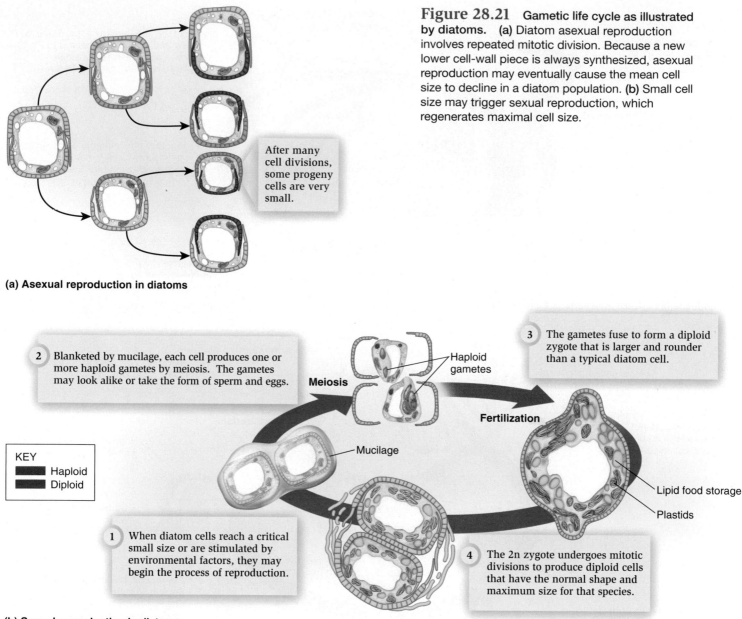

(a) Asexual reproduction in diatoms

After many cell divisions, some progeny cells are very small.

Figure 28.21   Gametic life cycle as illustrated by diatoms. **(a)** Diatom asexual reproduction involves repeated mitotic division. Because a new lower cell-wall piece is always synthesized, asexual reproduction may eventually cause the mean cell size to decline in a diatom population. **(b)** Small cell size may trigger sexual reproduction, which regenerates maximal cell size.

2 Blanketed by mucilage, each cell produces one or more haploid gametes by meiosis. The gametes may look alike or take the form of sperm and eggs.

Meiosis

Haploid gametes

3 The gametes fuse to form a diploid zygote that is larger and rounder than a typical diatom cell.

Fertilization

KEY
Haploid
Diploid

Mucilage

Lipid food storage

Plastids

1 When diatom cells reach a critical small size or are stimulated by environmental factors, they may begin the process of reproduction.

4 The 2n zygote undergoes mitotic divisions to produce diploid cells that have the normal shape and maximum size for that species.

(b) Sexual reproduction in diatoms

Parasitic Protists May Use Alternate Hosts for Different Life Stages

Parasitic protists are notable for often using more than one host organism, in which different life stages occur. The malarial parasite genus *Plasmodium* is a prominent example. About 40% of humans live in tropical regions of the world where malaria occurs, and as noted earlier, millions of infections and human deaths result each year. Malaria is particularly deadly for young children. The malarial parasite's alternate hosts are mosquitoes classified in the genus *Anopheles*. Though insecticides can be used to control mosquito populations, and antimalarial drugs exist, malarial parasites can develop drug resistance. Experts are concerned that cases may double in the next 20 years.

Plasmodium enters the human bloodstream via a mosquito bite as life stages known as sporozoites (**Figure 28.23**). These eventually reach the victim's liver and enter the liver cells. Following several cycles of cell division, life stages known as merozoites develop. Merozoites have protein complexes at their front ends, or apices, that allow them to invade human red blood cells. (The presence of these apical complexes gives rise to the phylum name Apicomplexa.) The merozoites consume the hemoglobin in red blood cells. While living within red blood cells, they form "ring stages," which can be visualized by staining and use of a microscope, allowing diagnosis. Merozoites reproduce asexually, generating large numbers of new merozoites that synchronously break out of red blood cells at intervals of 48 or 72 hours. These merozoite reproduction cycles correspond to cycles of chills and fever experienced by the infected person.

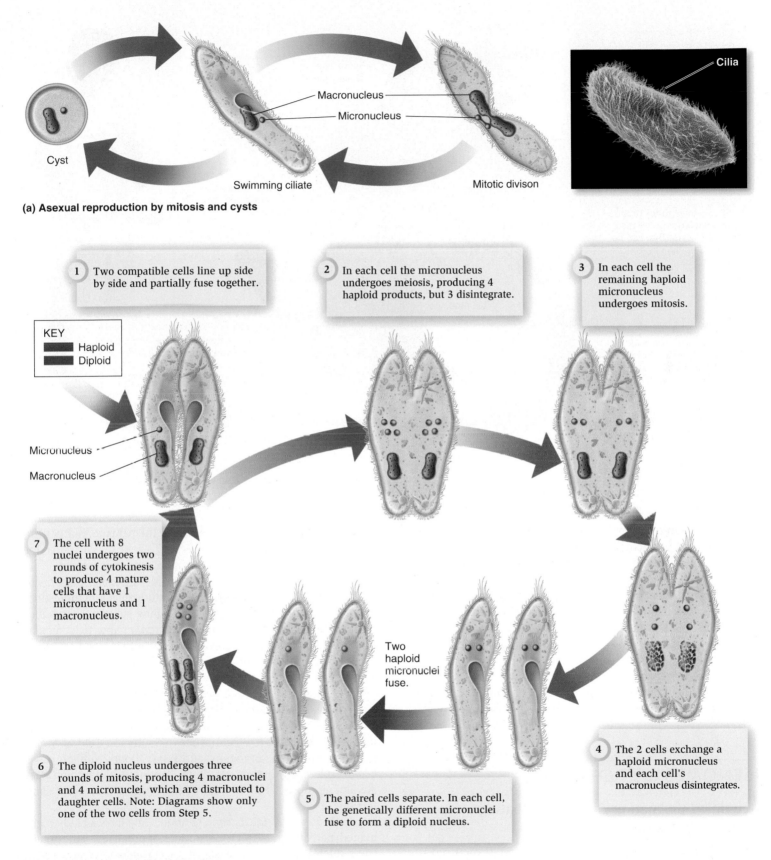

Macronucleus
Micronucleus
Cilia

Cyst

Swimming ciliate

Mitotic divison

(a) Asexual reproduction by mitosis and cysts

1 Two compatible cells line up side by side and partially fuse together.

2 In each cell the micronucleus undergoes meiosis, producing 4 haploid products, but 3 disintegrate.

3 In each cell the remaining haploid micronucleus undergoes mitosis.

KEY
Haploid
Diploid

Micronucleus
Macronucleus

Two haploid micronuclei fuse.

7 The cell with 8 nuclei undergoes two rounds of cytokinesis to produce 4 mature cells that have 1 micronucleus and 1 macronucleus.

6 The diploid nucleus undergoes three rounds of mitosis, producing 4 macronuclei and 4 micronuclei, which are distributed to daughter cells. Note: Diagrams show only one of the two cells from Step 5.

5 The paired cells separate. In each cell, the genetically different micronuclei fuse to form a diploid nucleus.

4 The 2 cells exchange a haploid micronucleus and each cell's macronucleus disintegrates.

(b) Sexual reproduction by conjugation

Figure 28.22   Ciliate reproduction.   (a) The asexual reproduction process in ciliates. **(b)** The sexual reproduction process in the ciliate *Paramecium caudatum*, which involves conjugation.

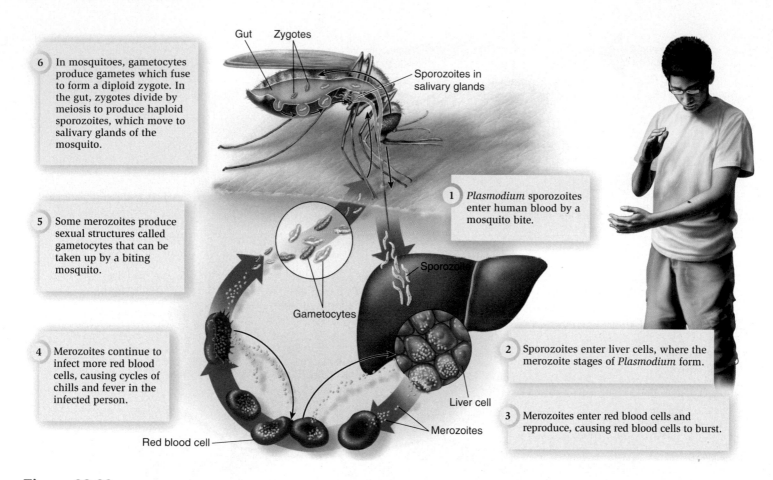

Figure 28.23 Diagram of the life cycle of *Plasmodium falciparum*, the agent of malaria. This life cycle requires two alternate hosts, humans and *Anopheles* mosquitoes.

6 In mosquitoes, gametocytes produce gametes which fuse to form a diploid zygote. In the gut, zygotes divide by meiosis to produce haploid sporozoites, which move to salivary glands of the mosquito.

5 Some merozoites produce sexual structures called gametocytes that can be taken up by a biting mosquito.

4 Merozoites continue to infect more red blood cells, causing cycles of chills and fever in the infected person.

1 *Plasmodium* sporozoites enter human blood by a mosquito bite.

2 Sporozoites enter liver cells, where the merozoite stages of *Plasmodium* form.

3 Merozoites enter red blood cells and reproduce, causing red blood cells to burst.

Some merozoites produce sexual structures, gametocytes, which are transmitted to a female mosquito as she bites.

Within the mosquito's body, the gametocytes produce gametes and fertilization occurs, yielding a zygote, the only diploid cell in *Plasmodium*'s life cycle. Within the mosquito gut, the zygote undergoes meiosis, generating structures filled with many sporozoites, the stage that can be transmitted to a new human host. Sporozoites move to the mosquito's salivary glands, where they remain until injected into a human host when the mosquito feeds. In recent years, genomic information bearing on life stages is helping medical scientists to develop new ways to prevent or treat malaria and other diseases caused by protists.

GENOMES & PROTEOMES

Genomic Information Aids in Combating Protistan Parasites

Genome sequences have been determined for several protists that cause human disease. This has been done as a way to find new cellular targets for drug treatments that will kill the parasites without harming the human host. The goal is to identify metabolic processes that are present in the parasite, but not the host. With such information, medical scientists can look for drugs that affect only the distinctive parasite process.

Genomic data are now available for the malarial parasite *Plasmodium falciparum* and three pathogenic kinetoplastids: *Trypanosoma brucei* (the agent of sleeping sickness), *Trypanosoma cruzi* (Chagas disease), and *Leishmania major* (leishmaniasis). Sleeping sickness is transmitted to humans and livestock by tsetse flies in equatorial Africa as well as in parts of South America and Asia. Nearly half a million people become infected with sleeping sickness each year, and the disease is fatal if untreated. Chagas disease is primarily a hazard in South America and can be transmitted via blood and organ donations. Leishmaniasis is endemic to 88 countries in South and Central America, the Mediterranean region, and Asia; about 2 million cases occur each year. Military personnel serving in the Middle East are among those at risk of contracting leishmaniasis from sand flea bites.

In the case of *P. falciparum*, genomic data have already highlighted potential new pharmaceutical approaches. The organism's nuclear genome consists of 14 chromosomes and about 5,300 genes. A large proportion of *P. falciparum*'s genes are related to evasion of the host's immune system and interactions between host and parasite. Two-thirds of the predicted 5,268 proteins appear to be unique to this organism, an unusual situation among eukaryotes. About 550 (some 10%) of the nuclear-encoded proteins are likely imported into a modified plastid known as an apicoplast, where they are needed for fatty acid metabolism and other processes. Because plastids are not present in the cells of mammals, enzymes in apicoplast pathways are possible targets for development of drug therapy. Mammals also lack calcium-dependent protein kinases (CDPKs), enzymes

that are essential to *P. falciparum*'s sexual development, offering another potential drug target. By contrast, genomic data reveal that kinetoplastids lack plastids, so drug strategies focused on this organelle will not affect these parasites. However, comparative study of their genomes reveals that *T. brucei*, *T. cruzi*, and *L. major* share some distinctive DNA transcription features that might prove useful in developing drug treatments.

CHAPTER SUMMARY

28.1 An Introduction to Protists

- Protists are eukaryotes that are abundant in moist habitats, and most are microscopic in size. Modern phylogenetic analysis has revealed that protists do not form a monophyletic group; instead, they are classified into several eukaryotic supergroups. (Figure 28.1, Table 28.1)

- Protists are often classified according to ecological roles into three major groups. The term algae is used informally for protist phyla that include photosynthetic protists. The term protozoa describes diverse heterotrophic protists that are often mobile. The term fungus-like protists describes protists having a structure and nutrition that resemble those of true fungi. (Figures 28.2, 28.3)

- Protists are particularly diverse in aquatic environments. The floating or swimming plankton are relatively small and simple in structure; the phytoplankton are adapted to life in well-illuminated near-surface waters; and the periphyton are attached to underwater surfaces. The multicellular protists known as seaweeds often display complex structure. (Figure 28.4)

- Microscopic protists propel themselves in several ways. Diverse small protists that swim by means of eukaryotic flagella are known as flagellates. The ciliates are a phylum of protists that move by means of many short cilia on their surfaces. Amoebae are diverse protists that move by means of pseudopodia. Other protists are able to glide across surfaces by secreting slime. (Figure 28.5)

28.2 Evolution and Relationships

- New discoveries are constantly changing concepts of protist evolution and relationships, and the relationships of some protists are unknown or controversial.

- The supergroup Excavata includes flagellate protists characterized by a feeding groove (such as the genus *Jakoba*), as well as the specialized parasites *Trichomonas vaginalis* and *Giardia lamblia*, which have modified mitochondria. (Figures 28.6, 28.7)

- The supergroup Euglenozoa consists of flagellates having disk-shaped mitochondrial cristae. Phyla include the kinetoplastids, which include some human parasites, and euglenoids, some of which are photosynthetic. (Figures 28.8, 28.9)

- The supergroup Archaeplastida includes three algal phyla (green algae, red algae, and glaucophytes), as well as the plant kingdom, all of which display primary plastids with two envelope membranes. (Figure 28.10)

- The supergroup Alveolata includes the ciliates, dinoflagellates, and apicomplexans, all known for the presence of saclike membranous vesicles called alveoli. (Figure 28.11)

- The supergroup Stramenopila includes a diverse group of protists that produce flagella with strawlike hairs that aid swimming. Stramenopiles include the fungus-like plant parasite *Phytophthora*, as well as diatoms and diverse algal classes whose secondary

plastids originated from a red alga by secondary endosymbiosis. Such plastids also occur in species of three other algal phyla: Haptophyta, Cryptophyta, and Dinozoa. (Figure 28.12)

- Tertiary plastids arose in protists on more than one occasion, and many heterotrophic protists and aquatic animals contain endosymbiotic algal cells. (Figure 28.13)

- The supergroup Rhizaria consists of flagellates and amoebae having thin cellular processes known as filose pseudopodia. Three prominent phyla are Chlorarachniophyta, which have secondary green plastids; mineral-shelled Radiolaria; and Foraminifera, which have shells composed of calcium carbonate. (Figure 28.14)

- The supergroup Amoebozoa is composed of many types of amoebae that move by means of pseudopodia, and includes the parasite *Entamoeba histolytica* and slime molds such as *Dictyostelium discoideum*.

- The supergroup Opisthokonta includes organisms that produce swimming cells that bear a single posterior flagellum. It includes the fungal and animal kingdoms. The protists known as choanoflagellates are considered the modern protists most closely related to the common ancestor of animals. (Figure 28.15)

28.3 Nutritional and Defensive Adaptations

- Protists display four basic types of nutrition. Heterotroph protists called phagotrophs rely on particle feeding, whereas heterotroph protists called osmotrophs absorb small organic molecules. Autotroph protists make their own organic food, and protists that use both autotrophic and heterotrophic means to obtain nutrients are known as mixotrophs. (Figure 28.16)

- Protists that break down wastes and dead organisms are called decomposers, whereas those that feed on living cells are termed parasites.

- Protists defend themselves from attackers by means of defensive adaptations that include protective cell coverings, sharp projectiles, light flashes, and toxic compounds. Dinoflagellates are particularly important as toxin producers. Aquatic ecologists have established that populations of the unicellular dinoflagellate genus *Pfiesteria* kill fish by means of a toxin that can also harm human cells. (Figure 28.17)

28.4 Reproductive Adaptations

- Protist populations grow by means of asexual reproduction involving mitosis, and many are able to persist through unfavorable conditions in the form of tough-walled cysts. (Figure 28.18)

- Protists having zygotic life cycles often use tough-walled zygotes to persist through unfavorable conditions. (Figure 28.19)

- Protists displaying sporic life cycles are able to occupy more types of habitats because they produce two or more alternating life stages having differing environmental preferences. (Figure 28.20)

- Diatoms, which have a gametic life cycle, use sexual reproduction to solve a cell size problem that originates from their unique mode of asexual cell division. (Figure 28.21)
- Ciliates display a unique type of sexual reproduction (conjugation) that involves the exchange of genetic material between a mating pair of cells. (Figure 28.22)
- Parasitic protists may have life cycles involving alternate hosts. One example is *Plasmodium*, the malarial agent, whose alternate hosts are humans and mosquitoes. Genomic sequence data have been obtained for several parasitic protists, with the goal of developing new treatments that will kill the parasites without harming hosts. (Figure 28.23)

TEST YOURSELF

1. Which protist phylum is most closely related to the animal kingdom?
   a. Rhodophyta
   d. Radiolaria
   b. Euglenida
   e. None of the above
   c. Choanomonada
2. Which protist phylum is most closely related to the plant kingdom?
   a. Chlorophyta
   d. Bacillariophyta
   b. Chlorarachniophyta
   e. Radiolaria
   c. Choanomonada
3. Which informal ecological group of protists includes autotrophs?
   a. protozoa
   d. all of the above
   b. algae
   e. none of the above
   c. fungus-like protists
4. How would you recognize a secondary plastid?
   a. It would have one envelope membrane.
   b. It would have two envelope membranes.
   c. It would have more than two envelope membranes.
   d. It would have pigments characteristic of primary plastids of red or green algal cells.
   e. Both c and d are correct.
5. What organisms have tertiary plastids?
   a. certain stramenopiles
   d. certain opisthokonts
   b. certain euglenoids
   e. none of the above
   c. certain cryptomonads
6. What is unusual about mixotrophs?
   a. They have no plastids, but they occur mixed in communities with autotrophs.
   b. They have mixed heterotrophic and autotrophic nutrition.
   c. Their cells contain a mixture of red and green plastids.
   d. Their cells contain a mixture of mitochondria having differently shaped cristae.
   e. They consume a mixed diet of algae.
7. What advantages do diatoms obtain from sexual reproduction?
   a. increased genetic variability
   b. increased ability of populations to respond to environmental change
   c. evolutionary potential
   d. regeneration of maximal cell size for the species
   e. all of the above
8. What are trichocysts?
   a. hairs that occur on flagella
   b. membrane sacs that occur beneath the cell surface
   c. tough-walled asexual cells that are able to withstand unfavorable conditions
   d. spearlike defensive structures that are shot from cells under attack
   e. none of the above

9. How do autotrophic protists use accessory pigments?
   a. Accessory pigments provide camouflage, so herbivores cannot see algae.
   b. Accessory pigments are able to absorb underwater light and transfer the energy to chlorophyll *a* for use in photosynthesis.
   c. Accessory pigments attract aquatic animals that carry gametes from one seaweed to another.
   d. All of the above.
   e. None of the above.
10. What are the two alternate hosts of the malarial parasite *Plasmodium falciparum*?
    a. humans and ticks
    b. ticks and mosquitoes
    c. humans and *Anopheles* mosquitoes
    d. humans and all types of mosquitoes
    e. none of the above

CONCEPTUAL QUESTIONS

1. Explain why protists are classified into multiple supergroups, rather than a single kingdom or phylum.
2. Why have molecular biologists sequenced the genomes of several parasitic protists?
3. Why are protistan cysts important to epidemiologists, biologists who study the spread of disease?

EXPERIMENTAL QUESTIONS

1. Why did the Burkholder team test two different strains of *Pfiesteria shumwayae*?
2. Why did the Burkholder team grow *Pfiesteria shumwayae* with algae or fish as food?
3. Why did the Burkholder team use a biohazard containment system?

COLLABORATIVE QUESTIONS

1. Imagine that you are studying an insect species, and you discover that the insects are dying of a disease that results in the production of cysts of the type that protists often generate. Thinking that the cysts might have been produced by a parasitic protist that could be used as an insect control agent, how would you go about identifying the disease agent?
2. Imagine that you are part of a marine biology team seeking to catalogue the organisms inhabiting a threatened coral reef. The team has found two new seaweeds, each of which occurs during a particular time of the year when the water temperature differs. You suspect that the two seaweeds might be different generations of the same species that have differing optimal temperature conditions. How would you go about testing your hypothesis?

www.brookerbiology.com
This website includes answers to the Biological Inquiry questions found in the figure legends and all end-of-chapter questions.

29

THE KINGDOM FUNGI

CHAPTER OUTLINE

29.1 Distinctive Features of Fungi

29.2 Fungal Sexual and Asexual Reproduction

29.3 Fungal Ecology and Biotechnology

29.4 Evolution and Diversity of Fungi

The orange peel fungus, *Aleuria aurantia*.

You might think that the largest organism in the world is a whale or perhaps a giant redwood tree. Amazingly, giant fungi would also be good candidates. For example, an individual of the fungus *Armillaria ostoyae* weighs hundreds of tons, is more than 2,000 years old, and spreads over 2,200 acres of Oregon forest soil! Scientists discovered the extent of this enormous fungus when they found identical DNA sequences in samples taken over this wide area. Other examples of such huge fungi have been found, and mycologists—scientists who study fungi—suspect that they may be fairly common, underfoot yet largely unseen.

Regardless of their size, fungi typically occur within soil or other materials, becoming conspicuous only when reproductive portions such as mushrooms extend above the surface. Even though fungi can be inconspicuous, they play essential roles in the Earth's environment, are associated in diverse ways with other organisms, and have many technological applications. In this chapter, we will explore the distinctive features of fungal structure, growth, nutrition, reproduction, and diversity. In the process you will learn how fungi are connected to forest growth, food production and food toxins, sick building syndrome, and other topics of great importance to humans.

29.1 Distinctive Features of Fungi

The eukaryotes known as fungi are so distinct from other organisms that they are placed in their own kingdom, the kingdom Fungi (**Figure 29.1**). The monophyletic kingdom Fungi includes early-diverging phyla such as chytrids, zygomycetes, and AM fungi, and higher fungi ascomycetes and basidiomycetes. Because fungi are closely related to the animal kingdom, fungi and animals display some common features. For example, both are **heterotrophic**,

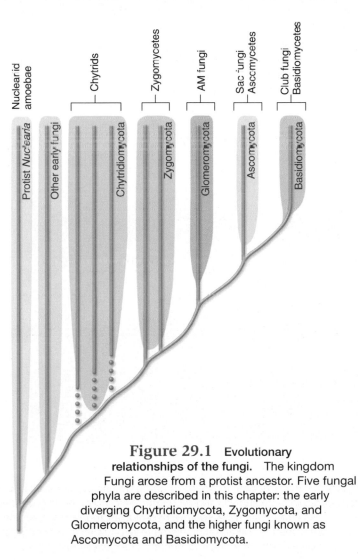

Figure 29.1 Evolutionary relationships of the fungi. The kingdom Fungi arose from a protist ancestor. Five fungal phyla are described in this chapter: the early diverging Chytridiomycota, Zygomycota, and Glomeromycota, and the higher fungi known as Ascomycota and Basidiomycota.

meaning that they cannot produce their own food but must obtain it from the environment. Fungi use an amazing array of organic compounds as food, which is termed their **substrate**. The substrate could be the soil, a rotting log, a piece of bread, a living tissue, or a wide array of other materials. Fungi are also like animals in having **absorptive nutrition**. Both fungi and animals secrete enzymes that digest organic materials and absorb the resulting small organic food molecules, as well as minerals and water, into their cells. In addition, both fungi and animals store surplus food as the carbohydrate glycogen in their cells. Despite these nutritional commonalities fungal structure is quite distinctive.

Fungi Have a Unique Cell-Wall Chemistry and Body Form

Unlike animal cells, which lack rigid cell walls, fungal cells are enclosed by tough cell walls composed of **chitin**, a polysaccharide that contains nitrogen. Chitin, which also forms the exoskeletons of arthropods such as insects, resists bacterial attack, thereby helping to protect fungal cells. Fungal cell walls influence two other features that distinguish fungal cells from animal cells: nutrition and motility. Because they have cell walls, fungal cells cannot engulf food particles by phagotrophy. In contrast, some animal cells are capable of phagotrophy, a feature inherited from ancestral protists (see Chapter 28). Additionally, the rigid walls of fungal cells, like those of plant cells, restrict the mobility of nonflagellate cells.

Most fungi have distinctive bodies known as **mycelia** (singular, *mycelium*) that are composed of microscopic, branched filaments known as **hyphae** (singular, hypha) (**Figure 29.2**). As discussed later, hyphae may be septate or aseptate, and cells may contain one nucleus or multiple nuclei. Most of the fungal mycelium is diffuse, or spread out, within food substrates and thus inconspicuous. In contrast, **fruiting bodies** are the visible fungal reproductive structures, which are composed of more densely packed hyphae that typically grow out of the substrate (**Figure 29.3**). Fruiting bodies are thus the only parts of fungal bodies that most people see. Mushrooms are one type of fungal fruiting body. Fungal fruiting bodies are specialized to produce the walled reproductive cells known as **spores**. Fungal fruiting bodies are amazingly diverse in form, color, and odor because they are adapted in ways that foster spore dispersal by wind, water, or animals. When spores settle in places where conditions are favorable for growth, they produce new mycelia. When the new mycelia undergo sexual reproduction, they produce new fruiting bodies.

If you remove the bark from rotting logs or look under dead, wet leaves in a forest, you may find what look like white or colored strings. The strings are **rhizomorphs**, fungal mycelia that have the shape of roots (**Figure 29.4**). Rhizomorphs transport water to other parts of the same mycelium living in low-moisture habitats.

Fungi Have Distinctive Growth Processes

If you have ever watched food become increasingly moldy over the course of several days, you have observed fungal growth. When a food source is plentiful, fungal mycelia can grow rapidly, adding as much as a kilometer of new hyphae per day. The mycelia grow at their edges as the fungal hyphae extend their tips through the undigested substrate. The narrow dimen-

(a) Hyphae on hair

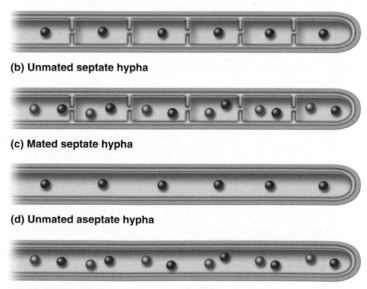

(b) Unmated septate hypha

(c) Mated septate hypha

(d) Unmated aseptate hypha

(e) Mated aseptate hypha

Figure 29.2   **Fungal hyphae.**   Fungal bodies are composed of delicate filaments known as hyphae. **(a)** Hyphae of *Trichophyton mentagrophites* are shown lying on human hair. This fungus produces enzymes that break down keratin, a protein that occurs abundantly in skin, nails, and hair. **(b)** A septate hypha prior to mating. Each cell contains only one nucleus. **(c)** A septate, dikaryotic hypha that occurs after mating. Dikaryotic cells have two nuclei of differing genetic types. **(d)** An aseptate hypha prior to mating lacks septa and contains only one genetic type of nuclei. **(e)** An aseptate hypha after mating contains two genetic types of nuclei.

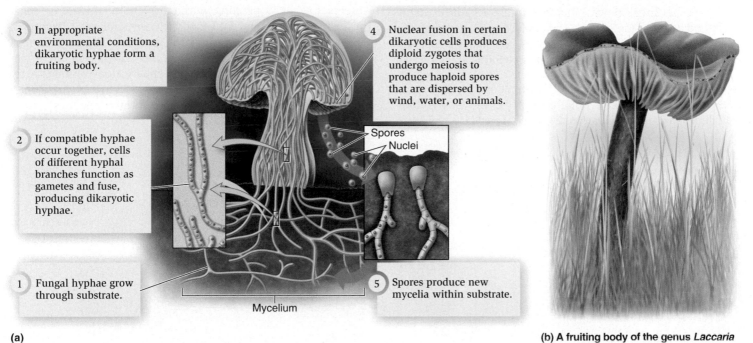

3 In appropriate environmental conditions, dikaryotic hyphae form a fruiting body.

4 Nuclear fusion in certain dikaryotic cells produces diploid zygotes that undergo meiosis to produce haploid spores that are dispersed by wind, water, or animals.

2 If compatible hyphae occur together, cells of different hyphal branches function as gametes and fuse, producing dikaryotic hyphae.

Spores
Nuclei

1 Fungal hyphae grow through substrate.

5 Spores produce new mycelia within substrate.

Mycelium

(a)

(b) A fruiting body of the genus *Laccaria*

Figure 29.3   **The reproductive fungal body.**   (a) Most of a fungus consists of hyphae that grow and branch from a central point to form a diffuse mycelium within a food substrate, such as soil. After mating, fungal hyphae may aggregate to form fleshy masses known as fruiting bodies that extend above the substrate surface. Fruiting bodies produce spores that are dispersed by wind, water, or animals. In suitable sites, spores may germinate, producing new mycelia. (b) The aboveground portion of the genus *Laccaria* and other mushrooms are examples of fruiting bodies.

Rhizomorphs

Underside of log

Figure 29.4   **Rhizomorphs.**   In moist forests, hyphae of fungi such as the honey fungus (*Armillaria mellea*) clump to form conspicuous string-shaped rhizomorphs.

sions and extensive branching of hyphae provide a very high surface area for absorption of organic molecules, water, and minerals.

Hyphal Tip Growth   How do hyphae grow? Cytoplasmic streaming and osmosis are important cellular processes in hyphal growth. Osmosis (see Chapter 5) is the diffusion of water through a membrane from an area with a low solute concentra-

tion into an area with a high solute concentration. Water enters fungal hyphae by means of osmosis because their cytoplasm is rich in sugars, ions, and other solutes. Water entry swells the hyphal tip, producing the force necessary for tip extension. Masses of tiny vesicles carrying enzymes and cell-wall materials made in the Golgi apparatus collect in the hyphal tip (**Figure 29.5**). The vesicles then fuse with the plasma membrane. Some vesicles deliver cell-wall materials to the hyphal tip, allowing it to extend. Other vesicles release enzymes that digest materials in the environment, releasing small organic molecules that are absorbed as food.

Hyphal Structure   The hyphae of most fungi are divided into many small cells by cross walls known as **septa** (see Figure 29.2b,c). Each of these cells has one or two nuclei. In such fungi, known as septate fungi, each round of nuclear division is followed by cross-wall formation. A central pore, which may be simple or more complex, perforates hyphal septa (**Figure 29.6**). Complex pores may have caps of modified endoplasmic reticulum. Septal pores are large enough to allow cytoplasmic structures and materials to pass through the hyphae. The hyphae of some fungi are not partitioned into smaller cells; rather, these hyphae are **aseptate** and multinucleate (see Figure 29.2d,e), a condition that results when nuclei repeatedly divide without intervening cytokinesis.

Fungal Nuclear Division   When the nuclei of fungi divide, in most cases a spindle forms within the nuclear envelope, which does not break down. This **intranuclear spindle** distinguishes nuclear division of fungi from that of animals and plants. (By contrast, in land plants and animals, the nuclear envelope vesiculates during prometaphase and then re-forms at telophase.)

Variations in Mycelium Growth Form Fungal hyphae grow rapidly through a substrate from areas where the food has become depleted to food-rich areas. In nature, mycelia may take an irregular shape, depending on the availability of food. A fungal mycelium may extend into food-rich areas for great distances, even spreading over thousands of acres like the giant fungus *A. ostoyae*, noted at the beginning of the chapter. In liquid laboratory media, fungi will grow as a spherical mycelium that resembles a cotton ball floating in water (**Figure 29.7a**). When grown in flat laboratory dishes, the mycelium assumes a more two-dimensional growth form (**Figure 29.7b**).

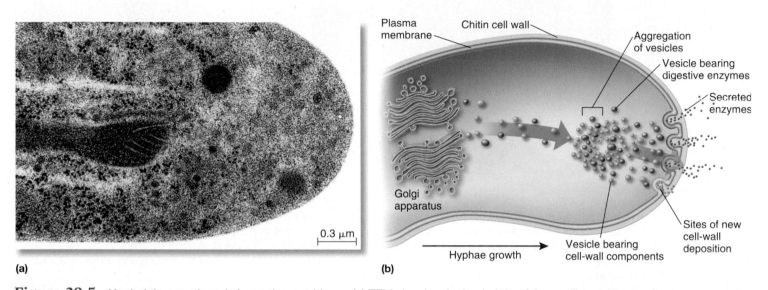

(a)

(b)

Figure 29.5 **Hyphal tip growth and absorptive nutrition.** (a) TEM showing the hyphal tip of *Aspergillus nidulans*, a fungus commonly used as a genetic model system. (b) Diagram of a hyphal tip, showing vesicles of two types: those containing cell-wall materials, and those bearing digestive enzymes. Upon fusion of these vesicles with the plasma membrane, both types of vesicle contents are released to the cell exterior. Digestive enzymes degrade extracellular organic polymers into smaller components that can be taken into hyphae by means of plasma membrane transporter proteins. The uptake of minerals and organic molecules causes osmotic water uptake, with the result that cells enlarge and hyphal tips extend. New cell-wall materials are added to the hyphal tip as it extends.

Biological inquiry: What do you think would happen to fungal hyphae that begin to grow into a substrate having higher solute concentration? How might your answer be related to food preservation techniques such as drying or salting?

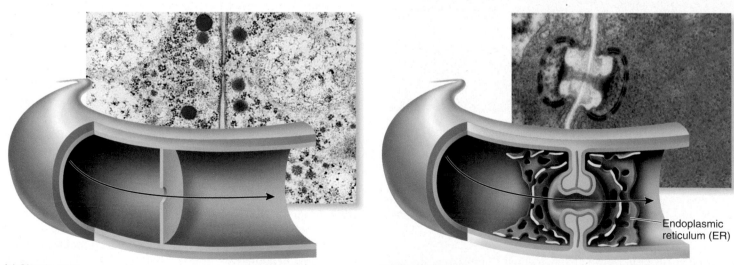

(a) Simple pore

(b) Complex pore

Figure 29.6 **Septa divide hyphae of higher fungi into compartments.** (a) The septa of ascomycetes have simple pores at the centers. (b) More complex pores distinguish the septa of most types of basidiomycetes.

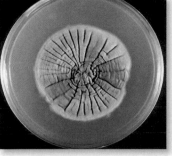

(a)   **(b)**

Figure 29.7 **Fungal shape shifting.** **(a)** When a mycelium, such as that of *Rhizoctonia solani*, is surrounded by food substrate in a liquid medium, it will grow into a spherical form. **(b)** When the food supply is limited to a two-dimensional supply, as shown by *Neotestudina rosatii* in the laboratory dish shown here, the mycelium will form a disk. Likewise, distribution of the food substrate determines the mycelium shape in nature.

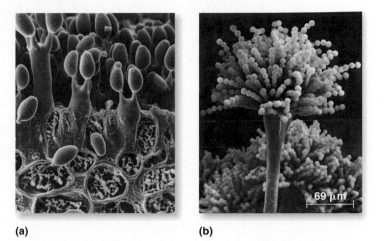

(a)   **(b)**

Figure 29.8 **Sexual and asexual reproductive cells of fungi.** **(a)** SEM showing spores produced after a sexual mating, zygote formation, and subsequent meiosis in the forest mushroom *Coprinus disseminatus*. When mature, these genetically variable spores detach and are dispersed into the environment, where they may find suitable conditions for growth. **(b)** SEM of the asexual conidia of *Aspergillus versicolor*, which causes skin infections in burn victims and lung infections in AIDS patients. Each of these small cells is able to detach and grow into an individual that is genetically identical to the parent fungus and able to grow in similar conditions.

29.2 Fungal Sexual and Asexual Reproduction

Many fungi are able to reproduce both sexually and asexually by means of microscopic spores, each of which can grow into a new adult (**Figure 29.8**). Sexual reproduction generates new allele combinations that may allow fungi to colonize new types of habitats. In contrast, asexual reproduction is a natural cloning process; it produces genetically identical organisms. Production of asexual spores allows fungi that are well adapted to a particular environment to disperse to similar, favorable places.

Fungi Have Distinctive Sexual Reproductive Processes

As is typical for eukaryotes, the fungal sexual reproductive cycle involves the union of gametes, the formation of zygotes, and the process of meiosis. However, many aspects of fungal sexual reproduction are unique, including the function of hyphal branches as gametes and the development of fruiting bodies (see Figure 29.3).

Fungal Gametes and Mating The gametes of most fungi are inconspicuous hyphal branches. During sexual reproduction, these hyphal branches fuse with those of a different mycelium of compatible mating type. Particular genes control mating compatibility. In contrast, some primitive fungi that live in the water produce flagellate sperm and nonmotile eggs.

The actual mating process in fungi is also remarkable. In most sexual organisms, gametes undergo fusion of their cytoplasms—a process known as **plasmogamy** and then the nuclei fuse—a process known as **karyogamy**. In contrast, the gamete nuclei of many fungi behave differently. After plasmogamy

occurs, the haploid gamete nuclei may remain separate for a long time. During this time period, the gamete nuclei both divide at each cell division, producing a **dikaryotic** (meaning "two nuclei") mycelium. Each cell of a dikaryotic mycelium possesses two unfused gamete nuclei (see Figure 29.2c). Dikaryotic mycelia are also known as **heterokaryons** (meaning "different nuclei"), reflecting the fact that the two nuclei of each cell are genetically distinct. Some fungi normally persist as heterokaryons, producing clones that can live for hundreds of years. Although the nuclei of dikaryotic mycelia remain haploid, alternate copies of many alleles occur in the separate nuclei. Thus, dikaryotic mycelia are functionally diploid. Eventually, dikaryotic mycelia produce fruiting bodies, the next stage of reproduction.

Fruiting Bodies Under appropriate environmental conditions such as seasonal change, a heterokaryotic mycelium may produce a fleshy fruiting body that emerges from the substrate (see Figure 29.3). All the cells of the fruiting body are dikaryotic. When the fruiting body is mature, the two nuclei in cells at the surface undergo nuclear fusion. This process produces many zygotes, which are the only cells in the fungal life cycle that have a diploid nucleus. In most cases, the fungal zygotes soon undergo meiosis to produce haploid spores. Each spore acquires a tough wall that protects it from drying and other stresses. Wind, rain, or animals disperse the mature spores, which grow into haploid mycelia. If a haploid mycelium encounters hyphae of an appropriate mating type, hyphal branches will fuse and start the sexual cycle over again.

(a)   (b)

Figure 29.9 Fruiting body adaptations that foster spore dispersal. **(a)** When touched by wind gusts or animal movements, spores puff from fruiting bodies of the puffball fungus (*Lycoperdon perlatum*). **(b)** Insects visit fruiting bodies of stinkhorn fungi such as *Phallus impudicus*, attracted by foul odors and colors like those of dung or rotting meat. Spores, produced in a sticky matrix, attach to the insects and are thereby carried away.

Figure 29.10 Toxic fruiting body of *Amanita muscaria*. Common in conifer forests, *A. muscaria* is both toxic and hallucinogenic. Ancient people used this fungus to induce spiritual visions and to reduce fear during raids. This fungus produces a toxin, amanitin, which specifically inhibits RNA polymerase II of eukaryotes (but not prokaryotes).

Biological inquiry: What effect would the amanitin toxin have on human cells?

By now, it should be clear why fruiting bodies usually emerge from the substrate, while most of the fungal mycelium lies inconspicuously within it. Mycelium growth requires organic molecules, minerals, and water provided by the substrate, but spores are more easily dispersed if released outside of the substrate. The structure of fruiting bodies varies in ways that reflect different adaptations that foster spore dispersal by wind, rain, or animals. For example, mature puffballs have delicate surfaces upon which just a slight pressure causes the spores to puff out into wind currents (**Figure 29.9a**). Birds' nest fungi form characteristic egg-shaped spore clusters. Raindrops splash on these clusters and disperse the spores. The fruiting bodies of stinkhorn fungi smell and look like rotting meat, which attracts carrion flies (**Figure 29.9b**). The flies land on the fungi to investigate the potential meal and then fly away, in the process dispersing spores that stick to their bodies. The fruiting bodies of fungal truffles are also specialized for spore dispersal by animals. However, they are unusual in being produced underground. Mature truffles emit odors that attract wild pigs and dogs, which break up the fruiting structures while digging for them, thereby dispersing the spores. When collectors seek to harvest valuable truffles from forests for the market, they use trained leashed pigs or dogs to locate these fungi.

Though fungi such as truffles are edible, the fruiting bodies of many fungi produce toxic substances which may deter animals from consuming them (**Figure 29.10**). For example, several fungi that attack stored grains, fruits, and spices produce **aflatoxins**; these cause liver cancer and are a major health concern worldwide. When people consume the forest mushroom *Amanita virosa*, known as the "destroying angel," they are also ingesting a powerful toxin that may cause liver failure so severe that a transplant may be required. Each year, thousands of people in North America alone are poisoned when they consume similarly toxic mushrooms gathered in the wild. There is no

reliable way for nonexperts to distinguish poisonous from nontoxic fungi; it is essential to receive instruction from an expert before foraging for mushrooms in the woods. Because several species of edible fungi are cultivated for market sale, many experts recommend that it is better to forage for mushrooms in the grocery store than in the wild (**Figure 29.11**).

Several types of fungal fruiting structures produce hallucinogenic or psychoactive substances. As in the case of fungal toxins, fungal hallucinogens may have evolved as antiherbivore adaptations. Humans have inadvertently experienced their effects. For example, *Claviceps purpurea*, which causes a disease of rye crops and other grasses known as ergot, produces a psychogenic compound related to LSD (lysergic acid diethylamide) (**Figure 29.12**). Some experts speculate that cases of hysteria, convulsions, infertility, and a burning sensation of the skin that occurred in Europe during the Middle Ages resulted from ergot-contaminated rye used in foods. Ergot poisoning may also have caused the behavioral changes that were at the heart of the Salem witch trials in the late 17th century. Another example of a hallucinogenic fungus is the "magic mushroom" (*Psilocybe*), which is used in traditional rituals in some cultures. Like ergot, the magic mushroom produces a compound similar to LSD, a controlled substance. Consuming hallucinogenic fungi is risky because the amount used to achieve psychoactive affects is dangerously close to a poisonous dose.

Asexual Reproduction Does Not Involve Mating and Meiosis

Asexual reproduction is particularly important to fungi, allowing them to spread rapidly through favorable environments by means of asexual spores. To reproduce asexually, fungi do not need to find compatible mates or expend resources on fruiting-body formation and meiosis. Though many fungi reproduce both sexually and asexually, more than 17,000 species reproduce primarily or exclusively by asexual means. DNA-sequencing studies have revealed that many types of modern fungi that reproduce only asexually have evolved from ancestors that had both sexual and asexual reproduction.

Figure 29.11 Several types of edible, delicious fungi are grown for the market.

Figure 29.13
An indoor mold fungus. The black mold *Stachybotrys chartarum* is one of several types of conidial fungi that grow on indoor paper and wood that have been wet for an extended period.

58 μm

Ergot

Figure 29.12 **Ergot of rye.** The fungus *Claviceps purpurea* infects rye and other grasses, producing hard masses of mycelia known as ergots in place of some of the grains (fruits). Ergots produce alkaloids related to LSD and thus cause psychotic delusions in humans and animals that consume products made with infected rye. Ergots were used in folk medicine to treat migraine or hasten childbirth.

Most fungi reproduce asexually by generating chains of tiny spores from the tips of hyphae. Many fungi produce asexual spores known as **conidia** (from the Greek word for "dust") (see Figure 29.8b). When they land on a favorable substrate, conidia germinate into a new mycelium that produces many more conidia. The green molds that form on citrus fruits are familiar examples of conidial fungi. A single fungus can produce as many as 40 million conidia per hour over a period of 2 days.

Because they can spread so rapidly, asexual fungi are responsible for costly fungal food spoilage and allergies. *Aspergillus fumigatus* is a common conidial mold that causes a potentially fatal lung disease. Such harmful conidia-producing molds growing in poorly ventilated, moist places may be the cause of "sick building syndrome," a term used to describe situations in which occupants of a building experience acute health effects that appear to be linked to time spent inside, but where no specific cause can be identified. However, some black molds that also grow on moist building materials are blamed for human illness, even though there is little evidence that they are harmful to adults (**Figure 29.13**). Medically important fungi that reproduce primarily by asexual means include the athlete's foot fungus (*Epidermophyton floccosum*) and the infectious yeast (*Candida albicans*).

29.3 Fungal Ecology and Biotechnology

The ability of fungi to degrade diverse materials is important in ecology, medicine, and human biotechnology applications. Fungal decomposers, also known as saprobes, are able to decompose nonliving organic materials. Recycling of materials in ecosystems depends on the activities of fungal decomposers. Fungal predators consume whole, though tiny, living organisms. Fungi that attack living animal or plant tissues are pathogens. As hyphae of pathogenic fungi grow through the tissues of plants and animals, food and minerals are absorbed from the host, causing disease symptoms. Mutualistic fungi have formed partnerships with photosynthetic organisms that provide the fungi with organic food. In return, fungi contribute mineral nutrients and water to their photosynthetic partners. In this section, we focus more closely on the ecological roles of decomposer, predatory, and mutualistic fungi.

Figure 29.14 **A predatory fungus.** The fungus *Arthrobotrys anchonia* traps nematode worms in hyphal loops that suddenly swell in response to the animal's presence. Fungal hyphae then grow into the worm's body and digest it.

Decomposer and Predatory Fungi Play Important Ecological Roles

Decomposer fungi are essential components of the Earth's ecosystems. Together with bacteria, they decompose dead organisms and organic wastes, preventing litter buildup. For example, only certain bacteria and fungi can break down cellulose, and a few fungi are the major decomposers of lignin, the decay-resistant component of wood. In the absence of decomposers, ecosystems would become clogged with organic debris. Decomposers are needed to break down organic compounds into carbon dioxide, which is used by algae and plants for photosynthesis. Decomposer fungi and bacteria are also Earth's recycling engineers. They release minerals to the soil and water, where plants and algae take up the minerals for growth. In the absence of fungi and bacteria, such minerals would remain forever bound up in dead organisms, unavailable for the growth of forests, coral reefs, and other biotic communities.

Some soil fungi are predators—they trap tiny soil animals, such as nematodes, in nooselike hyphae and then proceed to digest their bodies (**Figure 29.14**). Such fungi help control populations of nematodes, some of which attack plant roots. Some fungi attack and kill insects, and certain of these species are used as biological control agents to kill black field crickets, red-legged earth mites, and other pests.

Some Fungal Species Cause Plant and Human Diseases

Five thousand fungal species cause serious crop diseases and recent results show that new diseases can arise by horizontal gene transfer. Wheat rust is an example of a common crop disease caused by fungi (**Figure 29.15**). Rusts are named for reddish spores that emerge from the surfaces of infected plants. Many types of plants can be attacked by rust fungi, but rusts are of particular concern when new strains attack crops. For example, in late 2004, agricultural scientists discovered that a devastating rust, *Phakopsora pachyrhizi*, had begun to spread in the U.S. soybean crop. This rust kills soybean plants by attacking the leaves, causing complete leaf drop in less than 2 weeks. The disease had apparently spread to U.S. farms by means of

Figure 29.15 **Wheat rust.** (a) The plant pathogenic fungus *Puccinia graminis* grows within the tissues of wheat plants, using plant nutrients to produce rusty streaks of red spores that erupt at the stem and leaf surface where spores can be dispersed. (b) Red spore production is but one stage of a complex life cycle involving several types of spores. Rusts infect many other crops, causing immense economic damage.

spores blown on hurricane winds from South America. To control the spread of fungal diseases, agricultural experts work to identify effective fungicidal chemicals and develop resistant crop varieties.

In addition to being spread by wind, the spores of crop-disease fungi can also be introduced on travelers' clothing and other belongings. To reduce the entry of new crop disease fungi—as well as crop viruses and insect pests—agricultural customs inspectors closely monitor the entry of plants, soil, foods, and other materials that might harbor these organisms.

Fungi cause several types of disease in humans. For example, athlete's foot and ringworm are common skin diseases caused by fungi that are known as dermatophytes because they colonize the human epidermis (see Figure 29.2a). *Pneumocystis carinii* is a fungal pathogen that infects individuals with weakened immune systems, such as AIDS patients, sometimes causing death by pneumonia. *Coccidioides immitis*, *Cryptococcus neoformans*, and *Histoplasma capsulatum* are examples of fungi that cause lung and other diseases. These diseases are of special concern when they occur in cancer patients whose immune systems are weakened.

Though fungal diseases that attack humans are of medical concern, in nature, fungal pathogens often help to control populations of other organisms, which is an important ecological role. Some fungi play important ecological roles as beneficial partners in symbiotic associations with other organisms.

Fungi Form Mutually Beneficial Ecological Associations

Fungi form several types of associations that appear to benefit both partners; these are known as mutualistic interactions (see Chapter 57). For example, leaf-cutting ants, certain termites and beetles, and the salt marsh snail (*Littoraria irrorata*) cultivate particular fungi for food, much as do human mushroom growers. Other fungi obtain organic food molecules from photosynthetic organisms: plants, green algae, or cyanobacteria. We focus next on three types of fungi—mycorrhizal fungi, endophytes, and lichen fungi—that are beneficially associated with other organisms.

Mycorrhizae Associations between the hyphae of certain fungi and the roots of most seed plants are known as **mycorrhizae** (literally, "fungus roots") (see Chapter 37). Mycorrhizal associations are very important in ecology and agriculture; more than 80% of terrestrial plants form mycorrhizae. Plants that have mycorrhizal partners receive an increased supply of water and mineral nutrients, primarily phosphate, copper, and zinc. They do so because an extensive fungal mycelium is able to absorb minerals from a much larger volume of soil than can plant roots lacking fungal associates. Added together, all the branches of a fungal mycelium in 1 m^3 of soil can reach 20,000 km in total length. Experiments have shown that mycorrhizae greatly enhance plant growth in comparison to plants lacking fungal partners. In return, plants provide fungi with organic food molecules, sometimes contributing as much as 20% of their photosynthetic products—a very good investment.

By binding soils, fungal hyphae also reduce water loss and erosion, and they help protect plants against pathogens and toxic wastes. Fungi thereby help plants adapt to and thrive in new sites, thus playing an important role in plant succession (see Chapter 58). For this reason, ecologists increasingly incorporate mycorrhizal fungi into plant-community restoration projects.

The two most common types of mycorrhizae are ectomycorrhizae and endomycorrhizae. **Ectomycorrhizae** (from the Greek *ecto*, outside) are beneficial interactions between temperate forest trees and soil fungi (**Figure 29.16a**) whose hyphae coat tree-root surfaces (**Figure 29.16b**) and grow into the spaces between root cells (**Figure 29.16c**). Some species of oak, beech, pine, and spruce trees will not grow unless their ectomycor-

rhizal partners are also present. Mycorrhizae are thus essential to the success of commercial nursery tree production and reforestation projects.

Endomycorrhizae (from the Greek *endo*, inside) are partnerships between plants and fungi in which the fungal hyphae penetrate the spaces between root cell walls and plasma membranes and grow along the surfaces of these membranes (**Figure 29.17**). In such spaces, endomycorrhizal fungi often form highly branched, bushy arbuscules (from the word "arbor," referring to tree shape). As the arbuscules develop, the root plasma membrane also expands. Consequently, the arbuscules and the root plasma membranes surrounding them have very high surface areas that facilitate rapid and efficient exchange of materials: Minerals flow from fungal hyphae to root cells, and organic food molecules move from root cells to hyphae. These fungus-root associations are known as **arbuscular mycorrhizae**, abbreviated AM. AM fungi are associated with apple trees, coffee shrubs, and many herbaceous plants, including legumes, grasses, tomatoes, strawberries, and peaches.

Fungal Endophytes Other mutualistic fungi, known as **endophytes**, live compatibly within the leaf and stem tissues of various types of plants. The endophytes obtain organic food molecules from plants, and in turn contribute toxins or antibiotics that deter foraging animals, insect pests, and microbial pathogens. Endophytic fungi also help some plants to tolerate higher temperatures. As a result, plants with endophytes often grow better than plants of the same species without endophytic fungi. For example, fungal endophytes grow throughout the leaves and stems of many cool-season grasses, such as tall fescue (*Festuca arundinacea*). When such grasses reproduce, fungal hyphae transmitted in seeds will grow in the progeny plants. Grasses with endophytic fungi are larger, more toxic to herbivores, and more drought resistant than those without. These advantages allow tall fescue to invade plant communities and displace other plant species, with deleterious effects on wildlife.

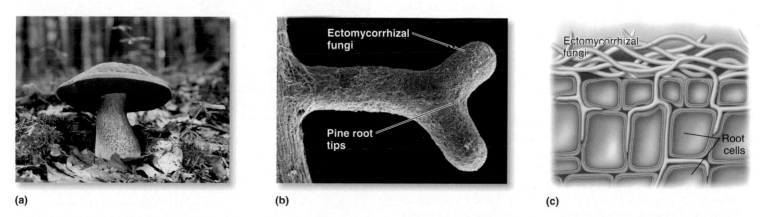

(a)   **(b)**   **(c)**

Figure 29.16 Ectomycorrhizae. (a) The fruiting body of the common forest fungus *Boletus*. This is an ectomycorrhizal fungus that is associated with tree roots. (b) Ectomycorrhizal fungal hyphae of *Laccaria bicolor* cover the surfaces of young *Pinus resinosa* root tips. (c) Diagram showing that the hyphae of ectomycorrhizal fungi do not penetrate root cell walls but grow within intercellular spaces. In this location, fungal hyphae are able to obtain organic food molecules produced by plant photosynthesis.

Biological inquiry: What benefits do plants obtain from this association?

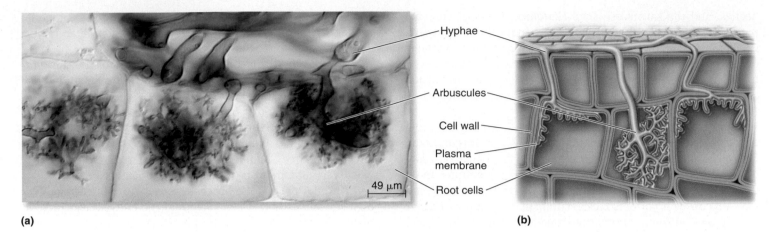

(a)

(b)

Figure 29.17 **Endomycorrhizae.** **(a)** Light micrograph showing black-stained AM fungi within the roots of the forest herb *Asarum canadensis*. Endomycorrhizal fungal hyphae enter plant roots via root hair cells, and then branches grow in the spaces between plant cell walls and plasma membranes. **(b)** Diagram showing the position of highly branched arbuscules. Hyphal branches of arbuscules are coated with plant plasma membrane, with the result that both hyphae and plant membrane have very high surface areas.

Fungal endophytes also occur in many tropical plants, producing compounds that help their hosts to repel attacks by herbivores and disease microbes. These compounds offer promise for the development of new antibiotics and anticancer drugs for use in human medicine.

Lichens Lichens are partnerships of particular fungi and certain photosynthetic green algae or cyanobacteria, and sometimes both. There are at least 25,000 lichen species, but these did not all descend from a common ancestor. DNA-sequencing studies suggest that lichens evolved independently in at least five separate fungal lineages. Molecular studies also show that some fungi have lost their ancestral ability to form lichen associations.

Lichen bodies take one of three major forms: (1) crustose—flat bodies that are tightly adherent to an underlying surface (**Figure 29.18a**); (2) foliose—flat, leaflike bodies (**Figure 29.18b**); or (3) fruticose—bodies that grow upright (**Figure 29.18c**) or hang down from tree branches. The photosynthetic green algae or cyanobacteria typically occur in a distinct layer close to the lichen's surface (**Figure 29.18d**). Lichen structure differs dramatically from that of the fungal components grown separately, demonstrating that the photosynthetic components influence lichen form.

The photosynthetic partner provides lichen fungi with organic food molecules and oxygen, and in turn it receives carbon dioxide, water, and minerals from the fungal partner. Lichen fungi also protect their photosynthetic partners from environmental stress. For example, lichens that occupy exposed habitats of high light intensity often produce bright yellow, orange, or red-colored compounds that help prevent damage to the photosynthetic apparatus (see Figure 29.18a). Lichen fungi also produce distinctive organic acids and other compounds that deter animal and microbial attacks.

(a)

(b)

(c)

(d)

Figure 29.18 **Lichen structure.** **(a)** An orange-colored crustose lichen grows tightly pressed to the substrate. **(b)** The flattened, leaf-shaped genus *Umbilicaria* is a common foliose lichen. **(c)** The highly branched genus *Cladonia* is a common fruticose lichen. **(d)** A handmade thin slice of *Umbilicaria* viewed with a compound microscope reveals that the photosynthetic algae occur in a thin upper layer. Fungal hyphae make up the rest of the lichen.

The fungal partners of many lichens can undergo sexual reproduction, producing fruiting bodies and sexual spores much like those of related fungi that do not form lichens. DNA studies have shown that some lichen fungi can self-fertilize, which is advantageous in harsh environments where potential mates may be lacking. To produce new lichens, hyphae that grow from sexual spores must acquire new photosynthetic partners, but only particular green algae or cyanobacteria are suitable. Even if compatible fungi and algae or cyanobacteria occur together, lichens may not form. In the laboratory, lichen formation has been accomplished only when the photosynthetic partner and fungus are exposed to stressful conditions. These results suggest that lichens form in nature under similarly harsh conditions.

The fungi associated with about one-third of lichen species can only reproduce asexually. Asexual reproductive structures include **soredia**, small clumps of hyphae surrounding a few algal cells that can disperse in wind currents (**Figure 29.19**). Soredia are lichen clones. By forming soredia, lichen fungi can disperse along with their appropriate photosynthetic partners.

Lichens often grow on rocks, buildings, tombstones, tree bark, soil, or other surfaces that easily become dry. When water is not available, the lichens are dormant until moisture returns. Thus lichens may spend much of their time in an inactive state, and for this reason they often grow very slowly. However, because they can persist for long periods, lichens can be very old; some are estimated to be more than 4,500 years old. Lichens occur in diverse types of habitats, and a number grow in some of the most extreme, forbidding terrestrial sites on Earth—deserts, mountaintops, and the Arctic and Antarctic—places where most plants cannot survive. In these locations, lichens serve as a food source for reindeer and other hardy organisms. Though unpalatable, lichens are not toxic to humans and have also served as survival foods for aboriginal peoples in times of shortages.

Lichens are recognized for their soil-building activities, which occur over very long timescales. Lichen acids help to break up the surfaces of rocks, beginning the process of soil

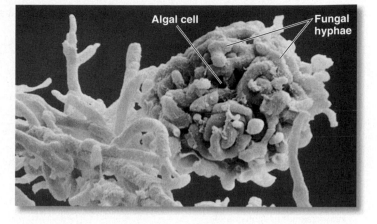

Figure 29.19 Lichen asexual reproduction. Soredia are examples of lichen asexual structures that contain algal or cyanobacterial cells wrapped by fungal hyphae. SEM of a soredium of the lichen *Cladonia coniocraea* that will break off from the parent lichen for dispersal in the environment.

development. Lichens having cyanobacterial partners can also increase soil fertility by adding fixed nitrogen. One study showed that such lichens released 20% of the nitrogen they fixed into the environment, where it is available for uptake by plants.

Lichens are useful as air-quality monitors because they are particularly sensitive to air pollutants such as sulfur dioxide. Air pollutants severely injure the photosynthetic components, causing death of the lichens. The disappearance of lichens serves as an early warning system of air-pollution levels that are also likely to affect humans. Lichens can also be used to monitor atmospheric radiation levels because they accumulate radioactive substances from the air. After the Chernobyl nuclear power plant accident, lichens in nearby countries became so radioactive that reindeer, which consume lichens as a major food source, became unfit for use as human food or in milk production.

FEATURE INVESTIGATION

Piercey-Normore and DePriest Discovered That Some Lichens Readily Change Partners

One of the most interesting current issues regarding symbiotic relationships is the extent to which partners have influenced each other's evolution and diversification. When symbiotic partners have influenced each other's evolution, they are said to have cospeciated (coevolved). In the past it was unclear whether or not lichen fungi and their photosynthetic partners coevolved. The fairly recent development of molecular techniques and computer software for analyzing relationships among microorganisms has

allowed researchers to attempt to answer this and many other fundamental questions about lichen biology.

Researchers Michelle Piercey-Normore and Paula DePriest devised a way to look for evidence of cospeciation of the fungal and algal partners in a large and diverse lichen group known as the Cladoniaceae (**Figure 29.20**). This group of lichens includes the common "reindeer lichen" (*Cladonia cristatella*). The investigators wanted to know if the pattern of speciation for the fungal partners matched that for the algal partner. The investigators first extracted DNA from the fungal and algal partners in 33 lichens, including diverse species of the genus *Cladonia*.

Figure 29.20   Piercey-Normore and DePriest discovered that some lichens readily change partners.

GOAL   To determine if lichen fungal and algal phylogenies are consistent, as evidence for or against cospeciation.

STARTING MATERIALS   Fungal and algal components of 33 lichen species in the family Cladoniaceae.

1   Isolate DNA from fungal and algal partners of 33 lichens.

2   Amplify a region of the rRNA gene using the technique of polymerase chain reaction (PCR) (described in Chapter 20). Use DNA-sequencing techniques (described in Chapter 20) to determine DNA sequences of the rRNA gene region from the algal and fungal partners of 33 lichens.

3   Use DNA-sequence information and computer programs to infer the phylogenies of the algae and the fungi from 33 lichens.

4   **THE DATA**

Connect lines between partners from the same lichen. Look for cases of cospeciation—when phylogenies of the algal and fungal partners match.

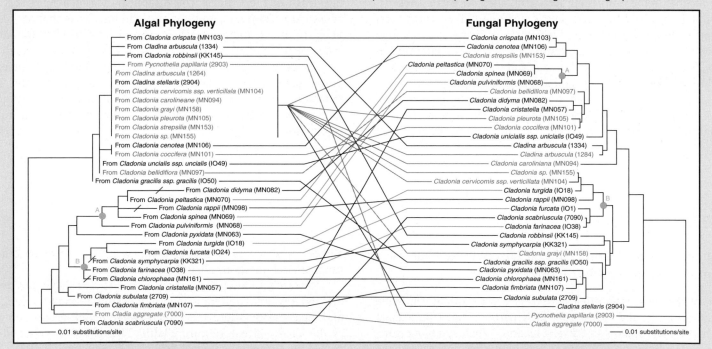

Results from step 4: The green dots marked A and B indicate lichen lineages showing some cospeciation. The partners of these lichens are connected by green lines. Other lineages show little evidence of cospeciation, shown by extensive crossed lines. These results imply a high degree of partner switching in this group of lichens. Partner switching even occurred in lineages A and B (marked with red slashes). These data show that in the lichen family Cladoniaceae, partner switching has been a more frequent occurrence than cospeciation.

Next, they amplified a region of the rRNA gene for all samples and determined the sequences of these DNAs. Finally, Piercey-Normore and DePriest used the DNA sequences to produce separate phylogenetic trees for the algal and fungal partners, and they used statistical procedures to test the possibility that the evolutionary divergence patterns of the fungi and algae were linked.

The results revealed that the fungal and algal phylogenies did not match, indicating that the patterns of their diversifica-tion were not congruent. These data indicated that lichen fungi and algae had not generally influenced each other's speciation patterns. If the lichen fungi and algae had undergone species diversification in tandem, their phylogenies should match, that is, be congruent. Instead, the phylogenetic evidence suggested that lichens often switch algal partners, "trading-up" for better algae. The results of this study, published in 2001, revealed that the lichen symbiosis is more complex than had previously been realized.

Fungi Have Many Applications in Biotechnology

The ability of fungi to grow on many types of substrates and produce many types of organic compounds represents their diverse ecological adaptations. Humans have harnessed fungal biochemistry in many types of biotechnology applications. Fungal biochemistry is a valuable asset to the chemical, food processing, and waste-treatment industries. A variety of industrial processes use fungi to convert inexpensive organic compounds into valuable materials such as citric acid used in the soft drink industry, glycerol, antibiotics such as penicillin, and cyclosporine, a drug widely used to prevent rejection of organ transplants. In the food industry, fungi are used to produce the distinctive flavors of blue cheese and other cheeses. Other fungi secrete enzymes that are used in the manufacture of protein-rich tempeh and other food products from soybeans. The baking industry depends on the yeast *Saccharomyces cerevisiae* for bread production, and the brewing and winemaking industries also find yeasts essential.

Fungi are increasingly being used in industrial processes to replace chemical procedures that generate harmful waste materials. For example, during the production of paper, wood pulp is chemically bleached to remove lignin, but this bleaching process generates harmful compounds, including dioxin, a carcinogenic compound. The forest fungus *Phanerochaete chrysosporium* produces several enzymes that allow it to break down lignin in wood (**Figure 29.21**). Removal of lignin allows the fungi to gain access to cell-wall polysaccharides, which they break down, thereby releasing sugars that the fungus absorbs as food. Fungal lignolytic ("lignin-breaking") enzymes such as those produced by *P. chrysosporium* can be used to bleach paper without producing dioxin, thereby reducing a harmful by-product of paper production. In addition, Khadar Valli, Hiroyuki Wariishi, and Michael Gold demonstrated that *P. chrysosporium* is able to decompose dioxin into nontoxic products, suggesting additional potential uses of the fungus in preventing pollution.

29.4 Evolution and Diversity of Fungi

The kingdom Fungi arose from a protist ancestor (see Figure 29.1). The modern fungi have been classified into five phyla, listed in **Table 29.1** by both their common and formal names:

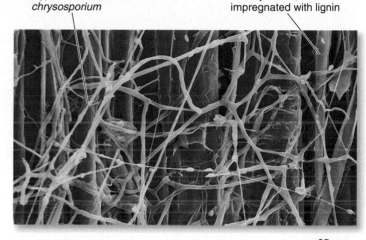

Hyphae of *Phanerochaete chrysosporium*

Woody tissue; cell walls impregnated with lignin

25 μm

Figure 29.21 **Ecological features of fungi can be harnessed in biotechnology.** The forest fungus *Phanerochaete chrysosporium* produces a cobweb-like mycelium on the surfaces of rotting wood. This fungus releases enzymes that break down lignin, thereby making plant cell-wall carbohydrates such as cellulose more accessible.

chytrids (Chytridiomycota), zygomycetes (Zygomycota), AM fungi (Glomeromycota), ascomycetes (Ascomycota), and basidiomycetes (Basidiomycota) (Table 29.1). (The suffixes *mycota* and *mycetes* derive from a Greek word meaning fungus.) DNA and other data strongly indicate that these phyla all descended from a single common ancestor. Several types of slime molds and fungus-like protists—though often studied with fungi—are classified with protists rather than true fungi (see Chapter 28). In this section, we will survey the characteristics of five phyla of true fungi and examine their distinctive reproductive cycles.

Chytrids Primarily Live in Aquatic Environments

Chytrids are the simplest fungi, and molecular evidence indicates that chytrids were among the earliest fungi to appear. Some chytrids occur as single, spherical cells that may produce hyphae (**Figure 29.22**), while others exist mainly as branched, aseptate hyphae. Chytrids are the only fungi that produce flagellate cells; these are used for spore or gamete dispersal.

Table 29.1	Distinguishing Features of Fungal Phyla			
Common name (Formal name)	**Habitat**	**Ecological role**	**Reproduction**	**Examples cited in this chapter**
Chytrids (Chytridiomycota)	Water and soil	Mostly decomposers; some pathogens	Flagellate spores	*Batrachochytrium dendrobatidis*
Zygomycetes (Zygomycota)	Mostly terrestrial	Decomposers and pathogens	Nonflagellate asexual spores produced in sporangia; resistant sexual zygospores	*Rhizopus stolonifer*
AM Fungi (Glomeromycota)	Terrestrial	Form mutually beneficial mycorrhizal associations with plants	Distinctively large, nonflagellate, multinucleate asexual spores	The genus *Glomus*
Ascomycetes (Ascomycota)	Mostly terrestrial	Many form lichens; some are mycorrhizal	Nonflagellate sexual spores (ascospores) in sacs (asci) on fruiting bodies (ascocarps); asexual conidia	*Venturia inaequalis, Aleuria aurantia, Saccharomyces cerevisiae*
Basidiomycetes (Basidiomycota)	Terrestrial	Less commonly form lichens; many are mycorrhizal; decomposers	Nonflagellate sexual spores (basidiospores) on club-shaped basidia on fruiting bodies (basidiocarps); several types of asexual spores	*Coprinus disseminatus, Rhizoctonia solani, Armillaria mellea, Puccinia graminis, Ustilago maydis, Phanerochaete chrysosporium, Laccaria bicolor, Amanita muscaria, Phallus impudicus, Lycoperdon perlatum*

The presence of a single, posterior flagellum on chytrid spores or gametes links fungi with the ancestry of choanoflagellates and animals (see Chapter 28). Since species within the phylum Chytridiomycota do not have a single, common ancestor, this phylum is not a monophyletic group. Using molecular data, systematists are now assigning some chytrids into new phyla.

Chytrids live in aquatic habitats or in moist soil. Most chytrids are decomposers, but some are parasites of protists (Figure 29.22), plants, or animals. For example, the chytrid *Batracho-chytrium dendrobatidis* has been associated with declining harlequin frog populations (look ahead to Figure 54.1).

Zygomycetes Produce Distinctive Zygospores

The **zygomycetes** feature a mycelium that is mostly composed of aseptate hyphae (those lacking cross walls) and distinctive reproductive structures. For example, like most zygomycetes, the black bread mold *Rhizopus stolonifer* produces asexual spores in enclosures known as **sporangia** (singular, sporangium) (**Figure 29.23a**). Bread mold sporangia form at hyphal tips in such large numbers that they make moldy bread appear black. Zygomycete sporangia may each release up to 100,000 spores into the air! The great abundance of such spores means that bread easily molds unless retardant chemicals are added.

Zygomycetes are named for the **zygospore**, a distinctive feature of sexual reproduction (**Figure 29.23b**). Zygospore production begins with the development of **gametangia** ("gamete bearers"). In the zygomycete fungi, gametangia are hyphal branches whose cytoplasm is isolated from the rest of the mycelium by cross walls. These gametangia enclose gametes that are basically a mass of cytoplasm containing several haploid nuclei. When food supplies run low, and if compatible mating strains are present, the gametangia of compatible mating types fuse, as do the gamete cytoplasms. The resulting cell, known as a zygosporangium, contains many haploid parental nuclei that fuse, producing many diploid nuclei. Eventually, a dark-pigmented, thick-walled spore known as a **zygospore** matures within the zygosporangium. Each zygospore contains many diploid nuclei and is capable of surviving stressful conditions.

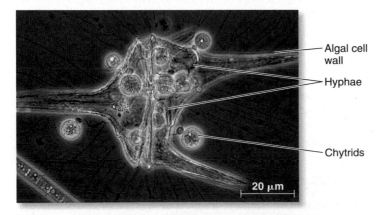

Algal cell wall

Hyphae

Chytrids

20 μm

Figure 29.22   Chytrids growing on a freshwater protist. The colorless chytrids produce hyphae that penetrate the cellulose cell walls of the dinoflagellate *Ceratium hirundinella*, absorbing organic materials from the alga. Chytrids use these materials to produce spherical flagellate spores that swim away to attack other algal cells.

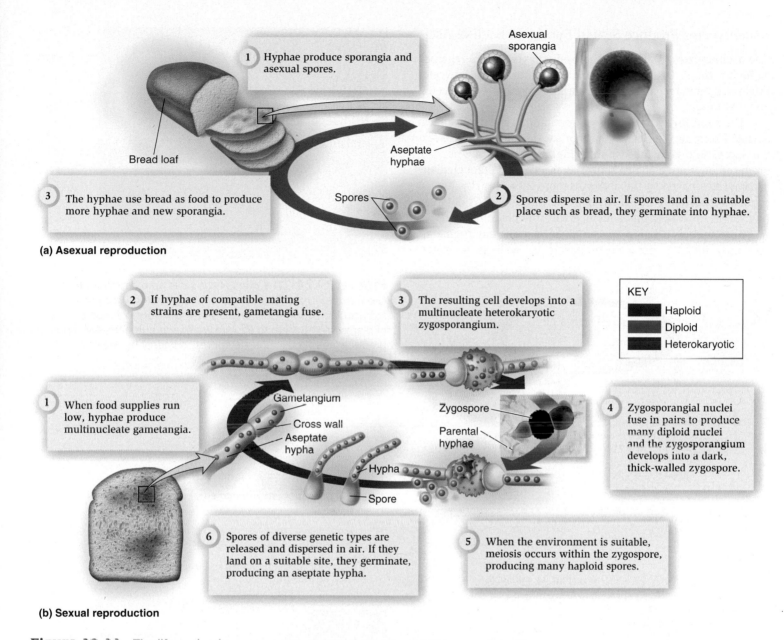

(a) Asexual reproduction

1 Hyphae produce sporangia and asexual spores.

Asexual sporangia

Bread loaf

Aseptate hyphae

3 The hyphae use bread as food to produce more hyphae and new sporangia.

Spores

2 Spores disperse in air. If spores land in a suitable place such as bread, they germinate into hyphae.

(b) Sexual reproduction

2 If hyphae of compatible mating strains are present, gametangia fuse.

3 The resulting cell develops into a multinucleate heterokaryotic zygosporangium.

KEY
■ Haploid
■ Diploid
■ Heterokaryotic

1 When food supplies run low, hyphae produce multinucleate gametangia.

Gametangium

Cross wall
Aseptate hypha

Hypha

Spore

Zygospore

Parental hypha

4 Zygosporangial nuclei fuse in pairs to produce many diploid nuclei and the zygosporangium develops into a dark, thick-walled zygospore.

6 Spores of diverse genetic types are released and dispersed in air. If they land on a suitable site, they germinate, producing an aseptate hypha.

5 When the environment is suitable, meiosis occurs within the zygospore, producing many haploid spores.

Figure 29.23 The life cycle of a zygomycete, the black bread mold *Rhizopus stolonifer*.

When the environment is suitable, the zygospore may undergo meiosis and germinate, dispersing many haploid spores. If the spores land in a suitable place, they germinate to form aseptate hyphae containing many haploid nuclei produced by mitosis.

Most zygomycetes are saprobes in soil, living on decaying materials, but some are parasites of plants or animals. The phylum Zygomycetes is not a monophyletic group and thus is likely to be split as more information becomes available.

AM Fungi Live with Plant Partners

The phylum Glomeromycota—commonly known as the **AM** (for <u>a</u>rbuscular <u>m</u>ycorrhizal) **fungi**—has only recently been defined as a distinct group. AM fungi were previously classified with

the Zygomycota. Glomeromycota have aseptate hyphae and reproduce only asexually by means of unusually large spores with many nuclei (**Figure 29.24**). Many vascular plants depend on AM fungi, and these fungi are not known to grow separately from plants or cyanobacterial partners.

Molecular evidence suggests that Glomeromycota originated around 600 million years ago. Fossils having aseptate hyphae and large spores similar to those of modern Glomeromycota are known from the time when land plants first became common and widespread, about 460 million years ago (see Chapter 30). This and other fossil evidence suggests that the ability of early plants to live successfully on land may have depended on help from fungal associates, as is common today.

Ascomycetes Produce Sexual Spores in Saclike Asci

The **ascomycetes**, along with the basidiomycetes, are regarded as higher fungi. DNA evidence indicates that the ascomycetes originated more recently than the earliest chytrids, zygomycetes, and AM fungi.

The name *ascomycetes* derives from unique sporangia known as **asci** (from the Greek *asco*, bags or sacs), which produce sexual spores known as **ascospores** (**Figure 29.25**). The asci are produced on fruiting bodies known as **ascocarps**. Although many ascomycetes have lost the ability to reproduce sexually, hyphal septa with simple pores (see Figure 29.6a) and DNA data can be used to identify them as members of this phylum.

Ascomycetes occur in terrestrial and aquatic environments, and they include many decomposers as well as parasites.

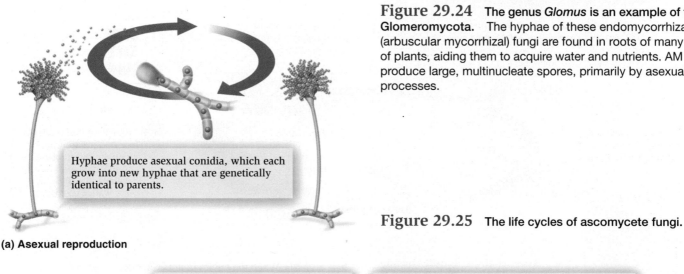

Hypha

Spore

70 µm

Figure 29.24 **The genus *Glomus* is an example of the Glomeromycota.** The hyphae of these endomycorrhizal (arbuscular mycorrhizal) fungi are found in roots of many types of plants, aiding them to acquire water and nutrients. AM fungi produce large, multinucleate spores, primarily by asexual processes.

Hyphae produce asexual conidia, which each grow into new hyphae that are genetically identical to parents.

(a) Asexual reproduction

Figure 29.25 The life cycles of ascomycete fungi.

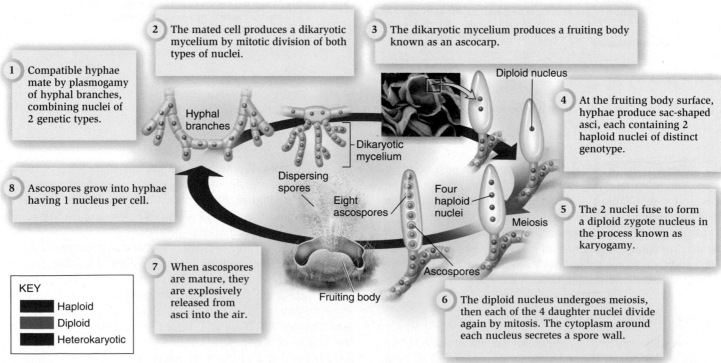

1 Compatible hyphae mate by plasmogamy of hyphal branches, combining nuclei of 2 genetic types.

2 The mated cell produces a dikaryotic mycelium by mitotic division of both types of nuclei.

3 The dikaryotic mycelium produces a fruiting body known as an ascocarp.

4 At the fruiting body surface, hyphae produce sac-shaped asci, each containing 2 haploid nuclei of distinct genotype.

5 The 2 nuclei fuse to form a diploid zygote nucleus in the process known as karyogamy.

6 The diploid nucleus undergoes meiosis, then each of the 4 daughter nuclei divide again by mitosis. The cytoplasm around each nucleus secretes a spore wall.

7 When ascospores are mature, they are explosively released from asci into the air.

8 Ascospores grow into hyphae having 1 nucleus per cell.

Hyphal branches

Dikaryotic mycelium

Dispersing spores

Eight ascospores

Fruiting body

Four haploid nuclei

Diploid nucleus

Meiosis

Ascospores

KEY
- Haploid
- Diploid
- Heterokaryotic

(b) Sexual reproduction

Important ascomycete plant pathogens include powdery mildews, chestnut blight (*Cryphonectria parasitica*), Dutch elm disease (the genus *Ophiostoma*), and apple scab (*Venturia inaequalis*) (**Figure 29.26**). Cup fungi (see chapter-opening photo) are common examples of ascomycetes. Edible truffles and morels are the fruiting bodies of particular ascomycetes whose mycelia form mycorrhizal partnerships with plants. Ascomycetes are the most common fungal components of lichens. Members of the ascomycetes include most **yeasts**, fungi that can occur as unicells and that reproduce by budding (**Figure 29.27**). (Some yeasts are classified in the Zygomycota or the Basidiomycota.) Ascomycete yeasts are important to the baking and brewing industries, and some are medically significant, as agents of disease. For example, the ascomycete yeast *Candida albicans* causes yeast infections and thrush (oral candidiasis). Yeasts are also widely used for fundamental biological studies.

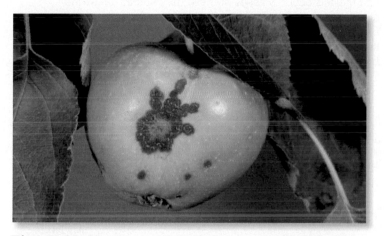

Figure 29.26   Apple infected with the ascomycete fungus *Venturia inaequalis*, which causes apple scab disease.   This fungus grows on leaves, flowers, and fruits, leaving harmless but unsightly scabs on their surfaces. Growers usually try to control apple scab by spraying trees with fungicides.

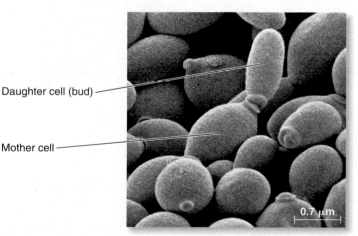

Daughter cell (bud)

Mother cell

0.7 μm

Figure 29.27   The budding yeast *Saccharomyces cerevisiae.*

Biological inquiry: Are all yeasts classified in the ascomycetes?

GENOMES & PROTEOMES

Yeast Is Used as a Model System in Genomics, Proteomics, and Metabolomics

Ascomycete yeasts such as *Saccharomyces cerevisiae* (Figure 29.27) have long served as model systems for cellular, biochemical, genetic, and molecular research. Because of *S. cerevisiae*'s scientific importance, this organism's genome was completely sequenced by an international coalition of more than 600 scientists in more than 100 laboratories, led by molecular biologist André Goffeau. This team found that *S. cerevisiae* has about 6,000 genes on 16 chromosomes and exhibits many cases of gene duplication and divergence, as do most organisms.

Genome sequences have recently been obtained for two of *S. cerevisiae*'s close relatives, the yeast *Schizosaccharomyces pombe* and *Ashbya gossypii*, a filamentous fungus used in industry for producing vitamin B₂. These data allow biologists to perform comparative studies that shed light on the evolution of entire genomes. Fred Dietrich, Peter Philippsen, and colleagues found that by comparison with other eukaryotes, *A. gossypii* has a comparatively small genome of only 9.2 million base pairs, which includes around 4,700 protein-coding genes on seven chromosomes. The genome of *A. gossypii* has no transposons, relatively few introns, and only a few gene duplications. A similarly "stripped-down" genome is present in *S. pombe*, with fewer than 5,000 protein-coding genes. These two fungi may illustrate the minimum genome size possible for free-living (nonparasitic) fungi and reveal the genetic mechanisms that result in small genome size.

Based on genomic data, the *S. cerevisiae* proteome has been estimated to include over 5,000 soluble proteins and more than 1,000 transmembrane proteins. In 2006, Anne-Claude Gavin, Giulio Superti-Furga, and many associates reported the existence of about 500 protein complexes in *S. cerevisiae*. These protein complexes combine in various ways with other proteins, in modular fashion, to form many types of cellular metabolic "machines."

Yeast research has also led the way in metabolomics, the study of all of the small molecules produced in an organism. For example, in 2005 Michael Snyder and his associates treated glass chips bearing an array of 1,300 yeast proteins with 87 different purified protein kinases plus radiolabeled ATP. The labeled ATP allowed researchers to determine which kinases phosphorylate particular proteins during signal transduction (see Chapter 9). This experiment revealed over 4,000 different signal transduction interactions. Such metabolomic studies have important applications in human medicine, because protein kinases are important drug targets.

Basidiomycetes Produce Diverse Fruiting Bodies

DNA-sequencing comparisons indicate that **basidiomycetes** are the most recently evolved group of fungi. Even so, they have a long fossil history, extending back nearly 300 million years.

Today, basidiomycetes are very important as decomposers and mycorrhizal partners in forests, producing diverse fruiting bodies commonly known as mushrooms, puffballs, stinkhorns, shelf fungi, rusts, and smuts (**Figure 29.28**; see also Figures 29.3b, 29.9, 29.10, and 29.16a).

The name given to the basidiomycetes derives from **basidia**, the club-shaped cells that produce sexual spores known as **basidiospores** on their surfaces (**Figure 29.29**). Basidia are typically located on the undersides of fruiting bodies, which are gener-

ally known as **basidiocarps**. Though some basidiomycetes have lost sexual reproduction, they can be identified as members of this phylum by unique hyphal structures known as **clamp connections** that help distribute nuclei during cell division. Basidiomycetes can also be identified by distinctive septa having complex pores (see Figure 29.6b) and by DNA methods. Basidiomycetes reproduce asexually by various types of spores. Rust fungi, for example, can produce up to four different types of asexual spores during their life cycle.

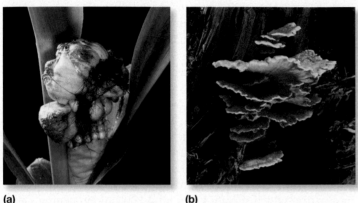

(a) (b)

Figure 29.28 Basidiomycetes produce diverse fruiting bodies. **(a)** Corn smut (*Ustilago maydis*) produces dikaryotic mycelial masses within the kernels (fruits) of infected corn plants. These mycelia produce many dark spores in which karyogamy and meiosis occur. Masses of these dark spores cause the smutty appearance. When the spores germinate, they produce basidiospores that can infect other corn plants. **(b)** Shelf fungi such as this sulfur shelf fungus are the fruiting bodies of basidiomycete fungi that have infected trees.

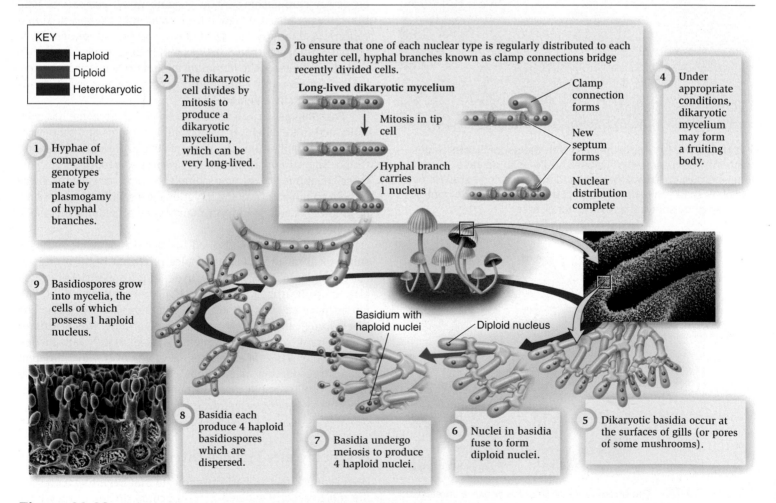

Figure 29.29 The sexual cycle of the basidiomycete fungus *Coprinus disseminatus*.

CHAPTER SUMMARY

29.1 Distinctive Features of Fungi

- Fungi form a monophyletic kingdom of heterotrophs that, like animals, display absorptive nutrition. Fungi secrete enzymes into food substrates, breaking down complex organic molecules into small organic molecules that are absorbed as food. (Figure 29.1)

- Fungal cells possess cell walls composed of chitin, a polysaccharide resistant to microbial attack. Fungal bodies, known as mycelia, are composed of microscopic branched filaments known as hyphae. (Figure 29.2)

- During sexual reproduction, fungi produce fruiting bodies, formed by aggregated hyphae, that usually extend out of the substrate. Fruiting bodies produce sexual spores and aid their dispersal by means of wind, water, or animals. (Figure 29.3)

- In most cases, the fungal body is inconspicuously dispersed within the substrate, but sometimes fungal hyphae aggregate to form stringlike rhizomorphs that aid fungal water transport. (Figure 29.4)

- Fungal mycelia grow at their edges, by tip growth of hyphae. Septate hyphae are divided into cells by cross walls, or septa, perforated by simple or complex pores. Some fungi have aseptate hyphae that are not subdivided into cells. (Figures 29.5, 29.6)

- Mycelial shape depends on the location of nutrients in the environment, which determines the direction in which cell division and hyphal growth will occur. (Figure 29.7)

29.2 Fungal Sexual and Asexual Reproduction

- Fungi disperse in their environments by means of spores produced by asexual or sexual reproduction. (Figure 29.8)

- The gametes of most fungi are relatively unspecialized hyphal branches. During sexual reproduction, hyphal branches fuse with those of a different mycelium of compatible mating type. In many fungi, dikaryotic hyphae (having two nuclei per cell) that result from mating often persist for long periods before nuclear fusion occurs. Nuclear fusion generates zygotes. Fungal zygotes are the only cells in the life cycle that possess a diploid nucleus.

- Zygotes undergo meiosis to produce haploid spores, which disperse from fruiting bodies. Spores germinate to produce haploid fungal mycelia. Fungi produce diverse types of fruiting bodies that foster spore dispersal by wind, water, or animals. Many fungal fruiting bodies produce defensive toxins or hallucinogens. (Figures 29.9, 29.10, 29.11, 29.12)

- Asexual reproduction does not involve mating or meiosis, and it occurs by means of asexual spores such as conidia. Many common and destructive molds spread by means of conidia. (Figure 29.13)

29.3 Fungal Ecology and Biotechnology

- Fungi play important roles in nature as decomposers, predators, and pathogens. Several types of fungi cause disease. (Figures 29.14, 29.15)

- Fungi form important mutualistic symbioses with other organisms. Mycorrhizae are common associations between fungi and plant roots. Ectomycorrhizae coat root surfaces, extending into root intercellular spaces. Endomycorrhizae commonly form highly branched arbuscules in the spaces between root cell walls and plasma membranes. (Figures 29.16, 29.17)

- Lichens are partnerships between fungi and photosynthetic green algae and/or cyanobacteria. Lichens can reproduce asexually by means of structures such as soredia, which consist of fungal hyphae wrapped around a few algal cells. When lichen fungi reproduce sexually, the hyphae arising from spore germination must find new algal partners. (Figures 29.18, 29.19)

- Lichens occur in diverse habitats, including harsh environments, and often grow slowly and to great age. Lichens help to build soils and are useful air-quality monitors.

- In at least one lichen family, the Cladoniaceae, the fungal partners have not evolved in tandem with symbiotic algae, and data suggest that such lichen fungi commonly switch algal partners. (Figure 29.20)

- Fungal biochemistry is useful in chemical, food processing, and waste-treatment industries, and fungi are increasingly used to replace chemical procedures that generate harmful waste materials. (Figure 29.21)

29.4 Evolution and Diversity of Fungi

- Currently, fungi are classified into five phyla commonly known as chytrids, zygomycetes, AM fungi, ascomycetes, and basidiomycetes. (Table 29.1)

- Chytrids are among the simplest and earliest-divergent fungi. They commonly occur in aquatic habitats and moist soil, where they produce flagellate reproductive cells. (Figure 29.22)

- Zygomycetes are named for their distinctive, large zygospores, the result of sexual reproduction. Common black bread mold and other zygomycetes reproduce asexually by means of many small spores. (Figure 29.23)

- The AM fungi produce distinctive large, multinucleate spores, and they form mutualistic arbuscular mycorrhizal relationships with many types of plants. (Figure 29.24)

- Ascomycetes produce sexual ascospores in saclike asci located at the surfaces of fruiting bodies known as ascocarps. Many are lichen symbionts. The ascomycete yeast *Saccharomyces cerevisiae* is widely used as a laboratory model system for genomic, proteomic, and metabolomic studies. (Figures 29.25, 29.26, 29.27)

- Basidiomycetes produce sexual basidiospores on club-shaped basidia located on the surfaces of fruiting bodies known as basidiocarps. Such fruiting bodies take a wide variety of forms, including mushrooms, puffballs, stinkhorns, shelf fungi, rust, and smuts. The hyphae of basidiomycete fungi are characterized by clamp connections, structures that aid in distributing nuclei of two types after cell division occurs in dikaryotic hyphae. (Figures 29.28, 29.29)

TEST YOURSELF

1. Fungal cells differ from animal cells in that fungal cells
   a. lack ribosomes, though these are present in animal cells.
   b. lack mitochondria, though these occur in animal cells.
   c. have cell walls, whereas animal cells lack rigid walls.
   d. lack cell walls, whereas animal cells possess walls.
   e. none of the above.

2. Conidia are
   a. cells produced by some fungi as the result of sexual reproduction.
   b. fungal asexual reproductive cells produced by the process of mitosis.
   c. structures that occur in septal pores.
   d. the unspecialized gametes of fungi.
   e. none of the above.

3. What are mycorrhizae?
   a. the bodies of fungi, composed of hyphae
   b. fungi that attack plant roots, causing disease
   c. fungal hyphae that are massed together into stringlike structures
   d. fungi that have symbiotic partnerships with algae or cyanobacteria
   e. mutually beneficial associations of particular fungi and plant roots

4. Where could you find diploid nuclei in an ascomycete or basidiomycete fungus?
   a. in spores
   b. in cells at the surfaces of fruiting bodies
   c. in conidia
   d. in soredia
   e. all of the above

5. Which fungi are examples of hallucinogens?
   a. *Claviceps* and *Psilocybe*
   b. *Epidermophyton* and *Candida*
   c. *Pneumocystis carinii* and *Histoplasma capsulatum*
   d. *Saccharomyces cerevisiae* and *Phanerochaete chrysosporium*
   e. *Cryphoenectria parasitica* and *Ventura inaequalis*

6. What role do fungal endophytes play in nature?
   a. They are decomposers.
   b. They are human parasites that cause skin diseases.
   c. They are plant parasites that cause serious crop diseases.
   d. They live within the tissues of grasses and other plants, helping to protect plants from herbivores and pathogens.
   e. All of the above.

7. What forms do lichens take?
   a. crustose, flat bodies
   b. foliose, leaf-shaped bodies
   c. fruticose, erect or dangling bodies
   d. single cells
   e. a, b, and c

8. Lichens consist of a partnership between fungi and what other organisms?
   a. red algae and green algae
   b. green algae and cyanobacteria
   c. heterotrophic bacteria and archaea
   d. choanoflagellates and *Nuclearia*
   e. none of the above

9. How can ascomycetes be distinguished from basidiomycetes?
   a. Ascomycete hyphae have simple pores in their septa and lack clamp connections, whereas basidiomycete hyphae display complex septal pores and clamp connections.
   b. Ascomycetes produce sexual spores in sacs, whereas basidiomycetes produce sexual spores on the surfaces of club-shaped structures.
   c. Ascomycetes are commonly found in lichens, whereas basidiomycetes are less commonly partners in lichen associations.
   d. Ascomycetes are not commonly mycorrhizal partners, but basidiomycetes are commonly present in mycorrhizal associations.
   e. All of the above are correct.

10. Which group of organisms listed is most closely related to the kingdom Fungi?
    a. the animal kingdom
    b. the green algae
    c. the land plants
    d. the bacteria
    e. the archaea

CONCEPTUAL QUESTIONS

1. Explain three ways that fungi are like animals, and two ways in which fungi resemble plants.
2. Explain why some fungi produce toxic or hallucinogenic compounds.
3. Explain three ways in which fungi function as mutualistic symbionts and what benefit the fungi receive from the partnerships.

EXPERIMENTAL QUESTIONS

1. What is meant by the term cospeciation?
2. How did Piercey-Normore and DePriest determine whether or not cospeciation had occurred in the Cladoniaceae?
3. What results did Piercey-Normore and DePriest find?

COLLABORATIVE QUESTIONS

1. Thinking about the natural habitats closest to you, where could you find fungi, and what roles do these fungi play?
2. Imagine that you are helping to restore the natural vegetation on a piece of land that had long been used to grow crops. You are placed in charge of planting pine seedlings (*Pinus resinosa*) and fostering their growth. In what way could you consider using fungi?

---

www.brookerbiology.com

This website includes answers to the Biological Inquiry questions found in the figure legends and all end-of-chapter questions.

30

PLANTS AND THE CONQUEST OF LAND

CHAPTER OUTLINE

A temperate rain forest in Olympic National Park, Washington State, containing diverse plant phyla.

W hen thinking about plants, people envision lush green lawns, shady street trees, garden flowers, or leafy fields of valuable crops. On a broader scale, they might imagine dense jungles, vast grassy plains, or tough desert vegetation. Shopping in the produce section of the local grocery store may remind us that plant photosynthesis is the basic source of our food. Just breathing crisp fresh air might bring to mind the role of plants as oxygen producers—the ultimate air fresheners. Do you start your day with a "wake-up" cup of coffee, tea, or hot chocolate? Then you may appreciate the plants that produce these and many other materials we use in daily life: medicines, cotton, linen, wood, bamboo, cork, and even the paper on which this textbook has been printed.

In addition to their importance to humans and modern ecosystems, plants have played dramatic roles in Earth's past. Throughout their evolutionary history, diverse plants have influenced Earth's atmospheric chemistry, climate, and soils and the evolution of many other groups of organisms, including humans. In this chapter, we will survey the diversity of modern plant phyla and their distinctive features. This chapter also explains how early plants adapted to land and how plants have continued to adapt to changing terrestrial environments. During this process, we will continue to gain insight into principles of diversity and evolution such as descent with modification.

30.1 Ancestry and Diversity of Modern Plants

Several hundred thousand modern species are formally classified into the kingdom Plantae, informally known as the plants or land plants (**Figure 30.1**). **Plants** can be defined as eukaryotic, primarily photosynthetic organisms that mostly live on land and display many adaptations to life in terrestrial habitats (**Table 30.1**). Molecular and other evidence indicate that land plants most likely evolved from aquatic algal ancestors. In this section, we will describe the modern algae that are most closely related to plants and survey the diverse phyla of living land plants.

Modern Green Algae Are Closely Related to the Ancestry of Land Plants

Molecular, biochemical, and structural data reveal that the plant kingdom is monophyletic, having probably originated from a single common protist ancestor (see Chapter 28). This common ancestor could probably be classified with modern freshwater green algae that are informally known as **charophyceans**. Such algae are named for the genus *Chara*, a multicellular alga that displays relatively complex structure and reproduction (**Figure 30.2a**), as does the related genus *Coleochaete* (**Figure 30.2b**). Either *Chara* and its close relatives or *Coleochaete* are the modern protists most closely related to the ancestry of land plants, depending on which genes are used in phylogenetic analyses. Thus, it is not surprising that these complex charophyceans share many features of biochemistry, cell structure, cell division, and sexual reproduction with land plants.

For example, complex charophyceans accomplish sexual reproduction by means of flagellate sperm and larger, nonmotile eggs, as do land plants. In contrast, more simply structured, early diverging charophyceans, such as the beautiful desmids (**Figure 30.2c**) and the single-celled flagellate *Mesostigma viride* (**Figure 30.2d**), reproduce differently. Such reproductive differences suggest that plant eggs and sperm evolved prior to divergence of the complex charophyceans. Diverse modern charophyceans are a vast reservoir of information about the origin of land plant sexual reproduction and other plant features.

Although charophyceans by themselves do not form a mono-phyletic group, what is known as a **clade**, some experts classify charophyceans together with land plants to form a clade known informally as streptophytes (see Figure 30.1). Charophycean algae have also been classified into larger groups in a variety of ways based on photosynthetic features. For example, charophyceans and other green algae are classified together with other protists having primary plastids, and plants into the eukaryote supergroup Archaeplastida (see Chapter 28). In addition, some experts classify plants and green algae together and name the combination in various ways (Viridiplantae, Chlorobionta, and Chloroplastida) that reflect the common feature of green plastids. The plastids of green algae and plants possess the same photosynthetic pigments: the main photosynthetic pigment chlorophyll a, the green accessory pigment chlorophyll b, and orange accessory pigments such as β-carotene. You may recall that accessory pigments absorb light energy and transfer it to chlorophyll a for use in photosynthesis. Because chlorophylls a and b are so abundant in plastids, the plastids of green algae and plants are generally green. By contrast, other algal groups have different colored plastids whose abundant amounts of red, gold, or brown accessory pigments hide chlorophyll a.

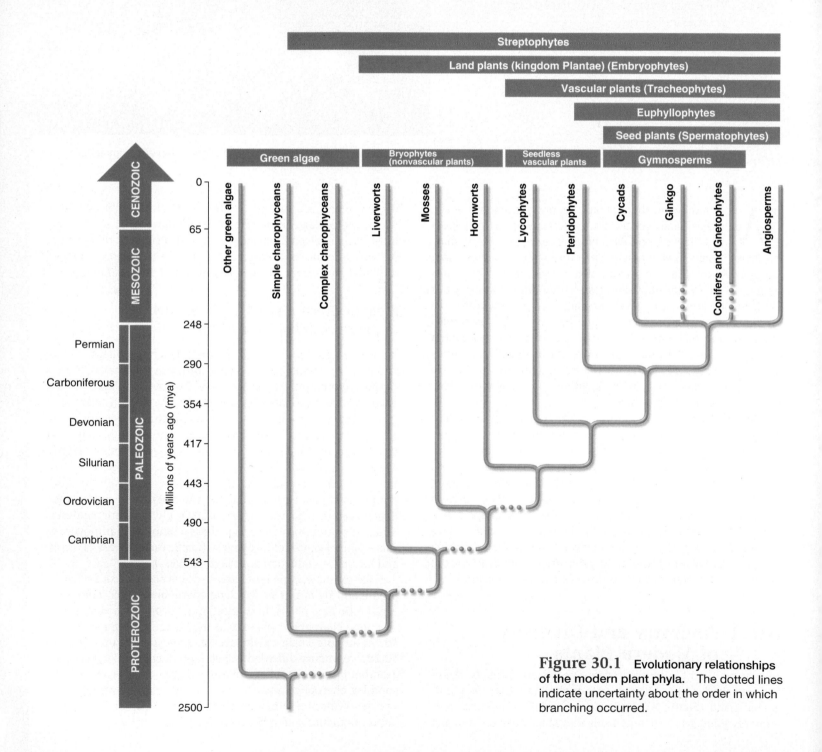

Figure 30.1 Evolutionary relationships of the modern plant phyla. The dotted lines indicate uncertainty about the order in which branching occurred.

Gre... ...or photosynthetic
feature... ...tids. By
contra... ...ynthetic
prod... ...al forms
(see... ...atures of

[Handwritten note overlapping the page:]

$$\left(p + q + y\right)^2$$

$$= p^2 + pq + py + pq + q^2 + qy + py + qy + y^2$$

aa → I^aI^a
ab → I^aI^b
ao → I^aI^o
bb → I^bI^b
bo → I^bI^o
oo → I^oI^o

...odern
...ants*

...e cycle; rosette
...porangia absent;
...glucans absent

...ric life cycle
...multicellular
...te and haploid
...bryos are
...least some time
...roducing sporangia;
...ia; sporopollenin
...ucans

...worts, **hornworts**,

...generation; supportive,
...ar tissue absent; true
...s absent; sporophytes
...e to grow independently

...YTES) (**lycophytes**,

...yte generation; lignified water
conducting ti... —xylem; specialized organic
food conducting tissue—phloem; sporophytes
branched; sporophytes eventually become
independent of gametophytes

SEEDLESS VASCULAR PLANTS (**lycophytes**,
pteridophytes)

Lycophytes   Leaves generally small with a
single, unbranched vein (lycophylls
or microphylls); sporangia borne on
sides of stems

EUPHYLLOPHYTES (**pteridophytes**, SPERMATOPHYTES)

Pteridophytes Leaves relatively large with
extensively branched vein system
(euphylls or megaphylls);
sporangia borne on leaves

SEED PLANTS (SPERMATOPHYTES) Seeds
present; leaves are euphylls

Gymnosperms (**cycads, ginkgos, conifers,
gnetophytes**)

Flowers and fruits absent; seed food stored
before fertilization in female gametophyte,
endosperm absent

Angiosperms (flowering plants)

Flowers and fruits present; seed food stored
after fertilization in endosperm formed by
double fertilization

*Key: **Phyla**; LARGER MONOPHYLETIC CLADES (synonyms). All other classification terms
are not clades.

ancestral green algae were inherited by diverse phyla of fossil
and living land plants. We next focus on the phyla of modern
plants and their distinguishing features.

Modern Land Plants Can Be Classified into 10 Phyla

Plant systematists use molecular and structural information
from living and fossil plants to classify plants into phyla. However, many controversies exist regarding the time of first appearance and species composition of some phyla, and the relationships among plant phyla are not completely clear. As a result of
our incomplete understanding of plant relationships, biologists
have classified modern plants in different ways, and classifications continue to change as new information becomes available.

1 cm

(a) *Chara zeylanica*

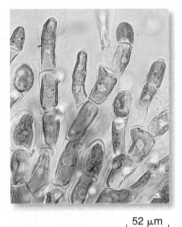

52 μm

(b) *Coleochaete pulvinata*

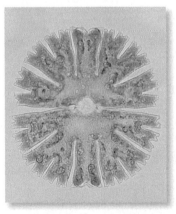

37 μm

(c) *Micrasterias radiosa*

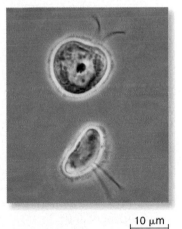

10 μm

(d) *Mesostigma viride*

Figure 30.2   Charophycean green algal relatives of the
land plants.   (a) *Chara zeylanica*. (b) *Coleochaete pulvinata*.
(c) The desmid *Micrasterias radiosa*. (d) *Mesostigma viride*,
named for its central red spot and green plastid color. All of
these algae inhabit freshwater lakes and ponds, and they
display structural, reproductive, biochemical, and molecular
features in common with land plants.

In this textbook, 10 phyla of living land plants are described: (1) the plants informally known as **liverworts** (formally called Hepatophyta), (2) **hornworts** (Anthocerophyta), (3) **mosses** (Bryophyta), (4) **lycophytes** (Lycopodiophyta), (5) **pteridophytes** (Pteridophyta), (6) **cycads** (Cycadophyta), (7) **ginkgos** (Ginkgophyta), (8) **conifers** (Coniferophyta), (9) **gnetophytes** (Gnetophyta), and (10) **angiosperms** (Anthophyta), also known as the **flowering plants** (see Table 30.1). Fossils reveal that additional plant phyla once lived but have since become extinct.

Phylogenetic information suggests that the modern plant phyla arose in a particular sequence. Liverworts, mosses, and hornworts originated relatively early; lycophytes and pteridophytes arose later; cycads, ginkgos, conifers, and gnetophytes came next; and flowering plants appeared most recently (see Figure 30.1). Each of these modern plant phyla displays features that reveal how plants became increasingly better adapted to life on land.

Bryophytes Include the Liverworts, Hornworts, and Mosses

Liverworts, hornworts, and mosses are Earth's simplest land plants (**Figures 30.3, 30.4,** and **30.5**). There are about 6,500 species of modern liverworts, about 100 species of hornworts, and 12,000 or more species of mosses. These monophyletic phyla are together known informally as **bryophytes**. Although the term bryophyte does not reflect a clade, it is useful for expressing common structural, reproductive, and ecological features of liverworts, mosses, and hornworts. For example, bryophytes possess simpler structure and reproduction than other plants, and bryophytes are most common and diverse in moist habitats. Because they diverged early in the evolutionary history of land plants (see Figure 30.1), bryophytes serve as models of the earliest terrestrial plants. They display several features that are absent from charophycean algae but are present in all other land plants. Such bryophyte features thus likely reflect early adaptations to life on land. The bryophyte life cycle, shared with other plants, is a good example. A comparison between the life cycle of aquatic charophyceans and that of bryophytes reveals the bryophyte life cycle's adaptive value on land.

The charophycean algae display a zygotic life cycle in which the diploid generation consists of only one cell, the zygote (**Figure 30.6a**). By contrast, sexual reproduction in bryophytes and all other plants involves a **sporic life cycle**, also known as **alternation of generations** (**Figure 30.6b**). A sporic life cycle originated independently in early plants and several other eukaryotic lineages that are not related to land plants (see Chapter 28). The plant life cycle is believed to have originated by a delay in meiosis, with the result that the diploid generation became multicellular before undergoing meiosis. For the earliest land plants, this multicellular diploid generation provided advantages in coping with terrestrial conditions. A closer look at the process of sexual reproduction in bryophytes reveals these advantages and also highlights ways in which bryophytes differ from other plants.

(a) *Marchantia polymorpha*

(b) *Marchantia polymorpha*   **(c)** A species of leafy liverwort

Figure 30.3 **Liverworts.** **(a)** The common liverwort *Marchantia polymorpha*, with raised, umbrella-shaped structures that bear sexual reproductive structures on the undersides. **(b)** A close-up of *M. polymorpha* showing surface cups that contain multicellular, frisbee-shaped asexual structures called propagules that are dispersed by wind and grow into new liverworts. **(c)** A species of liverwort having leaflike structures and thus known as a leafy liverwort.

Biological inquiry: Why do you think liverworts produce their spores on raised structures?

Figure 30.4 **Hornworts.** Sporophytes are the diploid, spore-producing structures that grow on gametophytes, the haploid, gamete-producing structures growing close to the ground. Hornwort sporophytes mature and open at the top, dispersing spores.

Figure 30.5 **Moss.** The common moss genus *Mnium* has leafy green gametophytes and unbranched, dependent sporophytes each bearing a sporangium at its tip. When mature, these sporophytes sprinkle spores into the wind.

Bryophyte Reproduction Illustrates Adaptation to Life on Land

Bryophytes—in common with other organisms having sporic life cycles—produce two multicellular life stages (**Figure 30.7**). These two stages are the sporophyte and the gametophyte, each stage having a distinctive reproductive role. The diploid **sporophyte** generation produces haploid spores by the process of meiosis, and multicellular plant sporophytes can produce many spores. Plant **spores** are single-celled reproductive structures that are dispersed into the air and are able to grow into gametophytes if they find suitable habitats. In contrast to the sporophyte generation, the haploid **gametophyte** generation produces gametes. Bryophytes may have separate male and female gametophytes. As we have earlier noted, plant **gametes** (from the Greek word meaning "wife") are nonflagellate eggs and smaller flagellate sperm that may fuse in pairs to form single-celled, diploid **zygotes**. Let's take a closer look at how sporophytes and gametophytes function together during the life cycle of a bryophyte, starting with gamete production.

Gametophytes The role of plant gametophytes is to produce haploid gametes, but meiosis is not involved in this process. The gametophytes of bryophytes and many other land plants produce gametes in specialized structures known as **gametangia** ("gamete containers"), in which developing gametes are protected by a jacket of tissue. The gametangial jacket protects delicate gametes from drying and microbial attack while they develop. Round or elongate gametangia that produce sperm are known as **antheridia**, whereas flask-shaped gametangia that enclose an egg cell are known as **archegonia** (Figure 30.7). When mature, if moist conditions exist, plant sperm are released from antheridia into

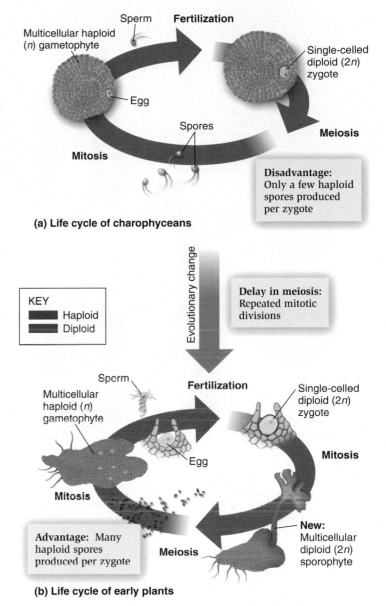

(a) Life cycle of charophyceans

Disadvantage: Only a few haploid spores produced per zygote

KEY
■ Haploid
■ Diploid

Delay in meiosis: Repeated mitotic divisions

Advantage: Many haploid spores produced per zygote

New: Multicellular diploid (2n) sporophyte

(b) Life cycle of early plants

Figure 30.6 Origin of the plant life cycle.

films of water. Under the influence of sex-attractant molecules secreted from archegonia, sperm swim toward eggs, twisting their way down the tubular archegonial necks. Sperm then fuse with egg cells to form diploid zygotes, which grow into new sporophytes. However, fertilization cannot occur in bryophytes unless water is present. Conditions of uncertain moisture, common in the land habitat, can thus limit plant reproductive fitness. The plant life cycle, featuring several reproductive advantages, is an adaptive response to this environmental challenge.

Sporophytes One reproductive advantage of the plant life cycle is that zygotes remain enclosed within gametophyte tissues, where they are sheltered and fed (a process described in more detail in Section 30.3). This critical innovation gives zygotes a good start as they begin to grow into young sporophytes, which are

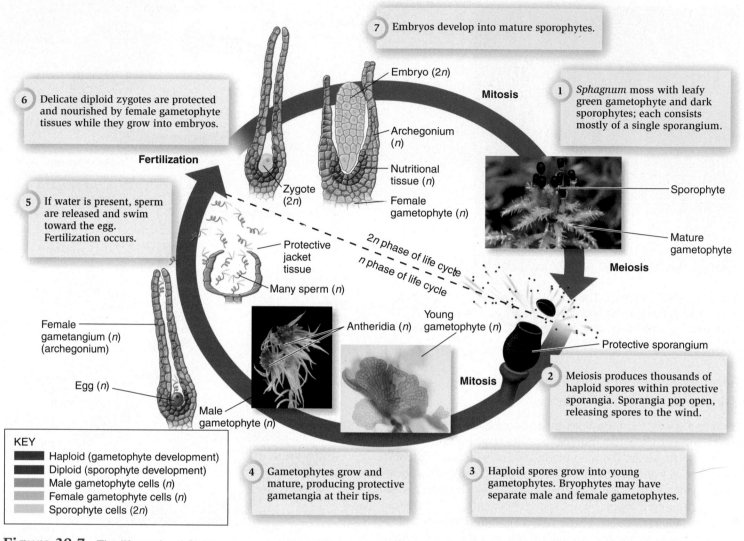

7 Embryos develop into mature sporophytes.

Embryo (2*n*)

Mitosis

6 Delicate diploid zygotes are protected and nourished by female gametophyte tissues while they grow into embryos.

1 *Sphagnum* moss with leafy green gametophyte and dark sporophytes; each consists mostly of a single sporangium.

Archegonium (*n*)

Fertilization

Nutritional tissue (*n*)

Zygote (2*n*)

Female gametophyte (*n*)

Sporophyte

5 If water is present, sperm are released and swim toward the egg. Fertilization occurs.

2*n* phase of life cycle

n phase of life cycle

Mature gametophyte

Protective jacket tissue

Meiosis

Many sperm (*n*)

Young gametophyte (*n*)

Female gametangium (*n*) (archegonium)

Antheridia (*n*)

Protective sporangium

Mitosis

Egg (*n*)

2 Meiosis produces thousands of haploid spores within protective sporangia. Sporangia pop open, releasing spores to the wind.

Male gametophyte (*n*)

KEY
▮	Haploid (gametophyte development)
▮	Diploid (sporophyte development)
▮	Male gametophyte cells (*n*)
▮	Female gametophyte cells (*n*)
▮	Sporophyte cells (2*n*)

4 Gametophytes grow and mature, producing protective gametangia at their tips.

3 Haploid spores grow into young gametophytes. Bryophytes may have separate male and female gametophytes.

Figure 30.7   The life cycle of *Sphagnum* moss, illustrating reproductive adaptations that likely helped early plants to reproduce on land.

known as **embryos**. For this reason, all land plants are known as **embryophytes**. Sheltering and feeding embryos is particularly important when embryo production is limited by water availability, as is the case for bryophytes. Another reproductive advantage is that, when mature, the multicellular sporophytes produce many **spores** in protective enclosures known as **sporangia** ("spore containers"). Bryophyte sporangia open in specialized ways that foster dispersal of spores into the air, allowing spore transport by wind. A third reproductive advantage is that plant spores have cell walls containing a tough material known as **sporopollenin** that helps to prevent cellular damage during transport in air. If spores reach habitats favorable for growth, their walls crack open, and new gametophytes develop by mitotic divisions, completing the life cycle.

Spore production is a measure of plant fitness, because plants can better disperse progeny throughout the environment when they produce more spores. The larger the diploid generation, the more spores a plant can produce. As a result, during plant evolution, the plant sporophyte has become larger and more complex (**Figure 30.8**).

Bryophytes Display Several Distinguishing Features

Bryophytes can be distinguished from other land plants by several features. First, bryophyte gametophytes are more common in nature, larger, and longer-lived than bryophyte sporophytes. Patches of moss in the woods are primarily gametophytes. In order to see bryophyte sporophytes, you would have to look very closely, because this life stage is quite small and attached to female gametophytes throughout its short lifetime (see Figures 30.5 and 30.7). Plant biologists say that bryophyte gametophytes are the dominant generation in their life cycle. By contrast, other plants have dominant sporophyte generations (see Figure 30.8).

Bryophyte sporophytes are small because they remain attached to parental gametophytes, are unable to branch, and have short lifetimes. At their tips, bryophyte sporophytes produce only a single sporangium containing a limited number of spores. By contrast, the sporophytes of other land plants become independent, produce branches with many sporangia, and are often able to produce spores for many years (see Figure 30.8).

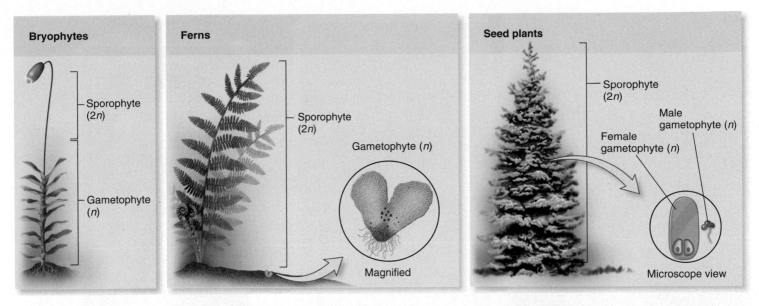

Figure 30.8 Relative sizes of the sporophyte generations of bryophytes, ferns, and seed plants.

In all land plants except bryophytes, the sporophyte generation is the larger, more complex, and longer-lived generation.

Yet another distinguishing feature of bryophytes is that they lack tissues that are specialized for both structural support and conduction. Plant tissues that provide both structural support and conduction of water, minerals, and organic compounds are known as **vascular tissues**. Although the gametophytes of some bryophytes display simple conducting tissues, these do not provide structural support. Thus, bryophytes are informally known as **nonvascular plants**. Other modern plant phyla are known as **vascular plants** because they produce vascular tissues that function in both conduction and support.

Lycophytes and Pteridophytes Are Vascular Plants That Do Not Produce Seeds

If you take a look outside, most of the plants in view are probably vascular plants. The presence of vascular tissue allows most vascular plants to grow much taller than bryophytes and to reproduce more effectively. As a result, vascular plants are more prominent than bryophytes in most modern plant communities. Vascular plants have been important to Earth's ecology for a long time. Fossils tell us that the first vascular plants appeared later than the earliest bryophytes (see Section 30.2), and that several early vascular plant lineages, such as the trimerophytes and psilophytes, once existed but became extinct. Molecular data indicate that the lycophytes are the oldest phylum of living vascular plants and that pteridophytes are the next oldest living plant phylum (see Figure 30.1). In the past, lycophytes were very diverse and included tall trees, but the tree lycophytes became extinct and now about 1,000 relatively small species exist (**Figure 30.9**). Pteridophytes have diversified more recently, and there are about 12,000 species of modern pteridophytes, including horsetails, whisk ferns, and other ferns (**Figure 30.10**).

Figure 30.9 The lycophyte *Lycopodium obscurum* growing in a forest. The stems bear many tiny leaves, known as lycophylls, and sporangia generally occur in club-shaped clusters. For this reason, lycophytes are informally known as club mosses or spike mosses, though they are not true mosses. The gametophytes of lycophytes are small structures that often occur underground.

The lycophytes and pteridophytes diverged prior to the origin of seeds, and they are informally known as seedless vascular plants. In contrast to bryophytes, and in common with the seed plants, the sporophytes of lycophytes and pteridophytes possess vascular tissues providing both structural support and conduction. Consequently, lycophytes, pteridophytes, and seed-producing plants are together known as the vascular plants or **tracheophytes**. The latter term takes its name from **tracheids**, a type of specialized vascular cell that conducts water and minerals and provides structural support.

(a) A whisk fern (*Psilotum nudum*)

(b) A horsetail (*Equisetum telmateia*)

(c) An early-diverging fern
(*Botrychium lunaria*)

Figure 30.10 Pteridophyte diversity. (a) The leafless, rootless green stems of the whisk fern (*Psilotum nudum*) branch by forking and bear many clusters of yellow sporangia that disperse spores via wind. The gametophyte of this plant is a tiny white structure that lives underground in a partnership with fungi. (b) The giant horsetail (*Equisetum telmateia*) displays branches in whorls around the green stems. The leaves of this plant are tiny, light brown structures that encircle branches at intervals. This plant produces spores in cone-shaped structures, and the wind-dispersed spores grow into small green gametophytes. (c) The fern *Botrychium lunaria*, showing a green photosynthetic leaf with leaflets and a modified leaf that bears many round sporangia. (d) The fern *Blechnum capense*, viewed from above, showing a whorl of young leaves that are in the process of unrolling from the bases to the tips. The surrounding older leaves have many leaflets. The stem of this fern grows parallel to the ground and is thus not illustrated. Most ferns produce spores in sporangia on the undersides of leaves. (e) The life cycle of a typical fern.

(d) A later-diverging fern (*Blechnum capense*)

Stems, Roots, and Leaves In common with other tracheophytes, the sporophytes of lycophytes and pteridophytes produce specialized organs—stems, roots, and leaves—that contain vascular tissue. **Stems** are branching structures that contain vascular tissue and also produce leaves and reproductive structures. At their centers, stems contain the specialized conducting tissues known as **phloem** and **xylem**, the latter of which contain tracheids. Such conducting tissues enable vascular plants to conduct organic compounds, water, and minerals throughout the plant body. The xylem also provides structural support, allowing vascular plants to grow taller than nonvascular plants. This support function arises from the presence of a compression-

and decay-resistant waterproofing material known as **lignin** that occurs in the cell walls of tracheids and some other types of plant cells. Most vascular plants also produce **roots**—organs specialized for uptake of water and minerals from the soil—and **leaves**, which generally have a photosynthetic function. Roots and leaves possess vascular systems that connect to the vascular tissue of stems, forming a continuous conduction system in the plant body. Vascular plant roots efficiently take up water and minerals from the soil, conduct these materials to stems via xylem, and receive sugar from phloem. The leaves of vascular plants obtain water and minerals needed for photosynthesis from the xylem, and they export sugar via the phloem.

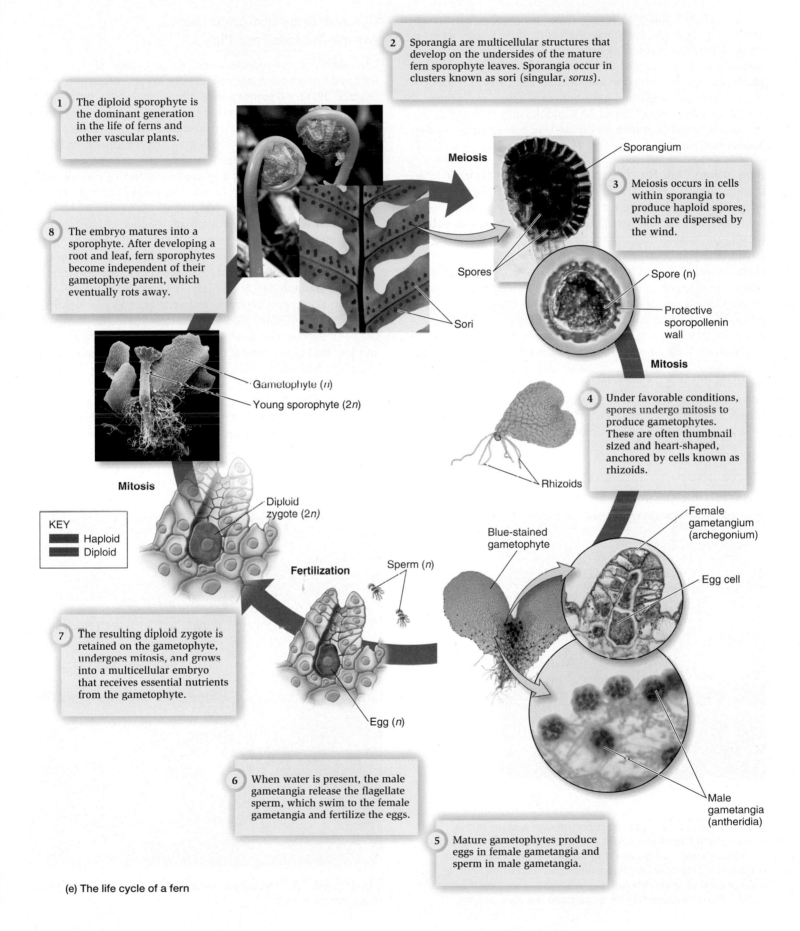

2 Sporangia are multicellular structures that develop on the undersides of the mature fern sporophyte leaves. Sporangia occur in clusters known as sori (singular, *sorus*).

1 The diploid sporophyte is the dominant generation in the life of ferns and other vascular plants.

Meiosis

Sporangium

3 Meiosis occurs in cells within sporangia to produce haploid spores, which are dispersed by the wind.

8 The embryo matures into a sporophyte. After developing a root and leaf, fern sporophytes become independent of their gametophyte parent, which eventually rots away.

Spores

Sori

Spore (*n*)

Protective sporopollenin wall

Mitosis

Gametophyte (*n*)
Young sporophyte (2*n*)

4 Under favorable conditions, spores undergo mitosis to produce gametophytes. These are often thumbnail sized and heart-shaped, anchored by cells known as rhizoids.

Rhizoids

Mitosis

Diploid zygote (2*n*)

KEY	
	Haploid
	Diploid

Female gametangium (archegonium)

Blue-stained gametophyte

Egg cell

Fertilization

Sperm (*n*)

7 The resulting diploid zygote is retained on the gametophyte, undergoes mitosis, and grows into a multicellular embryo that receives essential nutrients from the gametophyte.

Egg (*n*)

Male gametangia (antheridia)

6 When water is present, the male gametangia release the flagellate sperm, which swim to the female gametangia and fertilize the eggs.

5 Mature gametophytes produce eggs in female gametangia and sperm in male gametangia.

(e) The life cycle of a fern

Lycophyte roots and leaves differ from those of pterido-phytes. For example, lycophyte roots fork at their tips, whereas roots of pteridophytes branch from the inside like the roots of seed plants (see Chapter 35). Lycophyte leaves are relatively small and possess only one unbranched vein, whereas pterido-phyte leaves are larger and have branched veins, as do those of seed plants (compare Figures 30.9 and 30.10d). The evolution-ary origin of pteridophyte leaves will be discussed in Section 30.4 of this chapter.

Adaptations That Foster Stable Internal Water Content Lyco-phytes, pteridophytes, and other vascular plants display several adaptations that increase their ability to maintain stable inter-nal water content. For example, a protective **waxy cuticle** is present on most surfaces of vascular plant sporophytes (**Figure 30.11**). The plant cuticle contains a polyester polymer known as **cutin**, which helps to prevent attack by pathogens, and wax, which helps to prevent desiccation. The surface tissue of vascu-lar plant stems and leaves contains **stomata** (singular, stomate), pores that are able to open and close (see Figure 30.11). Stomata allow vascular plants to take in carbon dioxide needed for photo-synthesis and release oxygen to the air, while conserving water. When the soil is very dry, stomata close, which reduces water loss from plants via these pores. When the soil is moist, stomata open, allowing photosynthetic gas exchange to occur. These water-conserving features have allowed vascular plants to ex-ploit a wide spectrum of land habitats.

Gymnosperms and Angiosperms Are the Modern Seed Plants

The modern phyla commonly known as cycads, ginkgos, coni-fers, and gnetophytes are collectively known as gymnosperms (**Figure 30.12**). **Gymnosperms** reproduce using both spores and seeds, as do the flowering plants or angiosperms (**Figure 30.13**). Gymnosperms and angiosperms are thus known infor-mally as the **seed plants**. **Seeds** are complex structures having specialized tissues that protectively enclose embryos, that is, the young sporophytes. Seeds also contain stores of carbohydrate, lipid, and protein that can be used to provide energy for seed germination and seedling development.

Several additional seed plant phyla once existed and left fossils, but they are now extinct. Collectively, all of the living and fossil seed plant phyla are formally known as **spermatophytes** (see Figure 30.1). Modern and fossil seed plants are also formally known as **lignophytes**. Such plants are descended from seedless ancestors that produced wood, a specialized vascular tissue that is rich in lignin. Wood production enables plants to increase in girth and become tall. Many modern seed plants produce wood and are thus known informally as woody plants. By contrast, seed-less plants lack wood, as do many modern seed plants. Plants that produce little or no wood are informally known as herbaceous plants. Woody and herbaceous plants do not represent clades.

The **angiosperms** are distinguished by the presence of flow-ers, fruits, and a specialized seed tissue known as endosperm.

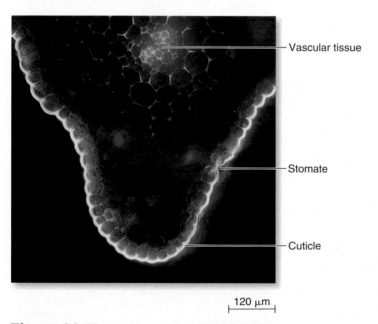

— Vascular tissue

— Stomate

— Cuticle

|— 120 μm —|

Figure 30.11 **A cross section through a stem of the pteridophyte *Psilotum nudum* shows tracheophyte adaptations for transporting and conserving water.** When viewed with fluorescence microscopy and illuminated with violet light, an internal core of xylem tracheids glows yellow, as does the surface cuticle. Surface pores known as stomata are also present.

Figure 30.12 **The pinyon pine (*Pinus edulis*) is an example of a gymnosperm.**

Figure 30.13 This bleeding heart plant (the genus *Dicentra*) is an example of an angiosperm.

Flowers are short stems bearing organs that are specialized in ways that enhance seed production (see Figure 30.13). **Fruits** are structures that develop from flower organs, enclose seeds, and foster seed dispersal in the environment. The term angiosperm means "enclosed seeds," reflecting the fact that the flowering plants produce seeds within fruits. **Endosperm** is a nutritive seed tissue that increases the efficiency by which food is stored in the seeds of flowering plants (see Section 30.4). Flowers, fruits, and endosperm are defining features of the angiosperms, and they are integral components of animal nutrition.

Though gymnosperms produce seeds, and many are woody plants, they lack flowers, fruits, or endosperm. The term gymnosperm means "naked seeds," reflecting the fact that gymnosperm seeds are not enclosed within fruits (see Table 30.1). Despite their lack of flowers, fruits, and seed endosperm, the modern gymnosperms are diverse and abundant in many places. Like the other phyla of plants that we have just surveyed, gymnosperms played significant roles in the past evolutionary history of plants, our next topic.

30.2 An Evolutionary History of Land Plants

A billion years ago, Earth's terrestrial surface was comparatively bare of life. Some green or brown crusts of cyanobacteria most likely grew in moist places, but there would have been very little soil, no plants, and no animal life. The origin of the first land plants was essential to development of the first substantial soils, the evolution of modern plants, and the ability of animals to colonize land. It is safe to say that if land plants had not appeared hundreds of millions of years ago, and become so marvelously well adapted to life on land, humans and the modern ecosystems on which we depend would not have evolved in their present form.

How can we know about events such as the origin and diversification of land plants? Biologists are able to infer the past by comparing molecular and other features of modern plants and by examining plant remains in the fossil record. Thus, we will begin our survey of plant history by taking a closer look at these techniques.

Modern and Fossil Plants Are Helpful in Inferring Evolutionary History

The distinctive features of living plant phyla (summarized in Table 30.1) reveal the order in which they first appeared—bryophytes earliest, lycophytes and pteridophytes later, gymnosperms next, and angiosperms most recently (see Figure 30.1). Molecular and fossil data support and provide additional information for this inference.

Molecular Approaches to Plant Evolutionary History Plant evolutionary biologists often use computers and special software to compare gene sequences from diverse plants. Such analyses aid in understanding relationships among plants and the time when different clades originated. Hypotheses of organism relationships are represented as diagrams known as phylogenetic trees (see Figure 30.1). The arrangement of branches on phylogenetic trees often changes as new data become available.

Sequence data often reveal past changes in genome structure, such as chromosomal rearrangements and the addition of introns to genes. You may recall that many eukaryotic genes possess noncoding regions known as introns and that transcribed intron sequences are removed during mRNA processing before translation occurs. Introns can be inserted into genes, but such events are relatively rare. As a result, introns can reveal ancient phylogenetic divergences. For example, in 1998, evolutionary biologist Y. L. Qiu and associates reported that liverworts lack three introns generally present in mitochondrial genes of other plants. More recently, Milena Groth-Malonek and coworkers identified additional introns that are missing from liverworts but present in other living plant phyla. Plant evolutionary biologists suspect that these introns first appeared within plant genes after liverworts had diverged from the mainstream of plant evolution, indicating that liverworts are the most ancient modern plant phylum. This conclusion has been supported by an analysis of sequence data for several genes, reported in 2005 by bryophyte expert Barbara Crandall-Stotler and associates.

Another example of the use of sequence data to infer plant phylogeny is an analysis of pteridophyte relationships published in 2001 by Kathleen Pryer and associates. Based on sequences of small subunit (SSU) rRNA and three chloroplast genes, Pryer's group concluded that pteridophytes are monophyletic and are the closest modern relatives of seed plants. The Pryer phylogeny contained a surprise—that plants commonly known as horsetails and whisk ferns were part of a clade that also contained the ferns. Plant biologists had earlier used structural features to classify horsetails and whisk ferns into phyla distinct from the ferns.

The Use of Fossils to Infer Plant Evolutionary History Fossils are the remains of dead organisms that have survived decay and other destructive processes long enough to become buried in materials that harden into rocks. The enclosed organic remains become stacked in layers, with the oldest at the bottom and the most recent fossils closer to the rock surface. This layering, together with chemical features, allows biologists to infer the relative ages of fossil remains (refer back to Figure 22.9).

The tough plant compounds lignin, cutin, and sporopollenin help to preserve the structure of plants as they fossilize (**Figure 30.14**). Such preservation allows plant scientists to deduce how fossil plants looked when alive. Plant biologists compare fossils to other fossils and to living plants to deduce the kinds of evolutionary changes that occurred during plant history. For example, several types of fossil plants were noted to have different types of early vascular tissues, but their relationships to the vascular tissue of modern plants were uncertain. Evolutionary biologists Martha Cook and Ned Friedman helped to resolve this issue by comparing vascular tissues of a modern lycophyte with those of fossilized early vascular plants. These investigators treated the stems and roots of a modern lycophyte with enzymes that would degrade all but the most resistant plant materials—those most likely to fossilize. In 2004, Cook and Friedman reported that the treated stem conducting tissues closely resembled particular fossils, enabling them to better interpret the fossils and understand the evolution of plant vascular tissue.

Another example of the use of fossils to deduce plant history was reported by evolutionary biologists Gar Rothwell and Kevin Nixon in 2006. These investigators performed a phylogenetic analysis of the structural features of many fossils, as well as living pteridophytes. Their study suggested the possibility that plants currently classified within the pteridophytes might actually belong to more than one clade, thus questioning the strength of conclusions based primarily on molecular data. Rothwell and Nixon have emphasized that current concepts of plant evolutionary history are incomplete and that studies of plant fossils are needed to complement molecular analyses.

The study of fossils, as well as molecular and other features of modern plants, has revealed an amazing story—how plants conquered the land. This story can be conceptualized as three dramatic episodes: (1) aquatic charophycean algae give rise to the first land-adapted plants; (2) seedless plants transform Earth's atmosphere and climate; and (3) an ancient cataclysm marks the rise of angiosperms.

Aquatic Charophycean Algae Give Rise to the First Land-Adapted Plants

Land plants got their start in the water. As we have earlier noted, plants most likely evolved from an aquatic ancestor similar to modern, complex charophycean algae. During their evolution in the water, charophyceans acquired several cellular and reproductive adaptations that aided survival in aquatic habitats. Some of these traits also proved useful on land, helping the first land plants to survive there. As an example, let's consider cytokinesis, the process by which the cytoplasm divides after nuclear division.

A distinctive feature of plant cytokinesis, the **phragmoplast**, first appeared during the diversification of charophyceans. Phragmoplasts are constructed largely of microtubules arranged perpendicularly to a developing cell plate (refer back to Figure 15.14b) and microfilaments. Phragmoplasts also promote the development of intercellular connections known as **plasmodesmata**, which are important modes of plant cell-to-cell communication (see Chapter 10). The earliest land plants inherited phragmoplasts and plasmodesmata from complex charophyceans, making use of them to produce cohesive tissues in which component cells could effectively communicate. Subsequent land plants used these traits to build increasingly more complex bodies that were even better adapted to the stresses of terrestrial life.

In addition to traits inherited from charophycean algae, early land plants acquired other features in response to stresses present on land, but not in the water. This fact explains why all land plants possess several features in common that are absent from charophyceans (see Table 30.1). For example, plant biologists Zoe Popper and Stephen Fry discovered that the cell walls of all land plants possess xyloglucan carbohydrates, which cross-link cellulose microfibrils, but that charophycean algae lack this feature. Cell-wall xyloglucans are thus among the new features that appeared as early plants began to adapt to land, probably aiding in the development of more complex bodies. The terrestrial environment also influenced the evolution of genes that control another essential plant cell-wall polysaccharide, cellulose.

Figure 30.14 **Plant fossils.** Fossil of *Pseudosalix handleyi*, an angiosperm.

Biological inquiry: What biochemical components of plants favor the formation of fossils?

GENOMES & PROTEOMES

The Number of Genes That Controls Cellulose Production Increased During Plant Evolutionary History

Land plant cell walls possess an essential component—the polymer of glucose known as cellulose. In fact, cellulose-rich cell walls, which strengthen and protect plant cells, are a hallmark of the plants and many of their green algal relatives. In plant cell walls, cellulose occurs as 3.5- to 10-nm-diameter cylindrical microfibrils containing 36 to 90 closely packed glucan chains (refer back to Figure 10.5). Extensive hydrogen bonding among the glucan polymers confers great strength, explaining the role of cellulose in cell walls, and its many useful applications, such as the paper pages of this textbook. The manner in which plant cellulose polymers are produced is key to comprehending how plants achieve this high level of biochemical organization. Understanding how plants produce cellulose is also fundamental to human efforts to genetically engineer plant cellulose, which has many industrial and medical applications.

Cellulose is spun outward from highly organized arrays of proteins known as terminal complexes, which are located in the plasma membrane (**Figure 30.15**). The terminal complexes of charophycean algae and plants have a unique rose shape and are thus known as rosettes. Cell biologists have determined that the rosettes contain cellulose synthase, an enzyme that synthesizes cellulose from sugars, and geneticists have discovered that *CesA* genes encode cellulose synthase.

Plant molecular biologist Alison Roberts and her associates compared the *CesA* genes of charophycean algae, seedless plants, and seed plants. They found that *Mesotaenium caldariorum*, a single-celled charophycean desmid, has at least two *CesA* genes.

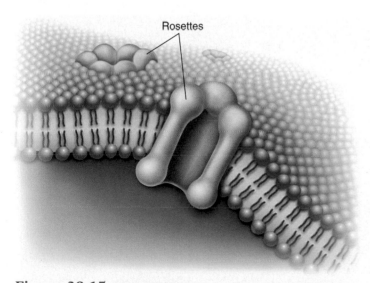

Rosettes

Figure 30.15 Rosette-shaped structures in plant plasma membranes contain cellulose synthase, which produces the cellulose-rich plant cell wall.

By contrast, the moss *Physcomitrella patens* has at least four *CesA* genes, and seed plants have numerous *CesA* genes. For example, the flowering plant *Arabidopsis thaliana* has a 10-member *CesA* gene family, and different *CesA* genes are expressed in particular tissues. Roberts and associates concluded that land plants had inherited cellulose synthase genes from ancestral charophycean algae and that the *CesA* gene family subsequently diversified by gene duplication and divergence. These investigators noted that cellulose synthase gene diversification is correlated with the evolution of greater plant structural complexity. As plants became larger and more complex, they exerted increasingly greater effects on their environments, as we will see next.

Seedless Plants Transformed Earth's Ecology

The spread of early plants dramatically changed the Earth's landscapes, fostering soil buildup, changes in atmospheric chemistry, and colonization of land by fungi and animals. For this reason, evolutionary biologists are interested in learning when land plants first appeared, and what ecological effects resulted.

The First Land Plants When did the earliest land plants first appear? Some evolutionary biologists use molecular clock methods to attempt to answer this question, and others use fossils. Molecular clock methods use rates of molecular sequence changes in modern organisms to infer the point in time when lineages diverged from a common ancestor. Based on molecular clock data, Daniel Heckman and colleagues have suggested that land plants might have been present as early as 700 million years ago. Although fossils of such a great age have not been identified, Heckman and associates hypothesized that populations of the earliest land plants were very small, inhibited by cold conditions that prevailed 750–580 million years ago (see Chapter 22). If so, the earliest plants would not have left many fossils. Only when warmer conditions allowed plant populations to expand do fossils become more abundant.

An early fossil record of presumed land plants consists of tiny sporelike structures reported by paleobiologists Wilson Taylor, Paul Strother, and colleagues, who studied rocks of the Cambrian period that are more than 500 million years old. Early plants whose bodies were too delicate to fossilize well presumably produced these tougher structures. By about 450 million years ago, during the Ordovician period, spores similar to those of modern bryophytes had become abundant and geographically widespread. This fact led paleobiologist Jane Gray to propose that bryophyte-like land plants had become well established by this point in time, millions of years prior to the appearance of vascular plant fossils. In 2003, paleobiologist Charles Wellman and associates reported that they had found fossil sporangia with spores, similar to those of modern bryophytes, in 470-million-year-old rocks.

Ecological Effects of Ancient and Modern Bryophytes Later-diverging modern liverworts and mosses typify a later phase of plant evolution. In addition to possessing tough spores and sporangia, these plants produce decay-resistant body tissues similar to fossils left by ancient land plants. Such tissues help modern bryophytes to avoid attack by decay bacteria and fungi. Plant evolutionary biologists Linda Graham and David Hanson and colleagues determined the amount of decay-resistant mass produced by several modern bryophytes and used these data to estimate the ecological impact of early nonvascular plants. The results suggested that nonvascular plants likely contributed organic substances to early soils, thereby helping to enrich them. The results also indicated that ancient nonvascular plants could have begun a process by which atmospheric carbon dioxide (CO_2) was converted to organic carbon that was buried and thus not respired back to carbon dioxide. Burial of organic carbon is a process that helps to reduce the amount of the greenhouse gas CO_2 in Earth's atmosphere, thereby influencing temperature and precipitation. The investigators calculated that such effects on soil, atmospheric chemistry, and climate might have been significant because they could have occurred over large geographic areas, and for millions of years before vascular plants became dominant.

Modern bryophytes likewise play important roles by storing CO_2 as decay-resistant organic compounds. The abundant modern moss genus *Sphagnum* contains so much decay-resistant mass that dead moss has accumulated over thousand of years into deep peat deposits. By storing organic carbon for long periods, *Sphagnum* moss functions as a giant global thermostat that helps to keep Earth's climate steady, to the benefit of humans and other life-forms. How does this biological thermostat work? Under cooler than normal conditions, *Sphagnum* grows more slowly and thus absorbs less CO_2, allowing atmospheric CO_2 to

rise a bit. Since atmospheric CO_2 helps to warm Earth's climate, increasing CO_2 warms the climate a little. When the climate warms sufficiently, *Sphagnum* grows faster, thereby sponging up more CO_2 as peat deposits. Reducing atmospheric CO_2 returns the climate to slightly cooler conditions. The world's peat deposits currently store an estimated 400 gigatons (billion tons) of organic carbon. Ecologists have expressed concern that modern global warming could increase the rate at which decay organism populations increase and break down organic peat into CO_2. Such processes have the potential to increase global temperatures sufficiently to harm human and other life.

Ecological Effects of Ancient Vascular Plants Vascular plants originated from extinct plants called protracheophytes such as the fossil *Aglaophyton major*. Protracheophytes had branched sporophytes and produced numerous sporangia, but their water-conducting cells lacked lignin and thus did not provide structural support, as does the xylem of vascular plants. Vascular plant fossils first appear in rocks deposited 420–430 million years ago and rapidly became diverse and abundant. These fossils reveal that the earliest vascular plants had no leaves or roots, but they did have stems with a central core of lignin-coated water-conducting cells, a tough outer cuticle, and stomata, much like modern pteridophytes (see Figure 30.11). These features suggest that early vascular plants had achieved the ability to maintain a stable internal water level. The presence of lignin and cutin also fostered the ability of vascular plant bodies to resist decaying long enough to fossilize, explaining why vascular plants have left a more extensive fossil record than did the earliest plants.

Fossils tell us that extensive forests dominated by tree-sized lycophytes, pteridophytes, and early lignophytes occurred in widespread swampy regions during the warm, moist Carboniferous period (354–290 million years ago) (**Figure 30.16**). Much

Giant lycophytes   Giant dragonfly   Giant horsetail (pteridophyte)

Figure 30.16 Reconstruction of a Carboniferous (Coal Age) forest, showing tree-sized lycophytes and pteridophytes.

Biological inquiry: Why did giant dragonflies occur during this time, but not now?

of today's coal derives from the abundant remains of these ancient plants, explaining why the Carboniferous is commonly known as the Coal Age. Carboniferous plants converted huge amounts of atmospheric CO_2 into decay-resistant organic materials such as lignin. Long-term burial of these materials, compressed into coal, together with chemical interactions between soil and the roots of vascular plants, dramatically changed Earth's atmosphere and climate.

Mathematical models of ancient atmospheric chemistry, supported by measurements of natural carbon isotopes, led paleoclimatologist Robert Berner to propose that the Carboniferous proliferation of vascular plants was correlated with a dramatic decrease in atmospheric carbon dioxide, which reached a his-

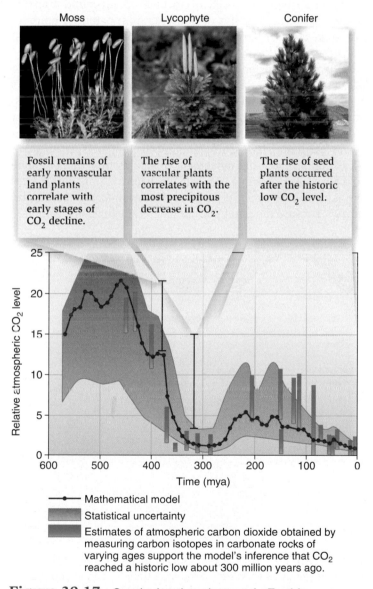

Moss | Lycophyte | Conifer

Fossil remains of early nonvascular land plants correlate with early stages of CO_2 decline.

The rise of vascular plants correlates with the most precipitous decrease in CO_2.

The rise of seed plants occurred after the historic low CO_2 level.

— Mathematical model

Statistical uncertainty

Estimates of atmospheric carbon dioxide obtained by measuring carbon isotopes in carbonate rocks of varying ages support the model's inference that CO_2 reached a historic low about 300 million years ago.

Figure 30.17 Graph showing changes in Earth's atmospheric carbon dioxide levels over geological time. Geological evidence indicates that carbon dioxide levels in Earth's atmosphere were once higher than they are now, but that the rise of land plants caused CO_2 to reach a historic low about 300 million years ago.

toric low about 300 million years ago (**Figure 30.17**). During this period of very low CO_2, atmospheric oxygen levels rose to historic high levels, because less O_2 was being used to break down organic carbon into CO_2. High atmospheric oxygen content has been invoked to explain the occurrence of giant Carboniferous dragonflies and other huge insects, which obtain their air by diffusion. Because atmospheric CO_2 is a greenhouse gas—allowing Earth's atmosphere to retain heat—the great Carboniferous decline in CO_2 level caused cool, dry conditions to prevail in the late Carboniferous and early Permian periods. As a result of this relatively abrupt global climate change, many of the giant seedless lycophytes and pteridophytes that had dominated Carboniferous forests became extinct, as did organisms such as the giant dragonflies. Cooler, drier Permian conditions favored extensive diversification of the first seed plants, the gymnosperms. Seed plants were better able than nonseed plants to reproduce in cooler, drier habitats (as we will see in Section 30.4). As a result, seed plants came to dominate Earth's terrestrial communities, as they continue to do.

An Ancient Cataclysm Marks the Rise of Angiosperms

Diverse phyla of gymnosperms dominated Earth's vegetation through the Mesozoic era (248–65 million years ago), what is sometimes called the Age of Dinosaurs (**Figure 30.18**). In fact, gymnosperms were probably a major source of food for terrestrial herbivorous dinosaurs, some of which grew to enormous sizes.

Figure 30.18 Gymnosperms dominated Earth's flora and dinosaurs dominated animal life during the Mesozoic era.

Though both early flowering plants and early mammals were present, they primarily existed in the shadows of gymnosperms and dinosaurs. One fateful day about 65 million years ago, disaster struck from the sky.

That day, at least one large meteorite or comet crashed into the Earth near the present-day Yucatan Peninsula in Mexico. This episode is known as the **K/T event** because it marks the end of the Cretaceous (sometimes spelled with a K) period and the beginning of the Tertiary (T) period. The impact, together with substantial volcanic activity that also occurred at this time, is thought to have produced huge amounts of ash, smoke, and haze that dimmed the sun's light long enough to kill many of the world's plants. Many groups of plants were forever lost, though some survived and persist to the present time. With a severely reduced food supply, the dinosaurs were also doomed, the only exceptions being their descendants, the birds.

After the K/T event, weedy ferns dominated long enough to leave huge numbers of fossil spores, and then surviving groups of flowering plants began to diversify into the space left by the demise of previous plants. The rise of angiosperms fostered the diversification of beetles (see Chapter 33) and other types of insects that associate with plants. For example, a large molecular analysis of ant relationships allowed Corrie Moreau, Naomi Pierce, and their associates to infer that the development of forests dominated by flowering plants had fostered ant diversification.

Our brief survey of plant evolutionary history reveals some important diversity principles. While environment certainly influenced the diversification of plants, plant diversification has also changed Earth's environment in ways that affected the evolution of other organisms. Plant evolutionary history also serves as essential background for a closer focus on the evolution of **critical innovations**, new features that foster the diversification of phyla. Among the critical innovations that appeared during plant evolutionary history, embryos, leaves, and seeds were particularly important, as we will discuss next.

30.3 The Origin and Evolutionary Importance of the Plant Embryo

The embryo, absent from charophyceans, was probably one of the first distinctive traits acquired by land plants. Recall that plant embryos are young sporophytes that develop from zygotes, enclosed by maternal tissues that provide sustenance. This feature is critical to plant reproduction in terrestrial environments. Drought, heat, ultraviolet light, and microbial attack could kill delicate plant egg cells, zygotes, and embryos if these were not protected and nourished by enclosing parental tissues. For land plants, the evolutionary origin of a dependent plant embryo was a critical innovation. The first embryo-producing plants diversified into hundreds of thousands of diverse modern species, as well as many species that have become extinct. The embryo is such a defining feature of plants that plant biologists often use

the term embryophytes as a synonym for plants (see Figure 30.1). A closer look at embryos reveals why their origin and evolution is so important to all land plants.

A plant embryo has several characteristic features. First, as we have previously noted, plant embryos are multicellular and diploid (see Section 30.1). Plant embryos develop by repeated mitosis from a single-celled zygote resulting from fertilization (see Figure 30.6b). In addition, we have also learned that plant eggs are fertilized while still attached to the maternal plant body, with the result that embryos begin their development within the protective confines of maternal tissues (see Figure 30.7). Plant biologists say that plants retain their zygotes and embryos. Third, plant embryo development depends on organic and mineral materials supplied by the mother plant. Nutritive tissues composed of specialized **placental transfer tissues** aid in the transfer of nutrients from parent to embryo. Let's take a closer look at the valuable role played by placental transfer tissues.

Placental transfer tissues function similarly to the placenta of mammals, which fosters nutrient movement from the mother's bloodstream to the developing fetus. Plant placental transfer tissues often occur in haploid gametophyte tissues that lie closest to embryos and in the diploid tissues of young embryos themselves. Such transfer tissues contain cells that are specialized in ways that promote the movement of solutes from gametophyte to embryo. For example, the cells of placental transport tissues display complex arrays of finger-like cell-wall ingrowths (**Figure 30.19**). Because the plant plasma membrane lines this elaborate

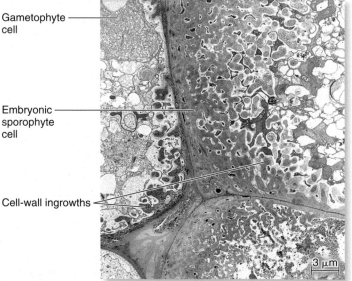

Figure 30.19   TEM of placental transfer tissue in the liverwort genus *Monoclea*.   Placental transfer tissues contain specialized cells having extensive finger-shaped cell-wall ingrowths. Such cells help nutrients to move rapidly from parental gametophytes to embryonic sporophytes, a process that fosters plant reproductive success.

Biological inquiry: How does increasing plasma membrane surface area foster rapid movement of nutrients in placental transfer tissue?

plant cell wall, the ingrowths vastly increase the surface area of plasma membrane. This increase provides the space needed for abundant membrane transport proteins, which move solutes into and out of cells. With more transport proteins present, materials can move at a faster rate from one cell to another. Experiments have revealed that dissolved sugars, amino acids,

and minerals first move from maternal cells into the intercellular space between parent tissues and embryo. Then, transporter proteins in the membranes of nearby embryo cells efficiently import materials into the embryo. Classic experiments have revealed the adaptive value of this process in land plant reproduction.

FEATURE INVESTIGATION

Browning and Gunning Demonstrated That Placental Transfer Tissues Increase the Rate at Which Organic Molecules Move from Gametophytes to Sporophytes

In the 1970s, plant cell biologists Adrian Browning and Brian Gunning explored placental transfer tissue function. Using a simple moss experimental system, they determined that placental transfer tissues increase the rate at which radioactively labeled carbon moves through placental transfer tissues from green gametophytes into young sporophytes. This process is expected to increase the ability of mature sporophytes to produce progeny

spores, thereby increasing reproductive success. Recall that embryos are very young, few-celled sporophytes, and that in mosses and other bryophytes all stages of sporophyte development are nutritionally dependent on gametophyte tissues. Browning and Gunning investigated nutrient flow into young sporophytes because these slightly older and larger developmental stages were easier to manipulate in the laboratory than were tiny embryos.

In a first step, the investigators grew many gametophytes of the moss *Funaria hygrometrica* in a greenhouse until young sporophytes developed as the result of sexual reproduction (**Figure 30.20**). In a second step, they placed black glass sleeves over young sporophytes as a shade to prevent photosynthesis,

Figure 30.20 Browning and Gunning demonstrated that placental transfer tissues increase plant reproductive success.

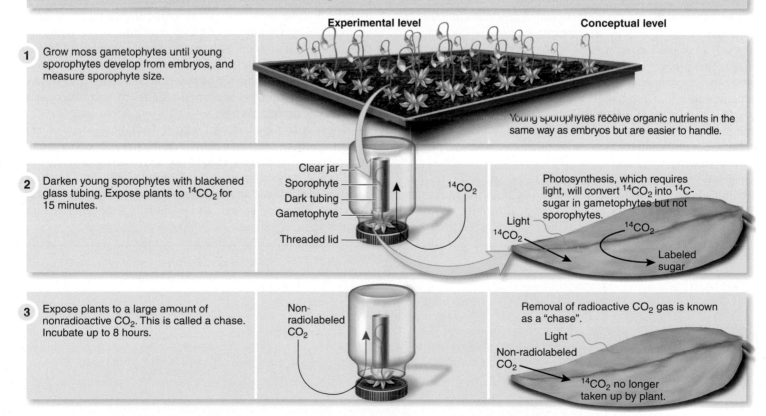

4 Pluck young sporophytes from gametophytes. Assay ^{14}C in both sporophytes and gametophytes using a scintillation counter. This was done immediately following the chase, or 2 or 8 hours after the chase.

Scintillation counter

Determine how much organic carbon flowed into sporophytes during each chase time.

5 THE DATA I

Carbon transfer from gametophyte to sporophyte:

Mean ^{14}C content of 5 gametophytes at 0 chase time	Mean ^{14}C lost from gametophytes after 8-hour chase	Mean ^{14}C gained by sporophytes after 8-hour chase
228 units	145 units	51 units

6 THE DATA II

Sporophyte size effect:

Sporophyte size	Mean ^{14}C content of 8 sporophytes at 0-hour chase	Mean ^{14}C content of 8 sporophytes after 2-hour chase
5–7 mm	$1.52 + 0.48$ units	8.47 ± 4.29 units
11–13 mm	0.22 ± 0.26 units	9.93 ± 3.94 units
23–25 mm	0.13 ± 0.05 units	24.97 ± 5.30 units

enclosed moss gametophytes and their attached sporophytes within transparent jars, and supplied the plants with radioactively labeled carbon dioxide for measured time periods known as pulses. Because the moss gametophytes were not shaded, their photosynthetic cells were able to convert the radioactively labeled carbon dioxide into labeled organic compounds, such as sugars and amino acids. Shading prevented the young sporophytes, which possess some photosynthetic tissue, from using labeled CO_2 to produce organic compounds.

In a third step, the researchers added an excess amount of nonradioactive CO_2 to prevent the further uptake of the radioactive CO_2 from their experimental system, a process known as a chase. This process stopped the radiolabeling of photosynthetic products. (Experiments such as these are known as pulse-chase experiments.) In a final step, Browning and Gunning plucked young sporophytes of different sizes (ages) from gametophytes and measured the amount of radioactive organic carbon present in the separated gametophyte and sporophyte tissues at various times following the chase.

From these data, they were able to calculate the relative amount of organic carbon that had moved from the photosynthetic moss gametophytes to sporophytes. Browning and Gunning discovered that about 22% of the organic carbon produced by gametophyte photosynthesis was transferred to the young sporophytes during an 8-hour chase period. They also calculated the rate of nutrient transfer between generations and compared this rate to the rate at which organic carbon moves in several other plant tissues that lack specialized transfer cells (determined in other studies). By so doing, Browning and Gunning discovered that organic carbon moved from moss gametophytes to young sporophytes nine times faster than organic carbon moves within other plant tissues. These investigators inferred that the increased rate of nutrient movement could be attributed to placental transfer cell structure, namely, the fact that cell-wall ingrowths enhanced plasma membrane surface area. By comparing the amount of radioactive carbon accumulated by young sporophytes of differing ages, they also learned that larger sporophytes absorbed labeled carbon about three times as fast as smaller ones.

These data are consistent with the hypothesis that placental transfer tissues increase plant reproductive success by providing embryos and older sporophytes with more nutrients than they would otherwise receive. Supplied with these greater amounts of nutrients, sporophytes are better able to grow larger than they otherwise would, and eventually they produce more progeny spores.

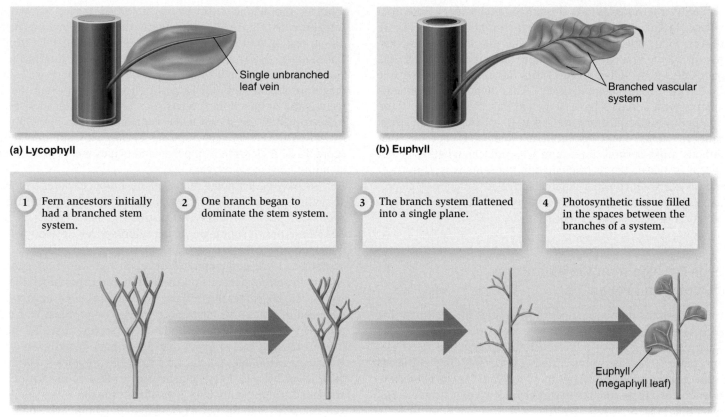

(a) Lycophyll — Single unbranched leaf vein

(b) Euphyll — Branched vascular system

1. Fern ancestors initially had a branched stem system.
2. One branch began to dominate the stem system.
3. The branch system flattened into a single plane.
4. Photosynthetic tissue filled in the spaces between the branches of a system.

Euphyll (megaphyll leaf)

(c) Euphyll evolution process

Figure 30.21   Lycophylls and euphylls.   (a) Most lycophylls possess only a single unbranched leaf vein having limited conduction capacity, explaining why lycophylls are generally quite small. **(b)** Euphylls possess branched vascular systems having greater conduction capacity, explaining why many euphylls are relatively large. **(c)** Fossil evidence suggests how pteridophyte euphylls might have evolved from branched stem systems.

30.4 The Origin and Evolutionary Importance of Leaves and Seeds

Like plant embryos, leaves and seeds are critical innovations that allowed plant phyla to diversify extensively. However, unlike the plant embryo, which likely originated just once at the birth of the plant kingdom, leaves and seeds probably evolved several times during plant evolutionary history. Comparative studies of diverse types of leaves and seeds in fossil and living plants suggest how these critical innovations might have originated.

The Leaves of Lycophytes Differ in Structure and Origin from Those of Other Vascular Plants

Leaves are the solar panels of the plant world. Their flat structure enables leaves to effectively capture sunlight for use in photosynthesis. This explains why leaflike structures evolved even on the gametophytes of leafy liverworts and mosses (see Figures 30.3 and 30.5), and also why they occur on the sporophytes of most vascular plants. Among the vascular plants, lycophytes produce the simplest and most ancient leaves. Modern lycophytes have tiny leaves, known as **lycophylls** or microphylls, that typically have only a single unbranched vein (**Figure 30.21a**). These small leaves are thought to have evolved from sporangia, which are borne along the stems of lycophytes.

In contrast, the leaves of other vascular plant phyla have extensively branched veins. Leaves with branched veins are known as **euphylls** (meaning "true leaves") (**Figure 30.21b**). This explains why the clade that includes pteridophytes and seed plants is known as the **euphyllophytes** (see Figure 30.1). The branched veins of euphylls are able to supply relatively large areas of photosynthetic tissue with water and minerals. Thus, euphylls are typically much larger than lycophylls, explaining why euphylls are also known as megaphylls (meaning "large leaves"). Euphylls provide considerable photosynthetic advantage to ferns and seed plants, because they provide more surface for solar energy capture than do small leaves. Hence, the evolution of relatively large leaves allowed plants to more effectively accomplish photosynthesis, enabling them to grow larger and produce more progeny.

Study of fern fossils indicates that euphylls likely arose from leafless, branched stem systems by a series of steps (**Figure 30.21c**). First, one branch assumed the role of the main axis, then the whole branch system became flattened, and finally the spaces between the branches of this flattened system became filled with photosynthetic tissue. This hypothetical process explains why euphylls have branched vascular systems; these apparently originated from the vascular system of an ancestral branched stem. Plant evolutionary biologists suspect that euphylls arose several times, and it is unclear whether or not the leaves of seed plants originated in the same way as those of ferns. Next, we'll consider what adaptive advantages seeds provide and how seeds might have evolved, beginning with a consideration of seed development.

Seeds Develop from the Interaction of Ovules and Pollen

The seed plants dominate modern ecosystems, suggesting that seeds offer reproductive advantages. Seed plants are also the plants with the greatest importance to humans. For these reasons, plant biologists are interested in understanding why seeds are so advantageous and how they evolved. To consider these questions, we must first take a closer look at seed structure and development.

Plants produce seeds by means of reproductive structures known as ovules and pollen, which are structures unique to seed plants. An **ovule** is a sporangium that contained only a single spore that developed into a very small egg-producing game-

tophyte, the whole enclosed by modified leaves known as **integuments** (**Figure 30.22a**). You can think of an ovule as being like a nesting doll with four increasingly smaller dolls inside. The smallest doll would correspond to an egg cell; intermediate-sized dolls would stand for the gametophyte, spore wall, and megasporangium; and the largest doll would represent the integuments. Fertilization converts such layered ovules into seeds. In seed plants, the sperm needed for fertilization are supplied by **pollen**, tiny male gametophytes enclosed by sporopollen in spore walls. A closer look at pollen and ovules will help in understanding how seeds develop.

We have earlier noted that all plants produce spores by meiosis within sporangia, and seed plants are no exception. However, seed plants produce two distinct types of spores in two different types of sporangia. Small **microspores** develop within microsporangia, and larger **megaspores** develop within megasporangia. Male gametophytes develop from the microspores, and the resulting pollen is released from microsporangia. Meanwhile, female gametophytes develop and produce eggs while enclosed by protective megaspore walls. The problem with this process is that female gametophytes are so tiny that they need help in feeding the embryos that develop from fertilized eggs. Female gametophytes get this help from the previous sporophyte generation by remaining attached to it. This is an advantage because the large sporophyte generation is thereby able to provide gametophytes with the nutrients needed for embryo development.

Embryos develop as the result of fertilization, which cannot occur until after **pollination**, the process by which pollen of the same species reaches ovules. Pollination typically occurs by means of wind or animal transport (see Chapter 31). Fertiliza-

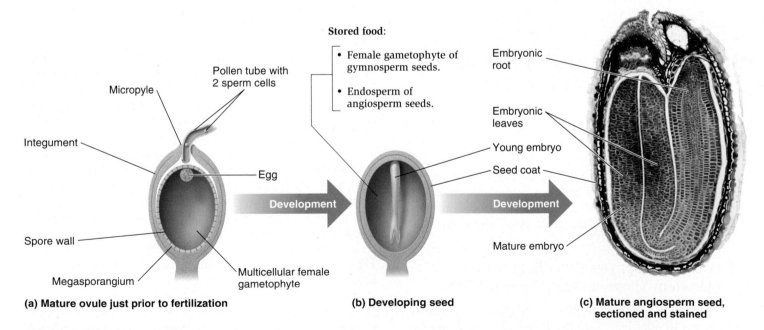

Figure 30.22   Ovule and seed structure.   (a) Ovule structure. (b) Seed structure. (c) Light micrograph of an angiosperm seed.

Biological inquiry: Why does the seed photo show little or no endosperm?

tion occurs when a male gametophyte extends a slender pollen tube that carries two sperm toward an egg. After a pollen tube enters the micropyle and releases the sperm, fertilization may occur. The fertilized egg becomes an embryo, and the ovule's integument develops into a protective, often hard and tough **seed coat** (**Figure 30.22b,c**).

Gymnosperm seeds contain female gametophyte tissue that has accumulated large amounts of protein, lipids, and carbohydrates prior to fertilization. These nutrients are used during seed germination to help support growth of the seedling. Angiosperm seeds also contain this useful food supply, but angiosperm ovules do not store food materials prior to fertilization. Instead, angiosperm seeds store food only after fertilization occurs, ensuring that the food is not wasted if an embryo does not form. How is this accomplished? The answer is a process known as **double fertilization**. This process produces both a zygote and a food storage tissue known as **endosperm**, which is a tissue unique to angiosperms. One of the two sperm delivered by each pollen tube fuses with the egg, producing a diploid zygote, as you might expect. The other sperm nucleus fuses with different gametophyte nuclei to form an unusual cell that has more than the diploid number of chromosomes; this cell generates the endosperm food tissue (see Chapter 39).

Seeds allow embryos access to food supplied by the older sporophyte generation, an option not available to seedless plants. The layered structure of ovules explains why seeds are also layered like nesting dolls, with a protective seed coat (developed from the ovule integuments) enclosing the embryo and also stored food, in many cases. These seed features improve the chances of embryo and seedling survival, thereby increasing seed plant fitness.

Seeds Confer Important Ecological Advantages

Seeds provide plants with numerous ecological advantages, which explains why seed plants have dominated Earth's ecosystems since the end of the Coal Age. First, many seeds are able to remain dormant in the soil for long periods, until conditions become favorable for germination and seedling growth. In contrast, most single-celled spores are unable to survive for long; if conditions are not suitable for germination, they die. In addition, seeds are larger and more complex than spores, which improves the ability of seeds to resist mechanical damage and pathogen attack. Further, seed coats have evolutionary adaptations that improve dispersal in diverse habitats. For example, many plants produce winged seeds that are effectively dispersed by wind. Other plants produce seeds with fleshy coverings that attract birds, which consume the seeds, digest their fleshy covering, and eliminate them at some distance from the originating plants.

Another advantage of seeds is that they can store considerable amounts of food, which helps plant seedlings grow large enough to compete for light, water, and minerals. This is especially important for seeds that must germinate in shady forests.

Finally, the sperm of seed plants can reach eggs without having to swim through water, because pollen tubes deliver sperm directly to ovules. Consequently, seed plant fertilization is not typically limited by lack of water, in contrast to that of seedless plants. Therefore, seed plants are better able to reproduce in arid and seasonally dry habitats. For these reasons, seeds are considered to be a particularly significant plant reproductive adaptation to reproduction in a land habitat.

Ovule and Seed Evolution Illustrate Descent with Modification

As we have seen, seed plants reproduce using both spores and seeds, and it is important to recognize that seed plants have not replaced spores with seeds. Instead, during seed plant evolution, ovules and seeds were added to an ancestral life history that begins with spores. Ovules and seeds thus originated by a sequence of changes to pre-existing structures and processes. Some clues about ovule and seed evolution arise from comparing reproduction in living lycophytes, pteridophytes, and spermatophytes. Fossils provide additional information.

Let's first consider reproduction of lycophytes and pteridophytes. Most modern lycophytes and pteridophytes release one type of spore and one type of gametophyte, which lives in the open environment and produces both male and female gametangia (see Figure 30.10e). However, some lycophytes and pteridophytes produce separate microspores and megaspores, which grow into separate male and female gametophytes, a process known as **heterospory** (meaning "different spores"). Such gametophytes also grow within the confines of microspore and megaspore walls and thus are known as **endosporic gametophytes**. The advantage of heterospory is that it has the potential to increase cross-fertilization, the fusion of eggs and sperm from gametophytes of distinct genotypes. Cross-fertilization thus increases the potential for genetic variation, which aids evolutionary flexibility.

The advantage of endosporic gametophytes is that they are protected to some degree by surrounding spore walls from environmental damage. As we have earlier noted, seed plants also produce distinct megaspores and microspores, which produce internal female and male gametophytes. From these observations, we can infer that heterospory and endosporic gametophytes were probably also features of seed plant ancestors, and constitute early steps toward seed evolution. Because seed plant ancestors are extinct, lycophytes and pteridophytes are useful sources of information about the origin of heterospory and endosporic gametophytes.

A next step in seed evolution may have been retention of megaspores within megasporangia, rather than releasing them. A further step may have been production of only one megaspore per sporangium rather than multiple spores per sporangium, which is common in nonseed plants. Reduction of megaspore numbers would have allowed plants to channel more nutrients into each megaspore. A final step might have been the retention of megasporangia on parental sporophytes. As we have noted, this adaptation would allow food materials to flow from photosynthetic plants to their dependent gametophytes and young embryos.

Fossils provide information about when the process of seed evolution first occurred. Very early fossil seeds such as *Elkinsia polymorpha* and *Archaeosperma arnoldii* were present 365 million years ago. Thus, paleobiologists think that the first seeds arose prior to the Coal Age, during the Devonian period (see Figure 30.1). A newly discovered Devonian fossil known as *Runcaria heinzelinii* is thought to represent a seed precursor (**Figure 30.23**). It had a lacy integument that did not completely enclose the megasporangium.

These hypothesized stages of seed evolution illustrate the stepwise way in which plants probably achieved an increased ability to cope with the stresses of life on land. The evolutionary journey illustrated by the transition from aquatic charophycean algae to bryophytes, to seedless plants, and finally to seed plants reveals how adaptation is related to environmental change, as well as ways in which plants themselves shaped Earth's ecosystems.

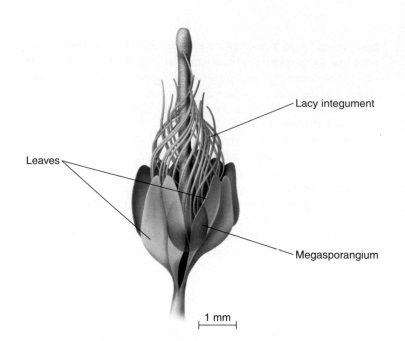

Figure 30.23   The fossil *Runcaria heinzelinii*.

CHAPTER SUMMARY

30.1 Ancestry and Diversity of Modern Plants

- Plants are multicellular organisms composed of eukaryotic cells having plastids, which display many adaptations to life on land, and primarily live on land. The modern plant kingdom consists of several hundred thousand species that can be classified into 10 phyla, informally known as the liverworts, mosses, hornworts, lycophytes, pteridophytes, cycads, ginkgos, conifers, gnetophytes, and angiosperms. (Figure 30.1, Table 30.1)

- The land plants evolved from ancestors that probably would be classified with modern, complex charophycean algae of freshwater habitats. (Figure 30.2)

- The monophyletic liverwort, hornwort, and moss phyla are together known informally as the bryophytes. Bryophytes illustrate early evolved features of land plants, such as a sporic life cycle involving embryos that develop within protective gametophytic tissues. (Figures 30.3, 30.4, 30.5, 30.6)

- Bryophytes are distinguished by a dominant gametophyte generation and a dependent, nonbranching, short-lived sporophyte generation. By contrast, other plant phyla possess dominant sporophyte generations that become independent, branch, and have increased reproductive capacity. Bryophytes also lack supportive vascular tissues, in contrast to other modern plant phyla, which are known as the vascular plants or tracheophytes. (Figures 30.7, 30.8)

- Lycophytes and pteridophytes possess stems, leaves, and roots having vascular tissues composed of xylem and phloem, but the roots and leaves of these phyla differ in distinctive ways. (Figures 30.9, 30.10, 30.11)

- Cycads, ginkgos, conifers, and gnetophytes are collectively known as gymnosperms. Gymnosperms inherited an ancestral capacity to produce wood. Angiosperms, also known as the flowering plants, produce seeds and many also produce wood. Flowers, fruits, and seed endosperm are distinctive features of the angiosperms. (Figures 30.12, 30.13)

30.2 An Evolutionary History of Land Plants

- Paleobiologists and plant evolutionary biologists infer the history of land plants by analyzing the molecular features of modern plants, and by comparing the structural features of fossil and modern plants. (Figures 30.14, 30.15)

- Seedless plants transformed Earth's ecology by fostering soil buildup and altering atmospheric chemistry and climate. Such plants also influenced the evolutionary history of terrestrial animal life. (Figures 30.16, 30.17, 30.18)

- The K/T meteorite or comet impact event that occurred 65 million years ago helped cause the extinction of previously dominant dinosaurs and many types of gymnosperms, leaving space into which angiosperms, insects, birds, and mammals diversified.

30.3 The Origin and Evolutionary Importance of the Plant Embryo

- Origin of the plant embryo was a critical innovation that fostered diversification of the land plants. Plant embryos are supported by nutrients supplied by female gametophytes with the aid of specialized placental transfer tissues. (Figure 30.19)

- In a classic experiment, Browning and Gunning inferred that placental transfer tissues were responsible for an enhanced flow rate of nutrients from parental gametophytes to embryos. (Figure 30.20)

30.4 The Origin and Evolutionary Importance of Leaves and Seeds

- Leaves are specialized photosynthetic organs that evolved more than once during plant evolutionary history. Lycophylls, which occur in lycophytes, are relatively small leaves having a single unbranched vein. Pteridophyte leaves are larger and have an extensively branched vascular system; they are known as euphylls. Fossils indicate that fern euphylls evolved from branched stem systems by dominance by one branch, flattening of the whole branch system, and development of photosynthetic tissue between the branches. (Figure 30.21)

- Seeds develop from ovules, megasporangia enclosed by leaflike integuments. Ovules develop into seeds after fertilization, following pollination. Pollen produces thin cellular tubes that deliver sperm cells to eggs produced by female gametophytes. Mature seeds contain an embryo sporophyte that develops from the zygote. Seeds also contain food stored within female gametophytes of gymnosperms or within endosperm tissues arising from double fertilization in angiosperms. (Figure 30.22)

- Seeds confer many reproductive advantages, including dormancy through unfavorable conditions, greater protection for embryos from mechanical and pathogen damage, seed coat modifications that enhance seed dispersal, and reduction of plant dependence on water for fertilization. Fossil seeds display stages in the evolution of seeds. (Figure 30.23)

TEST YOURSELF

1. The simplest and most ancient phylum of modern land plants is probably
   a. the pteridophytes.
   b. the cycads.
   c. the liverworts.
   d. the angiosperms.
   e. none of the above.

2. An important feature of land plants that originated during the diversification of charophycean algae is
   a. the sporophyte.
   b. spores, which are dispersed in air and coated with sporopollenin.
   c. tracheids.
   d. plasmodesmata.
   e. fruits.

3. A phylum whose members are also known as bryophytes is commonly known as
   a. liverworts.
   b. hornworts.
   c. mosses.
   d. all of the above.
   e. none of the above.

4. Plants possess a life cycle that involves alternation of two multicellular generations: the gametophyte and
   a. the lycophyte.
   b. the bryophyte.
   c. the pteridophyte.
   d. the lignophyte.
   e. the sporophyte.

5. The seed plants are also known as
   a. bryophytes.
   b. spermatophytes.
   c. pteridophytes.
   d. lycophytes.
   e. euphyllophytes.

6. A waxy cuticle is an adaptation that
   a. helps to prevent water loss from tracheophytes.
   b. helps to prevent water loss from charophyceans.
   c. helps to prevent water loss from bryophytes.
   d. aids in water transport within the bodies of vascular plants.
   e. all of the above.

7. Plant photosynthesis transformed a very large amount of carbon dioxide into decay-resistant organic compounds, thereby causing a historic low in atmospheric carbon dioxide levels during the geological period known as
   a. the Cambrian.
   b. the Ordovician.
   c. the Carboniferous.
   d. the Permian.
   e. none of the above.

8. Which phylum among the plants listed is likely to have the largest leaves?
   a. liverworts
   b. hornworts
   c. mosses
   d. lycophytes
   e. pteridophytes

9. Euphylls, also known as megaphylls, probably evolved from
   a. the leaves of mosses.
   b. lycophylls.
   c. branched stem systems.
   d. modified roots.
   e. none of the above.

10. A seed develops from
    a. a spore.
    b. a fertilized ovule.
    c. a microsporangium covered by integuments.
    d. endosperm.
    e. none of the above.

CONCEPTUAL QUESTIONS

1. Why do evolutionary biologists think that land plants evolved from ancestors related to modern charophycean algae?

2. Why have bryophytes such as mosses been able to diversify into so many species even though they have relatively small, dependent sporophytes?

3. What features help vascular plants to maintain stable internal water content?

EXPERIMENTAL QUESTIONS

1. What were the goals of the Browning and Gunning investigation?

2. How did Browning and Gunning prevent photosynthesis from occurring in moss sporophytes during the experiment (shown in the Feature Investigation in Figure 30.20), and why did they do this?

3. What measurements did Browning and Gunning make after adding an excess amount of unlabeled CO_2?

COLLABORATIVE QUESTIONS

1. Discuss at least one difference in environmental conditions experienced by early land plants and ancestral complex charophycean algae.

2. Discuss as many plant adaptations to land as you can.

---

www.brookerbiology.com

This website includes answers to the Biological Inquiry questions found in the figure legends and all end-of-chapter questions.

31

THE DIVERSITY OF MODERN GYMNOSPERMS AND ANGIOSPERMS

CHAPTER OUTLINE

The Madagascar periwinkle (*Catharanthus roseus*).

H umans have long recognized that plant extracts are useful in treating illnesses. For example, Native Americans used extracts of the flowering mayapple plant (*Podophyllum peltatum*) to treat skin tumors. Modern medicine has validated this practice. Compounds such as podophyllotoxin, obtained from the mayapple plant, are now used to treat leukemia and other medical conditions. Leukemia is also treated with vincristine, a drug extracted from another angiosperm, the beautiful Madagascar periwinkle (*Catharanthus roseus*). Vinblastine—another extract from *C. roseus*—is used to treat lymphatic cancers. Taxol, a compound used in the treatment of breast and ovarian cancers, was first discovered in extracts of the Pacific yew tree, a gymnosperm known as *Taxus brevifolia*.

Podophyllotoxin, vincristine, vinblastine, taxol, and many other plant-derived medicines are examples of plant secondary metabolites, which are distinct from the products of primary metabolism (carbohydrates, lipids, proteins, and nucleic acids). Secondary metabolites play essential roles in protecting plants from disease organisms and plant-eating animals, and they also aid plant growth and reproduction. Though all plants produce secondary metabolites, these natural products are exceptionally diverse in seed plants: the nonflowering gymnosperms and the flowering angiosperms.

In this chapter, we will learn how the hundreds of thousands of modern seed plants play many additional important roles in modern ecosystems and the lives of humans. This chapter also illustrates many diversity principles, including descent with modification, horizontal gene transfer, critical innovation, coevolution, biological services, and human impacts on biodiversity.

31.1 The Diversity of Modern Gymnosperms

Gymnosperms are plants that produce seeds that are exposed rather than enclosed in fruits, as is the case for angiosperms. The word gymnosperm, which means "naked seed," comes from the Greek *gymnos*, meaning naked (referring to the unclothed state of ancient athletes), and *sperma*, meaning seed. Gymnosperms inherited many adaptations to life on land from their ancestors (**Figure 31.1**) (see also Chapter 30). In this section, we will first consider some fossil plants that help to explain important gymnosperm traits. Then we will survey the structure, reproduction, and ecological roles of modern gymnosperm phyla.

Modern Gymnosperms Arose from Woody Ancestors

Most modern gymnosperms are woody shrubs or trees, such as the famous giant sequoias (*Sequoiadendron giganteum*) native to the Sierra Nevada of the western U.S. Giant sequoias are among Earth's largest organisms, weighing as much as 6,000 tons and reaching an amazing 100 m in height. The large size of sequoias and other trees is based on the presence of **wood**, a tissue composed of numerous pipelike arrays of empty, water-conducting cells whose walls are strengthened by an exceptionally tough secondary metabolite known as lignin. These properties enable woody tissues to transport water upward for great distances and also to provide the structural support needed for trees to grow tall and produce many branches.

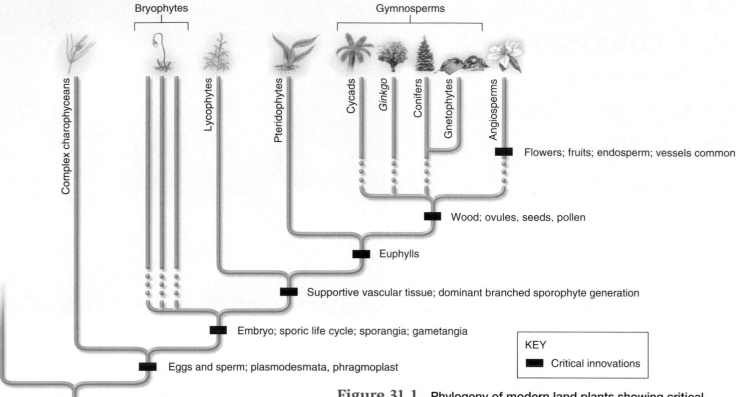

Figure 31.1 **Phylogeny of modern land plants showing critical innovations.** The dotted lines indicate uncertainty about branching order.

In modern seed plants, a special tissue known as the **vascular cambium** produces both thick layers of wood and thinner layers of inner bark. The inner bark transports watery solutions of organic compounds. (The structure and function of the vascular cambium, wood, and bark are described in more detail in Chapter 35.) Vascular cambium, wood, and inner bark are critical innovations that helped gymnosperms and other seed plants to compete effectively for light and other resources needed for photosynthesis.

Wood first appeared in a group of plants known as the **progymnosperms** (meaning "before gymnosperms"). Woody progymnosperms, such as the fossil plant genus *Archaeopteris*, which lived 370 million years ago, were the first true trees (**Figure 31.2**). Progymnosperms were able to produce a vascular cambium and wood because their vascular tissue was arranged in a ring around a central pith of nonvascular tissue. (In contrast, the vascular tissue of earlier tracheophytes was arranged differently.) This ring of vascular tissue, known as a **eustele**, contained cells that were able to develop into the vascular cambium as seedlings grew into saplings. The vascular cambium then produced wood, allowing saplings to grow into tall trees. Modern seed plants inherited the eustele, explaining why many gymnosperms and angiosperms are also able to produce vascular cambia and wood. Despite the fact that progymnosperms were woody plants, they did not produce seeds. This fact reveals that wood originated before seeds evolved.

Figure 31.2 **The progymnosperm *Archaeopteris*, an early tree.** This illustration was reconstructed from fossil data.

Table 31.1	Distinguishing Characteristics of Modern Seed Plant Phyla			
Seed plant group	Vessels in wood	Flowers, fruit, seed endosperm	Flagellate sperm	Female gametangia
Gymnosperms				
Cycads	No	No	Yes	Yes
Ginkgo biloba	No	No	Yes	Yes
Conifers	No	No	No	Yes
Gnetophytes	Yes	No	No	No
Angiosperms	Yes	Yes	No	No

Progymnosperms and diverse early gymnosperms were the major vegetation present during the Mesozoic era, also known as the Age of Dinosaurs. Some groups of gymnosperms became extinct before or as a result of the K/T event at the end of the Cretaceous period 65 million years ago (see Chapter 30). Only a few gymnosperm phyla have survived to modern times: cycads (the Cycadophyta); *Ginkgo biloba*, the only surviving member of a once large phylum termed Ginkgophyta; conifers (the Coniferophyta); and gnetophytes (Gnetophyta). These phyla display distinctive reproductive features and play important roles in ecology and human affairs (**Table 31.1**).

Cycads Are Endangered in the Wild but Widely Used as Ornamentals

Nearly 300 cycad species occur today, primarily in tropical and subtropical regions. However, many species of cycads are rare, and their tropical forest homes are increasingly threatened by human activities. Consequently, all cycads are listed as endangered species, and commercial trade in cycads is regulated by CITES (Convention on International Trade in Endangered Species of Wild Fauna and Flora), a voluntary international agreement.

The structure of cycads is so interesting and attractive that many species are cultivated for use in outdoor plantings or as houseplants. The nonwoody stems of some cycads emerge from the ground much like tree trunks, some reaching 15 m in height, while other cycads produce subterranean stems (**Figure 31.3**). Cycads display spreading, palmlike leaves (*cycad* comes from a Greek word meaning "palm"). Mature leaves of the African cycad *Encephalartos laurentianus* can reach an astounding 8.8 m in length!

In addition to underground roots, which provide anchorage and take up water and minerals, many cycads produce coralloid roots. Such roots extend aboveground and have branching shapes resembling corals (**Figure 31.4**). Coralloid root tissues harbor a bright blue-green ring of symbiotic cyanobacteria (see Figure 31.4 inset). These cyanobacteria use light to produce ATP by cyclic electron flow. The ATP helps to fuel cyanobacterial nitrogen fixation, which produces nitrogen minerals used in host growth (see Chapters 27, 37).

(a) Emergent cycad stem   **(b) Submergent cycad stem**

Figure 31.3   Cycads.   Palmlike foliage and conspicuous seed-producing cones are features of most cycads. **(a)** The stems of some cycads emerge from the ground. **(b)** The stems of other cycads are submerged in the ground.

Root surface

Cyanobacteria

(a) Coralloid roots

(b) Coralloid root cross section

Figure 31.4   Coralloid roots of cycads.   (a) Many cycads produce aboveground branching roots that resemble branched corals. **(b)** This magnified cross section of a coralloid root shows a ring of symbiotic blue-green cyanobacteria, which provide the plant with fixed nitrogen.

Biological inquiry: Why do the coralloid roots grow aboveground?

Recent studies have revealed that the cyanobacteria in *Cycas micronesica* produce an unusual amino acid that is distributed to the host plant's leaves and seeds. The distinctive amino acid, known as BMAA (β-N-methylamino-L-alanine), is harmful to the health of humans who consume flour made from cycad seeds or the meat of bats that have fed on *C. micronesica*. Botanist Paul Alan Cox and associates linked this toxin to the unusually high occurrence of a dementia resembling Alzheimer's disease among the Chamorro people of Guam in the Mariana Islands chain.

These researchers also found BMAA in the brain tissues of dementia patients who had not consumed foods originating from cycads. In an effort to understand these cases, Cox and associates examined many types of cyanobacteria from diverse habitats. In 2005, these investigators reported that most cyanobacteria produce BMAA in nature, suggesting that the toxic amino acid could be more widely present in the environment than previously thought. Studies of cycad toxicity have thus revealed a potential human health hazard of widespread concern. Because cycads generally produce a variety of toxins that likely deter herbivorous animals, experts recommend that humans should not consume food products made from these plants.

Cycad reproduction is distinctive in several ways. Individual cycad plants produce conspicuous conelike structures that bear either ovules and seeds, or pollen (see Figure 31.3). When mature, both types of reproductive structure emit odors that attract beetles. These beetles carry pollen to ovules, where the pollen produces tubes that deliver flagellate sperm to eggs.

Ginkgo Biloba Is the Last Survivor of a Once Diverse Group

The beautiful tree *Ginkgo biloba* is the single remaining species of a once diverse phylum that existed during the Age of Dinosaurs (**Figure 31.5a**). *G. biloba* takes its species name from the lobed shape of its leaves, which have unusual forked veins (**Figure 31.5b**). Today, *G. biloba* may be nearly extinct in the wild; widely cultivated modern *Ginkgo* trees are descended from seeds produced by a tree found in a remote Japanese temple garden and brought to Europe by 17th-century explorers.

G. biloba trees are widely planted along city streets because they are ornamental and also tolerate cold, heat, and pollution better than most trees. These trees are also long-lived. Individuals can live for more than a thousand years and grow to 30 m in height. Individual trees produce either ovules and seeds or pollen, based on a sex chromosome system much like that of humans. Ovule-producing trees have two X chromosomes; pollen-producing trees have one X and one Y chromosome. Wind disperses pollen to ovules, where pollen grains germinate to produce pollen tubes. These tubes grow through ovule tissues for several months, absorbing nutrients that are used for sperm development. Eventually the pollen tubes burst, delivering flagellate sperm to egg cells. After fertilization, zygotes develop into embryos, and the ovule integument develops into a fleshy, bad-smelling outer seed coat and a hard, inner seed coat. For streetside or garden plantings, people usually select pollen-producing trees to avoid the stinky seeds (**Figure 31.5c**).

Conifers Are the Most Diverse Modern Gymnosperm Lineage

The conifers are a lineage of trees named for their seed cones, of which pinecones are familiar examples (**Figure 31.6**). Conifers have been important components of terrestrial vegetation since the end of the Carboniferous period some 300 million years ago. Modern conifer families have existed for about 200 million years, survived the K/T event, and today include more than 500 species in 50 genera. Conifers are particularly common in mountain and high-latitude forests and are important sources of wood and paper pulp.

Conifers produce simple pollen cones and more complex ovule-bearing cones (**Figure 31.7**). The pollen cones of conifers bear many leaflike structures, each bearing a microsporangium in which meiosis occurs and pollen grains develop. By contrast, the ovule cones of conifers are composed of many short branch systems that bear ovules. No one could explain why conifer pollen and ovule cones were so different in structure until a new Coal Age fossil pollen cone came to light. In 2001, evolutionary botanist Genaro Hernandez-Castillo and colleagues reported that this early conifer pollen cone had the same complex structure as ovulate cones. This evidence suggests that modern, simple pollen cones probably evolved from more complex structures.

(a) *Ginkgo biloba* tree   **(b) *Ginkgo biloba* leaf**   **(c) *Ginkgo biloba* seed**

Figure 31.5 *Ginkgo biloba.* (a) A *Ginkgo biloba* tree; (b) fan-shaped leaves with forked veins; and (c) seeds having foul-smelling, fleshy seed coats.

(a) Pine (*Pinus ponderosa*)

(b) Dawn redwood (*Metasequoia glyptostroboides*)

Figure 31.6
Representative conifers.

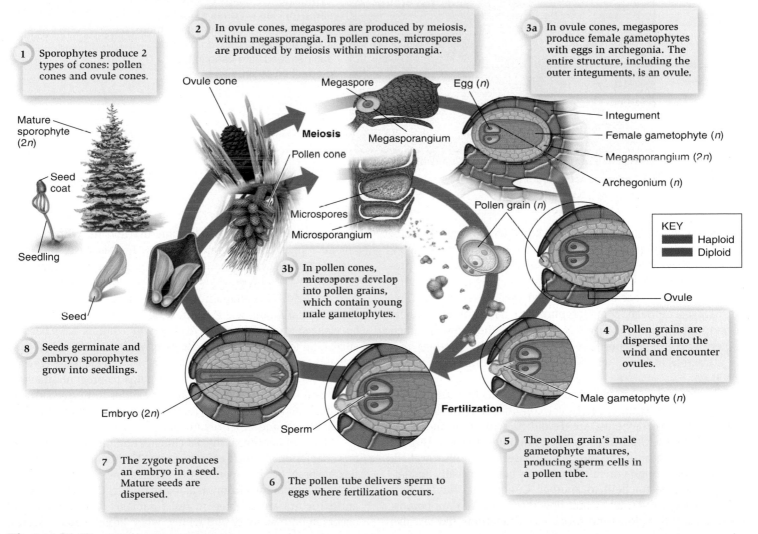

1 Sporophytes produce 2 types of cones: pollen cones and ovule cones.

2 In ovule cones, megaspores are produced by meiosis, within megasporangia. In pollen cones, microspores are produced by meiosis within microsporangia.

3a In ovule cones, megaspores produce female gametophytes with eggs in archegonia. The entire structure, including the outer integuments, is an ovule.

3b In pollen cones, microspores develop into pollen grains, which contain young male gametophytes.

4 Pollen grains are dispersed into the wind and encounter ovules.

5 The pollen grain's male gametophyte matures, producing sperm cells in a pollen tube.

6 The pollen tube delivers sperm to eggs where fertilization occurs.

7 The zygote produces an embryo in a seed. Mature seeds are dispersed.

8 Seeds germinate and embryo sporophytes grow into seedlings.

Ovule cone

Megaspore

Egg (*n*)

Meiosis

Megasporangium

Pollen cone

Microspores

Microsporangium

Integument

Female gametophyte (*n*)

Megasporangium (*2n*)

Archegonium (*n*)

Pollen grain (*n*)

Ovule

Male gametophyte (*n*)

Fertilization

Sperm

Embryo (*2n*)

Seed

Seedling

Seed coat

Mature sporophyte (*2n*)

KEY
Haploid
Diploid

Figure 31.7 The life cycle of *Pinus* spp.

(a) *Pinus* **spp. seed**   **(b)** *Taxus baccata* **seeds**

(c) *Juniperus scopularum* **cones**

Figure 31.8 Conifer seeds. (a) Winged, wind-dispersed seed of *Pinus* spp. (b) Fleshy-coated, bird-dispersed seeds of yew (*Taxus* spp.). (c) Fleshy cones of juniper (*Juniperus* spp.) contain one or more seeds and are dispersed by birds. Juniper seeds are used in the production of gin.

When conifer pollen is mature, it is released into the wind, which transports pollen to ovules. All together, it takes more than two years for pine (the genus *Pinus*) to complete the processes of male and female gamete development, fertilization by means of nonflagellate sperm, and seed development (Figure 31.7). The seed coats of pine and some other conifers develop wings that aid in wind dispersal. Other conifers, such as yew and juniper, produce seeds with fleshy coatings that are attractive to birds, which help to disperse the seeds (**Figure 31.8**).

Conifer wood contains water transport cells that are adapted for efficient conduction even in dry conditions. These transport cells, known as tracheids, are devoid of cytoplasm and occur in long columns that function like plumbing pipelines (**Figure 31.9a**). Tracheid side and end walls possess many thin-walled, circular **pits** through which water moves both vertically and laterally from one tracheid to another. Conifer pits have a porous outer region that lets water flow through and a nonporous, flexible central region—the **torus**—that functions like a valve

(**Figure 31.9b**). If tracheids become dry and fill with air, they are no longer able to conduct water. In this case, the torus presses against the pit opening, sealing it (**Figure 31.9c**). The torus valve thereby prevents air bubbles from spreading to the next tracheid. This adaptation localizes air bubbles, preventing them from stopping water conduction in other tracheids. These specialized tracheids help to explain why conifers have been so successful for hundreds of millions of years. Conifer wood (and leaves) may also display conspicuous resin ducts, passageways for the flow of syrup-like resin that helps to prevent attack by pathogens and herbivores. Resin that exudes from tree surfaces and hardens in the air may form amber, which traps and preserves insects and other organisms.

Many conifers occur in cold climates and thus display numerous adaptations to such environments. Their conical shapes and flexible branches help conifer trees shed snow, preventing heavy snow accumulations from breaking branches. Conifer leaf shape and structure are adapted to resist damage from drought that occurs in both summer and winter, when liquid water is scarce.

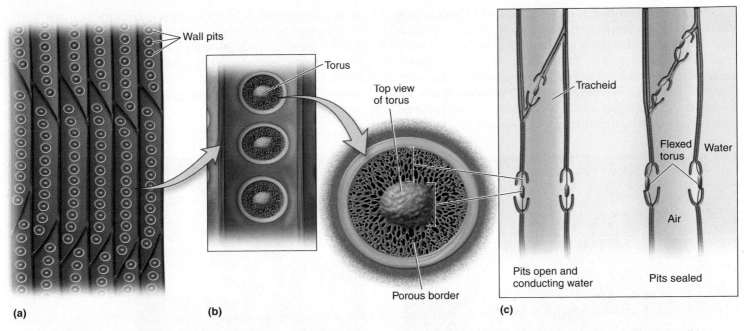

Figure 31.9 **Tracheids in conifer wood.** **(a)** The lignin-rich cell walls of tracheids, the major type of water-conducting cell in gymnosperms and seedless vascular plants. **(b)** Detailed view of a portion of a tracheid that shows the thin-walled areas known as pits, each with a torus. **(c)** A water-filled tracheid with open pit and an air-filled tracheid with pit sealed by the flexed torus.

Conifer leaves are often scalelike (**Figure 31.10a**) or needle-shaped (**Figure 31.10b**); these shapes reduce the area of leaf surface from which water can evaporate. In addition, a thick, waxy cuticle coats conifer leaf surfaces (**Figure 31.10c**), retarding water loss and attack by disease organisms.

Many conifers are evergreen; that is, their leaves live for more than one year before being shed and are not all shed during the same season. Retaining leaves through winter helps conifers start up photosynthesis earlier than deciduous trees, which in spring must replace leaves lost during the previous autumn. Evergreen leaves thus provide an advantage in the short growth season of alpine or high-latitude environments. However, some conifers do lose all their leaves in the autumn. The bald cypress (*Taxodium distichum*) of southern U.S. floodplains, tamarack (*Larix laricina*) of northern bogs, and dawn redwood (*Metasequoia glyptostroboides*) are examples of deciduous conifers. Fossils indicate that dawn redwoods once grew abundantly across wide areas of the Northern Hemisphere until a few million years ago and then disappeared. However, in the 1940s, a forester found a living dawn redwood growing in a remote Chinese village, and subsequent expeditions located forests of the conifers. Since then, dawn redwood trees have been widely planted as ornamentals, prized for their attractive foliage and cones. As recently as 1994, botanists found a previously unknown conifer species, *Wollemia nobilis*, in an Australian national park. Like the dawn redwood, *Wollemia* is an attractive tree that is likely to become more widely distributed as the result of human cultivation.

Gnetophytes Are of Evolutionary Interest

The modern gnetophytes consist of three unusual genera, *Gnetum*, *Ephedra*, and *Welwitschia*. *Gnetum* is unusual among modern gymnosperms in having broad leaves similar to those of many tropical plants (**Figure 31.11a**). More than 30 species of the genus *Gnetum* occur as vines, shrubs, or trees in tropical Africa or Asia. *Ephedra*, a gnetophyte genus native to arid regions of the southwestern U.S., has tiny brown scale-like leaves and green, photosynthetic stems (**Figure 31.11b**). *Ephedra* produces secondary metabolites that aid in plant protection but also affect human physiology. Early settlers of the western U.S. used *Ephedra* to treat colds and other medical conditions. In fact, the modern decongestant drug pseudoephedrine is based on the chemical structure of ephedrine, which was named for and originally obtained from *Ephedra*. Pseudoephedrine sales are now restricted in many places because this compound can be used as a starting point for the synthesis of illegal drugs. Ephedrine has also been used to enhance sports performance, a practice that has elicited medical concern.

Welwitschia, the third gnetophyte genus, has only one living representative species. *Welwitschia mirabilis* is a strange-looking plant that grows in the coastal Namib Desert of southwestern Africa, one of the driest places on Earth (**Figure 31.11c**). A long taproot anchors a stubby stem that barely emerges from the ground. Two very long leaves grow from the stem but rapidly become wind-shredded into many strips. The plant is thought to obtain most of its water from coastal fog, explaining how *W. mirabilis* can grow and reproduce in such a dry place.

Evolutionary plant biologists are interested in gnetophytes because their relationships to other seed plants are unclear. Some morphological features suggest that gnetophytes might be closely related to the angiosperms. For example, female gametangia (archegonia; refer back to Figure 30.7) are present in the ovules of most gymnosperms but have been lost from gnetophytes and angiosperms. Gnetophytes also possess vessels—a type of large-diameter water-conducting feature—in their vascular tissue, as do most angiosperms but not other gymnosperms (see Table 31.1). However, the vessels of gnetophytes are thought to have evolved independently of angiosperm vessels. In addition, recent molecular evidence links gnetophytes more closely to the conifers or specifically with the conifer genus *Pinus* (pine). The identity of the gymnosperm group from which the angiosperms arose remains unresolved.

(a) Scale-shaped leaves

(b) Needle-shaped leaves of pine

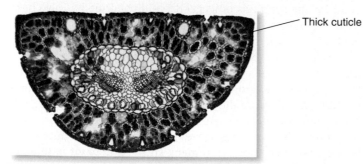

Thick cuticle

(c) Stained cross section of pine needle, showing the thick cuticle

Figure 31.10   Conifer leaves.

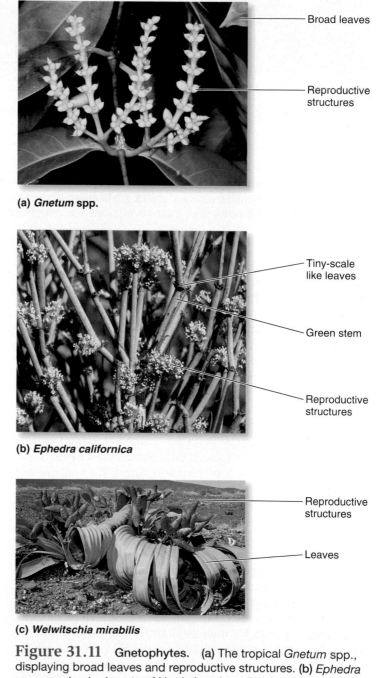

Broad leaves

Reproductive structures

(a) *Gnetum* spp.

Tiny-scale like leaves

Green stem

Reproductive structures

(b) *Ephedra californica*

Reproductive structures

Leaves

(c) *Welwitschia mirabilis*

Figure 31.11   Gnetophytes.   **(a)** The tropical *Gnetum* spp., displaying broad leaves and reproductive structures. **(b)** *Ephedra* spp. growing in deserts of North America, showing miniscule brown leaves on green, photosynthetic stems and reproductive structures. **(c)** *Welwitschia mirabilis* growing in the Namib Desert of southwestern Africa, showing long, wind-shredded leaves and reproductive structures.

31.2 The Diversity of Modern Angiosperms

Angiosperms, the flowering plants, retained many structural and reproductive features from ancestral plants. In addition, flowering plants evolved several traits not found or seldom found among gymnosperms and other land plants. Flowers and fruits are two of the defining features of angiosperms (**Figure 31.12**), because these features do not occur in other modern plants (see Table 31.1). The term **angiosperm** means "enclosed seed," which reflects the presence of seeds within fruits. Seed endosperm is another defining feature of the flowering plants (see Chapters 30 and 39).

Although humans obtain wood, medicines, and other valuable products from gymnosperms, we depend even more on the angiosperms. Our food, beverages, and spices—flavored by an amazing variety of secondary metabolites—primarily come from flowering plants. People surround themselves with ornamental flowering plants and decorative items displaying flowers or fruit. We also commonly use flowers and fruit in ceremonies. In this section, we focus on how flowers, fruits, and secondary metabolites played key roles in angiosperm diversification. We will also learn that features of flowers, fruits, and secondary metabolites are used to classify and identify angiosperm species.

Diverse Flower Types Are Adaptations That Foster Seed Production

Flowers are complex reproductive structures that are specialized for the efficient production of pollen and seeds. The sexual reproduction process of angiosperms depends on flowers. Thus, as the flowering plants diversified, flowers of varied types evolved as reproductive adaptations to differing environmental conditions. To understand this process, let's start by considering the basic flower parts and their roles in reproduction.

Flower Parts and Their Reproductive Roles Flowers are produced at stem tips and contain four types of organs: **sepals**, **petals**, pollen-producing **stamens**, and ovule-producing **carpels**

(**Figure 31.13**). These flower organs are supported by tissue known as a **receptacle**, located at the tip of a flower stalk—a **peduncle**. The functioning of several genes that control flower organ development explains why carpels are the central-most flower organs, why stamens surround carpels, and why petals and sepals are the outermost flower organs (refer back to Figure 19.26).

Many flowers produce attractive petals that play a role in **pollination**, the transfer of pollen among flowers. Sepals of many flowers are green and form the outer, protective layer of flower buds. By contrast, the sepals of other flowers resemble attractive petals. All of a flower's petals and sepals are collectively known as the **perianth**. Most flowers produce one or more stamens, the structures that produce and disperse pollen. Most flowers also contain carpels, structures that produce ovules. Some flowers lack perianths, stamens, or carpels. Flowers that possess all four types of flower organs are known as **complete flowers**, while flowers lacking one or more organ types are known as **incomplete flowers**. Flowers that contain both stamens and carpels are described as **perfect flowers**, while flowers lacking either stamens or carpels are **imperfect flowers**.

Flowers also differ in the numbers of organs they produce. Some flowers produce only a single carpel, others display several separate carpels, and many possess several carpels that are fused together into a compound structure. Both a single carpel and compound carpels are referred to as a **pistil** (from the Latin word *pistillum*, pestle), because it resembles the device people use to grind materials to powder in a mortar (see Figure 31.13). Only one pistil is present in flowers that have only one carpel and in flowers with fused carpels. By contrast, flowers possessing several separate carpels display multiple pistils.

Pistils are usually differentiated into three regions having distinct functions. A topmost portion of the pistil, known as the **stigma**, receives and recognizes pollen of the appropriate species or genotype. The stigma allows pollen of appropriate genetic type to germinate, producing a long pollen tube that grows through the elongate **style**. The pollen tube thereby delivers non-flagellate sperm cells to ovules and the eggs inside, allowing fertilization (**Figure 31.14**). If fertilization occurs, the ovule

Figure 31.12 **Angiosperm flowers and fruits.** Citrus plants display the defining features of flowering plants: flowers and fruits shown here and seed endosperm.

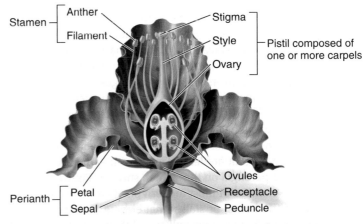

Figure 31.13 Generalized flower structure.

(a) Cherry (*Prunus* spp.)   **(b) Grape (*Vitis* spp.)**   **(c) Lemon (*Citrus* spp.)**

(d) Strawberry (*Fragaria* spp.)   **(e) Pineapple (*Ananas* spp.)**   **(f) Pea (*Pisum* spp.)**   **(g) Maple (*Acer* spp.)**

Figure 31.21   **Representative fruit types.**   **(a–c)** Cherry, grape, and lemon are fleshy, sweet fruits adapted to attract animals that consume the fruits and excrete the seeds. **(d)** Strawberry is an aggregate fruit, consisting of many tiny, single-seeded fruits produced by a single flower. The fruits are embedded in the surface of a fleshy receptacle that is adapted to attract animal seed dispersal agents. **(e)** Pineapple is a large multiple fruit formed by the aggregation of smaller fruits, each produced by one of the flowers in an inflorescence. **(f)** Peas produce legumes, fruits that open on two sides to release seeds. **(g)** Maple trees produce dry fruits with wings adapted for wind dispersal.

Flowers have also diversified by fusing different types of flower organs together, and orchids provide an interesting example. Orchid stamens and carpels are fused together into a single reproductive column that is surrounded by attractive petals and sepals (see Figure 31.18b). This arrangement of flower organs fosters orchid pollination by particular insects and is a distinctive feature of the orchid family. The sunflower family features unique inflorescences, groups of small flowers gathered into a head (see Figure 31.19c). Heads allow pollinators to transfer pollen among a large number of flowers at the same time (see Figure 31.20a). The grass family features flowers having few or no perianths, explaining why grass flowers are not showy (see Figure 31.18c). This adaptation fosters wind pollination, because petals would only get in the way of pollen transfer by wind. Diversification of fruit types, which we will consider next, has likewise contributed to the vast number of modern flowering plants.

Diverse Types of Fruits Function in Seed Dispersal

Fruits are structures that develop from ovary walls in diverse ways that aid the dispersal of enclosed seeds. Seed dispersal helps to prevent seedlings from competing with their larger par-

ents for scarce resources such as water and light. Dispersal of seeds also allows plants to colonize new habitats. Diverse fruit types illustrate many ways in which plants have become adapted for effective seed dispersal. Like flower types, fruit types are useful in classifying and identifying angiosperms.

Many mature angiosperm fruits, such as cherries, grapes, and citrus, are attractively colored, soft, juicy, and tasty (**Figure 31.21a–c**). Such fruits are adapted to attract animals that consume the fruits, digest the outer portion as food, and eliminate the seeds, thereby dispersing them. Hard seed coats prevent such seeds from being destroyed by the animal's digestive system. Strawberries are actually an aggregation of many fruits that all develop from a single flower having multiple pistils (**Figure 31.21d**). The ovaries of these pistils develop into the tiny, single-seeded yellow fruits on a strawberry surface; the fleshy, red, sweet portion of a strawberry develops from the flower receptacle. The strawberry system of aggregate fruits allows a single animal consumer, such as a bird, to disperse many seeds at the same time.

Pineapples (**Figure 31.21e**) are juicy multiple fruits that develop when many ovaries of an inflorescence fuse together. Multiple fruits have the advantage of being larger than individual fruits of the same species. Larger fruits attract relatively large animals that have the ability to disperse seeds for long distances.

Mulberries and figs are additional examples of multiple fruits that provide similar benefits.

The plant family informally known as **legumes** is named for its distinctive fruits, dry pods that open down both sides when seeds are mature, thereby releasing them (**Figure 31.21f**). Nuts and grains are additional examples of dry fruits, which have the advantage of rotting less rapidly than juicy fruits. **Grains** are the characteristic single-seeded fruits of cereal grasses such as rice, corn (maize), barley, and wheat. Maple trees produce dry and thus lightweight fruits having wings, features that foster effective wind dispersal (**Figure 31.21g**). Other plants produce dry fruits having surface burrs that attach to animal fur or possess floating fruits that disperse in water. Thus, flowering plants have diverse mechanisms for dispersing their seeds.

Angiosperms Produce Diverse Secondary Metabolites That Play Important Roles in Structure, Reproduction, and Protection

As we discussed in Chapter 7, secondary metabolism involves the synthesis of molecules that are not essential for cell structure and growth. These molecules, called **secondary metabolites**, are produced by various prokaryotes, fungi, protists, and plants but are most diverse in the angiosperms. About 100,000 different types of secondary metabolites are known, most of these produced by flowering plants. Because secondary metabolites play essential roles in plant structure, reproduction, and protection, diversification of these compounds has influenced flowering plant evolution. Three major classes of plant secondary metabolites occur: (1) terpenes and terpenoids; (2) phenolics, which include flavonoids and related compounds; and (3) alkaloids (**Figure 31.22**).

About 25,000 types of plant terpenes and terpenoids are derived from units of the hydrocarbon gas isoprene. Taxol, widely used in the treatment of cancer, is a terpene, as are citronella and a variety of other compounds that repel insects. Rubber, turpentine, rosin, and amber are complex terpenoids that likewise serve important roles in plant biology as well as having useful human applications.

Phenolic compounds are responsible for some flower and fruit colors as well as the distinctive flavors of cinnamon, nutmeg, ginger, cloves, chilies, and vanilla. Phenolics absorb ultraviolet radiation, thereby preventing damage to cellular DNA. They also help to defend plants against insects and disease microbes. Some phenolic compounds found in tea, red wine, grape juice, and blueberries have antioxidant properties that prevent the formation of damaging free radicals.

Alkaloids are nitrogen-containing secondary metabolites that often have potent effects on the animal nervous system. Plants produce at least 12,000 types of alkaloids, and some plants produce many alkaloids. Caffeine, nicotine, morphine, ephedrine, cocaine, heroin, and codeine are examples of alkaloids that influence the physiology and behavior of humans and are thus of societal concern. Like flower and fruit structure, secondary metabolites are useful in distinguishing among Earth's hundreds of thousands of flowering plant species.

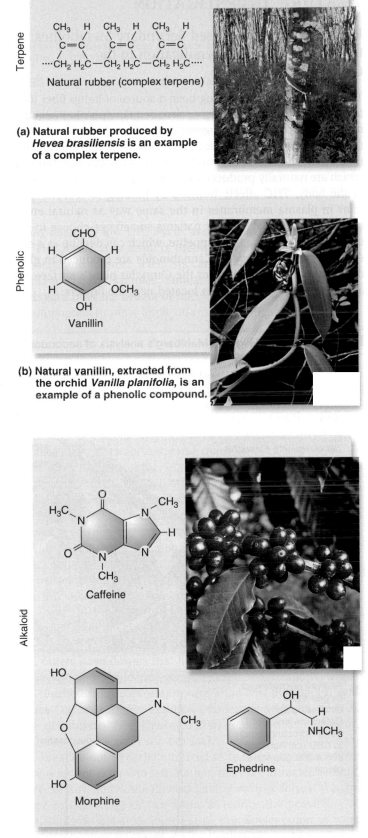

(a) **Natural rubber produced by** *Hevea brasiliensis* **is an example of a complex terpene.**

(b) **Natural vanillin, extracted from the orchid** *Vanilla planifolia*, **is an example of a phenolic compound.**

(c) **Alkaloids include caffeine produced by** *Coffea arabica*, **morphine, and ephedrine.**

Figure 31.22 Major types of plant secondary metabolites.

Table 31.2	Pollination Syndromes	
Animal features	**Coevolved flower features**
Bees	
Color vision includes UV, not red	Often blue, purple, yellow, white (not red) colors
Good sense of smell	Fragrant
Require nectar and pollen	Provide nectar and abundant pollen
Butterflies	
Good color vision	Blue, purple, deep pink, orange, red colors
Sense odors with feet	Light floral scent
Need landing place	Provide landing place
Feed with long tubular tongue	Nectar in deep, narrow floral tubes
Moths	
Active at night	Open at night; white or bright colors
Good sense of smell	Heavy, musky odors
Feed with long, thin tongue	Nectar in deep, narrow floral tubes
Birds	
Color vision, includes red	Often colored red
Often require perch	Strong, damage-resistant structure
Poor sense of smell	No fragrance
Feed in daytime	Open in daytime
High nectar requirement	Copious nectar in floral tubes
Hover (hummingbirds)	Pendulous (dangling) flowers
Bats	
Color blind	Light, reflective colors
Good sense of smell	Strong odors that attract bats
Active at night	Open at night
High food requirements	Copious nectar and pollen provided
Navigate by echolocation	Pendulous or borne on tree trunks

Figure 31.24 *Brighamia insignis*, an endangered plant. The pollinator that coevolved with *B. insignis* has become extinct, with the result that the plant is unable to produce seed unless artificially pollinated by humans.

Biological inquiry: What kind of animal likely pollinated B. insignis?

Pollination syndromes are also of practical importance in agriculture and in conservation biology. Fruit growers often import bees to pollinate flowers of fruit crops, thus increasing crop yields. Some plants have become so specialized to particular pollinators that if the pollinator becomes extinct, the plant becomes endangered. An example is the Hawaiian cliff-dwelling *Brighamia insignis* (**Figure 31.24**), whose presumed moth pollinator has become extinct. Humans that hand-pollinate *B. insignis* are all that stand between this plant and extinction.

Seed-Dispersal Coevolution Influences the Evolution of Fruits and Particular Animals

As in the case of pollination, coevolution between plants and their animal seed-dispersal agents has influenced both plant fruit characteristics and those of seed-dispersing animals. In addition, flowering plant fruits provide food for animals, an important biological service. For example, many of the plants of temperate forests produce fruits that are attractive to resident birds. Such juicy, sweet fruits have small seeds that readily pass

through bird guts. Many plants signal fruit ripeness by undergoing color changes from unripe green fruits to red, orange, yellow, blue, or black (**Figure 31.25**). Because birds have good color vision, they are able to detect the presence of ripe fruits and consume them before the fruits drop from plants and rot. Apples, strawberries, cherries, blueberries, and blackberries are examples of fruits whose seed dispersal adaptations have made them attractive food for humans as well. By contrast, the lipid-rich fruits of Virginia creeper (*Parthenocissus quinquefolia*) and some other autumn-fruiting plants energize migratory birds but are not tasty to humans. The Virginia creeper's leaves often turn fall colors earlier than surrounding plants, thereby signaling the availability of nutritious, ripe fruit to high-flying birds. It is important to such plants that lipid-rich fruits be consumed expeditiously because they rot easily, in which case seed dispersal cannot occur.

Primates of tropical Asia and Africa, but not generally those of Central and South America, have **trichromatic color vision**—the ability to distinguish blue, green, and red colors. Trichromatic color vision in Old World primates is the result of a gene duplication event that occurred early in their evolution, but after their divergence from New World primates. Primate specialists have speculated that trichromatic vision may enable Old World tree-dwelling primates to better detect the young, tender leaves of tropical trees, which are often slightly redder in pigmentation than mature leaves. Trichromatic vision also allows primates to more easily detect yellow, red, or orange ripe fruits against a background of leafy green foliage. Since humans evolved from primates native to Africa, we inherited their trichromatic color vision, helping to explain why humans find brightly colored flowers and fruits so attractive.

Figure 31.25 Fruits attractive to animal seed dispersal agents.   Color and odor signals alert coevolved animal species that fruits are ripe, thus favoring the dispersal of mature seeds.

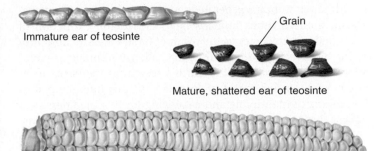

Immature ear of teosinte

Grain

Mature, shattered ear of teosinte

Nonshattering ear of *Z. mays*

Figure 31.26 Ears and grains of modern corn and its ancestor, teosinte.   Domesticated corn ears are larger than those of the ancestral grass teosinte. Corn fruits are softer and more edible than are grains of teosinte.

Biological inquiry: In what other way do corn ears differ from those of teosinte?

31.4 Human Influences on Angiosperm Diversification

By means of the process known as **domestication**, which involves artificial selection for traits desirable to humans, ancient humans transformed wild plant species into new crop species. More recently, human populations have increased so much that larger areas of natural habitat are being transformed for human use. Deforestation, for example, often results from the conversion of forests to agricultural land. Such habitat destruction is a leading cause of the extinction of plant and other species. In this section, we will consider how humans have produced new crop species and influenced the loss of wild species.

Humans Produced New Crop Plants by Domesticating Wild Species

Between 10,000 and 5,000 years ago, agriculture originated independently in eight different locations around the world. One of the earliest domesticated crops was an African plant commonly known as the bottle gourd (*Lagenaria siceraria*). The bottle gourd was grown for use as containers, musical instruments, and floats for fishing. For planting crops, humans selected seeds from gourds that had thicker rinds because these resisted breakage better than wild gourds. These differences in rind thickness can be detected in fossil remains that are about 10,000 years old. The fossils indicate that bottle gourds were grown as a crop in Asia and from there were transported to the Americas by ancient human colonists.

Cultivated bread wheat (*Triticum aestivum*) was probably among the earliest food crops, having originated more than 8,000 years ago, in what is now southeastern Turkey and northern Syria. Bread wheat originated by a series of steps from wild ancestors (*Triticum boeoticum* and *Triticum dicoccoides*). Among the earliest changes that occurred during wheat domestication was the loss of **shattering**, the process by which ears of wild wheat break apart and disperse the single-seeded fruits known as grains. A mutation probably caused the ears of some wheat plants to remain intact, a trait that is disadvantageous in nature but beneficial to humans. Nonshattering ears would have been easier for humans to harvest than normal ears. Early farmers probably selected plants as seed stock having nonshattering ears and other favorable traits such as larger grains. These ancient human selection processes, together with modern breeding efforts, explain why cultivated wheat differs from its wild relatives in shattering and other properties. The accumulation of these trait differences explains why cultivated and wild wheat plants are classified as different species.

About 6,500 years ago, people living in what is now Mexico domesticated a native grass known as teosinte (*Zea* spp.), producing a new species, *Zea mays*, known as corn or maize. The evidence for this pivotal event includes ancient ears that were larger than wild ones and distinctive fossil pollen. Modern ears of corn are much larger than those of teosinte, corn grains are larger and softer, and modern corn ears do not shatter, as do those of ancestral teosinte (**Figure 31.26**). These and other trait changes reflect artificial selection accomplished by humans. An analysis of the corn genome, reported in 2005 by Stephen Wright, Brandon Gaut, and coworkers, suggests that human selection has influenced about 1,200 genes.

Molecular analyses indicate that domesticated rice (*Oryza sativa*) originated from ancestral wild species of grasses (*Oryza nivara* and/or *Oryza rifipogon*). As in the cases of wheat and corn, domestication of rice involved loss of ear shattering. In 2006, Changbao Li, Ailing Zhou, and Tao Sang reported that a key amino acid substitution was primarily responsible for the loss of ear shattering in rice. As a result of this mutation, rice ears remain intact. Ancient humans might have unconsciously selected for this mutation while gathering rice from wild populations, because the mutants would not so easily have shed grains during the harvesting process. Eventually, the nonshattering mutant became a widely planted crop throughout Asia, and today it is the food staple for millions of people.

Although humans generated these and other new plant species, in modern times humans have caused the extinction of plants and other species as the result of habitat destruction. Protecting biodiversity will continue to challenge humans as populations and demands on the Earth's resources increase. Techniques such as the use of molecular barcodes have helped in the process of identifying and cataloging Earth's biodiversity, a process that aids in habitat conservation and restoration efforts.

GENOMES & PROTEOMES

DNA Barcodes Help in Identifying and Cataloging Plants

In view of the fact that flowering plants provide many essential biological services, ecologists have become concerned about current and future extinctions caused by human alterations of the environment. Plant extinction is a tragedy because so many plants produce useful secondary metabolites (including those having medicinal value) or serve as important food sources for other organisms. Botanists have responded by attempting to catalog the world's 300,000 species of plants, most of which are angiosperms. A globally useful cataloging system can help biologists keep track of all these plants, identify new species, and spot threats to plant biodiversity. One such system is **molecular barcodes**, DNA sequences that are used to identify and catalog Earth's biodiversity.

Sequences of a short region of the mitochondrial DNA have proved useful in barcoding animal species. Because each species has a distinctive genetic code for this region, species identifications can be made from tissue samples. Biologists accomplish this by extracting DNA from the samples and using polymerase chain reaction (PCR) techniques to make many copies of the barcode region, which can then be sequenced (see Chapter 20). Biologists then compare barcode sequences to those of other organisms in genetic databases, using computer methods to determine the specimen's closest relatives. Future miniaturization of the DNA processing and sequencing steps could allow field biologists to transmit information to databases and receive results, thereby identifying known or new organisms without removing them from their natural habitats.

Barcoding plant species has been a problem, because plant mitochondrial DNA lacks sufficient variation to distinguish closely related plant species. In other words, two closely related plant species may have identical mitochondrial gene sequences. For this reason, plant scientist W. John Kress and colleagues looked for a plastid DNA sequence that would parallel the use of the mitochondrial barcode in animals. By checking the variability of several plastid genes from a wide range of angiosperm species, including pairs of close relatives, these investigators identified a specific plastid gene for cataloging flowering plants. This sequence region can be easily amplified, even from dried specimens as old as 20 years. This work, reported in 2005, provides the information needed to barcode even larger arrays of plants, such as the flowering plants of entire countries. Such extensive barcoding efforts are expected to foster conservation and restoration projects, thereby helping to protect Earth's biodiversity.

CHAPTER SUMMARY

31.1 The Diversity of Modern Gymnosperms

- The seed plants inherited features of ancestral nonseed plants but display distinctive adaptations. The major phyla of seed plants mainly differ in reproductive features. (Figure 31.1)

- Gymnosperms are plants that produce exposed seeds rather than seeds enclosed in fruits. Many gymnosperms produce wood by means of a special tissue called vascular cambium.

- Several phyla of gymnosperms once existed but have become extinct and are known only from fossils. The diversity of modern gymnosperms includes four modern phyla: cycads, *Ginkgo biloba*, the conifers, and the gnetophytes. (Figure 31.2, Table 31.1)

- Cycads primarily live in tropical and subtropical regions. Features of cycads include palmlike leaves, nonwoody stems, coralloid roots with cyanobacterial endosymbionts, toxins, and large conelike seed-producing structures. (Figures 31.3, 31.4)

- *Ginkgo biloba* is the last surviving species of a phylum that was diverse during the Age of Dinosaurs. (Figure 31.5)

- Conifers have been widespread and diverse members of plant communities for the past 300 million years, and they are important sources of wood and paper pulp. Reproduction involves simple pollen cones and complex ovule-producing cones. Conifer wood contains water-conducting tracheids with thin-walled pits that allow water to move from one tracheid to another. Many conifers display additional adaptations that help them to survive in cold climates. (Figures 31.6, 31.7, 31.8, 31.9, 31.10)

- The gnetophytes are of particular interest to plant evolutionary biologists because their relationships to other seed plants are unclear. (Figure 31.11)

31.2 The Diversity of Modern Angiosperms

- Angiosperms inherited seeds and other features from ancestors but display distinctive features, such as flowers and fruits. Diversification of flower structure, fruit structure, and secondary metabolites has played important roles in angiosperm evolution. (Figure 31.12)

- Flowers foster seed production and are adapted in various ways that aid pollination in varying circumstances. The major flower organs are sepals, petals, stamens, and carpels, but some flowers lack one or more of these organs. Pollination is the transfer of pollen from a stamen to a pistil. Pistils display regions of specialized function:

the stigma is a receptive surface for pollen, pollen tubes grow through the style, and ovules develop within the ovary. If pollen tubes successfully deposit sperm near eggs in ovules, and fertilization occurs, ovules develop into seeds, and ovaries develop into fruits. (Figures 31.13, 31.14)

- Stamens and carpels may have evolved from leaflike structures bearing sporangia. The two largest lineages of flowering plants, the monocots and eudicots, diverged more than 120 million years ago. (Figures 31.15, 31.16, 31.17, 31.18, 31.19)

- Horizontal gene transfer occurs between modern monocots and eudicots, and between these plants and earlier-diverging lineages, with potential effects on evolution and diversification.

- Flower diversification involved evolutionary changes such as fusion of parts, changes in symmetry, loss of parts, and aggregation into inflorescences. (Figure 31.20)

- Fruits are structures that enclose seeds and aid in their dispersal. Fruits occur in many types that foster seed dispersal in varying circumstances. (Figure 31.21)

- Angiosperms produce three main groups of secondary metabolites: (1) terpenes and terpenoids; (2) phenolics, flavonoids, and related compounds; and (3) alkaloids. Secondary metabolites play essential roles in plant structure, reproduction, and defense. (Figure 31.22)

- Hillig and Mahlberg demonstrated the use of particular secondary metabolites in distinguishing species of the genus *Cannabis*. (Figure 31.23)

31.3 The Role of Coevolution in Angiosperm Diversification

- Coevolutionary interactions with animals that serve as pollen- and seed-dispersal agents played a powerful role in the diversification of both flowering plants and animals. (Figures 31.24, 31.25)

- Some flowers form specialized relationships with specific pollinators. These interdependent relationships are known as pollination syndromes. (Table 31.2)

- Human appreciation of flowers and fruits is based on sensory systems similar to those present in the animals with which angiosperms coevolved.

31.4 Human Influences on Angiosperm Diversification

- Humans have produced new crop species by domesticating wild plants. The process of domestication involved artificial selection for traits such as nonshattering ears of wheat, corn, and rice. (Figure 31.26)

- Large human populations are causing habitat destruction leading to species extinction. Genetic barcodes allow biologists to identify and catalog plants, thus aiding biodiversity conservation and habitat restoration efforts.

TEST YOURSELF

1. What feature must be present for a plant to produce wood?
   a. a type of conducting system in which vascular bundles occur in a ring around pith
   b. a eustele
   c. a vascular cambium
   d. all of the above
   e. none of the above

2. What is the correct order of evolution for these critical adaptations?
   a. embryos, vascular tissue, wood, seeds, flowers
   b. vascular tissue, embryos, wood, flowers, seeds
   c. vascular tissue, wood, seeds, embryos, flowers
   d. wood, seeds, embryos, flowers, vascular tissue
   e. seeds, vascular tissue, wood, embryos, flowers

3. How long have gymnosperms been important members of plant communities?
   a. 10,000 years, since the dawn of agriculture
   b. 100,000 years
   c. 300,000 years
   d. 65 million years, since the K/T event
   e. 300 million years, since the Coal Age

4. What similar features do gnetophytes and angiosperms possess that differ from other modern seed plants?
   a. Gnetophytes and angiosperms both produce flagellate sperm.
   b. Gnetophytes and angiosperms both produce flowers.
   c. Gnetophytes and angiosperms both produce tracheids, but not vessels, in their vascular tissues.
   d. Gnetophytes and angiosperms both produce fruits.
   e. None of the above.

5. Which part of a flower receives pollen from the wind or a pollinating animal?
   a. perianth
   b. stigma
   c. filament
   d. peduncle
   e. ovary

6. The primary function of a fruit is to
   a. provide food for the developing seed.
   b. provide food for the developing seedling.
   c. disperse pollen.
   d. disperse seeds.
   e. none of the above.

7. What are some ways in which flowers have diversified?
   a. color
   b. symmetry
   c. fusion of organs
   d. aggregation into inflorescences
   e. all of the above

8. Flowers of the genus *Fuchsia* produce deep pink to red flowers that dangle from plants, produce nectar in floral tubes, and have no scent. Based on these features, which animal is most likely to be a coevolved pollinator?
   a. bee
   b. bat
   c. hummingbird
   d. butterfly
   e. moth

9. Which type of plant secondary metabolite is best known for antioxidant properties of human foods such as blueberries, tea, and grape juice?
   a. alkaloids
   b. cannabinoids
   c. carotenoids
   d. phenolics
   e. terpenoids

10. What features of domesticated grain crops might differ from those of wild ancestors?
    a. the degree to which ears shatter, allowing for seed dispersal
    b. grain size
    c. number of grains per ear
    d. softness and edibility of grains
    e. all of the above

CONCEPTUAL QUESTIONS

1. Explain why humans should not consume food products made from cycads.

2. Explain why fruits such as apples, strawberries, and cherries are attractive and harmless foods for humans.

3. Is a sunflower really a flower?

EXPERIMENTAL QUESTIONS

1. Why did the investigators in the Feature Investigation in Figure 31.23 obtain nearly a hundred *Cannabis* fruit samples from around the world?

2. Why did Hillig and Mahlberg grow plants in a greenhouse before conducting the cannabinoid analysis?

3. Why did Hillig and Mahlberg collect samples from the leaves growing nearest the flowers?

COLLABORATIVE QUESTIONS

1. Where in the world would you have to travel to find wild plants representing all of the gymnosperm phyla, including the three types of gnetophytes?

2. How would you go about trying to solve what Darwin called an "abominable mystery," that is, the identity of the seed plant group that was ancestral to the flowering plants?

www.brookerbiology.com

This website includes answers to the Biological Inquiry questions found in the figure legends and all end-of-chapter questions.

32

AN INTRODUCTION TO ANIMAL DIVERSITY

CHAPTER OUTLINE

32.1 Characteristics of Animals

32.2 Traditional Classification of Animals

32.3 Molecular Views of Animal Diversity

There is a staggering variety of animal species on Earth, including thousands of vertebrates and hundreds of thousands of invertebrates.

Animals constitute a very species-rich kingdom. Well over a million species have been found and described by biologists, with many more species likely awaiting discovery and classification. Beyond being members of this kingdom ourselves, humans are dependent on animals. We eat many different kinds of animals, use a diverse array of animal products, and employ animals such as horses and oxen as a source of labor. We enjoy many animal species as companions and are dependent on other species to test lifesaving drugs. On the other hand, we are in competition with animals such as insects that threaten our food supply and are in turn parasitized by others. With such a huge number and diversity of existing animals, and with animals featuring so prominently in our lives, understanding animal diversity is of paramount importance. Researchers have spent great effort in determining the unique characteristics of different taxonomic groups and identifying their evolutionary relationships.

Since the time of Carl Linnaeus in the 1700s, scientists have classified animals based on their morphology, that is, on their physical structure. Then, as now, a lively debate has surrounded the question of what constitutes the "correct" animal phylogeny. In the 1990s, animal classifications based on similarities in DNA and rRNA became more common. Quite often, classifications based on morphology and those based on molecular data were similar, but some important differences arose. In this chapter, we will begin by defining the key characteristics of animals and then take a look at the major features of the animal body plans that form the basis of the traditional view of animal classification. We will explore how new molecular evidence has made significant alterations to this categorization of the animal kingdom and examine some of the similarities and differences

between the morphological and molecular-based phylogenies. As more molecular-based evidence becomes available, systematists will likely continue to redraw the tree of animal life. Thus, as you read this chapter, keep in mind that animal classification is not set in stone but rather is a work in progress.

32.1 Characteristics of Animals

The Earth contains over a million known animal species living in environments from the deep sea to the desert and exhibiting an amazing array of characteristics. Most animals move and eat multicellular prey, and thus they are loosely differentiated from species in other kingdoms. However, coming up with a firm definition of an animal is tricky, because animals are so diverse that biologists can find exceptions to nearly any given characteristic. Even so, a number of key features exist that can help us broadly characterize the group we call animals (Table 32.1).

In brief, animals are multicellular heterotrophs that lack cell walls, and most have nerves, muscles, the capacity to move at some point in their life cycle, and the ability to reproduce sexually. Unlike plants, animals cannot synthesize their essential organic molecules from inorganic sources and so must feed on other organisms. Many, if not most, animals are capable of some type of movement or locomotion in order to acquire food. This ability has led to the development of specialized systems of sensory structures and a nervous system to coordinate movement and prey capture. Sessile species, such as barnacles, use bristled appendages to obtain food. In many such sessile species, while adults are immobile, the larvae can swim.

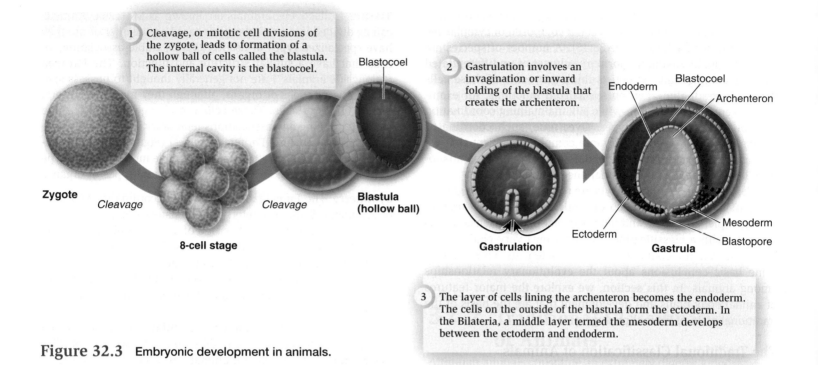

1 Cleavage, or mitotic cell divisions of the zygote, leads to formation of a hollow ball of cells called the blastula. The internal cavity is the blastocoel.

2 Gastrulation involves an invagination or inward folding of the blastula that creates the archenteron.

Blastocoel

Zygote

Cleavage   *Cleavage*   **Blastula (hollow ball)**

8-cell stage

Endoderm   Blastocoel   Archenteron

Gastrulation   **Gastrula**

Ectoderm   Mesoderm   Blastopore

3 The layer of cells lining the archenteron becomes the endoderm. The cells on the outside of the blastula form the ectoderm. In the Bilateria, a middle layer termed the mesoderm develops between the ectoderm and endoderm.

Figure 32.3   Embryonic development in animals.

called **germ layers**, while bilateral animals have three germ layers. In all animals except the sponges, the growing embryo develops different layers of cells through a process known as **gastrulation** to produce a **gastrula** (**Figure 32.3**). Fertilization of an egg by a sperm creates a diploid zygote. The zygote then undergoes **cleavage**, a succession of rapid cell divisions with no significant growth that produces a hollow sphere of cells called a **blastula**. In gastrulation, an area in the blastula invaginates and folds inward, creating in the process the primary germ layers. The inner layer of cells becomes the **endoderm**, which lines the **archenteron**, or primitive digestive tract. The outer layer, or **ectoderm**, covers the surface of the embryo and differentiates into the epidermis and nervous system.

The Bilateria develop a third layer of cells, termed the **mesoderm**, which develops between the ectoderm and endoderm. Mesoderm forms the muscles and most other organs between the digestive tract and the ectoderm. Because the Bilateria have these three distinct germ layers, they are often referred to as **triploblastic**, while the Radiata, which have only ectoderm and endoderm, are termed **diploblastic**.

Body Cavity   The next three significant divisions in the classification of animals concern the development of a fluid-filled body cavity called a **coelom** (**Figure 32.4**). In many animals, the body cavity is completely lined with mesoderm and is called a true coelom. Animals with a true coelom are termed **coelomates**. If the coelom is not completely lined by tissue derived from mesoderm, it is known as a **pseudocoelom**. Animals with a pseudocoelom are termed **pseudocoelomates** and include rotifers and roundworms. Some animals, such as flatworms, lack a

fluid-filled body cavity and are termed **acoelomates**. Instead of fluid, this region contains mesenchyme tissue. In some coelomate animals, such as mollusks and arthropods, the coelom is reduced to small pockets around the heart and excretory organs, and the main body cavity is known as the hemocoel. In these animals, a fluid called hemolymph bathes the tissues directly.

A body cavity has many important functions, perhaps the most important being that its fluid is relatively incompressible and thus cushions internal organs such as the heart and intestinal tract, helping to prevent injury from external forces. A body cavity also enables internal organs to move and grow independently of the outer body wall. Furthermore, in some soft-bodied invertebrates such as earthworms, the coelom functions as a **hydrostatic skeleton**, a fluid-filled body cavity surrounded by muscles that gives support and shape to the body of organisms. Because liquid is relatively incompressible, muscle contractions at one part of the body can push fluid toward another part of the body. This type of movement can best be observed in an earthworm (see Chapter 33). Finally, in some organisms, the fluid in the body cavity also acts as a simple circulatory system.

Embryonic Development   In the developing zygote, cleavage may occur by two mechanisms (**Figure 32.5a**). In **spiral cleavage**, the planes of cell cleavage are oblique to the axis of the embryo, resulting in an arrangement in which newly formed upper cells lie centered between the underlying cells. Animals that exhibit spiral cleavage are called **protostomes** and include mollusks, annelid worms, and arthropods. In **radial cleavage**, the cleavage planes are either parallel or perpendicular to the

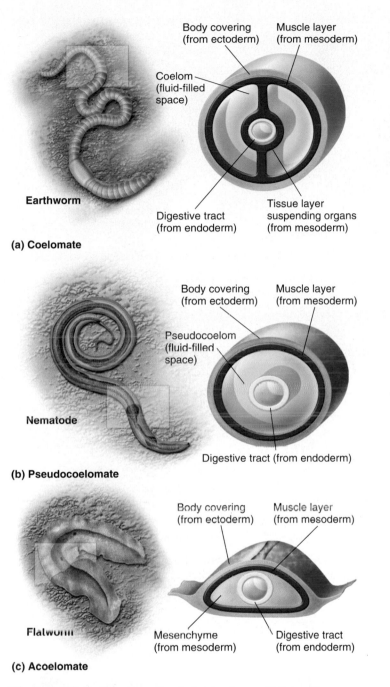

Earthworm

Body covering
(from ectoderm)

Muscle layer
(from mesoderm)

Coelom
(fluid-filled
space)

Digestive tract
(from endoderm)

Tissue layer
suspending organs
(from mesoderm)

(a) Coelomate

Nematode

Body covering
(from ectoderm)

Muscle layer
(from mesoderm)

Pseudocoelom
(fluid-filled
space)

Digestive tract (from endoderm)

(b) Pseudocoelomate

Flatworm

Body covering
(from ectoderm)

Muscle layer
(from mesoderm)

Mesenchyme
(from mesoderm)

Digestive tract
(from endoderm)

(c) Acoelomate

Figure 32.4 **The three basic body plans of bilaterally symmetrical animals.**

vertical axis of the egg. This results in tiers of cells, one directly above the other. Animals exhibiting radial cleavage are called **deuterostomes** and include echinoderms and vertebrates.

Protostome development is also characterized by so-called **determinate cleavage**, in which the fate of each embryonic cell is determined very early (**Figure 32.5b**). If we remove one of the cells from a four-cell mollusk embryo, neither the single cell nor the remaining three-cell mass can form viable embryos, and

development is halted. In contrast, deuterostome development is characterized by **indeterminate cleavage**, in which each cell produced by early cleavage retains the ability to develop into a complete embryo. For example, when one cell is excised from a four-cell sea urchin embryo, both the single cell and the remaining three can go on to form viable embryos. Other embryonic cells compensate for the missing cells. In human embryos, if individual embryonic cells separate from one another early in development, identical twins can result. Indeterminate embryonic cells are also known as stem cells and are found not just in the four-cell embryo but also in later embryonic stages. They are called pluripotent, because they can develop into almost every cell type of the body.

In addition to differences in cleavage patterns, protostomes and deuterostomes differ in two other embryonic features. The most fundamental of these concerns the development of a mouth and anus (**Figure 32.5c**). In gastrulation, the endoderm forms an indentation, the **blastopore**, which is the opening of the archenteron to the outside. In protostomes (from the Greek *protos*, first, and *stoma*, mouth), the blastopore becomes the mouth. If an anus is formed in a protostome, it develops from a secondary opening. In contrast, in the deuterostomes (Greek *deuteros*, second), the blastopore becomes the anus, and the mouth is formed from the secondary opening.

The last difference between protostomes and deuterostomes concerns coelom formation. In protostomes, a solid mass of mesoderm cells splits to form the cavity that becomes the coelom. This pattern is called **schizocoelous** development (from the Greek *schizo*, to split). In contrast, in deuterostomes, a layer of mesoderm cells form outpockets that bud off from the developing gut to form the coelom. This pattern is known as **enterocoelous** development. The resultant coeloms in both protostomes and deuterostomes are similar; they are just formed in different ways.

Other Methods of Classification In the traditional phylogenetic tree of animal life, further branches are based on features such as the possession of an exoskeleton (arthropods) or the development of a notochord (chordates). One other key feature of the animal body plan is the presence or absence of segmentation. In segmentation, also called **metamerism**, the body is divided into nearly identical subunits called segments. It is most obvious in the annelids, or segmented worms, in which each segment contains the same set of blood vessels, nerves, and muscles, but it is also evident in arthropods and chordates (**Figure 32.6**). Some segments may differ, such as those containing the brain or the sex organs, but most segments are very similar.

The advantage of segmentation is that it allows specialization of body regions. For example, in arthropods, some segments are specialized to form wings for flight. Recent studies have shown that changes in specialization among arthropod body segments can be traced to relatively simple changes in homeotic, or *Hox*, genes. In chordates, we can see segmentation in the backbone and muscles.

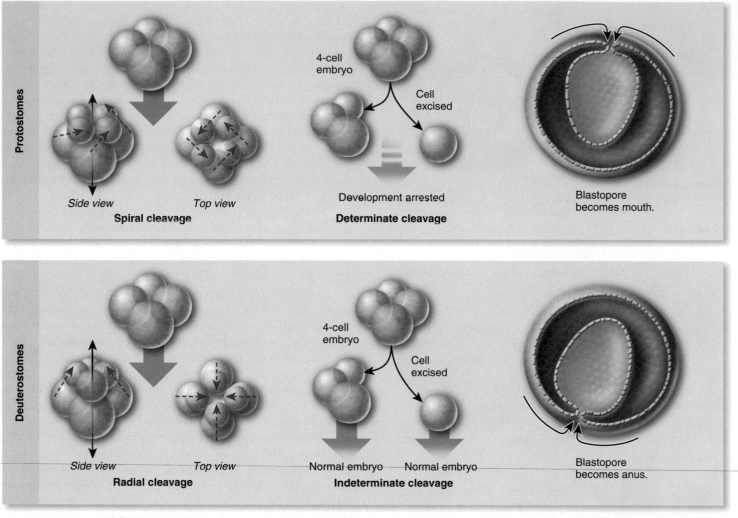

Figure 32.5 **Differences in embryonic development between protostomes and deuterostomes.** **(a)** Many protostomes have spiral cleavage, while most deuterostomes have radial cleavage. The dotted arrows indicate the direction of cleavage. **(b)** Protostomes have determinate cleavage, whereas deuterostomes have indeterminate cleavage. **(c)** In protostomes, the blastopore becomes the mouth. In deuterostomes, the blastopore becomes the anus.

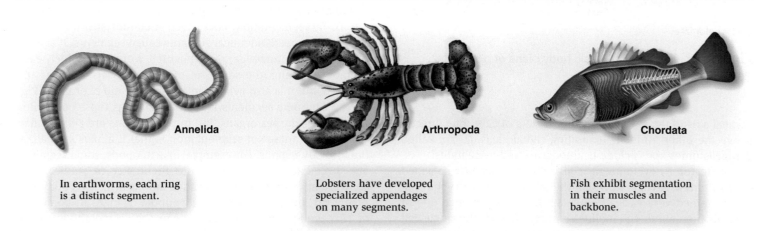

Figure 32.6 **Segmentation.** Annelids, arthropods, and chordates all exhibit segmentation.

GENOMES & PROTEOMES

Changes in *Hox* Gene Expression Control Body Segment Specialization

As we will see in Chapter 33, arthropods exhibit a vast degree of specialization of their segments. For example, many insects have wings and only three pairs of legs, whereas centipedes have no wings and many legs. Crabs, lobsters, and shrimps have highly specialized thoracic appendages called maxillipeds that aid in feeding. In the 1990s, Michalis Averof and coworkers elegantly showed how relatively simple changes in the expression patterns of **Hox genes**, genes involved in patterning the body axis, can account for this large variation in appendage types. As described in Chapter 19, animals have several *Hox* genes that are expressed in particular regions of the body. Some are expressed in anterior segments, while others are expressed in posterior segments (refer back to Figure 19.18). The *Hox* genes are designated with numbers 1 through 13.

Shifts in patterns of expression of *Hox* genes in the embryo along the anteroposterior axis are prominent in evolution. In vertebrates, the transition from one type of vertebra to another, for example, from cervical (neck) to thoracic (chest) vertebrae, is also controlled by particular *Hox* genes (**Figure 32.7**). The site of the cervical/thoracic boundary appears to be influenced by the *HoxC-6* gene. Differences in its relative position of expression, which occurs prior to vertebrae development, control neck length in vertebrates. In mice, which have a relatively short neck, the expression of *HoxC-6* begins between vertebrae numbers 7 and 8. In chickens and geese, which have longer necks, the expression begins further back, between vertebrae 14 and 15, or 17 and 18, respectively. The forelimbs also arise at this boundary in all vertebrates. Interestingly, snakes, which essentially have no neck or forelimbs, do not exhibit this boundary, and *HoxC-6* expression occurs toward their heads. This in effect means that snakes got longer by losing their neck and lengthening their chest.

Evolutionary and development biologist Sean Carroll has remarked that it is very satisfying to find that the evolution of body forms and novel structures in two of the most successful and diverse animal phyla, arthropods and vertebrates, is shaped by the shifting of *Hox* genes. It also reminds us of one of the basic tenets of diversity, that modern organisms illustrate Darwin's concept of descent with modification. Much of the diversity in animal phyla can be seen as variations on a common theme.

32.3 Molecular Views of Animal Diversity

While what we have called the traditional view of animal phylogeny has been accepted by most biologists for over a century, biologists are now using new molecular techniques to classify animals by comparing similarities in the DNA and the ribosomal RNA of animals, especially sequences of nucleotides in the gene that encodes RNA of the small ribosomal subunit (SSU rRNA) (see Chapter 26). The advantage of the molecular approach is that the data are generally more objective and subject to more rigorous testing than morphological data. The DNA sequence contains four easily identified and mutually exclusive character states: A, T, G, and C (RNA has A, U, G, and C).

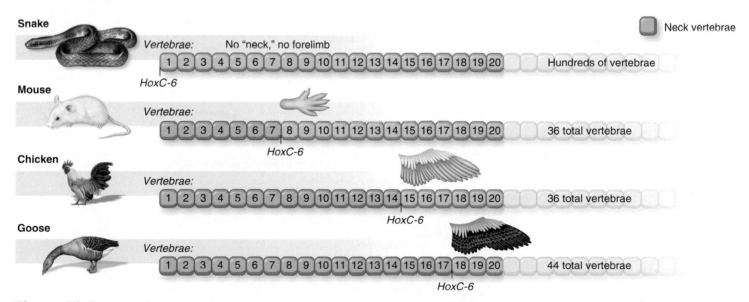

Figure 32.7 **Shifting zones of *Hox* gene expression control segment specialization.** In vertebrates, the transition between neck and trunk vertebrae is controlled by the position of the *HoxC-6* gene. In snakes, the expression of this gene is shifted so far forward that a neck does not develop.

Contrast this with morphological and embryological data, where characters are scored more subjectively, often based on the qualitative assessment of many traits. One of the most influential of the modern studies was the 1997 paper of Anna Marie Aguinaldo and colleagues, which established evidence for a new monophyletic group (clade) of molting animals, the Ecdysozoa (see Feature Investigation).

In addition, researchers have also studied the *Hox* genes that control early development and are present in all animals. *Hox* genes are important because many branches in the traditional phylogeny are based on early developmental differences such as cleavage, so examination of the genes that regulate these differences should provide insight into the evolution of animal development and to understanding how, when, and why animal body plans diversified.

While the use of molecular techniques in taxonomy is relatively new, the techniques have had a dramatic impact on traditional classification schemes. Phylogenies based on SSU rRNA and *Hox* genes are similar and, as we will explore, in many cases agree with the structure of the traditional phylogenetic tree (**Figure 32.8**). However, there are some important differences. Recently, the journal *Science* brought together numerous molecular taxonomists to discuss the current available data. In this section, we summarize their consensus on the new molecular phylogeny and examine the similarities and differences between its hypotheses and those of traditional phylogeny.

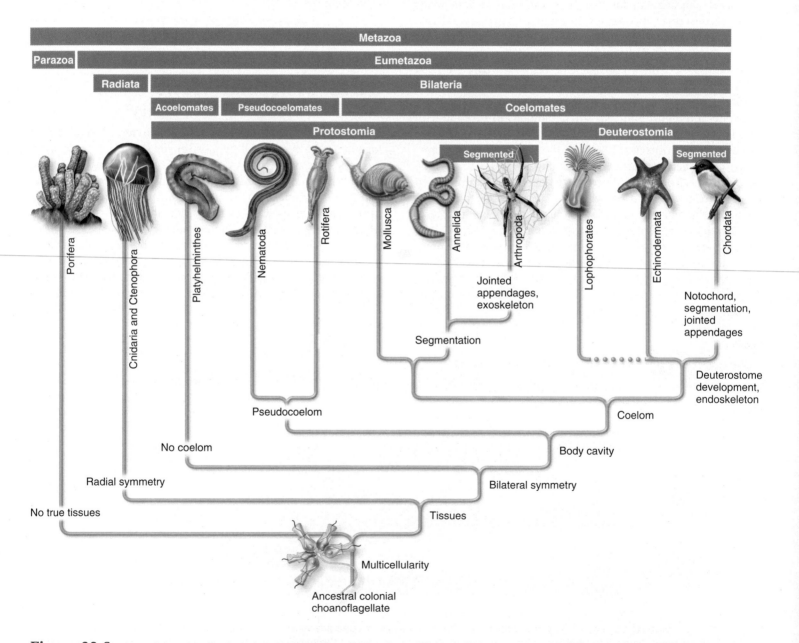

Figure 32.8 **A traditional animal phylogeny based on body plan.**   Though there are about 35 different animal phyla, we will focus our discussions here and in the next two chapters on the 11 groups with the greatest numbers of species. The dotted line represents the uncertainty of including the lophophorates with the deuterostomes.

FEATURE INVESTIGATION

Aguinaldo and Colleagues Used 18S rRNA to Analyze the Taxonomic Relationships of Arthropods to Other Taxa

In 1997, Anna Marie Aguinaldo, James Lake, and colleagues analyzed the relationships of arthropods to other taxa by sequencing the complete gene that encodes SSU rRNA from a variety of representative taxa (**Figure 32.9**). Total genomic DNA was isolated using standard techniques and amplified by the polymerase chain reaction (PCR). PCR fragments were then subjected to DNA sequencing, a technique described in Chapter 20. Using approaches that we have described in Chapter 26, the evolutionary relationships among 50 species were determined. The data indicated the existence of a monophyletic clade—the Ecdysozoa—containing the arthropods and nematodes plus a number of smaller phyla (see data in Figure 32.9). The name Ecdysozoa means molting animals, referring to the fact that all organisms in the clade undergo shedding of their exoskeleton.

The hypothesis that nematodes are more closely related to arthropods than previously thought has important ramifications. In particular, it implies that two well-researched model organisms, *Caenorhabditis elegans* (a nematode) and the fruit fly, *Drosophila melanogaster* (an arthropod), are more closely related than had been believed (see their positions in the data of Figure 32.9). Traditional classification assumed that developmental features common to both of these organisms had arisen early in metazoan evolution and thus might be broadly relevant to development in other coelomates, and in particular, humans. This new classification brings into question the applicability of studies of these organisms to human biology, since commonalities between these organisms might have evolved after the Ecdysozoans diverged. In this case, both nematodes and arthropods would be more closely related to each other than to humans.

Figure 32.9   A molecular animal phylogeny based on sequencing of SSU rRNA.

Biological inquiry: What is the major difference between molecular phylogeny and traditional phylogeny?

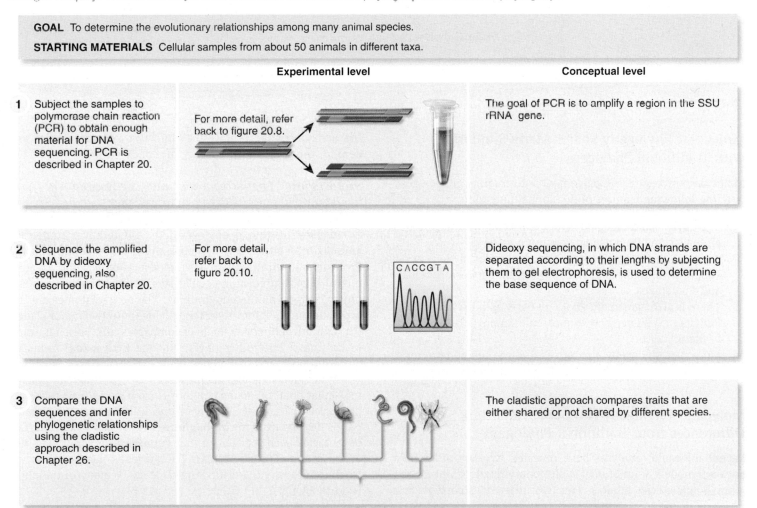

GOAL   To determine the evolutionary relationships among many animal species.

STARTING MATERIALS   Cellular samples from about 50 animals in different taxa.

	Experimental level	Conceptual level
1 Subject the samples to polymerase chain reaction (PCR) to obtain enough material for DNA sequencing. PCR is described in Chapter 20.	For more detail, refer back to figure 20.8.	The goal of PCR is to amplify a region in the SSU rRNA gene.
2 Sequence the amplified DNA by dideoxy sequencing, also described in Chapter 20.	For more detail, refer back to figure 20.10. CACCGTA	Dideoxy sequencing, in which DNA strands are separated according to their lengths by subjecting them to gel electrophoresis, is used to determine the base sequence of DNA.
3 Compare the DNA sequences and infer phylogenetic relationships using the cladistic approach described in Chapter 26.		The cladistic approach compares traits that are either shared or not shared by different species.

4 THE DATA

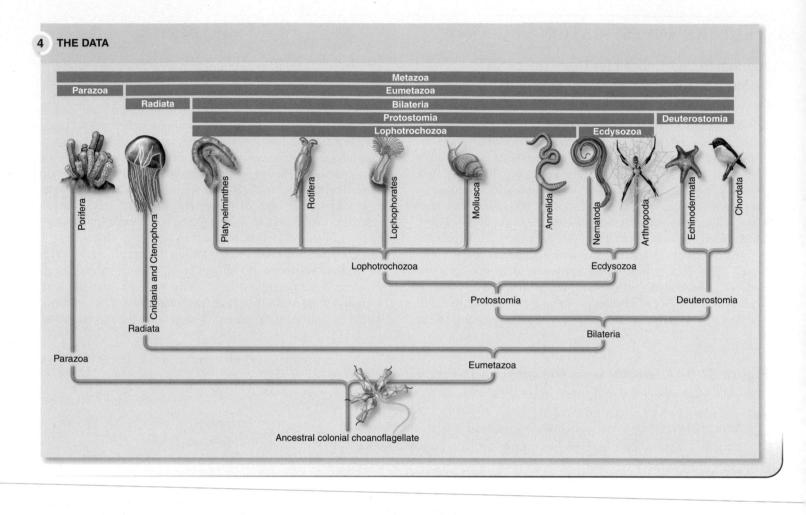

Molecular Phylogeny Shares Many Similarities with Traditional Phylogeny

Molecular phylogeny is in agreement with traditional phylogeny over the following features of the animal kingdom:

1. The clade called Metazoa is monophyletic, meaning all animals came from a single common ancestor.
2. At the earliest stages of evolution, molecular phylogeny supports the traditional view of the split between Parazoa and Eumetazoa.
3. There is also agreement about an early split between Radiata and Bilateria, with most animal phyla belonging to the Bilateria.
4. Molecular phylogeny also agrees that the echinoderms and chordates belong to a clade called the Deuterostomia.

Molecular Phylogeny Has Some Important Differences from Traditional Phylogeny

Recent molecular analyses have revealed two key differences between models of traditional evolutionary relationships among animals and newer models. The two differences concern first the division of the protostomes into two separate clades and second, the presence or absence of a body cavity.

Protostomes: Lophotrochozoa and Ecdysozoa The most important difference between molecular and traditional phylogenies involves relationships among the Bilateria. In the traditional view of animal phylogeny, the bilaterally symmetrical animals are split into two clades, the Deuterostomia and the Protostomia, reflecting two basic modes of embryonic development. However, recent molecular studies, from James Lake and others, suggest a different grouping. The deuterostomes are still separate, but the protostomes are divided into two major clades: the **Lophotrochozoa**, which encompasses the annelids, mollusks, and several other phyla, and the **Ecdysozoa**, primarily the arthropods and nematodes (Figure 32.9).

While the clade was organized primarily through analysis of molecular data, the name Lophotrochozoa stems from two morphological features seen in organisms of this clade. The "lopho" part is derived from the **lophophore**, a horseshoe-shaped crown of tentacles used for feeding that is present on some members of the group (**Figure 32.10a**). The "trocho" part refers to the **trochophore larva**, a distinct larval stage of many of the phyla (**Figure 32.10b**).

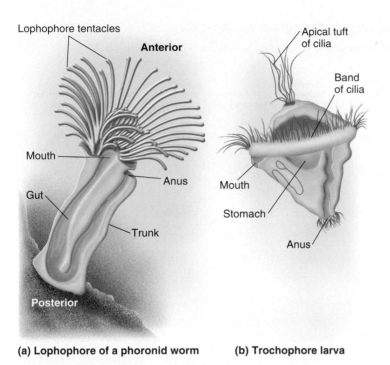

(a) **Lophophore of a phoronid worm**   (b) **Trochophore larva**

Figure 32.10   Characteristics of the Lophotrochozoa.
(a) A lophophore, a crown of ciliated tentacles, generates a current to bring food particles into the mouth. (b) The trochophore ("wheel-bearer") larval form is found in several animal lineages, such as polychaete worms and mollusks, indicating that both may have similar ancestry.

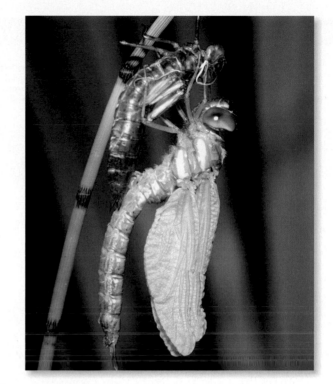

Figure 32.11   Ecdysis.   The Ecdysozoa are a clade of animals exhibiting ecdysis, the periodic shedding (molting) and re-formation of the exoskeleton.

Biological inquiry: What are the main members of the Ecdysozoa?

In traditional phylogeny, much debate surrounded the classification of the three minor phyla that possess lophophores—the Bryozoa, Phoronida, and Brachiopoda (collectively called the lophophorates). While these taxa exhibited some characteristics of protostomes and some of deuterostomes, they were often classified as the Deuterostomia (see Figure 32.8). However, molecular data support their inclusion within the Lophotrochozoa, along with annelids and mollusks.

The **Ecdysozoa** are so named because its members secrete a nonliving cuticle, typically an external skeleton (exoskeleton); think of the hard shell of a beetle or that of a crab. As these animals grow, the exoskeleton becomes too small and the animal molts, or breaks out of its old exoskeleton, and secretes a newer, larger one (**Figure 32.11**). This molting process is called **ecdysis**; hence the name Ecdysozoa. While named for this morphological characteristic, the clade is strongly supported by molecular evidence such as similarities in DNA.

Body Cavity The second most important difference between molecular and traditional phylogenies involves the presence or absence of a body cavity. In the traditional view of animal phylogeny, the bilaterally symmetrical animals are divided into those lacking a coelom (acoelomates), those with a pseudocoelom (pseudocoelomates), and those possessing a coelom (coelomates) (see Figure 32.4). The Platyhelminthes, or flatworms,

are classified as acoelomate and are thus seen as separate from coelomate phyla. Molecular evidence, however, now suggests that the flatworms should be included with the Lophotrochozoa. In this view, flatworms are not primitive acoelomate animals. Rather, they evolved from an ancestor that possessed a coelom but lost it during evolutionary modification. Similarly, molecular-based phylogeny places the rotifers and nematodes, the pseudocoelomate phyla, within the Lophotrochozoa and Ecdysozoa, respectively. Thus, molecular data suggest that the presence or absence of a coelom, a distinction traditionally used in the construction of animal phylogenies, may not be a useful way to classify animals.

It is important to stress, however, that more similarities are found between the molecular and traditional phylogenies than differences. Most of the major branch points in the phylogenies are in agreement. As a reference, **Table 32.2** summarizes the characteristics of the major animal phyla.

In the following two chapters, the discussion of animal phylogeny is based primarily on findings of molecular data. In Chapter 33, we will discuss the diversity of the Parazoa, Radiata, Lophotrochozoa, and Ecdysozoa phyla, and indeed all animals without a backbone. In Chapter 34, we will turn our attention to one phylum, Chordata, and in particular the vertebrates: fish, amphibians, reptiles, birds, and mammals.

Table 32.2 Summary of Characteristics of the Major Animal Phyla

Feature	*Porifera* Sponges	*Cnidaria* and *Ctenophora* Hydra, anemones, jellyfish	*Platyhelminthes* Flatworms	*Rotifera* Rotifers	*Lophophorates* Bryozoans and others	*Mollusca* Snails, clams, squid	*Annelida* Segmented worms	*Nematoda* Roundworms	*Arthropoda* Insects, arachnids, crustaceans	*Echinodermata* Sea stars, sea urchins	*Chordata* Vertebrates and others
Estimated number of species	8,000	11,000	20,000	2,000	4,000+	110,000	15,000	20,000	1,000,000+	6,000+	47,000+
Level of organization	Cellular; lack tissues and organs	Tissue; lack organs	Organs	Organs	Organs	Organs	Organs	Organs	Organs	Organs	Organs
Symmetry	Absent	Radial	Bilateral	Bilateral	Bilateral	Bilateral	Bilateral	Bilateral	Bilateral	Bilateral larvae, radial adults	Bilateral
Cephalization	Absent	Absent	Present	Present	Reduced	Present	Present	Present	Present	Absent	Present
Germ layers	Absent	Two	Three	Three	Three	Three	Three	Three	Three	Three	Three
Body cavity	Absent	Absent	Absent	Pseudo-coelom	Coelom	Reduced coelom	Coelom	Pseudo-coelom	Reduced coelom	Coelom	Coelom
Segmentation	Absent	Absent	Absent	Absent	Absent	Absent	Present	Absent	Present	Absent	Present (but reduced)
Digestive system	Absent	Gastro-vascular cavity; Ctenophores have complete gut	Incomplete gut	Complete gut (usually)	Complete gut	Complete gut	Complete gut	Complete gut	Complete gut	Usually complete gut	Complete gut
Circulatory system	Absent	Absent	Absent	Absent	Absent, open, or closed	Open	Closed	Absent	Open	Absent	Closed
Respiratory system	Absent	Absent	Absent	Absent	Absent	Gills, lungs	Absent	Absent	Trachae, gills, or book lungs	Tube feet, gills, respiratory tree	Gills, lungs
Excretory system	Absent	Absent	Protonephridia with flame cells	Protonephridia	Meta-nephridia	Meta-nephridia	Meta-nephridia	Excretory gland cells	Excretory glands resembling metanephridia	Absent	Kidneys
Nervous system	Absent	Nerve net	Brain, nerve net	Brain, nerve cords	No brain, nerve ring	Ganglia, nerve cords	Brain, ventral nerve cord	Brain, ventral nerve cord	Brain, ventral nerve cord	No brain, nerve ring and radial nerves	Well-developed brain; dorsal hollow nerve cord
Reproduction	Sexual; asexual (budding)	Sexual; asexual (budding)	Sexual (most hermaphroditic); asexual (body splits)	Mostly partheno-genetic; males appear only rarely	Sexual (some hermaphroditic); asexual (budding)	Sexual (some hermaphroditic)	Sexual (some hermaphroditic)	Sexual (some hermaphroditic)	Usually sexual (some hermaphroditic)	Sexual (some hermaphroditic); parthenogenetic; asexual by regeneration (rare)	Sexual, rarely partheno-genetic
Support	Endo-skeleton of spicules and collagen	Hydrostatic skeleton	Hydrostatic skeleton	Hydrostatic skeleton	Exoskeleton	Hydrostatic skeleton and shell	Hydrostatic skeleton	Hydrostatic skeleton	Exoskeleton	Endoskeleton of plates beneath outer skin	Endoskeleton of cartilage or bone

CHAPTER SUMMARY

32.1 Characteristics of Animals

- Animals constitute a very species-rich kingdom. They share a number of key characteristics, including multicellularity, heterotrophic feeding, the possession of nervous and muscle tissues, and sexual reproduction. (Table 32.1)

32.2 Traditional Classification of Animals

- The animal kingdom is monophyletic, meaning that all taxa have evolved from a single common ancestor. This can be observed in comparing gene sequences from animals and a protist. (Figure 32.1)

- The traditional classification of animals is based on four morphological and developmental features of animal body plans. (Table 32.2)

- Animals can be categorized according to the absence of different types of tissues (the Parazoa or sponges) and the presence of tissues (Eumetazoa or all other animals). The Eumetazoa can also be divided according to their type of symmetry, whether radial (Radiata, the cnidarians and ctenophores) or bilateral (Bilateria, all other animals). (Figure 32.2)

- The Radiata have two layers of embryonic cell layers (germ layers) called the endoderm and the ectoderm. The Bilateria develop a third germ layer termed the mesoderm, which develops between the ectoderm and the endoderm. (Figure 32.3)

- Animals can be classified according to the presence or absence of a coelom, or true body cavity. Animals with a coelom are termed coelomates. Animals that possess a pseudocoelom, or coelom that is not completely lined by tissue derived from mesoderm, are called pseudocoelomates. Those animals lacking a fluid-filled body cavity are termed acoelomates. (Figure 32.4)

- Animals are also classified according to patterns of embryonic development. Animals with spiral cleavage are called protostomes, and those exhibiting radial cleavage are considered deuterostomes. In protostomes, the blastopore, or opening of the gut to the outside, becomes the mouth; in deuterostomes, the blastopore becomes the anus. (Figure 32.5)

- Metamerism, the division of the body into identical subunits called segments, is another key feature of the animal body plan. (Figure 32.6)

- Shifts in the pattern of expression of *Hox* genes are prominent in evolution. In vertebrates, the transition from one type of vertebra to another is controlled by certain *Hox* genes. (Figure 32.7)

32.3 Molecular Views of Animal Diversity

- New molecular techniques that compare similarities in DNA and ribosomal RNA of animals are having a dramatic effect on traditional classification schemes. (Figure 32.8)

- In many cases, phylogenies based on these techniques are similar to those of traditional approaches; however, some important differences exist. Recent molecular studies propose a division of the protostomes into two major clades: the Lophotrochozoa and the Ecdysozoa. (Figure 32.9)

- The Lophotrochozoa are grouped primarily through analysis of molecular data, but they are distinguished by two morphological features—the lophophore, a crown of tentacles used for feeding, and the trochophore larva, a distinct larval stage. The lophophorates were often classified as deuterostomes, but molecular data support their inclusion within the Lophotrochozoa. (Figure 32.10)

- The Ecdysozoa are so named because its members secrete a nonliving cuticle, typically an exoskeleton or external skeleton. Ecdysis is the periodic shedding and re-formation of the exoskeleton. (Figure 32.11)

- Molecular evidence suggests that flatworms evolved from a coelomate ancestor and should be considered to be within the Lophotrochozoa, and not as primitive acoelomate animals.

TEST YOURSELF

1. Which of the following is *not* a distinguishing characteristic of animals?
   a. the capacity to move at some point in their life cycle
   b. possession of cell walls
   c. multicellularity
   d. heterotrophy
   e. all of the above are characteristics of animals

2. Terrestrial adaptations seen in animals include
   a. internal fertilization.
   b. tough, protective shells around eggs.
   c. a waxy cuticle covering exposed tissue.
   d. a and b only.
   e. all of the above.

3. Eumetazoa are animals that have
   a. true tissues.
   b. more than one tissue type.
   c. only one tissue type.
   d. radial symmetry.
   e. spiral cleavage.

4. The localization of sensory structures at the anterior end of the body is
   a. bilateral symmetry.
   b. spiral cleavage.
   c. cephalization.
   d. radial symmetry.
   e. gastrulation.

5. The germ layer that is present in triploblastic animals but is absent in diploblastic animals is
   a. the ectoderm.
   b. the mesoderm.
   c. the endoderm.
   d. the pseudocoelom.
   e. the coelom.

6. Pseudocoelomates
   a. lack a fluid-filled cavity.
   b. have a fluid-filled cavity that is completely lined with mesoderm.
   c. have a fluid filled cavity that is partially lined with mesoderm.
   d. have a fluid-filled cavity that is not lined with mesoderm.
   e. have an air-filled cavity that is partially lined with mesoderm.

7. Protostomes and deuterostomes can be classified based on
   a. cleavage pattern.
   b. destiny of the blastopore.
   c. whether the fate of the embryonic cells is fixed early during development.
   d. how the coelom is formed.
   e. all of the above.

8. Naturally occurring identical twins are possible only in animals that
   a. have spiral cleavage.
   b. have determinate cleavage.
   c. are protostomes.
   d. have indeterminate cleavage.
   e. a, b, and c.

9. Genes involved in the patterning of the body axis, that is, in determining characteristics such as neck length and appendage formation, are called
   a. small subunit (SSU) rRNA genes.
   b. *Hox* genes.
   c. metameric genes.
   d. determinate genes.
   e. none of the above.

10. A major difference between the molecular phylogeny of animals and traditional phylogeny of animals is that
    a. the presence or absence of the mesoderm is not important in molecular phylogeny.
    b. molecular phylogeny suggests that all animals do not share a single common ancestor.
    c. body symmetry, whether radial or bilateral, is not an important determinant in molecular phylogeny.
    d. molecular phylogeny does not include the echinoderms in the deuterostome clade.
    e. molecular phylogeny suggests that the presence or absence of a coelom is not important for classification.

CONCEPTUAL QUESTIONS

1. The traditional classification is based on what four features of animal body plans?

2. Distinguish between radial and bilateral symmetry.

3. Define cleavage and gastrulation.

EXPERIMENTAL QUESTIONS

1. What was the purpose of the study conducted by Aguinaldo and colleagues?

2. What was the major finding of this particular study?

3. What impact does the new view of nematode and arthropod phylogeny have on other areas of research?

COLLABORATIVE QUESTIONS

1. Discuss the different types of body cavities.

2. Discuss some of the characteristics of animal life.

www.brookerbiology.com

This website includes answers to the Biological Inquiry questions found in the figure legends and all end-of-chapter questions.

33

THE INVERTEBRATES

CHAPTER OUTLINE

This plantlike organism is actually a feather star, an echinoderm. The feather star's tube feet grip the ocean substrate while its arms filter feed on drifting microorganisms.

As we saw in Chapter 22, the history of animal life on Earth has evolved over hundreds of millions of years. While many date the divergence of animals to the early Paleozoic era, over 500 million years ago, biologist Blair Hedges and others have shown that molecular data point to a divergence among animals that may be as long as 800–1,200 million years ago. Some scientists suggest that changing environmental conditions, such as a buildup of dissolved oxygen and minerals in the ocean or an increase in atmospheric oxygen, eventually permitted higher metabolic rates and increased the activity of a wide range of animals. Others suggest that with the development of sophisticated locomotor skills, a wide range of predators and prey evolved, leading to an evolutionary arms race in which predators evolved powerful weapons and prey evolved more powerful defenses against them. Such adaptations and counteradaptations would have led to a proliferation of different lifestyles and taxa.

Over the next two chapters, we will survey the wondrous array of animal life on Earth (see chapter-opening photo). In this chapter, we examine the **invertebrates**, or animals without a backbone, a category that makes up more than 95% of all animal species. We begin by exploring some of the earliest animal lineages, the Parazoa and Radiata. We then turn to the Lophotrochozoa and Ecdysozoa, the two sister groups of protostomes introduced in Chapter 32. Finally, we turn to the deuterostomes, focusing here on the echinoderms and the invertebrate members of the phylum Chordata.

While the more modern molecular classification outlined in Chapter 32 will serve as the basis of our discussion of animal lineages, we will not ignore the concept of body plans, because it still can provide clues about how different phyla have evolved. The newer molecular phylogeny is still in its infancy, and many refinements will undoubtedly be made as increasing numbers of genes from more species are sequenced and compared. For this reason, many biologists are not yet ready to totally set aside the older body plan–based phylogeny.

33.1 Parazoa: Sponges, the First Multicellular Animals

The Parazoa consist of one phylum, Porifera (from the Latin, "pore bearers"), whose members are commonly referred to as sponges. Sponges are loosely organized and lack true tissues, groups of cells that have a similar structure and function. However, sponges are multicellular and possess several types of cells that perform different functions. Biologists have identified approximately 8,000 species of sponges, the vast majority of which are marine. Sponges range in size from only a few millimeters across to more than 2 m in diameter. The smaller sponges may be radially symmetrical, but most have no apparent symmetry. Some sponges have a low encrusting growth form, while others grow tall and erect (**Figure 33.1a**). While adult sponges are sessile, that is, anchored in place, the larvae are free-swimming.

The body of a sponge looks similar to a vase pierced with small holes or pores (**Figure 33.1b**). Water is drawn through these pores (ostia, singular, ostium) into a central cavity, the **spongocoel**, and flows out through the large opening at the top called the **osculum**. The water enters the pores by the beating action of the flagella of the **choanocytes**, or collar cells, that line the spongocoel (**Figure 33.1c**). In the process, the choanocytes trap and eat small particulate matter and tiny plankton.

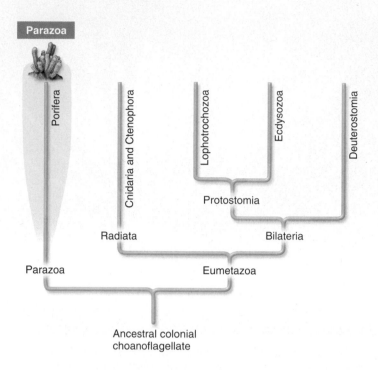

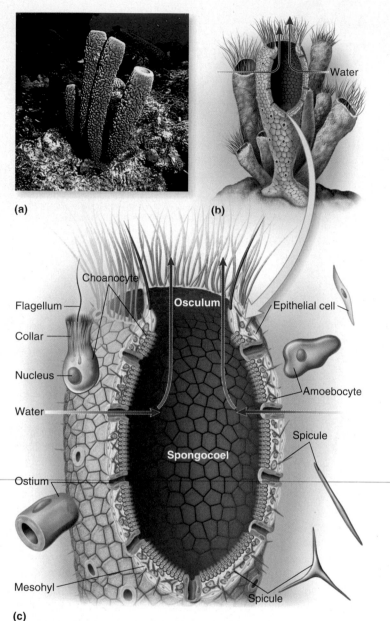

Figure 33.1 **Sponge morphology.** **(a)** The stovepipe sponge (*Aplysina archeri*) is a common sponge found on Caribbean reefs. **(b)** Many sponges have a vaselike shape. **(c)** A cross section reveals that sponges are truly multicellular animals, having various cell types but no distinct tissues.

Biological inquiry: If sponges are soft and sessile, why aren't they eaten by other organisms?

As we noted in Chapter 32, because of striking morphological and molecular similarities between choanocytes and choanoflagellates, a group of modern protists having a single flagellum, it is believed that sponges originated from a common choanoflagellate ancestor.

A layer of flattened epithelial cells similar to those making up the outer layer of other phyla protects the sponge body. In between the choanocytes and the epithelial cells lies a gelatinous, protein-rich matrix called the **mesohyl**. Within this matrix are mobile cells called **amoebocytes** that absorb food from choanocytes, digest it, and carry the nutrients to other cells. Thus, considerable cell-to-cell contact and communication exists in sponges.

Some amoebocytes can also form tough skeletal fibers that support the body. In many sponges, this skeleton consists of sharp spicules formed of calcium carbonate or silica. For example, some deep-ocean species, called glass sponges, are distinguished by needle-like silica spicules that form elaborate lattice-like skeletons. The presence of such tough spicules may help explain why there is not much predation of sponges. Sponge spicules come in a diverse array of shapes and sizes, and they are valuable taxonomic tools by which to distinguish different types of sponges. In a small family of carnivorous sponges, the spicules are sticky and capture small crustaceans. In these sponges, other cells migrate around the immobilized crustaceans and digest them extracellularly. Not all sponges have **spicules**, however. Others have fibers of a tough protein called **spongin** that lend skeletal support. Spongin skeletons are still commercially harvested and sold as bath sponges. Many species produce toxic defensive chemicals, some of which are thought to have possible antibiotic and anti-inflammatory effects in humans.

Around the turn of the 20th century, biologist Henry V. Wilson made the incredible discovery that if a sponge is dissociated into its individual cells after being passed through a sieve, its cells can reaggregate into a functional sponge within a short time. Wilson concluded that in order to do this, individual cells recognized and reaggregated with other cells of their own kind. Researchers have since discovered that the cells of other multicellular organisms also recognize cells of their own kind and tend to adhere when mixed with other cells. For example, in mammals, liver cells recognize and stick better to other liver cells, and brain cells recognize and adhere to brain cells.

Sponges reproduce through both sexual and asexual means. Most sponges are **hermaphrodites** (from the Greek god Hermes and the goddess Aphrodite), individuals that can produce both sperm and eggs. Gametes are formed in the mesohyl by amoebocytes or choanocytes. While eggs remain in the mesohyl, the sperm are released into the water and carried by water currents to fertilize the eggs of neighboring sponges. Zygotes develop into flagellated swimming larvae that eventually settle on a suitable substrate to become sessile adults. In asexual reproduction, a small fragment or bud may detach and form a new sponge.

33.2 Radiata: Jellyfish, Radially Symmetrical Animals

The Radiata consists of two closely related phyla: the Cnidaria (from the Greek *knide*, nettle, and *aria*, related to; pronounced nid-air′-e-ah) and the Ctenophora (from the Greek *ktenos*, comb, and *phora*, bearing; pronounced teen-o-for′-ah). Members of the Radiata phyla, or radiates, are mostly found in marine environments, although a few, primarily hydra, are freshwater species. The Cnidaria includes hydra, jellyfish, box jellies, sea anemones, and corals, and the Ctenophora consists of the comb jellies. The Radiata have only two embryonic germ layers: the ectoderm and the endoderm, which give rise to the epidermis and the gastrodermis, respectively. A gelatinous substance called the **mesoglea** connects the two layers. In jellyfish, the mesoglea is enlarged and forms the buoyant, transparent jelly, whereas in coral, the mesoglea is very thin.

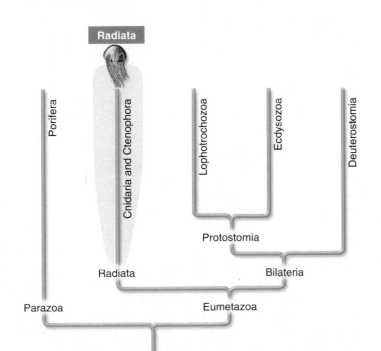

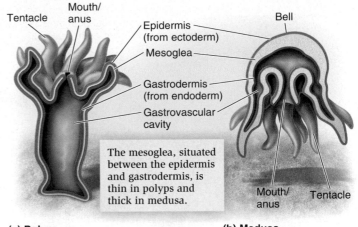

Figure 33.2 Cnidarians exist as a polyp or medusa form. Both **(a)** polyp and **(b)** medusa forms have two layers of cells, an outer epidermis (from ectoderm) and an inner layer of gastrodermis (from endoderm). In between is a layer of mesoglea, which is thick in jellyfish and thin in corals.

Both cnidarians and ctenophores possess a **gastrovascular cavity**, where extracellular digestion takes place (**Figure 33.2**). This feature allows the ingestion of larger food particles and represents a major advance over the sponges, which utilize only intracellular digestion. Most radiates have tentacles around the mouth that aid in food detection and capture. Radiates also have true nerve cells arranged as a **nerve net** consisting of interconnected neurons with no central control organ. In this section, we will provide an overview of the biology and diversity of the cnidarians and ctenophores.

The Cnidarians Have Specialized Stinging Cells

Most cnidarians exist as two different body forms and associated lifestyles: the sessile **polyp** or the motile **medusa** (Figure 33.2). For example, jellyfish exist predominantly in the medusa form, and corals exhibit only the polyp form. Many cnidarians, such as *Obelia*, have a life cycle that prominently features both polyp and medusa stages (**Figure 33.3**).

The polyp form has a tubular body with an opening at the oral end that is surrounded by tentacles and functions as both mouth and anus. The aboral end is attached to the substrate. In the 18th century, the Swiss naturalist Abraham Trembley discovered that when a freshwater hydra was cut in two, each part not only survived but could also regenerate the missing half. Polyps exist colonially, as they do in corals, or alone, as in sea anemones. Corals take dissolved calcium and carbonate ions from seawater and precipitate them as limestone underneath their bodies. In some species, this leads to a buildup of limestone deposits. As each successive generation of polyps dies, the limestone remains in place and new polyps grow on top. Thus, huge underwater limestone deposits called coral reefs are formed (look ahead to Figure 54.27b). The largest of these is Australia's Great Barrier Reef, which stretches over 2,300 km.

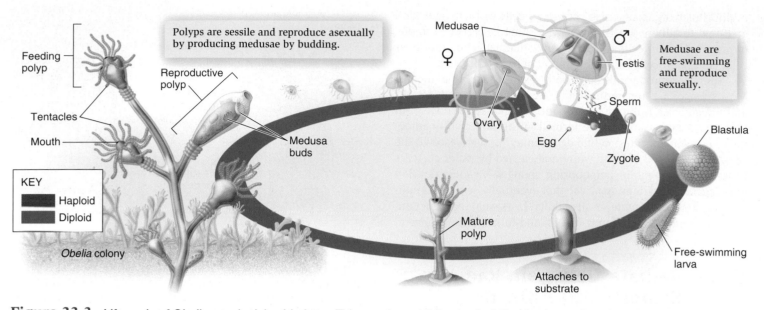

Polyps are sessile and reproduce asexually by producing medusae by budding.

Feeding polyp

Reproductive polyp

Tentacles

Mouth

Medusa buds

KEY

Haploid

Diploid

Obelia colony

Medusae

♀

Ovary

♂

Testis

Sperm

Egg

Zygote

Medusae are free-swimming and reproduce sexually.

Blastula

Mature polyp

Attaches to substrate

Free-swimming larva

Figure 33.3 Life cycle of *Obelia*, a colonial cnidarian. This species exhibits clearly defined polyp and medusa stages.

Biological inquiry: What are the dominant life stages of the following types of cnidarians: jellyfish, sea anemone, and Portuguese Man-of-War?

Many other extensive coral reefs are known, including the reef system along the Florida Keys, all of which occur in warm water, generally between 18°C and 30°C.

The free-swimming medusa form has an umbrella-shaped body with a mouth on the concave underside that is surrounded by tentacles. More mobile medusae possess simple sense organs near the bell margin, including organs of equilibrium called **statocysts** and photosensitive organs known as **ocelli**. When one side of the bell tips upward, the statocysts on that side are stimulated and muscle contraction is initiated to right the medusa. The ocelli allow medusae to position themselves in particular light levels.

One of the unique and characteristic features of the cnidarians is the existence of stinging cells called **cnidocytes**, which function in defense or the capture of prey (**Figure 33.4a**). Cnidocytes contain **nematocysts**, powerful capsules with an inverted coiled and barbed thread. While several different types of nematocysts are recognized, they all have the same general structure and function. Each cnidocyte has a hairlike trigger called a **cnidocil** on its surface. When the cnidocil is touched or a chemical stimulus is detected, the nematocyst fires the thread, which penetrates the prey and injects a small amount of toxin. Small prey are immobilized and passed into the mouth by the tentacles. Alternatively, some nematocyst threads can be sticky rather than stinging. After discharge, the cnidocyte is absorbed and a new one grows to replace it. The nematocysts of most cnidarians are not harmful to humans, but those on the tentacles of the larger jellyfish and the Portuguese Man-of-War (**Figure 33.4b**) can be extremely painful and even fatal. Tentacles of the largest jellyfish, *Cyanea arctica*, may be over 40 m long.

Muscles and nerves exist in their simplest forms in cnidarians. Contractile muscle fibers are found in both the epidermis and gastrodermis. While not true muscles, which only arise

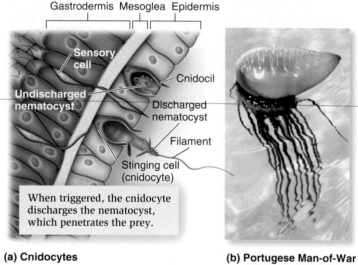

Gastrodermis Mesoglea Epidermis

Sensory cell

Cnidocil

Undischarged nematocyst

Discharged nematocyst

Filament

Stinging cell (cnidocyte)

When triggered, the cnidocyte discharges the nematocyst, which penetrates the prey.

(a) Cnidocytes

(b) Portugese Man-of-War (*Physalia physalis*)

Figure 33.4 Cnidarians have specialized stinging cells called cnidocytes. (a) Cnidocytes, which contain stinging capsules called nematocysts, are situated in the tentacles. (b) Portuguese Man-of-War (*Physalia physalis*) is an example of the medusa body form and lifestyle. A gas-filled float allows the animal to stay on the surface of the water.

from the mesoderm and therefore do not appear in diploblastic animals, these muscle fibers can contract to change the shape of the animal. For example, in the presence of a predator, an anemone can expel water very quickly through its open mouth and shrink down to a very small body form. The muscle fibers work against the fluid contained in the body, which thus acts as a hydrostatic skeleton. A nerve net that conducts signals from

Table 33.1	Main Classes and Characteristics of the Cnidaria	
	Class and examples (est. # of species)	**Class characteristics**
	Hydrozoa: Portuguese Man-of-War, *Hydra*, *Obelia*, some corals (2,700)	Mostly marine; most have both polyp and medusa stages with polyp stage colonial
	Scyphozoa: Jellyfish (200)	All marine; medusa stage dominant and large (up to 2 m); reduced polyp stage
	Anthozoa: Sea anemones, sea fans, most corals (6,000)	All marine; polyp stage dominant; medusa stage absent; many are colonial
	Cubozoa: Box jellies, sea wasps (20)	All marine; medusa stage dominant; box-shaped

Figure 33.5 **A ctenophore.** *Mnemiopsis leidyi*, commonly called the sea walnut, was accidentally introduced via the ballast water of ships to the Black and Caspian Seas, where it fed on plankton and decreased the food base for local fish, whose populations dramatically declined.

sensory nerves to muscle cells allows coordination of simple movements and shape changes.

The phylum Cnidaria consists of four classes—Hydrozoa (including *Obelia* and Portuguese Man-of-War), Scyphozoa (jellyfish), Anthozoa (sea anemones and corals), and Cubozoa (box jellies)—whose distinguishing characteristics are shown in **Table 33.1**.

Ctenophores Have a Complete Gut

Ctenophores, also known as comb jellies, are a small phylum of fewer than 100 species, all of which are marine and look very much like jellyfish (**Figure 33.5**). They have eight rows of cilia on their surface that resemble combs. The coordinated beating of the cilia, rather than muscular contractions, propels the ctenophores. Averaging about 1–10 cm in length, comb jellies are probably the largest animals to use cilia for locomotion. There are even a few ribbon-like species up to 1 m long.

Comb jellies possess two long tentacles but lack stinging cells. Instead, they have colloblasts, cells that secrete a sticky substance onto which small prey adhere. The tentacles are then drawn over the mouth. As with cnidarians, digestion occurs in the gastrovascular cavity, but waste and water are eliminated through two anal pores. Thus, the comb jellies possess the first complete gut. Prey are generally small and may include tiny crustaceans called copepods and small fish. Comb jellies are often transported around the world in ships' ballast water. *Mnemiopsis leidyi*, a ctenophore species native to the Atlantic coast of North and South America, was accidentally introduced into the Caspian and Black Seas in the 1980s. With a plentiful food supply and a lack of predators, *Mnemiopsis* underwent a population explosion and caused devastation to the local fishing industries.

All ctenophores are hermaphroditic, possessing both ovaries and testes, and gametes are shed into the water to eventually form a free-swimming larva that grows into an adult. There is no polyp stage. Nearly all ctenophores exhibit **bioluminescence**, a phenomenon that results from chemical reactions that give off light rather than heat. Thus, individuals can be particularly evident at night, and ctenophores that wash up onshore can make the sand or mud appear luminescent.

33.3 Lophotrochozoa: The Flatworms, Rotifers, Lophophorates, Mollusks, and Annelids

In the traditional view of animal phylogeny (refer back to Figure 32.8), the bilaterally symmetrical animals are split into those with no coelom (the platyhelminthes), those with a pseudocoelom (the nematodes and rotifers), and those with a coelom (the remaining phyla). However, as we explored in Chapter 32, molecular data suggest a different grouping in which the deuterostomes are still separate, but the protostomes are divided into two major lineages: the Lophotrochozoa and the Ecdysozoa (refer back to Figure 32.9). The Lophotrochozoa are a diverse group that generally includes taxa that possess either a lophophore (a crown of ciliated tentacles) or a distinct larval stage called a trochophore. In this grouping are seven major Lophotrochozoa phyla: the Platyhelminthes (flatworms), Rotifera (rotifers), Lophophorata (the lophophorates, a group of three phyla), Mollusca (mollusks), and Annelida (segmented worms). In this section, we explore the distinguishing characteristics of these phyla, beginning with some of the simplest lophotrochozoans, the Platyhelminthes.

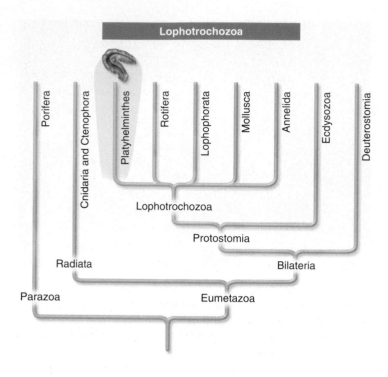

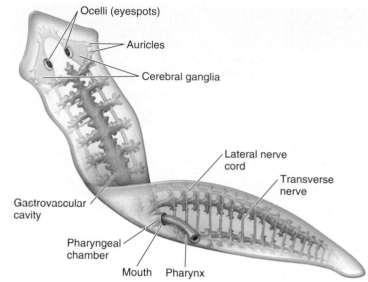

Figure 33.6   **Body plan of a flatworm.**   Flatworm morphology as represented by a planarian, a member of the class Turbellaria.

The Phylum Platyhelminthes Consists of Flatworms with No Coelom

Platyhelminthes (from the Greek *platy*, flat, and *helminth*, worm), or flatworms, lack a specialized respiratory or circulatory system to transport gases. They must obtain oxygen by diffusion, which makes a flattened shape necessary, in that no cell can be too far from the surface. Flatworms were among the first animals to develop an active predatory lifestyle. Platyhelminthes, and indeed most animals, are bilaterally symmetrical, with a head bearing sensory appendages.

The flatworms are also believed to be the first animals to develop three embryonic germ layers—ectoderm, endoderm, and mesoderm—with mesoderm replacing the simpler gelatinous mesoglea of cnidarians. As such, they are said to be triploblastic. The muscles in flatworms, which are derived from mesoderm, are well developed. The development of mesoderm was thus a critical evolutionary innovation in animals, because it also led to the development of more sophisticated organs. Flatworms are sometimes regarded as the first animals to reach the organ-system level of organization. The mesoderm fills the body spaces apart from the gastrovascular cavity; thus, the flatworms are acoelomate, lacking a fluid-filled body cavity in which the gut is suspended (**Figure 33.6**).

The digestive system of flatworms is incomplete, with only one opening, which serves as both mouth and anus, as in cnidarians. Most flatworms possess a muscular pharynx that may be extended through the mouth. The pharynx opens to a gastrovascular cavity, where food is digested. In large flatworms, the gastrovascular cavity is branched enough to distribute nutrients to all parts of the body. Any undigested material is egested back through the pharynx. The incomplete digestive system of flatworms thus prevents continuous feeding. Some flatworms

are predators, but many species have invaded other animals as parasites.

Flatworms have a distinct excretory system, consisting of **protonephridia**, two lateral canals with branches capped by **flame cells**. The flame cells, which are ciliated and waft water through the lateral canals to the outside (look ahead to Figure 49.8), exist primarily to maintain osmotic balance between the flatworm's body and the surrounding fluids. Simple though this system is, its development was key to permitting the invasion of freshwater habitats and even moist terrestrial areas.

Some free-living flatworms in the class Turbellaria possess light-sensitive eyespots or ocelli at the anterior end and chemoreceptive and sensory cells that are concentrated in organs called auricles. A pair of **cerebral ganglia** receives input from photoreceptors in eyespots and sensory cells. From the ganglia, a pair of lateral nerve cords running the length of the body allows rapid movement of information from anterior to posterior. In addition, transverse nerves form a nerve net on the ventral surface similar to that of cnidarians. Thus, the flatworms have retained the cnidarian-style nervous system, while possessing the beginnings of a more centralized type of nervous system seen throughout much of the rest of the animal kingdom. In all the Platyhelminthes, reproduction is either sexual or asexual. Most species are hermaphroditic but do not fertilize their own eggs. Flatworms can also reproduce asexually by splitting into two parts, with each half regenerating the missing fragment.

The four classes of flatworms are the Turbellaria, Monogenea, Trematoda (flukes), and Cestoda (tapeworms) (**Table 33.2**). Both cestodes and trematodes are internally parasitic and hence are of great medical and veterinary importance. They possess a variety of organs of attachment, such as hooks and suckers, that enable them to remain embedded within their hosts (**Figure 33.7**). Cestodes often require two separate vertebrate host

Table 33.2	Main Classes and Characteristics of Platyhelminthes
Class and examples (est. # of species)	**Class characteristics**
Turbellaria: Planarian (3,000)	Free-living flatworms; mostly marine; predatory or scavengers
Monogenea: Fish flukes (1,000)	Marine and freshwater; usually external parasites of fish; simple life cycle (no intermediate host)
Trematoda: Flukes (11,000)	Internal parasites of vertebrates; complex life cycle with several intermediate hosts
Cestoda: Tapeworms (5,000)	Internal parasites of vertebrates; no digestive system, nutrients absorbed across epidermis; complex life cycle, usually with one intermediate host

Figure 33.7 A tapeworm, *Taenia pisiformis*, a member of the class Cestoda. Note the scolex, the organ of attachment at the head end, complete with tiny hooks and suckers.

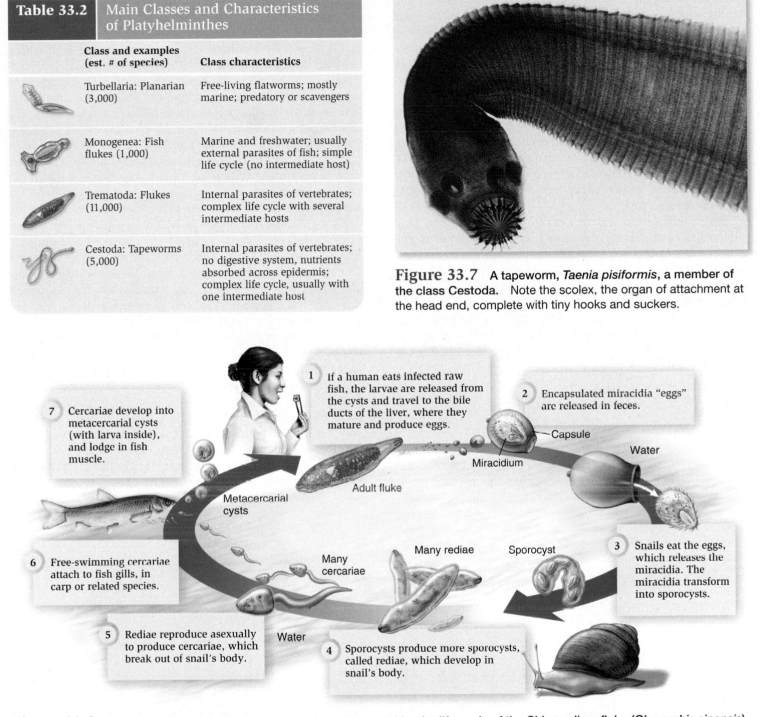

1 If a human eats infected raw fish, the larvae are released from the cysts and travel to the bile ducts of the liver, where they mature and produce eggs.

2 Encapsulated miracidia "eggs" are released in feces.

Capsule

Miracidium

Water

3 Snails eat the eggs, which releases the miracidia. The miracidia transform into sporocysts.

4 Sporocysts produce more sporocysts, called rediae, which develop in snail's body.

5 Rediae reproduce asexually to produce cercariae, which break out of snail's body.

6 Free-swimming cercariae attach to fish gills, in carp or related species.

7 Cercariae develop into metacercarial cysts (with larva inside), and lodge in fish muscle.

Adult fluke

Metacercarial cysts

Many cercariae

Many rediae

Sporocyst

Water

Figure 33.8 The complete life cycle of a trematode, as illustrated by the life cycle of the Chinese liver fluke (*Clonorchis sinensis*).

species, such as pigs or cattle to begin their life cycle and humans to complete their development. Many tapeworms can live inside humans who consume undercooked, infected meat—hence the value of thoroughly cooking meat to kill any parasites inside.

The life history of trematodes is even more complex than that of cestodes, involving multiple hosts. The first host, called the intermediate host, is usually a mollusk, and the final host, or definitive host, is usually a vertebrate, but often a second or even

a third intermediate host is involved. In the case of the Chinese liver fluke (*Clonorchis sinensis*), the adult parasite lives and reproduces in the definitive host, and the resultant "eggs" (encapsulated miracidia) pass from the host via the feces (**Figure 33.8**). An intermediate host, such as a snail, eats the eggs. The miracidia are released and transform into sporocysts. The sporocysts produce more sporocysts asexually called rediae. The rediae reproduce asexually to produce cercariae. Cercariae bore their way out of the snail and infect their definitive hosts directly by boring into their feet when in water or, for species with a second intermediate host (as in the Chinese liver fluke), by entering fish. Here, the cercariae develop into metacercarial cysts (juvenile flukes) and lodge in fish muscle, which the definitive host will eat. In the definitive host, the cyst protects the metacercaria from the host's gastric juices. In the small intestine, the metacercariae travel to the liver and grow into adult flukes and the life cycle begins anew. The life cycle of a trematode can thus involve at least seven stages: adult, egg (encapsulated embryo), miracidium, sporocyst, rediae, cercaria, and metacercaria. Because of the low probability of each larva reaching a suitable host, trematodes must produce large numbers of offspring to ensure that some survive.

Blood flukes, *Schistosoma* spp., are the most common parasitic trematodes infecting humans and cause the disease known as schistosomiasis. Over 200 million people worldwide, primarily in tropical Asia, Africa, and South America, are infected with schistosomiasis. The inch-long adult flukes can live for years in human hosts, and the release of eggs may cause chronic inflammation and blockage in many organs. Untreated schistosomiasis can lead to severe damage to the liver, intestines, and lungs and can eventually lead to death. Access to clean water can greatly reduce infection rates.

Members of the Phylum Rotifera Have a Pseudocoelom and a Ciliated Crown

Members of the phylum Rotifera (from the Latin *rota*, wheel, and *fera*, to bear) get their name from their ciliated crown or **corona**, which, when beating, looks similar to a rotating wheel (**Figure 33.9**). Most rotifers are microscopic animals, usually less than 1 mm long, and some have beautiful colors. There are about 2,000 species of rotifers, most of which inhabit fresh water, with a few marine or terrestrial species. Most often they are bottom-dwelling organisms, living on the pond floor or along lakeside vegetation.

Rotifers have an alimentary canal, a digestive tract with a separate mouth and anus, which means that they can feed continuously. The corona creates water currents that propel the animal through the water and that waft small planktonic organisms or decomposing organic material toward the mouth. The mouth opens into a circular muscular pharynx called a **mastax**, which has jaws for grasping and chewing. The mastax, which in some species can protrude through the mouth to seize small prey, is a structure unique to rotifers. The body of the rotifer bears a jointed foot with one to four toes. **Pedal glands** in the foot secrete a sticky substance that aids in attachment to the substrate. The internal organs lie within a pseudocoelom, a fluid-filled body cavity that is not completely lined with mesoderm. The pseudocoelom serves as a hydrostatic skeleton and as a medium for the internal transport of nutrients and wastes. Rotifers also have a pair of protonephridia with flame bulbs that collect excretory and digestive waste and drain into a cloacal bladder, which passes waste to the anus.

Reproduction in rotifers is unique. In some species, unfertilized diploid eggs that have not undergone meiotic division, called amictic eggs, develop into females through a process known as **parthenogenesis**. In other species, some eggs undergo meiosis and become haploid. These so-called mictic eggs, if unfertilized, develop into degenerate males that cannot feed and only live long enough to produce and release sperm that fertilize other mictic eggs. These fertilized eggs form zygotes, which have a thick shell and can survive for long periods of harsh conditions, such as if a water supply dries up, before developing into

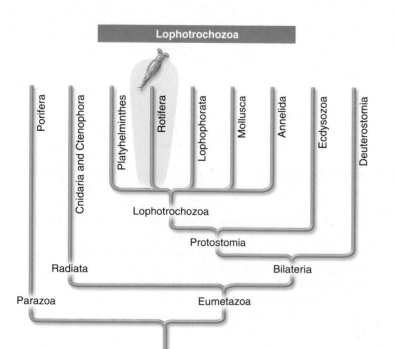

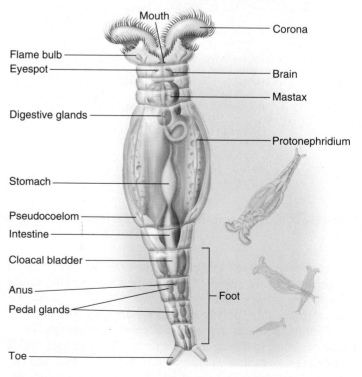

Figure 33.9   Body plan of a common rotifer, *Philodina* genus.

new females. Because the tiny zygotes are easily transported, rotifers show up in the smallest of aquatic environments, such as roof gutters or birdbaths.

The Lophophorata Includes Three Closely Related Phyla: Phoronida, Bryozoa, and Brachiopoda

The Lophophorates consist of three distinct phyla: the Phoronida, the Bryozoa, and the Brachiopoda. They all possess a lophophore, a ciliary feeding device (refer back to Figure 32.10a), and a true coelom (refer back to Figure 32.4a). The lophophore is a circular fold of the body wall that bears tentacles that draw water towards the mouth. Because a thin extension of the coelom penetrates each tentacle, the tentacles also serve as a respiratory device. Gases diffuse across the tentacles and into or out of the coelomic fluid and are carried throughout the body.

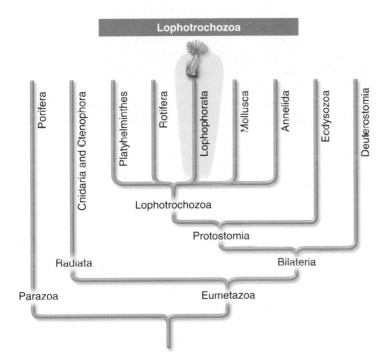

All three phyla have a U-shaped alimentary canal, with the anus located near the mouth but outside of the lophophore.

Phylum Phoronida Members of the Phoronida (from the Greek *phoros*, bearer, and the Latin *nidus*, nest) are elongated, tube-dwelling marine worms ranging in size from 1 mm to 50 cm in length. They live in a tough, leather-like chitinous tube that they secrete and that is often buried in the ground so only the lophophore sticks out (**Figure 33.10a**). The lophophore can be retracted quickly in the presence of danger. Only about 15 species of phoronids are found worldwide.

Phylum Bryozoa The bryozoans (from the Greek *bryon*, moss, and *zoon*, animal) are small colonial animals, most of which are less than 0.5 mm long, that can be found encrusted on rocks in shallow aquatic environments. They look very much like plants. There are about 4,000 species, many of which are fouling organisms that encrust boat hulls and have to be scraped off periodically. Within the colony, each animal secretes and lives inside a nonliving case called a **zoecium** (**Figure 33.10b**). The walls of the zoecium may be composed of chitin or calcium carbonate. For this reason, bryozoans have been important reef-builders, and since they date back to the Ordovician era, many fossil forms have been discovered and identified.

Phylum Brachiopoda Brachiopods (from the Greek *brachio*, arm, and *podos*, foot) are marine organisms with two shell halves, much like modern clams (**Figure 33.10c**). Unlike bivalve mollusks, however, which have a left and right valve (side) of the shell, brachiopods have a dorsal and ventral valve. Brachiopods are bottom dwelling species that attach to the substrate via a muscular pedicle. While they are a relatively small group, with about 300 living species, brachiopods flourished in the Paleozoic and Mesozoic eras, and about 30,000 fossil species have been identified. Some of these fossil forms tell of organisms that reached 30 cm in length, although their current relatives are only 0.5–8.0 cm long.

(a) A phoronid worm (*Phoronis californica*), buried in the sand with the lophophores extended.

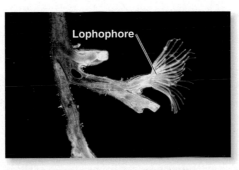

(b) Bryozoans, such as this *Plumatella repens*, are colonial lophophorates that reside in a nonliving case called a zoecium.

(c) Brachiopods including this northern lamp shell (*Terebratulina septentrionalis*) have a dorsal and ventral shell.

Figure 33.10 Lophophorates.

Biological inquiry: What are the two main functions of the lophophore?

The Mollusca Is a Large Phylum Containing Snails, Slugs, Oysters, Clams, Octopuses, and Squids

Mollusks (from the Latin *mollis*, soft) constitute a very large phylum, with over 100,000 living species, including organisms as diverse as snails, clams and oysters, cephalopods, and chitons. They are an ancient group, as evidenced by the classification of about 35,000 fossil species. Mollusks have a considerable economic, aesthetic, and ecological importance to humans. Many serve as sources of food, including scallops, oysters, clams, and squid. A significant industry involves the farming of oysters to produce cultured pearls, and rare and beautiful mollusk shells are extremely valuable to collectors. Snails and slugs can damage vegetables and ornamental plants, and boring mollusks can penetrate wooden ships and wharfs. Mollusks are intermediate hosts to many parasites, and several exotic species have become serious pests. For example, populations of the zebra mussel (*Dreissena polymorpha*) were introduced into North America from Asia via ballast water from transoceanic ships. Since their introduction, they have spread rapidly throughout the Great Lakes and an increasing number of inland waterways, significantly impacting native organisms and clogging water intake valves to municipal water treatment plants around the lakes.

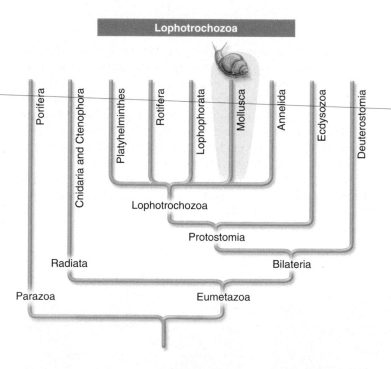

One common feature of the mollusks is their soft body, which exists, in many species, under a protective external shell. Most mollusks are marine, although some have colonized fresh water. Many snails and slugs have even moved onto land, but they survive only in humid areas and where the calcium necessary for shell formation is abundant in the soil. The ability to colonize freshwater and terrestrial habitats has led to a diversification of mollusk body plans. Thus, we again see how organismal diversity is related to environmental diversity.

While great variation in morphology occurs between classes, mollusks have a basic body plan consisting of three parts (**Figure 33.11**). A muscular **foot** is usually used for movement, and a **visceral mass** containing the internal organs rests atop the foot. The **mantle**, a fold of skin draped over the visceral mass, secretes a shell in those species that form shells. The mantle often extends beyond the visceral mass, creating a chamber called the **mantle cavity**, which houses delicate **gills**, specialized filamentous organs that are rich in blood vessels. A continuous current of water, often induced by cilia present on the gills or by muscular pumping, flushes out the wastes from the mantle cavity and brings in new oxygen-rich water.

Mollusks are coelomate organisms, but the coelom is confined to a small area around the heart. The mollusks' organs are served with oxygen and nutrients via a circulatory system. Mollusks have an **open circulatory system** with a heart that pumps body fluid called hemolymph through vessels and into sinuses, which are open, fluid-filled cavities. In comparison, in a closed circulatory system, the fluid called blood is always contained within vessels. The organs and tissues are thus continually bathed in hemolymph. From these sinuses, the hemolymph drains into vessels that take it to the gills and then back to the heart. The anus and pores from organs called **metanephridia**, which extract nitrogenous and other wastes, discharge into the mantle cavity. The metanephridial ducts may also serve to discharge sperm or eggs from the gonads.

The mollusk's mouth may contain a **radula**, a unique, protrusible, tonguelike organ that has many teeth and is used to eat plants, scrape food particles off rocks, or, if the mollusk is predatory, bore into shells of other species and tear flesh. In the cone shells (*Conus* spp.), the radula is reduced to a few poison-injecting teeth on the end of a long proboscis that is cast about in search of prey, such as a worm or even a fish (**Figure 33.12**). Some Indo-

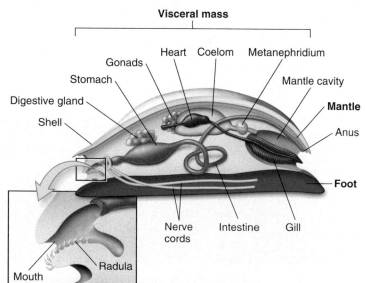

Figure 33.11 **The mollusk body plan.** The generalized body plan of a mollusk includes the characteristic foot, mantle, and visceral mass.

Figure 33.12 **Some gastropods are predators.** The cone shell, *Conus* spp., uses its long proboscis, equipped with poison-injecting teeth, to paralyze its prey—usually a fish.

Figure 33.13 **Bivalve shells have growth rings.** Quahog clams (*Mercenaria mercenaria*), also known as cherrystones or littlenecks (depending on their size), can live over 20 years.

Figure 33.14 **A veliger larva.** The immature free-swimming stage of molluscans develops from the trochophore larva.

Pacific cone shell species produce a neuromuscular toxin that can kill humans. Other mollusks, particularly bivalves, are suspension feeders that filter water brought in by ciliary currents.

Most shells are complex three-layered structures secreted by the mantle that continue to grow as the mollusk grows. Shell growth is often seasonal, resulting in distinct growth lines on the shell, much the same as tree rings (**Figure 33.13**). Using shell growth patterns, biologists have discovered some bivalves that are over 100 years old. The innermost layer of the shells of oysters, mussels, abalone, and other mollusks is a smooth, iridescent lining called **nacre**, which is commonly known as mother-of-pearl and is often collected from abalone shells for jewelry. Actual pearl production in mollusks, primarily oysters, occurs when a foreign object, such as a grain of sand, becomes lodged between the shell and the mantle, and layers of nacre are laid down around it to reduce the irritation.

Most mollusks have separate sexes, although some are hermaphroditic. Gametes are usually released into the water, where they mix and fertilization occurs. In some snails, however, fertilization is internal, with the male inserting sperm directly into the female. Internal fertilization was a key evolutionary development, enabling some snails to colonize land, and can be considered a critical innovation that fostered extensive adaptive radiation. In many species, reproduction involves the production of a trochophore larva that develops into a **veliger**, a free-swimming larva that has a rudimentary foot, shell, and mantle (**Figure 33.14**).

Of the eight molluscan classes, the four most common are the polyplacophora (chitons), gastropoda (snails and slugs), bivalvia (clams and mussels), and cephalopoda (octopuses, squid, and nautiluses) (**Table 33.3**). The class Gastropoda (from the Greek *gaster*, stomach, and *podos*, foot) is the largest group of mollusks and encompasses about 75,000 living species, including snails, periwinkles, limpets, and other shelled members. The class also includes species such as slugs and nudibranchs, whose shells have been greatly reduced or completely lost during their evolution (**Figure 33.15**). Most are marine or freshwater species, but some species, including snails and slugs, have

Table 33.3	Main Classes and Characteristics of Mollusks	
	Class and examples (est. # of species)	**Class characteristics**
	Polyplacophora: Chitons (860)	Marine; eight-plated shell
	Gastropoda: Snails, slugs, nudibranchs (75,000)	Marine, freshwater, or terrestrial; most with coiled shell, but shell absent in slugs and nudibranchs; radula present
	Bivalvia: Clams, mussels, oysters (30,000)	Marine or freshwater; shell with two halves or valves; primarily filter feeders with siphons
	Cephalopoda: Octopuses, squids, nautiluses (780)	Marine; predatory, with tentacles around mouth, often with suckers; shell often absent or reduced; closed circulatory system; jet propulsion via siphon

also colonized land. Most gastropods are slow-moving animals that are weighed down by their shell. Unlike bivalves, gastropods have a one-piece shell, into which the animal can withdraw to escape predators.

The 780 species of Cephalopoda (from the Greek *kephale*, head, and *podos*, foot) are the most morphologically complex of the mollusks and indeed among the most complex of all invertebrates. They include the octopuses, squids, cuttlefish, and nautiluses. Most are fast-swimming marine predators that range from organisms just a few centimeters in size to the giant squid (*Architeuthis*) which is known to reach over 17 m in length and 2 tons in weight. A cephalopod's mouth is surrounded by many long tentacles commonly armed with suckers. Octopuses have eight arms with suckers, and squids and cuttlefish have ten arms—eight with suckers and two long tentacles with suckers limited to their ends. Nautiluses have between 60 to 90 tentacles around the mouth.

All cephalopods have a beaklike jaw that allows them to bite their prey, and some, such as the blue-ringed octopus (*Hapalochlaena lunulata*), deliver a deadly poison through their saliva (**Figure 33.16**). Only one group, the nautiluses, has retained its external shell. In octopuses, the shell is not present, and in squid and cuttlefish, it is greatly reduced and internal. However, the fossil record is full of shelled cephalopods, called ammonites, some of which were as big as truck tires (**Figure 33.17**). They became extinct at the end of the Cretaceous period, although the reasons for this are not well understood.

The foot of some cephalopods has become modified into a muscular siphon. Water drawn into the mantle cavity is quickly expelled through the siphon, propelling the organisms forward or backward in a kind of jet propulsion. Such vigorous movement requires powerful muscles and a very efficient circulatory system to deliver oxygen and nutrients to the muscles. Cephalopods are the only mollusks with a **closed circulatory**

system, in which blood flows throughout an animal entirely within a series of vessels. One of the advantages of this type of system is that the heart can pump blood through the tissues rapidly. The blood of cephalopods contains the copper-rich protein hemocyanin for transporting oxygen. Less efficient than the iron-rich hemoglobin of vertebrates, hemocyanin gives the blood a blue color.

The nautiluses are impeded by a coiled, chambered shell and do not move as fast as the jet-propelled squids and octopuses (**Figure 33.18**). As it grows, the nautilus secretes a new chamber and seals off the old one with a **septum**. The older

Figure 33.17   A fossil ammonite.   These shelled cephalopods were abundant in the Cretaceous period.

Figure 33.15   A sea slug (*Phyllidia ocellata*).   The sea slugs, or nudibranchs, are a gastropod subclass whose members have lost their shell altogether.

Figure 33.16   The blue-ringed octopus (*Hapalochlaena lunulata*) is highly poisonous.

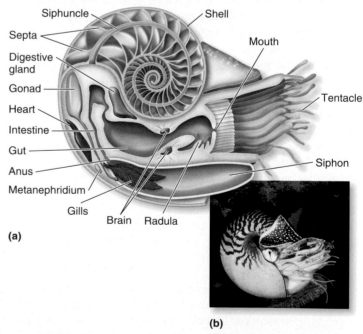

Siphuncle — Shell
Septa
Digestive gland — Mouth
Gonad
Heart — Tentacle
Intestine
Gut
Anus — Siphon
Metanephridium
Gills
Brain   Radula
(a)

(b)

Figure 33.18   The nautilus.   (a) A longitudinal section of a nautilus, showing the coiled shell with many chambers. The animal secretes a new chamber each year and lives only in the new one. (b) The chambered nautilus (*Nautilus pompilius*).

chambers are gas filled and act as buoyancy chambers. A thin strip of living tissue called the siphuncle removes liquid from the old chamber and replaces it with gas. The gas pressure within the chambers is only 1 atmosphere, despite the fact that nautiluses may be swimming at 400 m depths at a pressure of about 40 atmospheres. The shell's structure is strong enough to withstand this amount of pressure differential.

Cephalopods have a well-developed nervous system and brain that support their active lifestyle. Their sense organs, especially their eyes, are also very well developed. Many cephalopods (with the exception of nautiluses) have an ink sac that contains the pigment melanin; the sac can be emptied to provide a "smokescreen" to confuse predators. In many species, melanin is also distributed in special pigment cells in the skin that can produce color changes. Octopuses often change color when alarmed or during courtship, and they can rapidly change color to blend in with their background and escape detection. The central nervous system of the octopus is among the most complex in the invertebrate world. Behavioral biologists have demonstrated that octopuses can behave in sophisticated ways, and scientists are currently debating to what degree they are capable of learning by observation.

FEATURE INVESTIGATION

Fiorito and Scotto's Experiments Showed Invertebrates Can Exhibit Sophisticated Observational Learning Behavior

We tend to think of the ability to learn from others as a vertebrate phenomenon, especially among species that live in social groups. However, in 1992, Italian researchers Graziano Fiorito and Pietro Scotto demonstrated that octopuses can learn by observing the behavior of other octopuses (**Figure 33.19**). This was a surprising finding, in part because *Octopus vulgaris*, the species they studied, lives a solitary existence for most of its life.

In their experiments, octopuses were trained to attack either a red ball or a white ball by use of a reward (a small piece of fish placed behind the ball that it could not see) and a punishment (a small electric shock for choosing the wrong ball). This type of learning is called classical conditioning (see Chapter 55). Because octopuses are color blind, they must have been distinguishing between the relative brightness of the balls. Octopuses were considered to be trained when they made no mistakes in five trials. Observer octopuses in adjacent tanks were then allowed to watch the trained octopuses attacking the balls. In the third part of the experiment, the observer octopuses were themselves tested. In these cases, observers nearly always attacked the same color ball as they had observed the demonstrators attacking. In addition, learning by observation was achieved more quickly than the original training. This remarkable behavior is considered by some as the precursor to more complex forms of learning, including problem solving.

Figure 33.19   Observational learning in octopuses.

HYPOTHESIS   Octopuses can learn by observing another's behavior.

STUDY LOCATION   Laboratory setting with *Octopus vulgaris* collected from the Bay of Naples, Italy.

Experimental level   **Conceptual level**

1. Train 2 groups of octopuses, one to attack white balls, one to attack red.

Reward choice of correct ball (with fish) and punish choice of incorrect ball (with electric shock). Training complete when octopus makes no "mistakes" in 5 trials.

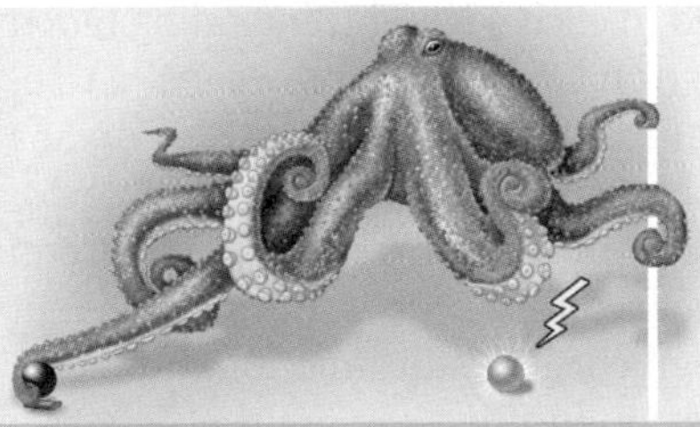

Conditions a demonstrator octopus to attack a particular color of ball.

2. In adjacent tanks, allow observer octopus to watch trained demonstrator octopus.

Observer octopus may be learning the correct ball by watching the demonstrator octopus.

Observer   Demonstrator

3 Drop balls into the tank of the observer octopuses. Test the observer octopus to see if it makes the same decisions as the demonstrator octopus.

Observer

If the observer octopus is learning from demonstator octopus, the observer octopus should attack the ball of the same color as the demonstrator octopus was trained to attack.

4 **THE DATA**

Participant	Color of ball chosen in 5 trials*	
	Red	**White**
Observers (watched demonstrator attack red)	4.31	0.31
Observers (watched demonstrator attack white)	0.40	4.10
Untrained (did not watch demonstrations)	2.11	1.94

*Average of 5 trials, data do not always sum to 5, because some trials resulted in no balls being chosen.

The Phylum Annelida Consists of the Segmented Worms

If you look at an earthworm, you will see little rings all down its body. Indeed, the phylum name Annelida is derived from the Latin, *annulus*, meaning little ring. Each ring is a distinct segment of the annelid's body, with each segment separated from the one in front and the one behind by a septum (**Figure 33.20a**). Segmentation, the division of the body into nearly identical subunits, is a critical evolutionary innovation in the annelids and confers at least three major advantages. First, many components of the body are repeated in each segment, including blood vessels, nerves, and excretory and reproductive organs. Excretion is accomplished by metanephridia, paired excretory organs in every segment that extract waste from the blood and coelomic fluid, emptying it to the exterior via pores in the skin (look ahead to Figure 49.9). If the excretory organs in one segment fail, the organs of another segment will still function.

Second, annelids possess a fluid-filled coelom that acts as a hydrostatic skeleton. In unsegmented coelomate animals, muscle contractions can distort the entire body during movement. However, such distortion is minimized in segmented animals, which allows for more effective locomotion over solid surfaces. In an earthworm, when the circular muscles around a segment contract against the hydrostatic skeleton, that segment becomes elongated. When the longitudinal muscles contract, the segment becomes compact. Waves of muscular contraction ripple down the segments, which elongate or contract independently.

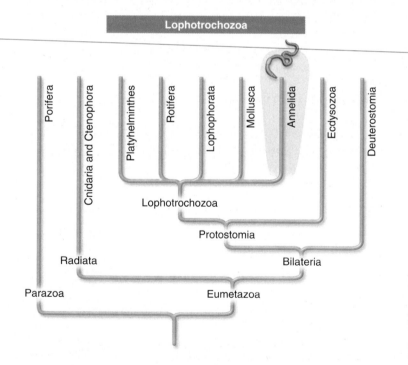

Third, segmentation also permits specialization of some segments, although such specialization is only minimally present at the annelid's anterior end. Annelids have a relatively sophisticated nervous system involving a pair of cerebral ganglia that connect to a subpharyngeal ganglion (**Figure 33.20b**).

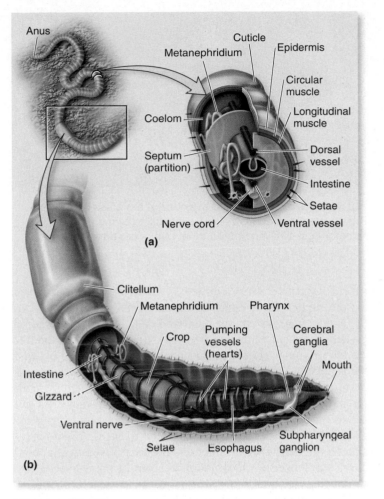

Figure 33.20 The segmented body plan of an annelid, as illustrated by an earthworm. (a) Individual segments are separated by a septum. (b) The segmented nature of the worm is apparent externally as well as internally.

Biological inquiry: What are some of the advantages of segmentation?

Table 33.4	**Main Classes and Characteristics of Annelids**

	Class and examples (est. # of species)	Class characteristics
	Polychaeta: Marine worms (10,000)	Well-developed head; usually free living; parapodia present
	Oligochaeta: Terrestrial and freshwater worms such as earthworms (3,500)	Undeveloped head; setae; no parapodia
	Hirudinea: Leeches (630)	Mostly ectoparasites; suckers present at both ends; flattened body; reduced coelom; no setae

From there, a large ventral nerve cord runs down the entire length of the body. The ventral nerve chord is unusual because it contains a few very large nerve cells called **giant axons** that facilitate high-speed nerve conduction and rapid responses to stimuli. Such axons are found in other invertebrates, the best known of which is the squid.

Annelids essentially have a double transport system, since both the circulatory system and the coelomic fluid carry nutrients, wastes, and respiratory gases, to some degree. The circulatory system is usually closed, with dorsal and ventral vessels connected by five pairs of pumping vessels that serve as muscular hearts. The blood of most annelid species contains the respiratory pigment hemoglobin, an iron-containing protein involved in oxygen transport in annelids as well as in vertebrates. Respiration occurs directly through the permeable skin surface, which restricts annelids to moist environments. The digestive system is complete and unsegmented, with many specialized regions: mouth, pharynx, esophagus, crop, gizzard, intestine, and

anus. Sexual reproduction involves two individuals, often of separate sexes, but sometimes hermaphrodites, which exchange sperm via internal fertilization. In some species, asexual reproduction by fission occurs, in which the posterior part of the body breaks off and forms a new individual.

Annelids are a large phylum with about 15,000 described species. Its members include the familiar earthworm, marine polychaete worms, and leeches, and range in size from less than 1 mm to enormous Australian earthworms that can reach a size of 3 m. All annelids except the leeches have chitinous bristles called **setae** on each segment. In one class, the polychaetes, these are situated on fleshy, footlike **parapodia** ("almost feet") that are pushed into the substrate to provide traction during movement. Many annelid species burrow into the Earth or into muddy marine sediments and extract nutrients from ingested soil or mud. Some annelids also feed on dead or living vegetation, while others are predatory or parasitic.

The phylum consists of three main classes: the Polychaeta, Oligochaeta, and Hirudinea (Table 33.4). Some biologists have recently suggested Oligochaeta and Hirudinea be combined into one larger class called the Clitellata, because they share a common structure called a clitellum, a glandular region of the body that has a role in reproduction. However, this newer taxonomic grouping is still being debated.

Class Polychaeta With over 10,000 species, the Polychaeta is the most species-rich class of the annelids. Many polychaetes are brightly colored, and all have many long setae bristling out of their body (polychaete means "many bristles"). Most of them are marine organisms, living in burrows in the mud or sand, or in rock crevices, and are often abundant in the intertidal mudflats. They are important prey for predators such as fish and crustaceans. The polychaete head is well developed and, in predatory species, may exhibit powerful jaws. Some species are filter feeders and have a crown of tentacles that sticks up out of the mud while the bulk of the worm remains hidden.

Class Oligochaeta The Oligochaeta (meaning "few bristles") includes the common earthworms and many species of freshwater worms. Earthworms play a unique and beneficial role in conditioning the soil, primarily due to the effects of their burrows and castings. Earthworms ingest soil and leaf tissue to extract nutrients and in the process create burrows in the Earth. As plant material and soil passes through the earthworm's digestive system, it is finely ground in the gizzard into smaller fragments. Once excreted, this material—called castings—enriches the soil. Because a worm can eat is own weight in soil every day, worm castings on the soil surface can be extensive. The biologist Charles Darwin was interested in earthworm activity, and his last work, *The Formation of Vegetable Mould, through the Actions of Worms, with Observations on Their Habits*, was the first detailed study of earthworm ecology. In it, he wrote, "All the fertile areas of this planet have at least once passed through the bodies of earthworms."

Class Hirudinea Leeches (class Hirudinea) are primarily found in freshwater environments, but there are also some marine species as well as terrestrial species that inhabit warm, moist areas such as tropical forests. Leeches have a fixed number of segments, usually 34, though the septa have disappeared in most species. All leeches feed on other organisms, often as blood-sucking parasites of vertebrates. They have powerful suckers at both ends of the body, and the anterior sucker is equipped with razor-sharp jaws that can bore or slice into the host's tissues. The salivary secretion of leeches (hirudin) acts as an anticoagulant to stop blood clotting. Leeches can suck up to several times their own weight in blood. As such, they were once used in the medical field in the practice of bloodletting, the withdrawal of often considerable quantities of blood from a patient in the belief that this would prevent or cure illness and disease. Even today, leeches may be used after surgeries, particularly those involving the reattachment of digits (**Figure 33.21**). In these cases, the blood vessels are not fully reconnected and much excess blood accumulates, causing swelling. This excess blood switches off the delivery of new blood and stops the formation of new arteries. If leeches remove the accumulated blood, new capillaries will be more likely to form, and the tissues will become healthy.

Unlike cestode and trematode flatworms, which are internally parasitic and quite host specific, leeches are generally external parasites that feed on a broad range of hosts, including fish, amphibians, and mammals. However, there are always exceptions. *Placobdelloides jaegerskioeldi* is a parasitic leech that lives only in the rectum of hippopotamuses.

33.4 Ecdysozoa: The Nematodes and Arthropods

The Ecdysozoa is the sister group to the Lophotrochozoa. While the separation is supported by molecular evidence, the Ecdysozoa is named for a morphological characteristic, the physical phenomenon of ecdysis, or molting (refer back to Figure 32.11).

Figure 33.21   **A leech, a member of the class Hirudinea.** This species, *Hirudo medicinalis*, is sucking blood from a hematoma, a swelling of blood that can occur after surgery.

All ecdysozoans possess a **cuticle**, a nonliving cover that serves to both support and protect the animal. Once formed, however, the cuticle typically cannot increase in size, which restricts the growth of the animal inside. The solution for growth is the formation of a new, softer cuticle under the old one. The old one then splits open and is sloughed off, allowing the new, soft cuticle to expand to a bigger size before it hardens. Where the cuticle is thick, as in arthropods, it impedes the diffusion of oxygen across the skin. Such species acquire oxygen by lungs, gills, or a set of branching, air-filled tubes called tracheae. There are no cilia on the cuticle for locomotion, and thus a variety of appendages specialized for locomotion evolved in many species, including legs for walking or swimming, and wings for flying.

The ability to shed the cuticle opened up developmental options for the ecdysozoans. For example, many species undergo a complete metamorphosis, changing from a wormlike larva into a winged adult. Animals with internal skeletons cannot do this because growth only occurs by adding more minerals to the existing skeleton. Another significant adaptation is the development of internal fertilization, in which the male deposits sperm directly into the female, where fertilization takes place. This trait, which allows animals to breed on dry land, evolved independently in the vertebrates.

Because of these innovations, ecdysozoans are an incredibly successful group. Of the eight ecdysozoan phyla, we will consider the most common two: the nematodes and arthropods. The grouping of nematodes and arthropods is a relatively new idea and implies that the process of molting arose only once in animal evolution. In support of this, certain hormones that stimulate molting have been discovered to exist in nematodes as well as arthropods.

The Phylum Nematoda Consists of Small Pseudocoelomate Worms Covered by a Tough Cuticle

The nematodes (from the Greek *nematos*, thread), also called roundworms, are small, thin worms that range from less than

1 mm to about 5 cm (**Figure 33.22**), although some parasitic species measuring 1 m or more have been found in the placenta of sperm whales. Nematodes are ubiquitous organisms that exist in nearly all habitats, from the poles to the tropics. They are found in the soil, in both freshwater and marine environments, and inside plants and animals as parasites. A shovelful of soil may contain a million nematodes. Over 20,000 species are known, but there are probably at least five times as many undiscovered species.

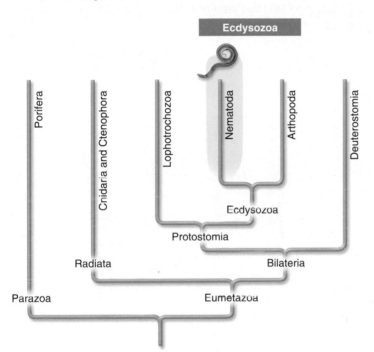

Figure 33.22 Scanning electron micrograph of a nematode within a plant leaf.

Nematodes have several distinguishing characteristics. A tough cuticle covers the body. The cuticle is secreted by the epidermis and is made primarily of **collagen**, a structural protein also present in vertebrates. The cuticle is shed periodically as the nematode grows. Beneath the epidermis are longitudinal muscles but no circular muscles, which means that muscle contraction results in more thrashing of the body than smoother wormlike movement. Nematodes possess a pseudocoelom that acts as a hydrostatic skeleton and a circulatory system. Diffusion of gases occurs through the cuticle. Roundworms have a complete digestive tract composed of a mouth, pharynx, intestine, and anus. The mouth often contains sharp, piercing organs called **stylets**, and the muscular pharynx functions to suck in food.

Nematode reproduction is usually sexual, with separate males and females, and fertilization takes place internally. Females are generally larger than males and can produce prodigious numbers of eggs, in some cases over 100,000 per day. Development in nematodes is easily observed because the organism is transparent and the generation time is short. For these reasons, the small, free-living nematode *Caenorhabditis elegans* has become a model organism for researchers to study (refer back to Figure 19.1b). In fact, the 2002 Nobel Prize in Medicine or Physiology was shared by Sydney Brenner, Robert Horvitz,

and John Sulston for their studies of the genetic regulation of development and programmed cell death in *C. elegans*. This nematode has 1,090 cells but 131 die, leaving exactly 959 cells. The cells die via a genetically controlled cell death. Many diseases in humans, including acquired immunodeficiency syndrome (AIDS), cause extensive cell death, while others, such as cancer and autoimmune diseases, reduce cell death so that cells that should die do not. Researchers are studying the process of programmed cell death in *C. elegans* in the hope of finding treatments for these and other human diseases.

A large number of nematodes are parasitic in humans and other vertebrates. The large roundworm (*Ascaris lumbricoides*) is a parasite of the small intestine that can reach up to 30 cm in length. Over a billion people worldwide carry this parasite. While infections are most prevalent in tropical or developing countries, the prevalence of *A. lumbricoides* is relatively high in rural areas of the southeastern U.S. Eggs pass out in feces and can remain viable in the soil for years. Eggs require ingestion before hatching into an infective stage. Hookworms (*Necator americanus*), so named because their anterior end curves dorsally like a hook, are also parasites of the human intestine. The eggs pass out in feces, and recently hatched hookworms can penetrate the skin of a host's foot to establish a new infection. In areas with modern plumbing, these diseases are uncommon.

Pinworms (*Enterobius vermicularis*), while a nuisance, have relatively benign effects on their hosts. The rate of infection in the U.S., however, is staggering: 30% of children and 16% of adults are believed to be hosts. Adult pinworms live in the large intestine and migrate to the anal region at night to lay their eggs, which causes intense itching. The resultant scratching spreads the eggs. In the tropics, some 250 million people are infected with *Wuchereria bancrofti*, a fairly large (100 mm) worm that lives in the lymphatic system, blocking the flow of lymph, and, in extreme cases, causing elephantiasis or extreme swelling of the legs and other body parts (**Figure 33.23**). Females release tiny, live young called microfilariae, which are transmitted to new hosts via mosquitoes.

men of an insect (**Figure 33.24**). Cephalization is extensive, and arthropods have well-developed sensory organs, including organs of sight, touch, smell, hearing, and balance. Arthropods have compound eyes composed of many independent visual units called **ommatidia** (singular, ommatidium) (look ahead to Figure 45.12). Each ommatidium functions as a separate photoreceptor capable of forming an independent image. Together, these lenses render a mosaic-like image of the environment. Some species, particularly some insects, possess additional simple eyes, or ocelli, that are probably only capable of distinguishing light from dark.

Figure 33.23 Elephantiasis in a human leg. The disease is caused by the nematode parasite *Wuchereria bancrofti*, which lives in the lymphatic system and blocks the flow of lymph.

The Phylum Arthropoda Contains the Insects, Crustaceans, and Spiders, All with Jointed Appendages

The arthropods (from the Greek *arthron*, joint, and *podos*, foot) constitute perhaps the most successful phylum on Earth. About three-quarters of all described living species are arthropods, and scientists have estimated that they are also numerically common, with an estimated 10^{18}, or a billion billion, individual organisms present on Earth. The huge success of the arthropods, in terms of their sheer numbers and diversity, is related to a body plan that permitted conquest of the major biomes on Earth, from the poles to the tropics, and from marine and freshwater habitats to dry land.

The body of a typical arthropod is covered by a hard cuticle, an **exoskeleton** (external skeleton) made of layers of chitin and protein. The cuticle can be extremely tough in some parts, as in the shells of crabs, lobsters, and even beetles, yet be soft and flexible in other parts, between body segments and segments of appendages, to allow for movement. In the class of arthropods called crustaceans, the exoskeleton is reinforced with calcium carbonate to make it extra hard. The exoskeleton provides protection and also a point of attachment for muscles, all of which are internal. It is also relatively impermeable to water, a feature that may have enabled many arthropods to conserve water and colonize land, in much the same way as a tough seed coat allowed plants to colonize land (see Chapter 30). From this point of view, the development of a hard cuticle was a critical innovation. It also reminds us that the ability to adapt to diverse environmental conditions can itself lead to increased organismal diversity.

Arthropods are segmented, and many of the segments bear appendages for locomotion, food handling, or reproduction. In many orders, the body segments have become fused into functional units, or **tagmata**, such as the head, thorax, and abdo-

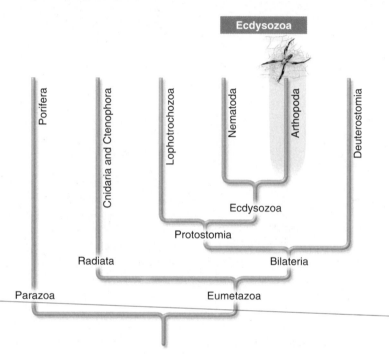

The arthropod brain is quite sophisticated, consisting of two or three ganglia connected to several smaller ventral nerve ganglia. Like most mollusks, arthropods have an open circulatory system (look ahead to Figure 47.2), in which hemolymph is pumped from the heart into the aorta or short arteries and then into sinuses, open fluid-filled cavities surrounding the major organs that coalesce to form the hemocoel. From the sinuses, gases and nutrients diffuse into tissues. The hemolymph flows back into the heart via pores, called ostia, that are equipped with valves.

Because the cuticle impedes the diffusion of gases through the body surface, arthropods possess special organs that permit gas exchange. In aquatic arthropods, these consist of feathery gills that have an extensive surface area in contact with the surrounding water. Terrestrial species have a highly developed **tracheal system**, a series of finely branched air tubes called tracheae that lead into the body from pores called **spiracles** (look ahead to Figure 48.9). The tracheal system delivers oxygen directly to tissues and cells. Some spiders have book lungs, consisting of a series of sheetlike structures extending into a hemolymph-filled chamber on the underside of the abdomen. Gases also diffuse across thin areas of the cuticle.

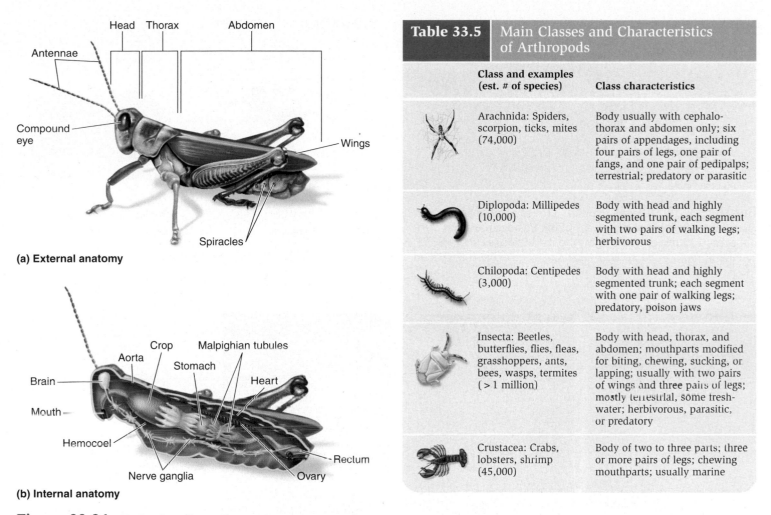

(a) External anatomy

(b) Internal anatomy

Figure 33.24 Body plan of an arthropod, as represented by a grasshopper.

Table 33.5	Main Classes and Characteristics of Arthropods

	Class and examples (est. # of species)	Class characteristics
	Arachnida: Spiders, scorpion, ticks, mites (74,000)	Body usually with cephalo-thorax and abdomen only; six pairs of appendages, including four pairs of legs, one pair of fangs, and one pair of pedipalps; terrestrial; predatory or parasitic
	Diplopoda: Millipedes (10,000)	Body with head and highly segmented trunk, each segment with two pairs of walking legs; herbivorous
	Chilopoda: Centipedes (3,000)	Body with head and highly segmented trunk; each segment with one pair of walking legs; predatory, poison jaws
	Insecta: Beetles, butterflies, flies, fleas, grasshoppers, ants, bees, wasps, termites (> 1 million)	Body with head, thorax, and abdomen; mouthparts modified for biting, chewing, sucking, or lapping; usually with two pairs of wings and three pairs of legs; mostly terrestrial, some fresh-water; herbivorous, parasitic, or predatory
	Crustacea: Crabs, lobsters, shrimp (45,000)	Body of two to three parts; three or more pairs of legs; chewing mouthparts; usually marine

The digestive system is complex and often includes a mouth, crop, stomach, intestine, and rectum. Excretion is accomplished by a specialized metanephridia or, in insects and some other taxa, by **Malpighian tubules**, delicate projections from the digestive tract that protrude into the hemolymph (look ahead to Figure 49.10). Nitrogenous wastes are absorbed by the tubules and emptied into the gut. The intestine and rectum reabsorb water and salts. This excretory system, allowing the retention of water, was another critical innovation that permitted the colonization of land by arthropods.

There are six main classes of arthropods, one now-extinct class, Trilobita (the trilobites), and five living classes, Arachnida (spiders and scorpions), Diplopoda (millipedes), Chilopoda (centipedes), Insecta (insects), and Crustacea (crabs and relatives) (**Table 33.5**).

Class Trilobita: Extinct Early Arthropods The trilobites were among the earliest arthropods, flourishing in shallow seas of the Paleozoic era, some 500 million years ago, and dying out about 250 million years ago. Most trilobites were bottom feed-ers and were generally 3–10 cm in size, although some reached almost 1 m in length (**Figure 33.25**). Like many arthropods, they had three main tagmata: the head, thorax, and abdomen. Trilobites also had two dorsal grooves that divided the body longitudinally into three lobes—a median lobe and two anterior lobes—a structural characteristic giving the class its name. Most of the body segments showed little specialization. In contrast, as we will explore, more advanced arthropods developed specialized appendages on many segments, including appendages for grasping, walking, and swimming.

Class Arachnida: The Spiders, Scorpions, Ticks, and Mites The class Arachnida contains predatory spiders and scorpions as well as the ticks and mites, some of which are blood-sucking parasites that feed on vertebrates. All species have a body consisting of two tagmata: a fused head and thorax called a **cephalothorax**, and an abdomen (**Figure 33.26a**).

In spiders (order Araneae), the two body parts are joined by a **pedicel**, a narrow, waistlike point of attachment. Spiders have six pairs of appendages: the chelicerae, or fangs (**Figure 33.26b**); a pair of **pedipalps**, which have various sensory, predatory, or reproductive functions; and four pairs of walking legs. The fangs are supplied with venom from poison glands.

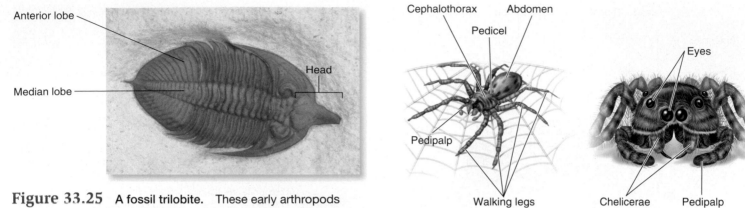

Figure 33.25 **A fossil trilobite.** These early arthropods were common in the shallow seas of the Paleozoic era but died out some 250 million years ago. About 4,000 fossil species, including *Huntonia huntonesis* shown here, have been described.

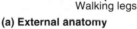

(a) External anatomy   **(b) Close-up of head**

Figure 33.26 Spider morphology.

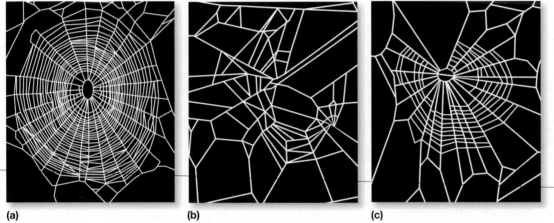

(a)   (b)   (c)

Figure 33.27 **Spider webs.** (a) Normal web. (b) Web spun by spider fed with prey containing caffeine. (c) Web spun by spider fed with prey containing marijuana.

Most spider bites are harmless to humans, although they are very effective in immobilizing and/or killing their insect prey. Venom from some species, including the black widow (*Latrodectus mactans*) and the brown recluse (*Loxosceles reclusa*), are potentially, although rarely, fatal to humans. The toxin of the black widow is a neurotoxin, which interferes with the functioning of the nervous system, while that of the brown recluse is hemolytic, meaning it destroys tissue around the bite. After the spider has subdued its prey, it pumps digestive fluid into the tissues via the fangs and sucks out the partially digested meal.

Spiders have abdominal silk glands called **spinnerets**, and many spin webs to catch prey (**Figure 33.27a**). The silk is a protein that stiffens after extrusion from the body because the mechanical shearing causes a change in the organization of the amino acids. Silk is stronger than steel of the same diameter. Each species constructs a characteristic size and style of web and can do it perfectly on its first attempt, indicating that web spinning is an innate (inherited) behavior (see Chapter 55). Spiders also use silk to wrap up prey and to construct egg sacs. Interestingly, spiders that are fed drugged food (flies) spin their webs differently than undrugged spiders (**Figure 33.27b,c**). Some scientists have suggested that web-spinning spiders be

used to test substances for the presence of drugs or even to indicate environmental contamination. Not all spiders use silk extensively. Other spiders, including the wolf spider (**Figure 33.28a**), actively pursue their prey.

Scorpions (order Scorpionida) are generally tropical or subtropical animals that feed primarily on insects, though they may eat spiders and other arthropods as well as smaller reptiles and mice. Their pedipalps are modified into large claws, and their abdomen tapers into a stinger, which is used to inject venom. While the venom of most North American species is generally not fatal to humans, that of the *Centruroides* genus from deserts in the U.S. Southwest and Mexico can be deadly. Fatal species are also found in India, Africa, and other countries. Unlike spiders, which lay eggs, scorpions bear live young that the mother subsequently carries around on her back until they have their first molt (**Figure 33.28b**).

In mites and ticks (order Acari), the two main body segments (cephalothorax and abdomen) are fused and appear as one large segment. Many mite species are free-living scavengers that feed on dead plant or animal material. Other mites are serious pests on crops, and some, like chiggers (*Trombicula alfreddugesi*), are parasites of humans that can spread diseases such as

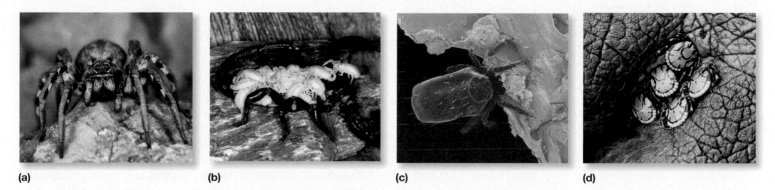

(a)   (b)   (c)   (d)

Figure 33.28 Common arachnids. **(a)** This wolf spider (*Lycosa tarantula*) does not spin a web but instead runs after its prey. Note the pedipalps, which look like short legs. **(b)** The Cape thick-tailed scorpion (*Parabuthus capensis*) is highly venomous and carries its white young on its back. **(c)** The chigger mite (*Trombicula alfreddugesi*) can cause irritation to human skin. **(d)** These South African bont ticks (*Amblyomma hebraeum*) are feeding on a white rhinoceros.

Biological inquiry: What is one of the main characteristics distinguishing arachnids from insects?

typhus (**Figure 33.28c**). Chiggers are parasites only in their larval stage. It is a myth that chiggers bore into the skin. Rather, it is their bite and salivary secretions that cause skin irritation. *Demodex brevis* is a hair-follicle mite that is common in animals and humans. The mite is estimated to be present in over 90% of adult humans. While the mite causes no irritation in most humans, *Demodex canis* causes the skin disease known as mange in domestic animals, particularly dogs.

Ticks are larger organisms than mites, and all are ectoparasitic, feeding on the body surface, on vertebrates. Their life cycle includes attachment to a host, sucking blood until they are replete, and dropping off the host to molt (**Figure 33.28d**). Ticks can carry a huge variety of viral and bacterial diseases, including Lyme disease, a bacterial disease so named because it was first found in the town of Lyme, Connecticut, in the 1970s.

Classes Diplopoda and Chilopoda: The Millipedes and Centipedes The millipedes and centipedes are both wormlike arthropods that are among the earliest terrestrial phyla known. Millipedes (class Diplopoda) have two pairs of legs per segment, as their Latin name denotes (*diplo*, two, and *podos*, feet), not 1,000 legs, as their common name suggests (**Figure 33.29a**). They are slow-moving herbivorous creatures that eat decaying leaves and other plant material. When threatened, the millipede's response is to roll up into a protective coil. Many millipede species also have repugnatorial glands on their underside that can eject a variety of toxic, repellent secretions. Some millipedes are brightly colored, warning potential predators that they can protect themselves.

Class Chilopoda (from the Latin *chilo*, lip, and *podos*, feet), or centipedes, are fast-moving carnivores that have one pair of walking legs per segment (**Figure 33.29b**). The head has many sensory appendages, including a pair of antennae and three pairs of appendages modified as mouthparts, including powerful claws connected to poison glands. The toxin from venom of some of the larger species, such as *Scolopendra heros*, is powerful enough to cause pain in humans. Most species do not have a waxy waterproofing layer on their cuticle and are restricted to

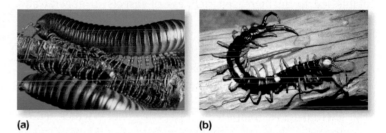

(a)   (b)

Figure 33.29 Millipedes and centipedes. **(a)** Millipedes have two pairs of legs per segment. **(b)** The venom of the giant centipede (*Scolopendra heros*) is known to produce significant swelling and pain to humans.

moist environments under leaf litter or in decaying logs, usually coming out at night to actively hunt their prey.

Class Insecta: A Diverse Array of Insects Insects are in a class by themselves, literally and figuratively. There are more species of insects than all other species of animal life combined. One million species of insects have been described and, according to best estimates, 9 million more species are awaiting description. At least 90,000 species of insects have been identified in the U.S. and Canada alone. It is believed that the explosion of diversity in the flowering plants (angiosperms) greatly promoted the huge diversification of insects relative to other classes.

GENOMES & PROTEOMES

Insect Diversity May Be Explained by the Appearance of the Angiosperms

More species are found in the class Insecta, over 1 million, than in any other class. Of these, over half a million are beetles, order Coleoptera. To understand what factors influence biodiversity, we need to understand why there are so many species of insects and beetles. In 1998, biologist Brian Farrell used

Table 33.6	Five Phylogenetic Comparisons of the Species Richness of Related Beetle Families and Subfamilies			
Comparison*	Gymnosperm-associated family/subfamily	# of Species	Angiosperm-associated family/subfamily	# of Species
1	Nemonychidae	85	Curculionidae	44,002
2	Oxycoryninae	30	Belinae	150
3	Aseminae	78	Cerambycinae	25,000
4	Palophaginae	3	Megalopodinae	400
5	Orsodacnidae	26	Chrysomelidae	33,400

* In each comparison, the families and subfamilies are closely related. For example, the *Nemonychidae* are closely related to the *Curculionidae*, but the latter has many more species.

a genomic study to show that the appearance of angiosperm plants led to an increased richness of insects in general and leaf-eating beetles in particular.

For about the first 50 million years of their order's existence, beetles fed on detritus, dead plant and animal material, and fungi. Plant feeding arose later in beetle history, during the Permian period, but relatively few fossilized beetle genera exist from that time. Jurassic fossils showed the existence of about 150 phytophagous (plant-feeding) beetle genera, which fed on gymnosperm plants: the conifers and cycads. However, many lineages of these ancestral beetles colonized angiosperms independently in the Cretaceous and Tertiary, after the diversification of angiosperms. In each case, the lineages that fed on angiosperms have become much more diverse than those that fed on gymnosperms.

In seeking to explain the huge diversity of beetles, Farrell analyzed DNA sequences for the entire 18S ribosomal subunit genes for 115 beetle species, drawn from all beetle subfamilies. In addition, he constructed a data set of 212 morphological characters of the same species. From both sets of data, Farrell constructed a phylogenetic tree. The oldest of the lineages for each of the subfamilies are gymnosperm-affiliated, while the more recently derived lineages are angiosperm-associated species.

Farrell also made five specific phylogenetic comparisons of the species richness of beetles in the same subfamilies but that either fed on gymnosperms or angiosperms (Table 33.6). In each case, the angiosperm-feeding group speciated frequently and became very diverse. In contrast to the gymnosperm-associated lineages, which generally feed on the pollen-rich reproductive parts of conifers and cycads, the angiosperm-associated subfamilies diversified in feeding habit to chew and mine leaves and feed on seeds and roots as well as reproductive parts. This illustrates how critical innovations can lead to extensive adaptive radiation, one of the principles of diversity noted at the beginning of this unit.

Insects are the subject of an entire field of scientific study, **entomology**. They are studied in large part because of their significance as pests of the world's agricultural crops. Insects live in all terrestrial habitats, and virtually all species of plants are fed upon by at least one, usually tens, and sometimes, in the case of large trees, hundreds of insect species. Because insects eat approximately one-quarter of the world's crops, we are constantly trying to find ways to reduce insect pest densities. Pest reduction often involves chemical control, the use of pesticides, or biological control, the use of living organisms, to reduce pest populations. Many species of insects are also important pests or parasites of humans and livestock, both by their own actions and as vectors of diseases such as malaria and sleeping sickness.

On the other hand, insects provide many types of essential biological services. We depend on insects such as honeybees to pollinate our crops. Bees also produce honey, and silkworms are the source of silk fiber. Despite the revulsion they provoke, fly larvae (maggots) are important in the decomposition process of both dead plants and animals. Insect parasites and predators are used in biological control to reduce densities of pest insects on crops.

Of paramount importance to the success of insects was the development of wings, a feature possessed by no other arthropod and indeed no other living animal except birds and bats. Unlike vertebrate wings, however, insect wings are actually outgrowths of the body wall cuticle and are not true segmental appendages. This means that insects still have all their walking legs. Insects are thus like the mythological horse Pegasus, which sprouted wings out of its back while retaining all four legs. In contrast, birds and bats have one pair of appendages (arms) modified for flight, which leaves them considerably less agile on the ground.

The great diversity of insects is illustrated by the fact that there are 35 different orders, some of which have over 100,000 species. The most common of the orders are discussed in **Table 33.7**. Different orders of insects have slightly different wing structures, and many of the orders are based on wing type (*pteron* is a Greek word meaning wing). Wasps and bees (Hymenoptera) have two pairs of wings hooked together that move as one wing. Butterflies (Lopidoptera) have wings that are covered in scales (from the Greek, *lepido*, scale), while other insects generally have clear, membranous wings. Flies (Diptera) possess only one pair of wings (the front pair); the back pair has been modified to a small pair of balancing organs, called halteres, that act like miniature gyroscopes. On the other hand, in beetles (Coleoptera) only the back pair of wings is functional, as the front wings have been hardened into protective shell-like coverings (elytra), under which the back pair folds when not in use. In ant and termite colonies, female individuals called workers have lost their wings, while the queen and the drones (males) have retained theirs. Other species, such as fleas and lice, are completely wingless.

Table 33.7 | Main Orders and Characteristics of Insects

Order and examples (est. # of species)		Order characteristics
Coleoptera: Beetles, weevils (500,000)		Two pairs of wings (front pair thick and leathery, acting as wing cases, back pair membranous); armored exoskeleton; biting and chewing mouthparts; complete metamorphosis; largest order of insects
Hymenoptera: Ants, bees, wasps (190,000)		Two pairs of membranous wings; chewing or sucking mouthparts; many have posterior stinging organ on females; complete metamorphosis; many species social; important pollinators
Diptera: Flies, mosquitoes (190,000)		One pair of wings with hindwings modified into halteres (balancing organs); sucking, piercing, or lapping mouthparts; complete metamorphosis; larvae are grublike maggots in various food sources; some adults are disease vectors
Lepidoptera: Butterflies, moths (140,000)		Two pairs of colorful wings covered with tiny scales; long tubelike tongue for sucking; complete metamorphosis; larvae are plant-feeding caterpillars
Hemiptera: True bugs; assassin bug, bedbug, chinch bug, cicada (100,000)		Two pairs of membranous wings; piercing or sucking mouthparts; incomplete metamorphosis; many plant feeders; some predatory or blood feeders; vectors of plant diseases
Orthoptera: Crickets, roaches, grasshoppers, mantids (30,000)		Two pairs of wings (front pair leathery, back pair membranous); chewing mouthparts; mostly herbivorous; incomplete metamorphosis; powerful hind legs for jumping
Odonata: Damselflies, dragonflies (6,000)		Two pairs of long, membranous wings; chewing mouthparts; large eyes; predatory on other insects; incomplete metamorphosis; larvae aquatic; considered primitive insects
Anoplura: Sucking lice (2,400)		Wingless ectoparasites; sucking mouthparts; flattened body; reduced eyes; legs with clawlike tarsi for clinging to skin; incomplete metamorphosis; very host specific; vectors of typhus
Siphonoptera: Fleas (2,000)		Wingless, laterally flattened; piercing and sucking mouthparts; adults are bloodsuckers on birds and mammals; jumping legs; complete metamorphosis; vectors of plague
Isoptera: Termites (2,000)		Two pairs of membranous wings when present; some stages wingless; chewing mouthparts; social species; incomplete metamorphosis

Insects in different orders have also evolved a variety of mouthparts (**Figure 33.30**). Grasshoppers, beetles, dragonflies, and many others have mouthparts adapted for chewing. Mosquitoes and many plant pests have mouthparts adapted for piercing and sucking. Butterflies and moths have a coiled tongue (**proboscis**) that can be uncoiled, enabling them to drink nectar from flowers. Finally, some flies have lapping, spongelike mouthparts that sop up liquid food. Their varied mouthparts allow insects to specialize their feeding on virtually anything: plant matter, decaying organic matter, and other living animals. The biological diversity of insects is thus related to environmental diversity, in this case, the variety of foods that insects eat. Parasitic insects attach themselves to other species, and there are even insect parasites that feed on parasites (called hyperparasites), proving, as the 18th-century English poet Jonathan Swift noted:

Big fleas have little fleas
upon their backs to bite 'em;
little fleas have smaller fleas
and so *ad infinitum*.

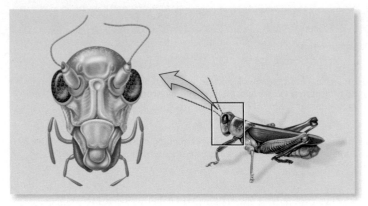

(a) Chewing (grasshopper)

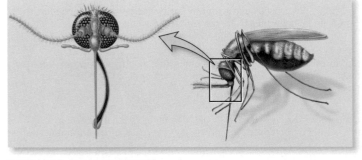

(b) Blood sucking (mosquito)

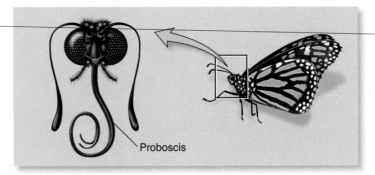

Proboscis

(c) Nectar sucking (butterfly)

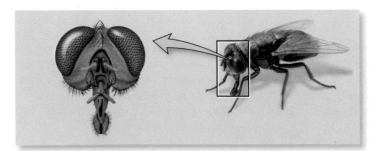

(d) Lapping up liquid (housefly)

Figure 33.30   A variety of insect mouthparts.   Insect mouthparts can be modified to allow insects to feed in a variety of ways, including **(a)** chewing (Orthoptera, Coleoptera, and others), **(b)** blood sucking (Diptera), **(c)** nectar sucking (Lepidoptera), and **(d)** lapping up liquid (Diptera).

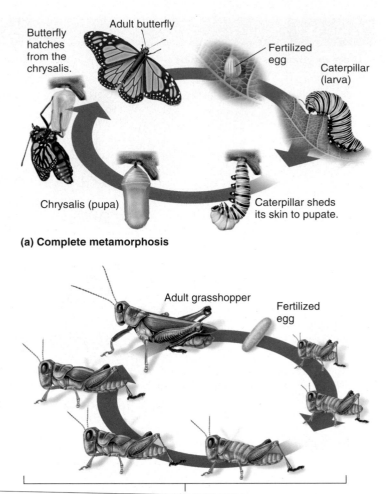

(a) Complete metamorphosis

(b) Incomplete metamorphosis

Figure 33.31   Metamorphosis.   (a) Complete metamorphosis, as illustrated by the life cycle of a monarch butterfly. The adult butterfly has a completely different appearance than the larval caterpillar. **(b)** Incomplete metamorphosis, as illustrated by the life cycle of a grasshopper. The eggs hatch into nymphs, essentially miniature versions of the adult.

All insects have separate sexes, and fertilization is internal. During development, the majority (approximately 85%) of insects undergo a change in body form known as **complete metamorphosis** (from the Greek *meta*, change, and *morph*, form) (**Figure 33.31a**). Complete metamorphosis has four stages: egg, larva, pupa, and adult. In these species, the larval stage is often spent in an entirely different habitat from that of the adult, and larval and adult forms utilize different food sources. They thus do not compete directly for the same resources. The dramatic body transformation from larva to adult occurs in the pupa stage. The remaining insects undergo **incomplete metamorphosis**, in which change is more gradual (**Figure 33.31b**). Incomplete metamorphosis has only three stages: egg, nymph, and adult. Young insects, called nymphs, look like miniature adults when they hatch from their eggs. As they grow and feed, they shed their skin several times, each time entering a new **instar**, or stage of growth.

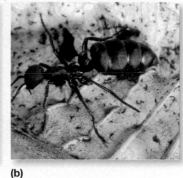

(a)   **(b)**

Figure 33.32 **The division of labor in insect societies.**
Individuals from the same insect colony may appear very different.
Among these army ants (*Eciton burchelli*) from Paraguay, there
are **(a)** workers that forage for the colony and soldiers that protect
the colony from predators, and **(b)** the queen, which reproduces
and lays eggs.

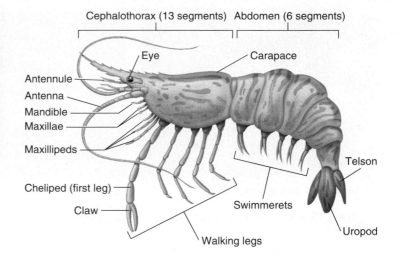

Figure 33.33   Body plan of a crustacean, as represented
by a shrimp.

Finally, some insects, such as bees, wasps, ants, and termites, have developed complex social behavior and live cooper-
atively in underground or aboveground nests. Such colonies
exhibit a division of labor, in that some individuals forage for
food and care for the brood (workers), others protect the nest
(soldiers), and some only reproduce (the queen and one or two
males) (**Figure 33.32**).

Class Crustacea: Crabs, Lobsters, Barnacles, and Shrimp
The crustaceans are common inhabitants of marine environments, although some species live in fresh water and a few are
terrestrial. Many are economically important food items for
humans, including crabs, lobsters, crayfish, and shrimp, and
smaller species are important food sources for other predators.

The crustaceans are unique among the arthropods in that
they possess two pairs of antennae at the anterior end of the
body—the antennule (first pair) and antenna (second pair)
(**Figure 33.33**). In addition, they have three or more sensory
and feeding appendages that are modified mouthparts, called
mandibles, maxillae, and maxillipeds. These are followed by
walking legs and, often, additional abdominal appendages called
swimmerets and a powerful tail consisting of a telson and uro-
pod. In some orders, the first pair of walking legs, or chelipeds,
is modified to form powerful claws. A lost crustacean appendage can regrow. The head and thorax are often fused together,
forming the cephalothorax. In many species, the cuticle covering the head extends over most of the cephalothorax, forming a
hard protective fold called the **carapace**. For growth to occur, a
crustacean must shed the entire exoskeleton.

Many crustaceans are predatory in nature, but others, such
as barnacles, are suspension feeders. Gas exchange typically
occurs via gills, and crustaceans, like all other arthropods, have
an open circulatory system. Crustaceans possess two excretory
organs: antennal glands and maxillary glands, both modified
metanephridia, which open at the bases of the antennae and
maxillae, respectively. Reproduction usually involves separate

Figure 33.34   **Crustacean larvae.**   The nauplius is a distinct
larval type possessed by most crustaceans, which molt several
times before reaching maturity.

sexes, and fertilization is internal. Most species carry their eggs
in brood pouches under the female's body. Eggs of most species
produce larvae that must go through many different molts prior
to assuming adult form. The first of these larval stages, called a
nauplius, is very different in appearance than the adult crustacean (**Figure 33.34**).

While there are many crustacean orders, most are small
and obscure, although many feature prominently in marine
food chains, a series of organisms in which each member of the
chain feeds on and derives energy from the member below it.
These include the Ostracoda, Copepoda, and Euphausiacea.
Ostracods are tiny creatures that superficially resemble clams,
and copepods are tiny and abundant planktonic crustaceans,
both of which are a food source of filter-feeding organisms and
small fish. Euphausiids are shrimplike krill that grow to about
3 cm and provide a large part of the diet of many whales.

(a) Barnacles—Order Cirripedia   **(b) Pill bug—Order Isopoda**   **(c) Coral crab—Order Decapoda**

Figure 33.35   Common crustaceans.   (a) Barnacles on intertidal rocks. (b) Pill bug or wood louse. (c) Coral crab (*Carpilius maculates*).

The order Cirripedia is composed of the barnacles, crustaceans whose carapace forms calcified plates that cover most of the body (**Figure 33.35a**). Their legs are modified into feathery filter-feeding structures. The order Isopoda (**Figure 33.35b**) contains many small species that are parasitic on marine fish. There are also terrestrial isopods, better known as pill bugs or wood lice, that retain a strong connection to water and need to live in moist environments such as leaf litter or decaying logs. When threatened, they curl up into a tight ball, making it difficult for predators to get a grip on them.

The most famous order, however, is the Decapoda, which includes the crabs and lobsters, the largest crustacean species (**Figure 33.35c**). As their name suggests, these decapods have 10 walking legs (five pairs), although the first pair is invariably modified to support large claws. Most decapods are marine, but there are many freshwater species, such as crayfish, and in hot, moist tropical countries, even some terrestrial species called land crabs. The larvae of many larger crustaceans are planktonic and grow to about 3 cm. These are abundant in some oceans and are a staple food source for many species.

33.5 Deuterostomia: The Echinoderms and Chordates

As we explored in Chapter 32, the deuterostomes are grouped together because they share similarities in patterns of development (refer back to Figure 32.5). Molecular evidence also supports a deuterostome clade. All animals in the phylum Chordata, which includes the vertebrates, are deuterostomes. Interestingly, so is one invertebrate group, the phylum Echinodermata, which includes the sea stars, sea urchins, and sea cucumbers. While there are far fewer phyla and species of deuterostomes than ecdysozoans, the species are generally much more familiar to us. After all, we humans are deuterostomes. While the majority of the deuterostome clade will be discussed in Chapter 34, we will conclude our discussion of invertebrate biology by turning our attention to the invertebrate deuterostomes. In this section, we will explore the phylum Echinodermata and then introduce the phylum Chordata, looking in particular at its distinguishing characteristics and at its two invertebrate subphyla: the urochordates, also known as the tunicates, and the cephalochordates, commonly referred to as the lancelets.

The Phylum Echinodermata Includes Sea Stars and Sea Urchins, Species with a Water Vascular System

The phylum Echinodermata (from the Greek *echinos*, spiny, and *derma*, skin) consists of a unique grouping of deuterostomes. A striking feature of all echinoderms is their modified radial symmetry. The body of most species can be divided into five parts pointing out from the center. As a consequence, cephalization is absent in most classes, and echinoderms move only very slowly. There is no brain and only a simple nervous system. The radial symmetry of echinoderms is secondary, however, because the free-swimming larvae have bilateral symmetry and metamorphose into the radially symmetrical adult form.

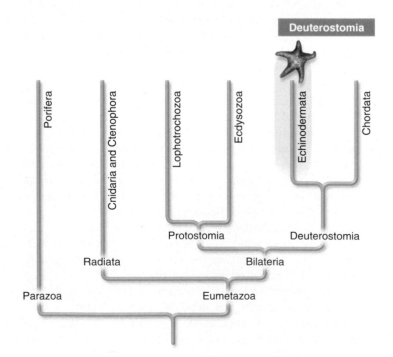

Most echinoderms have an **endoskeleton** (internal skeleton) composed of calcareous plates overlaid by a thin skin. The skeleton is covered with spines and jawlike pincers called **pedicellariae**, the primary purpose of which is to deter settling of animals such as barnacles (**Figure 33.36**). These structures can also possess poison glands.

A portion of the coelom has been adapted to serve as a unique **water vascular system**, a network of canals that branch into tiny **tube feet** that function in movement, gas exchange, and feeding. The water vascular system is powered by hydraulic power, that is, by water pressure generated by the contraction of muscles that enables the extension and contraction of the tube feet, allowing echinoderms to move slowly.

Water enters the water vascular system through the **madreporite**, a sieve-like plate on the animal's surface. From there it flows into a **ring canal** in the central disk, into five radial canals, and into the tube feet. At the base of each tube foot is a muscular sac called an **ampulla**, which stores water. Contractions of the ampullae force water into the tube feet, causing them to straighten and extend. When the foot contacts a solid surface, muscles in the foot contract, forcing water back into the ampulla. Sea stars also use their tube feet in feeding, where

they can exert a constant and strong pressure on bivalves, whose adductor muscles eventually tire, allowing the shell to open slightly. At this stage, the sea star everts its stomach and inserts it into the space, and then digests its prey using juices secreted from extensive digestive glands. Sea stars also feed on sea urchins, brittle stars, and sand dollars, prey that cannot easily escape them.

Echinoderms cannot osmoregulate, so no species have entered freshwater environments. No excretory organs are present, and both respiration and excretion of nitrogenous waste take place by diffusion across the tube feet. Coelomic fluid circulates around the body.

Most echinoderms exhibit **autotomy**, the ability to intentionally detach a body part, such as a limb, that will later regenerate. In some species, a broken limb can even regenerate into a whole animal. Some sea stars regularly reproduce by breaking in two. Most echinoderms reproduce sexually and have separate sexes. Fertilization is usually external, with gametes shed into the water. Fertilized eggs develop into free-swimming bilaterally symmetrical larvae, which undergo metamorphosis into sedentary adults.

While over 20 classes of echinoderms have been described from the fossil record, only 5 main classes of echinoderms exist today: the Asteroidea (sea stars), Ophiuroidea (brittle stars), Echinoidea (sea urchins and sand dollars), Crinoidea (sea lilies and feather stars), and Holothuroidea (sea cucumbers). The key features of the echinoderms and their classes are listed in Table 33.8.

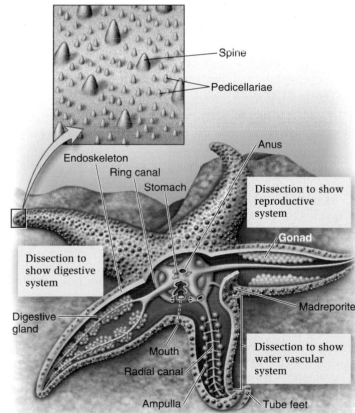

Figure 33.36 Body plan of an echinoderm, as represented by a sea star.   The arms of this sea star have been dissected to different degrees to show the echinoderm's various organs.

Biological inquiry: Echinoderms and chordates are both deuterostomes. What are three defining features of deuterostomes?

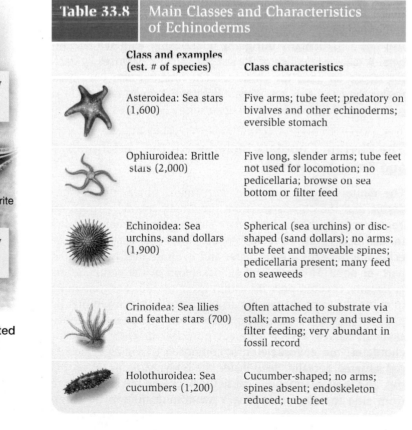

Table 33.8	Main Classes and Characteristics of Echinoderms	
	Class and examples (est. # of species)	**Class characteristics**
	Asteroidea: Sea stars (1,600)	Five arms; tube feet; predatory on bivalves and other echinoderms; eversible stomach
	Ophiuroidea: Brittle stars (2,000)	Five long, slender arms; tube feet not used for locomotion; no pedicellaria; browse on sea bottom or filter feed
	Echinoidea: Sea urchins, sand dollars (1,900)	Spherical (sea urchins) or disc-shaped (sand dollars); no arms; tube feet and moveable spines; pedicellaria present; many feed on seaweeds
	Crinoidea: Sea lilies and feather stars (700)	Often attached to substrate via stalk; arms feathery and used in filter feeding; very abundant in fossil record
	Holothuroidea: Sea cucumbers (1,200)	Cucumber-shaped; no arms; spines absent; endoskeleton reduced; tube feet

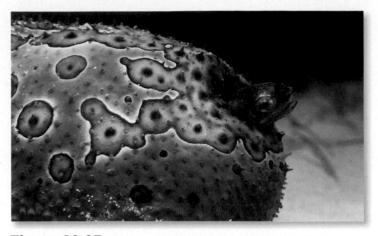

Figure 33.37 Sea cucumber with pearl fish.

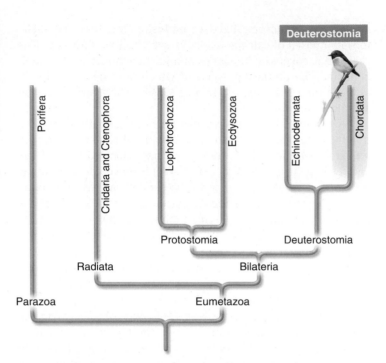

The most unusual of the echinoderms are members of the class Holothuroidea, the sea cucumbers. These animals really do look like a cucumber rather than a sea star or sea urchin (**Figure 33.37**). The hard plates of the endoskeleton are less extensive, so that the animal appears and feels fleshy. Sea cucumbers possess specialized respiratory structures called respiratory trees that pump water in and out of the anus. They are typically deposit feeders, ingesting sediment and extracting nutrients.

When threatened by a predator, a few tropical species of sea cucumber can eject sticky, toxic substances from their anus. If these do not serve to deter the predator, these species can undergo the process of evisceration, ejecting its digestive tract, respiratory structures, and gonads from the anus. If the sea cucumber survives, it can regenerate its organs later. The pearl fish has a particularly intriguing relationship with sea cucumbers. It takes advantage of the sea cucumber's defenses by darting inside its anus for protection. At first, the sea cucumber resists the pearl fish, but since it breathes through its anus, it is only a matter of time before an opportunity is afforded to the fish.

The Phylum Chordata Includes All the Vertebrates and Some Invertebrates

The deuterostomes consist of two major phyla, the echinoderms and the chordates (from the Greek *chorde*, string). As deuterostomes, both phyla share similar developmental traits. In addition, both have an endoskeleton, consisting in the echinoderm of calcareous plates and in chordates, for the most part, of bone. However, the endoskeleton of the echinoderms is usually covered only by a thin layer of skin and functions in much the same way as the arthropod exoskeleton, in that its primary function is providing protection. The chordate endoskeleton serves a very different purpose. In early divergent chordates, the endoskeleton is composed of a single flexible rod situated dorsally, deep inside the body. Muscles move this rod, and their contractions cause the back and tail end to move from side to side, permitting a swimming motion in water.

The endoskeleton becomes more complex in different lineages that develop limbs, as we will see in Chapter 34, but it is always internal, with muscles attached. This arrangement permits the possibility of complex movements, including the ability to move on land.

Let's take a look at the four critical innovations in the body design of chordates that distinguish them from all other animal life (**Figure 33.38**):

1. **Notochord.** Chordates are named for the **notochord**, a single flexible rod that lies between the digestive tract and the nerve cord. Composed of fibrous tissue encasing fluid-filled cells, the notochord is stiff yet flexible and provides skeletal support for all primitive chordates. In most chordates, such as vertebrates, a more complex jointed backbone usually replaces the notochord and only remnants exist as the soft material within the discs of vertebrae.

2. **Dorsal hollow nerve cord.** Many animals have a long nerve cord, but in nonchordate invertebrates, it is a solid tube that lies ventral to the alimentary canal. In contrast, the nerve cord in chordates is a hollow tube that develops dorsal to the alimentary canal. In 1822, the French naturalist Geoffroy Saint-Hilaire argued that this difference suggested that the ventral side of nonchordate invertebrates (as exemplified by the lobster) was homologous to the dorsal side of vertebrates. Some recent molecular work suggests that, as Hilaire first proposed, there was an inversion of the dorsoventral axis during animal evolution. In vertebrates, the dorsal hollow nerve cord develops into the brain and spinal cord.

3. **Pharyngeal slits.** Chordates, like many animals, have a complete gut, from mouth to anus. However, in chordates, slits develop in the pharyngeal region, close to the mouth,

that open to the outside. This permits water to enter through the mouth and exit via the slits, without having to go through the digestive tract. In early divergent chordates, **pharyngeal slits** function as a filter-feeding device, while in more advanced chordates, they develop into gills for gas exchange. In terrestrial chordates, the slits do not fully form and become modified for other purposes, such as the auditory (Eustachian) tubes in ears.

4. **Postanal tail.** Chordates possess a postanal tail of variable length that extends posterior to the anal opening. In aquatic chordates such as fish, the tail is used in locomotion. In terrestrial chordates, the tail may be used in a variety of functions or may be absent, as in humans. In virtually all other nonchordate phyla, the anus is terminal.

While few chordates apart from fishes possess all of these characteristics in their adult life, they all exhibit them at some time during development. For example, in adult humans the notochord becomes the spinal column and the dorsal hollow nerve cord becomes the central nervous system. However, humans only exhibit pharyngeal slits and a postanal tail during early embryonic development. All the pharyngeal slits, except one, which forms the Eustachian tubes in the ear, are eventually lost, and the postanal tail regresses to form the tailbone (the coccyx).

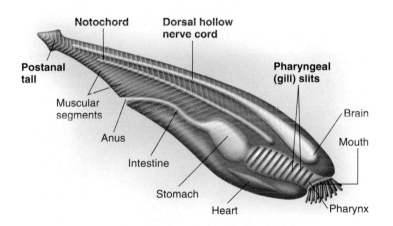

Figure 33.38 Chordate characteristics. The generalized chordate body plan has four main features: notochord, dorsal hollow nerve cord, pharyngeal slits, and postanal tail.

The phylum Chordata consists of the invertebrate chordates—the subphylum Urochordata (tunicates) and the subphylum Cephalochordata (lancelets)—along with the subphylum Vertebrata. While the Vertebrata is by far the largest of these subphyla, biologists have focused on the Urochordata and Cephalochordata for clues as to how the chordate phylum may have evolved. Comparisons of gene sequences coding for 18S rRNA show that these two subphyla in general, and the cephalochordates in particular, are our closest nonvertebrate relatives (**Figure 33.39**).

Subphylum Urochordata: The Tunicates The urochordates (from the Greek *oura*, tail) are a group of 3,000 marine species also known as tunicates. Looking at an adult tunicate, you might never guess that it is a relative of modern vertebrates. The only one of the four distinguishing chordate characteristics that it possesses is pharyngeal slits (**Figure 33.40a**). The larval tunicate, in contrast, looks like a tadpole and exhibits all four chordate hallmarks (**Figure 33.40b**). The larval tadpole swims for only a few days, usually without feeding, because the siphons are not functional in the larvae. Larvae settle on and attach to a rock surface via rootlike extensions called stolons. Here they metamorphose into adult tunicates and in the process lose most of their chordate characteristics. In 1928, the marine biologist Walter Garstang suggested modern vertebrates arose from a larval tunicate form that had somehow acquired the ability to reproduce. However, both the adult tunicate body plan and modern molecular data have led Billie Swalla and colleagues to propose that tunicates are so distantly related to other chordates that they should be considered a separate deuterostome phylum called Tunicata.

Adult tunicates are marine animals, some of which are colonial and others solitary, that superficially resemble sponges or cnidarians. Tunicates are filter feeders that draw water through the mouth through an **incurrent siphon**, using a ciliated pharynx, and filter it through extensive pharyngeal slits. The food is trapped on a mucus sheet secreted by what is known as an endostyle; passes via ciliary action to the stomach, intestine, and anus; and exits through the **excurrent siphon**. The whole animal is enclosed in a nonliving **tunic**, which it secretes, made of a protein and a cellulose-like material called tunicin.

Figure 33.39 Comparison of small subunit rRNA gene sequences of vertebrate, invertebrate chordate, and invertebrate species. The similarities between the invertebrate chordates (represented by the lancelet) and the vertebrates (represented by a human) suggest they are indeed our closest relatives.

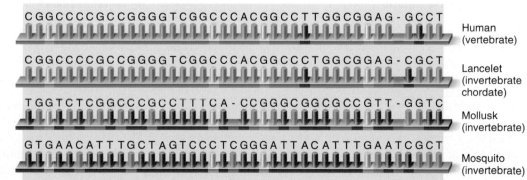

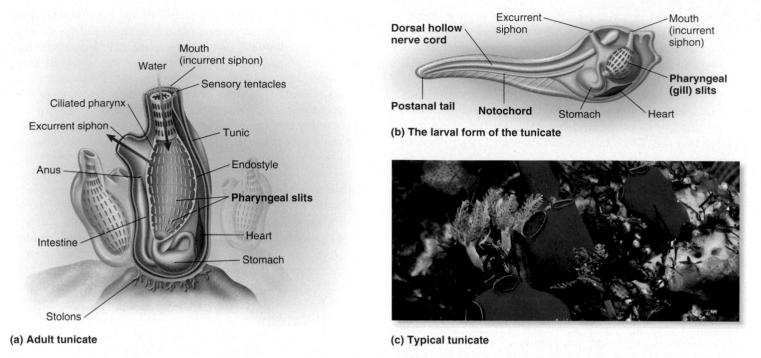

(a) Adult tunicate

(b) The larval form of the tunicate

(c) Typical tunicate

Figure 33.40   **Tunicates.**   **(a)** Body plan of the sessile, filter-feeding adult tunicate. **(b)** The larval form, which shows the four characteristic chordate features, has been proposed as a possible ancestor of modern vertebrates. **(c)** Blue tunicate, *Rhopalaea crassa*.

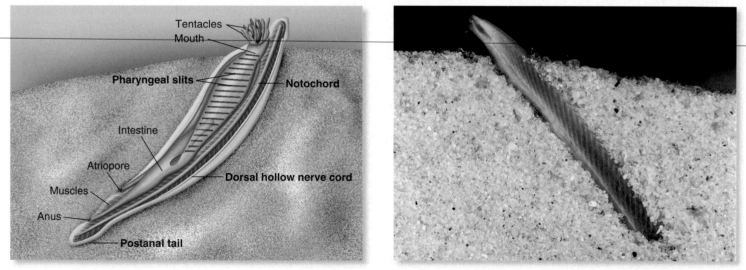

(a) Body plan of the lancelet

(b) Lancelet in the sand

Figure 33.41   **Lancelets.**   **(a)** The body plan of the lancelet clearly displays the four characteristic chordate features. **(b)** Lancelet in feeding position, showing the segmented body with chevron-shaped muscles in each segment.

Tunicates are also known as sea squirts for their ability to squirt out water from the excurrent siphon when disturbed. There is a rudimentary circulatory system with a heart and a simple nervous system of relatively few nerves connected to sensory tentacles around the incurrent siphon. The animals are mostly hermaphroditic.

Subphylum Cephalochordata: The Lancelets   The cephalochordates (from the Greek *cephalo,* head) look a lot more chordatelike than do tunicates. They are commonly referred to as lancelets, in reference to their bladelike shape and size, about 5–7 cm in length (**Figure 33.41a**). Lancelets are a small subphylum of 26 species, all marine filter feeders, with 4 species occurring in North American waters. Most of them belong to the genus *Branchiostoma*.

The lancelets live mostly buried in sand, with only the anterior end protruding into the water so that they can filter feed through the mouth (**Figure 33.41b**). Lancelets have the four distinguishing chordate characteristics: a clearly discernible notochord (extending well into the head), dorsal hollow nerve cord, pharyngeal slits, and postanal tail. In a fashion similar to tunicates, water enters the mouth and moves into the pharynx, where it is filtered through the pharyngeal slits. A mucus net across the pharyngeal slits traps food particles, and ciliary action takes the food into the digestive tract, while water exits via the excurrent siphon, here called the atriopore. Gas exchange generally takes place across the body surface. While the lancelet is usually sessile, it can leave its sandy burrow and swim to a new spot, using a series of serially arranged muscles that appear like chevrons (<<<<) along their sides. These muscles reflect the segmented nature of their body. With a body shape and a swimming motion characteristic of fish, it is not difficult to see how lancelets may have been the precursors to modern fish, especially the more ancient jawless fishes, which we will explore in Chapter 34.

CHAPTER SUMMARY

33.1 Parazoa: Sponges, the First Multicellular Animals

- Invertebrates, or animals without a backbone, make up more than 95% of all animal species. An early lineage—the Parazoa—consists of one phylum, the Porifera, or sponges. While sponges lack true tissues, they are multicellular animals possessing several types of cells. (Figure 33.1)

33.2 Radiata: Jellyfish, Radially Symmetrical Animals

- The Radiata consists of two phyla: the Cnidaria (hydra, jellyfish, sea anemones, corals, and box jellies) and the Ctenophora (comb jellies). Radiata have only two embryonic germ layers: the ectoderm and endoderm, with a gelatinous substance (mesoglea) connecting the two layers.

- Cnidarians exist as two forms: polyp and medusa. A characteristic feature of cnidarians is their stinging cells or cnidocytes, which function in defense and prey capture. Ctenophores possess the first complete gut and nearly all exhibit bioluminescence. (Figures 33.2, 33.3, 33.4, 33.5, Table 33.1)

33.3 Lophotrochozoa: The Flatworms, Rotifers, Lophophorates, Mollusks, and Annelids

- The Lophotrochozoa include taxa that possess either a lophophore or trochophore larva. Platyhelminthes, or flatworms, are regarded as the first animals to reach the organ-system level of organization. (Figure 33.6, Table 33.2)

- The four classes of flatworms are the Turbellaria, Monogenea, Trematoda (flukes), and Cestoda (tapeworms). Flukes and tapeworms are internally parasitic, with complex life cycles. (Figures 33.7, 33.8)

- Rotifers are microscopic animals that have a complete digestive tract with separate mouth and anus; the mastax, a muscular pharynx, is a structure unique to the rotifers. (Figure 33.9)

- The Lophophorates consist of the phoronids, bryozoa, and brachiopods, all of which possess a lophophore, a ciliary feeding structure. (Figure 33.10)

- The mollusks, which constitute a large phylum with over 100,000 diverse living species, have a basic body plan with three parts—a foot, a visceral mass, and a mantle—and an open circulatory system. (Figures 33.11, 33.12, 33.13, 33.14, 33.15)

- The four most common mollusk classes are the polyplacophora (chitons), gastropoda (snails and slugs), bivalvia (clams and mussels), and cephalopoda (octopuses, squid, and nautiluses). (Table 33.3)

- Cephalopods are among the most complex of all invertebrates. They are the only mollusks with a closed circulatory system, that have a well-developed nervous system and brain, and are believed to exhibit learning by observation. (Figures 33.16, 33.17, 33.18, 33.19)

- Segmentation, in which the body is divided into nearly identical subunits, is a critical evolutionary innovation in the annelids, although specialization is only minimally present at the anterior end. (Figure 33.20)

- Annelids are a large phylum with three main classes: Polychaeta (the most species-rich class), Oligochaeta (which includes the earthworms), and Hirudinea (leeches). (Figure 33.21, Table 33.4)

33.4 Ecdysozoa: The Nematodes and Arthropods

- The ecdysozoans are so named for their ability to shed their cuticle, a nonliving cover providing support and protection. The two most common ecdysozoan phyla are the nematodes and the arthropods.

- Nematodes, which exist in nearly all habitats, have a cuticle made of collagen, a structural protein. The small, free-living nematode *Caenorhabditis elegans* is a model organism. Many nematodes are parasitic in humans, including *Wuchereria bancrofti*, which causes elephantiasis. (Figures 33.22, 33.23)

- Arthropods are perhaps the most successful phylum on Earth. The arthropod body is covered by a cuticle made of layers of chitin and protein, and it is segmented, with segments fused into functional units called tagmata. (Figure 33.24)

- The six main classes of arthropods are Trilobita (the trilobites, now extinct), Arachnida (spiders and scorpions), Diplopoda (millipedes), Chilopoda (centipedes), Insecta (insects), and Crustacea (crabs and relatives). (Figures 33.25, 33.26, 33.27, 33.28, 33.29, Table 33.5)

- More insect species are known than all other animal species combined. Over half a million insect species are beetles. Researcher Brian Farrell found that the appearance of angiosperm plants led to an increased richness of leaf-eating beetles. (Tables 33.6, 33.7)

- The development of a variety of wing structures and mouthparts was a key to the success of insects. (Figure 33.30)

- Insects undergo a change in body form during development, either complete metamorphosis or incomplete metamorphosis, and have developed complex social behaviors. (Figures 33.31, 33.32)

- Most crustacean orders are small and feature prominently in marine food chains. The most well-known order of crustaceans is the Decapoda, which includes the crabs and lobsters. (Figures 33.33, 33.34, 33.35)

33.5 Deuterostomia: The Echinoderms and Chordates

- The Deuterostomia includes the phylums Echinodermata and Chordata. A striking feature of the echinoderms is their radial symmetry, which is secondary; the free-swimming larvae are bilaterally symmetrical. Most echinoderms possess an internal skeleton called an endoskeleton. (Figure 33.36)

- Five main classes of echinoderms exist today: the Asteroidea (sea stars), Ophiuroidea (brittle stars), Echinoidea (sea urchins and sand dollars), Crinoidea (sea lilies and feather stars), and Holothuroidea (sea cucumbers). (Figure 33.37, Table 33.8)

- The phylum Chordata is distinguished by four critical innovations: the notochord, dorsal hollow nerve chord, pharyngeal slits, and postanal tail. (Figure 33.38)

- The subphylum Urochordata (tunicates) and subphylum Cephalochordata (lancelets) are invertebrate chordates. Genetic studies have shown that cephalochordates are the closest nonvertebrate relatives of the vertebrate chordates (subphylum Vertebrata). (Figures 33.39, 33.40, 33.41)

TEST YOURSELF

1. Choanocytes are
   a. a group of protists that are believed to have given rise to animals.
   b. specialized cells of sponges that function to trap and eat small particles.
   c. cells that make up the gelatinous layer in sponges.
   d. cells of sponges that function to transfer nutrients to other cells.
   e. cells that form spicules in sponges.

2. Which of the following is *not* a characteristic of ctenophores?
   a. complete gut
   b. bioluminescence
   c. stinging cells
   d. ciliary locomotion
   e. all of the above are characteristics of ctenophores

3. Which of the following organisms can produce female offspring through parthenogenesis?
   a. cnidarians
   b. flukes
   c. choanocytes
   d. rotifers
   e. annelids

4. A lophophore is
   a. a contractile structure that aids in movement through the water column.
   b. a ciliated structure that aids in movement through the water column.
   c. a ciliated structure that functions primarily in feeding.
   d. a contractile structure that moves food into the coelom.
   e. a contractile structure important for defense.

5. In the annelids, the metanephridia function in
   a. transport of nutrients.
   b. movement.
   c. cellular communication.
   d. waste removal.
   e. reproduction.

6. A defining feature of the Ecdysozoa is
   a. a segmented body.
   b. a closed circulatory system.
   c. a cuticle.
   d. a complete gut.
   e. a lophophore.

7. In arthropods, the tracheal system is
   a. a unique set of structures that function in ingestion and digestion of food.
   b. a series of branching tubes extending into the body that allow for gas exchange.
   c. a series of tubules that allow waste products in the blood to be released into the digestive tract.
   d. the series of ommatidia that form the compound eye.
   e. none of the above.

8. Characteristics of the class Arachnida include
   a. two tagmata.
   b. six walking legs.
   c. an aquatic lifestyle.
   d. a lobed body.
   e. both b and d.

9. Incomplete metamorphosis
   a. is characterized by distinct larval and adult stages that do not compete for resources.
   b. is typically seen in arachnids.
   c. involves gradual changes in life stages where young resemble the adult stage.
   d. is characteristic of the majority of insects.
   e. always includes a pupal stage.

10. In the phylum Echinodermata, the tube feet function in
    a. movement.
    b. gas exchange.
    c. feeding.
    d. excretion.
    e. all of the above.

CONCEPTUAL QUESTIONS

1. Define hermaphrodite.

2. Define nematocyst and explain its function.

3. Explain the difference between complete metamorphosis and incomplete metamorphosis.

EXPERIMENTAL QUESTIONS

1. What was the hypothesis tested by Fiorito and Scotto?

2. What were the results of the experiment? Did these results support the hypothesis?

3. What is the significance of performing the experiment on both observer and untrained octopuses?

COLLABORATIVE QUESTIONS

1. Discuss some of the characteristics of the phylum Mollusca.

2. Discuss the four defining characteristics of chordates.

www.brookerbiology.com
This website includes answers to the Biological Inquiry questions found in the figure legends and all end-of-chapter questions.

34

THE VERTEBRATES

CHAPTER OUTLINE

Giant pandas (*Ailuropoda melanoleuca*). This species, shown here in Sichuan, China, is one of many charismatic vertebrate species.

In Chapter 33, we discussed two chordate subphyla, the Urochordates (tunicates) and Cephalochordates (lancelets). The third subphylum of chordates, the Vertebrata, or vertebrates, with about 48,000 species, is by far the largest and most dominant group of chordates. Species range in size from tiny fish weighing 0.1 g to huge whales of over 100,000 kg. They occupy nearly all of Earth's habitats, from the deepest depths of the oceans to mountaintops and the sky beyond. Throughout history, humans have depended on many vertebrate species for their welfare and have domesticated species such as horses, cattle, pigs, sheep, and chickens and had countless species, including cats and dogs, as pets. Many other vertebrate species are the subjects of conservation efforts, as we will see in Chapter 60 (see also chapter-opening photo). For example, some vertebrate species require large areas for their survival and as such are considered to be umbrella species, in that their preservation will help ensure conservation of others.

In this chapter, we begin by outlining the characteristics of **craniates**, chordates that have a brain encased in a skull. We will take a brief look at the hagfish, a craniate that has some, but not all, vertebrate characteristics. In the remainder of the chapter, we will explore the characteristics of vertebrates (**Figure 34.1**), discussing in some depth the evolutionary development of the major vertebrate classes, including fishes, amphibians, reptiles, birds, and mammals.

34.1 The Craniates

Although the hagfish, an eel-like, eyeless scavenger, is often called a primitive fish, its resemblance to a fish is misleading (**Figure 34.2**). The hagfish is not a true vertebrate and thus cannot be considered a true fish. At the same time, it is different from the tunicates and lancelets, the nonvertebrate chordates we discussed in Chapter 33. Although hagfish do not have a vertebral column—one of the vertebrate characteristics—they do have a skull, in this case, one made of cartilage and not bone. Paleontologists in China have recently discovered similar fishlike fossils that date to around 530 million years ago, the time of the Cambrian explosion. These are believed to be the earliest known craniates, of which the hagfish is the only living relative.

Craniates Are Distinguished by a Cranium and Neural Crest

Craniates have two defining characteristics that distinguish them from other nonvertebrate chordates:

1. **Cranium.** In craniates, the anterior end of the nerve cord elaborates to form a more developed brain that is encased in a protective bony or cartilaginous housing called the **cranium**. This continues the trend of cephalization—the development of the head end in animals.
2. **Neural crest.** The **neural crest** is a group of embryonic cells found on either side of the neural tube as it develops. The cells disperse throughout the embryo, where they contribute to the development of the skeleton, especially the cranium and other structures, including nerves, jaws, and teeth.

While these are the main distinguishing characteristics of craniates, there are others. For example, craniates have at least two clusters of *Hox* genes, compared to the single cluster of *Hox* genes in tunicates and lancelets. This additional gene cluster is believed to have permitted increasingly complex morphologies than those possessed by other nonvertebrate chordates.

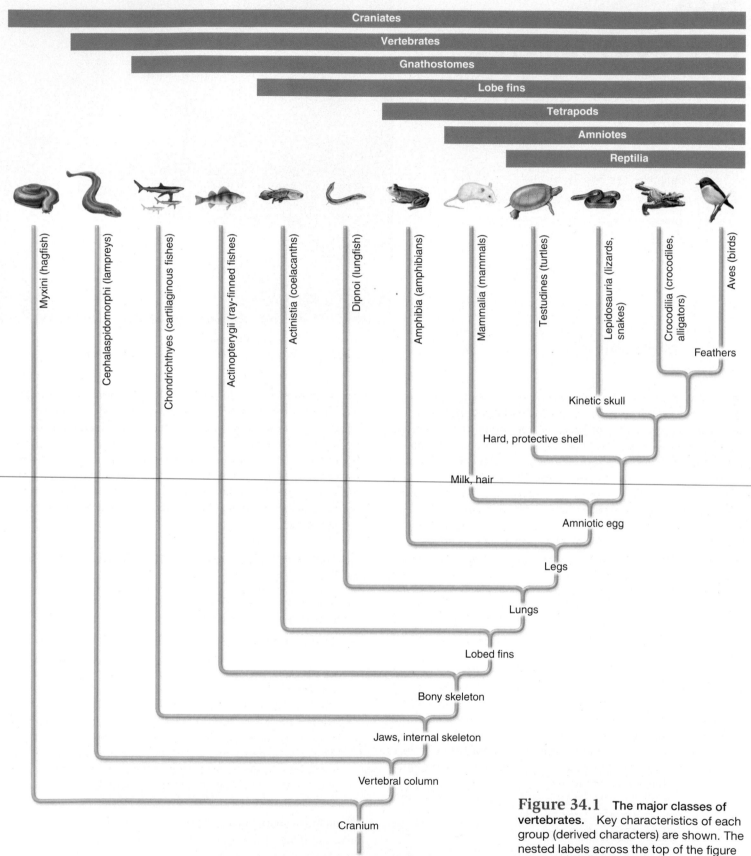

Figure 34.1 **The major classes of vertebrates.** Key characteristics of each group (derived characters) are shown. The nested labels across the top of the figure refer to groupings within the vertebrates.

Figure 34.2 Craniates. The hagfish possesses craniate characteristics (a skull and neural crest) but is not a true fish or vertebrate.

Biological inquiry: Why isn't the hagfish a true fish?

The Hagfish, Class Myxini, Are the Most Primitive Living Craniates

The hagfish are entirely marine, jawless, finless craniates that lack vertebrae. The hagfish skeleton consists largely of a skull that encloses the brain and a cartilaginous notochord. The lack of a vertebral column leads to extensive flexibility. Hagfish live in

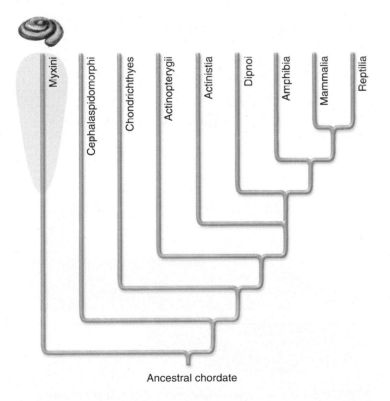

Ancestral chordate

the cold waters of northern oceans, close to the muddy bottom. Essentially blind, hagfish have a very keen sense of smell and are attracted to dead and dying fish, which they attach themselves to via toothed plates on the mouth. The powerful tongue then rasps off pieces of tissue. However, most of their diet consists of dead or disabled marine worms and other invertebrates. Though the hagfish cannot see approaching predators, they have special glands that produce copious amounts of slime, which may deter predatory attacks. When provoked, the hagfish's slime production increases dramatically, enough to potentially distract predators or coat their gills and interfere with breathing. Hagfish can sneeze to free their nostrils of their own slime.

34.2 Characteristics of Vertebrates

The **vertebrates** (from the Latin, *vertebratus*, back boned) retain all chordate characteristics that we outlined in Chapter 33 and all craniate characteristics noted previously, as well as possessing several additional traits, including:

1. **Vertebral column.** During development in vertebrates, the notochord is replaced by a bony or cartilaginous column of interlocking **vertebrae** that provides support and also protects the nerve cord, which lies within its tubelike structure.
2. **Endoskeleton of cartilage or bone.** The cranium and vertebral column are parts of the endoskeleton, the living skeleton of vertebrates that forms within the animal's body. Most vertebrates also have two pairs of appendages, whether fins, legs, or arms. The endoskeleton is composed of either bone or cartilage, both of which are very strong materials, yet they are more flexible than the chitin found in insects and other arthropods. The endoskeleton also contains living cells that secrete the skeleton, which grows with the animal, unlike the nonliving exoskeleton of arthropods.
3. **Internal organs.** Vertebrates possess a great diversity of internal organs, including a liver, kidneys, endocrine glands, and a heart with at least two chambers. The liver is unique to vertebrates, while the vertebrate heart, kidneys, and endocrine system are more complex than analogous structures in other taxa.

While these features are exhibited in all vertebrate classes, some classes developed innovations that helped them succeed in specific environments such as on land or in the air. For example, birds developed feathers and wings, structures that enable most species to fly. In fact, each of the vertebrate classes is distinctly different from one another, as outlined in **Table 34.1.** Of the 11 different classes of vertebrates, 5 of them are classes of fish, reflecting the dominance of fish within the vertebrates.

Table 34.1	The Main Classes and Characteristics of Living Vertebrates		
Class		**Examples (approx. # of species)**	**Main characteristics**
Cephalaspidomorphi		Lampreys (41)	Early divergent fish with no appendages, that is, fins; jawless sucking mouth; parasitic on other fish
Chondrichthyes		Sharks, skates, rays (850)	Fish with cartilaginous skeleton; teeth not fused to jaw; no swim bladder; well-developed fins; internal fertilization; single blood circulation
Actinopterygii		Ray-finned fish, most bony fish (24,600)	Fish with ossified skeleton; single gill opening covered by operculum; fins supported by rays, fin muscles within body; swim bladder often present; mucus glands in skin
Actinistia		Lobe-finned fish, of which coelacanth is only living member (1)	Fish with ossified skeleton; bony extensions, together with muscles, project into pectoral and pelvic fins; swim bladder filled with oil
Dipnoi		Lungfish (6)	Fish with ossified skeleton; rudimentary lungs allow fish to come to the surface to gulp air; limblike appendages
Amphibia		Frogs, toads, salamanders (4,000)	Tetrapods; adults able to live on land; fresh water needed for reproduction; development usually involving metamorphosis from tadpoles; adults with lungs and double blood circulation; moist skin; shell-less eggs
Mammalia		Mammals (5,500)	Mammary glands, hair, specialized teeth, enlarged skull, external ears, endothermic, highly developed brains, diversity of body forms
Testudines		Turtles (330)	Body encased in hard shell; no teeth; head and neck retractable into shell; eggs laid on land
Lepidosauria		Lizards, snakes (7,800)	Lower jaw not attached to skull; skin covered in scales
Crocodilia		Crocodiles, alligators (23)	Four-chambered heart; large aquatic predators; parental care of young
Aves		Birds (9,600)	Feathers, hollow bones, air sacs, reduced internal organs, endothermic, four-chambered heart

34.3 The Fishes

A critical innovation in vertebrate evolution is the hinged jaw, which first developed in fish. However, not all fish developed hinged jaws. The **ostracoderms**, an umbrella term for several classes of early divergent, heavily armored fish, now extinct, lacked a jaw (**Figure 34.3**). These fish most likely swam slowly above the substrate, using a strong pharyngeal pump as a way of vacuuming up particles. The ostracoderms were common in the Silurian and Devonian periods and many had fins, suggesting they were accomplished swimmers. However, their heavy armor also suggests that predatory fish, with hinged jaws, were also common at that time. A class of jawless fish—the lampreys—exists today, although these evolved separately from the ostracoderms.

Figure 34.3 **Ancient jawless fish.** *Cephalaspis* is one of an extinct group of primitive fish called ostracoderms that were partially armored and lacked a hinged jaw. Jawless species like this were particularly common in the Silurian and Devonian periods.

The jawed mouth was a significant evolutionary development. It enabled an animal to grip its prey more firmly, which may have increased its rate of capture, and to attack larger prey species, thus increasing its potential food supply. All vertebrate species that possess jaws are called **gnathostomes** (meaning "jaw mouth") (see Figure 34.1). In contrast, the jawless species are termed agnathans (meaning "without jaws"). Accompanying the jawed mouth was the development of more sophisticated head and body structures, including two pairs of appendages called fins. Gnathostomes also possess two additional *Hox* gene clusters (bringing their total to four), which permitted increased morphological complexity.

Biologists have identified about 25,000 species of living fish, more than all other species of vertebrates combined. All are aquatic, gill-breathing ectotherms that possess fins and a scaly skin. There are five separate classes of fish, each of which has distinguishing characteristics (see Table 34.1). We will begin by examining the most primitive class, the Cephalaspidomorphi (lampreys), which lack jaws and appendages. Most of our attention, however, will focus on the jawed fish: the Chondrichthyes (cartilaginous fish), Actinopterygii (ray-finned fish), Actinistia (coelacanths), and Dipnoi (lungfish).

The Lampreys, Class Cephalaspidomorphi, Are Eel-like Fish That Lack Jaws

Lampreys are unlike members of other classes of fish because they lack both a hinged jaw and true appendages. However, lampreys do possess a notochord surrounded by a cartilaginous rod that represents a rudimentary vertebral column, and thus they represent one of the most early divergent groups of vertebrates. Lampreys can be found in both marine and freshwater environments. Marine lampreys are parasitic as adults. They grasp other fish with their circular mouth (**Figure 34.4a**) and rasp a hole in the fish's side, sucking blood, tissue, and fluids until they are full (**Figure 34.4b**). Reproduction of all species is similar, whether they live in marine or freshwater environments. Males and females spawn in freshwater streams, and the resultant larval lampreys bury into the sand or mud, much like lancelets (refer back to Figure 33.41h), emerging to feed on small invertebrates or detritus at night. This stage can last for three to seven years, at which time the larvae metamorphose into an adult.

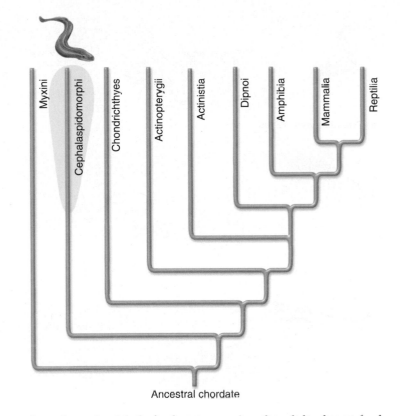

Ancestral chordate

Sometimes, in strictly freshwater species, the adults do not feed at all but quickly mate and die. Marine species migrate from freshwater back to the ocean for the remainder of their life. For parasitic species, the adults then begin their parasitic lifestyle.

Jawed Fish Include the Sharks and Rays and the Bony Fish

As we noted previously, one of the main innovations in fish was the development of the hinged jaw. The hinged jaw developed from the pharyngeal arches, cartilaginous rods that function to help support the respiratory tissue. Primitive jawless fish had nine gill arches surrounding the eight gill slits (**Figure 34.5a**). During the late Silurian period, about 417 million years ago, some of these gill arches became modified in jawless fish. The first and second gill arches were lost, while the third and fourth pairs became modified to form the jaws (**Figure 34.5b,c**).

(a) The sea lamprey (*Petromyzon marinus*) has a circular, jawless mouth.

(b) A sea lamprey feeding on a fish.

Figure 34.4 Modern jawless fish.

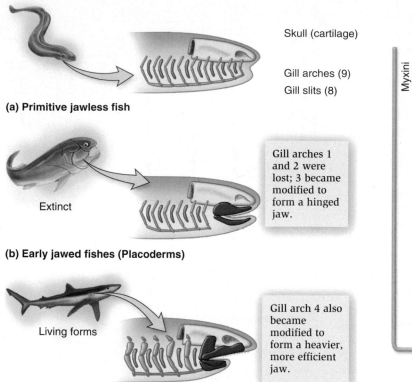

(a) Primitive jawless fish

Skull (cartilage)

Gill arches (9)

Gill slits (8)

(b) Early jawed fishes (Placoderms)

Extinct

Gill arches 1 and 2 were lost; 3 became modified to form a hinged jaw.

(c) Modern jawed fishes (cartilaginous and bony fishes)

Living forms

Gill arch 4 also became modified to form a heavier, more efficient jaw.

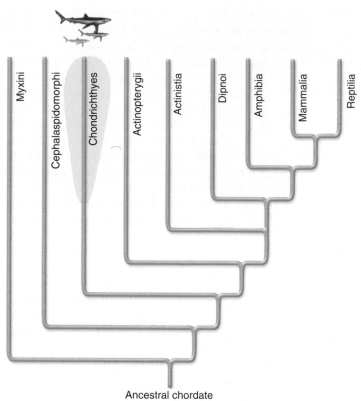

Ancestral chordate

Figure 34.5   The evolution of the vertebrate jaw. (a) Primitive fish and extant jawless fish such as lampreys have nine cartilaginous gill arches that support eight gill slits. **(b)** In early jawed fish such as the placoderms, the first two pairs of gill arches were lost, and the third pair became modified to form a hinged jaw. This left six gill arches to support the remaining five gill slits, which were still used in breathing. **(c)** In modern jawed fish, the fourth gill arch is also incorporated into the jaw, allowing stronger, more powerful bites to be delivered.

This is how evolution usually works; body features do not appear *de novo* but instead become modified to serve other functions.

By the mid-Devonian period, two classes of jawed fish, the Acanthodii (spiny fish) and Placodermi (armored fish) were common. Some of the placoderms were huge individuals, over 9 m long. Both classes died out by the end of the Devonian as part of one of several mass extinctions that occurred in the Earth's geological and biological history. The reasons for this extinction are not well understood, but other types of jawed fish—the cartilaginous and bony fish—did not go extinct. We will discuss the fish in these classes next.

Chondrichthyans: The Cartilaginous Fish Members of the class Chondrichthyes (the **chondrichthyans**)—sharks, skates, and rays—are also called cartilaginous fish because their skeleton is composed of flexible cartilage rather than bone. The cartilaginous skeleton is not considered an ancestral character but rather a derived character. This means that the ancestors of the chondrichthyans had bony skeletons, but that members of this class subsequently lost this feature. This hypothesis

is reinforced by the fact that during development, the skeleton of most vertebrates is cartilaginous, and then it becomes bony (ossified) as a hard calcium-phosphate matrix replaces the softer cartilage. A change in the developmental sequence of the cartilaginous fishes is believed to prevent the ossification process.

In the Carboniferous period, 354–290 million years ago, sharks were the great predators of the ocean. Aided by appendages called fins, sharks became fast, extremely efficient swimmers (**Figure 34.6a**). Perhaps the most important fin for propulsion is the large and powerful caudal fin, or tail fin, which, when swept from side to side, thrusts the fish forward at great speed. For example, great white sharks (*Carcharodon carcharias*) can swim at over 40 km per hour, and Mako sharks (*Isurus oxyrinchus*) have been clocked at nearly 50 km per hour. The paired pelvic fins (at the back) and pectoral fins (at the front) act like flaps on airplane wings, allowing the shark to dive deeper or rise to the surface. They also aid in steering. In addition, the dorsal fin (on the back) acts as a stabilizer to prevent the shark from rolling in the water as the tail fin pushes it forward.

Sharks were among the earliest fish to develop teeth. Their teeth evolved from rough scales on the skin that also contain dentin and enamel. While shark's teeth are very sharp and hard, they are not set into the jaw, as are human teeth, so they break off easily. To offset this, the teeth are continually replaced, row by row (**Figure 34.6b**). Sharks may have 20 rows of teeth, with the front pair in active use and the ones behind ready to grow as replacements when needed. Tooth replacement time varies from 9 days in the cookie-cutter shark (named for its

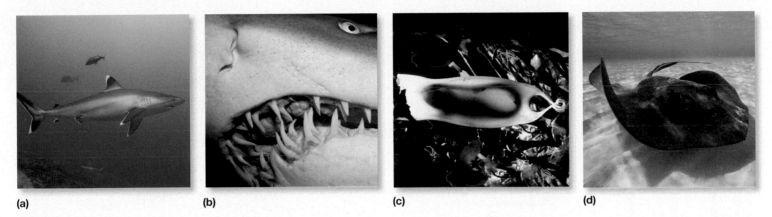

(a)   **(b)**   **(c)**   **(d)**

Figure 34.6 **Cartilaginous fish.** **(a)** The silvertip shark (*Carcharhinus albimarginatus*) is one of the ocean's most powerful predators. **(b)** Close-up of the mouth of a sand tiger shark (*Carcharias taurus*), showing rows of teeth. **(c)** This mermaid's purse (egg pouch) of a dogfish shark (*Scyliorhinus canicula*) is entwined in vegetation to keep it stationary. **(d)** Stingrays are essentially flattened sharks with very large pectoral fins.

characteristic of biting round plugs of flesh from its prey) to 242 days in great whites. Some experts estimate that certain sharks can use up to 20,000 teeth in a lifetime.

All chondrichthyans are denser than water, which theoretically means that they would sink if they stopped swimming. Many sharks never stop swimming and maintain buoyancy via the use of their fins and a large oil-filled liver. Another advantage of swimming is that water continually enters the mouth and is forced over the gills, allowing the sharks to extract oxygen and breathe. How then do skates and rays breathe when they rest on the ocean floor? These species, and a few sharks like the nurse shark, use a muscular pharynx and jaw muscles to pump water over the gills. In these and indeed all species of fish, the heart consists of two chambers, an atrium and a ventricle, that contract in sequence. They employ what is known as a single circulation, in which blood is pumped from the heart to capillaries in the gills to collect oxygen, and then it flows through arteries to the tissues of the body, before returning to the heart (look ahead to Figure 47.4a).

Many active predators possess a variety of acute senses, and sharks are no exception. They have a powerful sense of smell, facilitated by sense organs in the nostrils (sharks and other fish do not use nostrils for breathing). They can see well but cannot distinguish colors. While sharks have no eardrum, they can detect pressure waves generated by moving objects. All jawed fish have a row of microscopic organs in the skin, arranged in a line that runs laterally down each side of the body, that can detect movements in the surrounding water. This system of sense organs, known as the **lateral line**, picks up pressure waves and sends nervous signals to the inner ear and then on to the brain.

Fertilization is internal in chondrichthyans, with the male transferring sperm to the female via a pair of **claspers**, extensions of the pelvic fins. Some shark species are **oviparous**, that is, they lay eggs, often inside a protective pouch called a mermaid's purse (**Figure 34.6c**). The eggs, which are not guarded by either parent, then hatch into tiny sharks. In **ovoviparous** species, the eggs are retained within the female's body, but there

is no placenta to nourish the young. A few species are **viviparous**; the eggs develop within the uterus, receiving nourishment from the mother via a placenta. Both ovoviparous and viviparous sharks give birth to live young.

The sharks have been a very successful vertebrate group, with many species identified in the fossil record. While many species died out in the mass extinction at the end of the Permian period (290–248 million years ago), there was a further period of speciation of the survivors in the Mesozoic era, when most of the 375 modern-day species appeared. Skates and rays are essentially flattened sharks that cruise along the ocean floor using hugely expanded pectoral fins. In addition, their thin and whiplike tails are often equipped with a venomous barb that is used in defense (**Figure 34.6d**). Most of the 475 or so species of skates and rays feed on bottom-dwelling crustaceans and mollusks.

The Bony Fish Bony fish are the most numerous types of fish, with more individuals and more species (about 24,600) than any other. Most authorities now recognize three living classes: the Actinopterygii (ray-finned fish), the Actinistia, (coelacanths), and the Dipnoi (lungfish). Fish in all three classes possess a bony skeleton and scale-covered skin. The skin of bony fish, unlike the rough skin of sharks, is slippery and slimy because of glands that produce mucus, an adaptation that reduces drag during swimming. Just as in the cartilaginous fish, water is drawn over the gills for breathing, but in bony fish, a protective flap called an **operculum** covers the gills (**Figure 34.7**). Some early bony fish lived in shallow, oxygen-poor waters and developed lungs as an embryological offshoot of the pharynx. These fish could rise to the water surface and gulp air. As we will see, modern lungfish operate in much the same fashion. In most bony fish, these lungs evolved into a **swim bladder**, a gas-filled, balloon-like structure that helps the fish remain buoyant in the water even when it is completely stationary. The swim bladder is connected to the circulatory system, and gases can diffuse in and out of the blood, allowing the fish to change the volume of the swim bladder and to rise and sink.

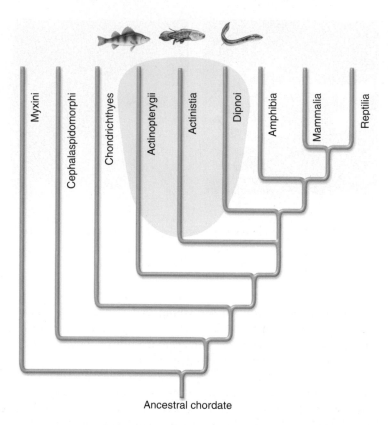

Ancestral chordate

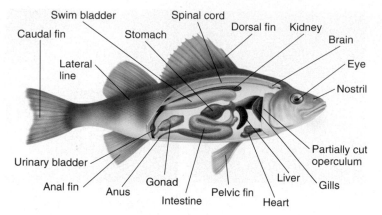

Figure 34.7   Generalized body plan of a bony fish.

logical adaptations to the different osmotic problems seawater presents compared to fresh water (see Chapter 49).

The most species-rich class of bony fish is the Actinopterygii, or **ray-finned fish**, which includes all bony fish except the coelacanths and lungfish. In Actinopterygii, the fins are supported by thin, bony, flexible rays and are moved by muscles on the interior of the body. The class has a diversity of forms, from large predatory moray eels to delicate sea dragons (**Figure 34.8**). Whole fisheries are built around the harvest of species such as cod, anchovies, and salmon.

The Actinistia (coelacanths) and Dipnoi (lungfish) are both considered Sarcopterygii or **lobe fins**. The name Sarcopterygii used to refer solely to the lobe-finned fish, but since it has become clear that terrestrial vertebrates (tetrapods) evolved from such fish, the definition of the group has been expanded to include both lobe-finned fish and tetrapods (see Figure 34.1). In the **lobe-finned fish**, the fins are part of the body, and they are supported by skeletal extensions of the pectoral and pelvic areas that are moved by muscles residing in the fins.

The fossil record revealed that the Actinistia, or coelacanths, were a very successful group in the Devonian period, but all fish of the class were believed to have died off at the end of the Mesozoic era, some 65 million years ago. You can therefore imagine the scientific excitement in 1938, when a modern coelacanth was

Thus, unlike the sharks, many bony fish can remain motionless and utilize a "sit-and-wait" ambush style. These three features—bony skeleton, operculum, and swim bladder—distinguish bony fish from cartilaginous fish.

Reproductive strategies of bony fish vary tremendously, but most species reproduce via external fertilization, with the female shedding her eggs and the male depositing sperm on top of them. While adult bony fish can maintain their buoyancy, their eggs tend to sink. This is why many species spawn in shallow, more oxygen- and food-rich waters, and why coastal areas are such important fish nurseries.

Bony fish have colonized all aquatic habitats. Most cartilaginous fish are marine, but bony fish probably evolved in freshwater habitats and secondarily returned to marine environments. This, of course, required the redevelopment of physio-

(a)   (b)   (c)

Figure 34.8   **The diversity of ray-finned fish.**   (a) Lionfish (*Pterois volitans*). (b) Whitemouth moray eel (*Gymnothorax meleagris*). (c) Leafy sea dragon (*Phycodurus eques*).

Figure 34.9   A lobefin fish, the coelacanth (*Latimeria chalumnae*).

Figure 34.10   An Australian lungfish (*Neoceratodus forsteri*).

Biological inquiry: How are lungfish similar to coelacanths?

discovered as part of the catch of a boat fishing near the Chalumna River in South Africa (**Figure 34.9**). Intensive searches in the area revealed that coelacanths were living off the southern African coast and especially off a group of islands near the coast of Madagascar called the Comoros Islands. Another species was more recently found in Indonesian waters.

Early diverging lobe-finned fish probably evolved in fresh water and had lungs, but the coelacanth lost them and returned to the sea. One distinctive feature is a special joint in the skull that allows the jaws to open extremely wide and gives the coelacanth a powerful bite. As further evidence of the coelacanth's unusual body plan, its swim bladder is filled with oil rather than gas, although it serves a similar purpose—to increase buoyancy.

The Dipnoi, or **lungfish**, like the coelacanths, are also not currently a very species-rich class, having just three genera and six species (**Figure 34.10**). Lungfish live in oxygen-poor freshwater swamps and ponds. They have both gills and lungs, the latter of which enable them to come to the surface and gulp air. In fact, lungfish will drown if they are unable to breathe air. When ponds dry out, some species of lungfish can dig a burrow and survive in it until the next rain. Because they also have muscular lobe fins, they are often able to successfully traverse quite long distances over shallow-bottomed lakes that may be drying out.

The morphological features of coelacanths, lungfish, and early divergent land animals, together with the similarity of coelacanth and lungfish nuclear genes to those of primitive terrestrial vertebrates, suggest to many scientists that lobe-fin ancestors gave rise to three lineages: the coelacanths, the lungfish, and the tetrapods.

34.4 Amphibia: The First Tetrapods

During the Devonian period, from about 417 to 354 million years ago, a diversity of plants colonized the land. The presence of plants served as both a source of oxygen and an extensive food source for animals that ventured out of the aquatic environment.

Terrestrial arthropods, especially insects, had evolved to feed on these plants and provided an additional food source for any vertebrate that could colonize the land.

The transition to life on land involved a large number of adaptations. Paramount among these were adaptations preventing desiccation and making locomotion and reproduction on land possible. We have seen that the lungfish evolved the ability to breathe air. In this section, we begin by outlining the development of the **tetrapods**, vertebrate animals having four legs or leglike appendages. We will discuss the first terrestrial vertebrates and their immediate descendants, the amphibians. We will then explore the characteristic features and diversity of modern amphibians.

Myxini

Cephalaspidomorphi

Chondrichthyes

Actinopterygii

Actinistia

Dipnoi

Amphibia

Mammalia

Reptilia

Ancestral chordate

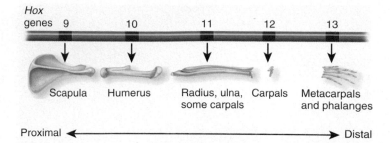

Hox
genes

Proximal ◄——————————————► Distal

Figure 34.11   *Hox* genes 9–13 work together to specify limb formation from the proximal to distal direction.   The axes of limb development in mice are shown, together with their associated genes.

(a) *Acanthostega*

(b) *Cacops*

The Origin of Tetrapods Involved the Development of Four Limbs

Over the Devonian period, fossils record the evolution of sturdy lobe-finned fishes that became fishes with four limbs. The abundance of light and nutrients in shallow waters encouraged a profusion of plant life and the invertebrates that fed on them. The development of lungs enabled lungfish to colonize these productive yet often oxygen-poor waters. Here, the ability to move in shallow water chock full of plants and debris was more vital than the ability to swim swiftly through open water and may have favored the progressive development of sturdy limbs. As an animal's weight began to be borne more by the limbs, the vertebral column strengthened, and hip bones and shoulder bones were braced against the backbone for added strength. Such modifications, while seemingly dramatic, are believed to be caused by relatively simple changes in the expression of genes, especially *Hox* genes (see Feature Investigation). In particular, *Hox* genes 9–13 work together to specify limb formation from the proximal to the distal direction (**Figure 34.11**).

Some of the transitional taxa between fish and amphibians possessed gills, like fish, but they had four limbs and free digits, characteristics of amphibians. *Acanthostega*, a very early diverging tetrapod, had forelimbs and hindlimbs, and eight digits on each hand, but it also retained many adaptations for aquatic life because it probably only rarely ventured onto land (**Figure 34.12a**). This is an important species, for it may represent a so-called stem species, a common tetrapod ancestor located at the beginning of the tetrapod branch on the tree of life. Eventually, however, species more like modern amphibians evolved, species that were still tied to water for reproduction but increasingly fed on land. In these species, the vertebral column, hip bones, and shoulder bones grew even sturdier (**Figure 34.12b**). Such changes were needed as the animal's weight was no longer supported by water but was borne entirely on the limbs. The evolution of a rib cage provided protection for the internal organs, especially the lungs and heart.

Figure 34.12   The development of primitive tetrapods. **(a)** *Acanthostega* represents a transitional form between a lobe-finned fish and a true amphibian. It possessed gills and was aquatic but had the tetrapod skeletal structure. **(b)** *Cacops* was a large, early amphibian of the Permian period.

By the middle of the Carboniferous period, about 320 million years ago, species similar to modern amphibians had become common in the terrestrial environment. For example, *Cacops* was a large amphibian, as big as a pony (Figure 34.12b). Its skin was heavy and tough, an adaptation that helped prevent water loss; its breathing was accomplished more by lungs than by skin; and it possessed **pentadactyl limbs** (limbs ending in five digits). With a bonanza of terrestrial arthropods to feast on, the amphibians became very numerous and species rich, and the mid-Permian period, some 260 million years ago, is sometimes known as the Age of Amphibians. However, most of the large amphibians became extinct at the end of the Permian period, coincident with the radiation of the reptiles, though whether the rise of reptiles was responsible for the extinction of amphibians is not known. Most surviving amphibians were smaller organisms resembling modern-day species.

FEATURE INVESTIGATION

Davis, Capecchi, and Colleagues Showed a Genetic-Developmental Explanation for Limb Length in Tetrapods

The development of limbs in tetrapods was a vital step in enabling animals to colonize land. The diversity of vertebrate limb types is amazing, from fins in fish and marine mammals to legs and arms in primates to different wing types in bats and birds.

In 1995, PhD student A. P. Davis, professor M. R. Capecchi, and colleagues analyzed the effects of mutations in specific *Hox* genes that are responsible for determining limb formation in mice. As described in **Figure 34.13**, they began with strains of mice carrying loss-of-function mutations in *HoxA-11* or *HoxD-11*. They bred the mice to obtain offspring carrying one, two, three, or four mutations.

Figure 34.13 Relatively simple changes in *Hox* genes control limb formation in tetrapods.

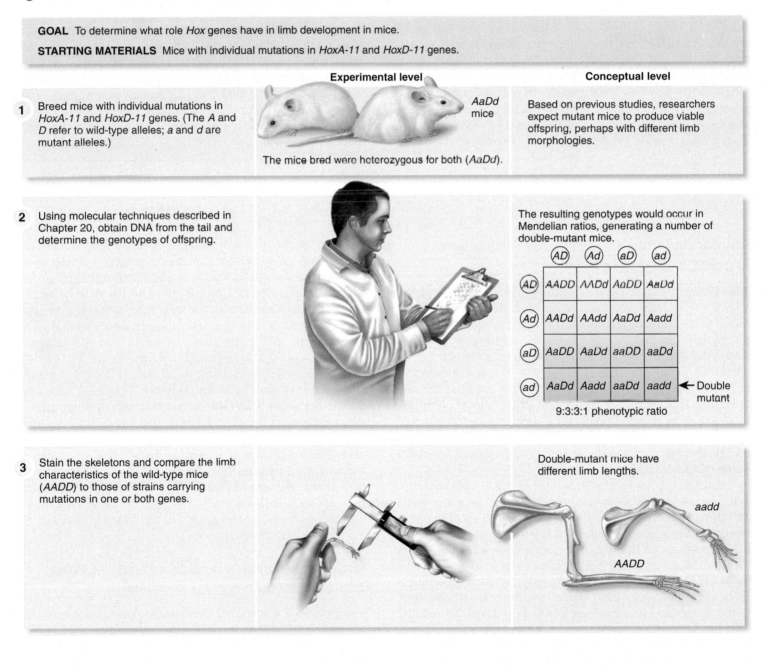

GOAL To determine what role *Hox* genes have in limb development in mice.

STARTING MATERIALS Mice with individual mutations in *HoxA-11* and *HoxD-11* genes.

Experimental level **Conceptual level**

1. Breed mice with individual mutations in *HoxA-11* and *HoxD-11* genes. (The *A* and *D* refer to wild-type alleles; *a* and *d* are mutant alleles.)

AaDd mice

The mice bred were heterozygous for both (*AaDd*).

Based on previous studies, researchers expect mutant mice to produce viable offspring, perhaps with different limb morphologies.

2. Using molecular techniques described in Chapter 20, obtain DNA from the tail and determine the genotypes of offspring.

The resulting genotypes would occur in Mendelian ratios, generating a number of double-mutant mice.

	AD	Ad	aD	ad
AD	AADD	AADd	AaDD	AaDd
Ad	AADd	AAdd	AaDd	Aadd
aD	AaDD	AaDd	aaDD	aaDd
ad	AaDd	Aadd	aaDd	aadd ← Double mutant

9:3:3:1 phenotypic ratio

3. Stain the skeletons and compare the limb characteristics of the wild-type mice (*AADD*) to those of strains carrying mutations in one or both genes.

Double-mutant mice have different limb lengths.

aadd

AADD

4 THE DATA

Genotype	Carpal bone fusions (% of mice showing the fusion)			
	Normal (none fused)	NL-T	T-P	NL-T-P
AADD	100			
AaDD	100			
aaDD	33	17	50	
AADd	100			
AAdd		17	17	67
AaDd	17	17	33	33

As seen in the data, the mutations affected the formation of limbs. For example, the wrist contains seven bones: three proximal carpals—called pisiform (P), triangular (T), and navicular lunate (NL)—and four distal carpals (d1–d4). In mice with the genotypes aaDD and AAdd, the proximal carpal bones are usually fused together. Individual heterozygotes (AADd and AaDD) do not show this defect, but compound heterozygotes (AaDd)

often do. Therefore, any two mutant alleles (either from both HoxA-11 and HoxD-11 or one from each locus) will cause carpal fusions. Deformities became even more severe with three mutant alleles (Aadd or aaDd) or four mutant alleles (aadd) (data not shown in the figure). Thus, scientists have shown that relatively simple mutations can control relatively large changes in limb development.

Amphibian Lungs and Limbs Are Adaptations to a Semiterrestrial Lifestyle

Amphibians (from the Greek, meaning "two lives") live in two worlds: They have successfully invaded the land but must return to the water to reproduce. One of the first challenges terrestrial animals had to overcome was breathing air when on land. Amphibians use the same technique as lungfish: They open their mouths to let in air then close and raise the floor of the mouth creating a positive pressure, thus pumping air into the lungs. This method of breathing is called buccal pumping. In addition, the skin of amphibians is much thinner than that of fish, and many amphibians can absorb oxygen from the air directly through their moist skin or the skin lining of the inside of the mouth or pharynx.

Amphibians have a three-chambered heart, with two atria and one ventricle. One atrium receives blood from the body, and the other receives blood from the lungs. Both atria pump blood into the single ventricle, which pumps some blood to the lungs and some to the rest of the body (look ahead to Figure 47.4b). This form of circulation allows the tissues to receive well-oxygenated blood at a higher pressure than is possible via single circulation because blood returns to the heart to be pumped to the tissues and is not slowed down by passage through the lung capillaries. This development enhances the delivery of nutrients and oxygen to the tissues, but it is still not maximally efficient because some oxygenated and deoxygenated blood are mixed in the ventricle.

Because the skin of amphibians is so thin, the animals face the problem of desiccation, or drying out. As a consequence, even amphibian adults are more abundant in damp habitats, such as swamps or rain forests, than in dry areas. Also, amphibians cannot venture too far from water because their larval stages are still aquatic. In frogs and toads, fertilization is generally external, with males shedding sperm over the gelatinous egg masses laid by the females in water (Figure 34.14a). The fertilized eggs lack a shell and would quickly dry out if exposed to the air. They soon hatch into tadpoles (Figure 34.14b), small fishlike herbivores that lack limbs and breathe through gills. As the tadpole nears the adult stage, the tail and gills are resorbed, and limbs and lungs appear (Figure 34.14c). Such a dramatic change in body form is known as metamorphosis, a process regulated by hormones from the thyroid gland. Reproduction is the link that ties amphibians to water, though a few species circumvent this need. Relatively few species are ovoviviparous or viviparous, retaining the eggs in the reproductive tract and giving birth to live young.

Modern Amphibians Include a Variety of Frogs, Toads, Salamanders, and Caecilians

Approximately 4,800 living amphibian species are known, and the vast majority of these, nearly 90%, are frogs and toads of the order Anura, meaning tail-less ones (Figure 34.15a). The other two orders are the Caudata ("tail visible"), the salamanders, and Gymnophiona ("naked snake"), the wormlike caecilians.

(a)   (b)   (c)

Figure 34.14 Amphibian development in the wood frog (*Rana sylvatica*).   (a) Amphibian eggs are laid in gelatinous masses in water. (b) The eggs develop into tadpoles, aquatic herbivores with a fishlike tail that breathe through gills. (c) During metamorphosis, the tadpole loses its gills and tail and develops limbs and lungs.

Adult anurans are carnivores, eating a variety of invertebrates by catching them on a long, sticky tongue. In contrast, the aquatic larvae (tadpoles) are primarily herbivores. Frogs generally have smooth, moist skin and long hind legs, making them excellent jumpers and swimmers. In addition to secreting mucus, which keeps their skin moist, some frogs can also secrete poisonous chemicals that deter would-be predators. Some amphibians advertise the poisonous nature of their skin with warning coloration (look ahead to Figure 57.10b). Others use camouflage as a way of avoiding detection by predators. Toads have a drier, bumpier skin and shorter legs than frogs. While they are less impressive leapers, toads can tolerate drier conditions than can frogs.

The salamanders (order Caudata) possess a tail and have a more elongate body than anurans (**Figure 34.15b**). During locomotion, they seem to sway from side to side, perhaps reminiscent of how the earliest tetrapods may have walked. Like frogs, salamanders often have colorful skin patterns that advertise their distastefulness to predators. Salamanders retain their moist skin by living in damp areas under leaves or logs or beneath lush vegetation. They generally range in size from 10 to 30 cm. Fertilization is usually internal, with females using their cloaca to pick up sperm packets deposited by males. A very few salamander species do not undergo metamorphosis, and the newly hatched young resemble tiny adults. On the other hand, some species, such as the axolotl, retain the gills and tail fins characteristic of the larval stage into adulthood, a phenomenon known as paedomorphosis.

Caecilians (order Gymnophiona) are a small order of about 160 species of legless, nearly blind amphibians (**Figure 34.15c**). Most are tropical and burrow in forest soils, while a few live in ponds and streams. They are secondarily legless, which means that they developed from legged ancestors. Caecilians eat worms and other soil invertebrates and have tiny jaws equipped with teeth. In this order, fertilization is internal, and females usually bear live young. The young are nourished inside the mother's body by a thick, creamy secretion known as uterine milk. In most caecilian species, the young grow into adults about 30 cm long, though some species up to 1.3 m in length are known.

(a)

(b)   (c)

Figure 34.15 Amphibians   (a) Most amphibians are frogs and toads of the order Anura, including this red-eyed tree frog (*Agalychnis callidryas*). (b) The order Gymnophiona includes wormlike caecilians such as this species from Colombia, *Caecilia nigricans*. (c) The order Caudata includes species such as this mud salamander (*Pseudotriton montanus*).

Biological inquiry: Do all amphibians produce tadpoles?

34.5 Amniotes: The Appearance of Reptiles

While successfully living in a terrestrial environment, adult amphibians remain tied to water for reproduction. Amphibians lay their eggs in water or in a very moist place, so the shell-less eggs do not dry out on exposure to air. Thus, a critical innovation in animal evolution was the development of a shelled egg that sheltered the embryo from desiccating conditions on land.

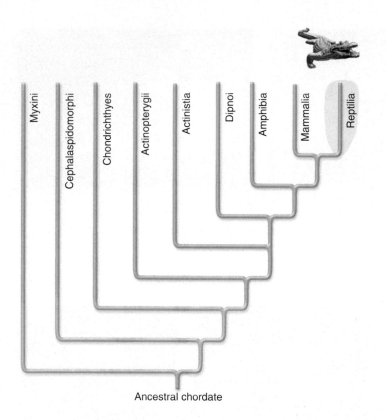

Ancestral chordate

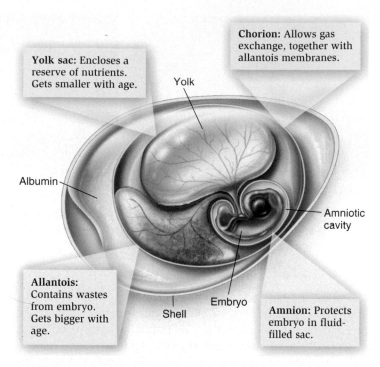

Yolk sac: Encloses a reserve of nutrients. Gets smaller with age.

Chorion: Allows gas exchange, together with allantois membranes.

Yolk

Albumin

Amniotic cavity

Allantois: Contains wastes from embryo. Gets bigger with age.

Shell

Embryo

Amnion: Protects embryo in fluid-filled sac.

Figure 34.16 The amniotic egg.

A shelled egg containing fluids was like a personal enclosed pond for each developing individual. Such an egg evolved in the common ancestor of turtles, lizards, snakes, crocodiles, birds, and mammals—a group of tetrapods collectively known as the **amniotes**. The amniotic egg permitted animals to lay their eggs in a dry place so that reproduction was no longer tied to water. It was truly a critical innovation. The amniotic egg untethered animals from water in a similar way as the development of seeds liberated plants from water (see Chapter 31).

In time, the amniotes came to dominate the Earth. Mammals are considered amniotes, too, because even though most of them do not lay eggs, they retain other features of amniotic reproduction. In this section, we begin by discussing in detail the morphology of the amniotic egg and other adaptations that permitted species to become fully terrestrial. We then discuss the biology of the reptiles, the first group of vertebrates to fully exploit land.

The Amniotic Egg and Other Innovations Permitted Life on Land

The **amniotic egg (Figure 34.16)** contains the developing embryo and the four separate extraembryonic membranes that it produces:

1. The innermost membrane is the **amnion**, which protects the developing embryo in a fluid-filled sac called the amniotic cavity.
2. The **yolk sac** encloses a stockpile of nutrients, in the form of yolk, for the developing embryo.

3. The **allantois** functions as a disposal sac for metabolic wastes.
4. The **chorion**, along with the allantois, provides gas exchange between the embryo and the surrounding air.

Surrounding the chorion is the albumin, or egg white, which also stores nutrients. The **shell** provides a tough, protective covering that is impermeable to water and prevents the embryo from drying out. However, the shell remains permeable to oxygen and carbon dioxide, so the embryo can breathe. In birds, this shell is hard and calcareous, while in reptiles and primitive mammals like the platypus and echidna, it is soft and leathery. Most mammals, however, do not have shelled eggs. Instead, the embryos embed into the wall of the uterus and receive their nutrients directly from the mother.

Along with the amniotic egg, other critical innovations that enabled the conquest of land included:

- **Desiccation-resistant skin.** While the skin of amphibians is moist and aids in respiration, the skin of amniotes is thicker and watertight and contains keratin, a tough protein. This requires that gas exchange take place through the lungs.

- **Thoracic breathing.** Amphibians breathe by contracting the mouth to force air into the lungs. In contrast, amniotes use **thoracic breathing**, in which coordinated contractions of muscles expand the rib cage, creating a negative pressure to suck air in and then forcing it out later. This results in a greater volume of air being displaced with each breath.

- **Water-conserving kidneys.** The ability to concentrate wastes prior to elimination and thus conserve water is an important role of the amniote kidneys.

- **Internal fertilization**. Because sperm cannot penetrate a shelled egg, fertilization occurs internally, within the female's body before the shell is secreted. In this process, the male of the species often uses a copulatory organ (penis) to transfer sperm into the female reproductive tract. However, birds transfer sperm from cloaca to cloaca.

Classification of the Reptiles Is Currently Under Revision

Early amniote ancestors gave rise to all modern amniotes we know today, from lizards and snakes to birds and mammals. The traditional view of amniotes involved three living classes: the reptiles (turtles, lizards, snakes, and crocodilians), birds, and mammals. As we will see later in the chapter, modern systematists have argued that enough similarities exist between birds and the classic reptiles that birds should be considered part of the reptilian lineage. Other systematists believe that enough differences are found between the living reptilian taxa that each should be considered a distinct class: the Testudines (turtles), Lepidosauria (lizards and snakes), Crocodilia (crocodiles), and Aves (birds), as illustrated in Figure 34.1. This is the classification scheme that we will follow in this chapter. The fossil record includes other reptilian classes, all of which are extinct, including two classes of dinosaurs (ornithischian and saurischian dinosaurs), flying reptiles (pterosaurs), and two classes of ancient aquatic reptiles (icthyosaurs and plesiosaurs).

Class Testudines: The Turtles
Turtles is an umbrella term for terrestrial species, also called tortoises, and aquatic species, also known as terrapins. The turtle lineage is ancient and has remained virtually unchanged for 200 million years. The major distinguishing characteristic of the turtle is a hard protective shell into which the animal can withdraw its head and limbs. In most species, the vertebrae and ribs are fused to this shell. All turtles lack teeth but have sharp beaks for biting.

The majority of turtles are aquatic and have webbed feet. The forelimbs of marine species have evolved to become large flippers. All turtles, even the aquatic species, lay their eggs on land, usually in soft sand. The gender of hatchlings is temperature dependent, with high temperatures producing more females. Marine species often make long migrations to sandy beaches to lay their eggs (**Figure 34.17a**). Most land tortoises are quite slow movers, possibly due to a low metabolic rate and a heavy shell. On the other hand, they are very long-lived species, often surviving for 120 years or more. Many turtle species are in danger of extinction, due to egg hunting, destruction of habitat and nesting sites, and death from entanglement in fishing nets.

Class Lepidosauria: Lizards and Snakes
The class Lepidosauria is the largest class within the traditional reptiles, with about 3,000 species of snakes (order Serpentes) and 4,800 species of lizards (order Sauria). Many species have an elongated body form. One of the defining characteristics of the orders is a **kinetic skull**, in which the joints between various parts of the skull are extremely mobile. The lower jaw does not join directly to the skull but rather is connected by a multi-jointed hinge, and the upper jaw is hinged and moveable from the rest of the head. This allows the jaws to open relatively wider than other vertebrate jaws, with the result that lizards, and especially snakes, can swallow large prey (**Figure 34.18**). Nearly all species are carnivores.

A main difference between lizards and snakes is that lizards generally have limbs, whereas snakes do not. Also, snakes may be venomous, whereas lizards are generally not. However, there are exceptions to these general rules. Many legless lizard species exist (see **Figure 34.17b**), and two lizards are poisonous: the Gila monster (*Heloderma suspectum*) of the U.S. Southwest (see **Figure 34.17c**) and the Mexican beaded lizard (*Heloderma horridum*). A more reliable distinguishing characteristic is that lizards have moveable eyelids and external ears (at least ear canals), whereas snakes do not.

(a) (b) (c)

Figure 34.17 A variety of reptiles. (a) A green turtle (*Chelonia mydas*) laying eggs in the sand in Malaysia. (b) The Florida worm lizard (*Rhineura floridana*) is an amphisbaenian, a type of legless soil-burrowing lizard. (c) The Gila monster (*Heloderma suspectum*), one of only two poisonous lizards, is an inhabitant of the desert Southwest of the U.S. and of Mexico.

Class Crocodilia: The Crocodiles and Alligators The Crocodilia is a small class of large aquatic animals that have remained essentially unchanged for nearly 200 million years (**Figure 34.19**). Indeed, these animals existed at the same time as the dinosaurs. Most of the 23 recognized species live in tropical or subtropical regions. There are only two extant species of alligators: one living in the southeastern U.S. and one found in China.

While the class is small, it is evolutionarily very important. Crocodiles have a four-chambered heart, a feature they share with birds and mammals (look ahead to Figure 47.4c). In this regard, crocodiles are more closely related to birds than any other living reptile class. Their teeth are set in sockets, a feature typical of the dinosaurs and the earliest birds. Similarly, crocodiles care for their young, another trait they have in common with birds. These and other features suggest that crocodiles and birds are more closely related than crocodiles and lizards. As with turtles, nest temperature influences the sex ratio of offspring. With crocodiles, however, when temperatures are warm, more males are produced. Biologists are concerned that a scenario of global warming may reduce the number of breeding females, leading to extinction of some species.

The Dinosaurs: Classes Ornithischia and Saurischia In 1841, the English paleontologist Richard Owen coined the term **dinosaur**, meaning terrible lizard, to describe some of the wondrous fossil animals discovered in the 19th century. About 215 million years ago, dinosaurs were the dominant tetrapods on Earth and remained so for 150 million years, far longer than any other vertebrate. The two main classes were the ornithischian or bird-hipped dinosaurs, which were herbivores such as *Stegosaurus*, and the saurischian or lizard-hipped dinosaurs, which were fast, bipedal carnivores such as *Tyrannosaurus* (**Figure 34.20**). In contrast to the limbs of lizards, amphibians, and crocodiles, which splayed out to the side, the legs of dinosaurs were positioned directly under the body, like pillars, a position that could help support their heavy body. Because less energy was devoted in lifting the body from the ground, some dinosaurs are believed to have been fast runners. Members of different but closely related classes—the pterosaurs (the first vertebrates to fly), and plesiosaurs and ichthyosaurs (marine reptiles)—were also common at this time.

Figure 34.18 **The kinetic skull.** In snakes and lizards, the top of the jaw, as well as the bottom of the jaw, is hinged on the skull, thus permitting large prey to be swallowed. This Halloween snake (*Pliocercus euryzonus*) is swallowing a Costa Rican rain frog.

(a)

(b)

Figure 34.19 **Crocodilians.** The Crocodilia is an ancient class that has existed unchanged for millions of years. (a) Alligators, such as this American alligator (*Alligator mississippiensis*), have a broad snout, and the lower jaw teeth close on the inside of the upper jaw (and thus are almost completely hidden when the mouth is closed). (b) Crocodiles, including this American crocodile (*Crocodylus acutus*), have a longer, thinner snout, and the lower jaw teeth close on the outside of the upper jaw (and thus are visible when the mouth is closed).

Biological inquiry: In what ways are crocodilians similar to birds?

(a) Ornithischian (*Stegosaurus*)   (b) Saurischian (*Tyrannosaurus*)

Figure 34.20   Dominant terrestrial dinosaur classes.
(a) Ornithischians included *Stegosaurus*, and (b) saurischians included bipedal species such as *Tyrannosaurus*.

Dinosaurs were the biggest animals ever to walk on the planet, with some animals weighing up to 50 tonnes (metric tons) or over 100,000 pounds. The variety of the thousands of dinosaur species found in fossil form around the world is staggering. However, perhaps not surprisingly for such long-extinct species, the details of their lives are still being hotly debated. For example, an issue still unresolved is whether some dinosaur species were **endothermic**, that is, capable of generating and retaining body heat through their metabolism, just as birds and mammals are. Another issue is whether dinosaurs exhibited parental care of their young.

All dinosaurs, and many other animals, went extinct quite abruptly at the end of the Cretaceous period, about 65 million years ago. Although widely attributed to climatic change brought about by the impact of an asteroid, there is ongoing debate over the cause of this mass extinction. It is not known why all the dinosaurs died out, while many other animals, including small mammals, survived.

34.6 Birds

The defining characteristics of birds (class Aves, plural of the Latin *avis*, bird) are that they have feathers and nearly all species can fly. As we will see, the ability to fly has shaped nearly every feature of the bird body. The other vertebrates that have evolved the ability to fly, the bats and the now-extinct pterosaurs, used skin stretched tight over elongated limbs to fly. Such a surface can be irreparably damaged, though some holes may heal remarkably quickly. In contrast, birds utilize feathers, epidermal outgrowths that can be replaced if damaged. Recent studies in cell signaling show that the developmental changes necessary for animals to grow feathers instead of scales may not be that complex at the molecular level.

In this section, we will begin by discussing the likely evolution of birds from dinosaur ancestors, outline the key characteristics of birds, and provide a brief overview of the various bird orders.

GENOMES & PROTEOMES

The Differentiation of Scales, Feathers, and Fur May Be Caused by Simple Changes in Developmental Pathways

The skin of reptiles is protected by tough scales containing keratin, a tough waterproof protein. In birds, the skin is covered in feathers, and in mammals, it is covered by hair or fur. These epithelial appendages look very different but share similarities in initial development. All originate as **epithelial placodes**, regions of slightly thickened epithelial cells. Their spatial distribution on the body is determined by many developmental signaling pathways, including those controlled by proteins designated Wnt, Notch, and BMP that regulate the properties of the epithelial cells (or epidermis) and the dermis. The timing and level of expression of these regulatory molecules determine the actual appendage that develops in the dermal layer. Wnt and Notch promote cell proliferation, while BMP programs cell death. Scales are formed as an outgrowth of the skin and are turned on by low levels of Wnt and Notch and high levels of BMP, while feathers develop with inverse levels of such regulatory molecules (**Figure 34.21**).

Placode — Epidermis
— Dermis
Signals

| Epithelial placode formation | Outgrowth | Follicle formation (feather, hair) |

(a)

Morphology	Scale	Feather
Wnt levels (cell proliferation)	Low	High
Notch levels (cell proliferation)	Low	High
BMP levels (cell death)	High	Low

(b)

Figure 34.21   Development of epithelial appendages.
(a) Epithelial appendages begin as an outgrowth of the epidermis and dermis. In the development of feathers and hair, this outgrowth gives way to the formation of a follicle. (b) The difference between the identities of different epithelial appendages, such as scales and feathers, are caused, in part, by the timing and level of early signaling events.

Different regulatory genes acting in later stages of development can alter the properties of the scales, feathers, or hair. For example, the mouse *HoxC-13* gene is necessary for normal development of mouse hair. Mice with a mutation in this gene have fragile, brittle hair that breaks easily. Other features of the body that are hardened by keratin, the structural component of hair, may also be affected, such as nails and even filiform papillae, which are small projections on the surface of the tongue. The seemingly radical differences in body coverings of different vertebrate classes are actually caused by relatively simple changes in the levels of regulatory molecules.

Modern Birds Probably Evolved from Small, Feather-Covered Dinosaurs

To trace the evolution of birds, we must look for the earliest type of animals that had feathers. One of the first known fossils exhibiting the faint impression of feathers was *Archaeopteryx lithographica* (meaning "ancient wings" and "stone picture"), found in a limestone quarry in Germany in 1861. The fossil was dated at 150 million years old. Except for the presence of feathers, *Archaeopteryx* appears to have had features similar to those of dinosaurs (**Figure 34.22a**). First, the fossil had an impression of a long tail with many vertebrae, a dinosaur feature. Some modern birds have long tails, but they are made of feathers, with the actual tailbone being much reduced. Second, the wings had claws

halfway down the leading edge, another dinosaur-like character. Among modern birds, only the hoatzin, a South American swamp-inhabiting bird, has claws on its wings, which enable the chicks to climb back into the nest if they fall out. A third dinosaur-like feature is *Archaeopteryx*'s toothed beak. Fourth, the fossils show that *Archaeopteryx* lacked an enlarged breastbone, a feature that modern birds possess to anchor their large flight muscles, so it likely could not fly.

Similarities between the structure of the skull, feet, and hind leg bones have led scientists to conclude that *Archaeopteryx* is closely related to **theropods**, a group of bipedal saurischian dinosaurs. The wings and feathers of *Archaeopteryx* may have enabled it to glide from tree to tree, to help it keep warm, or to fold over its head when hunting, in much the same way as some herons fold their wings over their heads when they are fishing. Later, the wings and feathers may have taken on functions of flight. We often find that evolution proceeds when previously evolved structures are co-opted into different uses, a diversity principle known as descent with modification.

In 1997, paleontologists unearthed fossils of about the same age of *Archaeopteryx* in China that similarly suggest a close kinship between dinosaurs and modern birds. *Caudipteryx zoui* was a dinosaur-like animal with feathers on its wings and tail and a toothed beak (**Figure 34.22b**). *Confuciusornis sanctus* was a small, flightless but completely feathered dinosaur lacking the long bony tail and toothed jaw found in other theropod dinosaurs. Its large tail feathers may have functioned in courtship displays (**Figure 34.22c**).

(a) *Archaeopteryx lithographica*   **(b) *Caudipteryx zoui***   **(c) *Confuciusornis sanctus***

Figure 34.22 **Missing links between dinosaurs and birds.** (a) *Archaeopteryx lithographica* was a Jurassic animal with dinosaur-like features as well as wings and feathers. (b) *Caudipteryx zoui* was a dinosaur with feathers on its tail and wings. (c) *Confuciusornis sanctus* was a birdlike animal with a horny bill.

These three species—*Archaeopteryx*, *Caudipteryx*, and *Confuciusornis*—help trace a lineage from dinosaurs to birds. By the early Cretaceous period, and only a relatively short period after *Archaeopteryx* evolved, the fossil record shows the existence of a huge array of bird types resembling modern species. These were to share the skies with pterosaurs for 70 million years, before eventually having the airways to themselves.

Birds Display Four Key Characteristics

Modern birds possess many characteristics that reveal their reptilian ancestry. For example, they have scales on their feet and legs, and they lay shelled eggs. In addition, however, among living animals they have four features unique to birds, all of which are associated with flight.

1. **Feathers.** Feathers are modified scales that keep birds warm, enable flight, and are supported on a modified forelimb (**Figure 34.23a**). Soft, downy feathers maintain heat, while stiffer contour feathers give the wing the airfoil shape it needs to generate lift. Each contour feather develops from a follicle, a tiny pit in the skin. If a feather is lost, a new one can be regrown. The contour feathers consist of many paired barbs, each of which supports barbules that hook together with barbules from neighboring barbs to give the feather its shape (**Figure 34.23b**).

2. **Lightweight skeleton.** Most bird bones are thin and hollow and are crisscrossed internally by tiny pieces of bone to give them a honeycomb structure (**Figure 34.23c**). The keeled breastbone, or **sternum**, provides an anchor on which a bird's powerful flight muscles attach. These muscles may contribute up to 30% of the bird's body weight. The bird skull lacks teeth, but the keratin of bird beaks is tough and malleable, and it takes very different shapes in different species (**Figure 34.24**).

3. **Air sacs.** Flight requires a great deal of energy generated from an active metabolism that requires abundant oxygen. Birds have nine air sacs, large, hollow sacs that may extend into the bones (look ahead to Figure 48.13). On inhalation, air passes into the posterior air sacs. On exhalation, half of this air is forced back out through the lungs, where oxygen is extracted, and then goes into the anterior air sacs. On the next inhalation, air is sucked from the posterior air sacs into the anterior air sacs via the lungs. Air is therefore being constantly moved across the lungs during inhalation and exhalation. While making bird breathing very efficient, this process also makes birds especially susceptible to airborne toxins (hence, the utility of the canary in the coal mine; the bird's death signaled the presence of harmful carbon dioxide or methane gas that was otherwise unnoticed by miners).

4. **Reduction of organs.** To decrease the total mass the bird must carry, some organs are reduced in size or are lacking altogether. For example, birds have only one ovary and can carry relatively few eggs. As a result, they lay fewer eggs than most other reptile species. In fact, the gonads of both males and females are reduced, except during the breeding season, when they increase in size. Most birds also lack a urinary bladder.

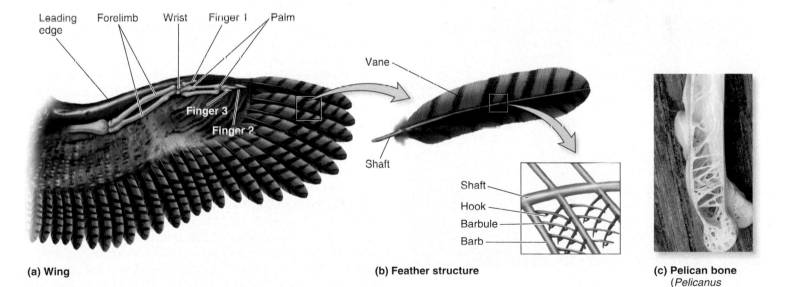

(a) Wing   **(b) Feather structure**   **(c) Pelican bone**
(*Pelicanus occidentalis*)

Figure 34.23   Features of the bird wing and feather.   (a) The wing is supported by an elongated and modified forelimb with extended fingers. **(b)** Each feather has a hollow shaft that supports many barbs, which in turn support barbules that interlock with hooks to give the feather its form. **(c)** The bones of a pelican (*Pelicanus occidentalis*) are hollow but crisscrossed with a honeycomb structure that provides added strength.

Biological inquiry: What adaptations in birds help reduce their body weight to enable flight?

Birds also have other distinct features, though mammals also possess some of these. For example, birds have a warm body temperature, which ensures rapid metabolism and the quick production of ATP that these active organisms need to fuel flight and other activities. In fact, birds' body temperatures are generally 40–42°C, considerably warmer than the human body's average of 37°C. Birds have a double circulation and a four-chambered heart to ensure rapid blood circulation. Rapid flight also requires acute vision, and bird vision is among the best in the vertebrate world. Generally, plant material is not energy-rich enough to supply the dietary needs of birds, so most birds are carnivores, eating insects and other invertebrates, though some birds like parrots specialize on more nutrient-rich fruits and seeds. Bird eggs also have to be kept warm for successful development, which entails brooding by an adult bird. Often, the males and females take turns brooding so that one parent can feed and maintain its strength. Picking successful partners is therefore an important task, and birds often engage in complex courtship rituals.

Birds Have Many Orders, All with the Same Body Plan

Birds are the most species-rich class of terrestrial vertebrates, with 28 orders, 166 families, and about 9,600 species (**Table 34.2**). Despite this diversity, birds lack the variety of body shapes that exist in the other endothermic class of vertebrates, the mammals, where some species swim, others fly, others walk on four legs, and yet others walk only on two legs. Most birds fly and thus most have the same general body shape. The biggest departures from this body shape are the flightless birds, including the kiwis, ostriches, cassowaries, emus, and rheas. These birds have smaller wing bones and the keel on the breastbone is greatly reduced or absent. Penguins are also flightless birds whose upper limbs are modified as flippers for swimming.

(a) **(b)** **(c)** **(d)** **(e)** **(f)**

Figure 34.24 **A variety of bird beaks.** Birds have evolved a variety of beak shapes for different types of food gathering. **(a)** Hyacinthe macaw (*Anodorhynchus hyacinthinus*)—cracking. **(b)** White pelican (*Pelecanus onocrotalus*)—scooping. **(c)** Verreaux's eagle (*Aquila verreauxii*)—tearing. **(d)** American avocet (*Recurvirostra americana*)—probing. **(e)** Lucifer hummingbird (*Calothorax lucifer*)—nectar feeding. **(f)** Roseate spoonbill (*Ajaia ajaia*)—sieving.

34.7 Mammals

Mammals evolved from amniote ancestors earlier than birds. About 225 million years ago, the first mammals appeared in the mid-Triassic period. They evolved from small mammal-like lizards that went extinct about 170 million years ago. Mammals survived, although most are believed to have been small, insect-eating species that lived in the shadows of dinosaurs. However, in January 2005, two fossils of a 130-million-year-old mammalian species called *Repenomamus* were discovered that challenge the notion of mammals as small insect eaters. One fossil was of an animal estimated to weigh about 13 kg (30 lb), while the other had the remains of a baby dinosaur in its stomach.

After extinction of the dinosaurs in the Cretaceous period, some 65 million years ago, mammals flourished. Today, biologists have identified about 5,500 species of mammals with a diverse array of lifestyles, from fishlike dolphins to birdlike bats, and from small insectivores such as shrews to large herbivores such as giraffes and elephants. The range of sizes and body forms of mammals is unmatched by any other group, and mammals are prime illustrations of the concept that organismal diversity is related to environmental diversity. In this section, we will outline the features that distinguish mammals from other taxa. We will also examine the diversity of mammals that exists on Earth and will end by turning our attention to primates and in particular to the evolution of modern humans.

Table 34.2	The Main Orders of Birds, in Order of Species Richness

Order		Examples (approx. # of species)	Main characteristics
Passeriformes		Robins, starlings, sparrows, warblers (5,300)	Perching birds with perching feet; songbirds
Apodiformes		Hummingbirds, swifts (430)	Fast fliers with rapidly beating wings; small bodies
Piciformes		Woodpeckers, toucans (380)	Large with specialized bills; two toes pointing forward and two backward
Psittaciformes		Parrots, cockatoos (340)	Large, powerful beaks
Chadradriiformes		Seagulls, wading birds (330)	Shorebirds
Columbiformes		Doves, pigeons (300)	Round bodies; short legs
Falconiformes		Eagles, hawks, kestrels, vultures (290)	Diurnal carnivores; birds of prey; powerful talons, and strong beaks
Galliformes		Chickens, pheasants, quail (270)	Often large birds; weak flyers; ground nesters
Coraciiformes		Hornbills, kingfishers (200)	Large bills; cavity nesters
Anseriformes		Ducks, swans, geese (150)	Able to swim; webbed feet; broad bills
Strigiformes		Owls (150)	Nocturnal carnivores; powerful talons, and bills
Pelecaniformes		Pelicans, frigate birds, cormorants (55)	Large, colonial fish eaters; often tropical
Sphenisciformes		Penguins (18)	Flightless, wings modified into flippers for swimming; marine; Southern Hemisphere
Casuariformes		Cassowaries, emus (3)	Large, flightless; Australia and New Guinea
Struthioniformes		Ostrich (1)	Large, flightless; only two toes; Africa only

Mammals Have Mammary Glands, Hair, Specialized Teeth, and an Enlarged Skull

The distinguishing characteristics of mammals are the posses sion of mammary glands, hair, specialized teeth, and an enlarged skull.

- **Mammary glands.** Mammals, or the class Mammalia (from the Latin, *mamma*, breast), are named after the female's distinctive **mammary glands**, which secrete milk.

Milk is a fluid rich in fat, sugar, protein, and vital minerals, especially calcium. Newborn mammals suckle this fluid, which helps promote rapid growth.

- **Hair.** All mammals have hair, although some have more than others. Whales retain only a few hairs on their snout, and humans are relatively hairless. In some animals, the hair is dense and referred to as fur. In some aquatic species such as beavers, the fur is so dense it cannot be thoroughly wetted, so the hair underneath remains dry.

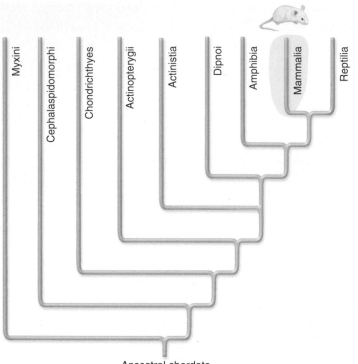

Ancestral chordate

Mammals are endothermic, and their fur is an efficient insulator. Hair can also take on functions other than insulation. Many animals, including cats, dogs, walruses, and whales, have sensory hairs called vibrissae (**Figure 34.25a**). Hair can be of many colors, to allow the mammals to blend into their background (**Figure 34.25b**). In some cases, as in porcupines and hedgehogs, the hairs become long, stiffened, and sharp (quills) and serve as a defense mechanism (**Figure 34.25c**).

- **Specialized teeth.** Mammals are the only vertebrates with teeth that are adapted for different types of diets (**Figure 34.26**). While teeth are generally present in all species, different teeth are larger, smaller, lost, or reduced, depending

on diet. Of particular importance to carnivores are the piercing canine teeth, while herbivorous species like antelopes depend on their chisel-like incisors to snip off vegetation and on their many molars to grind plant material. Only mammals chew their food in this fashion. Rodent incisors grow continuously throughout life, and species such as beavers wear them down by gnawing tough plant material such as wood.

- **Enlarged skull.** The mammalian skull differs from other amniote skulls in several ways. First, mammals have a single lower jawbone—the dentary—unlike reptiles, whose lower jaw is composed of multiple bones. Second, mammals have three bones in the inner ear, as opposed to reptiles, which have one bone in the inner ear. Third, mammals have pinnae, or external ears. Fourth, the brain is enlarged and is contained within a relatively large skull.

In addition to those uniquely mammalian characteristics, some but not all mammals possess these additional features:

- **The ability to digest plants.** Apart from tortoises and marine iguanas, certain species of mammals are the only large vertebrates that can exist on a steady diet of grasses or tree leaves; indeed, most large mammals are herbivores. Though mammals cannot digest cellulose, the principal constituent of the cell wall of many plants, some species have a large four-chambered stomach containing cellulose-digesting bacteria. These bacteria can break down the cellulose and make the plant cell contents available to the animal.

- **Horns and antlers.** Mammals are the only living class of vertebrates to possess horns or antlers. Many mammals, especially antelopes, cattle, and sheep, have horns, typically consisting of a bony core that is a permanent outgrowth of the skull surrounded by a hairlike keratin sheath (**Figure 34.27a**). Rhinoceros horns are outgrowths of the epidermis consisting of very tightly matted hair (**Figure 34.27b**).

Figure 34.25 **Mammalian hair.** **(a)** The sensory hairs (vibrissae) of the walrus (*Odobenus rosmarus*). **(b)** The camouflaged coats of Burchell's zebra (*Equus burchelli*). **(c)** The defensive quills of the crested porcupine (*Hystrix africaeaustralis*).

(a)

(b)

(c)

In contrast, deer antlers are made entirely out of bone (**Figure 34.27c**). Deer grow a new set of antlers each year and shed them after the mating season. Hooves and claws are also made of keratin and protect an animal's toes from the impact of its feet striking the ground and aid in prey capture, respectively.

Mammals Are the Most Diverse Group of Vertebrates Living on Earth

Modern mammals are incredibly diverse (**Table 34.3**). They vary in size from tiny insect-eating bats, weighing in at only 2 g, to leviathans such as the blue whale, the largest animal ever known, which tips the scales at 100 tonnes (over 200,000 lbs.). The 26 different mammalian orders are divided into three dis-

tinct subclasses. The subclass Prototheria contains only the order Monotremata, or **monotremes**, which are found in Australia and New Guinea. There are only three species: the duck-billed platypus (**Figure 34.28a**) and two species of echidna, a spiny animal resembling a hedgehog. Monotremes are considered primitive mammals because they lay eggs rather than bear live young, lack a placenta, and have mammary glands with poorly developed nipples. The mothers incubate the eggs, and upon hatching, the young simply lap up the milk as it oozes onto the fur.

The subclass Metatheria, or the **marsupials**, is a group of seven orders, with about 280 species (**Figure 34.28b**). Once widespread, members of this order are now largely confined to Australia, although some marsupials exist in South America and one species—the opossum—is found in North America.

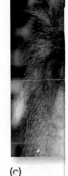

(a)

(c)

several de
dwelling i

- **Grasp**
  chara
  (see F
  an **op**
  the fi
  grip a
  prima

- **Large**
  the al
  efficie
  As a i
  In tur

- **At lea**
  This f
  objec

- **Binoc**
  are pc
  other
  branc
  This i
  vision

There are
onomists
anthropoi
species su
generally i

(a)   (b)   (c)

(d)   (e)

Figure 34.26 Mammalian teeth. Mammals have different types of teeth, according to their diet. **(a)** The wolf has long canine teeth for biting its prey. **(b)** The deer has a long row of front molars for grinding plant material. **(c)** The beaver, a rodent, has long, continually growing incisors. **(d)** The elephant's incisors are extremely modified into tusks. **(e)** Dolphins and other fish or plankton feeders have numerous small teeth for prey capture.

(a)   (b)   (c)

Figure 34.27 Horns and antlers. Mammals have a variety of outgrowths that are used for defense or by males as weapons in contests over females. **(a)** The horns of this male kudu (*Tragelaphus strepsiceros*) are bony outgrowths of the skull covered in a keratin sheath. **(b)** The horns of the black rhinoceros (*Diceros bicornis*) are outgrowths of the epidermis and made of tightly matted hair. **(c)** The antlers of the caribou (*Rangifer tarandus*) are grown and shed each year.

(a)   (b)   (c)

Figure 34.30   **Primate classification.**   Many authorities divide the primates into two groups: **(a)** the prosimians, smaller, nocturnal species such as this tarsier (*Tarsius syrichta)*, and the anthropoids, larger diurnal species. **(b)** Anthropoids comprise the monkeys, such as this Capuchin monkey, *Cebus capucinus*, and **(c)** the hominoids, species such as this white-handed gibbon (*Hylobates lar*).

Biological inquiry: What are the defining features of primates?

The second group consists of the larger-brained and diurnal **anthropoids**: the monkeys (**Figure 34.30b**) and the **hominoids** (gibbons, orangutans, gorillas, chimpanzees, and humans) (**Figure 34.30c**).

What differentiates monkeys from hominoids? Most monkeys have tails, whereas hominoids do not. In addition, the shoulder bones in the two groups have a different structure; while hominoids can swing from branch to branch, monkeys cannot. Instead, monkeys run along the tops of branches and their movements are more like other four-footed mammals. The 13 species of hominoids are split into two groups: the lesser apes (family Hylobatidae), or the gibbons; and the greater apes (family Hominidae), or the orangutans, gorillas, chimpanzees, and humans. The lesser apes are strictly arboreal, while the greater apes often descend to the ground to feed.

Humans Evolved from Ancestral Primates

Although humans are closely related to chimpanzees and gorillas, they did not evolve directly from them. Rather, all hominoid species shared a common ancestor. Recent molecular studies show that gorillas, chimpanzees, and humans are closer to one another than gibbons and orangutans, so scientists have split the hominoids into groups, including the Ponginae (orangutans) and the Homininae (gorillas, chimpanzees, and humans and their ancestors). In turn, the Homininae are split into three tribes: the Gorillini (gorillas), the Panini (chimpanzees), and the Hominini (humans and their ancestors).

About 6 million years ago in Africa, a lineage that led to humans began to separate from other primate lineages. The evolution of humans should not be viewed as a neat, stepwise progression from one species to another. Rather, human evolution, like the evolution of most species, can be visualized more like a tree, with one or two **hominin** species (extinct and modern forms of humans) likely coexisting at the same point in time, with some eventually going extinct and some giving rise to other species (**Figure 34.31**).

One of the key characteristics differentiating hominins from other apes is that hominins walked on two feet, that is, they are **bipedal**. At about the time when hominins diverged from other ape lineages, the Earth's climate had cooled and the forests of Africa had given way to grassy savannas. Now that the hands were not engaged in climbing, bipedalism could have been selected for, as it is a more energetically economical form of walking. A bipedal method of locomotion and upright stance may also have been advantageous in allowing hominins to peer over the tall grass of the savanna.

In any event, bipedalism resulted in many anatomical changes in hominins. First, the opening of the skull where the spinal cord enters shifted forward, allowing the spine to be more directly underneath the head. Second, the hominin pelvis became broader to support the additional weight, and third, the lower limbs, used for walking, became relatively larger than those in other apes. These are also the types of anatomical changes paleontologists look for in the fossil record to help determine whether fossil remains are hominin. The earliest group

Unlike
they are
ventral
tially, th
simply
marsup
long tim
   All
class Eu
mals (**F**
it is not
rians, fe
in mars
the you
fetus w

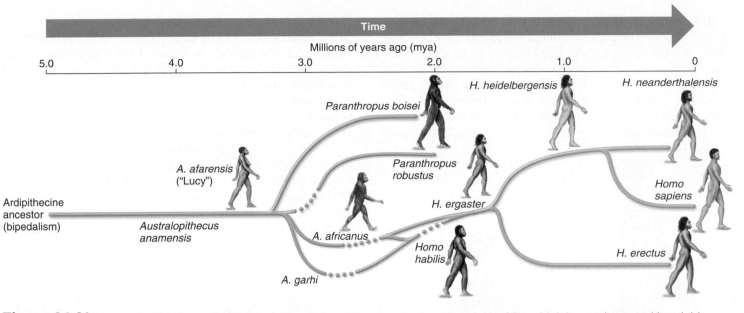

Figure 34.31 **Human evolution.** In this human family tree (based on the ongoing work of Donald Johanson), several hominid species lived contemporaneously with one another, but only one lineage gave rise to modern humans (*Homo sapiens*).

of hominins included several species of a smaller-brained genus, *Australopithecus*, one of which gave rise to the genus *Homo*, the humans.

Australopithecines Since 1924, when the first fossil australopithecine ("southern ape") was found in South Africa, hundreds of fossils of this group have been unearthed all over southern and eastern Africa, the areas where fossil deposits are best exposed to paleontologists. This was a widespread group, with at least six species. Compared to modern humans, all were relatively small, about 1–1.5 m in height, and possessed a facial structure and brain size similar to those of a chimp. Females were much smaller than males, a condition known as sexual dimorphism. In 1974, paleontologist Donald Johanson unearthed the skeleton of a female *Australopithecus afarensis*, dubbed Lucy (the Beatles' song "Lucy in the Sky with Diamonds" was playing in the camp the night when Johanson was sorting the unearthed bones). Over 40% of the skeleton had been preserved, enough to provide a good idea of the physical appearance of australopithecines. When compared to modern humans, it is striking how small australopithecines were (**Figure 34.32**). Examination of the bones revealed that *A. afarensis* walked on two legs. Two of the larger species now considered to be a separate genus, *Paranthropus*, weighed about 40 kg and lived contemporaneously with australopithecines and members of *Homo* species. Both *Paranthropus* species died out rather suddenly about 1.5 million years ago.

Modern Humans and the Genus **Homo** In the 1960s, paleontologist Louis Leakey found hominin fossils estimated to be about 2 million years old in Olduvai Gorge, Tanzania. Two particularly interesting facts stand out about these fossils. First, reconstruction of the skull showed an increased brain size, about 680 cubic centimeters (cc) compared to the 500 cc of *Australopithecus*.

Second, the fossils were found with a wealth of stone tools. As a result, Leakey assigned the fossils to a new species, *Homo habilis*, meaning "handyman." The discovery of several more *Homo* fossils followed, but there have been no extensive finds, as there were with Lucy. This makes it difficult to determine which *Australopithecus* lineage gave rise to the *Homo* lineage (see Figure 34.31), and scientists remain divided on this point.

One of the most important species of *Homo* is believed to be *Homo ergaster*, a hominin with a "human-looking" face and skull, with downward-facing nostrils. *H. ergaster* also evolved in Africa and is thought to have given rise to many species, including *Homo erectus*, *Homo heidelbergensis*, *Homo neanderthalensis*, and *Homo sapiens*. *H. ergaster* is believed to be a direct ancestor of modern humans, with *Homo heidelbergensis* viewed as an intermediary step.

H. erectus was a large hominin, as large as a modern human but with heavier bones and a smaller brain capacity of between 750 and 1,225 cc. Fossil evidence shows that *H. erectus* was a social species that used tools, hunted animals, and cooked over fires. *H. erectus* spread out of Africa soon after it appeared, over a million years ago, and fossils have been found as far away as eastern Asia.

Homo heidelbergensis gave rise to two species, *H. neanderthalensis* and *H. sapiens*. *H. neanderthalensis* was named for the Neander Valley of Germany, where the first fossils of its type were found. Neanderthals were a shorter, stockier species than modern humans, with a more massive skull and large brain size of about 1,450 cc, perhaps associated with their bulk. Males were about 168 cm (5 ft 6 in) in height and would have been very strong by modern standards. However, about 30,000 years ago, this species was replaced by another hominin species, *H. sapiens* ("wise man"), our own species. *H. sapiens* was a taller, lighter-weight species with a slightly smaller brain capacity of about 1,350 cc.

Paleontologists remain divided as to whether *H. sapiens* evolved in Africa and spread from there to other areas of the world, or whether premodern humans such as *H. ergaster* migrated from Africa and evolved to become modern humans in different parts of the world. The first model, the Out of Africa hypothesis, posits that the migration of hominins, after evolution in Africa, happened at least twice, once for *H. ergaster* and once for *H. sapiens*, with *H. sapiens* gradually replacing

H. ergaster and other intermediate species such as *H. erectus* and *H. neanderthalensis* in other parts of the world. Some scientists find this difficult to accept and suggest that human groups have evolved from *H. ergaster* populations in a number of different parts of the world, a model known as the multiregional hypothesis. According to this hypothesis, gene flow between neighboring populations prevented the formation of several different species.

Studies of human mitochondrial DNA, which occurs only in the cellular organelles called mitochondria, show that all modern people share a common ancestor that dates to about 170,000 years ago. This evidence is consistent with the Out of Africa model, since the common ancestor would have to be much older than that to support the multiregional hypothesis. Furthermore, recent analyses of DNA from Neanderthal bones show it to be quite distinct from the DNA of *H. sapiens*. This also suggests that there was no interbreeding of Neanderthals with *H. sapiens* that migrated into Europe.

Evidence overall appears to support the Out of Africa hypothesis, in which *H. ergaster* evolved in Africa and spread to Europe and Asia. *H. erectus* evolved in Asia and *H. neanderthalensis* evolved in Europe. Both species shared the same fate, extinction at the hands of another species, the "wise man," *H. sapiens*. Many, but not all researchers think that humans evolved in Africa about 170,000 years ago from *H. heidelbergensis* and then spread into the rest of the world. Recently, scientists have suggested that a genetic mutation in modern humans some 50,000 years ago enabled increased problem-solving skills and cognitive abilities, which may have allowed them to succeed at the expense of Neanderthals and others.

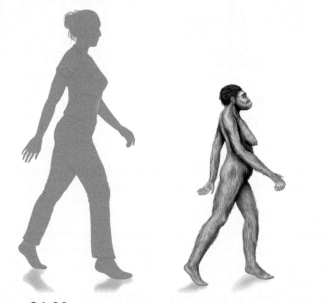

Figure 34.32   Australopithecines.   These ancestors of modern humans, as illustrated by this reconstruction of the famous fossil "Lucy," were only about 1 m tall and weighed around 18 kg.

Chapter Summary

34.1 The Craniates

- The subphylum Vertebrata is the largest and most dominant group of chordates, occupying nearly all of Earth's major habitats. (Figure 34.1)

- Craniates have two defining characteristics that distinguish them from other nonvertebrate chordates: a cranium and a neural crest. The hagfish (class Myxini) is considered the most primitive living craniate. (Figure 34.2)

34.2 Characteristics of Vertebrates

- Vertebrates have several characteristic features, including a vertebral column, endoskeleton of cartilage or bone, and internal organs. Of the 11 different vertebrate classes, 5 are classes of fish, reflecting the dominance of fish. (Table 34.1)

34.3 The Fishes

- A critical innovation in vertebrate evolution is the hinged jaw, which first developed in fish. Ostracoderms, a name for several extinct classes of fish, and lampreys, eel-like fish, lack a hinged jaw. The hinged jaw evolved from cartilaginous gill arches. (Figures 34.3, 34.4, 34.5)

- The chondrichthyans (sharks, skates, and rays) have a skeleton composed of flexible cartilage and powerful appendages called fins. They are active predators with acute senses and were among the earliest fish to develop teeth. (Figure 34.6)

- Bony fish consist of the Actinopterygii (ray-finned fish, the most species-rich class), Actinistia (coelacanths), and the Dipnoi (lungfish). In Actinopterygii, the fins are supported by thin, flexible rays and moved by muscles. (Figures 34.7, 34.8)

- The lobe fins comprise the lobe-finned fish (Actinistia and Dipnoi) and the tetrapods. In the lobe-finned fish, the fins are part of the body. (Figures 34.9, 34.10)

34.4 Amphibia: The First Tetrapods

- Fossils record the evolution of lobe-finned fishes to fishes with four limbs (*Acanthostega*). Recent research has shown that relatively simple mutations control large changes in limb development. (Figures 34.11, 34.12, 34.13)

- Amphibians live on land but return to the water to reproduce. The larval stage undergoes metamorphosis, losing gills and tail for lungs and limbs. (Figure 34.14)

- The majority of amphibians belong to the order Anura (frogs and toads). Other orders are the Caudata (salamanders) and Gymnophiona (caecilians). (Figure 34.15)

34.5 Amniotes: The Appearance of Reptiles

- The amniotic egg permitted animals to become fully terrestrial. Other critical innovations included desiccation-resistant skin, thoracic breathing, water-conserving kidneys, and internal fertilization. (Figure 34.16)

- Living reptilian classes include the Testudines (turtles), Lepidosauria (lizards and snakes), Crocodilia (crocodiles), and Aves (birds). Other reptilian classes, all of which are extinct, include two classes of dinosaurs (Ornithischia and Saurischia). (Figures 34.17, 34.18, 34.19, 34.20)

34.6 Birds

- The differences in body coverings of different vertebrate classes (scales, feathers, hair) are caused by relatively simple changes in the levels of regulatory molecules. (Figure 34.21)

- Three species—*Archaeopteryx*, *Caudipteryx*, and *Confuciusornis*—help trace a lineage from dinosaurs to birds. (Figure 34.22)

- The four key characteristics of birds are feathers, a lightweight skeleton, air sacs, and reduced organs. Birds are the most species-rich class of terrestrial vertebrates. (Figures 34.23, 34.24, Table 34.2)

34.7 Mammals

- The distinguishing characteristics of mammals are the possession of mammary glands, hair, specialized teeth, and an enlarged skull. Other unique characteristics of some mammals are the ability to digest plants and possession of horns or antlers (Figures 34.25, 34.26, 34.27)

- Three subclasses exist: the Prototheria (monotremes), Metatheria (marsupials), and Eutheria (placental mammals). (Figure 34.28, Table 34.3)

- Many defining characteristics of primates relate to their tree-dwelling nature and include grasping hands, large brain, nails instead of claws, and binocular vision. (Figures 34.29, 34.30)

- About 6 million years ago in Africa, a lineage that led to humans began to separate from other primate lineages. A key characteristic of hominins (extinct and modern humans) is bipedalism.

- Human evolution can be visualized like a tree, with one or two hominin species coexisting at the same point in time, with some eventually going extinct and some giving rise to other species. (Figures 34.31, 34.32)

- The Out of Africa hypothesis posits that the migration of hominins from Africa happened at least twice, with *Homo sapiens* gradually replacing other hominin species in other parts of the world. The multiregional hypothesis posits that human groups evolved in a number of different parts of the world, a model known as the multiregional hypothesis.

TEST YOURSELF

1. Which of the following is *not* a defining characteristic of craniates?
   a. cranium
   b. neural crest
   c. two clusters of *Hox* genes
   d. protective housing around the brain
   e. cephalization

2. The hinged jaw is an important adaptation in vertebrates that allowed for
   a. movement onto land.
   b. evolution of more complex body development.
   c. improvement in prey capture.
   d. evolution of speech.
   e. all of the above.

3. The presence of a bony skeleton, an operculum, and a swim bladder are all defining characteristics of
   a. Myxini.
   b. lampreys.
   c. Chondrichthyes.
   d. bony fishes.
   e. amphibians.

4. Organisms that lay eggs are said to be
   a. oviparous.
   b. ovoviparous.
   c. viviparous.
   d. placental.
   e. none of the above.

5. Adaptations in animals that allow for life on land include
   a. a waxy cuticle.
   b. strong limbs that provide movement.
   c. internal fertilization.
   d. b and c only.
   e. all of the above.

6. In some amphibians, the adult retains certain larval characteristics, which is known as
   a. metamorphosis.
   b. parthenogenesis.
   c. cephalization.
   d. paedomorphosis.
   e. hermaphrodism.

7. The membrane of the amniotic egg that serves as a site for waste storage is
   a. the amnion.
   b. the yolk sac.
   c. the allantois.
   d. the chorion.
   e. the albumin.

8. Which of the following is *not* a distinguishing characteristic of birds?
   a. amniotic egg
   b. feathers
   c. air sacs
   d. lack of certain organs
   e. lightweight skeletons

9. Placental mammals that gestate their young for a prolonged time are
   a. monotremes.
   b. marsupials.
   c. eutherians.
   d. therapsids.
   e. all of the above.

10. *Homo sapiens* are believed to have evolved most recently from
    a. *Homo erectus*.
    b. *Homo neanderthalensis*.
    c. *Homo habilis*.
    d. *Homo ergaster*.
    e. *Australopithecus africanus*.

CONCEPTUAL QUESTIONS

1. What are the two defining characteristics that distinguish craniates from other nonvertebrate chordates?

2. Explain the function of the lateral line and the operculum.

3. List the four extraembryonic membranes in the amniotic egg and explain the function of each.

EXPERIMENTAL QUESTIONS

1. What was the purpose of the study conducted by A. P. Davis and colleagues?

2. How were the researchers able to study the effects of individual genes?

3. Explain the results of the experiment shown in Figure 34.13 and how this relates to limb development in vertebrates.

COLLABORATIVE QUESTIONS

1. Discuss the characteristics of the vertebrates.

2. Discuss the three different ways mammals bring their offspring into the world.

www.brookerbiology.com

This website includes answers to the Biological Inquiry questions found in the figure legends and all end-of-chapter questions.

<u>Notes</u>

Notes

<u>Notes</u>

<u>Notes</u>

Notes

54

AN INTRODUCTION TO ECOLOGY AND BIOMES

CHAPTER OUTLINE

What controls the densities of these flowering plants at Glacier National Park, Montana? Is it temperature or rainfall? Or is it the availability of pollinators, herbivory by insects or vertebrates, or competition with other plant species for resources? Ecology seeks to answer questions such as these.

In 2006, a study led by J. Alan Pounds of the Monteverde Cloud Forest Preserve in Costa Rica reported that fully two thirds of the 110 species of harlequin frogs in mountainous areas of Central and South America had become extinct over the previous 20 years. The researchers noted that populations of other species, such the Panamanian golden frog (*Atelopus zeteki*), had been greatly reduced (**Figure 54.1**). The question was why. The culprit was identified as a disease-causing fungus, *Batrachochytrium dendrobatidis*, but this study implicated global warming as the agent causing outbreaks of the fungus.

Figure 54.1 Diminishing and disappearing populations. Population sizes of the Panamanian golden frog (*Atelopus zeteki*) have diminished greatly over the past 20 years, while populations of many other species of harlequin frogs have disappeared entirely. Ecologists are investigating the reasons for this decline.

One effect of global warming is to increase the cloud cover, which reduces daytime temperatures and raises nighttime temperatures. Researchers believe that this combination has created favorable conditions for the spread of *B. dendrobatidis*, which thrives in cooler temperatures. Pounds, the team's lead researcher and an ecologist, was quoted as saying, "Disease is the bullet killing frogs, but climate change is pulling the trigger."

Ecology is the study of interactions among organisms and between organisms and their environments. The interactions among organisms are called **biotic** interactions, while those between organisms and their nonliving environment are termed **abiotic** interactions. These interactions in turn govern the population densities of plants and animals and the numbers of species in an area. In this first chapter of the ecology unit, we will introduce you to the four broad areas of ecology: organismal, population, community, and ecosystems ecology. Next, we will explore how ecologists approach and conduct their work. We will then turn our focus to abiotic interactions and examine the effects of factors such as temperature, water, light, pH, and salt concentrations on the distributions of organisms. We conclude with a consideration of climate and its large influence on the major types of habitats, called biomes, where organisms are found.

Ecological studies have important implications in the real world, as will be amply illustrated by examples discussed throughout the unit. However, a distinction must be made between ecology and **environmental science**, the application of ecology to real-world problems. To use an analogy: Ecology is to environmental science as physics is to engineering. Both physics and ecology provide the theoretical framework on which to pursue more applied studies. Engineers rely on the principles of physics to build bridges. Environmental scientists rely on the principles of ecology to solve environmental problems.

Ecology describes the necessary framework for understanding how populations are affected by features of the physical environment, like temperature and moisture, and by other organisms. In this unit, we'll also learn how plants compete with one another, how herbivores affect plant abundance, and how natural enemies impact prey populations. We'll examine the effects of humans on the environment, including pollution, global warming, and the introduction of exotic species of plants and animals.

Ecologists are among the best equipped scientists to study these phenomena. Before 1960, the field of ecology was dominated by taxonomy, natural history, and speculation about observed patterns. An ecologist's tools of the trade might have included sweep nets, quadrats (small, measured plots of land used to sample living things), and specimen jars. Since that time, an explosion in the number of ecological studies has occurred, and ecologists have become active in investigating environmental change on regional and global scales. Ecologists have embraced experimentation and adapted concepts and methods derived from agriculture, physiology, biochemistry, genetics, physics, chemistry, and mathematics. Now an ecologist's equipment is just as likely to include portable computers, satellite-generated images, and chemical autoanalyzers.

54.1 The Scale of Ecology

Ecology ranges in scale from the study of an individual organism through the study of populations to the study of communities and ecosystems (**Figure 54.2**). In this section, we introduce each of the broad areas of organismal, population, community, and ecosystems ecology. We will provide an investigation that helps illuminate the field of population ecology

and conclude with an exploration of how ecologists conduct their experiments.

Organismal Ecology Investigates How Adaptations and Choices by Individuals Affect Their Reproduction and Survival

Organismal ecology can be divided into two subdisciplines. The first, **physiological ecology**, investigates how organisms are physiologically adapted to their environment and how the environment impacts the distribution of species. Much of this chapter discusses physiological ecology.

The second area, **behavioral ecology**, focuses on how the behavior of an individual organism contributes to its survival and reproductive success, which in turn eventually affects the population density of the species. If we see individuals constantly competing, we may view nature as involving a fierce struggle for survival. If we see individuals cooperating with one another and acting in selfless, even altruistic ways, we tend to believe nature is neat and harmonious.

Population Ecology Describes How Populations Grow and Interact with Other Species

Population ecology focuses on groups of interbreeding individuals, called populations. A primary goal is to understand the factors that affect a population's growth and determine its size and density. Although the attention of a population ecologist may be aimed at studying the population of a particular species, the relative abundance of that species is often influenced by its interactions with other species. Thus, population ecology includes the study of **species interactions** such as predation, competition, and parasitism. Much of the theory of ecology is built upon the ecology of populations. Knowing what factors impact populations can help us lessen species endangerment, stop extinctions, and control invasive species.

(a)   (b)   (c)   (d)

Figure 54.2 The scales of ecology. (a) Organismal ecology. What is the temperature tolerance of this desert locust, *Schistocerca gregaria*? (b) Population ecology. What factors influence the growth of desert locust populations in Mauritania, Africa? (c) Community ecology. What factors influence the interaction of species in functional communities? These desert locusts were killed by fungal sprays developed in England. (d) Ecosystems ecology. How does energy flow among organisms, and how do nutrients cycle between organisms and the environment in the desert ecosystem? These desert locusts have died over the ocean and have been washed ashore.

FEATURE INVESTIGATION

Callaway and Aschehoug's Experiments Showed That the Secretion of Chemicals Gives Exotic Plants a Competitive Edge over Native Species

One important topic in the area of population ecology concerns **exotic species**, species moved from a native location to another location, usually by humans. Such species sometimes spread or invade so aggressively that they crowd out native species. Of the 300 most invasive exotic plants in the U.S., over half were brought in for gardening, horticulture, or landscaping purposes. Invasive exotic plants have traditionally been thought to succeed because they have escaped their natural enemies, primarily insects that remained in the country of origin and were not transported to the new locale. One way of controlling exotic species,

therefore, has been to import the plant's natural enemies. This is known as **biological control**. However, new research on the population ecology of diffuse knapweed (*Centaurea diffusa*), an invasive Eurasian plant that has established itself in many areas of North America, suggests a different reason for their success.

Researchers Ragan Callaway and Erik Aschehoug hypothesized that this particular species secretes powerful root exudates called **allelochemicals** that kill the roots of other species, allowing *Centaurea* to proliferate. To test their hypothesis, Callaway and Aschehoug collected seeds of three native Montana grasses, *Koeleria cristata*, *Festuca idahoensis*, and *Agropyron spicata*, and grew them both without other species and with the exotic *Centaurea* species (**Figure 54.3**). As hypothesized, *Centaurea* depressed the biomass of the native grasses.

Figure 54.3   Experimental evidence of the effect of allelochemicals on plant production.

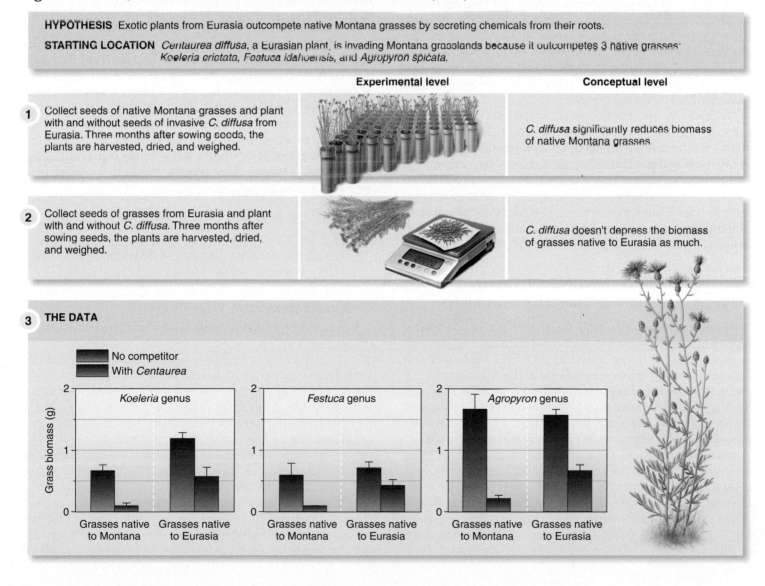

HYPOTHESIS Exotic plants from Eurasia outcompete native Montana grasses by secreting chemicals from their roots.

STARTING LOCATION *Centaurea diffusa*, a Eurasian plant, is invading Montana grasslands because it outcompetes 3 native grasses: *Koeleria cristata*, *Festuca idahoensis*, and *Agropyron spicata*.

Experimental level   **Conceptual level**

1   Collect seeds of native Montana grasses and plant with and without seeds of invasive *C. diffusa* from Eurasia. Three months after sowing seeds, the plants are harvested, dried, and weighed.

C. diffusa significantly reduces biomass of native Montana grasses

2   Collect seeds of grasses from Eurasia and plant with and without *C. diffusa*. Three months after sowing seeds, the plants are harvested, dried, and weighed.

C. diffusa doesn't depress the biomass of grasses native to Eurasia as much.

3   **THE DATA**

When the experiments were repeated with grasses native to Eurasia, *Koeleria laerssenii*, *Festuca ovina*, and *Agropyron cristatum*, the species were affected, but to a significantly lesser degree than the Montana species were.

In other experiments not described in Figure 54.3, Callaway and Aschehoug added activated carbon, which absorbs the chemical excreted by the *Centaurea* roots. With activated carbon added to the soil, the Montana grass species increased in biomass compared to the previous experiments. The researchers concluded that *C. diffusa* outcompetes Montana grasses by secreting

an allelochemical and that Eurasian grasses are not as susceptible to the chemical's effect, since they coevolved with it. If the reason for the success of invasive plants can be attributed to the chemicals they secrete, this calls into question the effectiveness of biological control of exotic weeds by importation of their natural enemies. This study on the population biology of an exotic plant has changed the way we think about why invasive exotic species succeed and could affect the way we attempt to control them in the future.

Community Ecology Focuses on What Factors Influence the Number of Species in a Given Area

We've just seen how a population of an exotic grass may produce allelochemicals that give it a growth advantage over populations of other species. On a larger scale, **community ecology** studies how populations of species interact and form functional communities. In a forest, there are many populations of trees, herbs, shrubs, grasses, the herbivores that eat them, and the carnivores that in turn prey on the herbivores. Community ecology focuses on why certain areas have high numbers of species (that is, are species rich), while other areas have low numbers of species (that is, are species poor).

While ecologists are interested in species richness for its own sake, a link also exists between species richness and community function. Ecologists generally believe that species-rich communities perform better than do species-poor communities. It has also been posited that more species make a community more stable, that is, more resistant to disturbances such as introduced species. Community ecology also considers how species composition and community structure change over time and in particular after a disturbance, a process called succession.

Ecosystems Ecology Describes the Passage of Energy and Nutrients Through Communities

Ecosystems ecology deals with the flow of energy and cycling of nutrients among organisms within a community and between organisms and the environment. As such, it is concerned with both the biotic and abiotic components of the environment. Following this flow of energy and nutrients necessitates an understanding of feeding relationships between species, called food chains. In food chains, each level is called a trophic level and many food chains interconnect to form complex food webs.

The second law of thermodynamics states that in every energy transformation, free energy is reduced because heat energy is lost from the ecosystem in the process, and the entropy of the universe increases. There is, therefore, a unidirectional flow of energy through an ecosystem, with energy dissipated at every step. An ecosystem needs a recurring input of energy from an external source—in most cases, the sun—to sustain itself.

In contrast, chemicals such as nitrogen do not dissipate and constantly cycle between abiotic and biotic components of the environment, often becoming more concentrated in organisms in higher trophic levels.

Ecological Methods Focus on Observation and Experimentation

How do ecologists go about studying their subject? Let's suppose you are employed by the United Nations' Food and Agricultural Organization (FAO), which operates internationally out of Rome, Italy. As an ecologist, you are charged with finding out what causes outbreaks of locusts, a type of grasshopper that periodically erupts in Africa and other parts of the world, destroying crops and other vegetation. First of all, you might draw up a possible web of interaction between the factors that could affect locust population size (**Figure 54.4**). These interactions are many and varied, and they include

- abiotic factors such as temperature, rainfall, wind, and soil pH;
- host plants, including increases or decreases in either the quality or quantity of the plants;
- predators, including bird predators, insect parasites, and bacterial parasites;
- competitors, including other insects and larger vertebrate grazers.

With such a vast array of factors to be investigated, where is the best place to start? As discussed in Chapter 1, hypothesis testing involves a five-stage process: (1) observations, (2) hypothesis formation, (3) experimentation, (4) data analysis, and (5) acceptance or rejection of the hypothesis. In our study of locusts, we begin by careful observation of the organism in its native environment. We can analyze the fluctuations of locusts and determine if the populations vary with fluctuations in the other phenomena such as levels of parasitism, numbers of predators, or food supply. Imagine we found that locust numbers are affected by bird predation levels, and that an inverse relationship exists between predation levels and locust numbers.

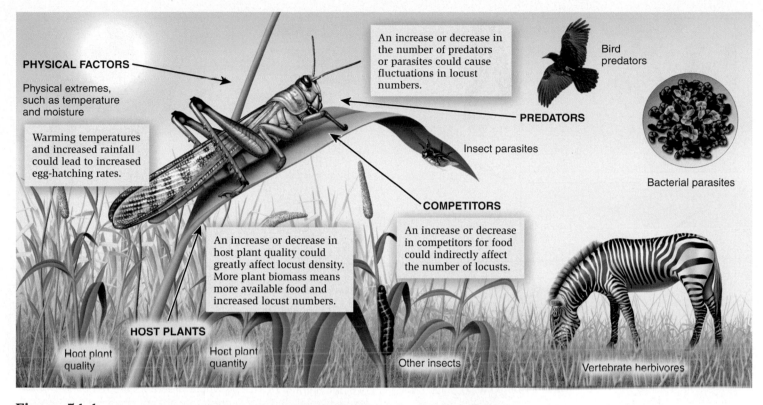

PHYSICAL FACTORS

Physical extremes, such as temperature and moisture

Warming temperatures and increased rainfall could lead to increased egg-hatching rates.

An increase or decrease in the number of predators or parasites could cause fluctuations in locust numbers.

Bird predators

PREDATORS

Insect parasites

Bacterial parasites

COMPETITORS

An increase or decrease in competitors for food could indirectly affect the number of locusts.

An increase or decrease in host plant quality could greatly affect locust density. More plant biomass means more available food and increased locust numbers.

HOST PLANTS

Host plant quality

Host plant quantity

Other insects

Vertebrate herbivores

Figure 54.4   Interaction web of factors that might influence locust population size.

As predation levels increase, locust numbers decrease. If we plotted this relationship graphically, the resulting graph would look like that depicted in **Figure 54.5a**. This result would give us some confidence that predation levels determined locust numbers and this would be our hypothesis. In fact, we would have so much confidence that we could create a "line of best fit" to represent a summary of the relationship between these two variables, as shown in **Figure 54.5b**.

However, if the points were not highly clustered, as in **Figure 54.5c**, we would have little confidence that predation affects locust density. Many statistical tests are used to determine whether or not two variables are significantly correlated. In the studies in this unit, unless otherwise stated, most graphs like Figure 54.5b imply that a meaningful relationship exists between the two variables. We call this type of relationship a significant **correlation**. If locust density shows a linear relationship with predation, we say that locust density is correlated with predation.

We have to be cautious when forming conclusions based on correlations. For example, large numbers of locusts could be associated with large, dense plants. We might conclude from this that food availability controls locust density. However, an alternative conclusion would be that large plants provide locusts refuge from bird predators, which cannot attack them in the dense interior. While it would appear that biomass affects locust density by providing abundant food, in actuality, predation would still be the most important factor affecting locust density.

Thus, correlation does not always mean causation. For this reason, after conducting observations, ecologists usually turn to experiments to test their hypotheses.

Continuing with our example, an experiment might involve removing predators from locust populations. If predators are having a significant effect, then removing them should cause an increase in locust numbers. Thus, we would have two groups: a group of locusts with predators removed (the experimental group) and a group of locusts with predators still intact (the control group), with equal numbers of locusts in both groups at the start of the experiment. Any differences in locust population density would be due solely to differences in predation. Reduced predation might be achieved by putting a cage made of chicken wire over and around bushes containing locusts, so that birds are denied access. We could look at locust survivorship over the course of one generation of locust and predator.

Performing the experiments several times is called **replication**. We might replicate the experiment 5 times, 10 times, or even more. At the end of the replications, we would add up the total number of living locusts and calculate the mean. In the experimental group, let's suppose that the surviving numbers of locusts in each replicate are 5, 4, 7, 8, 12, 15, 13, 6, 8, and 10; the mean number surviving would be 8.8. In the control group, which still allows predator access, the numbers surviving might be 2, 4, 7, 5, 3, 6, 11, 4, 1, and 3, with a mean of 4.6. Without predators, the mean number of locusts surviving would therefore be almost double the average number surviving with predators.

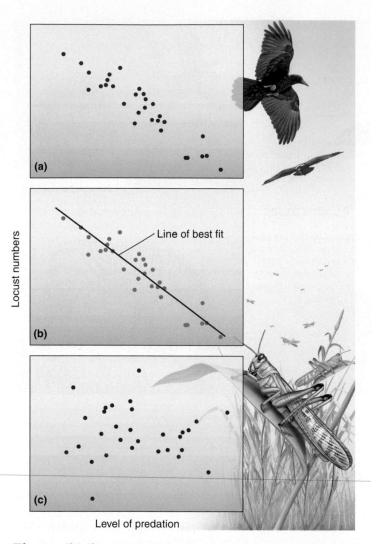

Figure 54.5 How locust numbers might be correlated with predation. (a) Higher locust numbers are found in nature where predation levels are lowest. We can draw a line of best fit (b) to represent this relationship. In (c), the relationship between locust numbers and predation levels might be so weak that we would not have much confidence in a linear relationship between the variables.

Biological inquiry: What would it mean if the line of best fit sloped in the opposite direction?

Our data analysis would give us confidence that predators were indeed the cause of our changes in locust numbers. The results of such experiments can be illustrated graphically by a bar graph (**Figure 54.6**). Ecologists can use a variety of tests to see if these differences are statistically significant. We won't look at the mechanics of these tests, but in this unit, when experimental and control groups are presented as differing, these are considered to be statistically significant differences unless stated otherwise.

By the way, it turns out that predation is not really the primary factor that controls locust populations. The results we have been discussing were hypothetical. Weather, in particular, rain, is the most important feature governing locust population size.

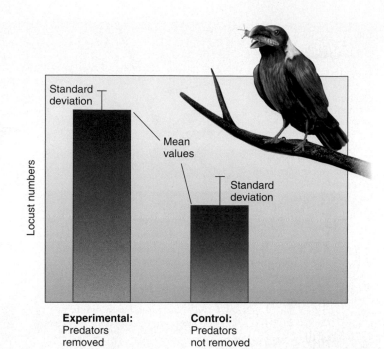

Figure 54.6 Graphic display of hypothetical results of a predator removal experiment. The two bars represent the average number of locusts where predators are removed (experimental) and where predators are not removed (control). The vertical lines (the standard deviations) give an indication of how tightly the individual replicate results are clustered around the mean. The shorter the lines, the tighter the cluster, and the more confidence we have in the result.

Moist soil allows eggs to hatch and provides water for germinating plants, allowing a ready source of food for the hatchling locusts. In fact, physical or abiotic factors such as amount of moisture usually have powerful effects in most ecological systems. In the next part of the chapter, we turn our attention to an examination of the effects of the physical environment on the distribution patterns of organisms.

54.2 The Environment's Impact on the Distribution of Organisms

Both the distribution patterns of organisms and their abundance are limited by physical features of the environment such as temperature, wind, availability of water and light, salinity, pH, and water currents. Some species can tolerate a relatively wide range of environmental conditions, and others only a narrow range, but each species usually functions best over only a limited part of the range known as a species' optimal range or **fundamental niche**. In this section, we will examine these features of the physical environment, focusing on the impact of each on the distribution of organisms.

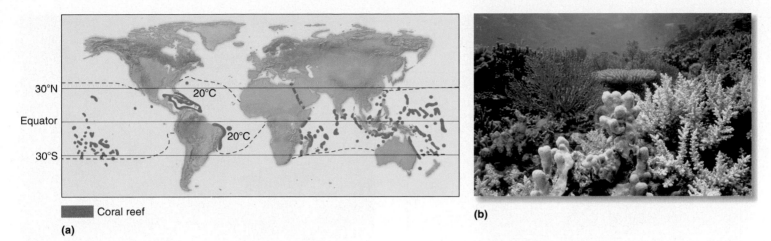

(a)

(b)

Figure 54.7 **Locations of coral reefs.** (a) Coral reef formation is limited to waters bounded by the 20°C isotherm (dashed line), a line where the average daily temperature is 20°C during the coldest month of the year. (b) Coral reef from the Pacific Ocean.

Temperature Has an Important Effect on the Distribution of Plants and Animals

Temperature is perhaps the most important factor in the distribution of organisms because of its effect on biological processes and because of the inability of most organisms to regulate their body temperature precisely. For example, the organisms that form coral reefs secrete a calcium carbonate shell. Shell formation and coral deposition are accelerated at high temperatures but are suppressed in cold water. Coral reefs are therefore abundant only in warm water, and a close correspondence is observed between the 20°C isotherm for the average daily temperature during the coldest month of the year and the limits of the distribution of coral reefs (**Figure 54.7**). An isotherm is a line on a map connecting points of equal temperature. Coral reefs are located between the two 20°C isotherm lines that are formed above and below the equator.

In plants, cold temperature can be lethal because cells may rupture if the water they contain freezes. Frost is probably the single most important factor limiting the geographic distribution of tropical and subtropical plants. In the Sonoran Desert in Arizona, saguaro cacti can easily withstand frost for one night as long as temperatures rise above freezing the following day, but they are killed when temperatures remain below freezing for 36 hours. This means that the cactus's distribution is limited to places where the temperature does not remain below freezing for more than one night (**Figure 54.8**).

The geographic range limits of endothermic animals are also affected by temperature. For example, the eastern phoebe (*Sayornis phoebe*), a small bird, has a northern winter range that coincides with an average minimum January temperature of above −4°C. Such limits are probably related to the energy demands associated with cold temperatures. Cold temperatures mean higher metabolic costs, which are in turn dependent on high feeding rates. Below −4°C, the eastern phoebe cannot feed fast enough or, more likely, find enough food to keep warm.

- - - Boundary of saguaro cactus range

● Temperatures remain below freezing for 1 or more days/year

● Temperatures remain below freezing for less than 0.5 days/year

● No days below freezing on record

Figure 54.8 **Saguaro cacti in freeze-free zones.** A close correspondence is seen between the range of the saguaro cactus (dashed line) and the area in which temperatures do not go below freezing (0°C).

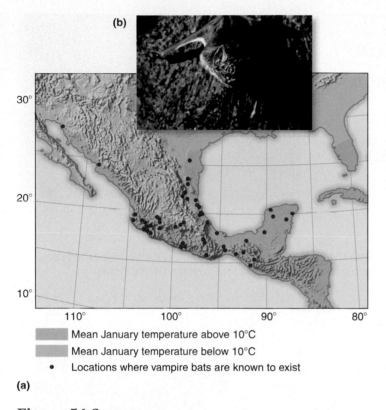

(a)

Mean January temperature above 10°C

Mean January temperature below 10°C

• Locations where vampire bats are known to exist

Figure 54.9 **Geographic range of vampire bats.** **(a)** The northern distribution of the vampire bat in Mexico is limited to areas warmer than the 10°C isotherm in January because of the animal's poor capacity to thermoregulate. Its southern distribution in Argentina and Chile is limited by the same isotherm. **(b)** The vampire bat, *Desmodus rotundus*.

Biological inquiry: Should customs officials in the U.S. be concerned about the entry of vampire bats into the U.S.?

Figure 54.10 Giant sequoia. A park ranger uses a drip torch to ignite a fire at Sequoia National Park in California. Periodic, controlled human-made fires mimic the sporadic wildfires that normally burn natural areas. Such fires are vital to the health of giant sequoia populations, since they serve to open the pine cones and release the seeds.

Similarly, cold temperatures limit the distribution of the vampire bat (*Desmodus rotundus*). The vampire bat is found in an area from central Mexico to northern Argentina. Its range in Mexico is limited to that area where the average minimum temperature in January is above 10°C. Because of the bat's poor capacity for thermal regulation, it cannot survive in areas below that temperature (**Figure 54.9**).

High temperatures are also limiting for many plants and animals because relatively few species can survive internal temperatures more than a few degrees above their metabolic optimum. While we have discussed how corals are sensitive to low temperatures, they are sensitive to high temperatures as well. When temperatures are too high, the symbiotic algae that live within coral die and are expelled, causing a phenomenon known as coral bleaching. Once bleaching occurs, the coral tissue loses its color and turns a pale white. El Niño is a weather phenomenon characterized by a major increase in the water temperature of the equatorial Pacific Ocean. In the winter of 1982–1983, an influx of warm water from the eastern Pacific raised temperatures just 2–3°C for six months, which was enough to kill many of the reef-building corals on the coast of Panama. By May 1983, just a few individuals of one species, *Millepora intricata*, were alive. Scientists are concerned that future climate changes may increase the frequency of coral bleaching.

The ultimate high temperatures that many terrestrial organisms face are brought about by fire. However, some species depend on frequent, low-intensity fires for their reproductive success. The longleaf pine (*Pinus palustris*) of the southeast U.S. produces serotinous cones, which remain sealed by pine resin until the heat of a fire melts them open and releases the seeds. In the west, giant sequoia trees are similarly dependent on periodic low-intensity fires for germination of their seeds. Such fires both enhance the release of seeds and clear out competing vegetation at the base of the tree so that seeds can germinate (**Figure 54.10**). Fire-suppression practices that attempt to protect forests from fires actually can have the opposite result by preventing the regeneration of fire-dependent species. Furthermore, fire prevention can result in more growth of vegetation beneath the canopy (the understory) that may later fuel hotter and more damaging fires.

Keep in mind that it is not mean average temperatures that usually limit the range of species but rather the frequency of occasional extremes, such as freezes for the saguaro cacti. Farmers know this only too well. The frequency and strength of periodic freezes limit the northern distribution of oranges in Florida and the southern distribution of coffee in Brazil, not average temperatures for the coldest months. Experimentally moving organisms outside their normal range and monitoring survivorship is a useful way to establish what factors control the natural limits of species. Of course, such movements should be done only in a carefully controlled manner and when there is no risk of species becoming invasive in their new habitats. Again, physical extremes, such as fires or severe freezes, may be apparent only in isolated years, so ecologists often have to wait many years to determine whether extremes are limiting the distribution and abundance of plants and animals in the field.

Despite the obvious relationships between species distributions and temperature, we need to be cautious about solely relating the two. The temperatures measured for constructing isotherm maps are not always the temperatures that the organisms experience. In nature, an organism may choose to lie in the sun or hide in the shade, both of which affect the temperatures it experiences. Such local variations of the climate within a given area, or **microclimate**, can be important for a particular species. For example, the rufous grasshopper (*Gomphocerippus rufus*) is distributed widely in Europe, but in Great Britain it reaches its northern limit only 150 m from the south coast, where it is restricted to steep, south-facing, and therefore relatively sun-drenched and warm grassy slopes.

Because so many species are limited in their distribution patterns by global temperatures, ecologists are concerned that if global temperatures rise, many species will be driven to extinction or that their geographic ranges will shrink and the location of centers of agriculture and forestry will be altered. The increase in the average temperature of Earth's atmosphere and oceans is called global warming, and it is caused by a process known as the greenhouse effect.

The Greenhouse Effect The Earth is warmed by the **greenhouse effect**. In a greenhouse, sunlight penetrates the glass and raises temperatures, with the glass acting to trap the resultant heat inside. Similarly, solar radiation in the form of short-wave energy passes through the atmosphere to heat the surface of the Earth. At night this energy is radiated from the Earth's warmed surface back into the atmosphere, but in the form of long-wave infrared radiation. Instead of letting it escape back into space, however, atmospheric gases absorb much of this infrared energy and reradiate it to the Earth's surface, causing its temperature to rise further (**Figure 54.11**). Without some type of greenhouse effect, life on Earth would not exist. Global temperatures would be much lower than they are, perhaps averaging only $-17°C$ compared with the existing average of $+15°C$.

The greenhouse effect is caused by a small group of gases, mainly water vapor, that together make up less than 1% of the total volume of the atmosphere. After water vapor, the four most significant greenhouse gases are carbon dioxide, methane, nitrous oxide, and chlorofluorocarbons (**Table 54.1**).

Global Warming Ecologists are concerned that human activities are increasing the greenhouse effect, causing **global warm-**

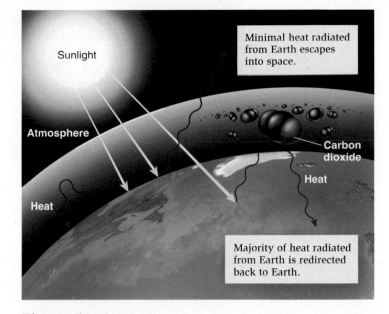

Figure 54.11 **The greenhouse effect.** Solar radiation, in the form of short-wave energy, passes through the atmosphere to heat the Earth's surface. Long-wave infrared energy is radiated back into the atmosphere. Most infrared energy is reflected by atmospheric gasses, including carbon dioxide molecules, back to Earth, causing global temperatures to rise.

ing, a gradual elevation of the Earth's surface temperature. All greenhouse gases have increased in atmospheric concentration since industrial times. The most important of these gases is carbon dioxide (CO_2). As Table 54.1 shows, CO_2 has a lower global warming potential per unit of gas (relative absorption) than any of the other major greenhouse gases, but its concentration in the atmosphere is much higher. Concentrations of atmospheric CO_2 have increased from about 280 ppm (parts per million) in the preindustrial 18th century to 379 ppm in 2005.

To predict the effect of global warming, most scientists focus on a future point, about 2100, when the concentration of atmospheric CO_2 will have doubled—that is, increased to about 700 ppm compared with the late-20th-century level of 350 ppm. Ecologists argue that at that time, average global temperatures will be about $1-6°C$ (about $2-10°F$) warmer than present and will increase an additional $0.5°C$ each decade. This increase in heat might not seem like much, but it is comparable to the warming that ended the last ice age.

Assuming this scenario of gradual global warming is accurate, we need to consider what the consequences on natural and human-made ecosystems might be. Although many species can adapt to slight changes in their environment, the anticipated changes in global climate are expected to occur too rapidly to be compensated for by normal evolutionary processes such as natural selection. Plant species cannot simply disperse and move north or south into the newly created climatic regions that will be suitable for them. Many tree species take hundreds, even thousands, of years for seed dispersal.

Table 54.1	Greenhouse Gases and Their Contribution to Global Warming			
	Carbon dioxide (CO_2)	**Methane (CH_4)**	**Nitrous oxide (N_2O)**	**Chlorofluorocarbons (CFCs)**
Relative absorption per ppm of increase*	1	32	150	>10,000
Atmospheric concentration†	379 ppm	1.78 ppm	315 ppb	0.28–0.54 ppb
Contribution to global warming	50%	19%	4%	15%
Percent from natural sources; type of source	20–30%; volcanoes	70–90%; swamps, gas from termites and ruminants	90–100%; soils	0%
Major human-made sources	Fossil fuel use, deforestation	Rice paddies, landfills, biomass burning, coal and gas exploitation	Cultivated soil, fossil-fuel use, automobiles, industry	Previously industrially manufactured products (for example, as aerosol propellants) but banned in U.S. and the E.U.

*Relative absorption is the warming potential per unit of gas.

†ppm = parts per million, ppb = parts per billion.

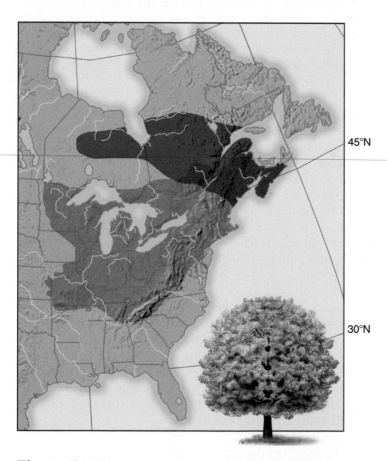

Figure 54.12 **The range of sugar maples could be reduced by global warming.** The present geographic range of the sugar maple (blue shading) and its potential range under doubled CO_2 levels (red shading) in North America. Purple shading indicates the region of overlap, which is the only area where the sugar maple would be found before it spread into its new potential range.

Paleobotanist Margaret Davis predicted that in the event of a CO_2 doubling, the sugar maple (*Acer saccharum*), which is presently distributed throughout the midwestern and northeastern U.S. and southeastern Canada, would die back in all areas except in northern Maine, northern New Brunswick, and southern Quebec (**Figure 54.12**). Of course, this contraction in the tree's distribution could be offset by the creation of new favorable habitats in central Quebec. However, most scientists believe that the climatic zones would shift toward the poles faster than trees could migrate via seed dispersal, hence extinctions would occur. Interestingly, scientists are beginning to be able to genetically modify organisms to change their temperature tolerances.

GENOMES & PROTEOMES

Temperature Tolerance May Be Manipulated by Genetic Engineering

Below-freezing temperatures can be very damaging to plant tissue, either killing the plant or greatly reducing its productivity. Frost injury causes losses to agriculture of more than $1 billion annually in the U.S. While frost has been considered an unavoidable result of subfreezing temperatures, genetic engineering is beginning to change this view.

Between 0°C and −40°C, pure water will be a liquid unless provided with an ice nucleus or template on which an ice crystal can be built. Researchers discovered that some bacteria commonly found on leaf surfaces act as ice nuclei, triggering the formation of ice crystals and eventually causing frost damage. The genes that confer ice nucleation have been identified, isolated, and prevented from working in an engineered strain of the bacteria *Pseudomonas syringae*. When this strain is allowed to colonize strawberries, frost damage is greatly reduced, and plants can withstand an additional 5°C drop in temperature before frost forms. The promise of this technique for increasing agricultural yields and altering normal plant-distribution patterns is staggering.

At the other end of the temperature spectrum, heat shock proteins (HSPs) function as "molecular chaperones" to help organisms cope with the stress of high temperatures. At high temperatures, proteins may either unfold or bind to other proteins to form misfolded protein aggregations. HSPs act to prevent this type of event from taking place or to help aggregated proteins regain their normal configuration. HSPs normally constitute only about 2% of the cell's soluble protein content, but this can increase to 20% when a cell is stressed, whether by heat, cold, drought, or other stressors. In fact, HSPs are extremely common and are found in all organisms, from bacteria to plants and animals.

In the tropics, high temperatures can substantially decrease the growth rates and productivity of many crop species. There is now substantial interest in identifying crop strains with naturally high HSP levels and identifying thermally tolerant varieties for use in crop-breeding programs. Given the seemingly inevitable prospect of global warming, such research seems particularly timely.

Wind Can Amplify the Effects of Temperature

Wind is created by temperature gradients. As air heats up, it becomes less dense and rises. As hot air rises, cooler air rushes in to take its place. Hot air rising in the tropics is replaced by cooler air flowing in from more temperate regions, thereby creating northerly or southerly winds.

Wind affects living organisms in a variety of ways. It increases heat loss by evaporation and convection (the windchill factor). Wind also contributes to water loss in organisms by increasing the rate of evaporation in animals and transpiration in plants. For example, the tree line in alpine areas is often determined by a combination of low temperatures and high winds such that transpiration exceeds water uptake.

Winds can also intensify oceanic wave action, with resulting effects for organisms. On the ocean's rocky shore, seaweeds survive heavy surf by a combination of holdfasts and flexible structures. The animals of this zone have powerful organic glues and muscular feet to hold them in place (**Figure 54.13**).

The Availability of Water Has Important Effects on the Abundance of Organisms

Water has an important effect on the distribution of organisms. Cytoplasm is 85–90% water, and without moisture there can be no life. As noted in Chapter 3, water performs crucial functions in all living organisms. It acts as a solvent for chemical reac-

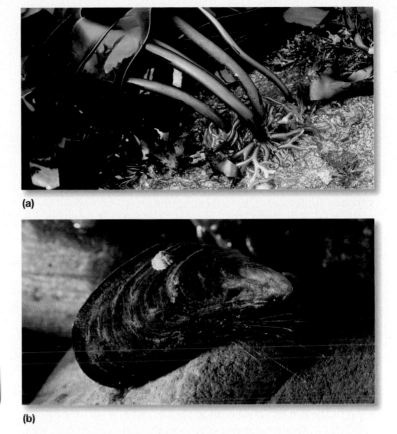

(a)

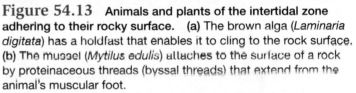

(b)

Figure 54.13 **Animals and plants of the intertidal zone adhering to their rocky surface.** (a) The brown alga (*Laminaria digitata*) has a holdfast that enables it to cling to the rock surface. (b) The mussel (*Mytilus edulis*) attaches to the surface of a rock by proteinaceous threads (byssal threads) that extend from the animal's muscular foot.

tions, takes part in the reactions of hydrolysis, is the means by which animals eliminate wastes, and is used for support in plants and in some invertebrates as part of a hydrostatic skeleton.

The distribution patterns of many plants are limited by available water. Some plants, such as the water tupelo tree (*Nyssa aquatica*) in the southeast U.S., do best when completely flooded and are thus found predominantly in swamps. Others, for example, coastal plants that grow on sand dunes, experience very little fresh water. Their roots penetrate deep into the sand to extract moisture. In cold climates, water can be present but locked up as permafrost and, therefore, unavailable—this is termed a frost-drought situation. Alpine trees can be affected by frost drought. The trees stop growing at a point on the mountainside where they cannot take up enough moisture to offset transpiration losses. This point, known as the timberline, is readily apparent on many mountainsides. Not surprisingly, the density of many plants is limited by the availability of water. For example, a significant correlation is observed between increased rainfall and increased creosote bush density in the Mojave Desert.

Animals face problems of water balance, too, and their distribution and population density can be strongly affected by water availability. Because most animals depend ultimately on plants as food, their distribution is intrinsically linked to those of their food sources. Such a phenomenon regulates the number of buffalo (*Syncerus caffer*) in the Serengeti area of Africa. In this area, grass productivity is related to the amount of rainfall in the previous month. Buffalo density is governed by food availability, so a significant correlation is found between buffalo density and rainfall (**Figure 54.14**). The only exception occurs in the vicinity of Lake Manyara, where groundwater promotes plant growth.

The importance of water in limiting animal population density was underscored by an extraordinary event caused by El Niño. During the 1982–1983 rainy season, rainfall on Isla Genovesa in the Galápagos Islands off of Ecuador increased from its normal 100–150 mm to 2,400 mm. Plants responded with prodigious growth, and Darwin's finches (*Geospiza* spp.) bred up to eight times rather than their normal maximum of three, probably because of increased abundance of fruits and seeds.

Light Can Be a Limiting Resource for Plants and Algae

Because light is necessary for photosynthesis, it can be a limiting resource for plants. However, what may be sufficient light to support the growth of one plant species may be insufficient for another. Many plant species grow best in shady conditions, such as eastern hemlock (*Tsuga canadensis*). Its saplings grow in the understory below the forest canopy, reaching maximal photosynthesis at one-quarter of full sunlight. Other plants, like sugarcane (*Saccharum officinarum*) or the desert shrub *Larrea*, continue to increase their photosynthetic rate as light intensity increases.

One reason photosynthetic rates vary among plants is related to three different biochemical pathways by which carbon

fixation can occur: C_3, C_4, and CAM (see Chapter 8). Because C_4 plants such as sugarcane grow faster in areas with high daytime temperatures and intense sunlight, those species outcompete C_3 plants and are more common in tropical areas than in temperate areas. On the other hand, in cooler, cloudier temperate areas, C_3 species can tolerate lower light and live in areas where C_4 plants cannot. Thus, C_3 plants are much more common in areas outside the tropics. CAM plants, such as some desert succulents, are the opposite of typical plants in that they open their stomata to take up CO_2 at night, presumably as an adaptation to minimize water loss in the day. The absorbed CO_2 is stored as malic acid, which is then used to complete photosynthesis during the day. These plants are adapted to live in very dry desert areas where little else can grow.

In aquatic environments, light may be an even more limiting factor because water absorbs light, preventing photosynthesis at depths greater than 100 m. Most aquatic plants and algae are limited to a fairly narrow zone close to the surface, where light is sufficient to allow photosynthesis to exceed respiration. This zone is known as the **euphotic zone**. In marine environments, seaweeds at greater depths have wider thalli (leaflike light-gathering structures) than those nearer the surface, because wide thalli can collect more light. In addition, in aquatic environments, plant color changes with depth. At the surface, plants and algae appear green, as they are in terrestrial conditions, because they absorb red and blue light, but not green. At greater depths, red light is mostly absorbed by water, leaving predominantly blue-green light. Red algae occur in deeper water because these possess pigments that enable them to utilize blue-green light efficiently (**Figure 54.15**).

The Concentration of Salts in Soil or Water Can Be Critical

Salt concentrations vary widely in aquatic environments and have a great impact on osmotic balance in animals. Oceans contain considerably more dissolved minerals than rivers. Oceans continually receive the nutrient-rich waters of rivers, and the sun evaporates pure water from ocean surfaces, making concentrations of minerals such as salt even higher.

The phenomenon of osmosis influences how living organisms cope with different environments. Freshwater fish cannot live in salt water, and saltwater fish cannot live in fresh water. Each employs different mechanisms to maintain an osmotic balance with their environment. Freshwater fish are hyperosmotic (having a higher concentration of ions) to their environment and tend to gain water by osmosis as it passes over the thin tissue of the gills and mouth. To counter this, the fish continually eliminate water in the urine. However, to avoid losing all dissolved ions, many ions are reabsorbed into the bloodstream at the kidneys. Many marine fish are hypoosmotic (having a lower concentration of ions) to their environment and tend to lose water as seawater passes over the mouth and gills. They drink more water to compensate for this loss, but the water contains a higher concentration of salts, which must then be excreted at the gills and kidneys (look back to Chapter 49).

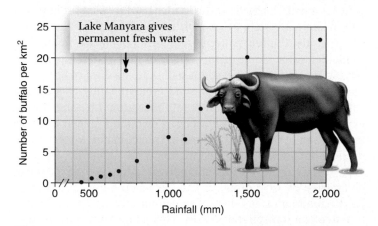

Figure 54.14 **The relationship between the amount of rainfall and the density of buffalo.** In the Serengeti area of Africa, buffalo density is very much dependent on grass availability, which itself is dependent on annual rainfall. The main exception is where there is permanent water, such as Lake Manyara, where greater water availability leads to greater grass growth and buffalo densities.

(a)

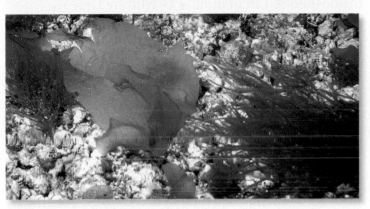

(b)

Figure 54.15   Green algae of surface marine waters compared to red algae of deeper marine waters.   (a) In the eastern Pacific Ocean, off the coast of California, these giant kelp floating at the ocean surface are green, just like terrestrial plants. (b) In contrast, at 75-m depth, in the McGrail Bank off of the Gulf of Mexico, most seaweeds are pink and red because the pigments can absorb the blue-green light that occurs at such depths.

Figure 54.16   Plant adaptations for salty conditions. Special salt glands in *Spartina* leaves exude salt, enabling this grass to exist in saline intertidal conditions.

(a)   (b)

Figure 54.17   Species-rich floras of chalk grassland compared to species-poor floras of acid soils.   (a) At Mount Caburn, in the lime-rich chalk hills of Sussex County, England, there is a much greater variety of plant and animal species than at (b) a heathland site in England. Heathlands are a product of thousands of years of human clearance of natural forest areas and are characterized by acidic, nutrient-poor soils.

Salt in the soil also affects the growth of plants. In arid terrestrial regions, crystalline salt accumulates in soil where water evaporates. This can be of great significance in agriculture, where continued watering in arid environments, together with the addition of salt-based fertilizers, greatly increases salt concentration in soil and reduces crop yields. A very few terrestrial plants are adapted to live in saline soil along seacoasts. Here the vegetation consists largely of **halophytes**, species that can tolerate higher salt concentrations in their cell sap than regular plants. Species such as mangroves and *Spartina* grasses have salt glands that excrete salt to the surface of the leaves, where it forms tiny white salt crystals (**Figure 54.16**).

The pH of Soil or Water Can Limit the Distribution of Organisms

As discussed in Chapter 2, the pH of water can be acidic, alkaline, or neutral. Variation in pH can have a major impact on the distribution of organisms. Normal rainwater has a pH of about

5.6, which is slightly acidic because the absorption of atmospheric CO_2 and SO_2 into rain droplets forms carbonic and sulphuric acids. However, most plants grow best at a soil water pH of about 6.5, a value at which soil nutrients are most readily available to plants. Only a few genera, such as rhododendrons (*Rhododendron*) and azaleas (*Azalea*), can live in soils with a pH of 4.0 or less. Furthermore, at a pH of 5.2 or less, nitrifying bacteria do not function properly, which prevents organic matter from breaking down. In general, soils containing chalk and limestone, the so-called high-lime soils, have a higher pH and sustain a much richer flora (and associated fauna) than do acidic soils (**Figure 54.17**).

Generally, the number of fish and other species also decreases in acidic waters. The optimal pH for most freshwater fish and bottom-dwelling invertebrates is between 6.0 and 9.0. Acidity in lakes increases the amount of toxic metals, such as aluminum, mercury, and lead, which can leach into the water from surrounding soil and rock. Both too much mercury and too much aluminum can interfere with gill function, causing fish to suffocate. Interestingly, Florida has a great number of acidic lakes, more than any other region in the U.S., but most of these are naturally acidic, and most fish found in these lakes are tolerant of the relatively low pH levels.

The susceptibility of both aquatic and terrestrial organisms to changes in pH explains why ecologists are so concerned about **acid rain**, precipitation with a pH of less than 5.6. Acid rain results from the burning of fossil fuels such as coal, natural gas, and oil, which releases sulfur dioxide and nitrogen oxide into the atmosphere. These react with oxygen in the air to form sulfuric acid and nitric acid, which falls to the surface in rain or snow. When this precipitation falls on rivers and especially lakes, it can turn them more acidic and they lose their ability to sustain fish and other aquatic life. For example, lake trout disappear from lakes in Ontario and the eastern U.S. when the pH falls below about 5.2. Although this low pH does not affect survival of the adult fish, it affects the survival of juveniles.

Acid rain is important in terrestrial systems, too. For example, acid rain can directly affect forests by killing leaves or pine needles, as has happened on some of the higher mountaintops in the Great Smoky Mountains. It can also greatly depress soil pH, which can result in a loss of essential nutrients such as calcium and nitrogen. Low calcium in the soil results in calcium deficiencies in plants, in the snails that consume the plants, and in the birds that eat the snails, ultimately causing weak eggshells that break before hatching. Decreased soil pH also kills certain soil microorganisms, preventing decomposition and recycling of nitrogen in the soil. Decreases in soil calcium and nitrogen weaken trees and other plants, and may make them more susceptible to insect attack.

Acid rain is a common problem in the northeastern U.S. and Scandinavia, where sulfur-rich air drifts over from the Midwest and the industrial areas of Britain, respectively, causing the deposition of highly acidic rain. The problem was particularly acute during the 1960s and 1970s, but decreased manufacturing and the use of low-sulfur coal and the introduction of sulfur-absorbing scrubbers on the smokestacks of coal-burning power plants have reduced the problem somewhat in recent years. Acid rain is clearly a problem with a wide-ranging impact on ecological systems.

54.3 Climate and Its Relationship to Biological Communities

Temperature, water, wind, and light are components of **climate**, the prevailing weather pattern in a given region. Because the distribution and abundance of organisms are often limited by the abiotic environment, to understand the patterns of abundance of life on Earth, ecologists need to study the global climate. We begin this section by examining global climate patterns, focusing on how temperature variation drives atmospheric circulation and how features such as elevation and landmass can alter these patterns. Knowing how and why climate changes around the world enables us to understand and predict the occurrence of different **biomes**, the major community types on Earth such as tropical forests and hot deserts. In this section, we also explore the main characteristics of Earth's major terrestrial and aquatic biomes.

Atmospheric Circulation Is Driven by Global Temperature Differentials

Substantial differences in temperature occur over the Earth, mainly due to variations in the incoming solar radiation. In higher latitudes, such as northern Canada and Russia, the sun's rays hit the Earth obliquely and are spread out over more of the planet's surface than they are in equatorial areas (**Figure 54.18**). More heat is also lost in the atmosphere of higher altitudes because the sun's rays travel a greater distance through the atmosphere, allowing more heat to be dissipated by cloud cover. The result is that a much smaller amount of solar energy (40% less) strikes polar latitudes than equatorial areas. Generally, temperatures increase as the amount of solar radiation increases (**Figure 54.19**). However, at the tropics both cloudiness and rain reduce average temperature, so that temperatures do not continue to increase toward the equator.

The English meteorologist George Hadley made the initial contribution to a model of general atmospheric circulation in 1735. Hadley proposed that solar energy drives winds, which in

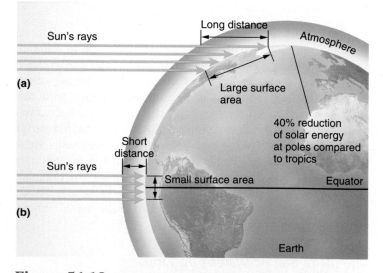

Figure 54.18 **The intensity of solar radiation at different latitudes.** In polar areas (a), the sun's rays strike the Earth at an oblique angle and deliver less energy than at tropical locations. In tropical areas (b), the energy is concentrated over a smaller surface and travels a shorter distance through the atmosphere.

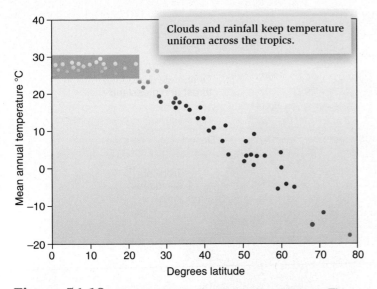

Figure 54.19 **Variation of the Earth's temperature.** The temperatures shown in this figure were measured at moderately moist continental locations of low elevation. Note the wide band of similar temperatures at the tropics.

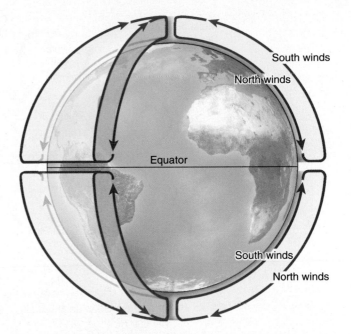

Figure 54.20 George Hadley's 1735 model of atmospheric circulation. In Hadley's model, simple convective circulation of air on a uniform, nonrotating Earth, heated at the equator and cooled at the poles, took the form of one large convection cell in each hemisphere. Winds are named according to the direction from which they blow, so the south wind blows from south to north.

turn influence the global circulation of the atmosphere. In his model, the warmth at the equator causes the surface equatorial air to heat up and rise vertically into the atmosphere. As the warm air rises away from its source of heat, it cools and becomes less buoyant, but the cool air does not sink back to the surface because of the warm air behind it. Instead, the rising air spreads north and south away from the equator, eventually returning to the surface at the poles. From there it flows back toward the equator to close the circulation loop. Hadley suggested that on a nonrotating Earth, this air movement would take the form of one large convection cell in each hemisphere, as shown in **Figure 54.20**.

When the effect of the Earth's rotation is added, however, the surface flow is deflected toward the west. This consequence is known as the **Coriolis effect**. Hadley's one-cell circulation has since been modified to account for the Coriolis effect and other more modern data. In the 1920s, a three-cell circulation in each hemisphere was proposed to fit the Earth's heat balance (**Figure 54.21**). The contribution of George Hadley is still recognized, in that the most prominent of the three cells, the one nearest the equator, is called the **Hadley cell**. In the Hadley cell, the warm air rising near the equator forms towers of cumulus clouds that provide rainfall, which in turn maintains the lush vegetation of the equatorial rain forests. As the upper flow in this cell moves toward the poles, it begins to subside, or fall back to Earth, at about 30° north and south of the equator. These **subsidence zones** are areas of high pressure and are the sites of the world's tropical deserts, because the subsiding air is relatively dry, having released all of its moisture over the equator. Winds are generally weak and variable near the center of this zone of descending air. Subsidence zones have popularly been called the horse latitudes. The name is said to have been coined by Spanish sailors crossing the Atlantic, whose ships were

sometimes rendered motionless in these waters and who reportedly were forced to throw horses overboard, or eat them, as they could no longer water or feed them.

From the center of the subsidence zones, the surface flow splits into the westerlies that flow toward the poles, and the equatorial flow, which is deflected by the Coriolis effect and forms the reliable trade winds. In the Northern Hemisphere, the trades are from the northeast, the direction from which they provided the sail power to explore the New World; in the Southern Hemisphere, the trades are from the southeast. The trade winds from both hemispheres meet near the equator in a region called the intertropical convergence zone (ITCZ), also known as the doldrums. Here the light winds and humid conditions provide the monotonous weather that may be the basis for the expression "in the doldrums."

In the three-cell model, the circulation between 30° and 60° latitude, called the **Ferrell cell**, is opposite that of the Hadley cell. The net surface flow is poleward, and because of the Coriolis effect, the winds have a strong westerly (flowing from west to east) component. These prevailing westerlies were known to Benjamin Franklin, perhaps the first American weather forecaster, who noted that storms migrated eastward across the colonies. In fact, the secondary zones of high precipitation can come anywhere from about 45° to 60°, with between 45° and 55° being most common. The final circulation cell is known as the **polar cell**. At the poles, the air has cooled and descends, but it has little moisture left, explaining why many high-latitude regions are actually desert-like in condition.

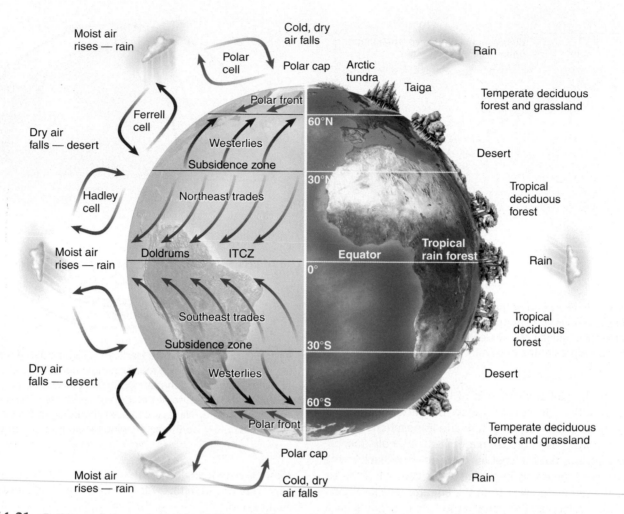

Figure 54.21 **Patterns of atmospheric circulation and biome positions.** Three-cell model of the atmospheric circulation on a uniform, rotating Earth heated at the equator and cooled at the poles. Tropical forests exist mainly in a band around the equator, where it is hot and rainy. At around 30° north and south, the air is hot and dry, and deserts exist. A secondary zone of precipitation exists at around 45° to 55° north and south, where temperate forests are located. The polar regions are generally cold and dry.

Thus, the distributions of the major biomes are determined by temperature differences and the wind patterns they generate. Hot, tropical forest blankets the tropics, where rainfall is high. At about 30° latitude, the air cools and descends, but it is without moisture, so the hot deserts occur around that latitude. The middle cell of the circulation model shows us that at about 45° to 55° latitude the air has warmed and gained moisture, so it ascends, dropping rainfall over the wet, temperature forests of this region in the Pacific Northwest and Western Europe in the Northern Hemisphere and over New Zealand and Chile in the Southern Hemisphere.

Elevation and Other Features of a Landmass Can Also Affect Climate

Thus far, we have considered how global temperatures and wind patterns affect climate. The geographic features of a landmass can also have an important impact. For example, the elevation of a region greatly influences its temperature range. On mountains, temperatures decrease with increasing elevation.

This decrease is a result of a process known as **adiabatic cooling**, in which increasing elevation leads to a decrease in air pressure. When air is blown across the Earth's surface and up over mountains, it expands because of the reduced pressure. As it expands, it cools at a rate of about 10°C for every 1,000 m in elevation, as long as no water vapor or cloud formation occurs. (Adiabatic cooling is also the principle behind the function of a refrigerator, in which Freon gas cools as it expands coming out of the compressor.) A vertical ascent of 600 m produces a temperature change roughly equivalent to that brought about by an increase in latitude of 1,000 km. This explains why mountaintop vegetation, even in tropical areas, can have the characteristics of tundra.

Mountains can also influence patterns of precipitation. For example, when warm, moist air encounters the windward side of a mountain, it flows upward and cools, releasing precipitation in the form of rain or snow. On the side of the mountain sheltered from the wind (the leeward side), drier air descends, producing what is called a **rain shadow**, an area where precipitation is noticeably less (**Figure 54.22a**). In this way, the

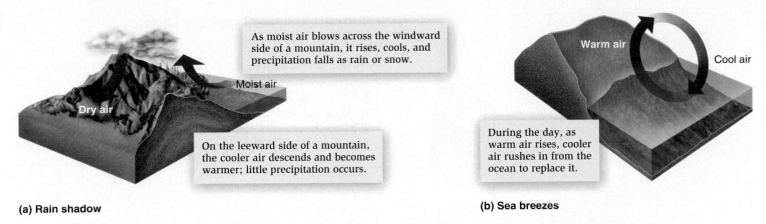

As moist air blows across the windward side of a mountain, it rises, cools, and precipitation falls as rain or snow.

On the leeward side of a mountain, the cooler air descends and becomes warmer; little precipitation occurs.

During the day, as warm air rises, cooler air rushes in from the ocean to replace it.

(a) Rain shadow

(b) Sea breezes

Figure 54.22 The influence of elevation and proximity to water on climate.

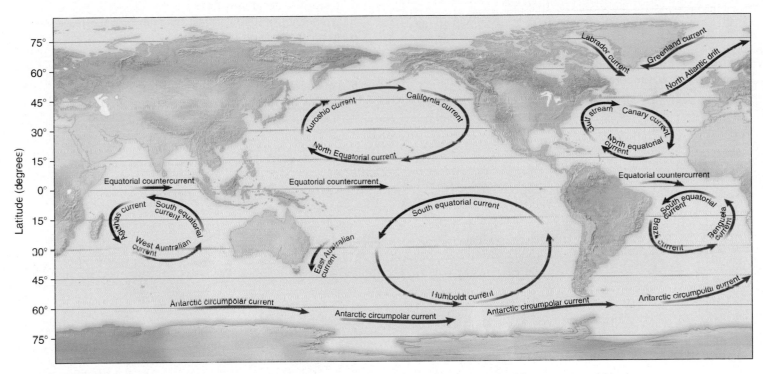

Figure 54.23 Ocean currents of the world. The red arrows represent warm water; the blue arrows, cold water.

western side of the Cascade Range in Washington receives more than 500 cm of annual precipitation, whereas the eastern side receives only 50 cm.

The proximity of a landmass to a large body of water can affect climate because land heats and cools more quickly than the sea does. The specific heat capacity of the land is much lower than that of the water, allowing the land to warm more quickly in the day. The warmed air rises and cooler air flows in to replace it. This pattern creates the familiar onshore sea breezes in coastal areas (**Figure 54.22b**). At night, the land cools quicker than the sea, and so the pattern is reversed, creating offshore breezes. The sea, therefore, has a moderating effect on the temperatures of coastal regions and especially islands. The climates of coastal regions may differ markedly from those of their climatic zones. Many never experience frost, and fog is

often evident. Thus, along coastal areas, different vegetation patterns may occur compared to those in areas farther inland. In fact, some areas of the U.S., including Florida, would be deserts were it not for the warm water of the sea and the moisture-laden clouds that form above them.

Together with the rotation of the Earth, winds also create ocean currents. The major ocean currents act as "pinwheels" between continents, running clockwise in the ocean basins of the Northern Hemisphere and counterclockwise in those of the Southern Hemisphere (**Figure 54.23**). Thus, the Gulf Stream, equivalent in flow to 50 times the world's major rivers combined, brings warm water from the Caribbean and the U.S. coasts across to Europe, the climate of which is correspondingly moderated. The Humboldt Current brings cool conditions almost to the equator along the western coast of South America.

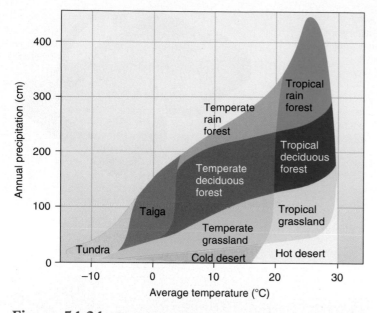

Figure 54.24 The relationship between the world's terrestrial biome types and temperature and precipitation patterns.

Biological inquiry: What other factors may influence biome types?

Terrestrial Biome Types Are Dictated by Climate Patterns

Differences in climate on Earth help to define its different terrestrial biomes. Many types of classification schemes are used for mapping the geographic extent of biomes, but one of the most useful was developed by the American ecologist Robert Whittaker, who classified biomes according to the physical factors of average annual precipitation and temperature (**Figure 54.24**). In this scheme, we can recognize 10 terrestrial biomes (**Figure 54.25**):

1. tropical rain forest;
2. tropical deciduous forest;
3. temperate rain forest;
4. temperate deciduous forest;
5. temperate coniferous forest (taiga or boreal forest);
6. tropical grassland (savanna);
7. temperate grassland (prairie);
8. hot desert;
9. cold desert;
10. tundra.

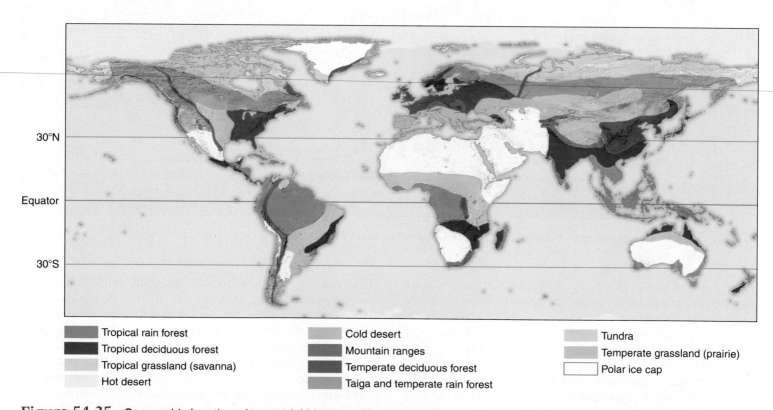

Figure 54.25 Geographic location of terrestrial biomes. The distribution patterns of taiga and temperate rain forest are combined because of their similarity in tree species and because temperate rain forest is actually limited to a very small area.

Figure 54.26a–k illustrates these 10 terrestrial biomes and identify their main characteristics.

While these broad terrestrial biomes are a useful way of defining the main types of communities on Earth, ecologists acknowledge that not all communities fit neatly into one of these 10 major biome types. Also, one biome type often grades into another, as seen on mountain ranges (**Figure 54.26k**). Soil conditions can also influence biome type. In California, serpentine soils, which are dry and nutrient-poor, support only sparse vegetation. In the eastern U.S., most of New Jersey's coastal plain, called the Pine Barrens, consists of sandy, nutrient-poor soil that cannot support the surrounding deciduous forest and instead contains grasses and low shrubs growing among open stands of pygmy pitch pine and oak trees.

Some ecologists recognize chaparral, a type of temperate grassland, as a distinct biome type. Winters are cool and wet, and summers warm and dry. Summer conditions are so dry that fires occur normally and frequently at this time. Such conditions are most commonly seen at around 30° latitude, where cool ocean waters moderate the climate, as along the coasts of California, South Africa, Chile, and southwest Australia and in countries surrounding the Mediterranean Sea.

TROPICAL RAIN FOREST

Figure 54.26a

Tropical rain forest in Fiji

Physical Environment: Rainfall exceeding 230 cm per year, temperature hot year round, averaging 25–29°C, soils often shallow and nutrient-poor.

Location: Equatorial, between the Tropics of Cancer and Capricorn. Tropical forests cover much of northern South America, Central America, western and central Africa, South east Asia, and various islands in the Indian and Pacific oceans.

Plant Life: The numbers of plant species found in tropical forests can be staggering, often reaching as many as 100 tree species per square kilometer. Leaves often narrow to "drip-tips" at the apex so that rainwater drains quickly. Many trees have large buttresses that help support their shallow root systems. Little light penetrates the **canopy**, the uppermost layer of tree foliage, and the ground cover is often sparse. Epiphytes, plants that live perched on trees and are not rooted in the ground, are common. Bromeliads are common epiphytes in North and South American forests. Lianas, or climbing vines, are also common.

Animal Life: Animal life in the tropical rain forests is diverse; insects, reptiles, amphibians, and mammals are well represented. Large mammals, however, are not common. Because many of the plant species are widely scattered in tropical forests, it is risky for plants to rely on wind for pollination or to disperse their seed. This means that animals are important in pollinating flowers and dispersing fruits and seeds. Mimicry and bright protective coloration, warning of bad taste or the existence of toxins, are rampant.

Effects of Humans: Humans are impacting tropical forests greatly by logging and by clearing the land for agriculture. Also, many South American tropical forests are cleared to create grasslands for cattle.

TROPICAL DECIDUOUS FOREST

Figure 54.26b

Tropical deciduous forest in Bandhavgarh National Park, India

Physical Environment: Rainfall is substantial, at around 130–280 cm a year, but the dry season is distinct, often two to three months or longer. Soil water shortages can occur in the dry season. Temperatures are hot year round, averaging 25–39°C.

Location: Equatorial, where rainfall is more seasonal. Much of India consists of tropical deciduous forest, containing teak trees. Brazil, Thailand, and Mexico also contain tropical deciduous forest. At the wet edges of this biome, it may grade into tropical forests; at the dry end, it may grade into tropical grasslands or savannas.

Plant Life: Because of the distinct dry season, many of the trees in tropical deciduous forests shed their leaves, just as they do in temperate forests, and an understory of herbs and grasses may grow during this time. Indeed, because the canopy is often more open than in the tropical rain forest, a denser closed forest—what we might think of as a "tropical jungle"—exists at the forest floor. Where the dry season is six to seven months long, tropical deciduous forests may contain shorter, thorny plants such as acacia trees and the forest is referred to as a tropical thorn forest.

Animal Life: The diversity of animal life is high, and species such as monkeys, antelopes, wild pigs, and tigers are present. However, as with plant diversity, animal diversity is less than in tropical rain forests. Tropical thorn forests may contain more browsing mammals; hence, the development of plant thorns as a defense.

Effects of Humans: Logging and agriculture have had a large impact on tropical deciduous forests. The litter layer of dead, decaying leaves is much thicker than in tropical rain forests and renders the soil more fertile.

TEMPERATE RAIN FOREST

Figure 54.26c

Hoh Rain Forest in Olympic National Park, Washington

Physical Environment: Temperatures are a little cooler, 5–25°C on average, and winters are mild, but there is abundant rainfall, usually exceeding 200 cm a year. The condensation of water from dense coastal fogs augments the normal rainfall.

Location: The coverage of this biome type is small, consisting of a thin strip along the northwest coast of North America from northern California through Washington State, British Columbia, and into southeast Alaska (here called tongass). It also exists in southwestern South America along the Chilean coast. Indeed, it is found only in coastal situations because of the moderating influence of the ocean on air temperature.

Plant Life: The dominant vegetation type, especially in North America, consists of large evergreen trees such as western hemlock, Douglas fir, and Sitka spruce. The high moisture content allows epiphytes to thrive. Cool temperatures slow the activity of decomposers, so that the litter layer is thick and spongy.

Animal Life: In North America, the temperate rain forest is rich in species such as mule deer, elk, squirrels, and numerous birds such as jays and nuthatches. Because of the abundant moisture and moderate temperatures, reptiles and amphibians are also common.

Effects of Humans: This biome is a prolific producer of wood and supplies much timber, although logging threatens the survival of the forest in some areas.

TEMPERATE DECIDUOUS FOREST

Figure 54.26d

Temperate deciduous forest in Minnesota

Physical Environment: Temperatures fall below freezing each winter but not usually below –12°C, and annual rainfall is generally between 75 and 200 cm.

Location: Large tracts of temperate deciduous forest are evident in the eastern U.S., eastern Asia, and western Europe. In the Southern Hemisphere, eucalyptus forests occur in Australia, and stands of southern beech are found in southern South America, New Zealand, and Australia.

Plant Life: Species diversity is much lower in temperate deciduous forests than in the tropical forests, with about only three to four species per square kilometer, and several tree genera may be dominant in a given locality—for example, oaks, hickories, and maples are usually dominant in the eastern U.S. Commonly, leaves are shed in the fall and reappear in the spring. Many herbaceous plants flower in spring before the trees leaf out and block the light, though even in the summer, the forest is not as dense as in tropical forests, so there is abundant ground cover. There are few epiphytes and lianas.

Animal Life: Animals are adapted to the vagaries of the climate; many mammals hibernate during the cold months, birds migrate, and insects enter diapause, a condition of dormancy passed usually as a pupa. Reptiles, which are dependent on solar radiation for heat, are relatively uncommon. Mammals include squirrels, wolves, bobcats, foxes, bears, and mountain lions.

Effects of Humans: Logging has eliminated much temperate forest from heavily populated portions of Europe and North America. Soils are rich because the annual leaf drop promotes high soil nutrient levels. With careful agricultural practices, soil richness can be conserved, and as a result agriculture can flourish.

TEMPERATE CONIFEROUS FOREST OR TAIGA

Figure 54.26e

Temperate coniferous forest in Canada

Physical Environment: Precipitation is generally between 30 and 70 cm, often occurring in the form of snow. Temperatures are very cold, below freezing for long periods of time.

Location: The biome of coniferous forests, known commonly by its Russian name, taiga, lies north of the temperate-zone forests and grasslands. Vast tracts of taiga exist in North America and Russia, and mountain taiga exists on mountainous areas. In the Southern Hemisphere, little land area occurs at latitudes at which one would expect extensive taiga to exist.

Plant Life: Most of the trees are evergreens or conifers with tough needles, hence its similarity to temperate rain forest. In this biome, spruces, firs, and pines generally dominate, and the number of tree species is relatively low. Many of the conifers have conical shapes to reduce bough breakage from heavy loads of snow. As in tropical forests, the understory is sparse because the dense year-round canopies prevent sunlight from penetrating. Soils are poor because the fallen needles decay so slowly in the cold temperatures that a layer of needles builds up. This layer of needles acidifies the soil, further reducing the numbers of understory species.

Animal Life: Reptiles and amphibians are rare because of the low temperatures. Insects are strongly periodic but may often reach outbreak proportions in times of warm temperatures. Mammals that inhabit this biome such as bears, lynxes, moose, beavers, and squirrels are heavily furred.

Effects of Humans: Humans have not extensively settled these areas, but they have been quite heavily logged.

TROPICAL GRASSLAND

Figure 54.26f

Tropical grassland of the Masai Mara Game Reserve in Kenya

Physical Environment: Hot, tropical areas, averaging 24–29°C with a low or seasonal rainfall of between 50 and 130 cm per year. There is often an extensive dry season.

Location: Extensive savannas occur in Africa, South America, and northern Australia.

Plant Life: Wide expanses of grasses dominate savannas but occasional thorny trees, such as acacias, may occur. Fire is prevalent in this biome, so most plants have well-developed root systems that enable them to resprout quickly after a fire.

Animal Life: The world's greatest assemblages of large mammals occur in the savanna biome. Herds of antelope, zebras, and wildebeest are found, together with their associated predators: cheetah, lion, leopard, and hyena. Termite mounds dot the landscape in some areas. The extensive herbivory of large grazers, together with frequent fires, may help maintain savannas and prevent their development into forests.

Effects of Humans: Savanna soils are often poor because the occasional rain leaches nutrients out. Nevertheless, conversion of this biome to agricultural land is rampant, especially in Africa. Overstocking of land for domestic animals can greatly reduce grass coverage through over-grazing, turning the area more desert-like. This process is known as **desertification**.

TEMPERATE GRASSLAND

Figure 54.26g

Temperate grassland in Wyoming State

Physical Environment: Annual rainfall generally between 25 and 100 cm, too low to support a forest but higher than that in deserts. Temperatures in the winter often fall below −10°C, while summers may be very hot, approaching 30°C.

Location: Temperate grasslands include the prairies of North America, the steppes of Russia, the pampas of Argentina, and the veldt of South Africa. In addition to the limiting amounts of rain, fire and grazing animals may also prevent the establishment of trees in the temperate grasslands. Where temperatures rarely fall below freezing and most of the rain falls in the winter, chaparral, a fire-adapted community featuring shrubs and small trees, occurs.

Plant Life: From east to west in North America and from north to south in Asia, grasslands show differentiation along moisture gradients. In Illinois, with an annual rainfall of 80 cm, tall prairie grasses such as big bluestem and switchgrass grow to about 2 m high. Along the eastern base of the Rockies, 1,300 km to the west, where rainfall is only 40 cm, prairie grasses such as buffalo grass and blue grama rarely exceed 0.5 m in height. Similar gradients occur in South Africa and Argentina.

Animal Life: Where the grasslands remain, large mammals are the most prominent members of the fauna: bison (buffalo) and pronghorn (antelope) in North America, wild horses in Eurasia, and large kangaroos in Australia. Burrowing animals such as North American gophers and African mole rats are also common.

Effects of Humans: Prairie soil is among the richest in the world, having 12 times the humus layer of a typical forest soil. Worldwide, most prairies have been converted to agricultural cropland, and original grassland habitats are among the rarest biomes in the world.

HOT DESERT

Figure 54.26h

Saguaro National Park in Arizona, part of the Sonoran Desert

Physical Environment: Temperatures are variable from below freezing at night to as much as 50°C in the day. Rainfall is less than 30 cm per year.

Location: Hot deserts are found around latitudes of 30° north and south. Prominent deserts include the Sahara of North Africa, the Kalahari of southern Africa, the Atacama of Chile, the Sonoran of northern Mexico and the southwest U.S., and the Simpson of Australia.grow crops there. However, salinization of soils is prevalent.

Plant Life: Three forms of plant life are adapted to deserts: annuals, succulents, and desert shrubs. Annuals circumvent drought by growing only when there is rain. Succulents, such as the saguaro cactus and other barrel cacti of the southwestern deserts, store water. Desert shrubs, such as the spraylike ocotillo, have short trunks, numerous branches, and small, thick leaves that can be shed in prolonged dry periods. In many plants, spines or volatile chemical compounds serve as a defense against water-seeking herbivores.

Animal Life: To conserve water, desert plants produce many small seeds, and animals that eat those seeds, such as ants, birds, and rodents, are common. Reptiles are numerous because high temperatures permit these ectothermic animals to maintain a warm body temperature. Lizards and snakes are important predators of seed-eating mammals.

Effects of Humans: Ambitious irrigation schemes and the prolific use of underground water have allowed humans to colonize deserts and grow crops there. However, salinization of soils is prevalent. Off-road vehicles can disturb the fragile desert communities.

COLD DESERT

Figure 54.26i

The Gobi Desert of Mongolia

Physical Environment: Precipitation is less than 25 cm a year, often in the form of snow. Rainfall usually comes in the spring. In the daytime, temperatures can be high in the summer, 21–26°C, but average around freezing, −2 to 4°C, in the winter.

Location: Cold deserts are found in dry regions at middle to high latitudes, especially in the interiors of continents and in the rain shadows of mountains. Cold deserts are found in North America (the Great Basin Desert), in eastern Argentina (the Patagonian Desert), and in central Asia (the Gobi Desert).

Plant Life: Cold deserts are relatively poor in terms of numbers of species of plants. Most plants are small in stature, being only between 15 and 120 cm tall. Many species are deciduous and spiny. The Great Basin Desert in Nevada, Utah, and bordering states is a cold desert dominated by sagebrush.

Animal Life: As in hot deserts, large numbers of plants produce small seeds on which numerous ants, birds, and rodents feed. Many species live in burrows to escape cold and to keep warm. In the Great Basin Desert, pocket mice, jackrabbits, kit fox, and coyote are common.

Effects of Humans: Agriculture is hampered because of low temperatures and low rainfall, and human populations are not extensive. If the top layer of soil is disturbed by human intrusions such as off-road vehicles, erosion occurs rapidly and even less vegetation is able to exist.

TUNDRA

Figure 54.26j

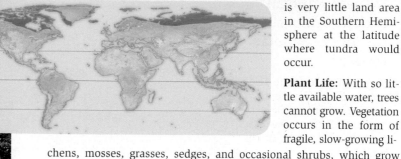

is very little land area in the Southern Hemisphere at the latitude where tundra would occur.

Plant Life: With so little available water, trees cannot grow. Vegetation occurs in the form of fragile, slow-growing lichens, mosses, grasses, sedges, and occasional shrubs, which grow close to the ground. Plant diversity is very low. In some places desert conditions prevail because so little moisture falls.

Animal Life: Animals of the arctic tundra have adapted to the cold by having good insulation. Many birds, especially shorebirds and waterfowl, migrate. The fauna is much richer in summer than in winter. Many insects spend the winter at immature stages of growth, which are more resistant to cold than the adult forms. The larger animals include such herbivores as musk oxen and caribou in North America, called reindeer in Europe and Asia, as well as the smaller hares and lemmings. Common predators include arctic fox, wolves, and snowy owls, and polar bears near the coast.

Effects of Humans: Though this area is sparsely populated, mineral extraction, especially of oil, has the potential to significantly impact this biome. Ecosystem recovery from such damage would be very slow.

Denali National Park in Alaska

Physical Environment: Precipitation is generally less than 25 cm per year and is often locked up as snow and unavailable for plants. Deeper water can be locked away for a large part of the year in **permafrost**, a layer of permanently frozen soil. The growing season here is short, only 50–60 days. Summer temperatures are only 3–12°C, and even during the long summer days, the ground thaws to less than 1 m in depth. Midwinter temperatures average −32°C.

Location: Tundra (from the Finnish *tunturia*, treeless plain) exists mainly in the Northern Hemisphere, north of the taiga, because there

---

MOUNTAIN RANGES

Figure 54.26k

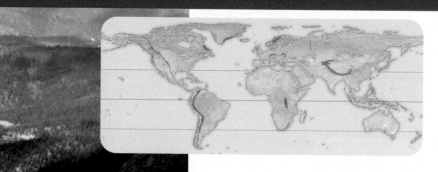

Rocky Mountains of Colorado

Physical Environment: Mountain ranges must be treated differently than other biomes. Biome type relies predominantly on climate. On mountains, temperature decreases with increasing elevation through adiabatic cooling, as discussed in the text. Thus, precipitation and temperature may change dramatically, depending on elevation and whether the mountainside is on the windward or leeward side.

Location: Mountain ranges exist in many areas of the world, but among the largest are the Himalayas in Asia, the Rockies in North America, and the Andes in South America.

Plant Life: Biome type may change from temperate forest through taiga and into tundra on an elevation gradient in the Rocky Mountains, and even from tropical forest to tundra on the highest peaks of the Andes in tropical South America. In tropical regions, daylight varies little from the 12 hours per day throughout the year. Instead of an intense period of productivity, vegetation in the tropical alpine tundra exhibits slow but steady rates of photosynthesis and growth all year.

Animal Life: The animals of this biome are as varied as the number of habitats they contain. Generally, more species of plants and animals are found at lower elevations than at higher ones. At higher elevations, animals such as bighorn sheep and mountain goats have to be very sure-footed to climb the craggy slopes and have skidproof pads on their hooves. Despite the often-strong winds, birds of prey, such as eagles, are frequent predators of the furry rodents found at higher elevations, including guinea pigs and marmots.

Effects of Humans: Logging and agriculture at lower elevations can cause habitat degradation. Because of the steep slopes, mountain soils are often well drained, thin, and especially susceptible to erosion.

Aquatic Biomes Consist of Marine and Freshwater Regions

Within aquatic environments, several different biome types are also recognized, including marine aquatic biomes (the intertidal zone, coral reef, and open ocean) and freshwater lakes, rivers, and wetlands. These biomes are distinguished by differences in parameters such as salinity, oxygen content, depth, current strength, and availability of light (**Figures 54.27a–f**).

INTERTIDAL ZONE

Figure 54.27a

Olympic Coast National Marine Sanctuary in Washington State

Physical Environment: The **intertidal zone**, the area where the land meets the sea, is alternately submerged and exposed by the daily cycle of tides. The resident organisms are subject to huge daily variation in temperature, light intensity, and availability of seawater, which makes life difficult.

Location: Throughout the world, the area where the land meets the sea consists of sandy shore, mudflats, or rocky shore. Commonly, there is a vertical zonation consisting of three broad zones, most evident on rocky shores. The upper littoral zone is submerged only during the highest tides. The midlittoral zone is submerged during the highest regular tide and exposed during the lowest tide each day. The lower littoral zone is exposed only during the lowest tide.

Plant Life: Plant life may be quite limited because the sand or mud is constantly shifted by the tide. Mangroves may colonize mudflats in tropical areas, and salt marsh grasses may colonize mudflats in temperate locations. On the rocky shore, plant life consists of green algae and seaweeds.

Animal Life: Animal life may be quite diverse. On the rocky shore, sea anemones, snails, hermit crabs, and small fishes live in tide pools. On the rock face, there may be a variety of limpets, mussels, sea stars, sea urchins, snails, sponges, tube worms, whelks, isopods, and chitons. At low tides, organisms may be dry and vulnerable to predation by a variety of animals, including birds and mammals. High tides bring predatory fish. In sandy or muddy shores, the biome may contain burrowing marine worms, crabs, and small isopods.

Effects of Humans: Urban development has greatly reduced the beach area available to breeding turtles and shorebirds. Oil spills have greatly impacted some rocky intertidal areas.

CORAL REEF

Figure 54.27b

Caribbean Coral Reef

Physical Environment: Corals need warm water of at least 20°C but less than 30°C. They are also limited to the euphotic zone, where light penetrates. Sunlight is important because many corals harbor symbiotic algae, or dinoflagellates, that contribute nutrients to the animals and that require light to live.

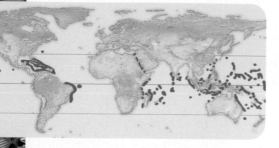

Location: Coral reefs exist in warm tropical waters where there are solid substrates for attachment and water clarity is good. The largest coral reef in the world is the Great Barrier Reef off the Australian coastline, but other coral reefs are found throughout the Caribbean Sea and Pacific Ocean.

Plant Life: Dinoflagellate algae live within the coral tissue, and a variety of red and green algae live on the coral reef surface.

Animal Life: An immense variety of microorganisms, invertebrates, and fish live among the coral, making the coral reef one of the most interesting and species rich biomes on Earth. Probably 30–40% of all fish species on Earth are found on coral reefs. Prominent herbivores include snails, sea urchins, and fish. These are in turn consumed by octopus, sea stars, and carnivorous fish. Many species are brightly colored, warning predators of their toxic nature.

Effects of Humans: Collectors have removed many corals and fish for the aquarium trade, and marine pollution threatens water clarity in some areas. Perhaps the greatest threat is from global warming and climate change. Water temperatures that are too high (over 30°C) can cause coral bleaching.

THE OPEN OCEAN

Figure 54.27c

Manta Ray in the open ocean

Physical Environment: In the open ocean, sometimes called the **pelagic zone**, water depth averages 4,000 m and nutrient concentrations are typically low, though the waters may be periodically enriched by ocean **upwellings**, which carry mineral nutrients from the bottom waters to the surface. Pelagic waters are mostly cold, only warming near the surface.

Location: Across the globe, covering 70% of the Earth's surface.

Plant Life: Where light levels are high at the surface, many microscopic, photosynthetic organisms (**phytoplankton**) grow and reproduce. Phytoplankton account for nearly half the photosynthetic activity on Earth and produce much of the world's oxygen.

Animal Life: Open-ocean organisms include **zooplankton**, minute animal organisms consisting of some worms, copepods (tiny shrimplike creatures), tiny jellyfish, and the small larvae of invertebrates and fish that graze on the phytoplankton. The open ocean also includes free-swimming animals collectively called **nekton**, which can swim against the currents to locate food. The nekton include large squids, fish, sea turtles, and marine mammals that feed on phytoplankton, zooplankton, or each other. Only a few of these organisms live at any great depth. In some areas, there exists a unique assemblage of animals associated with deep-sea hydrothermal vents that spew out hot (350°C) water rich in hydrogen sulfide. In this dark, oxygen-poor environment, large polychaete worms exist together with other organisms that are chemoautotrophs (see Figure 59.3).

Effects of Humans: Oil spills and a long history of garbage disposal have polluted the ocean floors of many areas. Overfishing has caused many fish populations to crash, and the whaling industry has greatly reduced the numbers of most species of whales.

FRESHWATER LENTIC HABITATS

Figure 54.27d

Everglades National Park, Florida

Physical Environment: Freshwater habitats are traditionally divided into **lentic**, or standing-water habitats (from the Latin *lenis*, calm), and **lotic**, or running-water habitats (from the Latin *lotus*, washed). The lentic habitat consists of still, often deep water. Its physical characteristics depend greatly on the surrounding land, which dictates what nutrients collect in the lake. Young lakes often start off clear and with little plant life. Such lakes are called **oligotrophic**. With age, the phytoplankton bloom and algae spread, reducing the water clarity. Such lakes are termed **eutrophic**. The process of eutrophication is discussed more thoroughly in Chapter 59.

Location: Throughout all the continents of the world.

Plant Life: In addition to free-floating phytoplankton and algae, lentic habitats may have rooted vegetation, which often emerges above the water surface (emergent vegetation), such as cattails, plus deeper dwelling aquatic plants and algae.

Animal Life: Animals include fish, frogs, turtles, crayfish, insect larvae, and many species of insects. In tropical and subtropical lakes, alligators and crocodiles commonly are seen.

Effects of Humans: Agricultural runoff, including fertilizers and sewage, can greatly increase lake nutrient levels and speed up the process of eutrophication. This may result in algal blooms and fish kills. In some areas, exotic species of invertebrates and fish are outcompeting native species.

FRESHWATER LOTIC HABITATS

Figure 54.27e

Fast-flowing river in the Pacific Northwest

Physical Environment: In lotic habitats, flowing water prevents nutrient accumulations and phytoplankton blooms. The current also mixes water thoroughly, providing a well-aerated habitat of relatively uniform temperature. However, current, oxygen level, and clarity are greater in headwaters than in the lower reaches of rivers. Nutrient levels are generally less in headwaters.

Location: On all continents except Antarctica.

Plant Life: In slow-moving streams and rivers, algae and rooted plants may be present; in swifter moving rivers, leaves from surrounding forests are the primary food source for animals.

Animal Life: Lotic habitats have a fauna completely different from that of lentic waters. Animals are adapted to stay in place despite an often-strong current. Many of the smaller organisms are flat and attach themselves to rocks to avoid being swept away. Others live on the underside of large boulders, where the current is much reduced. Fish such as trout may be present in rivers with cool temperatures, high oxygen, and clear water. In warmer, murkier waters, catfish and carp may be abundant.

Effects of Humans: Animals of lotic systems are not well adapted for low-oxygen environments and are particularly susceptible to oxygen-reducing pollutants such as sewage. Dams across rivers have prevented the passage of migratory species such as salmon.

WETLANDS

Figure 54.27f

Yellow Waters River, Kakadu National Park, Northern Territory, Australia

Physical Environment: At the margins of both lentic and lotic habitats, wetlands may develop. Wetlands are areas regularly saturated by surface water or groundwater. They can range from marshes and swamps to bogs. Many are seasonally flooded when rivers overflow their banks or lake levels rise. Some wetlands also develop along estuaries, where rivers merge with the ocean and high tides can flood the land. Here, salt marshes or, in tropical areas, stands of mangroves can develop. Because of generally high nutrient levels, oxygen levels are fairly low. Temperatures vary substantially with location.

Location: Worldwide, except in Antarctica.

Plant Life: Wetlands are among the most productive and species-rich areas in the world. In North America, floating plants such as lilies and rooted species such as sedges, cattails, cypress, and gum trees predominate.

Animal Life: Most wetlands are rich in animal species. Wetlands are a prime habitat for wading and diving birds. In addition, they are home to a profusion of insects, from mosquitoes to dragonflies. Vertebrate predators include many amphibians, reptiles, otters, and alligators.

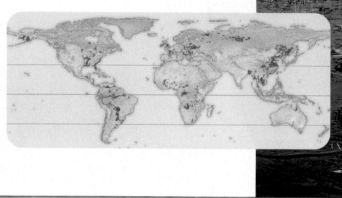

Effects of Humans: Long mistakenly regarded as wasteland by humans, many wetlands have been drained and developed for subdivisions and industry.

In this chapter we have discussed how the physical environment profoundly influences the distribution and abundance of life on Earth and the existence of different biome types. In subsequent chapters, we will learn how other factors influence the distribution of plants and animals. On a smaller scale, the presence of predators, parasites, or competitors can also influence where organisms are found. Part of an organism's fundamental niche may be occupied by a competitively superior species. We will examine the influences of such factors in Chapter 57. For animals, social interactions with other members of the same species such as fights over territory or mates can also influence population distributions. To discuss these issues, the next chapter examines animal behavior.

CHAPTER SUMMARY

54.1 The Scale of Ecology

- Ecologists study the interactions among organisms and between organisms and their environments. (Figure 54.1)

- The field of ecology can be subdivided into broad areas of organismal, population, community, and ecosystems ecology. (Figure 54.2)

- Organismal ecology considers how individuals are adapted to their environment and how the behavior of an individual organism contributes to its survival and reproductive success and the population density of the species.

- Population ecology explores those factors that influence a population's growth, size, and density, and community ecology studies how populations of species interact and form functional communities. (Figure 54.3)

- Ecosystems ecology examines the flow of energy and cycling of nutrients among organisms within a community and between organisms and the environment.

- Ecological methods focus on observation and experimentation. A variety of statistical tests exist that help determine whether two variables are related. (Figures 54.4, 54.5, 54.6)

54.2 The Environment's Impact on the Distribution of Organisms

- Abiotic factors such as temperature, wind, water, light, salinity, pH, and water currents can have powerful effects on ecological systems.

- Temperature exerts important effects on the distribution of organisms because of its effect on biological processes and the inability of most organisms to regulate their body temperature. (Figures 54.7, 54.8, 54.9, 54.10)

- Life on Earth is made possible by the greenhouse effect, in which short-wave solar radiation passes through the atmosphere to warm the Earth but is radiated back to space as long-wave infrared radiation. Much of this radiation is absorbed by atmospheric gases and reradiated back to Earth's surface, causing its temperature to rise. (Figure 54.11)

- The major atmospheric gases causing the greenhouse effect are water, carbon dioxide, methane, nitrous oxide, and chlorofluorocarbons. (Table 54.1)

- An increase in atmospheric gases is increasing the greenhouse effect, causing global warming, a gradual elevation of the Earth's surface temperature. Ecologists expect that global warming will have a large effect on the distribution of the world's organisms. (Figure 54.12)

- Wind can amplify the effects of temperature and modify wave action. (Figure 54.13)

- The availability of water has an important effect on the abundance of organisms. (Figure 54.14)

- Light can be a limiting resource for plants in both terrestrial and aquatic environments. (Figure 54.15)

- The concentration of salts and the pH of soil and water can limit the distribution of organisms. (Figures 54.16, 54.17)

54.3 Climate and Its Relationship to Biological Communities

- Global temperature differentials are caused by variations in incoming solar radiation and patterns of atmospheric circulation. (Figures 54.18, 54.19, 54.20, 54.21)

- Elevation and the proximity between a landmass and large bodies of water can similarly affect climate. (Figures 54.22, 54.23)

- Climate has a large effect on biomes, major types of habitats characterized by distinctive plant and animal life. (Figures 54.24, 54.25)

- Terrestrial biomes are generally named for their climate and vegetation type and include tropical rain forest, tropical deciduous forest, temperate rain forest, temperate deciduous forest, temperate coniferous forest (taiga), tropical grassland (savanna), temperate grassland (prairie), hot and cold deserts, and tundra. In mountain ranges, biome type may change on an elevation gradient. (Figure 54.26)

- Within aquatic environments, biomes include marine aquatic biomes (the intertidal zone, coral reef, and open ocean) and freshwater lakes, rivers, and wetlands. These are distinguished by differences in parameters including salinity, oxygen content, depth, current strength (lentic versus lotic), and availability of light. (Figure 54.27)

TEST YOURSELF

1. Which of the following is probably the most important factor in the distribution of organisms in the environment?
   a. light
   b. temperature
   c. salinity
   d. water availability
   e. pH

2. The greenhouse effect is
   a. a new phenomenon resulting from industrialization.
   b. due to the absorption of solar radiation by atmospheric gases.
   c. responsible for the natural warming of the Earth.
   d. all of the above.
   e. b and c only.

3. Which of the following is not a component of climate?
   a. temperature
   b. vegetation
   c. water
   d. wind
   e. light

4. The study of how groups of different species interact in their common environment is
   a. ecology.
   b. organismal ecology.
   c. community ecology.
   d. population ecology.
   e. environmental ecology.

5. In aquatic environments, plants and algae are usually found in the _____ zone near the surface of the water, where light is able to penetrate.
   a. aphotic
   b. littoral
   c. euphotic
   d. limnetic
   e. photosynthetic

6. What is the driving force that determines the circulation of the atmospheric air?
   a. temperature differences of the Earth
   b. winds
   c. ocean currents
   d. mountain ridges
   e. all of the above

7. The distribution of the major biomes of the Earth is determined by
   a. temperature differences.
   b. mountain ridges
   c. ocean currents.
   d. temperature and moisture differences.
   e. all of the above.

8. What characteristics are commonly used to identify the biomes of the Earth?
   a. temperature
   b. precipitation
   c. vegetation
   d. all of the above
   e. a and b only

9. Terrestrial areas that are regularly saturated by surface water or groundwater are called
   a. chaparral.
   b. lakes.
   c. wetlands.
   d. deserts.
   e. grasslands.

10. Which is the most important contribution to human-caused global warming?
    a. carbon dioxide
    b. nitrous oxide
    c. sulfur dioxide
    d. methane
    e. chlorofluorocarbons

CONCEPTUAL QUESTIONS

1. Define ecology.
2. Explain the greenhouse effect.
3. Define biome.

EXPERIMENTAL QUESTIONS

1. Prior to Callaway and Aschehoug's study, what was the prevailing hypothesis of why invasive species succeed in new environments?
2. Briefly describe the evidence collected to support the allelochemical hypothesis.
3. What was the function of the activated charcoal used in a subsequent test of the hypothesis?

COLLABORATIVE QUESTIONS

1. Discuss several ways in which the physical environment affects the distribution of organisms.
2. Based on your knowledge of biomes, identify the biome in which you live. In your discussion, list and describe the organisms that you have observed in your biome.

www.brookerbiology.com

This website includes answers to the Biological Inquiry questions found in the figure legends and all end-of-chapter questions.

55

BEHAVIORAL ECOLOGY

CHAPTER OUTLINE

Chimpanzee (*Pan troglodytes*) using a stick to catch food.

Behavior is the observable response of organisms to external or internal stimuli. Defined as such, behavior takes into account a very broad range of activities. In this chapter, we focus our attention on the field of **behavioral ecology**, the study of how behavior contributes to the differential survival and reproduction of organisms. For example, nesting black-headed gulls (*Larus ridibundus*) always pick up broken eggshells after a chick has hatched and carry them away from the nest. One might think that they are being neat and tidy, or are minimizing the risk of bacterial infection to the chicks, but there is more to the behavior than this. The chicks and unhatched eggs are well camouflaged in the nest, but the white color of the empty eggshell quickly attracts the attention of predators such as crows that would kill and eat the chicks or remaining eggs. By removing the old eggshells, the gull parents are increasing the chances that their offspring—and thus their genes—will survive.

Behavioral ecology builds upon earlier work that focused primarily on how organisms behave. In the early 20th century, scientific studies of animal behavior, termed **ethology** (from the Greek *ethos*, habit, manner), focused on the specific genetic and physiological mechanisms of behavior. These factors are called **proximate causes**. For example, we could hypothesize that male deer rut, or fight with other males, in the fall because a change in day length stimulates the eyes, brain, and pituitary gland and triggers hormonal changes in their bodies. The founders of ethology, Karl von Frisch, Konrad Lorenz, and Niko Tinbergen, shared the 1973 Nobel Prize in Physiology or Medicine for their pioneering discoveries concerning the proximate causes of behavior.

However, we could also hypothesize that male deer fight to determine which deer get to mate with a female deer and pass on their genes. This hypothesis leads to a different answer than the one that is concerned with changes in day length. This answer focuses on the adaptive significance of fighting to the deer, that is, on why a particular behavior evolved, in terms of its effect on reproductive success. These factors are called **ultimate causes** of behavior. Since the 1970s, behavioral ecologists have focused more on understanding the ultimate causes of behavior.

In this chapter, we will explore the role of both proximate and ultimate causes of behavior. We begin the chapter by investigating how behavior is achieved, that is, we look at the role of both genetics and the environment. In doing so, we will examine the important contributions of ethologists von Frisch, Lorenz, and Tinbergen. We consider how different behaviors are involved in movement, gathering food, and communication. Later, we investigate how organisms interact in groups, whether an organism can truly behave in a way that benefits others at a cost to itself, and how behavior shapes different mating systems. You will note that much of the chapter focuses on animal behavior, because the behavior of other organisms is more limited and less well understood.

55.1 The Impact of Genetics and Learning on Behavior

Behavior is rarely due solely to genes (nature) or solely to environmental influences (nurture), but usually to both. Determining to what degree a behavior is influenced by genes versus the environment will depend on the particular genes and environment examined. However, in a few cases, changes in behavior may be caused by variation in just one gene. Even if a given behavior is influenced by many genes, if one gene is altered, it is possible that the entire behavior can change. To use the analogy of baking a cake, a change in one ingredient of the recipe may change the whole taste of the cake, but that does not mean that the one ingredient is responsible for the entire cake.

In this section, we begin by examining the effect that a single gene can exert on behavior and consider several examples of simple, genetically programmed behaviors. Later, we explore several types of learned behavior, including classical and operant conditioning and cognitive learning, and conclude the section by exploring an example of the interaction of genetics and learning on behavior.

GENOMES & PROTEOMES

Some Behavior Results from Simple Genetic Influences

An excellent example of the effect of a single gene on complex behavior was demonstrated in W. C. Rothenbuhler's 1964 work on honeybees. Some strains of bees are termed hygienic, that is, they detect and remove diseased larvae from the nest. This behavior involves two distinct maneuvers: uncapping the wax cells and then discarding the dead larvae. Other strains are not hygienic and do not exhibit such behavior. Rothenbuhler demonstrated, by genetic crosses, that one gene (*u*) controlled cell uncapping and another gene (*r*) controlled larval removal. Double recessives (*uurr*) were hygienic strains, and double dominants (*UURR*) were nonhygienic strains. When the two strains were crossed, all the F$_1$ hybrids were nonhygienic (*UuRr*). When the F$_1$ hybrids were backcrossed with the pure hygienic strain (*uurr*), four different genotypes were produced, as Mendel's law of independent assortment predicts (Chapter 16): one-quarter of the offspring were hygienic (*uurr*), one-quarter were nonhygienic and showed neither behavior (*UuRr*), one-quarter uncapped the cells but failed to remove the larvae (*uuRr*), and one-quarter removed the larvae but only if the cells were uncapped for them (*Uurr*).

In 1996, Jennifer Brown and her colleagues showed how a single gene, *fosB*, controlled nurturing behavior in mice. Normal mice, which carry the gene, clean their newborn pups, nurse them, and crouch over them to keep them warm. Brown's group created mutant mice lacking the *fosB* gene. Despite fully functioning mammary glands, mutant mothers paid little attention to their newborn pups, which remained scattered around the cage and died within a few days. If the pups of the *fosB* mutants were placed with normal females, they thrived, but if normal pups were given to mutant mothers, they too were neglected.

Genes for behavior actually act on the development of nervous systems and musculature, physical traits that evolve through natural selection. Many genes are needed for the proper development and function of the nervous system and musculature. Even so, as described by Rothenbuhler's and Brown's work, variation in a single gene can have a dramatic impact on behavior.

Fixed Action Patterns Are Genetically Programmed

Behaviors that seem to be genetically programmed are referred to as **innate** (also called instinctual). While we recognize that expression of genes varies, often in response to environmental stimuli, some behavior patterns evidently are genetically quite fixed. Most individuals will exhibit the same behavior regardless of the environment. A spider will spin a specific web without ever seeing a member of its own species build one. The courtship behaviors of many bird species are so stereotyped as to be virtually identical.

A classic example of innate behavior is the egg-rolling response in geese (**Figure 55.1**). If an incubating goose notices an egg out of the nest, she will extend her neck toward the egg, get up, and then roll the egg back to the nest using her beak. Such behavior functions to improve fitness because it increases the survival of offspring. Eggs that roll out of the nest get cold and fail to hatch. Geese that fail to exhibit the egg-rolling response would pass on fewer of their genes to future generations. Egg-rolling behavior is an example of what ethologists term a **fixed action pattern (FAP)**, a behavior that, once initiated, will continue until completed. For example, if the egg is removed while the goose is in the process of rolling it back toward the nest, the goose still completes the FAP, as though she were rolling back

1   Female goose extends her neck toward egg.

2   Goose gets up from nest and approaches egg.

3   Goose places her neck above the egg.

4   Goose rolls egg back to nest with her beak and neck.

Figure 55.1   **A fixed action pattern as an example of innate behavior.**   Female geese retrieve eggs that have rolled outside the nest through a set sequence of movements. The goose will complete this entire sequence even if a researcher takes the egg away before the goose has rolled it back to the nest.

the now-absent egg to the nest. The stimulus to initiate this behavior is obviously a strong one, which ethologists term a **sign stimulus**. The sign stimulus for the goose is that an egg had rolled out of the nest. According to ethologists, this stimulus acts on the goose's central nervous system, which provides a neural stimulus to initiate the motor program or FAP. Interestingly, any round object will elicit the egg-rolling response, from a wooden egg to a volleyball. While sign stimuli usually have certain key components, they are not necessarily very specific.

Niko Tinbergen's study of male stickleback fish provides another classic example of FAPs and sign stimuli. Male sticklebacks, which have a characteristic red belly, will attack other male sticklebacks that invade their territory. Tinbergen found that sticklebacks attacked small, unrealistic model fish having a red ventral surface, while ignoring a realistic male stickleback model that lacked a red underside (**Figure 55.2**).

Conditioning Occurs When a Relationship Between a Stimulus and a Response Is Learned

Although many of the behavioral patterns exhibited by animals are largely innate, sometimes animals can make modifications to their behavior based on previous experience, a process called **learning**. Perhaps the simplest form of learning is **habituation**, in which an organism learns to ignore a repeated stimulus. For example, animals in African safari parks become habituated to the presence of vehicles containing tourists; these vehicles are neither a threat nor a benefit. After a while, birds can become habituated to the presence of a scarecrow, resulting in damage to crops. Habituation can be a problem at airfields, where birds eventually ignore the alarm calls designed to scare them away from the runways.

Habituation is a form of nonassociative learning in which there is a decrease in response to a stimulus due to repetition. Alternatively, an association may gradually develop between a stimulus and a response. Such a change in behavior is termed **associative learning**. In associative learning, a behavior is changed or conditioned through the association. The two main types of associative learning are termed classical conditioning and operant conditioning.

In **classical conditioning**, an involuntary response comes to be associated positively or negatively with a stimulus that did not originally elicit the response. This type of learning is generally associated with the Russian psychologist Ivan Pavlov. In his original experiments in the 1920s, Pavlov restrained a hungry dog in a harness and presented small portions of food at regular intervals. The dog would salivate whenever it smelled the food. Pavlov then began to sound a metronome when presenting the food. Eventually the dog would salivate at the sound of the metronome, whether or not the food was present. Pavlov termed the food the **unconditioned stimulus** for salivation, while the metronome was the **conditioned stimulus**. Likewise, he termed salivation in response to food as an **unconditioned response** and salivation in response to the metronome as a **conditioned response**. Classical conditioning is pervasive in human society and is widely observed in other animals. For example, many insects quickly learn to associate certain flower odors with nectar rewards and other flower odors with no rewards.

In **operant conditioning**, an animal's behavior is reinforced by a consequence, either a reward or a punishment. The classic example of operant conditioning is generally associated with the American psychologist B. F. Skinner, who placed laboratory animals, usually rats, in a specially devised cage that came to be known as a Skinner box. As the rat explored its cage, it could press on a small lever that would result in food being released. At the beginning of the experiment, the rat would often bump into the lever by accident, eat the food, and continue exploring its cage. Later, it would learn to associate the lever with getting food. Eventually, if it was hungry, the rat would almost continually press the lever. Operant conditioning, also called trial-and-error learning, is common in animals. Often it is associated with negative rather than positive reinforcement. For example, toads will eventually refuse to strike at insects that sting, such as wasps and bees, and birds will learn to avoid bad-tasting butterflies (**Figure 55.3**).

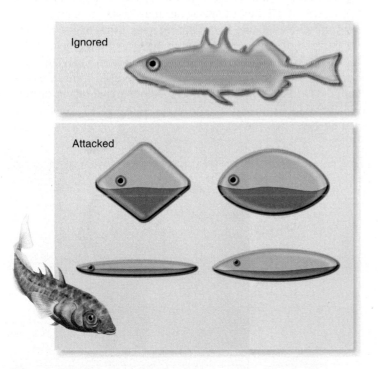

Figure 55.2 A fixed action pattern elicited by a sign stimulus. The sign stimulus for male sticklebacks to attack other males entering their territory is a red ventral surface. In experiments, male sticklebacks attacked all models that had a red underside, while ignoring a realistic model of a stickleback that lacked the red belly.

Cognitive Learning Involves Conscious Thought

Cognitive learning refers to the ability to solve problems with conscious thought and without direct environmental feedback.

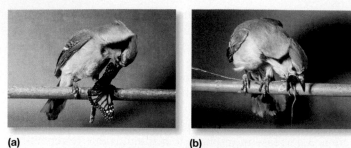

(a)   **(b)**

Figure 55.3 Operant conditioning, also known as trial-and-error learning. **(a)** A young blue jay will eat a monarch butterfly, not knowing that it is noxious. **(b)** After the first experience of vomiting after eating a monarch, a blue jay will avoid the insects in the future.

In the 1920s, psychologist Wolfgang Kohler conducted a series of classic experiments with chimpanzees that suggested that animals could exhibit cognitive learning. In the experiments, a chimpanzee was left in a room with a banana hanging from the ceiling and out of reach (**Figure 55.4**). Also present in the room were several wooden boxes. At first, the chimp tried in vain to jump up and grab the bananas. After a while, however, it began to arrange the boxes one on top of another underneath the fruit. Eventually, the chimp climbed the boxes and retrieved the fruit. This clearly looks like an example of problem solving that involves cognitive learning.

In fact, many other examples of such behavior exist. Chimps strip leaves off twigs and use the twigs to poke into ant nests, withdrawing the twig and licking the ants off (see chapter-opening photo). Captive ravens have been shown to retrieve meat suspended from a branch by a string, even though they have never encountered the problem before. They pull up on the string, step on it, and then pull up on the string again, repeating the process until the meat is within reach.

The Interaction of Genetics and Learning on Behavior Often Involves Imprinting

Much of the behavior we have discussed so far has been presented as either innate or learned, but the behavior we observe in nature is often a mixture of both. Bird songs present a good example. Many birds learn their songs as juveniles, when they hear their parents sing. If juvenile white-crowned sparrows are raised in isolation, their adult songs do not resemble the typical species-specific song (**Figure 55.5**). If they hear only the song of a different species, such as the song sparrow, they again sing a poorly developed adult song. However, if they hear the song of the white-crowned sparrow, they will learn to sing a fully developed white-crowned sparrow song. The birds are genetically programmed to learn, but they will sing the correct song only if the appropriate instructive program is in place to guide learning.

Another example of how innate behavior interacts with learning can occur during a limited time period of development, called a **critical period**. At this time, many animals develop species-specific patterns of behavior. This process is called **imprinting**. One of the best examples of imprinting was demonstrated by the Austrian ethologist Konrad Lorenz in the 1930s. Lorenz noted that young birds of some species imprint on their mother during a critical period, usually within a few hours after hatching. This behavior serves them well, because in many species of ducks and geese it would be hard for the mother to keep track of all her offspring as they walk or swim. After imprinting takes place, the offspring keep track of the mother. The survival of the young ducks requires that they quickly learn to follow their mother's movements. Lorenz raised greylag geese from eggs, and soon after they hatched, he used himself as the model for imprinting. As a result, the young goslings imprinted on Lorenz and followed him around (**Figure 55.6**). For the rest of their life, they preferred the company of Lorenz and other

Figure 55.4 Cognitive behavior involves problem-solving ability. This chimp has devised a solution to the problem of retrieving bananas that were initially out of its reach.

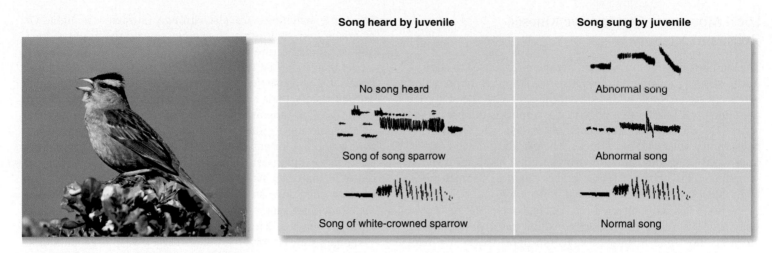

Song heard by juvenile

No song heard

Song of song sparrow

Song of white-crowned sparrow

Song sung by juvenile

Abnormal song

Abnormal song

Normal song

Figure 55.5   **The interaction between genetics and learning.**   The juvenile white-crowned sparrow will sing an abnormal song if it is kept in isolation or hears only the song of a different species. However, the juvenile will sing the normal white-crowned sparrow song if exposed to it.

Biological inquiry: Cuckoos lay their eggs in other birds' nests, so their young are reared by parent birds of a different species. However, unlike the white-crowned sparrow, adult cuckoos always sing their own distinctive song, not that of the host species they hear as juveniles. How is this possible?

humans to geese. Studies have shown that even an object as foreign as a black box, watering can, or flashing light will be imprinted on if it is the first moving object the chick sees during the critical period. In nature, if the young geese failed to be provided with any stimulus during the critical period, they would fail to imprint on anything, and without parental care, they would almost certainly die.

Other animals imprint in different ways. Newborn shrews imprint on the scent of their mother. Mothers also can imprint

Figure 55.6   **Konrad Lorenz being followed by his imprinted geese.**   Newborn geese follow the first object that they see on hatching and later will follow that particular object only. They normally will follow their mother but can be induced to imprint on humans. The first thing these young geese saw on hatching was ethologist Konrad Lorenz.

on their own young within a few hours. For example, if sheep mothers are kept apart from their offspring for only a few hours after birth, they will reject them. In these situations, the innate behavior is the ability to imprint soon after birth, and the factors in the environment are the stimulus to which the imprinting is directed.

Finally, innate behavior can interact with learning during animal migration. Inexperienced juvenile birds will migrate in a particular direction but will fail to correct for deviations if they are blown off course. Experienced adult birds, on the other hand, can often correct for storm-induced displacement, indicating they have more complex navigational skills. There are in fact many complex behaviors involved in movement, as we will explore in the next section.

55.2 Local Movement and Long-Range Migration

Organisms need to find their way, both locally and over what can be extremely long distances. Locally, organisms continually need to locate sources of food, water, and perhaps nesting sites. Migration involves the longer-distance seasonal movement of animals between overwintering areas and summer breeding sites; these are often hundreds or even thousands of kilometers apart. Several different types of behavior may be involved in these movements. In this section, we begin by exploring local movement and, in particular, how one species uses landmarks to guide its movements. We then consider the longer-range seasonal movement called migration and examine three possible mechanisms used by migrating animals to find their way.

Local Movement Can Involve Kinesis, Taxis, and Memory

The simplest forms of movement are mere responses to stimuli. A **kinesis** is a movement in response to a stimulus, but one that is not directed toward or away from the source of the stimulus. A simple experiment often done in classrooms is to observe the activity levels of woodlice, sometimes called sow bugs or pill bugs, in dry areas and moist areas. The woodlice move faster in drier areas, and they slow down when they reach moist environments. This behavior tends to keep them in damper areas, which they prefer in order to avoid desiccation.

A **taxis** is a more directed type of response either toward or away from a stimulus. Cockroaches exhibit negative phototaxis, meaning they tend to move away from light. Under low-light conditions, the photosynthetic unicellular flagellate *Euglena gracilis* shows positive phototaxis and moves toward a light source.

Sea turtle hatchings are also strongly attracted to light. On emerging from their nests, they crawl toward the brightest location, traditionally the reflected moonlight on the ocean's surface. Lighted houses on the shore can confuse the hatchlings, however, and cause them to lose their way, with sometimes disastrous results. Male silk moths orient themselves in relation to wind direction (anemotaxis). If the air current carries the scent of a female moth, they will move upwind to locate it. Some freshwater fish orient themselves to the currents of streams. Many fish exhibit positive rheotaxis (from the Greek *rheos*, current), in that they swim against the water current to prevent being washed downstream.

Sometimes memory and landmarks may be used to aid in local movements. The prominent Dutch-born ethologist Niko Tinbergen showed how the female digger wasp uses landmarks to relocate her nests as described next.

FEATURE INVESTIGATION

Tinbergen's Experiments Show That Digger Wasps Use Landmarks to Find Their Nests

In the sandy, dry soils of Europe, the solitary female digger wasp (*Philanthus triangulum*) digs four to five nests in which to lay her eggs. Each nest stretches obliquely down into the ground for 40–80 cm. The wasp, sometimes called a bee wolf, follows this by performing a sequence of apparently genetically programmed events. She catches and stings a honeybee, which paralyzes it; returns to the nest; drags the bee into the nest; and lays an egg on it. The egg hatches into a larva, which feeds on the paralyzed bee. However, the larva needs to ingest five to six bees before it is fully developed. This means the wasp must catch and sting four to five more bees for each larva. She can only carry one bee at a time. After each visit, the wasp must seal the nest, find a new bee, relocate the nest, open it, and add the bee. How does the wasp relocate the nest after spending considerable time away? Niko Tinbergen observed the wasps hover and

fly around the nest each time they took off. He hypothesized that they were learning the nest position by creating a mental map of the landmarks in the area.

To test his hypothesis, Tinbergen experimentally adjusted the landmarks around the burrow that the wasps might be using as cues (**Figure 55.7**). First, he put a ring of pinecones around the nest entrance to train the wasp to associate the pinecones with the nest. Then, when the wasp was out hunting, he moved the circle of pinecones a distance from the real nest and constructed a sham nest, making a slight depression in the sand and mimicking the covered entrance of the burrow. On returning, the wasp flew straight to the sham nest and tried to locate the entrance. Tinbergen chased it away. When it returned, it again flew to the sham nest. Tinbergen repeated this nine times, and every time the wasp chose the sham nest. Tinbergen got the same result with 16 other wasps and not once did they choose the real nest.

Figure 55.7   How Niko Tinbergen discovered the digger wasp's nest-locating behavior.

Biological inquiry: How would you test what type of spatial landmarks are used by female digger wasps?

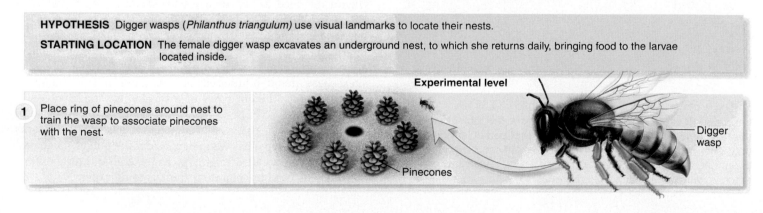

HYPOTHESIS Digger wasps (*Philanthus triangulum*) use visual landmarks to locate their nests.

STARTING LOCATION The female digger wasp excavates an underground nest, to which she returns daily, bringing food to the larvae located inside.

Experimental level

1   Place ring of pinecones around nest to train the wasp to associate pinecones with the nest.

Pinecones

Digger wasp

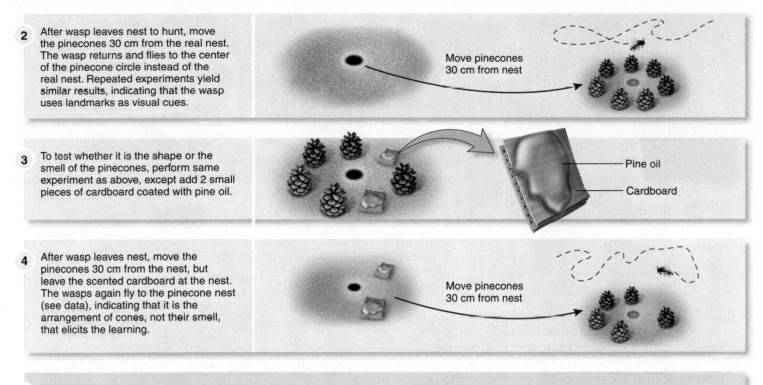

2 After wasp leaves nest to hunt, move the pinecones 30 cm from the real nest. The wasp returns and flies to the center of the pinecone circle instead of the real nest. Repeated experiments yield similar results, indicating that the wasp uses landmarks as visual cues.

Move pinecones 30 cm from nest

3 To test whether it is the shape or the smell of the pinecones, perform same experiment as above, except add 2 small pieces of cardboard coated with pine oil.

Pine oil

Cardboard

4 After wasp leaves nest, move the pinecones 30 cm from the nest, but leave the scented cardboard at the nest. The wasps again fly to the pinecone nest (see data), indicating that it is the arrangement of cones, not their smell, that elicits the learning.

Move pinecones 30 cm from nest

5 THE DATA

Wasp #	Number of return visits to real nest with scented cardboard	Number of return visits to sham nest with pinecones
18	0	5
19	0	5
20	0	6
21	0	0
22	0	5

*Seventeen wasps were studied as described in steps 1 and 2. Five wasps, numbered 18–22, were studied as described in steps 3 and 4.

Next Tinbergen experimented with the type of stimulus that might be eliciting the learning. He hypothesized that the wasps could be responding to the distinctive scent of the pinecones rather than their appearance. He trained the wasps by placing a circle of pinecones that had no scent and two small pieces of cardboard coated in pinecone oil around the real nest.

He then moved the cones to surround a sham nest and left the scented cardboard around the real nest. The returning wasps again ignored the real nest with the scented cardboard and flew to the sham. He concluded that for the wasps, sight was apparently more important than smell in determining landmarks.

Migration Involves Long-Range Movement and More Complex Spatial Navigation

As well as learning to navigate over short-range distances, many animals undergo **migration**, long-range seasonal movement. Migrations occur in many species and usually involve a movement away from a birth area to feed and a return to the birth area to breed, with the movement generally being linked to

seasonal availability of food. For example, nearly half the breeding birds of North America migrate to South America to escape the cold winters and feed, and then they return to North America in the spring to breed. Arctic terns that breed in Arctic Canada and Asia migrate to the Antarctic to feed in the winter and then return to breed. This staggering journey involves a 40,000-km (24,800-mi) round-trip, most of it over the open ocean, during which the birds must stay airborne for days at a time!

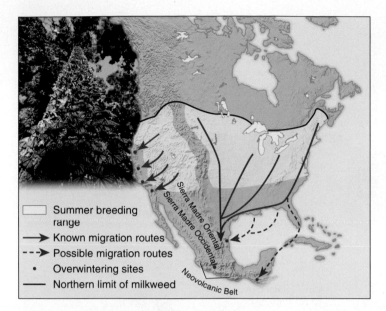

Figure 55.8 **Monarch butterfly migration.**   Many monarch butterflies east of the Rocky Mountains migrate to a small area in Mexico to avoid the cold northern weather. Here they roost together in large numbers in fir trees (inset). Some butterflies may stay in Florida and Cuba. Butterflies west of the Rockies overwinter in mild coastal California locations.

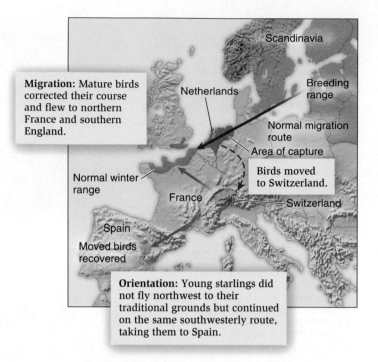

Figure 55.9 **Orientation versus navigation.**   Starlings normally migrate from breeding grounds in Scandinavia and northeastern Europe through the Netherlands and northern Germany to overwintering sites in northern France and southeast England. This involves a southwest flight. When juveniles were captured in the Netherlands and moved to Switzerland, they continued on in a southwestern direction and ended up in Spain. When experienced birds were captured and moved, they flew to the normal overwintering areas.

Many mammals, including wildebeest and caribou, make migrations that track the appearance of new vegetation on which they feed. The monarch butterfly of North America migrates to overwinter in California, Mexico, and probably south Florida and Cuba (**Figure 55.8**). This migration is unique in that none of the individuals has ever been to the migration sites before. An interesting point about the return journey of the monarch is that it involves several generations of butterflies to complete. On their way back to the northern U.S. and Canada, they lay eggs and die. The caterpillars develop on milkweed plants, and the resultant adults continue to journey farther north. This happens several times in the course of the return journey.

How do migrating animals find their way? Three mechanisms have been proposed: piloting, orientation, and navigation. In **piloting**, an animal moves from one familiar landmark to the next. For example, many whale species migrate between overwintering areas and summer calving grounds. Grey whales migrate between the Bering Sea near Alaska to coastal areas of Mexico. Features of the coastline, including mountain ranges, and rivers, may aid in navigation. In **orientation**, animals have the ability to follow a compass bearing and travel in a straight line. **Navigation** involves the ability not only to follow a compass bearing but also to set or adjust it.

An experiment with starlings helps illuminate the difference between orientation and navigation (**Figure 55.9**). European starlings breed in Scandinavia and northeastern Europe and migrate in a southwest direction toward coastal France and southern England to spend the winter. Migrating starlings were captured and tagged in the Netherlands and then transported south to Switzerland and released. Juvenile birds, which had never made the trip before, flew southwest in their migration and were later recaptured in Spain. Adult birds, with more experience, returned to their normal wintering range by adjusting their course by approximately 90°. This implies that the adult birds can actually navigate, whereas the juveniles rely on orientation.

Many species use a combination of navigational reference points, including the position of the sun, the stars (for nighttime travel), and the Earth's magnetic field. Homing pigeons have magnetite in their beaks that acts as a compass to tell direction (look back to Section 45.4, Electromagnetic Sensing). Navigation by the sun or the stars also requires the use of a timing device to compensate for the ever-changing position of these reference points. Many migrants, therefore, possess the equivalent of an internal clock. Pigeons integrate their internal clock with the position of the sun. Researchers have altered the internal clock of pigeons by keeping them under artificial lights for certain periods of time. When the pigeons are released, they display predictable deviations in their flight. For every hour that their internal clock is shifted, the orientation of the birds shifts about 15°.

Not all examples of animal migration are well understood. Green sea turtles feed off the coast of Brazil yet swim east for 2,300 km (1,429 mi) to Ascension Island, an 8-km-wide island in the center of the Atlantic Ocean between Brazil and Africa, to lay their eggs. It is not known why the turtles lay their eggs on this speck of an island or how they succeed in finding it. Perhaps fewer predators exist on Ascension than on other beaches. Thus, while scientists have made many discoveries about animal navigation, much remains to be learned about how animals acquire a "map sense."

To a large extent, local and long-distance movement involves searching for food. Organisms are constantly faced with a decision of how long to stay in a food patch and when to look for a new source of food. In the next section, we will investigate how such foraging decisions are made.

55.3 Foraging Behavior

Food gathering, or foraging, often involves decisions about whether to remain at a resource patch and look for more food or look for a completely new patch. The analysis of these decisions is often performed in terms of **optimality theory**, which predicts that an animal should behave in a way that maximizes the benefits of a behavior minus its costs. In this case, the benefits are the nutritional or calorific value of the food items, and the costs are the energetic or calorific costs of movement. When the difference between the energetic benefits of food gathering and the energetic costs of food gathering is maximized, an organism is said to be optimizing its foraging behavior. We should note that optimality theory can also be used to investigate other behavioral issues such as how large a territory to defend. Too small a territory would contain insufficient food, and too large a territory would be too energetically costly to defend. Theoretically, then, there is an optimal territory size for a given species.

Optimal Foraging Entails Maximizing the Benefits and Minimizing the Costs of Food Gathering

Optimal foraging proposes that in a given circumstance, an animal seeks to obtain the most energy possible with the least expenditure of energy. The underlying assumption of optimal foraging is that natural selection favors animals that are maximally efficient at propagating their genes and at performing all other functions that serve this purpose. In this model, the more net energy an individual gains in a limited time, the greater the reproductive success.

Studies show that animals tend to modify their behavior in a way that maximizes the difference between their energy uptake and their energy expenditure. For example, shore crabs (*Carcinus maenas*) will eat many different-sized mussels but tend to feed preferentially on intermediate-sized mussels, which give them the highest rate of energy return (**Figure 55.10**). Very

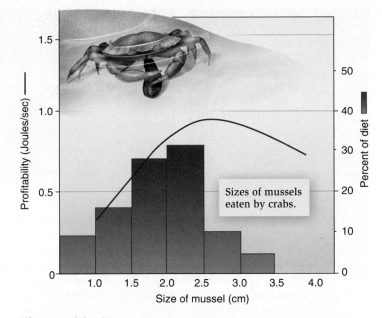

Figure 55.10 **Optimal foraging behavior in shore crabs.** In an aquarium setting, when offered a choice of equal numbers of each size mussel, the shore crab (*Carcinus maenas*) prefers intermediate-sized mussels that provides the highest rate of energy return. Profitability is the energy yield per second of time used in breaking open the shell.

large mussels yield more energy, but they take so long for the crab to open that they are actually less profitable, in terms of energy yield per unit time spent, than smaller sizes. Very small mussels are easy to crack open but contain so little flesh that they are not worth the effort. This leaves intermediate-sized mussels as the preferred size. Of course, the intermediate-sized mussels may take a longer time to locate, because more crabs are looking for them, so crabs eat some less profitable but more frequently encountered sizes. The result is that the diet consists of mussels in a range of sizes around the preferred optimal size.

In some cases, animals do not always forage optimally. For example, animals seek not only to maximize food intake but also to minimize the risk of predation. Some species may only dart out to take food from time to time. The risk of predation thus has an influence on foraging behavior. Many animals also maintain territories to minimize competition with other individuals and control resources, whether food, nesting sites, or mates. However, as we will see, defending these territories also has an energetic cost.

Defending Territories Has Costs and Benefits

Many animals, or groups of animals, such as a pride of lions, actively defend a **territory**, a fixed area in which an individual or group excludes other members of its own species, and sometimes other species, by aggressive behavior or territory marking.

(b) Cheetah

(c) Nesting gannets

(a) Golden-winged sunbird

Figure 55.11   Territory sizes differ among animals.   (a) The golden-winged sunbird of East Africa (*Nectarinia reichenowi*) has a medium territory size that is dependent on the number of flowers it can obtain resources from and defend. (b) Cheetahs (*Acinonyx jubatus*) hunt over large areas and can have extensive territories. This male is urine-marking part of his territory in the southern Serengeti, near Ndutu, Tanzania. (c) Nesting gannets (*Morus bassanus*) have much smaller territories, in which each bird is just beyond the pecking range of its neighbor.

Territory owners tend to optimize territory size according to the costs and benefits involved. The primary benefit of a territory is that it provides exclusive access to a particular resource, whether it be food, mates, or sheltered nesting sites. Large territories may provide more of a resource but may be costly to defend, while small territories that are less costly to defend may not provide enough of a resource.

In studies of the territorial behavior of the golden-winged sunbird (*Nectarinia reichenowi*) in East Africa, researchers Frank Gill and Larry Wolf measured the energy content of nectar as the benefit of maintaining a territory while the energy costs included activities such as perching, foraging, and defending (**Figure 55.11a**). Defending the territory ensured that other sunbirds did not take nectar from available flowers, thus increasing the amount of nectar in each flower. In defending a territory, the sunbird saved 780 calories a day in reduced foraging activity. However, the sunbird also spent 728 calories in defense of the territory, yielding a net gain of 52 calories a day.

Territory size differs considerably among species, and optimality theory predicts that it should evolve to maximize the difference between benefits and costs, thus maximizing the fitness "profit" to the territory holder. Because cheetahs need large areas to be able to hunt successfully, they establish large territories relative to their size (**Figure 55.11b**). In contrast, territories set up solely to defend areas for mating or nesting are often relatively small. For example, male sea lions defend small areas of beach. The preferred areas contain the largest amount of females and are controlled by the largest breeding bulls. The size of the territory of some nesting birds, such as gannets, is determined by how far the bird can reach to peck its neighbor without leaving its nest (**Figure 55.11c**).

Territories may be held for a season, a year, or the entire lifetime of the individual. Ownership of a territory needs to be periodically proclaimed, and thus communication between indi-viduals is necessary for territory owners. This may involve various types of signaling, which we discuss next.

55.4 Communication

Communication is the use of specially designed signals or displays to modify the behavior of others. It may be used for many purposes, including defining territories, maintaining contact with offspring, courtship, and contests between males. The use of different forms of communication between organisms depends on the environment in which they live. For example, visual communication plays little role in the signals of nocturnal animals. Similarly, animals in dense forests often cannot see each other, so sounds are of prime importance in mapping out territories. Sound, however, is a temporary signal. Scent can last longer and is often used to mark the large territories of some mammals. In this section, we outline the various types of communication—chemical, auditory, visual, and tactile—that occur among animals.

Chemical Communication Is Often Used to Mark Territories or Attract Mates

The chemical marking of territories is common among animals, especially among members of the canine and feline families (see Figure 55.11b). Scent trails are often used by social insects to recruit workers to help bring prey to the nest. Fire ants (*Solenopsis* spp.) attack large, living prey, and many ants are needed to drag the prey back to the nest. In this case, the scout that finds the prey lays down a scent trail from the prey back to the nest. The scent excites other workers, which follow the trail to the prey. The scent marker is very volatile, and the trail effectively disappears in a few minutes to avoid mass confusion over old trails.

Animals frequently use chemicals to attract mates. Female moths attract males by powerful chemical attractants called **pheromones**. Among social organisms, some individuals use pheromones to manipulate the behavior of others. A queen bee releases pheromones that suppress the reproductive system of workers, which ensures that she is the only reproductive female in the hive.

Auditory Communication Is Often Used to Attract Mates and to Deter Competitors

Many organisms communicate by making sound. Because the Earth itself can absorb sound waves, sound travels farther in the air, which is why many birds and insects use perches when singing. Air is on average 14 times less turbulent at dawn and dusk than during the rest of the day, which helps explain the preference of most animals for calling at these times. Some insects utilize the very plants on which they feed as a medium of song transmission. Many male leafhopper and planthopper insects vibrate their abdomens on leaves and create species-specific courtship songs that are transmitted by adjacent and touching vegetation and are picked up by nearby females of the same species.

While many males use auditory communication to attract females, some females use calls to attract the attention of males. Female elephant seals use this behavior to their advantage. When a nondominant male attempts to mate with a female, she screams loudly, attracting the attention of the dominant male, which drives the nondominant male away. In this way, she is guaranteed a mating with the strongest male. Sound production can attract predators as well as mates. Some bats listen for the mating calls of male frogs to find their prey. Parasitic flies detect and locate chirping male crickets and then deposit tiny maggots on or near them. The maggots latch onto and penetrate the cricket, and eventually kill it. Sound may also be used by males during competition over females. In many animals, lower-pitched sounds come from bigger males, so by calling to one another, males can gauge the size of their opponents and save fighting energy.

Visual Communication Is Often Used in Courtship and Aggressive Displays

In courtship, animals use a vast number of visual signals to identify and select potential mates. Competition among males for the most impressive displays to attract females has led to elaborate coloration and extensive ornamentation in some species. For example, peacocks and males of many bird species have developed elaborate plumage to attract females.

Male fireflies have developed light flashes that are species specific with regard to number and duration of flashes (**Figure 55.12**). Females respond with a flash of their own. Such bright flashes are also bound to attract predators. Some female fireflies use mimicry to their advantage. Female *Photuris versicolor* fireflies mimic the flashing responses normally given by females of other species, such as *Photinus tamytoxus*, in order to lure the males of those species close enough to eat them.

Visual signals are also used to resolve disputes over territories or mates. Deer and antelope have antlers or horns that they use to display and spar over territory and females. Most of these matches never develop into outright fights, since the males gauge their opponent's strength by the size of these ornaments. Among insects, the "horns" of rhinoceros beetles and the eye stems of stalk-eyed flies send similar signals.

Tactile Communication Is Used to Strengthen Social Bonds and to Convey Information About Food

Animals often use tactile communication to establish bonds between group members. Primates frequently groom one another, while canines and felines may nuzzle and lick each other. Many insects use tactile communication to convey information on the whereabouts of food. Members of the ant genus *Leptothorax* feed on immobile prey such as dead insects. When a scouting ant encounters such prey, it usually needs an additional worker to help bring it back to the nest.

(a)   (b)   (c)

Figure 55.12   Visual communication in fireflies. (a) Communication between fireflies is conducted by species-specific light flashes emitted by organs located on the underside of the abdomen. (b) At dusk, many different light flashes can be seen. (c) Sometimes females, such as this large *Photuris versicolor*, mimic the displays of other species, luring an unwitting male, such as this *Photinus tamytoxus*, and then eating him.

(a) **Bees clustering around a recently returned worker.**

(b) **Round dance**

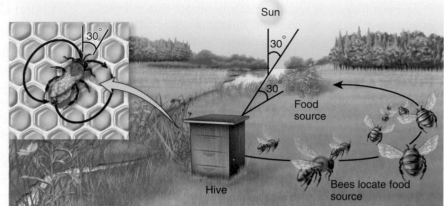

(c) **Waggle dance: The angle of the waggle to the vertical orientation of the honeycomb corresponds to the angle of the food source from the sun.**

Figure 55.13 Tactile communication among honeybees regarding food sources. (a) Bees gather around a newly returned scout to receive information about nearby food sources. (b) If the food is less than 50 m away, the scout performs a round dance. (c) If the food is more than 50 m away, the scout performs a waggle dance, which conveys information about its location. If the dance is performed at a 30° angle to the right of the hive's vertical plane, then the food source is located at a 30° angle to the right of the sun.

Rather than laying a scent trail, which is energetically costly, the scout ant recruits a helper and physically leads it to the food source. The helper runs in tandem with the scout, its antennae touching the scout's abdomen.

Perhaps the most fascinating example of tactile communication among animals is the dance of the honeybee, elegantly studied by German ethologist Karl von Frisch in the 1940s. Bees commonly live in large hives; in the case of the European honeybee (*Apis mellifera*), the hive consists of 30,000–40,000 individuals. The flowering plants on which the bees forage are sometimes located miles from the hive and are distributed in a patchy manner, with any given patch usually containing many flowers that store more nectar and pollen than an individual bee can carry back to the nest. The scout bee that locates the resource patch returns to the hive and recruits more workers to join it (**Figure 55.13a**). Because it is dark inside the hive, the bee uses a tactile signal. The scout dances on the vertical side of a honeycomb, and the dance is monitored by other bees, which follow and touch her to interpret the message. If the food is relatively close to the hive, less than 50 m away, the scout performs a round dance, rapidly moving in a circle, first in one direction and then the other. The other bees know the food is relatively close at hand, and the smell of the scout tells them what flower species to look for (**Figure 55.13b**).

If the food is more than 50 m away, the scout will perform a different type of dance, called a "waggle dance." In this dance, the scout traces a figure 8, in the middle of which she waggles her abdomen and produces bursts of sound. Again, the other bees maintain contact with her. Occasionally, the scout will regurgitate a small sample of nectar so the bees know the type of food source they are looking for. The truly amazing part of the waggle dance is that the angle at which the central part of the figure 8 deviates from the vertical direction of the comb represents the same angle at which the food source deviates from the

sun (**Figure 55.13c**). The direction is always up-to-date, because the bee adjusts the dance as the sun moves across the sky.

As we have seen, much communication occurs not only to defend territories but also to communicate information to other individuals in the population, including potential mates. While living on your own and maintaining a territory has advantages, living in a group also has its benefits, including ready availability of mates and increased protection from predators. In the following section, we examine group living and the behavior it engenders.

55.5 Living in Groups

Much animal behavior is directed at other animals. As such, some of the more complex behavior occurs when animals live together in groups such as flocks or herds. If a central concern of ecology is to explain the distribution patterns of organisms, then one of our most important tasks is to identify and understand the behavior that results from group living. While congregations promote competition for food, there are also benefits of group living that compensate for the costs involved. As we discuss in this section, many of these benefits relate to group defense against predators. Group living can reduce predator success in at least two ways: through increased vigilance and through protection in numbers.

Living in Large Groups May Reduce the Risk of Predation Because of Increased Vigilance

For many predators, success depends on surprise. If an individual is alerted to an attack, the predator's chance of success is lowered. A woodpigeon (*Columba palumbus*) will take to the air when it spots a goshawk (*Accipiter gentilis*). Once one pigeon

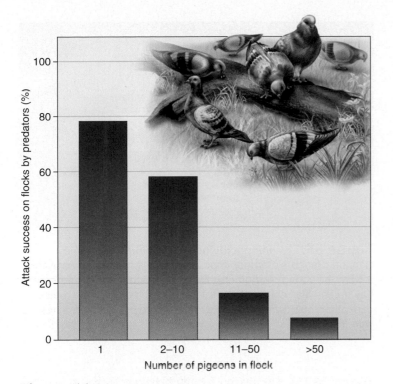

Figure 55.14 Living in groups and the many eyes hypothesis. The larger the number of woodpigeons, the less likely an attack will be successful.

takes flight, the other members of the flock are alerted and follow suit. If each pigeon occasionally looks up to scan for a hawk, the bigger the group, the more likely that one bird will spot the hawk early enough for the flock to take flight. This is referred to as the **many eyes hypothesis** (**Figure 55.14**). By living in groups, individuals may decrease the amount of time scanning for predators and increase the time they have to feed. Of course, cheating is a possibility, because some birds might never look up, relying on others to keep watch while they keep feeding. However, the individual that happens to be scanning when a predator approaches is most likely to escape, a fact that tends to discourage cheating.

Living in Groups Offers Protection by the "Selfish Herd"

Group living also provides protection in sheer numbers. Typically, predators take one prey animal per attack. In any given attack, an individual antelope in a herd of 100 has a 1 in 100 chance of being selected, whereas a single individual has a 1 in 1 chance. Large herds may be attacked more frequently than a solitary individual, but a herd is unlikely to attract 100 times more attacks than an individual, often because of the territorial nature of predators. Furthermore, large numbers of prey are able to defend themselves better than single individuals, which usually choose to flee. For example, groups of nesting black-headed gulls will mob a crow relentlessly, thereby reducing the crow's ability to steal the gulls' eggs.

Research has shown that within a group, each individual can minimize the danger to itself by choosing the location that is as close to the center of the group as possible. This was the subject of a famous paper, "The Geometry of the Selfish Herd," by the British evolutionary biologist W. D. Hamilton. The explanation of this type of defense is that predators are likely to attack prey on the periphery because they are easier to isolate visually. Many animals in herds tend to bunch close together when they are under attack, making it physically difficult for the predator to get to the center of the herd.

In the end, group size may be the result of a trade-off between the costs and benefits of group living. Although much group behavior serves to reduce predation, other complex behavior occurs in groups, including grooming behavior and behavior that appears to benefit the group at the expense of the individual. For example, a honeybee will sting a potential hive predator to discourage it. The bee's stinger is barbed, and once it has penetrated the predator's skin, the bee cannot withdraw it. The bee's only means of escape is to tear away part of its abdomen, leaving the stinger behind and dying in the process. In the next section, we explore the reasons for altruistic behavior, in which an individual risks its life for the benefit of others.

55.6 Altruism

In Chapter 23, we learned that a primary goal of an organism is to pass on its genes, yet we see many instances in which some individuals forego reproducing altogether, apparently to benefit the group. How do ecologists explain **altruism**, behavior that appears to benefit others at a cost to oneself. In this section, we begin by discussing whether such behavior evolved for the good of the group or for the good of the individual. As we will see, most altruistic acts serve to benefit the individual's close relatives. We explore the concept of kin selection, which argues that acts of self-sacrifice indirectly promote the spread of an organism's genes, and see how this plays out in an extreme form in the genetics of social insect colonies. We conclude by examining reciprocal altruism as an attempt to explain the evolution of altruism among nonkin.

In Nature, Individual Selfish Behavior Is More Likely Than Altruism

One of the first attempts to explain the existence of altruism was called **group selection**, the premise that natural selection produces outcomes beneficial for the whole group or species. In 1962, the British ecologist V. C. Wynne-Edwards argued that a group containing altruists, each willing to subordinate its interests for the good of the group, would have a survival advantage over a group composed of selfish individuals. In concept, the idea of group selection seemed straightforward and logical: a group that consisted of selfish individuals would overexploit its resources and die out, while the fitness of a group with altruists would be enhanced.

In the late 1960s, the idea of group selection came under severe attack. Leading the charge was the biologist G. C. Williams, who argued that evolution acts through **individual selection**, which proposes that adaptive traits generally are selected for because they benefit the survival and reproduction of the individual rather than the group. Williams' arguments against group selection follow.

Mutation Mutant individuals that readily use resources for themselves or their offspring will have an advantage in a population where individuals limit their resource use. Consider a species of bird in which a pair lays only two eggs, that is, it has a clutch size of two and there is no overexploitation of resources. Two eggs would ensure a replacement of the parent birds but would prevent a population explosion. Imagine a mutant bird arises that lays three eggs. If the population is not overexploiting its resources, sufficient food may be available for all three young to survive. If this happens, the three-egg genotype will eventually become more common than the two-egg genotype. This process would work for even larger brood sizes, such as four eggs or five eggs, and brood sizes would tend to increase until they became so large that the parents could not look after all their young. Field studies of great tits in Wytham Woods, England, show a median clutch size of eight to nine eggs, not because females couldn't incubate more, but because adult birds cannot reliably supply sufficient food for more than eight or nine chicks to survive.

Immigration Even in a population in which all pairs laid two eggs and no mutations occurred to increase clutch size, selfish individuals that laid more could still immigrate from other areas. In nature, populations are rarely sufficiently isolated to prevent immigration of "selfish" mutants from other populations.

Individual Selection For group selection to work, some groups must die out faster than others. In practice, groups do not become extinct very frequently. Individuals die off more frequently than groups, so individual selection will be the more powerful evolutionary force.

Resource Prediction Group selection assumes that individuals are able to assess and predict future food availability and population density within their own habitat. There is little evidence that they can. For example, it is difficult to imagine that songbirds would be able to predict the future supply of the caterpillars that they feed to their young and adjust their clutch size accordingly.

Most ecologists accept individual gain as a more plausible result of natural selection than group selection. Population size is more often controlled by competition in which individuals strive to command as much of a resource as they can. Such selfishness can cause some seemingly surprising behaviors. For example, male Hanuman langurs (*Semnopithecus entellus*) kill infants when they take over groups of females from other males

Figure 55.15   Infanticide as selfish behavior.   Male Hanuman langurs (*Semnopithecus entellus*) can act aggressively toward the young of another male, even killing them, hastening the day the females come into estrus and the time when the males can father their own offspring. Note that the mother is running with the infant.

(**Figure 55.15**). The reason for the behavior is that when they are not nursing their young, females become sexually receptive much sooner, hastening the day when the male can father his own offspring. Infanticide ensures that the male will father more offspring, and the genes governing this tendency spread by natural selection.

If individual selfishness is more common than group selection, how do we account for what appear to be examples of altruism in nature?

Apparent Altruistic Behavior in Nature Is Often Associated with Kin Selection

All offspring have copies of their parents' genes, so parents taking care of their young are actually caring for copies of their own genes. Genes for altruism toward one's young are favored by natural selection and will become more numerous in the next generation, because offspring have copies of those same genes.

The probability that any two individuals will share a copy of a particular gene is a quantity, r, called the **coefficient of relatedness**. During meiosis in a bi-allelic system, any given copy of a gene has a 50% chance of going into an egg or sperm. A mother and father are on average related to their children by an amount $r = 0.5$, because half of a child's genes come from its mother and half from its father. By similar reasoning, brothers or sisters are related by an amount $r = 0.5$ (they share half their mother's genes and half their father's); grandchildren and grandparents, by 0.25; and cousins, by 0.125 (**Figure 55.16**). In 1964, ecologist W. D. Hamilton realized the implication of the coefficient of relatedness for the evolution of altruism. An organism not only can pass on its genes through having offspring, it also can pass them on through ensuring the survival of siblings, nieces, nephews, and cousins. This means an organ-

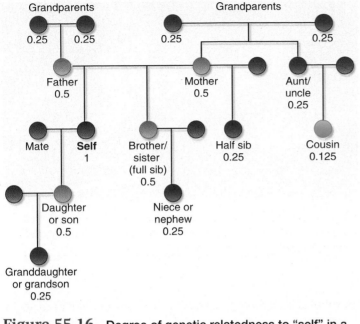

Figure 55.16 **Degree of genetic relatedness to "self" in a diploid organism.** Pink circles represent completely unrelated individuals.

Biological inquiry: In theory, should you sacrifice your life to save two sisters or nine cousins?

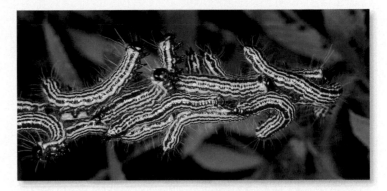

Figure 55.17 **Altruistic behavior or kin selection?** *Datana ministra* caterpillars exhibit a bright, striped warning pattern to advertise their bad taste to predators. All the larvae in the group are likely to be the progeny of one egg mass from one adult female moth. The death of the one caterpillar it takes to teach a predator to avoid the pattern benefits the caterpillar's close kin.

ism has a vested interest in protecting its brothers and sisters, and even their offspring.

The term **inclusive fitness** is used to designate the total number of copies of genes passed on through one's relatives, as well as one's own reproductive output. Selection for behavior that lowers an individual's own fitness but enhances the reproductive success of a relative is known as **kin selection**. Hamilton proposed that an altruistic gene will be favored by natural selection when

$$rB > C$$

where r is the coefficient of relatedness of donor (the altruist) to the recipient, B is the benefit received by the recipients of the altruism, and C is the cost incurred by the donor. This is known as **Hamilton's rule**.

Imagine two sisters who are not yet mothers. One has a rare kidney disease and needs a transplant from her sister. Let's assume both sisters will have two children of their own. The risk of the transplant to the donor involves a 1% chance of dying, but the benefit to the recipient involves a 90% chance of living and having children. In this example, $r = 0.5$, $B = 0.9 \times 2 = 1.8$, and $C = 0.01 \times 2 = 0.02$. Since the genetic benefit (rB) of 0.9 is much greater than the genetic cost (C) of 0.02, it makes evolutionary sense to proceed with the transplant. While humans are unlikely to do this type of calculation before deciding whether to risk their lives to save their brothers or sisters from a life-threatening event, this example shows how such behavior could arise and spread in nature.

Let's examine a situation involving altruism within a group of animals. Many insect larvae, especially caterpillars, are soft-bodied creatures. They rely on a bad taste or toxin to deter predators and advertise this condition with bright warning colors. For example, noxious *Datana ministra* caterpillars, which feed on oaks and other trees, have bright red and yellow stripes and adopt a specific posture with head and tail ends upturned when threatened (**Figure 55.17**). Unless it is born with an innate avoidance of this prey type, a predator has to kill and eat one of the caterpillars in order to learn to avoid similar individuals in the future. It is of no personal use to the unlucky caterpillar to be killed. However, animals with warning colors often aggregate in kin groups because they hatch from the same egg mass. In this case, the death of one individual is likely to benefit its siblings, which are less likely to be attacked in the future, and thus its genes will be preserved. This explains why the genes for bright color and a warning posture are successfully passed on from generation to generation. In a case where $r = 0.5$, B might be 50, and $C = 1$, the benefit of 25 > 1, so the genes for this behavior will spread.

A common example of altruism in social animals occurs when a sentry raises an alarm call in the presence of a predator. This behavior has been observed in Belding's ground squirrels (*Spermophilus beldingi*) (**Figure 55.18**). The squirrels feed in groups, with certain individuals acting as sentries and watching for predators. As a predator approaches, the sentry typically gives an alarm call and the group members retreat into their burrows. In drawing attention to itself, the caller is at a higher risk of being attacked by the predator. However, in many groups, those closest to the sentry are most likely to be offspring or brothers or sisters; thus, the altruistic act of alarm calling is reasoned to be favored by kin selection. Supporting this is the fact that most alarm calling is done by females, because they are more likely to stay in the colony where they were born and thus have kin nearby, whereas the males are more apt to disperse far from the colony.

Figure 55.18 **Alarm calling, a possible example of kin selection.** This Belding's ground squirrel sentry is emitting an alarm call to warn other individuals, which are often close kin, of the presence of a predator. It is believed that by doing so, the sentry draws attention away from the others but becomes an easier target itself.

Figure 55.19 **A naked mole rat colony.** In this mammal species, females do not reproduce and only the queen (shown resting on workers) has offspring.

Altruism in Social Insects Arises Partly from Genetics and Partly from Lifestyle

Perhaps the most extreme form of altruism is the evolution of sterile castes in social insects, in which the vast majority of females, known as workers, rarely reproduce themselves but instead help one reproductive female (the queen) to raise offspring, a phenomenon called **eusociality**. The explanation of eusociality lies partly in the particular genetics of most social insect reproduction. Females develop from fertilized eggs and are diploid, the product of fertilization of an egg by a sperm. Males develop from unfertilized eggs and are haploid.

Such a genetic system is called **haplodiploidy**. If they have the same parents, each daughter receives an identical set of genes from her haploid father. The other half of a female's genes come from her diploid mother, so the coefficient of relatedness (r) of sisters is 0.50 (from father) + 0.25 (from mother) = 0.75. The result is that females are more related to their sisters (0.75) than they would be to their own offspring (0.50). This suggests it is evolutionarily advantageous for females to stay in the nest or hive and care for other female offspring of the queen, which are their full sisters.

Elegant though these types of explanations are, they do not provide the whole picture. Large eusocial colonies of termites exist, but termites are diploid, not haplodiploid. In this case, how do we account for the existence of eusociality?

In the 1970s, Richard Alexander suggested that it was the particular lifestyle of these animals, rather than genetics, that promoted eusociality. He argued that in a normal diploid organism, females are related to their daughters by 0.5 and to their sisters by 0.5, so it should matter little to them whether they rear siblings or daughters of their own. He predicted, well before eusociality was discovered to occur in mammals, that a eusocial mammalian species could exist when certain conditions were met, including that the nests or burrows be enclosed and subterranean, and that the colony have a food supply such

as tubules and roots. In addition, the soil would need to be hard, dry clay to keep the colony safe from digging predators. He proposed that the colony would be defended by a few members of the colony willing to give their lives in defense of others, and he posited the existence of mechanisms by which a queen could manipulate other individuals.

At the time, Alexander had no idea that a mammal with these characteristics existed. Surprisingly, subsequent discoveries confirmed the existence of a eusocial mammal that satisfied all of the predictions of Alexander's model: the naked mole rat. Naked mole rats are diploid species that live in arid areas of Africa in large underground colonies where only one female, the queen, produces offspring (**Figure 55.19**). A renewable food supply is present in the form of tubers of the plant *Pyrenacantha kaurabassana*. These weigh up to 50 kg and provide food for a whole colony, though the food would be insufficient if all the mole rats reproduced. Because the burrows are as hard as cement, there are few ways to attack them, and a heroic effort by a mole rat blocking the entrance can effectively stop a predator (commonly a rufous-beaked snake). The queen mole rat does indeed manipulate the colony members; she suppresses reproduction in other females by producing a pheromone in her urine that is passed around the colony by grooming. Hence, the mole rats seem to have evolved the appropriate behavior to exploit this ecological niche. As Alexander argued, lifestyle characteristics can provide an explanation for the evolution of eusociality in species such as naked mole rats, in which both sexes are diploid.

Unrelated Individuals May Engage in Altruistic Acts if the Altruism Is Likely to Be Reciprocated

Even though we have argued that kin selection can explain instances of apparent altruism, cases of altruism are known to exist between unrelated individuals. What drives this type of behavior appears to be a "You scratch my back, I'll scratch yours" type of reciprocal altruism, in which the cost to the animal of behaving altruistically is offset by the likelihood of a return benefit. This occurs in nature, for example, when unrelated chimps groom each other.

Researcher Gerald Wilkinson has noted that female vampire bats exhibit reciprocal altruism via food sharing. Vampire bats can die after 60 hours without a blood meal, because they can no longer maintain their correct body temperature. Adult females will share their food with their young, the young of other females, and other unrelated females that have not fed. The females roost together in groups of 8 to 12 and their dependent young. A hungry female will solicit food from another female by approaching and grooming her. The female being groomed then regurgitates part of her blood meal for the other. The roles of blood donor and recipient are often reversed, and Wilkinson showed that unrelated females are more likely to share with those that had recently shared with them. The probability of a female getting a free lunch is increased because the roost consists of individuals that remain associated with each other for long periods of time.

55.7 Mating Systems

In nature, most males seem superfluous because one male could mate with many females in a local area. If one male can mate with many females, why are there few species with a sex ratio of, say, 1 male : 20 females? The sex ratio is more often about 1:1. The answer lies with natural selection. Let's consider a hypothetical population that contains 20 females for every male; each male mates, on average, with 20 females. A parent whose children were exclusively sons could expect to have 20 times the number of grandchildren compared to a parent with the same number of daughters. Under such conditions, natural selection would favor the production of sons, and males would become prevalent in the population. However, if the population were mainly males, females would be at a premium, and natural selection would favor their production. Such constraints operate on the numbers of both male and female offspring, keeping the sex ratio at about 1:1.

Even though the sex ratio is fairly even in most species, that doesn't mean that one female always mates with one male or vice versa. Several different types of mating systems occur among animals. In monogamy, each individual mates exclusively with one partner over at least a single breeding cycle and sometimes for longer. In contrast, polygamy, a system in which individuals mate with more than one partner in a breeding season, is much more common among animals. There are two types of polygamy. In polygyny ("many females"), one male mates with more than one female, but females mate only with one male. In polyandry ("many males"), one female mates with several males, but males mate with only one female.

In this section, we examine the characteristics of these mating systems and explore the role of sexual selection, a type of natural selection in which competition for mates drives the evolution of certain traits.

Sexual Selection Involves Mate Choice and Mate Competition

As we learned in Chapter 24, sexual selection promotes traits that will increase an organism's mating success. You will recall that sexual selection can take two forms. In intersexual selection, members of one sex, usually females, choose mates based on particular characteristics such as color of plumage or sound of courtship song. In intrasexual selection, members of one sex, usually males, compete over partners, and the winner performs most of the matings. Let's explore each of these in a little more detail.

Female Mate Choice Females have many different ways to choose their prospective partners. Female hangingflies (*Hylobittacus* spp.) demand a nuptial gift of a food package, an insect prey item that the male has caught (**Figure 55.20**). Such a nutrient-rich gift may permit females to produce more eggs. The bigger the gift, the longer it takes the female to eat it and the longer the male can copulate with her. Females will not mate with males that do not offer such a package. Female spiders and mantids will sometimes eat their mate during or after copulation, with the male's body constituting the ultimate nuptial gift.

Males may also have parenting skills that females desire. Among 15-spined sticklebacks (*Spinachia spinachia*), males perform nest fanning, cleaning, and guarding of the offspring. Females prefer to mate with males that shake their bodies most during courtship, apparently using this cue to assess the quality of the male as a potential father.

Often, females choose mates without the offering of obvious material benefits and make their choices based on plumage color or courtship display. The male African long-tailed widow bird (*Euplectes progne*) has long tail feathers that he displays to females via aerial flights. Researcher Malte Andersson experimentally shortened the tails of some birds by clipping their tail feathers, and he lengthened the tails of others by taking the clippings and sticking them onto other birds with superglue.

Figure 55.20 **Female choice of males based on nuptial gifts.** A male hangingfly presents a nuptial gift, a small moth, to a female.

Males with experimentally lengthened tails attracted four times as many females as males with shortened tails and they fathered more clutches of eggs (**Figure 55.21**).

Some researchers have suggested that males with excessively long tail feathers must be very healthy to bear this "handicap" and that these ornaments therefore advertise the male's genetic quality. However, other important benefits may be associated with plumage quality. Bright colors are often caused by pigments called carotenoids that help stimulate the immune system. In zebra finches and red jungle fowl, colorful plumage has been associated with heightened resistance to disease, suggesting that females that choose such males are choosing genetically healthier fathers.

On rare occasions, a sex role reversal occurs and the male discriminates among females. In the Mormon cricket (*Anabrus simplex*), males only mate once because they provide a nutrient-rich nuptial gift of a spermatophore to females, which is energetically costly to produce. Here, males choose heavier females to mate with because these females have more eggs and the males can father more offspring.

Mate Competition Between Individuals In many species, females do not actively choose their preferred mate; instead, they mate with competitively superior males. In such cases, dominance is determined by fighting or by ritualized sparring. Outcomes may be dictated by the size of weapons such as antlers or horns or by body size. In the southern elephant seal (*Mirounga leonina*), females haul up onto the beach to give birth and gain safe haven for their pups from marine predators. Following birth, they are ready to mate. In this situation, dominant males are able to command a substantial group and constantly lumber across the beach to fight other males and defend their harem. Such competition tends to promote increased body size, and in species with male-male competition, males are often substantially larger than females (**Figure 55.22**).

Large body size does not always guarantee paternity. Smaller male elephant seals may intercept females in the ocean and attempt to mate with them there, rather than on the beach, where the competitively dominant males patrol. Such "satellite" males, which move around the edge of the mating arena, are also visible in other species. For example, small male frogs hang around ponds waiting to intercept females headed toward the call of dominant males. Thus, even though competitively dominant males father most offspring, smaller males can have reproductive success.

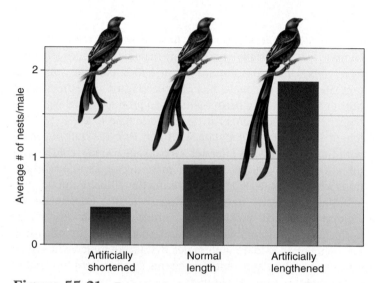

Figure 55.21 **Female choice based on male appearance.** Males with artificially lengthened tails mate with more females, and therefore have more nests, than males with a normal or artificially shortened tail.

Biological inquiry: Why do you think it is rarer for female birds or mammals to have more colorful plumage or elaborate adornments than males of the same species?

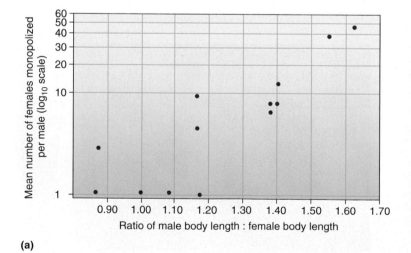

(a)

(b)

Figure 55.22 **Large male size and mating success.** (a) In southern elephant seals, the larger the male, in relation to female size, the greater the number of females that can be mated with and monopolized. (b) These male elephant seals are fighting to maintain control of their female harems.

In Monogamous Mating Systems, Males and Females Are Paired for at Least One Reproductive Season

Mating success is also dependent on the type of the particular species' mating system. In **monogamy**, one male mates with one female, and most individuals have mates. Males and females are generally similar in body size and appearance (**Figure 55.23a**). Several hypotheses explain the existence of monogamy. The first is the **mate-guarding hypothesis**, which suggests that males stay with a female to protect her from being fertilized by other males. Such a strategy may be advantageous when receptive females are widely scattered and difficult to find.

The **male assistance hypothesis** maintains that males remain with females to help them rear their offspring. Monogamy is common among birds, about 70% of which are socially monogamous, that is, the pairings remain intact during at least one breeding season. According to this hypothesis, monogamy is prevalent in birds because eggs and chicks take a considerable amount of parental care. Most eggs need to be incubated continuously if they are to hatch, and chicks require almost continual feeding. It is therefore in the male's best interest to help fledge his young, because he would have few surviving offspring if he did not.

The **female-enforced monogamy hypothesis** suggests that females stop their male partners from being polygynous. Male and female burying beetles (*Nicrophorus defodiens*) work together to bury small, dead animals, which will provide a food resource for their developing offspring. Males will release pheromones to attract other females to the site. However, while an additional female might increase the male's fitness, the additional developing offspring might compete with the offspring of the first female, decreasing her fitness. As a result, on smelling these pheromones, the first female will interfere with the male's attempts at signaling, preserving the monogamous relationship.

Recent research by Larry Young and Elizabeth Hammock has shown that fidelity may have a genetic basis. These researchers found that fidelity of male voles depends on the length of repeating, short genetic sequences, once considered junk DNA, in a gene that codes for a key hormone receptor. Adult male voles with the long version of the sequence were more apt to form pair bonds with female partners and nurture their offspring than were voles with the short version.

In Polygynous Mating Systems, One Male Mates with Many Females

In **polygyny**, one male mates with more than one female in a single breeding season, but females mate only with one male. Physiological constraints often dictate that female organisms must care for the young, because they are the ones most often left "holding the baby." Because of these constraints, at least in many organisms with internal fertilization, such as mammals and some fish, males are able to desert and mate with more females. Polygynous systems are therefore associated with uniparental care of young, with males contributing little. Sexual dimorphism is typical in polygynous mating systems, particularly when males engage in competition over mates (**Figure 55.23b**). Sexual maturity is often delayed in males that fight because of the considerable time it takes to reach a sufficiently large size to compete for females.

Polygyny is influenced by the spatial or temporal distribution of breeding females. In cases where all females are sexually receptive within the same narrow period of time, little opportunity exists for a male to garner all the females for himself. Where female reproductive receptivity is spread out over weeks or months, there is much more opportunity for males to mate with more than one female. For example, females of the common toad (*Bufo bufo*) all lay their eggs within a week, and males generally have time to mate only with one female. In contrast, female bullfrogs (*Rana catesbeiana*) have a breeding season of several weeks, and males may mate with as many as six females in a season.

(a)   (b)   (c)

Figure 55.23 **Sexual dimorphism in body size and mating system.** (a) In monogamous species, such as these Manchurian cranes, *Grus japonensis*, males and females appear very similar. (b) In polygynous species, like white-tailed deer, *Odocoileus virginianus*, males are bigger than females and have large horns with which they engage in combat over females. (c) In polyandrous species, females are usually bigger, as with these golden silk spiders, *Nephila clavipes*.

Resource-Based Polygyny Where some critical resource is patchily distributed and in short supply, certain males may dominate the resource and breed with more than one visiting female. In the lark bunting (*Calamospiza melanocorys*), which mates in North American grasslands, males arrive at the grasslands first, compete for territories, and then display with special courtship flight patterns and songs to attract females. The major source of nestling death in this species is overheating from too much exposure to the sun. Prime territories are therefore those with abundant shade, and some males with shaded territories attract two females, even though the second female can expect no help from the male in the process of rearing young. Males in some exposed territories remain bachelors for the season. From the dominant male's point of view, resource-based polygyny is advantageous; from the female's point of view, there may be costs. Although by choosing dominant males, a female may be gaining access to good resources, she may also have to share these resources with other females.

Harem Mating Structures Sometimes males defend a group of females without commanding a resource-based territory. This pattern is more common when females naturally congregate in groups or herds, perhaps to avoid predation, as with southern elephant seals (see Figure 55.22). Usually the largest and strongest males command most of the matings, but being a harem master is usually so exhausting that males may only manage to remain the dominant male for a year or two.

Communal Courting Polygynous mating can occur where neither resources nor harems are defended. In some instances, particularly in birds and mammals, males display in designated communal courting areas called **leks** (**Figure 55.24**). Females come to these areas specifically to find a mate, and they choose a prospective mate after the males have performed elaborate displays. Most females seek to mate with the best male, so a few successful males perform the vast majority of the matings. At a lek of the white-bearded manakin (*Manacus manacus*) of South America, one male accounted for 75% of the 438 matings where there were as many as 10 males. A second male mated 56 times (13% of matings), while six others mated only a total of 10 times.

In Polyandrous Mating Systems, One Female Mates with Many Males

In most systems in which one individual mates with more than one individual of the opposite sex, the polygamous sex is the male. The opposite condition, **polyandry**, in which one female mates with several males, is much more rare. Nevertheless, it is practiced by some species of birds, fish, and insects. Sexual dimorphism is present, with the females being the larger of the sexes (see **Figure 55.23c**). In the Arctic tundra, the summer season is short but very productive, providing a bonanza of insect

Figure 55.24 Some male birds and mammals congregate in communal courting grounds called leks. A male white-bearded manakin waits at a lek, where males have pulled the leaves off a branch to create a clear display area. Females visit the leks and males display to them, puffing up their white "beard" feathers.

food for two months. The productivity of the breeding grounds of the spotted sandpiper (*Actitis macularia*) is so high that the female becomes rather like an egg factory, laying up to five clutches of four eggs each in 40 days. Her reproductive success is limited not by food but by the number of males she can find to incubate the eggs, and females compete for males, defending territories where the males sit.

Polyandry is also seen in some species where egg predation is high and males are needed to guard the nests. For example, in the pipefish (*Syngnathus typhle*), males have brood pouches that provide eggs with safety and a supply of oxygen- and nutrient-rich water. Females produce enough eggs to fill the brood pouches of two males, and may mate with more than one male.

Ultimately, as we have seen, most behaviors have evolved to maximize an individual's reproductive output. In a successful group of individuals, this leads to population growth. But we are not knee-deep in sandpipers or elephant seals, so there must be some constraints on reproductive output. In Chapter 56, we next turn to the realm of population ecology to explore how populations grow and what factors limit their growth.

CHAPTER SUMMARY

55.1 The Impact of Genetics and Learning on Behavior

- Behavior is usually due to the interaction of an organism's genes and the environment.

- Genetically programmed behaviors are termed innate and often involve a sign stimulus that initiates a fixed action pattern. (Figures 55.1, 55.2)

- Organisms can often make modifications to their behavior based on previous experience, a process called learning. Some forms of learning include habituation, classical conditioning, operant conditioning, and cognitive learning. (Figures 55.3, 55.4)

- Much behavior is a mixture of innate and learned behaviors. A good example of this occurs in a process called imprinting, in which animals develop strong attachments that influence subsequent behavior. (Figures 55.5, 55.6)

55.2 Local Movement and Long-Range Migration

- The simplest forms of local movement involve kinesis, taxis, and memory. (Figure 55.7)

- Many animals undergo long-range seasonal movement called migration in order to feed or breed. They do this using three proposed mechanisms: piloting, the ability to move from one landmark to the next; orientation, the ability to follow a compass bearing; and navigation, the ability to set, follow, and adjust a compass bearing. (Figures 55.8, 55.9)

55.3 Foraging Behavior

- Animals use complex behavior in food gathering or foraging. Optimality theory views foraging behavior as a compromise between the costs and benefits involved.

- The theory of optimal foraging assumes that animals modify their behavior to keep the ratio of their energy uptake to energy expenditure high. (Figure 55.10)

- The size of a territory, a fixed area in which an individual or group excludes other members of its own species, tends to be optimized according to the costs and benefits involved. (Figure 55.11)

55.4 Communication

- Communication is a form of behavior. The use of different forms of communication between organisms depends on the environment in which they live.

- Chemical communication often involves marking territories; auditory and visual forms of communication are often used to attract mates. A fascinating form of tactile communication involves the dance of the honeybee. (Figures 55.12, 55.13)

55.5 Living in Groups

- Many benefits of group living relate to defense against predators, offering protection through sheer numbers and through what is called the many eyes hypothesis or the geometry of the selfish herd. (Figure 55.14)

55.6 Altruism

- Infanticide is an example of selfish behavior. (Figure 55.15)

- Altruism is behavior that benefits others at a cost to oneself. One of the first explanations of altruism, called group selection, posited that natural selection produced outcomes beneficial for the group. Biologists now believe that most apparently altruistic acts are often associated with outcomes beneficial to those most closely related to the individual, a concept termed kin selection. (Figures 55.16, 55.17, 55.18)

- Altruism among eusocial animals may arise partly from the unique genetics of the animals and partly from lifestyle. (Figure 55.19)

- Altruism is known to exist among nonrelated individuals that live in close proximity for long periods of time.

55.7 Mating Systems

- Sexual selection takes two forms: intersexual selection, in which the female chooses a mate based on particular characteristics, or intrasexual selection, in which males compete with one another for the opportunity to mate with a female. (Figures 55.20, 55.21, 55.22)

- Several types of mating systems are found among animals, including monogamy, polygyny, and polyandry. (Figure 55.23)

- Polygynous mating can often occur in situations where males dominate a resource, defend groups of females (harems), or display in common courting areas called leks. (Figure 55.24)

TEST YOURSELF

1. Behavioral ecology is the study of
   a. how organisms interact with their environment.
   b. courtship behavior.
   c. how an individual's behavior affects reproductive success.
   d. how an individual's behavior affects survival.
   e. both c and d.

2. Geotaxis is a response to the force of gravity. Fruit flies placed in a vial will move to the top of the vial. This is an example of _____ geotaxis.
   a. positive
   b. neutral
   c. innate
   d. negative
   e. learned

3. Certain behaviors seem to have very little environmental influence. Such behaviors are the same in all individuals regardless of the environment and are referred to as _____ behaviors.
   a. genetic
   b. instinctual
   c. innate
   d. b and c only
   e. all of the above

4. Patrick has decided to teach his new puppy a few new tricks. Each time the puppy responds correctly to Patrick's command, the puppy is given a treat. This is an example of
   a. habituation.
   b. classic conditioning.
   c. operant conditioning.
   d. imprinting.
   e. orientation.

5. An animal using landmarks to move from one area to another is exhibiting
   a. navigation.
   b. classic conditioning.
   c. migration.
   d. piloting.
   e. orientation.

6. For group living to evolve, the benefits of living in a group must be greater than the cost of group living. Which of the following is an example of a benefit of living in a group?
   a. reduced spread of disease and/or parasites
   b. increased food availability
   c. reduced competition for mates
   d. decreased risk of predation
   e. all of the above

7. The modification of behavior based on prior experience is called
   a. a fixed action pattern.
   b. learning.
   c. navigation.
   d. adjustment behavior.
   e. innate.

8. When an individual behaves in a way that reduces its own fitness but increases the fitness of others, the organism is exhibiting
   a. kin selection.
   b. group selection.
   c. altruism.
   d. selfishness.
   e. ignorance.

9. Hamilton's theory of kin selection suggests that altruism could evolve in a population if the altruistic behavior of one individual increased the reproductive behavior of
   a. all members of the group.
   b. only the females of the group.
   c. relatives.
   d. nonrelatives.
   e. the youngest individuals of the group.

10. When each female in the population mates with several males, but each male mates with only one female, the mating system is referred to as
    a. polygamy.
    b. polyandry.
    c. polygyny.
    d. monogamy.
    e. harem mating.

CONCEPTUAL QUESTIONS

1. Define ethology.
2. Define a fixed action pattern.
3. Describe the distinguishing features of the different mating systems.

EXPERIMENTAL QUESTIONS

1. What observations were important for the development of Niko Tinbergen's hypothesis explaining how digger wasps located their nests?

2. How did Tinbergen test the hypothesis that the wasps were using landmarks to relocate the nest? What were the results?

3. Did the Tinbergen experiment rule out any other cue the wasps may have been using besides the sight of pinecones?

COLLABORATIVE QUESTIONS

1. Discuss the two main types of associative learning and give an example of each.

2. Discuss several ways in which organisms communicate with each other.

www.brookerbiology.com

This website includes answers to the Biological Inquiry questions found in the figure legends and all end-of-chapter questions.

56

POPULATION ECOLOGY

CHAPTER OUTLINE

A population of king penguins (*Aptenodytes patagonicus*) in Salisbury Plain on South Georgia Island, Antarctica.

A **population** can be defined as a group of interbreeding individuals occupying the same area at the same time. In this way, we can think of a population of water lilies in a particular lake, the lion population in the Ngorogoro crater in Africa, or the human population of New York City. However, the boundaries of a population can often be difficult to define, though they may correspond to geographic features such as the boundaries of a lake or forest or be contained within a mountain valley or a certain island. Individuals may enter or leave a population, such as the human population of New York City or the deer population in North Carolina. Thus, populations are often fluid entities, with individuals moving into (immigrating) or out of (emigrating) an area. For the purposes of simplicity, in our discussion we will assume that immigration and emigration cancel each other out as factors.

This chapter explores **population ecology**, the study of how populations grow and what factors promote and limit growth. To study populations, we need to employ some of the tools of **demography**, the study of birth rates, death rates, age distributions, and the sizes of populations. We begin our discussion by exploring characteristics of populations, including density and how it is quantified, dispersion, reproductive strategies, and age classes. We will consider how life tables and survivorship curves help summarize demographic information such as birth and death rates. Similarly, growth rates will be examined by determining how many reproductive individuals are in the population and their fertility rate. The data are then used to construct simple mathematical models that allow us to analyze and predict population growth. We will also look at the factors that limit the growth of populations, and conclude the chapter by using the population concepts and models to explore the growth of human populations.

56.1 Understanding Populations

Within their areas of distribution, organisms occur in varying numbers. We recognize this pattern by saying a plant or animal is "rare" in one place and "common" in another. For more precision, ecologists quantify commonness further and talk in terms of population **density**, the numbers of organisms in a given unit area. Population growth affects population density, and knowledge of both can help us make decisions about the management of species. How long will it take for a population of an endangered species to recover to a healthy level if we protect it from its most serious threats? For example, in Florida, boat propellers kill dozens of manatees a year. A knowledge of manatee population growth rates and population densities would allow us to determine at what point populations can no longer recover from such losses and could help determine when and where to set speed limits for boaters. Since 1994, several parts of Georges Bank, once one of the world's richest fishing grounds, have been closed to commercial fishing because of overfishing. How many fish can we reasonably trawl from the sea and still ensure that an adequate population will exist for future use? Such information is vital in making determinations of size limits, catch quotas, and length of season for fisheries to ensure an adequate future population size.

In this section, we discuss density and other characteristics of populations within their habitats. We will also discuss the different reproductive strategies organisms use and assign individuals to different groups called age classes. Population growth can be predicted using knowledge of age classes, reproductive age, and reproductive strategies. We then analyze survivorship and fertility data, which can tell us at what rates populations may grow. Let's begin our exploration of populations by considering density and the various ways that ecologists attempt to quantify it.

Ecologists Use Many Different Methods to Quantify Population Density

The simplest method to measure population density is to visually count the number of organisms in a given area. We can reasonably do this only if the area is small and the organisms are relatively large. For example, we can determine the number of gumbo limbo trees (*Bursera simaruba*) on an island in the Florida Keys. Normally, however, population ecologists calculate the density of plants or animals in a small area and use this figure to estimate the total abundance over a larger area. Several different sampling methods exist for quantifying density in this way, including the use of traps to catch animals, from insects to mammals. Suction traps, like giant aerial vacuum cleaners, can suck flying insects from the sky. Pitfall traps set into the ground can catch species wandering over the surface, such as spiders, lizards, or beetles. Mist nets, consisting of very fine netting spread between trees, can entangle flying birds and bats. Simple baited snap traps, like mouse traps, or live traps, can catch small mammals. Population density can thus be estimated as the number of animals caught per unit area where the traps were set, for example, per 100 m² of habitat.

Sometimes population biologists will capture animals and then tag and release them (**Figure 56.1**). The rationale behind the **mark-recapture technique** is that after the tagged animals are released, they mix freely with unmarked individuals and within a short time are randomly mixed within the population. The population is resampled and the numbers of marked and unmarked individuals are recorded. We assume that the ratio of marked to unmarked individuals in the second sample is the same as the ratio of marked individuals in the first sample to the total population size. Thus:

$$\frac{\text{Number of marked individuals in first catch}}{\text{Total population size, } N} = \frac{\text{Number of marked recaptures in second catch}}{\text{Total number of second catch}}$$

Let's say we catch 50 largemouth bass in a lake and mark them with colored fin tags. A week later we return to the lake and catch 40 fish and 5 of them are previously tagged fish. If we assume that no immigration or emigration has occurred, which is quite likely in a closed system like a lake, and we assume there have been no births or deaths of fish, then the total population size is given by rearranging the equation:

$$\text{Total population size, } N = \frac{\begin{array}{c}\text{Number of marked individuals in first catch}\\ \times \text{ Total number of second catch}\end{array}}{\begin{array}{c}\text{Number of marked recaptures}\\ \text{in second catch}\end{array}}$$

Using our data:

$$N = \frac{50 \times 40}{5} = \frac{2,000}{5} = 400$$

Figure 56.1   **The mark-recapture technique is often used to estimate population size.**   An ear tag identifies this Rocky Mountain goat (*Oreamnos americanus*) in Olympic National Park, Washington. Recapture of such marked animals permits accurate estimates of population size.

Biological inquiry: If we mark 110 Rocky Mountain goats and recapture 100 goats, 20 of which have ear tags, what is the estimate of the total population size?

From this equation, we estimate that the lake has a total population size of 400 largemouth bass. This could be useful information for game and fish personnel who wish to know the total size of a fish population in order to set catch limits. However, the mark-recapture technique can have drawbacks. Some animals that have been marked may learn to avoid the traps. Recapture rates will then be low, resulting in an overestimate of population size. Imagine that instead of 5 tagged fish out of 40 recaptured fish, we only get 2 tagged fish. Now our population size estimate is 2,000/2 = 1,000, a dramatic increase in our population size estimate. On the other hand, some animals can become trap-happy, particularly if the traps are baited with food. This would result in an underestimate of the population size.

Because of the limitations of the mark-recapture technique, ecologists also use other, more novel methods to estimate population density. For many species with valuable pelts we can track population densities through time by examining pelt records taken from trading stations. We can also estimate relative population density by examining catch per unit effort, which is especially valuable in commercial fisheries. We can't easily expect to count the number of fish in an area of ocean, but we can count the number caught, say, per 100 hours of trawling. For some species that leave easily recognizable fecal pellets, like rabbits or deer, we can count pellet numbers. For frogs or birds, we can count chorusing or singing individuals. Many plant individuals are clonal, that is, they grow in patches of genetically identical individuals, so that rather than count individuals we can use the amount of ground covered by plants as an estimate of vegetation density. We can also count leaf scars or chewed leaves as an estimate of the density of the animals that damage them.

Patterns of Spacing Individuals within a population can show different patterns of spatial **dispersion**; that is, they can be clustered together or spread out to varying degrees. The three basic kinds of dispersion pattern are clumped, uniform, and random. We can visualize these patterns by imagining people in a meeting room. If some people know each other, they get together in small groups, creating a clumped pattern. If people do not know each other, and were perhaps even wary of each other, they might maintain a certain minimum personal distance between themselves to produce a uniform dispersion. If nobody thinks or cares about their position relative to anyone else, we would get a random dispersion.

The type of dispersion observed in nature can tell us a lot about what processes shape group structure. The most common dispersion pattern is **clumped**, likely because resources tend to be clustered in nature. For example, certain plants may do better in moist conditions, and moisture is greater in low-lying areas (**Figure 56.2a**). Social behavior between animals may also promote a clumped pattern. Many animals are clumped into flocks or herds.

On the other hand, competition may cause a **uniform** dispersion pattern between individuals, as between trees in a forest. At first, the pattern of trees and seedlings may appear random as seedlings develop from seeds dropped at random, but competition between roots may cause some trees to be outcompeted by others, causing a thinning out and resulting in a uniform distribution. Thus, the dispersion pattern starts out random but

ends up uniform. Uniform dispersions may also result from social interactions, as between some nesting birds, which tend to keep an even distance from each other (**Figure 56.2b**).

Perhaps the rarest dispersion pattern is **random** because resources in nature are rarely randomly spaced. Where resources are common and abundant, as in moist, fertile soil, the dispersion patterns of plants may be random and lacking a pattern (**Figure 56.2c**).

Reproductive Strategies To better understand how populations grow in size, let's consider their reproductive strategies. For example, some organisms can produce all of their offspring in a single reproductive event. This pattern, called **semelparity** (from the Latin *semel*, once, and *parere*, to bear), is common in insects and other invertebrates and also occurs in organisms such as salmon, bamboo grasses, and agave plants (**Figure 56.3a**). These individuals reproduce once only and die. Semelparous organisms may live for many years before reproducing, like the agaves, or they may be annual plants that develop from seed, flower, and drop their own seed within a year.

Other organisms reproduce in successive years or breeding seasons. The pattern of repeated reproduction at intervals throughout the life cycle is called **iteroparity** (from the Latin *itero*, to repeat), and it is common in most vertebrates and perennial plants such as trees. Among iteroparous organisms much variation occurs in the number of reproductive events and in the number of offspring per event. Many species, such as temperate birds or temperate forest trees, have distinct breeding seasons (seasonal iteroparity) that lead to distinct generations (**Figure 56.3b**). For a few species, individuals reproduce repeatedly and at any time of the year. This is termed continuous iteroparity and is exhibited by some tropical species, many parasites, and many mammals (**Figure 56.3c**).

(a) Clumped **(b) Uniform** **(c) Random**

Figure 56.2 **Three types of dispersion.** **(a)** A clumped distribution pattern, as in these plants clustered around an oasis, often results from the uneven distribution of a resource, in this case, water. **(b)** A uniform distribution pattern, as in these nesting black-browed albatrosses (*Diomedea melanophris*) on the Falkland Islands, may be a result of competition or social interactions. **(c)** A random distribution pattern, as in these bushes at Leirhnjukur Volcano in Iceland, is the least common form of spacing.

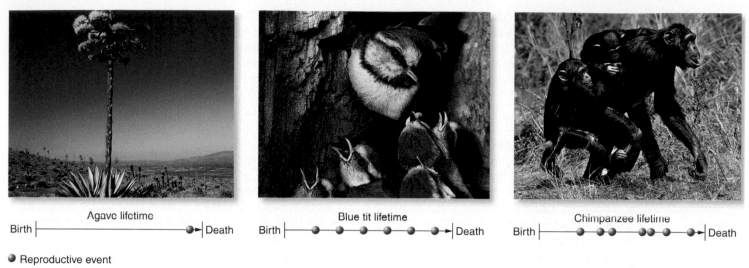

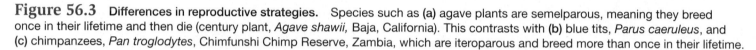

● Reproductive event

(a) Semelparity

(b) Iteroparity (seasonal)

(c) Iteroparity (continuous)

Figure 56.3 **Differences in reproductive strategies.** Species such as **(a)** agave plants are semelparous, meaning they breed once in their lifetime and then die (century plant, *Agave shawii,* Baja, California). This contrasts with **(b)** blue tits, *Parus caeruleus,* and **(c)** chimpanzees, *Pan troglodytes,* Chimfunshi Chimp Reserve, Zambia, which are iteroparous and breed more than once in their lifetime.

Why do species reproduce in a semelparous or iteroparous mode? The answer may lie in part in environmental uncertainty. If survival of juveniles is very poor and unpredictable, then selection favors repeated reproduction and long reproductive life to increase the chance that juveniles will survive in at least some years. This is often referred to as bet-hedging. If the environment is stable, then selection favors a single act of reproduction, because the organism can devote all its energy to making offspring, not maintaining its own body. Under favorable circumstances, annuals produce more seeds than trees, which have to invest a lot of energy in maintenance. However, when the environment becomes stressful, annuals run the risk of their seeds not germinating. They must rely on some seeds successfully lying dormant and germinating after the environmental stress has ended.

Age Classes The reproductive strategy employed by an organism has a strong effect on the subsequent age classes of a population. Semelparous organisms often produce groups of same-aged young called **cohorts**, which may grow at similar rates. Iteroparous organisms generally have many young of different ages because the parents reproduce frequently. The age classes of populations can be characterized by specific categories, such as years in mammals, stages (eggs, larvae, or pupae) in insects, or size in plants.

We expect that a population increasing in size should have a large number of young, whereas a decreasing population should have few young. An imbalance in age classes can have a profound influence on a population's future. For example, in an overexploited fish population, the bigger, older reproductive age classes are often removed. If the population experiences reproductive failure for one or two years, there will be no young fish to move into the reproductive age class to replace the removed fish, and the population may collapse. Other popula-

tions experience removal of younger age classes. Where populations of white-tailed deer are high, they overgraze the vegetation and eat many young trees, leaving only older trees, whose foliage is too tall for them to reach (**Figure 56.4**). This can have disastrous effects on the future population of trees, for while the forest might consist of healthy mature trees, when these die, there will be no replacements. Removal of deer predators such as panthers and wolves often allows deer numbers to skyrocket and survivorship of young trees in forests to plummet. To accurately examine how populations grow, we need to examine and understand the demography of the population.

Life Tables and Survival Curves Summarize Survival Patterns

One way to determine how a population will grow is to examine a cohort of individuals from birth to death. For most animals and plants this involves marking a group of individuals in a population as soon as they are born or germinate and following their fate through their lifetime. For some long-lived organisms such as tortoises, elephants, or trees, this is impractical, so a snapshot approach is used, in which researchers examine the age structure of a population at one point in time. Recording the presence of juveniles and mature individuals, researchers use this information to construct a life table. A **life table** provides data on the number of individuals alive in each particular age class. Age classes can be created for any time period, but they often represent one year. Males are not usually included in these tables, since they are typically not the limiting factor in population growth.

Let's examine a life table for the North American beaver (*Castor canadensis*). Prized for their pelts, by the mid-19th century these animals had been hunted and trapped to near extinction.

Figure 56.4 Theoretical age distribution of two forest populations. **(a)** Age distribution of an undisturbed forest with numerous young trees, many of which die as the trees age and compete with one another for resources, leaving relatively few big, older trees. **(b)** Age distribution of a forest where overgrazing has reduced the abundance of young trees, leaving mainly trees in the older age classes.

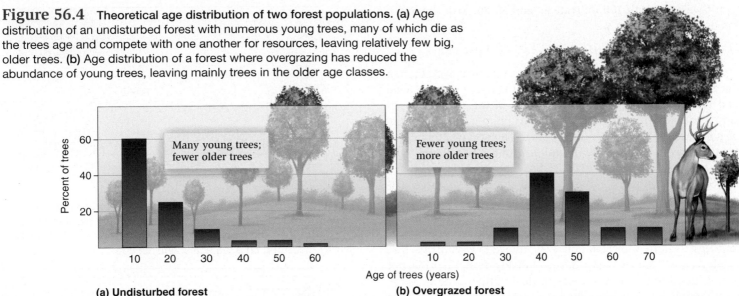

(a) Undisturbed forest **(b) Overgrazed forest**

| Table 56.1 | Life Table for the Beaver (*Castor canadensis*) in Newfoundland, Canada |

Age (years)	Number alive at start of year, n_x	Number dying during year, d_x	Proportion alive at start of year l_x	Age-specific fertility, m_x	$l_x m_x$
0–1	3,695	1,995	1.000	0.000	0
1–2	1,700	684	0.460	0.315	0.145
2–3	1,016	359	0.275	0.400	0.110
3–4	657	286	0.178	0.895	0.159
4–5	371	98	0.100	1.244	0.124
5–6	273	68	0.074	1.440	0.107
6–7	205	40	0.055	1.282	0.071
7–8	165	38	0.045	1.280	0.058
8–9	127	14	0.034	1.387	0.047
9–10	113	26	0.031	1.080	0.033
10–11	87	37	0.024	1.800	0.043
11–12	50	4	0.014	1.080	0.015
12–13	46	17	0.012	1.440	0.017
13–14	29	7	0.007	0.720	0.005
14+	22	22	0.006	0.720	0.004

Net reproductive rate, $\Sigma l_x m_x = 0.938$

Beavers began to be protected by laws in the 20th century, and populations recovered in many areas, often growing to what some considered to be nuisance status. In Newfoundland, Canada, legislation supported trapping as a management technique. From 1964 to 1971, trappers provided mandibles from which teeth were extracted for age classification. If many teeth were obtained from, say, 1-year-old beavers, then such animals were probably common in the population. If the number of teeth from 2-year-old beavers was low, then we know there was high mortality for the 1-year-old age class. From the mandible data, researchers constructed a life table (**Table 56.1**). The number of individuals alive at the start of the time period (in this case, a year) is referred to as n_x, where n is the number and x refers to the particular age class. By subtracting the value of n_x from the number alive at the start of the previous year, we can calculate the number dying in a given age class or year, d_x. Thus

$d_x = n_x - n_{x+1}$. For example, in Table 56.1, 273 beavers were alive at the start of their sixth year (n_5) and only 205 were alive at the start of the seventh year (n_6); thus, 68 died during the fifth year: $d_5 = n_5 - n_6$, or $d_5 = 273 - 205 = 68$.

A simple but informative exercise is to plot numbers of surviving individuals at each age, creating a **survivorship curve** (**Figure 56.5**). The value of n_x, the number of individuals, is typically expressed on a log scale. Ecologists use a log scale to examine rates of change with time, not change in absolute numbers. Although we could accomplish the same thing with a linear scale, the use of logs makes it easier to examine a wide range of population sizes. For example, if we start with 1,000 individuals and 500 are lost in year 1, the log of the decrease is

$$\log_{10} 1,000 - \log_{10} 500 = 3.0 - 2.7 = 0.3 \text{ per year}$$

If we start with 100 individuals and 50 are lost, the log of the decrease is similarly

$$\log_{10}100 - \log_{10}50 = 2.0 - 1.7 = 0.3 \text{ per year}$$

In both cases the rates of change are identical even though the absolute numbers are different. Plotting the n_x data on a log scale ensures that regardless of the size of the starting population, the rate of change of one survivorship curve can easily be compared to that of another species. The survivorship curve for the beaver follows a fairly uniform rate of death over the life span.

Survivorship curves generally fall into one of three patterns (**Figure 56.6**). In a type I curve, the rate of loss for juveniles is relatively low, and most individuals are lost later in life, as they become older and more prone to sickness and predators (see Feature Investigation that follows). Organisms that exhibit type I survivorship have relatively few offspring but invest much time and resources in raising their young. Many large mammals, including humans, exhibit type I curves. At the other end of the scale is a type III curve, in which the rate of loss for juveniles is relatively high, and the survivorship curve flattens out for those organisms that have survived early death. Many fish and marine invertebrates fit this pattern. Most of the juveniles die or are eaten, but a few reach a favorable habitat and thrive. For example, once they find a suitable rock face on which to attach themselves, barnacles grow and survive very well. Many insects and plants also fit the type III survivorship curve, because they lay many eggs or release hundreds of seeds, respectively. Type II curves represent a middle ground, with fairly uniform death rates over time. Species with type II survivorship curves include many birds, small mammals, reptiles, and some annual plants. The beaver population most closely resembles this survivorship curve. Keep in mind, however, that these are generalized curves and that few populations fit them exactly.

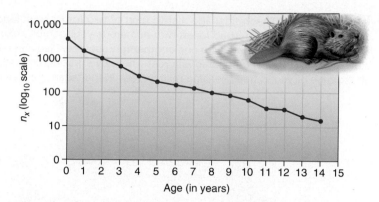

Figure 56.5 **Survivorship curve for the North American beaver.** The survivorship curve is generated by plotting the number of surviving beavers, n_x, from any given cohort of young, usually measured on a log scale, against age.

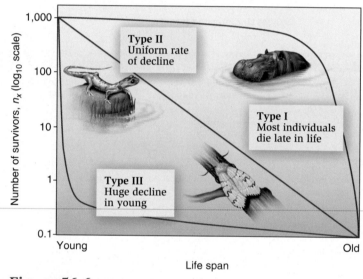

Figure 56.6 Idealized survivorship curves.

FEATURE INVESTIGATION

Murie's Collections of Dall Mountain Sheep Skulls Permitted Accurate Life Tables to Be Constructed

The Dall mountain sheep (*Ovis dalli*) lives in mountainous regions, including the Arctic and sub-Arctic regions of Alaska. In the late 1930s, the U.S. National Park Service was bombarded with public concerns that wolves were responsible for a sharp decline in the population of Dall mountain sheep in Denali National Park (then Mt. McKinley National Park). Shooting the wolves was advocated as a way of increasing the number of sheep. Because meaningful data on sheep mortality were nonexistent, the Park Service enlisted biologist Adolph Murie to collect relevant information. In addition to spending many hours observing interactions between wolves and sheep, Murie also

gathered data on sheep age at death. To do this, he collected sheep skulls in Denali National Park. He determined their age by counting annual growth rings on the horns. Thus, the analysis of skulls gave a snapshot of how old the animals were when they died.

In 1947, Edward Deevey put Murie's data in the form of a life table that listed each age class and the number of skulls in it (**Figure 56.7**). While Murie had collected 608 skulls, Deevey expressed the data per 1,000 individuals to allow for comparison with other life tables. From the data, Deevey constructed a survivorship curve. For the Dall mountain sheep, there was a slight initial decline in survivorship as young lambs were lost and then the survivorship curve flattened out, indicating that the sheep survived well through about age 7 or 8.

Figure 56.7 Examining the survivorship curve of a Dall mountain sheep population reveals information on the cause of death.

HYPOTHESIS Culling the wolf population would protect reproductively active adults in the Dall mountain sheep population.

STARTING LOCATION Wolf predation of sheep is commonly observed at Denali National Park (formerly known as Mt. McKinley National Park) in Alaska.

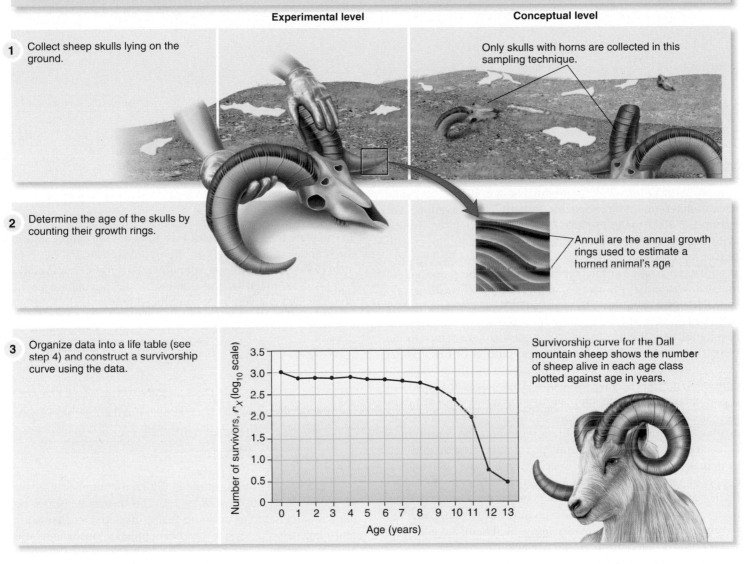

Experimental level

Conceptual level

1 Collect sheep skulls lying on the ground.

Only skulls with horns are collected in this sampling technique.

2 Determine the age of the skulls by counting their growth rings.

Annuli are the annual growth rings used to estimate a horned animal's age

3 Organize data into a life table (see step 4) and construct a survivorship curve using the data.

Survivorship curve for the Dall mountain sheep shows the number of sheep alive in each age class plotted against age in years.

4 THE DATA

Results used in step 3:

Age class	Number alive, n_x	$n_x \log_{10}$	Age class	Number alive, n_x	$n_x \log_{10}$
0–1	1,000	3.00	7–8	640	2.81
1–2	801	2.90	8–9	571	2.76
2–3	789	2.90	9–10	439	2.64
3–4	776	2.89	10–11	252	2.40
4–5	764	2.88	11–12	96	1.98
5–6	734	2.86	12–13	6	0.78
6–7	688	2.84	13–14	3	0.48

Then the number of sheep declined rapidly as they aged. These data underlined what Murie had previously observed, which was that wolves preyed primarily on the most vulnerable members of the sheep population, the youngest and the oldest. The Park Service ultimately ended a limited wolf control program that had been in effect since 1929. It also determined that the decline in the Dall mountain sheep had actually been precipitated by a series of cold winters that killed many sheep and weakened others, making them easier prey for the wolves, but that wolf predation *per se* was not to blame.

Age-Specific Fertility Data Can Tell Us When to Expect Growth to Occur

To calculate how a population grows, we need information on birth rates as well as mortality and survivorship rates. For any given age, we can determine the proportion of female offspring that are born to females of reproductive age. Using these data we can determine an **age-specific fertility rate**, called m_x. For example, if 100 females produce 75 female offspring, $m_x = 0.75$. With this additional information, we can calculate the growth rate of a population.

First, we use the survivorship data to find the proportion of individuals alive at the start of any given age class. This age-specific survivorship rate, termed l_x, equals n_x/n_0, where n_0 is the number alive at time 0, the start of the study, and n_x is the number alive at the beginning of age class x. Let's return to the beaver life table in Table 56.1. The proportion of the original beaver population still alive at the start of the sixth age class, l_5, equals $n_5/n_0 = 273/3{,}695$, or 0.074. This means that 7.4% of the original beaver population survived to age 5. Next we multiply the data in the two columns, l_x and m_x, for each row, to give us a column $l_x m_x$. This column represents the contribution of each age class to the overall population growth rate. An examination of the beaver age-specific fertility rates illustrates a couple of general points. First, for this beaver population in particular, and for many organisms in general, there are no babies born to young females. As females mature sexually, age-specific fertility goes up and it remains fairly high until later in life, when females reach postreproductive age.

The number of offspring born to females of any given age class depends on two things: the number of females in that age class and their age-specific fertility rate. Thus, although fertility of young beavers is very low, there are so many females in the age class that $l_x m_x$ for 1-year-olds is quite high. Age-specific fertility for older beavers is much higher, but the relatively few females in these age classes cause $l_x m_x$ to be low. Maximum values of $l_x m_x$ occur for females of an intermediate age, 3–4 years old in the case of the beaver. The overall growth rate per generation is then the number of offspring born to all females of all ages, where a generation is defined as the mean period between birth of females and birth of their offspring. Thus, to get the generational growth rate, we sum all the values of $l_x m_x$, that is, $\Sigma l_x m_x$, where the symbol Σ means "sum of." We term this summed value R_0 and refer to it as the **net reproductive rate**.

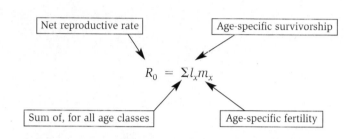

To calculate the future size of a population, we simply multiply the number of individuals in the population by the net reproductive rate. Thus, the population size in the next generation, N_{t+1}, is determined by the number in the population now, at time t, which is given by N_t, multiplied by R_0.

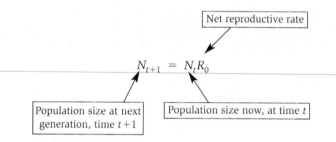

Let's consider an example in which the number of beavers alive now, N_t, is 1,000 and $R_0 = 1.1$. This means the beaver population is reproducing at a rate that is 10% greater than simply replacing itself. The size of the population next generation, N_{t+1}, is given by

$$N_{t+1} = N_t R_0$$
$$N_{t+1} = 1{,}000 \times 1.1$$
$$= 1{,}100$$

Thus, the number of beavers in the next generation is 1,100 and the population will have grown larger.

In determining population growth, much depends on the value of R_0. If $R_0 > 1$, then the population will grow. If $R_0 < 1$, the population is in decline. If $R_0 = 1$, then the population size stays the same and we say it is at **equilibrium**. In the case of the beavers, Table 56.1 reveals that $R_0 = 0.938$, which is less than 1, and therefore the population is in decline. This is valuable information, because it tells us that at that time, the beaver population in Newfoundland needed more protection (perhaps in the form of bans on trapping and hunting) in order to attain a population level at equilibrium.

56.2 How Populations Grow

Life tables can provide us with accurate information about how populations can grow from generation to generation. However, other population growth models can provide us with valuable insights into how populations grow over shorter time periods. The most simple of these assumes that populations grow if, for any time interval, the number of births is greater than the number of deaths. In this section, we will examine two different types of these simple models. The first assumes resources are not limiting, and it results in prodigious growth. The second, and perhaps more biologically realistic, assumes resources are limiting, and it results in limits to growth and eventual stable population sizes. We then consider what other factors might limit population growth, such as natural enemies, and discuss the overall life history strategies employed by different species to enable them to exist on Earth.

Knowing the Per Capita Growth Rate Helps Predict How Populations Will Grow

The change in population size over any time period can be written as the number of births per unit time interval minus the number of deaths per unit time interval.

For example, if in a population of 1,000 deer there were 100 births and 50 deaths over the course of one year, then the population would grow in size to 1,050 the next year. We can write this formula mathematically as

$$\frac{\text{Change in numbers}}{\text{Change in time}} = \text{Births} - \text{deaths}$$

or

$$\frac{\Delta N}{\Delta t} = B - D$$

The Greek letter Δ indicates change, so that ΔN is the change in number and Δt is the change in time; B is the number of births per time unit; and D is the number of deaths per time unit.

Often, the numbers of births and deaths are expressed per individual in the population, so the birth of 100 deer to a population of 1,000 would represent a birth rate, b, of 100/1,000, or 0.10 per individual. Similarly, the death of 50 deer in a population of 1,000 would be a death rate, d, of 50/1,000, or 0.05 per individual. Now we can rewrite our equation giving the rate of change in a population.

$$\frac{\Delta N}{\Delta t} = bN - dN$$

For our deer example,

$$\frac{\Delta N}{\Delta t} = 0.10 \times 1000 - 0.05 \times 1000 = 50$$

so if $\Delta t = 1$ year, the deer population would increase by 50 individuals in a year.

Ecologists often simplify this formula by representing $b - d$ as r, the **per capita growth rate**. Thus $bN - dN$ can be written as rN. Because they are also interested in population growth rates over very short time intervals, so-called instantaneous growth rates, instead of writing

$$\frac{\Delta N}{\Delta t}$$

ecologists write

$$\frac{dN}{dt}$$

which is the notation of differential calculus. The equations essentially mean the same thing, except that dN/dt reflects very short time intervals. Thus,

$$\frac{dN}{dt} = rN = (0.10 - 0.05)N = 50$$

Exponential Growth Occurs When the Per Capita Growth Rate Remains Above Zero

How do populations grow? Clearly, much depends on the value of r. When $r < 0$, the population decreases; when $r = 0$, the population remains constant; and when $r > 0$, the population increases. When $r = 0$ the population is often referred to as being at equilibrium, where no changes in population size will occur and there is **zero population growth**.

Even if r is only fractionally above 0, population increase is rapid and when plotted graphically, a characteristic J-shaped curve results (**Figure 56.8**). We refer to this type of population growth as geometric or **exponential growth**. When conditions are optimal for the population, r is at its maximum rate and is called the **intrinsic rate of increase** (denoted r_{max}). Thus, the rate of population growth under optimal conditions is $dN/dt = r_{max} N$. Again, the larger the value of r_{max}, the steeper the slope of the curve. Because population growth depends on the value of N as well as the value of r, the population increase is even greater as time passes.

How do field data fit this simple model for exponential growth? Clearly population growth cannot go on forever, as envisioned under exponential growth. But initially at least, in a new and expanding population where resources are not limited, exponential growth is often observed. Let's look at a few examples. Tule elk (*Cervus elaphus nannodes*) are a subspecies of elk that are native to California and were isolated from other herds during the Ice Ages. Hunted nearly to extinction in the 19th

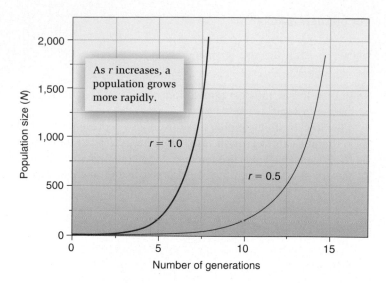

Figure 56.8 **Exponential population growth.** As the value of *r* increases, the slope of the curve gets steeper. In theory, a population with unlimited resources could grow indefinitely.

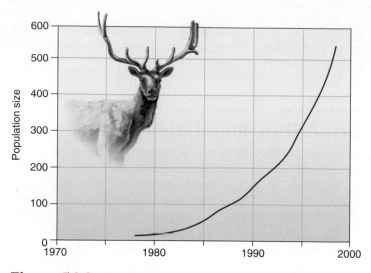

Figure 56.9 **Exponential growth following reintroduction of a population to a habitat.** A population of tule elk (*Cervus elaphus nannodes*) reintroduced to Point Reyes National Seashore in 1978 fits a pattern of exponential growth.

century, less than a dozen individuals survived on a private ranch. In the 20th century, reintroductions resulted in the recovery of tule elk to around 3,500 individuals. One reintroduction was made in March 1978 at Point Reyes National Seashore in California, where 10 animals—2 males and 8 females—were released. By 1993, the herd had reached 214 individuals and continued to grow in an exponential fashion until 1998, when the herd size stood at 549 (**Figure 56.9**). This was deemed an excessive number for the size of the available habitat, and animals were removed to begin herds in other locations. Since then, herd size at Point Reyes has been maintained at around 350.

The growth of some exotic species introduced into new habitats also seems to fit the pattern of exponential growth. The rapid expansion of rabbits after their introduction into South Australia in the late 19th century is a case in point. In 1859, Thomas Austin received two dozen European rabbits from England. Rabbit gestation lasts a mere 31 days, and in South Australia each doe could produce up to 10 litters of at least six young each year. The rabbits had essentially no enemies and ate the grass used by sheep and other grazing animals. Even when two-thirds of the population was shot for sport, which was the purpose of the initial introduction, the population grew into the millions in a few short years. By 1875, rabbits were reported on the west coast, having moved over 1,760 km across the continent, despite the deployment of huge, thousand-kilometer-long fences meant to contain them.

Finally, one of the most prominent examples of exponential growth is the growth of the global human population, which because of its large importance, we will examine separately later in the chapter.

Logistic Growth Occurs in Populations in Which Resources Are Limited

Despite its applicability to rapidly growing populations, the exponential growth model is not appropriate in many situations. The model assumes unlimited resources, which is not often the case in the real world. For most species, resources become limiting as populations grow. Thus the per capita growth rate decreases as resources are used up. The upper boundary for the population size is known as the **carrying capacity (K)**. Thus, a more realistic equation to explain population growth, one that takes into account the amount of available resources, is

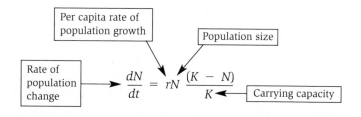

where $(K - N)/K$ represents the proportion of unused resources remaining. This equation is called the **logistic equation**.

In essence, this equation means that the larger the population size, N, the closer it becomes to the carrying capacity, K, and the fewer the available resources for population growth. At large values of N, $(K - N)/K$ becomes small and thus population growth is small. If $K = 1,000$, $N = 900$, and $r = 0.1$, then

$$\frac{dN}{dt} = (0.1)(900) \times \frac{(1,000 - 900)}{1,000}$$

$$\frac{dN}{dt} = 9$$

In this instance, population growth is nine individuals per unit of time.

At medium values of N, $(K - N)/K$ is closer to a value of 1, and population growth is larger. If $K = 1,000$, $N = 500$, and $r = 0.1$, then

$$\frac{dN}{dt} = (0.1)(500) \times \frac{(1,000 - 500)}{1,000}$$

$$\frac{dN}{dt} = 25$$

However, if population sizes are low ($N = 100$), even though $(K - N)/K$ is very close to 1, population sizes are so small that growth is again low.

$$\frac{dN}{dt} = (0.1)(100) \times \frac{(1,000 - 100)}{1,000}$$

$$\frac{dN}{dt} = 9$$

Thus, growth is small at high and low values of N and is greatest at immediate values of N. Growth is greatest when $N = K/2$ (try some calculations to verify this for yourself).

Let's consider how an ecologist would use the logistic equation. First, the value of K would come from intense field and laboratory work where researchers would determine the amount of resources, such as food, needed by each individual and then determine the amount of available food in the wild. Field censuses would determine N, and field censuses of births and deaths per unit time would provide r. When this type of population growth is plotted over time, an S-shaped growth curve results (**Figure 56.10**). This pattern, in which the growth of a population typically slows down as it approaches K, is called **logistic growth**.

Does the logistic growth model provide a better fit to growth patterns of plants and animals in the wild than the exponential model, which is also shown in Figure 56.10? In some instances, such as laboratory cultures of bacteria and yeasts, the logistic growth model provides a good fit (**Figure 56.11**). However, for many other populations, including those of the shrews, voles, and red squirrels shown in **Figure 56.12**, there is much variation. If the logistic model held true for these species, we would expect to see population growth leveling off, indicative of the populations' having reached their equilibrium density. However, in nature variations in temperature, rainfall, or resources cause changes in carrying capacity and thus in population densities. The uniform conditions of temperature and resource levels of the laboratory do not exist. In fact, when studying the population patterns of short-tailed shrews, researchers at the

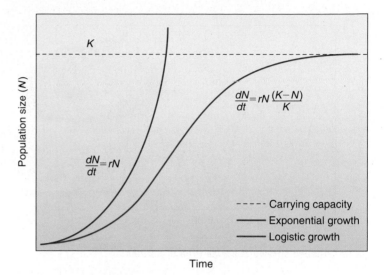

Figure 56.10 **Exponential versus logistic growth.** Exponential (J-shaped) growth occurs in an environment with unlimited resources, while logistic (S-shaped) growth occurs in an environment with limited resources.

Biological inquiry: What is the population growth per unit of time when r = 0.1, N = 200, and K = 500?

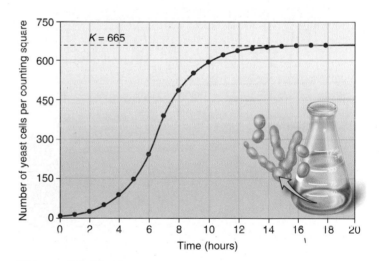

Figure 56.11 **Growth of yeast cells in culture fits the logistic growth model.** Early tests of the logistic growth curve were validated by growth of yeast cells in laboratory cultures. These populations showed the typical S-shaped growth curve.

Konza Prairie Biological Station in Kansas found a tight relationship between available soil moisture and the relative abundance of shrews (**Figure 56.13**). One hypothesis is that increased soil moisture provides more free water for drinking and an increased amount of invertebrate prey such as worms. This reminds us again of the influence of abiotic factors on population densities, as we explored in Chapter 54.

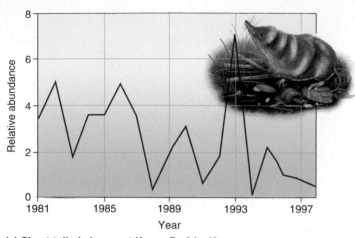

(a) Short-tailed shrews at Konza Prairie, Kansas

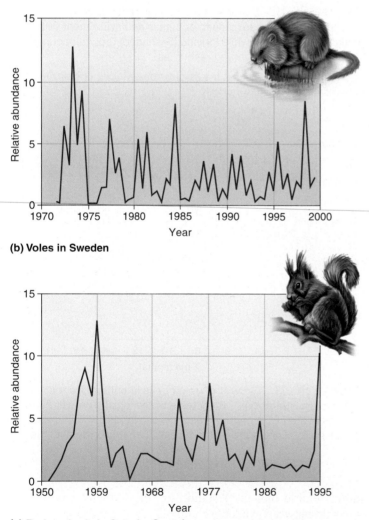

(b) Voles in Sweden

(c) Red squirrels in Ontario, Canada

Figure 56.12   **The logistic growth model does not describe all populations.**   Variation in abundance of different species of small mammal populations over many years of data collection shows great variability and a lack of fit to the idealized logistic growth curve. Data reflect relative abundance in a set number of traps at different sampling points.

Is the logistic model of little value because it fails to describe population growth accurately? Not really. It is a useful starting point for thinking about how populations grow, and it seems intuitively correct. However, the carrying capacity is a difficult feature of the environment to identify for most species, and it also varies temporally, according to climatic and local weather patterns; thus, logistic growth is difficult to measure accurately.

Also, as we will discover, populations are affected by predators, parasites, or competition with other species. In Chapter 57, we will examine how natural enemies and competitors affect population densities and explore whether species interactions commonly limit population growth. However, it is instructive to first see how these natural enemies can reduce population size. Such population reductions are often caused by a process known as density dependence.

Density-Dependent Factors May Regulate Population Sizes

Factors that influence birth and death rates, and thus regulate population size, can be either density dependent or density independent. A **density-dependent factor** is a mortality factor whose influence varies with the density of the population. Parasitism, predation, and competition are some of the many density-dependent factors that may reduce the population densities of living organisms and stabilize them at equilibrium levels. Such factors can be density dependent in that their impact depends on the density of the population; they kill relatively more of a population when densities are higher and less of a population when densities are lower. For example, many predators develop a visual search image for a particular prey. When a prey is rare, predators tend to ignore it and kill relatively few. When a prey is common, predators key in on it and kill relatively more.

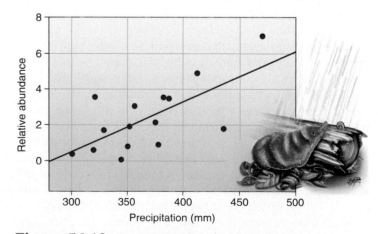

Figure 56.13   **Abiotic factors influence population densities.**   The number of short-tailed shrews at Konza Prairie, Kansas, increases as precipitation and thus soil moisture increases because of a greater availability of drinking water and invertebrate prey. Each point refers to a year from Figure 56.12a.

In England, for example, predatory shrews kill proportionately more moth pupae in leaf litter when the pupae are common compared to when they are rare.

Density dependence can be detected by plotting mortality, expressed as a percentage, against population density (**Figure 56.14**). If a positive slope results and mortality increases with density, the factor tends to have a greater effect on dense populations than on sparse ones and is clearly acting in a density-dependent manner.

A **density-independent factor** is a mortality factor whose influence is not affected by changes in population size or density. In general, density-independent factors are physical factors, including weather, drought, freezes, floods, and disturbances such as fire. For example, in hard freezes the same proportion of organisms such as birds or plants are usually killed, no matter how large the population size. However, even physical factors such as weather can act in a density-dependent manner. For example, in an environment where there were many beavers and a limited number of rivers to dam, some individuals would not have a lodge. In such a situation, a cold winter could kill a high percentage of beavers. If, on the other hand, there were few beavers, most would have a lodge to provide them with protection from a hard freeze. In this case, the cold would kill a lower percentage.

Determining which factors act in a density-dependent fashion has large practical implications. Foresters, game managers, and conservation biologists alike are interested in learning how to maintain populations at equilibrium levels. For example, if disease were to act in a density-dependent manner on white-tailed deer, there wouldn't be much point in game managers attempting to kill off predators such as mountain lions to increase herd sizes for hunters, because proportionately more deer would be killed by disease.

Finally, a source of mortality that decreases with increasing population size is considered an **inverse density-dependent factor**. For example, if a territorial predator such as a lion always ate the same number of wildebeest prey, regardless of wildebeest density, it would be acting in an inverse density-dependent manner, because it is taking a smaller proportion of the population at higher density. Some mammalian predators, being highly territorial, often act in this manner on herbivore density. Parasites that infect 10% of a host population regardless of host density are also exhibiting inverse density dependence.

Thus, natural enemies can act in a density-dependent manner and control prey populations, or they can act in a density-independent or inverse density-dependent manner and not regulate them. In order to make generalizations about which factors control populations in nature, we need to know the frequency of density dependence in natural systems.

Which factors tend to act in a density-dependent manner? In the 1980s, a broad review of many research studies, which considered 51 populations of insects, 82 populations of large mammals, and 36 populations of small mammals and birds, showed a wide variety of density-dependent factors. No single process such as parasitism, competition, or predation could be regarded as a regulatory factor of overriding importance. Even for individual taxa, such as insects, density dependence varied from parasitism and predation to competition and abiotic factors. This finding is disconcerting to ecologists interested in species management, because it means that generalizations about which factors are likely to act in a density-dependent manner are not easily made. Each species has to be analyzed on a case-by-case basis.

Life History Strategies Incorporate Traits Relating to Survival and Competitive Ability

The population parameters we have discussed—including iteroparity versus semelparity, continuous versus seasonal iteroparity, and density-dependent versus density-independent factors—have important implications for how populations grow and indeed for the reproductive success of populations and species. However, these reproductive strategies can be viewed in the context of a much bigger picture of life history strategies, sets of physiological and behavioral features that incorporate not only reproductive traits but also survivorship and length of life characteristics, habitat type, and competitive ability.

Life history strategies can be considered a continuum (**Figure 56.15**). At the one end are species, termed **r-selected species**, that have a high rate of per capita population growth, r, but

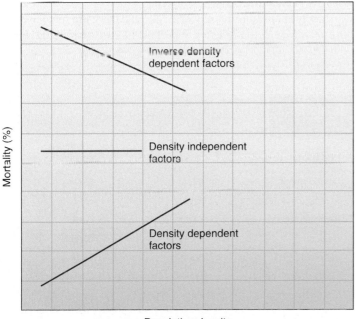

Mortality (%)

Inverse density dependent factors

Density independent factors

Density dependent factors

Population density

Figure 56.14 **Three ways that factors affect mortality in response to changes in population density.** For a density-dependent factor, mortality increases with population density, while for a density-independent factor, mortality remains unchanged. For an inverse-dependent factor, mortality decreases as a population increases in size.

poor competitive ability. An example is a weed that quickly colonizes vacant habitats (such as barren land), passes through several generations, and then is outcompeted. Weeds produce huge numbers of tiny seeds and therefore have high values of *r*. At the other end are species, termed ***K*-selected species**, that have more or less stable populations adapted to exist at or near the carrying capacity, *K*, of the environment. An example is a tree that exists in a mature forest. While trees do not have a high reproductive rate, they tend to compete well and eventually displace many species by overtopping them and shading them out. The *r*- and *K*-selection continuum brings together

various life history features including reproductive strategy, dispersal ability, growth rates, and population size (**Table 56.2**).

Let's return to our previous comparison of weeds versus trees. Weeds exist in disturbed habitats such as gaps in a forest canopy where trees have blown down, allowing light to penetrate to the forest floor. An *r*-selected species like a weed grows quickly and reaches reproductive age early, devoting much energy to producing a large number of seeds that disperse widely. These weed species remain small and do not live long, perhaps passing a few generations in the light gap before it closes.

A mature forest tree, on the other hand, often exists in undisturbed native habitats. Forest trees grow slowly and reach reproductive age late, having to devote much energy to growth and maintenance. A *K*-selected species like a tree grows large and shades out *r*-selected species like weeds, eventually outcompeting them. Such trees live a long time and produce seeds repeatedly every year when mature. These seeds are bigger than those of *r*-selected species; consider the acorns of oaks versus the seeds of dandelions. Acorns contain a large food reserve that helps them grow, whereas dandelion seeds must rely on whatever nutrients they can gather from the soil.

As you might expect, a trade-off exists between seed size and number—a plant can produce a few big seeds or lots of small ones. While the weed–tree example is a useful way to think about the *r*- and *K*-selection continuum, we must remember that other organisms can be *r*- or *K*-selected, too. For example, insects can be considered small, *r*-selected species that produce many young and have short life cycles. Mammals, such as elephants, that grow slowly, have few young, and reach large sizes are typical of *K*-selected species.

In a human-dominated world, almost every life history attribute of a *K*-selected species sets it at risk of extinction. First,

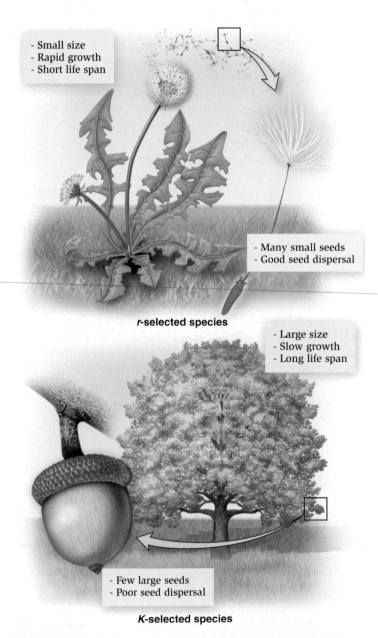

- Small size
- Rapid growth
- Short life span

- Many small seeds
- Good seed dispersal

***r*-selected species**

- Large size
- Slow growth
- Long life span

- Few large seeds
- Poor seed dispersal

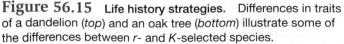

***K*-selected species**

Figure 56.15   Life history strategies.   Differences in traits of a dandelion (*top*) and an oak tree (*bottom*) illustrate some of the differences between *r*- and *K*-selected species.

Table 56.2	Characteristics of *r*- and *K*-Selected Species	
Life history feature	***r*-selected species**	***K*-selected species**
Development	Rapid	Slow
Reproductive rate	High	Low
Reproductive age	Early	Late
Body size	Small	Large
Length of life	Short	Long
Competitive ability	Weak	Strong
Survivorship	High mortality of young	Low mortality of young
Population size	Variable	Fairly constant
Dispersal ability	Good	Poor
Habitat type	Disturbed	Not disturbed
Parental care	Low	High

K-selected species tend to be bigger, so they need more habitat in which to live. Florida panthers need huge tracts of land to establish their territories and hunt for deer. There is only room for about 22 panthers on publicly owned land in South Florida. Privately owned land currently supports another 50 panthers. As the amount of land shrinks through development, so does the number of panthers. Second, *K*-selected species tend to have fewer offspring and so their populations cannot recover as fast from disturbances like fire or overhunting. California condors, for example, produce only a single chick every other year. Third, *K*-selected species breed at a later age, and their generation time and time to grow from a small population to a larger population is long. Gestation time in elephants is 22 months, and elephants take at least seven years to become sexually mature. Large trees such as the giant sequoia; large terrestrial mammals like elephants, rhinoceros, and grizzly bears; and large marine mammals like blue whales and sperm whales all run the risk of extinction. Interestingly, the coast redwood seems to be an exception, a fact perhaps attributable to its unusual genome (see the following Genomes & Proteomes section).

What are the advantages to being a *K*-selected species? In a world not perturbed by humans, *K*-selected species would fare well. Being a *K*-selected species is as viable an option as being *r*-selected. However, in a human-dominated world, many *K*-selected species are selectively logged or hunted or their habitat is altered, and the resulting small population sizes make extinction a real possibility.

GENOMES & PROTEOMES

Hexaploidy Increases the Growth of Coast Redwood Trees

Besides having the world's most massive tree, the giant sequoia (*Sequoiadendron giganteum*), California is also home to the world's tallest tree, the coast redwood (*Sequoia sempervirens*), a towering giant that can grow to over 90 m and can live for up to 2,000 years. (**Figure 56.16**). These trees are currently confined to a relatively small 700-km strip along the Pacific coast from California to southern Oregon, an area characterized by moderate year-long temperatures, heavy winter rains, and dense summer fog. Interestingly, because this climate was far more common in earlier eras, these trees were once dispersed throughout the Northern Hemisphere. As of today, however, over 95% of the old-growth coast redwoods are gone, the result of extensive logging.

How is this huge species different from other tree species? In 1948, researchers made the startling discovery that the tree is a hexaploid, that is, each of its cells contains six sets of chromosomes, with 66 chromosomes in total. While hexaploidy is not unknown in grasses and shrubs, it is unusual in trees. Of all the

Figure 56.16 **The coast redwood (*Sequoia sempervirens*) is a hexaploid conifer.** The coast redwood can grow to over 90 m, and the oldest living trees are over 2,000 years old. Their great genetic variation may help explain their incredible growth and longevity.

conifers on Earth, the coast redwood is the only hexaploid. Having this quality means each tree may have several different alleles for any given gene, which leads to a very genetically diverse population. Molecular biologist Chris Brinegar has found that hardly any two trees have exactly the same genetic constitution. Such genetic diversity allows greater adaptation to environmental conditions and more adaptations against insect or fungal pests. Indeed, living redwoods have no known lethal diseases, and pests do not cause significant damage. What's more, with six different sets of genes, trees also have the potential for great variety in their gene products, the proteins, which may help explain their prodigious growth.

56.3 Human Population Growth

In 2005, the world's population was estimated to be increasing at the rate of 153 people every minute: 2 in developed nations and 151 in less developed nations. The United Nations' 2005 projections pointed to a world population stabilizing at around 10 billion near the year 2150, as would happen with logistic growth. However, until now, human population growth has better fit an exponential growth pattern than a logistic one. In this section, we examine human population growth trends in more detail and discuss how knowledge of a population's age structure can help predict its future growth. We then investigate the carrying capacity of the Earth for humans and explore how the concept of an ecological footprint, which measures human resource use, can help us determine this carrying capacity.

Human Population Growth Fits an Exponential Pattern

Until the beginning of agriculture and the domestication of animals, about 10,000 B.C.E., the average rate of population growth was very low. With the establishment of agriculture, the world's population grew to about 300 million by 1 C.E. and to 800 million by the year 1750. Between 1750 and 1998, a relatively tiny period of human history, the world's human population surged from 800 million to 6 billion (**Figure 56.17**). In 2006, the number of humans was estimated at 6.5 billion, with two people added to the world's population every second. Considering this phenomenal increase in growth, the biggest questions remain, when will the human population level off and at what level?

Human populations can exist at equilibrium densities in one of two ways:

1. *High birth and high death rates.* Before 1750, this was often the case, with high birth rates offset by deaths from wars, famines, and epidemics.
2. *Low birth and low death rates.* In Europe, beginning in the 18th century, better health and living conditions reduced the death rate. Eventually, social changes such as increasing education for women and marriage at a later age reduced the birth rate.

The shift in birth and death rates with development is known as the **demographic transition** (**Figure 56.18**). In the first stage of the transition, birth and death rates are both high, and the population remains in equilibrium. In the second stage of this transition, which first occurred in Western Europe beginning in the late 18th century, the death rate declines first, while the birth rate remains high. High rates of population growth result. In the third stage, the birth rates drop and death rates stabilize, so that low population growth ensues. In the fourth stage, both birth and death rates are low, and the population is again at equilibrium.

The exact pace of the demographic transition between countries differs, depending on culture, economics, politics, and religion. This is illustrated by examining the demographic transition

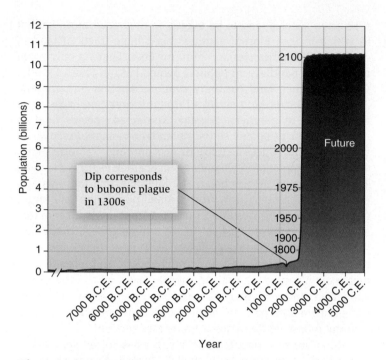

Figure 56.17 **World human population growth through history illustrates an exponential growth pattern.**   If and when human population growth will level off are issues of considerable debate.

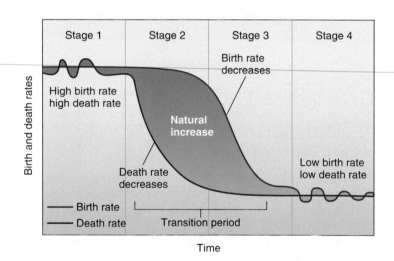

Figure 56.18   **The classic stages of the demographic transition.**   The difference in the birth rate and the death rate equals the rate of natural increase or decrease.

in Sweden and Mexico (**Figure 56.19**). In Mexico, the demographic transition occurred more recently and was typified by a faster decline in the death rate, reflecting rapid improvements in public health. A relatively longer lag occurred between the decline in the death rate and the decline in the birth rate, however, with the result that Mexico's population growth rate is still well above Sweden's, perhaps reflecting differences in culture or the fact that in Mexico the demographic transition is not yet complete.

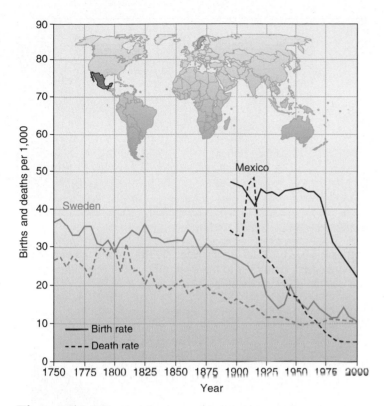

Figure 56.19   The demographic transition in Sweden and Mexico. While the transition began earlier in Sweden than it did in Mexico, the transition was more rapid in Mexico and the overall rate of population increase remains higher. (The spike in the death rate in Mexico prior to 1920 is attributed to the turbulence surrounding the Mexican revolution.)

Knowledge of a Population's Age Structure Can Help Predict Its Future Growth

Changes in the age structure of a population also characterize the demographic transition. In all populations, **age structure** refers to the relative numbers of individuals of each defined age group. This information is commonly displayed as a population pyramid (**Figure 56.20**). In West Africa, for example, children under the age of 15 make up nearly half of the population, creating a pyramid with a wide base and narrow top. Thus, even if fertility rates decline, there will still be a huge increase in the population as these children move into childbearing age. The age structure of Western Europe is much more balanced. Even if the fertility rate of young women in Western Europe increased to a level higher than that of their mothers, the annual numbers of births would still decline because of the low number of women of childbearing age.

Predicting the Earth's Carrying Capacity for Humans Is a Difficult Task

What is the Earth's carrying capacity for humans and when will it be reached? Estimates have been quite varied. Much of the speculation on the future size of the world's population centers on lifestyle. To use a simplistic example, if everyone on the planet ate meat extensively and drove large cars, then the carrying capacity would be a lot less than if people were vegetarians and used bicycles as their main means of transportation.

Most estimates propose that the human population will grow to between 10 to 15 billion people by the end of the 21st century.

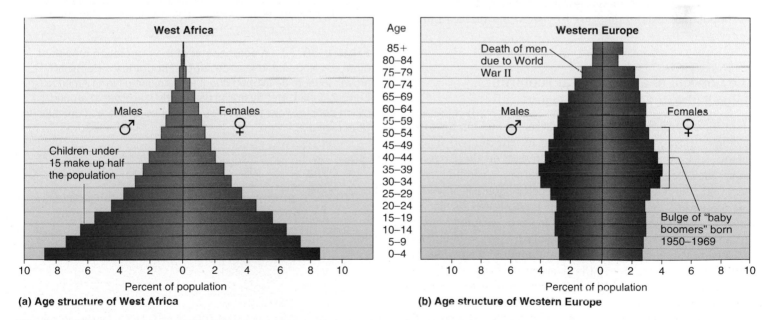

(a) **Age structure of West Africa**

(b) **Age structure of Western Europe**

Figure 56.20   The age structure of human populations in West Africa and Western Europe, as of 2000.   (a) In developing areas of the world such as West Africa there are far more children than any other age group. (b) In the developed countries of Western Europe, the age structure is more even. The bulge represents those born in the post–World War II baby boom, when birth rates climbed due to stabilization of political and economic conditions.

Global population growth can be examined by looking at **total fertility rates (TFR)**, the average number of live births a woman has during her lifetime (**Figure 56.21**). The global fertility rate has been declining, from almost 5.0 in the 1960s to 2.9 in the late 1990s. This is still greater than the 2.1 needed for zero population growth. (The fact that the replacement rate is slightly higher than 2.0, to replace mother and father, reflects mortality before individuals reach reproductive age.) The fertility rate differs considerably between geographic areas. In Africa, the total fertility rate of 5.2 in 2003 has declined relatively little since the 1950s, when it was around 6.7 children per woman. In Latin America and Southeast Asia, the rates have declined considerably from the 1950s and are now at around 2.7 and 2.6, respectively. Sweden and most other countries in Europe and North America have a TFR of less than 2.1; in Russia, fertility rates have dropped to a low of 1.1. In China, while the TFR is only 1.8, the population there will still continue to increase until at least 2025 because of the large number of women of reproductive age.

Total fertility rates of less than 2.1 will lead to population decline over the long term. The results illustrate that in highly industrialized or developed countries, the population has nearly stabilized at a little over 1 billion people, while in less developed countries, the population is still increasing dramatically (**Figure 56.22**).

A recent United Nations report shows world population projections to the year 2050 for three different growth scenarios: low, medium, and high (**Figure 56.23**). The three scenarios are based on three different assumptions about fertility rate. Using a low fertility rate estimate of only 1.5 children per woman, the population would reach a maximum of about 7.4 billion people by 2050. A more realistic assumption may be to use the fertility rate estimate of 2.0 or even 2.5, in which case the population would continue to rise to 8.9 or 10.6 billion, respectively.

The Concept of an Ecological Footprint Helps Estimate Carrying Capacity

In the 1990s, researcher Mathis Wackernagel and his coworkers calculated how much land is needed for the support of each person on Earth. Everybody has an impact on the Earth, because they consume the land's resources, including crops, wood, oil, and other supplies. Thus, each person has an **ecological footprint**, the aggregate total of land needed for survival in a sustainable world. The average footprint size for everyone on the planet is about 2 hectares or 5 acres (1 ha = 10,000 m²), but a wide variation is found around the globe (**Figure 56.24**). The ecological footprint of the average Canadian is 7.5 hectares versus about 10 hectares for the average American. In most developed countries, the largest component of land is for energy, followed by food and then forestry. Much of the land needed for energy serves to absorb the CO_2 emitted by the use of fossil fuels. If everyone required 10 hectares, as the average American does, we would need three Earths to provide us with the needed resources. Many people in less developed countries are much more frugal in their use of resources.

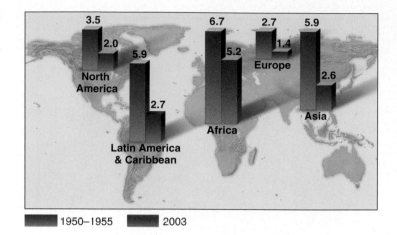

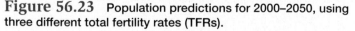
1950–1955   2003

Figure 56.21 Total fertility rates (TFRs) among major regions of the world. Data refer to the average number of children born to a woman during her lifetime.

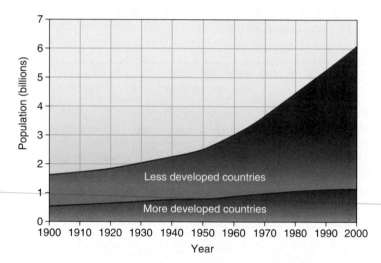

Figure 56.22 A comparison of the population growth of more developed and less developed countries. Population growth in more developed countries (including North America, Europe, Japan, and Australia) has nearly stabilized at around 1.1 billion people, while that of less developed countries continues to skyrocket.

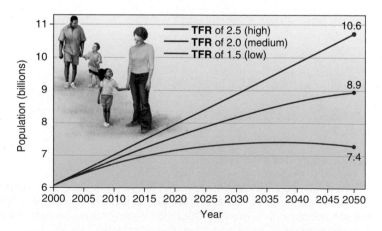

Figure 56.23 Population predictions for 2000–2050, using three different total fertility rates (TFRs).

However, globally we are already in an ecological deficit. This is possible because many people currently live in an unsustainable manner, using supplies of nonrenewable resources, such as groundwater and fossil fuels.

What's your personal ecological footprint? Several different calculations are available on the Internet that you can use to find out. Just type "ecological footprint" (in quotes) into a search engine. A rapidly growing human population combined with an increasingly large per capita ecological footprint makes it increasingly difficult to conserve other species on the planet, a subject we will examine further in our discussion of conservation biology (Chapter 60).

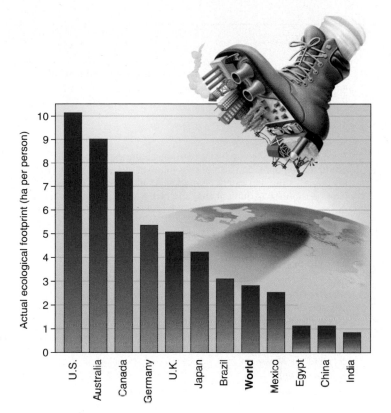

Figure 56.24 **Ecological footprints of different countries.** The term ecological footprint refers to the amount of productive land needed to support the average individual of that country.

Biological inquiry: What is your ecological footprint?

CHAPTER SUMMARY

56.1 Understanding Populations

- Population ecology studies how populations grow and what factors promote and limit growth.

- Ecologists measure population density, the numbers of organisms in a given unit area, in many ways, including the mark-capture technique. (Figure 56.1)

- Individuals within populations show different patterns of dispersion, including clumped (the most common), uniform, and random; different reproductive strategies; and different age classes. (Figures 56.2, 56.3, 56.4)

- Life tables summarize the survival pattern of a population. (Table 56.1)

- Survivorship curves illustrate life tables by plotting the numbers of surviving individuals at different ages. (Figures 56.5, 56.6, 56.7)

- Age-specific fertility and survivorship data help determine the overall growth rate per generation, or the net reproductive rate (R_0).

56.2 How Populations Grow

- The per capita growth rate (r) helps determine how populations grow over any time period.

- When r is > 0, exponential (J-shaped) growth occurs. Exponential growth can be observed in an environment where resources are not limited. (Figures 56.8, 56.9)

- Logistic (S-shaped) growth takes into account the upper boundary for a population, called carrying capacity, and occurs in an environment where resources are limited. (Figures 56.10, 56.11)

- Variations in temperature, rainfall, or resource quantity or quality cause changes in population densities, and thus the idealized logistic growth model does not describe all populations. (Figures 56.12, 56.13)

- Density-dependent factors are mortality factors whose influence varies with population density. (Figure 56.14)

- Life history strategies are a set of features including reproductive traits, survivorship and length of life characteristics, habitat type, and competitive ability.

- Life history strategies can be viewed as a continuum, with r-selected species (those with a high rate of population growth but poor competitive ability) at one end and K-selected species (those with a lower rate of population growth but better competitive ability) at the other. (Figures 56.15, 56.16, Table 56.2)

56.3 Human Population Growth

- Up to the present, human population growth has fit an exponential growth pattern. (Figure 56.17)

- Human populations have been moving from states of high birth and death rates to low birth and death rates, a shift called the demographic transition. (Figures 56.18, 56.19)

- Differences in the age structure of a population, the numbers of individuals in each age group, are also characteristic of the demographic transition. (Figure 56.20)

- Though they have been declining worldwide, total fertility rates (TFRs) differ markedly in less developed and more developed countries. Predicting the growth of the human population depends on the total fertility rate that is projected. (Figures 56.21, 56.22, 56.23)

- The ecological footprint refers to the amount of productive land needed to support each person on Earth. Because people in many countries live in a nonsustainable manner, globally we are already in an ecological deficit. (Figure 56.24)

TEST YOURSELF

1. The number of organisms in a given unit area is termed population
   a. dispersion.
   b. dispersal.
   c. density.
   d. ecology.
   e. growth.

2. A student decides to conduct a mark-recapture experiment to estimate the population size of mosquitofish in a small pond near his home. In the first catch, he marked 45 individuals. Two weeks later he captured 62 individuals, of which 8 were marked. What is the estimated size of the population based on these data?
   a. 134
   b. 349
   c. 558
   d. 1,016
   e. 22,320

3. Which of the following factors often results in a clumped pattern of dispersion?
   a. competition over limited resources
   b. random dispersal of seed
   c. social behavior among members of a population
   d. all of the above
   e. a and c only

4. A life table usually contains information about
   a. the number of surviving individuals of a particular age class.
   b. fertility for specific age classes.
   c. dispersal patterns of a population.
   d. all of the above.
   e. a and b only.

5. _____ survivorship curves are usually associated with organisms that have high mortality rates in the early stages of life.
   a. Type I
   b. Type II
   c. Type III
   d. Types I and II
   e. Types II and III

6. If the net reproductive rate (R_0) is equal to 0.5, what assumptions can we make about the population?
   a. This population is essentially not changing in numbers.
   b. This population is in decline.
   c. This population is growing.
   d. This population is in equilibrium.
   e. None of the above.

7. The maximum number of individuals a certain area can sustain is known as
   a. the intrinsic rate of growth.
   b. the resource limit.
   c. the carrying capacity.
   d. the logistic equation.
   e. the equilibrium size.

8. Which of the following factors may alter carrying capacity over time?
   a. weather pattern changes
   b. numbers of other species that are present in the habitat
   c. deforestation
   d. soil chemistry changes
   e. all of the above

9. A mortality factor whose influence does not vary with the density of the population is known as a _____ factor.
   a. limiting
   b. density-independent
   c. resource-partitioning
   d. density-dependent
   e. inverse density-dependent

10. The amount of land necessary for survival for each person in a sustainable world is known as
    a. the sustainability level.
    b. an ecological impact.
    c. an ecological footprint.
    d. survival needs.
    e. all of the above.

CONCEPTUAL QUESTIONS

1. Define population and population ecology.

2. Describe and list the assumptions of the mark-recapture technique.

3. Describe the three types of dispersion.

EXPERIMENTAL QUESTIONS

1. What problem led to the study conducted by Murie on the Dall mountain sheep population of Denali National Park?

2. Describe the survivorship curve developed by Deevey based on Murie's data.

3. How did the Murie and Deevey data affect the decision of the Park Service on the control of the wolf population?

COLLABORATIVE QUESTIONS

1. Discuss the two main types of life history strategies.

2. Discuss the concept of demographic transition.

www.brookerbiology.com

This website includes answers to the Biological Inquiry questions found in the figure legends and all end-of-chapter questions.

57

SPECIES INTERACTIONS

CHAPTER OUTLINE

Species interactions are numerous and varied. This larval seven-spot ladybird (*Coccinella septumpunctata*) is feeding on black bean aphids (*Aphis fabae*) which in turn feed on bean leaves (*Vicia faba*).

I n this chapter we turn from considering populations on their own to investigating how they interact with populations of other species that live in the same locality. Species interactions can take a variety of forms (**Table 57.1**). **Competition** is an interaction that affects both species negatively (/), as both species compete over food or other resources. Sometimes this interaction is quite one-sided, where it is detrimental to one species but not to the other, an interaction called **amensalism** (/0). **Predation**, **herbivory**, and **parasitism** all have a positive effect on one species and a negative effect on the other (+/−). However, while predators always kill their prey, the hosts of parasites and herbivores often survive their attacks. **Mutualism** is an interaction in which both species benefit (+/+), while **commensalism** benefits one species and leaves the other unaffected (+/0). Last is the interaction, or rather lack of interaction, termed **neutralism**, when two species occur together but do not interact in any measurable way (0/0).

Neutralism may be quite common, but few people have quantified its occurrence.

To illustrate how species interact in nature, let's consider a rabbit population in a woodland community (**Figure 57.1**). To determine what factors influence the size and density of the rabbit population, we need to understand the range of its possible species interactions. For example, the rabbit population could be limited by the quality of available food. It is also likely that other species, such as deer, use the same resource and thus compete with the rabbits. The rabbit population could be limited by predation from foxes or by parasitism in the form of the virus that causes the disease myxomatosis. It is also possible that other associations, such as mutualism or commensalism, may occur. This chapter examines each of these species interactions in turn, beginning with competition, an important interaction among species. We conclude with a discussion of conceptual models of species interactions that ecologists use when trying to determine which factors are most important in influencing population densities within ecological systems.

Table 57.1	Summary of the Types of Species Interactions	
Nature of interaction	**Species 1***	**Species 2***
Competition	−	−
Amensalism	−	0
Predation, herbivory, parasitism	+	−
Mutualism	+	+
Commensalism	+	0
Neutralism	0	0

* + = positive effect; − = negative effect; 0 = no effect.

57.1 Competition

In this section, we will see how ecologists have studied different types of competition and how they have shown that the competitive effects of one species on another can change as the environment changes or as different predators or parasites are present. Although species may compete, we will also learn how sufficient differences in lifestyle or morphology can exist that reduce the overlap in their ecological niches, thus allowing them to coexist.

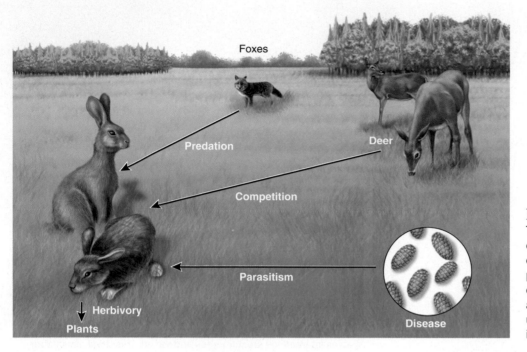

Figure 57.1 Species interactions. These rabbits can interact with a variety of species, experiencing competition with deer for food, predation by foxes, and parasitism from various disease-causing organisms. Understanding what factors affect the rabbit population size requires an understanding of each of these species interactions.

Several Different Types of Competition Occur in Nature

Several different types of competition are found in nature (**Figure 57.2**). Competition may be **intraspecific**, between individuals of the same species, or **interspecific**, between individuals of different species. Competition can also be characterized as resource competition or as interference competition. In **resource competition**, organisms compete indirectly through the consumption of a limited resource, with each obtaining as much as it can. For example, when fly maggots compete in a mouse carcass, not all the individuals can command enough of the resource to survive and become adult flies. In **interference competition**, individuals interact directly with one another by physical force or intimidation. Often this force is ritualized into aggressive behavior associated with territoriality (as discussed in Chapter 55). In these cases, strong individuals survive and take the best territory, and weaker ones perish or at best survive under suboptimal conditions.

Competition between species is not always equal. In fact, in many cases one species has a strong effect on another, but the reverse effect is negligible. Thus many interactions involving competition are a $-/0$ relationship, called **amensalism**, rather than a $-/-$ relationship. Such extreme asymmetric competition can also be observed between plants, where one species produces and secretes chemicals from its roots that inhibit the growth of another species. In Chapter 54's Feature Investigation, we saw how diffuse knapweed, an exotic species, secretes root chemicals called allelochemicals that kill the roots of native grass species. This phenomenon is termed **allelopathy**. The action of penicillin, which directly inhibits the growth of bacteria and other fungi, is a classic case of allelopathy.

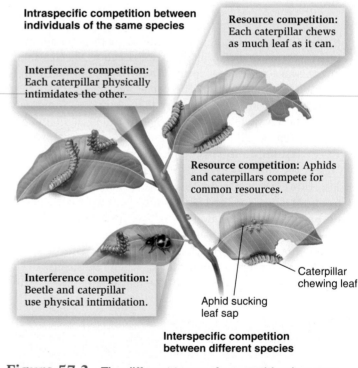

Intraspecific competition between individuals of the same species

Resource competition: Each caterpillar chews as much leaf as it can.

Interference competition: Each caterpillar physically intimidates the other.

Resource competition: Aphids and caterpillars compete for common resources.

Interference competition: Beetle and caterpillar use physical intimidation.

Caterpillar chewing leaf

Aphid sucking leaf sap

Interspecific competition between different species

Figure 57.2 The different types of competition in nature.

Researchers have established that one of the best methods of studying competition between two species is to temporarily remove one of them and examine the effect on the other species. A now-classic example of this method involved a study of the interactions between two species of barnacles conducted on the west coast of Scotland, as described next.

FEATURE INVESTIGATION

Connell's Experiments with Barnacle Species Showed That One Species Can Competitively Exclude Another in a Natural Setting

The most direct method of assessing the effect of competition is to remove individuals of species A and measure the response of species B. Often, however, such manipulations are difficult to conduct outside the laboratory. If individuals of species A are removed, what is to stop them from migrating back into the area of removal?

In 1954, ecologist Joseph Connell conducted an experiment that overcame this problem. *Chthamalus stellatus* and *Semibalanus balanoides* (formerly known as *Balanus balanoides*) are two species of barnacles that dominate the Scottish coastline. Each organism's **niche**, or physical distribution and ecological role, on the intertidal zone was well defined, with *Chthamalus* occurring in the upper intertidal zone, and *Semibalanus* restricted to the lower intertidal zone. Connell sought to determine what the range of *Chthamalus* adults might be in the absence of competition from *Semibalanus* (**Figure 57.3**).

To do this, Connell obtained rocks from high on the rock face, just below the high-tide level, where only *Chthamalus* grew. These rocks already contained young and mature *Chthamalus*. He then moved the rocks into the *Semibalanus* zone, fastened them down with screws, and allowed *Semibalanus* to also colonize them. Once *Semibalanus* had colonized these rocks, he took the rocks out, removed all the *Semibalanus* organisms from one side of the rocks with a needle, and then returned the rocks to the lower intertidal zone, screwing them down once again. As seen in the data, the mortality of *Chthamalus* on rock halves with *Semibalanus* was fairly high. On the *Semibalanus*-free halves, however, *Chthamalus* survived well.

Figure 57.3   Connell's experimental manipulation of species indicated the presence of competition.

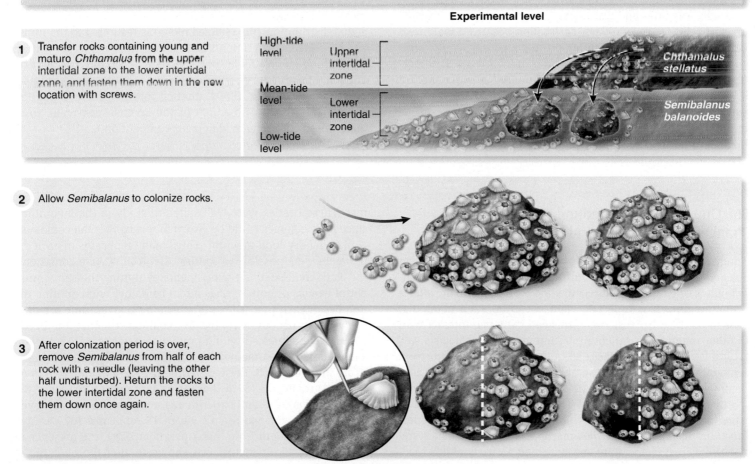

HYPOTHESIS Adult *Chthamalus stellatus* were being competitively excluded from the lower intertidal zone by the species *Semibalanus balanoides*.

STARTING LOCATION Two species of barnacles, *Chthamalus stellatus* and *Semibalanus balanoides*, grow in the intertidal zone of the rocky shores of the Scottish coast.

Experimental level

1 Transfer rocks containing young and mature *Chthamalus* from the upper intertidal zone to the lower intertidal zone, and fasten them down in the new location with screws.

High-tide level
Mean-tide level
Low-tide level
Upper intertidal zone
Lower intertidal zone
Chthamalus stellatus
Semibalanus balanoides

2 Allow *Semibalanus* to colonize rocks.

3 After colonization period is over, remove *Semibalanus* from half of each rock with a needle (leaving the other half undisturbed). Return the rocks to the lower intertidal zone and fasten them down once again.

4 Monitor survival of *Chthamalus* on both sides of rocks.

Chthamalus grows on the side where *Semibalanus* has been removed, indicating that *Semibalanus* may exclude *Chthamalus* from certain habitats.

Chthamalus realized niche

Chthamalus fundamental niche

Semibalanus fundamental and realized niches

5 THE DATA

Rock No.	Side of rock	% *Chthamalus* mortality over one year	
		Young barnacles	Mature barnacles
13b	*Semibalanus* removed	35	0
	Undisturbed	90	31
12a	*Semibalanus* removed	44	37
	Undisturbed	95	71
14a	*Semibalanus* removed	40	36
	Undisturbed	86	75

Connell also monitored survival of natural patches of both barnacle species where both occurred on the intertidal zone at the upper margin of the *Semibalanus* distribution. In a period of unusually low tides and warm weather, when no water reached any barnacles for several days, desiccation became a real threat to the barnacles' survival. During this time, young *Semibalanus* suffered a 92% mortality rate and older individuals a 51% mortality rate. At the same time, young *Chthamalus* experienced a 62% mortality rate compared with a rate of only 2% for more resistant older individuals. Clearly, *Semibalanus* is not as resistant to desiccation as *Chthamalus* and thus could not survive in the upper intertidal zone where *Chthamalus* occurs. *Chthamalus* is more resistant to desiccation than *Semibalanus* and thus can be found higher in the intertidal zone.

Thus, while the potential distribution (the fundamental niche) of *Chthamalus* extends over the entire intertidal zone, its actual distribution (the realized niche) is restricted to the upper zone. Connell's experiments were among the first to show that, in a natural environment, one species can actually outcompete another, affecting its distribution within a habitat. Interestingly, 30 years later, in the 1980s, Connell repeated his competition experiments in the same area of the Scottish coast and once more observed strong evidence for competition.

The Outcome of Competition Can Vary with Changes in the Abiotic and Biotic Environments

Using experiments to temporarily remove individuals of one species and examine the results on the remaining species, as Connell did, is often the most direct method to investigate the effects of competition. It is especially valuable to do such manipulations in the field because organisms can then also interact with all other organisms in their environment, or, as ecologists say, natural variation can be factored in. However, in laboratory experiments, the investigator can often control and vary important factors systematically.

In the late 1940s, biologist Thomas Park began a series of experiments examining competition between two flour beetles, *Tribolium castaneum* and *Tribolium confusum*, in which he systematically varied temperature, moisture, and a variety of other factors. These beetles were well suited to study in the laboratory, since large colonies could be grown in relatively small containers containing dry food medium. Thus, many replicates of each experiment were possible to confirm that results were consistent.

Park conducted the experiments by putting the same number of beetles of both species into a container and counting the number of each that were still alive after a given time interval. *T. confusum* usually won, but in initial experiments, the beetle cultures were infested with a protozoan parasite called *Adelina triboli* that killed some beetles and preferentially killed *T. castaneum* individuals. In these experiments, *T. confusum* won in 66 out of 74 replicates (89%) because it was more resistant to the parasite. In subsequent experiments, the parasite was removed and *T. castaneum* won in 12 out of 18 replicates (67%). Two things were evident from this experiment. First, the presence of a parasite (a biotic factor) was shown to alter the outcome of competition. Second, with or without the parasite, there was no absolute victor. For example, even with the parasite present, sometimes *T. castaneum* won. Thus, some random variation, which we call stochasticity, was evident.

Park then began to vary the abiotic environment and found that competitive ability was greatly influenced by climate (**Figure 57.4**). Generally, *T. confusum* did better in dry conditions, and *T. castaneum*, in moist conditions. However, *T. confusum* also won in cold-wet conditions. Once again, some stochasticity occurred, and victory was not always absolute. Later, it was found that the mechanism of competition was largely predation of eggs and pupae by larvae and adults, which, as they ate the flour medium, would also bite the stationary eggs and pupae, killing them. In general, the species were mutually antagonistic, that is, they bit more eggs and pupae of the other species than they did of their own.

In summary, Park's important series of experiments illustrated that the results of competition could vary as a function of at least three factors: parasites, temperature, and moisture. The experiments also showed how much stochasticity occurred, even in controlled laboratory conditions. But what of systems in nature, where far more variability exists? Is competition a common occurrence? One line of evidence shows that competition is frequent in nature and a strong enough force to cause the extinction of one of two competing species that share the same niche. Researchers have proposed several mechanisms by which two competing species can coexist. One states that similar species can coexist if they occupy different niches. Another is that species may occupy similar niches but undergo physical changes in form, termed character displacement, that allow them to utilize different resources. Let's explore each of these concepts in more detail.

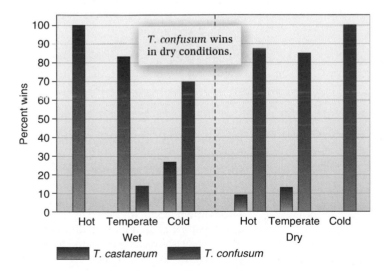

Figure 57.4 **Influence of abiotic factors on competition between *T. castaneum* and *T. confusum*.** Results of competition between the flour beetles *Tribolium castaneum* and *Tribolium confusum* show that each species usually performs better in a given habitat; for example, *T. confusum* does better in dry conditions.

Field Studies Show Competition Occurs Frequently in Nature

By reviewing studies that have investigated competition in nature, we can see how frequently it occurs and in what particular circumstances it is most important. In one 1983 review by Joseph Connell, competition was found in 55% of 215 species surveyed, demonstrating that it is indeed frequent in nature. Generally in studies of single pairs of species utilizing the same resource, competition is almost always reported (90%), whereas in studies involving more species, the frequency of competition drops to 50%. Why should this be the case? Imagine a resource such as a series of different-sized grains with four species—ants, beetles, mice, and birds—feeding on it. The ants feed on the smallest grain, the beetles and mice on the intermediate sizes, and the birds, on the largest. If only adjacent species competed with each other, competition would be expected only between the ant-beetle, beetle-mouse, and mouse-bird. Thus, competition would be found in only three out of the six possible species pairs (50%) (**Figure 57.5**). Naturally, the percent would vary according to the number of species on the axis. If there were only three species along the axis, we would expect competition in two of the three pairs (67%).

Some other general patterns were evident from Connell's review. Plants showed a high degree of competition, perhaps because they are rooted in the ground and cannot easily escape or perhaps because they are competing for the same set of limiting nutrients—water, light, and minerals. Marine organisms tended to compete more than terrestrial ones, perhaps because many of the species studied lived in the intertidal zone and were attached to the rock face, in a manner similar to that of plants. Because the area of the rock face is limited, competition for space is quite important. However, even though we see competition frequently in nature, we often see apparent competitors occupying the same general areas. How is this possible?

Species May Coexist if They Do Not Occupy Identical Niches

In 1934 the Russian microbiologist Georgyi Gause began to study competition between three protist species, *Paramecium aurelia*, *Paramecium bursaria*, and *Paramecium caudatum*, all of which fed on bacteria and yeast, which in turn fed on an oatmeal medium in a culture tube in the laboratory. The bacteria occurred more in the oxygen-rich upper part of the culture tube, and the yeast in the oxygen-poor lower part of the tube. Because each species was a slightly different size, Gause calculated population growth as a combination of numbers of individuals per milliliter of solution multiplied by their unit volume to give a population volume for each species. When grown separately, population volume of all three *Paramecium* species followed a logistic growth pattern (**Figure 57.6a**). When Gause cultured *P. caudatum* and *P. aurelia* together, *P. caudatum* went extinct (**Figure 57.6b**). Both species utilized bacteria as food but *P. aurelia* grew at a rate six times faster than *P. caudatum* and was better able to convert the food into offspring.

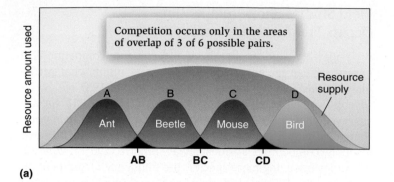

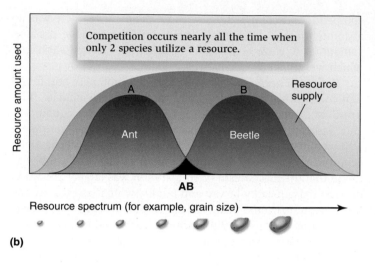

Figure 57.5 **The frequency of competition according to the number of species involved.** **(a)** Resource supply and utilization curves of four species, A, B, C, and D, along the spectrum of a hypothetical resource such as grain size. If competition occurs only between species with adjacent resource utilization curves, competition would be expected between three of the six possible pairings: A and B; B and C; and C and D. **(b)** When only two species utilize a resource set, competition would nearly always be expected between them.

Biological inquiry: If five species utilized the resource set in part (a), what would be the expected frequency of competition observed?

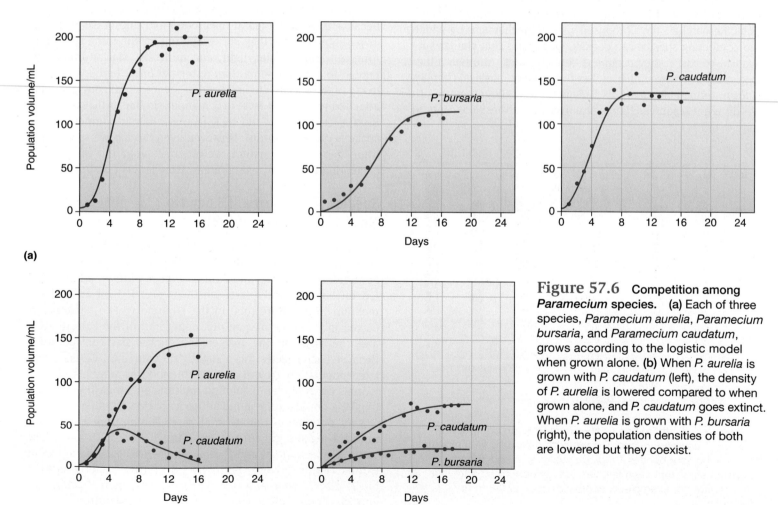

Figure 57.6 **Competition among** ***Paramecium*** **species.** **(a)** Each of three species, *Paramecium aurelia*, *Paramecium bursaria*, and *Paramecium caudatum*, grows according to the logistic model when grown alone. **(b)** When *P. aurelia* is grown with *P. caudatum* (left), the density of *P. aurelia* is lowered compared to when grown alone, and *P. caudatum* goes extinct. When *P. aurelia* is grown with *P. bursaria* (right), the population densities of both are lowered but they coexist.

Figure 57.7 Resource partitioning among five species of warblers feeding in North American spruce trees. Each warbler species prefers to feed at a different height and portion of the tree, thus reducing competition.

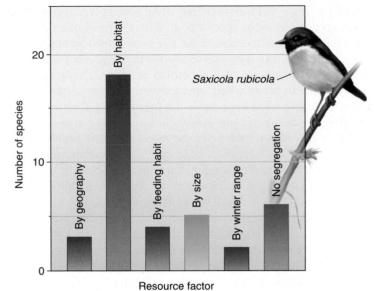

Figure 57.8 Segregation according to resource factor among 40 species of passerine birds. Most segregation is by habitat, followed by size, feeding habit, geography, and type of winter range they forage in. In about 15% of cases, no obvious segregation was observed. More than half of all bird species, including *Saxicola rubicola*, are passerines, also known as perching birds.

However, when Gause cultured *P. caudatum* and *P. bursaria* together, neither went extinct. The population volumes of each was much less compared to when they were grown alone, because some competition occurred between them. Gause discovered, however, that *P. bursaria* was better able to utilize the yeast in the lower part of the culture tubes. From these experiments Gause concluded that two species with exactly the same requirements cannot live together in the same place and use the same resources, that is, occupy the same niche. His conclusion, later termed the **competitive exclusion hypothesis**, essentially means that complete competitors cannot coexist.

If complete competitors drive one another to extinction, how different do they have to be to coexist and in what features do they usually differ? To address such questions, in 1958 ecologist Robert MacArthur examined coexistence between five species of warblers feeding within spruce trees in New England. All belonged to the genus *Dendroica*, so one would expect these closely related bird species to compete strongly, possibly sufficiently strongly to cause extinctions. MacArthur found that the species occupied different heights and portions in the tree and thus each probably fed on a different range of insects (**Figure 57.7**). In addition, the Cape May warbler exists usually at low densities but can respond quickly to periodic outbreaks of insects that occur. It thus gains a temporary advantage in numbers over the other species during these times.

The term **resource partitioning** describes the differentiation of niches, both in space and time, that enables similar species to coexist in a community, just as *Chthamalus* and *Semibalanus* live on different parts of the intertidal zone in Scotland. In a way, we can think of resource partitioning as reflecting the results of past competition. British ornithologist David Lack examined competition and coexistence among about 40 species

of British passerines, or perching birds (**Figure 57.8**). As a group, these perching birds had fairly similar lifestyles. Most segregated according to some resource factor, with habitat being the most common one. For example, while all the passerines fed on insects, some would feed exclusively in grasslands, others in forests, some low to the ground, and others high in trees, where the insects present would likely be different. Birds also segregated by size, so that bigger species would take different-sized food than smaller species, and by feeding habit, with some feeding on insects on trees, others on tree trunks, and so on. Some species also fed in different winter ranges, while others occurred in different parts of the country (separation by geography). About 15% of bird species showed no segregation at all.

What about the species that do not appear to live in different habitats or have different food habits? To answer this question, researchers have looked at differences in physical form, or morphology, between coexisting species.

Morphological Differences May Allow Species to Coexist

Although the competitive exclusion principle acknowledges that complete competitors cannot coexist, some partial level of competition may exist that is not severe enough to drive one of the competitors to extinction. In 1959, biologist G. Evelyn Hutchinson looked at size differences in feeding apparatus, such as mouthparts or other body parts important in feeding, between species when they were **sympatric** (occurring in the same geographic area) and **allopatric** (occurring in different geographic areas).

Hutchinson's hypothesis was that when species were sympatric, each species tended to specialize on different types of food. This was reflected by differences in the size of their feeding apparatus. The tendency for two species to diverge in morphology and thus resource use because of competition is called **character displacement**. In areas where species were allopatric, there was no need to specialize on a particular prey type, so the size of the feeding apparatus did not evolve to become larger or smaller; rather it retained a "middle of the road" size that allowed species to exploit the largest range of prey size distribution.

One of the classic cases of character displacement involves a study of Galápagos finches, several closely related species of finches Charles Darwin discovered on the Galápagos Islands (see also the Feature Investigation in Chapter 23). When two species, *Geospiza fuliginosa* and *Geospiza fortis*, are sympatric, their beak sizes (bill depths) are different. *G. fuliginosa* has a smaller bill depth, which enables it to crack small seeds more efficiently, while *G. fortis* has a larger bill depth, which enables it to feed on bigger seeds. However, when both species are allopatric, that is, existing on different islands, their bills are more similar in depth. Researchers studying *Geospiza* concluded that the bill depth differences evolved in ways that minimized competition. But how great do differences between characters have to be in order to permit coexistence?

Hutchinson noted that the ratio between feeding characters when species were sympatric (and thus competed) averaged about 1.3 (**Table 57.2**). In contrast, the ratio between feeding characters when species were allopatric (and did not compete) was closer to 1.0. Hutchinson proposed that the value of 1.3, a roughly 30% difference, could be used as an indication of the amount of difference necessary to permit two species to coexist. One problem with Hutchinson's ratio is that some differences of 1.3 between similar species might have evolved for reasons other than competition. Some ecologists have argued that we should not attach too much biological meaning to what may in fact be an arbitrary pattern. Nevertheless, while the actual ratio may be disputed, Hutchinson's findings have shown that competition in nature can cause character displacement.

In summary, competition occurs frequently in nature and several different competitive mechanisms are involved. Competition over the same limited resource can lead to the extinction of one of two species occupying the same niche. However, species may coexist if their niche overlap is reduced through resource partitioning or character displacement. We also saw that the outcome of competitive interactions may vary according to both abiotic and biotic factors in the environment. Biotic factors such as parasitism, herbivory, and predation may be powerful mortality factors in their own right and could be sufficient to reduce population densities to the extent that competition does not occur. In the following sections we will examine the strength of these factors, beginning with predation.

57.2 Predation, Herbivory, and Parasitism

Several interactions between species have a positive effect for one species and a negative effect for the other. The traditional view of predation is one of carnivory, in which a predator, typically an animal, feeds on another animal. However, herbivory and parasitism are also interactions in which one species is helped and the other is harmed, and while they can be viewed in the broad context of predation, each has special characteristics that set it apart. Herbivory usually involves nonlethal predation on plants, while predation generally results in the death of the prey. Parasitism, like herbivory, is typically nonlethal, and differs from predation in that the adult parasite typically lives and reproduces in the living host, as in Chinese liver flukes (refer back to Figure 33.8). Parasitoids, insects that lay eggs in living hosts, have features in common with both predators and parasites. They always kill their hosts, as predators do, but one host can support the development of one or more than one parasitoid, just as in parasitism. Parasitoids are common in the insect world and include parasitic wasps and flies that feed on many other insects or spiders. Omnivores are individuals that can feed on both plants and animals. For example, bears feed on both berries and fish. These varied categories of predation can be classified according to how lethal they are for the prey and how close is the association between predator and prey (**Figure 57.9**).

In this section we begin by looking at antipredator strategies and how, despite such strategies, predation remains a factor affecting the density of prey. We survey the strategies plants use to deter herbivores and how, in turn, herbivores overcome host plant defenses. Finally, we investigate parasitism, which may be the predominant lifestyle on Earth, and explore the growing role of genomics in the fight against parasites.

Table 57.2	Comparison of Feeding Characters of Sympatric and Allopatric Species				
Animal (character)	**Species**	**Measurement (mm) when**		**Ratio* when**	
		Sympatric	**Allopatric**	**Sympatric**	**Allopatric**
Weasels (skull)	*Mustela erminea*	50.4	46.0	1.28	1.07
	Mustela nivalis	39.3	42.9		
Mice (skull)	*Apodemus flavicollis*	27.0	26.7	1.09	1.04
	Apodemus sylvaticus	24.8	25.6		
Nuthatches (beak)	*Sitta tephronota*	29.0	25.5	1.23	1.02
	Sitta neumayer	23.5	26.0		
Galápagos finches (beak)	*Geospiza fortis*	12.0	10.5	1.43	1.13
	Geospiza fuliginosa	8.4	9.3		

*Ratio of the larger to smaller character.

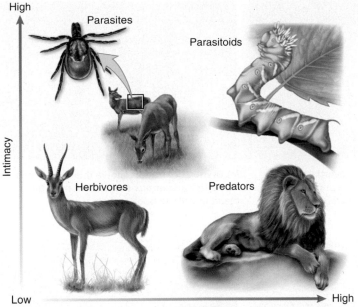

Figure 57.9 **Possible interactions between populations.**
Lethality represents the probability that an interaction results in
the death of the prey. Intimacy represents the closeness and
duration of the relationship between the consumer and the prey.

Animals Have Evolved Many Antipredator Strategies

The variety of strategies that animals have evolved to avoid
being eaten suggests that predation is a strong selective force.

Chemical Defense A great many species have evolved chemical defenses against predation. One of the classic examples of a
chemical defense involves the bombardier beetle (*Stenaptinus
insignis*), which has been studied by Tom Eisner and coworkers. These beetles possess reservoirs of hydroquinone and hydrogen peroxide in their abdomen. When threatened, they eject
the chemicals into an "explosion chamber," where the subsequent release of oxygen causes the whole mixture to be vio-
lently ejected as a hot spray (about 88°C, or 190°F) that can be
directed at the beetle's attackers (**Figure 57.10a**). Many other
arthropods, such as millipedes, have chemical sprays too, and
the phenomenon is also found in vertebrates, as anyone who
has had a close encounter with a skunk can testify.

Often associated with a chemical defense is an **aposematic
coloration**, or warning coloration, which advertises an organism's unpalatable taste. For instance, the ladybird beetle's bright
red color warns of the toxic defensive chemicals it exudes when
threatened, and many tropical frogs have bright warning coloration that calls attention to their skin's lethality (**Figure 57.10b**).
In the 1960s, entomologist Lincoln Brower and coworkers showed
that after inexperienced blue jays ate a monarch butterfly and
suffered a violent vomiting reaction, they learned to associate
the striking orange-and-black appearance of the butterfly with a
noxious reaction (refer back to Figure 55.3). While some ani-

mals synthesize toxins using their own metabolic processes such
as a rattlesnake, other animal poisons are actually acquired
from plants. For example, the monarch butterfly caterpillars
accumulate toxic chemicals called cardiac glycosides from milkweed species.

Cryptic Coloration **Cryptic coloration** is an aspect of camouflage, the blending of an organism with the background of its
habitat. Cryptic coloration is a common method of avoiding
detection by predators. For example, many grasshoppers are
green and blend in with the foliage on which they feed. Stick
insects mimic branches and twigs with their long, slender bodies. In most cases, these animals stay perfectly still when threatened, because movement alerts a predator. Behavior that allows
an animal to blend into its environment is referred to as **catalep-
sies**. Cryptic coloration is prevalent in the vertebrate world, too.
Many seahorses adopt a body shape and color pattern that are
similar to the habitats in which they are found (**Figure 57.10c**).

Mimicry **Mimicry**, the resemblance of an organism (the mimic)
to another organism (the model), also secures protection from
predators. There are two major types of mimicry. In **Müllerian
mimicry**, many noxious species converge to look the same,
thus reinforcing the basic distasteful design. One example is
the black and yellow striped bands of several different types of
bees and wasps. Müllerian mimicry is also found among noxious Amazonian butterflies. **Batesian mimicry** is the mimicry
of an unpalatable species (the model) by a palatable one (the
mimic). Some of the best examples involve flies, especially
hoverflies of the family Syrphidae, which are striped black and
yellow to resemble stinging bees and wasps but are themselves
harmless. The nonvenomous scarlet king snake (*Lampropeltis
triangulum*) mimics the venomous coral snake (*Micrurus nigrocinctus*), thereby gaining protection from would-be predators
(**Figure 57.10d**).

It is sometimes difficult to distinguish between Batesian
and Müllerian mimics. Until recently, the viceroy butterfly was
believed to be an example of Batesian mimicry because of its
striking resemblance to the unpalatable monarch. However, in
1991, researcher David Ritland, a graduate student at the time,
showed that the viceroy was equally as unpalatable as the monarch and is actually an example of Müllerian mimicry. In theory,
Batesian mimics should lose their protection when they are
common, because they need plenty of distasteful models whose
unpalatability maintains the validity of their warning colors. If
the edible mimic became too common, predators would learn to
ignore the coloration, which would also harm the model.

Displays of Intimidation Some animals put on displays of
intimidation in an attempt to discourage predators. For example,
a toad swallows air to make itself appear larger, frilled lizards
extend their collars when frightened to create this same effect,
and porcupine fish inflate themselves to large proportions when
threatened (**Figure 57.10e**). All of these animals use displays to
deceive potential predators about the ease with which they can
be eaten.

(a) As it is held by a tether attached to its back, this bombardier beetle (*Stenaptinus insignis*) directs its hot, stinging spray at a forceps "attacker."

(b) Aposematic coloration advertises the poisonous nature of this blue poison arrow frog (*Dendrobates azureus*) from South America.

(c) Cryptic coloration allows this Pygmy sea horse (*Hippocampus bargibanti*) from Bali, to blend in with its background.

(d) In this example of Batesian mimicry, an innocuous scarlet king snake (*Lampropeltis triangulum*) (*left*) mimics the poisonous coral snake (*Micrurus nigrocinctus*) (*right*).

(e) In a display of intimidation, this porcupine fish (*Diodon hystrix*) puffs itself up to look threatening to its predators.

Figure 57.10 Antipredator adaptations.

Biological inquiry: According to the classification of species interactions in Table 57.1, how would you classify Batesian and Müllerian mimicry?

Fighting Many animals have developed horns and antlers, which although primarily used in competition over mates can be used in defense against predators (refer back to Figure 34.26c). Invertebrate species often have powerful claws, pincers, or, in the case of scorpions, venomous stingers that can be used in defense as well as offense.

Agility Some groups of insects, such as grasshoppers, have a powerful jumping ability to escape the clutches of predators. Many frogs are prodigious jumpers, and flying fish can glide above the water to escape their pursuers.

Armor The shells of tortoises and turtles are a strong means of defense against most predators, as are the quills of porcupines (refer back to Figure 34.24c). Many beetles have a tough exoskeleton that protects them from attack from other arthropod predators such as spiders.

Predator satiation **Predator satiation** is the synchronous production of many progeny by all individuals in a population to satiate predators and thereby allow some progeny to survive. Predator satiation is commonly called masting when discussed in relation to seed herbivory in trees, which tend to have years of unusually high seed production that reduces predation. However,

a similar phenomenon is exhibited by the emergence of 13-year and 17-year periodical cicadas (*Magicicada* spp.). These insects are termed periodical because the emergence of adults is highly synchronized to occur once every 13 or 17 years. Adult cicadas live for only a few weeks, during which time females mate and deposit eggs on the twigs of trees. The eggs hatch 6–10 weeks later, and the nymphs drop to the ground and begin a long subterranean development, feeding on the contents of the xylem of roots. Because the xylem is low in nutrients, it takes many years for nymphs to develop, though there appears to be no physiological reason why some cicadas couldn't emerge after, say, 12 years of feeding, and others after 14. Their synchrony of emergence is thought to maximize predator satiation. Worth noting in this context is the fact that both 13 and 17 are prime numbers, and thus predators on a shorter multiannual cycle cannot repeatedly utilize this resource. For example, a predator that bred every three years could not rely on cicadas always being present as a food supply.

How common is each of these defense types? No one has done an extensive survey over the entire animal kingdom. However, in 1989, Brian Witz surveyed studies that documented antipredator mechanisms in arthropods, mainly insects. By far the most common antipredator mechanisms were chemical defenses and associated aposematic coloration, noted in 51% of the examples. All other types of defense mechanisms occurred with considerably less frequency.

Despite the Impressive Array of Defenses, Predators Can Still Affect Prey Densities

The importance of predation on prey populations may be dependent on whether the system is donor controlled or predator controlled. In a donor-controlled system, prey supply is determined by factors other than predation, such as food supply, so that removal of predators has no effect on prey density. Examples include predators that feed on intertidal communities in which space is the limiting factor that controls prey populations.

In a predator-controlled system, the action of predator feeding eventually reduces the supply of prey. The removal of predators in a donor-controlled system is thus likely to have little effect, whereas in a predator-controlled system, such an action would probably result in large increases in prey abundance. Research studies have shown that many predators can have a significant effect on many prey populations. Considerable data exist on the interaction of the Canada lynx (*Lynx canadensis*) and its prey, snowshoe hares (*Lepus americanus*), because of the value of the pelts of both animals. In 1942, British ecologist Charles Elton analyzed the records of furs traded by trappers to the Hudson's Bay Company in Canada over a 100-year period. Analysis of the records showed that a dramatic 9- to 11-year cycle existed for as long as records had been kept (**Figure 57.11**). This cycle appears to be an example of a stable predator-prey oscillation that is predator controlled: the hares go up in density, followed by an increase in density of the lynx, which then depresses hare numbers. This is followed by a decline in the number of lynx, and the cycle begins again.

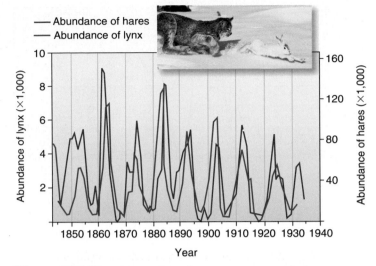

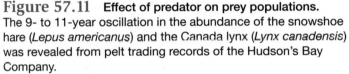

Figure 57.11   **Effect of predator on prey populations.** The 9- to 11-year oscillation in the abundance of the snowshoe hare (*Lepus americanus*) and the Canada lynx (*Lynx canadensis*) was revealed from pelt trading records of the Hudson's Bay Company.

Invasive species have provided striking examples of the effects of predators. The brown tree snake (*Boiga irregularis*) was inadvertently introduced by humans to the island of Guam, in Micronesia, shortly after World War II. The growth and spread of its population over the next 40 years closely coincided with a precipitous decline in the island's forest birds. The snake is a new predator with no natural predators on Guam to control it, and since the birds did not evolve with the snake, they have no defenses against it. Eight of the island's 11 native species of forest birds went extinct by the 1980s (**Figure 57.12**).

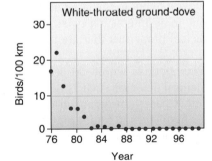

Figure 57.12   **Invasive species provide an example of predator pressure.**   Population trends for four native Guam birds, as indicated by 100 km roadside surveys conducted from 1976 to 1998, show a precipitous decline because of predation by the introduced brown tree snake.

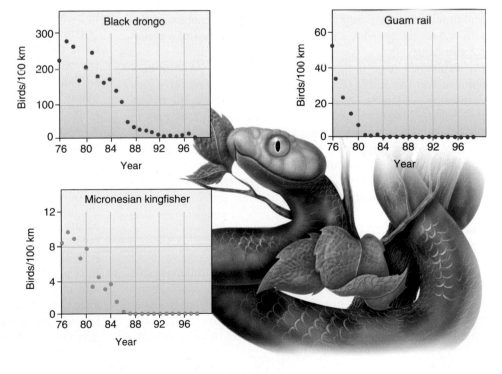

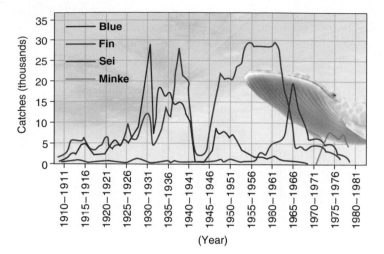

Figure 57.13 Sequential decline of whale catches in the Antarctic shows the strong effect of human predators. Whale catches are believed to be directly related to whale population sizes. The catches of blue whale, the first species to be strongly affected by human predation, started a precipitous decline in the 1940s, as the whale was hunted to very low levels. Humans then began hunting more abundant fin whales, and in turn sei and minke whales, as each species became depleted.

One of the biggest reductions in prey in response to predation has been the systematic decline of whales of various species in response to the human whaling industry. The history of whaling has been characterized by a progression from larger, more valuable or easily caught species to smaller, less valuable or easily caught ones, as numbers of the original targets have been depleted (**Figure 57.13**). Belatedly, the International Whaling Commission enacted a moratorium on all commercial whaling in 1986. Following the moratorium, the populations of some whales have increased. Blue whales are thought to have quadrupled their numbers off the California coast during the 1980s, and numbers of the California gray whales have recovered to pre-whaling levels, showing the impact that an absence of a predator can have.

Ecologists have found that in nearly 1,500 predator-prey studies, over two-thirds (72%) showed a large depression of prey density by predators. Thus, we can conclude that in the majority of cases, predators influence the abundance of their prey in their native environment. The variety of antipredator mechanisms discussed earlier also shows how predation is important enough to select for the evolution of chemical defenses, camouflage, and mimicry in prey. Taken together these data indicate that predation is a powerful force in nature.

Plants and Herbivores May Be Engaged in an Evolutionary Arms Race

Herbivory involves the predation of plants or similar species such as algae. Such herbivory can be lethal to the plants, especially for small species, but often it is nonlethal, since many plant species, particularly larger ones, can regrow. We can distinguish two types of herbivores: generalist herbivores and spe-

cialist herbivores. Generalist herbivores can feed on many different plant species and are usually mammals. Specialist herbivores are often restricted to one or two species of host plants and are usually insects. There are, however, exceptions. Pandas are specialists because they feed only on bamboo, and koalas specialize on eucalyptus trees. On the other hand, grasshoppers are generalists, because they feed on a wide variety of plant species, including crops.

Plants appear to present a luscious green world of food to any organism versatile enough to use it, so why don't we see more plants being eaten by herbivores? After all, unlike most animals, plants cannot move to escape being eaten. Two hypotheses have been proposed to answer the question of why more plant material is not eaten. First, predators and parasites might keep herbivore numbers low, thereby sparing the plants. The many examples of the strength of predation provide evidence for this view. Second, the plant world is not as helpless as it appears; the sea of green is in fact armed with defensive spines, tough cuticles, and noxious chemicals. Let's take a closer look at plant defenses against herbivores and the ways that herbivores attempt to overcome them.

Plant Defenses Against Herbivores As described in Chapter 7, an array of unusual and powerful chemicals is present in plants, including alkaloids (nicotine in tobacco, morphine in poppies, cocaine in coca, and caffeine in tea), phenolics (lignin in wood and tannin in leaves), and terpenoids (in peppermint) (**Figure 57.14a–c**). Two general chemical classes can be recognized, those containing nitrogen compounds (e.g., alkaloids) and those lacking nitrogen (e.g., phenolics and terpenoids). Such compounds are not part of the primary metabolic pathway that plants use to obtain energy. They are therefore referred to as **secondary metabolites**, or secondary chemicals. Most of these chemicals are bitter tasting or toxic, and they deter herbivores from feeding. The staggering variety of secondary metabolites in plants, over 25,000, may be testament to the large number of organisms that feed on plants. In an interesting twist, many of these compounds have medicinal properties that are beneficial to humans (**Table 57.3**).

In addition to chemical compounds within the leaf, many plants have an array of mechanical defenses such as thorns and spines (**Figure 57.14d**). In other cases, an organism may provide a plant protection against herbivores in return for resources such as light, water, nutrients, or nesting sites (as we will see in Section 57.3).

An understanding of plant defenses is of great use to agriculturalists, since the more crops that are defended against pests, the higher the crop yields. The ability of plants to prevent herbivory is also known as **host plant resistance**. Host plant resistance may be due to chemical or mechanical defenses. One serious problem associated with commercial development of host plant resistance is that it may take a long time to breed into plants—between 10 and 15 years. This is because of the long time it takes to identify the responsible chemicals and develop the resistant genetic lines. Also, resistance to one pest may come at the cost of increasing susceptibility to other pests.

(a)   **(b)**

(c)   **(d)**

Figure 57.14 **Defenses against herbivory.** Plants possess an array of unusual and powerful chemicals, including **(a)** alkaloids, such as nicotine in tobacco, **(b)** phenolics, such as tannins in tea leaves (near Mount Fuji, Japan), and **(c)** terpenoids in peppermint leaves. **(d)** Mechanical defenses include plant spines and thorns, as on this shrub, *Rosa multiflora*.

Biological inquiry: Which type of defense would be most effective in deterring invertebrate herbivores?

Finally, some pest strains can overcome the plant's mechanisms of resistance. Circumvention of resistance by pests evolves in much the same way as resistance to pesticides.

Despite these problems, host plant resistance is a good tactic for the farmer. After the initial development of resistant varieties, the cost is minimal. Perhaps more importantly, host plant resistance is environmentally benign, generally having few side effects on other species in the community. It is estimated that about 75% of cropland in the U.S. utilizes pest-resistant plant varieties, most of these being resistant to plant pathogens. Bt corn is a variety of corn that has been genetically modified to incorporate a gene from the soil bacterium *Bacillus thuringiensis* (Bt), which provides a natural pesticide that is toxic to some insects. Genetic engineers have also produced Bt cotton, Bt tomato, and genetically modified varieties of many other crop species.

Overcoming Plant Resistance Herbivores can often overcome plant defenses. They can detoxify many poisons, mainly by two chemical pathways: oxidation and conjugation. Oxidation, the most important of these mechanisms, occurs in mammals in the liver and in insects in the midgut. It involves catalysis of the secondary metabolite to a corresponding alcohol by a group of enzymes known as mixed-function oxidases (MFOs). Conjugation, often the next step in detoxification, occurs by the uniting of the harmful element, or the compound resulting from the oxidation, with another molecule to create an inactive and readily excreted product.

In addition, certain chemicals that are toxic to generalist herbivores actually increase the growth rates of adapted specialist species, which can circumvent the defense or put the chemicals to good use in their own metabolic pathways. The Brassicaceae, the plant family that includes mustard, cabbage, and other species, contains acrid-smelling mustard oils called glucosinolates, the most important one of which is sinigrin.

Table 57.3	Plant Species with Compounds Used as Drugs		
Plant name	**Compound name**	**Use in medicine**	**Country of origin or cultivation**
Atropa belladonna	Atropine	Dilation of pupil	Central and Southern Europe; cultivated in U.S., U.K., other countries
Cassia senna	Danthron	Laxative	Cultivated in Egypt
Catharanthus roseus	Vincristine	Antitumor agent	Pantropical; cultivated in U.S., India, other countries
Cinchona ledgeriana	Quinine	Antimalarial	Cultivated in Indonesia, Zaire
Datura metel	Scopolamine	Sedative	Cultivated in Asia
Ephedra sinica	Ephedrine	Chronic bronchitis	China
Digitalis spp.	Digitoxin	Cardiotonic	Cultivated in Europe and Asia
Papaver sominiferum	Codeine, morphine	Analgesic, sedative	Cultivated in Turkey, India, Burma, Thailand
Pausinystalia yohimbe	Yohimbine	Erectile dysfunction	Cameroon, Nigeria, Rwanda
Rauvolfia spp.	Reserpine	Tranquilizer	India, Bangladesh, Sri Lanka, Burma, Malaysia, Indonesia, Nepal
Silybum merianum	Silymarin	Liver disorders	Mediterranean region

Large white butterflies (*Pieris brassicae*) preferentially feed on cabbage over other plants. In fact, if newly hatched larvae are fed an artificial diet, they do much better when sinigrin is added to it. When larvae are fed cabbage leaves on hatching from eggs and are later switched to an artificial diet without sinigrin, they die rather than eat. In this case, the secondary metabolite has become an essential feeding stimulant.

Ultimately, a good method to estimate the effects of herbivory on plant populations is to remove the herbivores and examine subsequent growth and reproductive output. Analysis of the hundreds of such experiments that have been conducted have yielded several interesting generalizations. First, herbivory in aquatic systems is more extensive than in terrestrial ones, generally because aquatic systems contain species, such as algae, that are especially susceptible to herbivory. Second, invertebrate herbivores such as insects have a stronger effect on plants than vertebrate herbivores such as mammals, at least in terrestrial systems. Thus, while one might consider large grazers like bison in North America or antelopes in Africa to be of huge importance in grasslands, it is more likely that grasshoppers are the more significant herbivores. In forests, invertebrate grazers such as caterpillars have greater access to canopy leaves than vertebrates and are also likely to have a greater effect.

Which types of plants are most affected by herbivory? The effect of herbivores is usually greatest on algae, presumably because these organisms are the least sophisticated in terms of their ability to manufacture complex secondary metabolites. Grasses and shrubs are also significantly affected by herbivores, but woody plants such as trees are less so. This may be because large and long-lived trees can draw on large resource reserves to buffer the impact of herbivores.

Parasitism Might Be the Predominant Lifestyle on Earth

When one organism feeds on another, but does not normally kill it outright, the predatory organism is termed a **parasite** and the prey a **host**. Some parasites remain attached to their hosts for most of their lives; for example, tapeworms spend their entire adult life inside the host's alimentary canal and even reproduce within their host. Others, such as the lancet fluke, have more complex life cycles that require the use of multiple hosts (**Figure 57.15**). Some, such as ticks and leeches, drop off their hosts after prolonged periods of feeding. Others, like mosquitoes, remain attached for relatively short periods.

Some flowering plants are parasitic on other plants. **Holoparasites** lack chlorophyll and are totally dependent on the host plant for their water and nutrients. One famous holoparasite is the tropical *Rafflesia arnoldii*, which lives most of its life within the body of its host (**Figure 57.16**). Only the flower develops externally, and it is a massive flower, 1 m in diameter and the largest known in the world. **Hemiparasites** generally do photosynthesize, but they lack a root system to draw water and thus depend on their hosts for that function. Mistletoe (*Viscum album*) is a hemiparasite. Hemiparasites usually have a broader range of hosts than do holoparasites, which may be confined to a single or a few host species.

We can define parasites that feed on one species or just a few closely related hosts as **monophagous**. **Polyphagous** species, by contrast, feed on many host plants, often from more than one plant family. We can also distinguish parasites as **microparasites** (for example, pathogenic bacteria and viruses), which multiply within their hosts, usually within the cells, and

5 Some of the flukes migrate into the ant's brain, causing the ant to climb to the tip of a blade of grass, where it is more likely to be eaten by a cow.

Ant

Adult lancet fluke

1 Adult flukes produce eggs inside a cow. The eggs are passed in the cow's feces.

2 Snails eat the fluke eggs; which later hatch in the snail's intestine and asexually produce offspring.

4 Ants eat the slime balls.

Slime balls

3 Offspring are passed from the snail in slime balls.

Figure 57.15 **A parasite life cycle.** The life cycle of the lancet fluke (*Dicrocoelium dendriticum*) involves behavioral changes in ants, one of its three hosts, that increase its transmission rate.

Figure 57.16 *Rafflesia arnoldii*, the world's biggest flower, lives as a parasite in Indonesian rain forests.

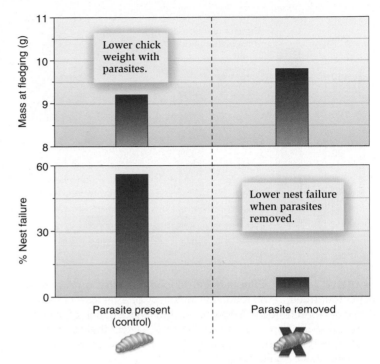

Figure 57.17 Parasite removal experiments. The left side shows the results when blowfly larva were present in the nests of young blue tits, while the right side shows the results when these parasites were removed.

macroparasites (such as schistosomes), which live in the host but release infective juvenile stages outside the host's body. Usually, the host has a strong immunological response to microparasitic infections. For macroparasitic infections, however, the response is short-lived, the infections tend to be persistent, and the hosts are subject to continual reinfection.

Last, we can distinguish **ectoparasites**, such as ticks and fleas, which live on the outside of the host's body, from **endoparasites**, such as pathogenic bacteria and tapeworms, which live inside the host's body. Problems of definition arise with regard to plant parasites, which seem to straddle both camps. For example, some parasitic plants, such as dodder, *Cuscuta* spp., an orange, stringlike plant, exist partly outside of the host's body and partly inside (refer back to Figure 37.22). Outgrowths called haustoria penetrate inside the host plant to tap into nutrient supplies. Being endoparasitic on a host seems to require greater specialization compared to ectoparasitism. Thus, ectoparasitic animals like leeches feed on a wider variety of hosts than do internal parasites such as liver flukes.

As we have seen throughout this textbook, there are vast numbers of species of parasites, including viruses, bacteria, protozoa, flatworms (flukes and tapeworms), nematodes, and various arthropods (ticks, mites, and fleas). Therefore, parasitism is a common way of life. Parasites may outnumber free-living species by four to one. Most plant and animal species harbor many parasites. For example, leopard frogs have nematodes in their ears and veins, and flukes in their bladders, kidneys, and intestines. A free-living organism that does not harbor parasitic individuals of a number of species is a rarity.

As with studies of other species interactions, a good method to determine the effect of parasites on their host populations is to remove the parasites and to re-examine the system. However, this is difficult to do, primarily because of the small size and unusual life histories of many parasites, which makes them difficult to remove from a host completely. The few cases of experimental removal confirm that parasites can reduce host population densities. The nests of birds such as blue tits are often infested with parasitic blowfly larvae that feed on the blood of nestlings. In 1997, Sylvie Hurtrez-Bousses and colleagues exper-

imentally reduced blowfly larval parasites of young blue tits in nests in Corsica. Parasite removal was cleverly achieved by taking the nests from 145 nest boxes, removing the young, microwaving the nests to kill the parasites, and then returning the nests and chicks to the wild. The success of chicks in microwaved nests was compared to that in nonmicrowaved (control) nests, and it was found that parasite-free blue tit chicks had greater body mass at fledging, the time when feathers first grow (**Figure 57.17**). Perhaps more important was that complete nest failure, that is, death of all chicks, was much higher in control nests than in treated nests.

Because parasite removal studies are difficult to do, ecologists have also examined the strength of parasitism as a mortality factor by studying introduced parasite species. Evidence from natural populations suggests that introduced parasites have substantial effects on their hosts. Chestnut blight, a fungus from Asia that was accidentally introduced to New York around 1904, virtually eliminated chestnut trees (*Castanea dentata*) in North America. Although the fungus took many years to spread throughout North America, by the 1950s, it had significantly reduced the density of American chestnut trees in North Carolina (**Figure 57.18**). In Europe and North America, Dutch elm disease has similarly devastated populations of elms. The disease wiped out 25 million of Britain's original 30 million elm trees between the 1960s and the 1990s. In Italy, canker has had similar severe effects on cypress. Fortunately, as we will discuss shortly, the field of genomics is helping greatly in the fight against plant diseases.

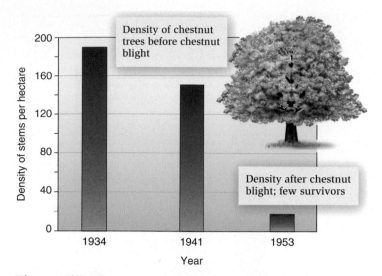

Figure 57.18 **Effects of introduced parasites on American chestnut trees.** The reduction in density of American chestnuts trees in North Carolina following the 1904 introduction of chestnut blight disease from Asia shows the severe effect that parasites can have on their hosts. By the 1950s, this once-prevalent species was virtually eliminated.

Introduced parasites have a strong effect on native animals as well as plants. In the early 1980s, canine distemper virus caused a large decline in the last remaining population of black-footed ferrets in the U.S. Conservationists are particularly concerned that many endangered animals are threatened by disease from domestic animals. In 1994, one-third of the population of lions in Serengeti National Park was wiped out by canine distemper virus, presumably contracted by domestic dogs. Some believe that the early 20th century extinction of the thylacine (*Thylacinus cynocephalus*), a Tasmanian marsupial that resembled a wolf, was caused by a distemper-like disease transmitted by dogs. To prevent the threat of disease among endangered species, some populations have been vaccinated; for example, the Ethiopian wolf (*Canis simensis*) has been vaccinated against rabies, and mountain gorillas (*Gorilla gorilla*) have been vaccinated against measles.

GENOMES & PROTEOMES

Transgenic Plants May Be Used in the Fight Against Plant Diseases

Many important native forest trees, which are also grown in urban landscapes, have been almost entirely wiped out by diseases spread by the importation of exotic plants. Sudden oak death is a recently recognized disease that is killing tens of thousands of oak trees and other plant species in California. The symptoms vary between species but include leaf spots, oozing of a dark sap through the bark, and twig dieback. Although sudden oak death is a forest disease, the organism causing this disease is known to infect many woody ornamental plants,

such as rhododendrons, that are commonly sold by nurseries. In March 2004, a California nursery was found to have unknowingly shipped plants infected with sudden oak death to all 50 states. Following this discovery, California nurseries halted shipments of trees to other states in an attempt to stop the spread of the disease, originally thought to have been imported on rhododendrons. In 2004, scientists mapped out the genome sequence of the disease-carrying fungus, *Phytophthora ramorum*. They are hoping that identifying the genes and their proteins will help them develop specific diagnostic tests to quickly detect the presence of sudden oak death in trees, which is currently impossible to detect until a year or more after the tree is infected.

However, scientists hope for much more from the field of genomics in their fight against plant diseases. Many scientists have suggested limited, cautious transfer of resistance genes from the original host species in the source regions of the disease to threatened species. Original host species have usually evolved over millions of years of exposure to these diseases and have acquired genes that provide resistance. In the regions of recent introduction of parasites, there has been no selection for resistance, so the host plants are often killed en masse. Transgenic trees that have received resistance genes could be produced and then be replanted in forests or urban areas. An advantage of this technique over traditional cross-breeding strategies involving two different species is that transgenic methods involve the introduction of fewer unnecessary genes of the nonnative species. Also, fewer tree generations would be required to develop resistance. For example, using traditional breeding technology, Asian chestnut trees (*Castanea mollissima*) are being bred with American chestnuts (*Castanea dentata*) to reduce the susceptibility of the latter to chestnut blight, but the resultant hybrid is often significantly altered in appearance from the traditional American chestnut and the process takes more than a decade to produce trees that are ready to plant. Transgenic technology could minimize these drawbacks. William Powell and colleagues are working to enhance the American chestnut's resistance by inserting a gene taken from wheat. The gene, oxalate oxidase, destroys a toxin produced by the fungus that causes chestnut blight.

57.3 Mutualism and Commensalism

Mutualism and commensalism are interactions that are beneficial to at least one of the species involved. The unique feature of mutualism is that both species benefit from the interaction. For example, in mutualistic pollination systems, both plant and pollinator benefit, the former by the transfer of pollen and the latter typically by a nectar meal. In commensalism, one species benefits and the other remains unaffected. For example, in some forms of seed dispersal, barbed seeds are transported to new germination sites in the fur of mammals. The seeds benefit, but the mammals are generally unaffected. In this section we will examine the major types of mutualism and commensalism.

It is interesting to note that humans have entered into mutualistic relationships with many species. For example, the mutualistic association of humans with plants has resulted in some of the most far-reaching ecological changes on Earth. Humans have planted huge areas of the Earth with crops, allowing these plant populations to reach densities they never would on their own. In return, the crops have led to expanded human populations because of the increased amounts of food they provide.

Mutualism Is an Association Between Two Species That Benefits Both Species

Many close associations are known between species in which both species benefit. For example, leaf-cutting ants of the tribe Attini, of which there are about 210 species, enter into a mutualistic relationship with a fungus. A typical colony of about 9 million ants has the collective biomass of a cow and harvests the equivalent of a cow's daily requirement of fresh vegetation. Instead of consuming it directly, the ants chew it into a pulp, which they store underground as a substrate on which the fungus grows. The ants shelter and tend the fungus, protecting it from competing fungi and helping it reproduce and grow. In turn, the fungus produces specialized structures known as gongylidia, which serve as food for the ants. In this way, the ants circumvent the chemical defenses of the leaves, which are digested by the fungus.

The ant-fungus mutualism permits both species to live in close association, utilizing a common resource. Other different types of mutualisms occur in nature. Some of these are **defensive mutualisms**, often involving an animal defending a plant or an herbivore. **Dispersive mutualisms** include plants and pollinators that disperse their pollen, and plants and fruit eaters that disperse the plant's seeds.

Defensive Mutualism One of the most commonly observed mutualisms occurs between ants and aphids. Aphids are fairly defenseless creatures and are easy prey for most predators. The aphids feed on plant sap and have to process a significant amount of it to get their required nutrients. In doing so, they excrete a lot of fluid, and some of the sugars still remain in the excreted fluid, which is called honeydew. The ants drink the honeydew and, in return, protect the aphids from an array of predators, such as ladybird beetle larvae, by driving the predators away. In some cases, the ants herd the aphids like cattle, moving them from plant to plant (**Figure 57.19a**).

In other cases, ants enter into a mutualistic relationship with a plant itself. One of the most famous cases involves acacia trees in Central America, whose large thorns provide food and nesting sites for ants (**Figure 57.19b**). In return, the ants bite and discourage both insect and vertebrate herbivores from feeding on the trees. They also trim away foliage from competing plants and kill neighboring plant shoots, ensuring more light, water, and nutrient supplies for the acacias. In this case, neither species can live without the other, a concept called **obligatory mutualism**. This contrasts with **facultative mutualism**, in which the interaction is beneficial but not essential to the survival and reproduction of either species. For example, the ant-aphid mutualisms are generally facultative. Both species benefit from the association, but each could live without the other.

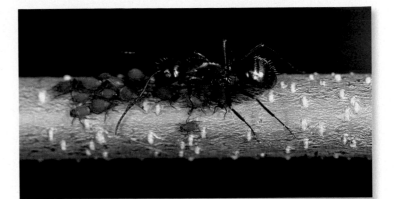

(a)

(b)

Figure 57.19 Defensive mutualism involves species that receive food or shelter in return for providing protection. (a) This red carpenter ant, *Camponotus pennsylvanicus*, tends aphids feeding on a twig. The ants receive sugar-rich honeydew produced by the aphids and in return they protect the aphids from predators. (b) Ants, usually *Pseudomyrmex ferruginea*, make nests inside the large, hornlike thorns of the bull's horn acacia and defend the plant against insects and mammals. In return, the acacia (*Acacia collinsii*) provide two forms of food to the ants: protein-rich granules called Beltian bodies and nectar from extrafloral nectaries (nectar-producing glands that are physically apart from the flower).

Biological inquiry: Is the relationship between red carpenter ants and aphids an example of facultative or obligatory mutualism?

Dispersive Mutualism Many examples of plant-animal mutualism involve pollination and seed dispersal. From the plant's perspective, an ideal pollinator would be a specialist, moving quickly among individuals but retaining a high fidelity to a plant species. One way that plant species in an area encourage the pollinator's species fidelity is by sequential flowering of different species through the year and by synchronized flowering within a species. The plant should provide just enough nectar to attract a pollinator's visit. From the pollinator's perspective, it would be best to be a generalist and obtain nectar and pollen from as many flowers as possible in a small area, thus minimizing the energy spent on flight between patches. This suggests that although mutualisms are beneficial to both species, their optimal needs are quite different.

Figure 57.20   **Dispersive mutualism.**   This blackbird (*Turdus merula*) is an effective seed disperser.

(a)   (b)

Figure 57.21   **Commensalisms that involve cheating.**
(a) Bee orchids (*Ophrys apifera*) mimic the shape of a female bee. Male bees copulate with the flowers, transferring pollen but getting no nectar reward. **(b)** Hooked seeds of burdock (*Arctium minus*) have lodged in the fur of a white-footed mouse (*Peromyscus leucopus*). The plant benefits from the relationship by the dispersal of its seeds, and the animal is not affected.

Mutualistic interactions are also highly prevalent in the seed-dispersal systems of plants. Fruits provide a balanced diet of proteins, fats, and vitamins. In return for this juicy meal, animals unwittingly disperse the enclosed seeds, which pass through the digestive tract unharmed. Fruits taken by birds and mammals often have attractive colors (**Figure 57.20**); those dispersed by nocturnal bats are not brightly colored but instead give off a pungent odor that attracts the bats.

Commensal Relationships Are Those in Which One Partner Receives a Benefit While the Other Is Unaffected

In commensalism, one member derives a benefit while the other neither benefits nor is harmed. Such is the case when orchids or other epiphytes grow in forks of tropical trees. The tree is unaffected, but the orchid gains support and increased exposure to sunlight and rain. Cattle egrets feed in pastures and fields among cattle, whose movements stir up insect prey for the birds. The egrets benefit from the association, but the cattle generally do not. One of the best examples of commensalism involves **phoresy**, in which one organism uses a second organism for transportation. Flower-inhabiting mites travel between flowers in the nostrils (nares) of hummingbirds. The flowers the mites inhabit live only a short while before dying, so the mite relocates to distant flowers by scuttling into the nares of visiting hummingbirds and hitching a ride to the next flower. Presumably the hummingbirds are unaffected.

Some commensalisms involve one species "cheating" on the other without harming it. In the bogs of Maine, the grass-pink orchid (*Calopogon pulchellus*) produces no nectar, but it mimics the nectar-producing rose pogonia (*Pogonia ophioglossoides*) and is therefore still visited by bees. Bee orchids mimic the appearance and scent of female bees; males pick up and transfer pollen while trying to copulate with the flowers (**Figure 57.21a**). The stimuli of flowers of the orchid genus *Ophrys* are so effective that male bees prefer to mate with them even in the presence of

actual female bees! Many plants have essentially cheated their potential mutualistic seed-dispersal agents out of a meal by developing seeds with barbs or hooks to lodge in the animals' fur or feathers rather than their stomachs (**Figure 57.21b**). In these cases, the plants receive free seed dispersal, and the animals receive nothing, except perhaps minor annoyance. This type of relationship is fairly common; most hikers and dogs have at some time gathered spiny or sticky seeds as they wandered through woods or fields.

57.4 Conceptual Models

In this chapter, we have seen that competition, predation, herbivory, parasitism, mutualism, and commensalism are important in nature. How can we determine which of these factors, along with abiotic factors such as temperature and moisture, is the most important in regulating population size? The question is one asked by many applied biologists, such as foresters, marine biologists, and conservation biologists, who are interested in reducing population mortality in order to maximize a population's size. Many different theoretical models have been proposed to describe which mortality factors are the most significant in limiting population size. Some models stress the importance of so-called bottom-up factors such as plant or prey quality and abundance in controlling the herbivores or predators that feed on them. Others stress the importance of top-down factors, such as predators and parasites, acting on their animal or plant prey (**Figure 57.22**). Still others incorporate components of both these models. In this section, we will briefly discuss some of the evidence for the existence of bottom-up versus top-down control.

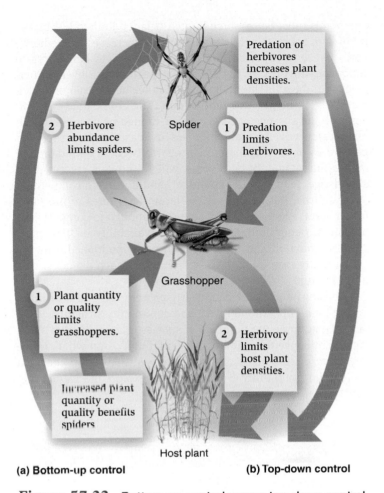

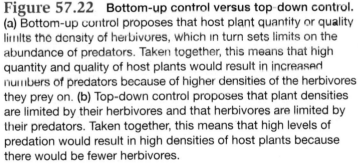

(a) Bottom-up control   **(b) Top-down control**

Figure 57.22 Bottom-up control versus top-down control. (a) Bottom-up control proposes that host plant quantity or quality limits the density of herbivores, which in turn sets limits on the abundance of predators. Taken together, this means that high quantity and quality of host plants would result in increased numbers of predators because of higher densities of the herbivores they prey on. (b) Top-down control proposes that plant densities are limited by their herbivores and that herbivores are limited by their predators. Taken together, this means that high levels of predation would result in high densities of host plants because there would be fewer herbivores.

Bottom-Up Models Suggest Food Limitation Controls Population Densities

At least two lines of evidence exist to suggest that bottom-up effects are important in limiting population sizes. First, we know there is a progressive lessening of available energy passing from plants through herbivores to carnivores and to the carnivores that eat carnivores (termed secondary carnivores). This line of evidence, based on the thermodynamic properties of energy transfer, suggests that plants usually regulate the population size of all other species that rely on them.

Second, much evidence supports the **nitrogen-limitation hypothesis** that organisms select food in terms of the nitrogen content of the tissue. This is largely due to the different propor-

tions of nitrogen in plants and animals. Animal tissue generally contains about 10 times as much nitrogen as plant tissue; thus, animals favor high-nitrogen plants. For example, red deer feed preferentially on grasses defecated upon by herring gulls (*Larus argentatus*). Where the number of gull droppings increases, so does the vegetation nitrogen content. Fertilization has repeatedly been shown to benefit herbivores. Nearly 60% of 186 studies investigating the effects of fertilization on herbivores reported that increasing a plant's tissue nitrogen concentration through fertilization had strong positive effects on herbivore population sizes, survivorship, growth, and fecundity.

Top-Down Models Suggest Natural Enemies Control Population Densities

Top-down models suggest that predators control populations of their prey and that herbivores control plant populations. Supporting evidence comes from the world of biological control, where natural enemies are released to control pests of agriculture. Many weeds are invaders that were accidentally introduced to an area from a different country, as seeds in ships' ballasts or in agricultural shipments. For example, over 50% of the 190 major weeds in the U.S. are invaders from outside the country. Many of these weeds have become separated from their native natural enemies; this is one reason the weeds become so prolific. Because chemical control is expensive and may have unwanted environmental side effects, many land managers have reverted to biological control, in which the invading weed is reunited with its native natural enemy.

Ecologists have noted many successes in the biological control of weeds. Klamath weed or St. John's wort (*Hypericum perforatum*), a pest in California pastures, was controlled by two beetles from its homeland in Europe. The Brazilian weevil, *Cyrtobagous salviniae*, is a proven biological control agent for the floating fern (*Salvinia molesta*), which has invaded lakes in Australia and New Guinea. Alligator weed has been controlled in Florida's rivers by the alligatorweed flea beetle (*Agasicles hygrophila*) from South America. The prickly pear cactus (*Opuntia stricta*) provides a prime example of effective biological control of a weed. The cactus was imported into Australia in the 19th century and quickly established itself as a major pest of rangeland. The small cactus moth (*Cactoblastis cactorum*) was introduced in the 1920s and within a short time successfully saved hundreds of thousands of acres of valuable rangeland from being overrun by the cacti (**Figure 57.23**). The numerous examples showing that these pest populations are controlled when reunited with their natural enemies provide strong evidence of top-down control in nature.

Modern Models Suggest Both Top-Down and Bottom-Up Effects Are Important

More recently, different models have been proposed that take into account the effects of both natural enemies and limited resources on species. In 1981, Laurie Oksanen and coworkers suggested that the strength of mortality factors varies with availability of plant biomass involved, a model they termed the ecosystem exploitation hypothesis (**Table 57.4**). Thus, for very simple systems where mainly plants exist, like Arctic tundra, not enough plant material is available to support herbivores, and plants must be resource limited (that is, limited by competition with each other). As plant biomass increases, some herbivores can be supported, but there are too few herbivores to support carnivores. In the absence of carnivores, levels of herbivory can be quite high. Plant abundance becomes limited primarily by herbivory, not competition. The abundance of herbivores, in the absence of carnivores, is limited by competition for plant resources. As plant biomass increases still further and herbivores become common, carnivores become abundant and reduce the number of herbivores, which in turn increases plant abundance. Plants, being abundant, endure severe competition for resources, but herbivores, suffering high rates of mortality from natural enemies, are not abundant enough to compete. The predators, being limited only by the availability of their prey, are also common and thus experience competition. Ecologists are currently examining the degree to which such a model holds true in nature.

Species interactions can clearly be very important in influencing both population growth and community structure. What factors determine the numbers of species in a given area? What factors influence the stability of a community, and what are the effects of disturbances on community structure? In the next chapter, on community ecology, we will explore these and other questions.

(a)   **(b)**

Figure 57.23 Successful biological control of prickly pear cactus. The prickly pear cactus (*Opuntia stricta*) in Chinchilla, Australia, before **(a)** and after **(b)** top-down control by the cactus moth (*Cactoblastis cactorum*).

Table 57.4	**Primary Mortality Factors According to Availability of Plant Biomass**			
	Low ⟶ Plant biomass ⟶ High			
Taxa	**Plants only**	**Plants and herbivores**	**Plants, herbivores, and carnivores**	**Plants, herbivores, carnivores, and secondary carnivores**
Plants	Competition	Herbivory	Competition	Herbivory
Herbivores		Competition	Predation	Competition
Carnivores			Competition	Predation
Secondary carnivores				Competition

CHAPTER SUMMARY

57.1 Competition

- Species interactions can take a variety of forms that differ based on their effect on the species involved. (Figure 57.1, Table 57.1)

- Competition can be categorized as intraspecific (between individuals of the same species) or interspecific (between individuals of different species). Competition can also be categorized as resource competition or interference competition. (Figure 57.2)

- One species can exclude the other in a natural environment, affecting its distribution within a habitat. (Figure 57.3)

- Laboratory and field experiments show that competition occurs frequently in nature, and varies as a function of both abiotic and biotic factors. (Figures 57.4, 57.5)

- The competitive exclusion hypothesis states that two species with the same resource requirements cannot occupy the same niche. (Figure 57.6)

- Resource partitioning and morphological differences between species allow them to coexist in a community. (Figures 57.7, 57.8, Table 57.2)

57.2 Predation, Herbivory, and Parasitism

- The many antipredator strategies that animals have evolved suggest that predation is a strong selective force. Of these strategies, chemical defense and aposematic coloration are the most common. (Figures 57.9, 57.10)

- Despite these defenses, oscillations in predator-prey cycles, the effect of introduced species, and examples of human predation illustrate that predators can have a large effect on prey densities. (Figures 57.11, 57.12, 57.13)

- Plants have also evolved an array of defenses against predators, termed host plant resistance, which includes chemical defenses such as secondary metabolites and mechanical defenses such as thorns and spines. (Figure 57.14, Table 57.3)

- Parasitism is a common lifestyle on Earth, and some parasites have complex life cycles involving multiple hosts. (Figures 57.15, 57.16)

- Evidence from experimental removal of parasites and from the study of introduced plant and animal parasites confirms that parasites can greatly reduce prey densities. (Figures 57.17, 57.18)

57.3 Mutualism and Commensalism

- Mutualism is an association between two species that benefits both. Defensive mutualisms typically involve an animal defending either a plant or herbivore; dispersive mutualisms involve plants and pollinators that disperse their pollen, and plants and fruit eaters that disperse the plant's seeds. (Figures 57.19, 57.20)

- In commensal relationships, one partner receives a benefit while the other is not affected. (Figure 57.21)

57.4 Conceptual Models

- Conceptual models of species interactions describe the importance of factors such as competition and predation. Bottom-up models propose that plant quality or quantity regulates the abundance of all herbivore and predator species; top-down models propose that the abundance of predators controls herbivore and plant densities. (Figures 57.22, 57.23)

- Modern models, which incorporate components of both bottom-up and top-down models, propose that mortality varies according to productivity of plant biomass. (Table 57.4)

TEST YOURSELF

1. Two species of birds feed on similar types of insects and nest in the same tree species. This is an example of
   a. intraspecific competition.
   b. interference competition.
   c. resource competition.
   d. mutualism.
   e. none of the above.

2. The experiments conducted by Thomas Park using flour beetles provided evidence that the results of competition are influenced by
   a. moisture.
   b. temperature.
   c. parasitism.
   d. stochasticity.
   e. all of the above.

3. According to the competitive exclusion hypothesis
   a. two species that use the exact same resource show very little competition.
   b. two species with the same niche cannot coexist.
   c. one species that competes with several different species for resources will be excluded from the community.
   d. all competition between species results in the extinction of at least one of the species.
   e. none of the above.

4. In Lack's study of British passerine birds, different species seem to segregate based on resource factors, such as location of prey items. This differentiation among the niches of these passerine birds is known as
   a. competitive exclusion.
   b. intraspecific competition.
   c. character displacement.
   d. resource partitioning.
   e. allelopathy.

5. Divergence in morphology that is a result of competition is termed
   a. competitive exclusion.
   b. resource partitioning.
   c. character displacement.
   d. amensalism.
   e. mutualism.

6. Some organisms have bright coloration that serves as a warning to predators of the organism's bad taste. This type of signal is called
   a. Batesian mimicry.
   b. aposematic coloration.
   c. allelopathy.
   d. cryptic coloration.
   e. masting.

7. Batesian mimicry differs from Müllerian mimicry in that
   a. in Batesian mimicry, both species possess the chemical defense.
   b. in Batesian mimicry, one species possesses the chemical defense.
   c. in Müllerian mimicry, one species has several different mimics.
   d. in Müllerian mimicry, aposematic coloration is always found.
   e. in Batesian mimicry, cryptic coloration is always found.

8. Secondary metabolites
   a. are produced as chemical defenses by plants.
   b. are by-products of cellular respiration.
   c. are typically produced by organisms that exhibit Batesian mimicry.
   d. play an important role in energy production in plants.
   e. can be used as alternative reactants in photosynthesis.

9. Parasitic plants that rely solely on their host for nutrients are called
   a. hemiparasites.
   b. fungi.
   c. holoparasites.
   d. monophagous.
   e. polyphagous.

10. A species interaction in which one species benefits but the other species is unharmed is called
    a. mutualism.
    b. amensalism.
    c. parasitism.
    d. commensalism.
    e. mimicry.

CONCEPTUAL QUESTIONS

1. Define the competitive exclusion hypothesis.
2. Distinguish between Müllerian and Batesian mimicry.
3. Describe some plant defenses against herbivory.

COLLABORATIVE QUESTIONS

1. Discuss different types of competition found in nature.
2. Discuss several antipredator strategies that animals have evolved.

EXPERIMENTAL QUESTIONS

1. Describe the realized niches for the two species of barnacles used in Connell's experiment.
2. Outline the procedure that Connell used in the experiments.
3. How did Connell explain the presence of *Chthamalus* in the upper intertidal zone if *Semibalanus* was shown to outcompete the species in the first experiment?

www.brookerbiology.com

This website includes answers to the Biological Inquiry questions found in the figure legends and all end-of-chapter questions.

58

COMMUNITY ECOLOGY

CHAPTER OUTLINE

Waterhole with lion pride and elephants, Savuti Reserve, Botswana.

So far in this unit, we have examined ecology in terms of the behavior of individual organisms, the growth of populations, and interactions between pairs of species. Most populations, however, exist not on their own but together with populations of many other species. This assemblage of many populations that live in the same place at the same time is known as a **community**. For example, a tropical forest community consists of not only the tree species, vines, and other vegetation but also of the insects that pollinate them, the herbivores that feed upon the insects, and the predators and parasites of the herbivores. Communities can occur on a wide range of scales and can be nested. The tropical forest community also encompasses smaller communities such as the water filled recesses of bromeliads, which form a microhabitat for different species of insect and their larvae. Both of these entities—the tropical forest and the bromeliad tank— are viable communities, depending on one's frame of reference with regard to scale.

Community ecology studies how groups of species interact and form functional communities. In Chapter 57, we looked in detail at interactions between individual species. In this chapter, we widen our focus to explore the factors that influence the number and abundance of species in a community. We begin by examining the nature of ecological communities. Are communities loose assemblages of species that happen to live in the same place at the same time, or are they more tightly organized groups of mutually dependent species? We explore why, on a global scale, the number of species is usually greatest in the tropics and declines towards the poles. Community ecology also addresses what factors act to stabilize species richness in a community. However, ecologists recognize that communities may change, for example, following a disturbance such as a fire.

This recovery tends to occur in a predictable way that ecologists have termed succession. Finally, in certain situations, for example, on islands recovering from physical disturbance, succession has been postulated to occur via waves of colonization of species from neighboring landmasses, followed by extinctions of some species. The equilibrium model of island biogeography holds that island size and distance from the mainland govern the process of succession.

58.1 Differing Views of Communities

Ecologists have long held differing views on the nature of a community and how it is structured and functions. Some of the initial work in the field of community ecology considered a community to be equivalent to a superorganism, in much the same way that the body of an animal is more than just a collection of organs. In this view, individuals, populations, and communities have a relationship to each other that resembles the associations found between cells, tissues, and organs. Indeed, American botanist Frederic Clements, the champion of this viewpoint, suggested in 1905 that ecology was to the study of communities what physiology was to the study of individual organisms. This view of community, with predictable and integrated associations of species separated by sharp boundaries, is termed the **organismic model**. Some modern-day ecologists still depict the community as a superorganism. As we will see later in this chapter, Clements's hypotheses on communities were also linked to his study of succession, the changes in the types of plant and animal species that occupy a given area through time.

Clements's ideas were challenged in 1926 by botanist Henry Allan Gleason. Gleason proposed an **individualistic model**, which viewed a community as an assemblage of species coexisting primarily because of similarities in their physiological requirements and tolerances. While acknowledging that some assemblages of species were fairly uniform and stable over a given region, Gleason suggested that distinctly structured ecological communities usually do not exist. Instead, species distribute independently along an environmental gradient. Viewed in this way, communities do not necessarily have sharp boundaries, and associations of species are much less predictable and integrated than in Clements's organismic model.

By the 1950s, many ecologists had abandoned Clements's view in favor of Gleason's. In particular, Robert Whittaker's studies asserted the **principle of species individuality**, which stated that each species is distributed according to its physiological needs and population dynamics, that most communities intergrade continuously, and that competition does not create distinct vegetational zones. For example, let's consider an environmental gradient such as a moisture gradient on an uninterrupted slope of a mountain. Whittaker proposed that four hypotheses could explain the distribution patterns of plants and animals on the gradient (**Figure 58.1**):

1. Competing species, including dominant plants, exclude one another along sharp boundaries. Other species evolve toward a close, perhaps mutually beneficial association with the dominant species. Communities thus develop along the gradient, each zone containing its own group of interacting species giving way at a sharp boundary to another assemblage of species. This corresponds to Clements's organismic model.

2. Competing species exclude one another along sharp boundaries but do not become organized into groups of species with parallel distributions.

3. Competition does not usually result in sharp boundaries between species. However, the adaptation of species to similar physical variables will result in the appearance of groups of species with similar distributions.

4. Competition does not usually produce sharp boundaries between species, and the adaptation of species to similar physical variables does not produce well-defined groups of species with similar distributions. The centers and boundaries of species populations are scattered along the environmental gradient. This corresponds to Gleason's individualistic model.

To test these possibilities, Whittaker examined the vegetation on various mountain ranges in the western U.S. He sampled plant populations along an elevation gradient from the tops of the mountains to the bases and collected data on various physical variables, such as soil moisture.

The results supported the fourth hypothesis, that competition does not produce sharp boundaries between species and that adaptation to physical variables does not result in defined groups of species. Whittaker concluded that his observations agreed with Gleason's predictions that (1) each species is dis-

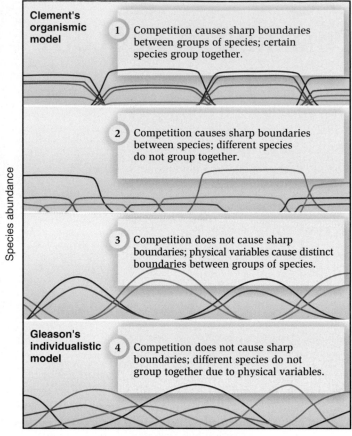

Figure 58.1 **Four hypotheses on how populations might relate to one another along an environmental gradient.** Each curve in each part of the figure represents one species and the way its population might be distributed along an environmental gradient such as a moisture gradient on a mountain slope.

tributed in its own way, according to its genetic, physiological, and life cycle characteristics; and (2) the broad overlap and scattered centers of species populations along a gradient implies that most communities grade into each other continuously rather than form distinct, clearly separated groups. Thus, Whittaker's observations showed that the composition of species at any one point in an environmental gradient was largely determined by factors such as temperature, water, light, pH, and salt concentrations (features discussed in Chapter 54).

58.2 Patterns of Species Richness

Even though most communities intergrade along environmental gradients such as a mountain slope, ecologists recognize distinct differences between communities. The community at the top of a mountain will be quite different from that at the bottom, so distinguishing between communities on a broad scale is useful. Also, some sharp boundaries between groups of species sometimes do exist, especially related to physical differences

Figure 58.2 Unique communities may result from distinct environmental conditions. In New Zealand's Dun Mountain area, we can contrast the sparse vegetation on the serpentine soil to the left to that of the beech forest on the nonserpentine soil on the right.

such as water quality and soil type, that cause distinct communities to develop. For example, serpentine soils are rich in metals, including magnesium, iron, and nickel, but poor in plant nutrients. The species that have adapted to these harsh conditions form a unique community restricted to this area (**Figure 58.2**).

One of the most straightforward means of distinguishing between communities is to determine the number of species in each community, or **species richness**. The number of species of most taxa varies according to geographic location, generally increasing from polar areas to temperate areas and reaching a maximum in the tropics. For example, the species richness of North American birds increases from Arctic Canada to Panama (**Figure 58.3**). A similar pattern exists for mammals and reptiles. Species richness is also increased by topographical variation. More mountains mean more hilltops, valleys, and differing habitats; thus, there is an increased number of birds in the mountainous western U.S. However, species richness is reduced by the peninsular effect, in which diversity decreases as a function of distance from the main body of land, and thus we see a decrease in the number of species in Florida and in Baja California.

Many hypotheses for the polar-equatorial gradient of species richness have been advanced. We will consider three of the most important, which propose that communities diversify with increased time, area, and productivity, respectively. Although they are treated separately here, it is important to note that these hypotheses are not mutually exclusive and that all can contribute to patterns of species richness.

The Time Hypothesis Suggests Communities Diversify with Age

Many ecologists argue that communities diversify, that is, gain species, with time and that temperate regions have less rich communities than tropical ones because they are younger and have only more recently (relatively speaking) recovered from re-

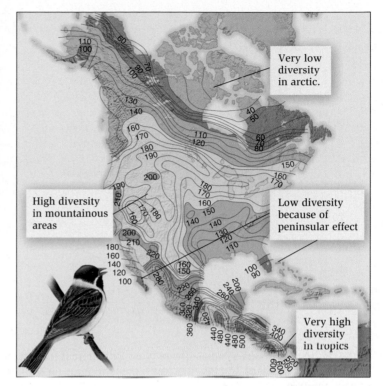

Figure 58.3 Species richness of birds in North America. The values indicate the numbers of different species in a given area. Contour lines show equal numbers of bird species. Note the pronounced latitudinal gradient heading south toward the tropics and the high diversity in California and northern Mexico, regions of considerable topographical variation and habitat diversity.

cent glaciations and severe climatic disruption. The time hypothesis has two variations. It proposes that, first, resident species have not yet evolved new forms to exploit vacant niches. And second, it proposes that species that could possibly live in temperate regions have not migrated back from the unglaciated areas into which the ice ages drove them.

In support of the time hypothesis, ecologists compared the species richness of bottom-dwelling invertebrates such as worms in historically glaciated (covered with ice) and unglaciated lakes in the Northern Hemisphere that occur at similar latitudes. Lake Baikal in Siberia is an ancient unglaciated temperate lake and contains a very diverse fauna; for example, there are 580 species of invertebrates in the bottom zone. Great Slave Lake, a comparable sized lake in northern Canada that was once glaciated, contains only four species in the same zone.

There are, however, drawbacks to the utility of the time hypothesis. For example, the time hypothesis may help explain variations in the species richness of terrestrial organisms, but it has limited applicability to marine organisms. While we might not expect terrestrial species to redistribute themselves quickly following a glaciation—especially if there is a barrier like the English Channel to overcome—there seems to be no reason that marine organisms couldn't relatively easily shift their distribution patterns during glaciations, yet the polar-equatorial gradient of species richness still exists in marine habitats.

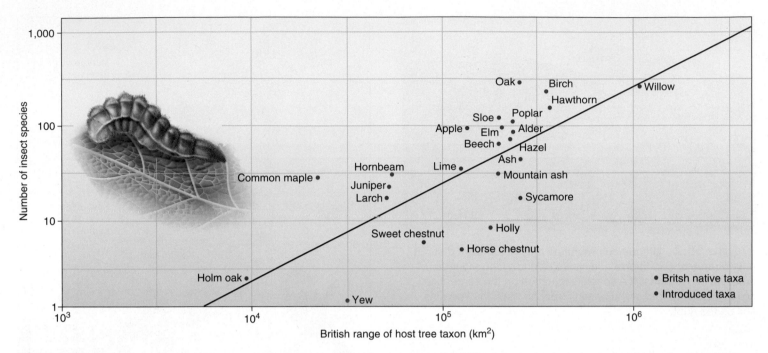

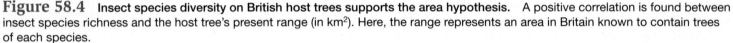

Figure 58.4 **Insect species diversity on British host trees supports the area hypothesis.** A positive correlation is found between insect species richness and the host tree's present range (in km²). Here, the range represents an area in Britain known to contain trees of each species.

The Area Hypothesis Suggests Large Areas Support More Species

The **area hypothesis** proposes that larger areas contain more species than smaller areas because they can support larger populations and a greater range of habitats. Much evidence supports the area hypothesis. For example, in 1974, Donald Strong showed that insect species richness on tree species in Britain was better correlated with the area over which a tree species could be found than with time of habitation since the last ice age. The relationship between the amount of available area and the number of species present is called the **species-area effect** (**Figure 58.4**). Some introduced tree species, such as apple and lime, were relatively new to Britain, but they bore many different insect species, a fact Strong argued did not support the time hypothesis.

As we saw in Chapter 54, the three-cell model of atmospheric circulation leads to a symmetry of climates between polar regions and temperate areas in the Northern and Southern Hemispheres (refer back to Figure 54.21). Symmetrically similar climates are only adjacent in the tropics, however, which creates one large area. This has been proposed as a reason why the tropics have greater species richness. However, the area hypothesis seems unable to explain why, if increased richness is linked to increased area, there are not more species in the vast contiguous landmass of Asia. Furthermore, while tundra may be the world's largest land biome, it has low species richness. Finally, the largest marine system, the open ocean, which has the greatest

volume of any habitat, has fewer species than tropical surface waters, which have a relatively small volume.

The Productivity Hypothesis Suggests That More Energy Permits the Existence of More Species

The **productivity hypothesis** proposes that greater production by plants results in greater overall species richness. An increase in plant biomass, the total weight of plant material produced, leads to an increase in the number of herbivores and hence an increase in the number of predator, parasite, and scavenger species. Production itself is influenced by factors such as temperature and rainfall, because many plants grow better where it is warm and wet. For example, in 1987, David Currie and colleagues showed that the species richness of trees in North America is best predicted by the **evapotranspiration rate**, the rate at which water moves into the atmosphere through the processes of evaporation from the soil and transpiration of plants (**Figure 58.5**).

Once again, however, there are exceptions to this rule. Some tropical seas, such as the southeast Pacific off of Colombia and Ecuador, have low productivity but high species richness. On the other hand, the sub-Antarctic Ocean has a high productivity but low species richness. Estuarine areas, where rivers empty into the sea, are similarly very productive yet low in species, presumably because they represent a stressful environment for many organisms that are alternately inundated by fresh water

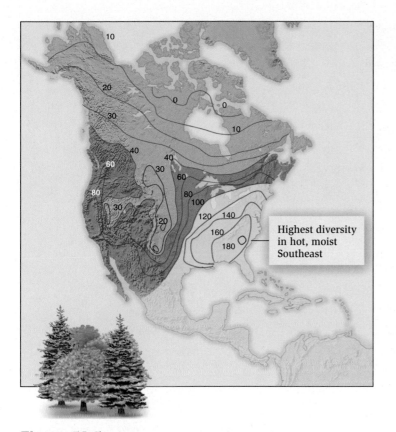

Figure 58.5 Tree species richness in North America is correlated with evapotranspiration rate. Contour lines show equal numbers of tree species.

Biological inquiry: Why doesn't the species richness of trees increase in mountainous areas of the West, as it does for birds?

and salt water with daily changes in the tide. Some lakes that are polluted with fertilizers have high productivity but low species richness.

Last, Robert Latham and Robert Ricklefs showed in 1993 that while patterns of tree diversity in North America support the productivity hypothesis, the pattern does not hold for broad comparisons between continents: the temperate forests of eastern Asia support substantially higher numbers of tree species (729) than do climatically similar areas of North America (253) or Europe (124). These three areas have different evolutionary histories and different neighboring areas from which species might have invaded.

Thus, in terms of accounting for the polar-equatorial gradient of species richness, each of the three hypotheses we have explored has some evidence to support it and some to contradict it. Different processes may occur over different scales. On a regional scale, we know that the time since the last glaciation has the potential to change patterns of species richness. On a more local scale, area and productivity of available habitat may be important. At any given point on the globe, species richness may be affected by the interaction of these different factors.

58.3 Calculating Species Diversity

So far, we have discussed communities in terms of variations in species richness. However, to measure species diversity, ecologists need to take into account not only the number of species in a community but also their frequency of occurrence, or **relative abundance**. For example, imagine two hypothetical communities, A and B, both with two species and 100 total individuals.

	Number of individuals of species 1	Number of individuals of species 2
Community A	99	1
Community B	50	50

The species richness of community B equals that of community A, because they both contain two species. However, community B is considered more diverse because the distribution of individuals between species is more even. One would be much more likely to encounter both species in community B than in community A, where one species dominates.

To measure the diversity of a community, therefore, ecologists calculate what is known as a diversity index. Although there are a great many of these indices, the most widely used is the **Shannon diversity index (H_S)**, which is calculated as

$$H_S = -\Sigma p_i \ln p_i$$

where p_i is the proportion of individuals belonging to species i in a community, ln is the natural logarithm, and the Σ is a summation sign. For example, for a species in which there are 50 individuals out of a total of 100 in the community, p_i is 50/100, or 0.5. The natural log of 0.5 is −0.693. For this species, $p_i \ln p_i$ is then $0.5 \times -0.693 = -0.347$. For a hypothetical community with 5 species and 100 total individuals, the Shannon diversity index would be calculated as follows:

Species	Abundance	p_i	$p_i \ln p_i$
1	50	0.5	−.347
2	30	0.3	−.361
3	10	0.1	−.230
4	9	0.09	−.217
5	1	0.01	−.046
Total 5	100	1.00	$\Sigma p_i \ln p_i$ −1.201

Note that in this example, even the rarest species, species 5, contributes some value to the index, so that if there were many such rare species, their contributions would accumulate. This makes the Shannon diversity index very valuable to conservation biologists, who often study rare species and their importance to the community. Remember too that the negative sign in front of the summation changes these values to positive, so the index actually becomes 1.201, not −1.201. (The minus sign in the equation is used to give us a positive number for the index, which is more appealing than an index with a negative number.)

Table 58.1	Shannon Diversity Index of Bird Species on Logged and Unlogged Sites in Indonesia					
	Unlogged			Logged		
Species	N	p_i	$p_i \ln p_i$	N	p_i	$p_i \ln p_i$
Nectarinia jugularis	410	.225	−.336	910	.386	−.367
Ducula bicolor	230	.126	−.261	220	.093	−.221
Philemon subcorniculatus	210	.115	−.249	240	.102	−.233
Nectarinia aspasia	190	.104	−.235	120	.051	−.152
Dicaeum vulneratum	185	.101	−.232	280	.119	−.253
Ducula perspicillata	170	.093	−.221	180	.076	−.196
Phylloscopus borealis	160	.088	−.214	140	.059	−.167
Eos bornea	88	.048	−.146	73	.031	−.108
Ixos affinis	76	.042	−.133	31	.013	−.056
Geoffroyus geoffroyi	44	.024	−.089	54	.023	−.087
Rhyticeros plicatus	24	.013	−.056	27	.011	−.050
Cacatua moluccensis	12	.007	−.035	1	.001	−.007
Tanygnathus megalorynchos	9	.005	−.026	11	.005	−.026
Electus roratus	7	.004	−.022	0	0	0
Macropygia amboinensis	6	.003	−.017	7	.003	−.017
Cacomantis sepulcralis	3	.002	−.012	0	0	0
Trichoglossus haematodus	0	0	0	64	.027	−.097
Total	**1,824**	**1.0**		**2,358**	**1.0**	
Shannon diversity index			**2.284**			**2.037**

Values of the Shannon diversity index for real communities often fall between 1.5 and 3.5, with the higher the value, the greater the diversity. **Table 58.1** calculates the diversity of two bird communities in Indonesia with similar species richness but differing species abundance. The bird communities were surveyed on lowland forest that had been selectively logged or on pristine, unlogged forest. To document diversity, biologist Stuart Marsden established census stations in the two types of forest and recorded the identity and density of all bird species for a number of 10-minute periods. Although a greater number of individual birds was seen in the logged areas (2,358) compared to unlogged ones (1,824), a high percentage of the individuals in the logged areas (38.6%) belonged to just one species, *Nectarinia jugularis*. While there was only one more bird species found in the unlogged area than in the logged area, calculation of the Shannon diversity index showed a significantly higher diversity of birds in the unlogged area, 2.284 versus 2.037, a difference of 12.1%.

Of course, an accurate determination of community diversity depends on detailed knowledge of which and how many of each species are present. While this is relatively easy to determine for communities of vertebrates and some invertebrates, it is much more difficult for microbial communities. Yet knowledge of microbial communities is of vital importance if only because of the decomposition functions they perform. As described next, with the advent of modern molecular tools, our knowledge of the diversity of microbial communities is beginning to expand.

GENOMES & PROTEOMES

Metagenomics May Be Used to Measure Community Diversity

Bacteria are abundant members of all communities and are vital to their functioning. They serve as food sources for other organisms and participate in the decomposition process. However, most microorganisms are taxonomically unknown, mainly because they cannot be cultivated on known culture media. The new field of **metagenomics** seeks to identify and analyze the collective microbial genomes contained in a community of organisms, including those that are not easily cultured in the laboratory.

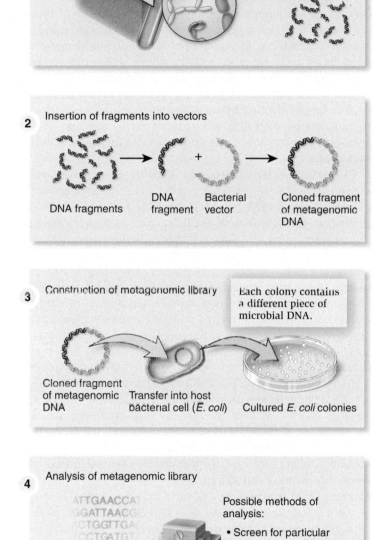

1 Isolation of many types of bacterial DNA from environmental sample

DNA isolated

Restriction endonucleases cleave DNA

2 Insertion of fragments into vectors

DNA fragments

DNA fragment

Bacterial vector

Cloned fragment of metagenomic DNA

3 Construction of metagenomic library

Each colony contains a different piece of microbial DNA.

Cloned fragment of metagenomic DNA

Transfer into host bacterial cell (*E. coli*)

Cultured *E. coli* colonies

4 Analysis of metagenomic library

ATTGAACCA
GGATTAACG
ACTGGTTGA
CCTGATGT
TTGAACC
ATTAAC

Possible methods of analysis:

- Screen for particular sequences by PCR or hybridization
- Random DNA sequencing
- Screen for expression of particular phenotypes

Figure 58.6 **The standard protocol of a metagenomics experiment.** (1) Isolation and fragmentation of DNA from the sample, (2) insertion of fragments into bacterial vectors, (3) insertion of cloned DNA into host cell, and culturing in selective growth media to create a library, and (4) analysis of DNA sequences and protein expression.

This technique has only been in existence since the early 1990s, but significant progress has already been made in providing libraries that have advanced our understanding of which bacteria are present in various communities and how they function.

The process involves four main steps (**Figure 58.6**). First, an environmental sample containing many bacterial species is collected and its DNA is isolated from the cells using chemical or physical methods. Because the genomic DNA of each species is relatively large, it is cut up into fragments with enzymes called restriction endonucleases. Second, the fragments are combined with vectors, small units of DNA that can be inserted into a model organism, usually a bacterium. The third step is transformation, the physical insertion of the foreign DNA into a host bacterial cell. Individual bacteria are then grown on a selective media so that only the transformed cells survive, and they grow into a colony of cloned cells. A collection of thousands of clones, each containing a different piece of microbial DNA, is called a metagenomic library. Lastly, the DNA from the metagenomic libraries is analyzed. In some cases, expression of the new DNA results in the synthesis of a new protein that changes the phenotype of the host, for example, a new enzyme that is detected by a chemical technique or an unusual color or shape in the model organism.

Metagenomics is helping determine the identity and activities of bacteria in diverse habitats, from the soil to deep-sea sediments to the oral cavity of humans. Recent studies have used shotgun DNA sequencing (see Chapter 21) to analyze bacterial genomes directly from soil and ocean samples. Sampling of deep-sea and ocean sediment samples has permitted sequencing of particular genes, most often the 16S ribosomal RNA (16S rRNA) that all bacteria possess. A sample of many different bacterial species will yield many different 16S rRNA gene sequences. These are then compared to a database containing all known sequences of this gene. Larger DNA segments will be compared to corresponding genomes of cultured bacteria to see if we can assign a tentative metabolic function to each segment. In this way, we can determine the community diversity of soil, sediment, or water samples and make better determinations of how the community is functioning.

In 2004, Jill Banfield and colleagues used metagenomics techniques to identify the five dominant species of bacteria living at temperatures of 42°C (107°F) and pH 0.8 in the acidic waste water (called acid mine drainage) from a mine in California. They detected 2,036 proteins from these species, including proteins from the most abundant bacterial species, a *Leptospirillum* group II bacteria. This represented the first large-scale proteomics-level expression of a natural microbial community. One of the proteins, a cytochrome, oxidizes iron and probably influences the rate of breakdown of acid mine drainage products. Many other proteins appear responsible for defending against free radicals, suggesting that this is a challenging task. The hope is that the team can now identify enzymes and metabolic pathways that will help in the cleanup of this and other contaminated sites in the future.

58.4 Species Richness and Community Stability

A community is often seen as stable when little to no change can be detected in the number of species and their abundance over a given time period. The community may then be said to be in equilibrium. The frame of reference for detecting change may encompass a study of a few years or, preferably, several decades. For example, long-term data from Bookham Common, England, revealed bird communities in which the number of species appeared stable for nearly 30 years. Community stability is an important consideration in conservation biology, as we will discuss further in Chapter 60. A decrease in the stability of a community over time may alert us to a possible problem. In the 1970s, the populations of many raptor species, including peregrine falcons, bald eagles, and osprey, declined precipitously. Eventually, the decline was traced to the pesticide DDT (dichlorodiphenyltrichloroethane), which caused eggshells to become thin and break before the birds could hatch. After DDT was banned later in the decade, raptor communities began to recover.

In this section, we consider the relationship between species richness and community stability. We begin by exploring the question of whether communities with more species are more stable than communities with fewer species. We then examine the link between diversity and stability, using evidence from the field. In a final consideration, we look at the relationship from a different angle and consider whether or not stable communities are more species rich than communities that have been disturbed.

The Diversity-Stability Hypothesis States That Species-Rich Communities Are More Stable Than Those with Fewer Species

Community stability may be viewed in several different ways. Some communities, such as extreme deserts, are considered stable because they are resistant to change by anything other than water. Other communities, such as river communities, are considered stable because they can recover quickly after a disturbance, such as pollution, being cleansed by the rapid flow of fresh water. Lake communities, on the other hand, may be less stable because there is no drainage outlet and pollutants can accumulate quickly.

Because maintaining community stability is seen as important, much research has gone into understanding the factors that enhance it. This work has produced the prevailing idea that species-rich communities are more stable than species-poor communities. There is much debate, for example, over whether species-rich communities are more resistant to invasion by undesirable exotic species, such as weeds, than species-poor communities. The link between species richness and stability was first explicitly proposed by the English ecologist Charles Elton

in the 1950s. He suggested that a disturbance in a diverse, or species-rich, community would be cushioned by large numbers of interacting species and would not produce as drastic an effect as it would on a less diverse community. Thus, an introduced predator or parasite could cause extinctions in a species-poor system but possibly not in a diverse system, where its effects would be buffered by interactions with more species in the community.

In support of this hypothesis, Elton suggested that species-poor island communities are much more vulnerable to invading species than are species-rich continental communities. For example, introduced mosquito species have caused many extinctions of native Hawaiian birds, which are extremely susceptible to the avian malaria spread by the insects. Elton also argued that outbreaks of pests are often found on cultivated land or land disturbed by humans, both of which are greatly simplified areas with few naturally occurring species. His arguments collectively became known as the **diversity-stability hypothesis**.

However, ecologists began to challenge Elton's association of diversity with stability. They pointed out that many examples of introduced species that have assumed pest proportions on continents, not just islands, including rabbits in Australia and pigs in North America. They noted that disturbed or cultivated land may suffer from pest outbreaks not because of its simple nature but because individual species, including exotic pest species and native species of natural enemies, often have no evolutionary history with one another, in contrast to the long associations between pests and natural enemies evident in natural biomes. For example, in Europe, coevolved predators such as foxes prevent rabbit populations from increasing to pest proportions. What was needed was an experiment to determine if a link existed between diversity and stability.

In 1996, ecologist David Tilman reported the relationship between species richness and stability from an 11-year study of 207 grassland plots in Minnesota. He measured the biomass of every species of plant, in each plot, at the end of every year and obtained the average species biomass. He then calculated how much this biomass varied from year to year through a statistical measure called the coefficient of variation. Year-to-year variation in plant community biomass was significantly lower in plots with greater plant species richness (**Figure 58.7**). Tilman's results showed that community stability is positively correlated with diversity.

Tilman suggested that diversity stabilizes communities because diverse plots were more likely to contain disturbance-resistant species that, in the event of a disturbance, could grow and compensate for the loss of disturbance-sensitive species. For example, when a change in climate such as drought harmed competitively dominant species that thrived in normal conditions, decreasing their abundance, unharmed drought-resistant species increased in mass and replaced them. Such decreases in susceptible species and compensatory increases in other species acted to stabilize total community biomass.

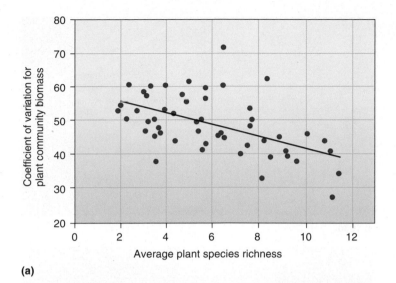

(a)

(b)

Figure 58.7 Relationships between biomass variation and species diversity. (a) Tilman's 11-year study of grassland plots in Minnesota revealed that year-to-year variability in community biomass was lower in species-rich plots. Each dot represents an individual plot, although only the plots from one field are graphed. (b) This aerial photograph shows the grassland plots.

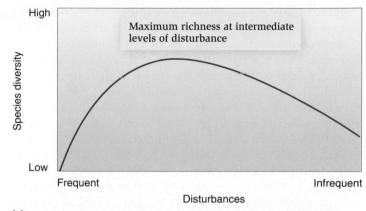

(a)

(b)

Figure 58.8 The intermediate-disturbance hypothesis of community organization. (a) This hypothesis proposes that species richness is highest at intermediate levels of disturbances caused by events such as fires or windstorms. (b) A light gap in a tropical rain forest in Costa Rica promotes the growth of small herbaceous species until trees colonize the light gap and gradually grow over and outcompete the species.

Biological inquiry: According to the intermediate-disturbance hypothesis, why are there so many species in the tropics?

The Intermediate-Disturbance Hypothesis Proposes That Moderately Disturbed Communities Are More Diverse

Up to this point, we have focused on the question of whether diverse communities are more stable than species-poor communities. Large-scale field studies have verified that stability is increased by diversity. What if we were to rephrase the question and ask whether stable communities are more diverse than unstable communities? In answer to this question, ecologist Joseph Connell has argued that the highest diversities are maintained not in stable communities but in communities with intermediate levels of disturbance, a concept called the **intermediate-disturbance hypothesis** (**Figure 58.8a**). Disturbance in communities may be brought about by many different phenomena such as droughts,

fires, floods, and hurricanes, or even herbivory, predation, or parasitism. Recall from Chapter 56 that some species, termed *r*-selected species, are better dispersers than other species, and that *K*-selected species are better competitors. Connell reasoned that at high levels of disturbance, only colonists that were *r*-selected species would survive, giving rise to low diversity. This is because these species would be the only ones able to disperse quickly to a highly disturbed area. At low rates of disturbance, competitively dominant *K*-selected species would outcompete all other species, which would also yield low diversity. The most diverse communities would lie somewhere in between.

Connell argued that natural communities fit into this model fairly well. Tropical rain forests and coral reefs are both examples of communities with high species diversity. Connell pointed out that coral reefs maintain their highest diversity in areas disturbed by hurricanes and that the richest tropical forests occur where disturbance by storms causes landslides and tree falls. For example, the fall of a tree creates a hole in the rain forest canopy known as a light gap, where direct sunlight is able to reach the rain forest floor. The light gap is rapidly colonized by *r*-selected species, such as small herbaceous plants, which are well adapted for rapid growth. While these pioneering species grow rapidly, they are overtaken by hardier *K*-selected species, such as mature trees, which fill in the gap in the canopy (**Figure 58.8b**). Although events such as hurricanes and tree falls are fairly frequent events in these communities, their occurrence in any one area is usually of intermediate frequency.

In 1979, Wayne Sousa provided an elegant experimental verification of the intermediate-disturbance hypothesis in the marine intertidal zone. He found that small boulders, which were easily disturbed by waves, carried a mean of 1.7 sessile plant and animal species. These frequently moving boulders crushed or dislodged most colonizing species. Large boulders, which were rarely moved by waves, had a mean of 2.5 species. On these boulders, competitively dominant species supplanted many other species. Sousa found that intermediate-sized boulders had the most species, an average of 3.7 species per boulder, because they contained a mix of *r*- and *K*-selected species. To test the hypothesis, Sousa cemented small boulders to the ocean floor and obtained an increase in species richness to near the value for large boulders, showing that the resulting number of species was a result of rock stability, not rock size.

To summarize, large-scale field experiments have shown a definite link between stability and diversity. However, if we turn the question around, we find that the most diverse communities are not necessarily the most stable, but rather that the most diverse communities exist at intermediate levels of disturbance. Finally, it is instructive to realize that many communities are not permanently stable and that change is a normal ecological phenomenon. Such change often proceeds in predictable ways, a process that ecologists term succession.

58.5 Succession: Community Change

At 8:32 a.m. on May 18, 1980, Mount St. Helens, a previously little studied peak in the Washington Cascades, erupted. The blast felled trees over a 600-km² area, and the landslide that followed—the largest in recorded history—destroyed everything in its path, killing nearly 60 people and thousands of animals. However, since the eruption, much of the area has experienced a relatively rapid recovery of plant and animal communities (**Figure 58.9**).

Ecologists have developed several terms to describe how community change occurs. The term **succession** describes the

(a) 1980

(b) 1997

Figure 58.9   Succession on Mount St. Helens.   (a) The initial blast occurred on May 18, 1980. (b) By 1997, 17 years later, many of the areas initially overrun by lava had developed low-lying vegetation and new trees sprouted up between the old dead tree trunks.

gradual and continuous change in species composition and community structure over time. **Primary succession** refers to succession on a newly exposed site that was not previously occupied by soil and vegetation, such as bare ground caused by a volcanic eruption or the rubble created by the retreat of glaciers. In primary succession on land, the plants must often build up the soil, and thus a long time—even hundreds of years—may be required for the process. Only a tiny proportion of the Earth's surface is currently undergoing primary succession, for example, around Mount St. Helens and the volcanoes in Hawaii and off the coast of Iceland, and behind retreating glaciers in Alaska and Canada.

Secondary succession refers to succession on a site that has already supported life but has undergone a disturbance, such as a fire, tornado, hurricane, or flood (as in the 2004 tsunami in Indonesia). Clearing a natural forest and farming the land for several years is an example of a severe forest disturbance that does not kill all native species. Some plants and many soil bacteria, nematodes, and insects are still present. Cessation of farming may lead to a distinct secondary succession. The secondary succession in abandoned farmlands (also called old fields) can lead to a pattern of vegetation quite different from one that develops after primary succession following glacial retreat. For example, the plowing and added fertilizers, herbicides, and pesticides may have caused substantial changes in the soil of an old field, allowing species that require a lot of nitrogen to colonize. These species would not be present for many years in newly created glacial soils.

Frederic Clements is often viewed as the founder of successional theory as well as community ecology in general. His work in the early 20th century emphasized succession as proceeding to a distinct end point or **climax community**. Each phase of succession was called a **sere**, or seral stage. The initial sere was known as the pioneer seral stage. While disturbance could return a community from a later seral stage to an earlier seral stage, generally the community headed towards climax.

A key assumption of Clements was that each colonizing species made the environment a little different—a little more shady or a little more rich in soil nitrogen—so that it became more suitable for other species, which then invaded and outcompeted the earlier residents. This process, known as **facilitation**, supposedly continued until the most competitively dominant species had colonized, when the community was said to be at climax. The composition of the climax community for any given region was thought to be determined by climate and soil conditions. While Clements's ideas on succession focused on the mechanism of facilitation, two other mechanisms affecting succession—inhibition and tolerance—have since been proposed. Next, we'll examine the evidence for each of them.

Facilitation Assumes Each Invading Species Creates a More Favorable Habitat for Succeeding Species

Succession following the gradual retreat of Alaskan glaciers is often used as a specific example of facilitation as a mechanism of succession. Over the past 200 years, the glaciers in Glacier Bay have undergone a dramatic retreat of nearly 100 km (**Figure 58.10**). This is one of the few instances where we can trace the chronology of physical change in an area, from 1794, when Captain George Vancouver visited the inlet and made notes on the positions of the glaciers, to the present.

Succession in Glacier Bay follows a distinct pattern of vegetation. As glaciers retreat, they leave moraines, deposits of stones, pulverized rock, and debris that serve as soil. In Alaska, the bare soil has a low nitrogen content and scant organic matter. In the pioneer stages, the soil is first colonized by a black crust of cyanobacteria, lichens, moss, horsetail (*Equisetum variegatum*), and the occasional river beauty (*Epilobium latifolium*) (**Figure 58.11a**). Because the cyanobacteria are nitrogen fixers, the soil nitrogen increases a little, but soil depth and litterfall (fallen leaves, twigs, and other plant material) are still minimal. At this stage there may be a few seeds and seedlings of dwarf shrubs of the rose family commonly called mountain avens (*Dryas drummondii*), alders, and spruce, but they are rare in the community. After about 40 years, *D. drummondii* dominates the

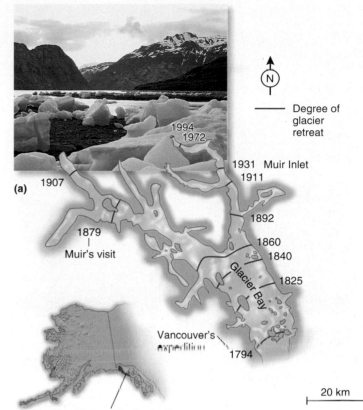

(a)

(b)

Figure 58.10 The degree of glacier retreat at Glacier Bay, Alaska, since 1794. (a) Primary succession begins on the bare rock and soil evident at the edges of the retreating glacier. (b) The lines reflect the position of the glacier in 1794 and its subsequent retreat.

landscape (**Figure 58.11b**). Soil nitrogen increases, as does soil depth and litterfall, and alder trees begin to invade.

At about 60 years, alders (*Alnus sinuata*) form dense, close thickets (**Figure 58.11c**). Alders have nitrogen-fixing bacteria that live mutualistically in their roots and convert nitrogen from the air into a biologically useful form. The excess nitrogen fixed by these bacteria accumulates in the soil. Soil nitrogen dramatically increases, as does litterfall. Spruce trees (*Picea sitchensis*) begin to invade at about this time. After about 75 to 100 years, the spruce trees begin to overtop the alders, shading them out. The litterfall is still high and the large volume of needles turns the soil acidic. The shade causes competitive exclusion of many of the original understory species, including alder, and only mosses carpet the ground. At this stage, seedlings of western hemlock (*Tsuga heterophylla*) and mountain hemlock (*Tsuga mertensiana*) may also occur, and after 200 years, a mixed spruce-hemlock climax forest results (**Figure 58.11d**).

What other evidence is there of facilitation? Experimental studies of early primary succession on Mount St. Helens, which show that decomposition of fungi allows mosses and other

Seral stage	Pioneer	*Dryas*	Alder	Spruce
Time (years) since glacial retreat	5	40	60	200
Soil depth (cm)	5.2	7.0	8.8	15.1
Soil N (g/m²)	3.8	5.3	21.8	53.3
Soil pH	7.2	7.3	6.8	3.6
Litterfall (g/m² /yr)	1.5	2.8	277	261

Cyanobacteria
Moss
Lichens

Mountain avens
(*Dryas drummondii*)

Alder
(*Alnus sinuata*)

Spruce
(*Picea sitchensis*)
Western hemlock
(*Tsuga heterophylla*)

(a)   (b)   (c)   (d)

Figure 58.11 **The pattern of primary succession at Glacier Bay, Alaska.** (a) The first species to colonize the bare Earth following retreat of the glaciers are small species such as cyanobacteria, moss, and lichens. (b) Mountain avens (*Dryas drummondii*) is a flower common in the *Dryas* seral stage. (c) Soil nitrogen and litterfall increase rapidly as alder (*Alnus sinuata*) invade. Note also the appearance of a few spruce trees higher up the valley. (d) Spruce (*Picea sitchensis*) and hemlock (*Tsuga heterophylla*) trees comprise a climax spruce-hemlock forest at Glacier Bay, with moss carpeting the ground. Two hundred years ago, glaciers occupied this spot.

fungi to colonize the soil, provide evidence of facilitation. In New England salt marshes, *Spartina* grass facilitates the establishment of beach plant communities by stabilizing the rocky substrate and reducing water velocity, which enables other seedlings to emerge. Succession on sand dunes also supports the facilitation model, in that pioneer plant species stabilize the sand dunes and facilitate the establishment of subsequent plant species. The foredunes, those nearest the shoreline, are the most frequently disturbed and are maintained in a state of early succession, while more stable communities develop farther away from the shoreline.

Succession also occurs in aquatic communities. Although soils do not develop in marine environments, facilitation may still be encountered when one species enhances the quality of settling and establishment sites for another species. When experimental test plates used to measure settling rates of marine organisms were placed in the Delaware Bay, researchers discovered that certain cnidarians enhanced the attachment of tunicates, and both facilitated the attachment of mussels, which were the dominant species in the community. In this experiment, the smooth surface of the test plates prevented many species from colonizing but once the surface became rougher, because of the presence of the cnidarians, many other species

were able to colonize. In a similar fashion, early colonizing bacteria, which create biofilms on rock surfaces, can facilitate succession of other organisms.

Inhibition Implies That Early Colonists Prevent Later Arrivals from Replacing Them

Although data on succession in some communities fit the facilitation model, researchers have proposed alternative hypotheses of how succession may operate. Another view is that possession of space is all-important, and that what gets there first determines subsequent community structure. In this process, known as **inhibition**, early colonists may exclude subsequent colonists. For example, removing the litter of *Setaria faberi*, an early successional plant species in New Jersey old fields, causes an increase in the biomass of a later species, *Erigeron annuus*. The release of toxic compounds from decomposing *Setaria* litter or physical obstruction by the litter itself contributes to the inhibition of *Erigeron*. Without the litter present, *Erigeron* dominates and reduces the biomass of *Setaria*. Plant species that grow in dense thickets, such as some grasses, ferns, vines, pine trees, and bamboo, can inhibit succession, as can many introduced plant species.

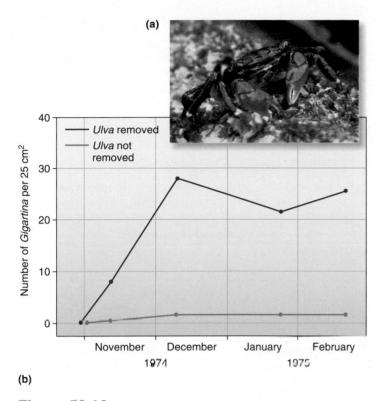

(a)

(b)

Figure 58.12 Inhibition as a primary method of succession in the marine intertidal zone. (a) *Gigartina* on rock face with the herbivorous crab, *Pachygrapsus crassipes*. (b) Removing *Ulva* from intertidal rock faces allowed colonization by *Gigartina*.

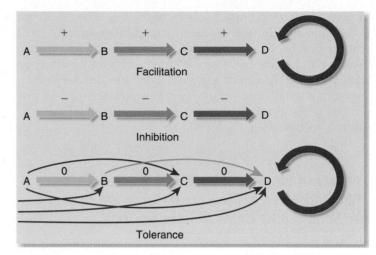

Figure 58.13 Three models of succession. A, B, C, and D represent four stages or seres. D represents the climax community. An arrow indicates "is replaced by," and + = facilitation, − = inhibition, and 0 = no effect. The facilitation model is the classic model of succession. In the inhibition model, much depends on which species gets there first. The tolerance model is similar to the facilitation model, in that later species may be facilitated by earlier species, but they can also invade in their absence.

Biological inquiry: Inhibition implies competition between species with early arriving species tending to outcompete later arrivals, at least for a while. Does competition or mutualism feature more prominently in facilitation?

Inhibition has been seen as the primary method of succession in the marine intertidal zone, where space is limited. In this habitat, early successional species are at a great advantage in maintaining possession of valuable space. In 1974, ecologist Wayne Sousa created an environment for testing how succession works in the intertidal zone by scraping rock faces clean of all algae or putting out fresh boulders or concrete blocks. The first colonists of these areas were the green algae *Ulva*. By removing *Ulva* from the substrate, Sousa showed that the large red alga *Gigartina canaliculata* was able to colonize more quickly (**Figure 58.12**). The results of Sousa's study indicate that early colonists can inhibit rather than facilitate the invasion of subsequent colonists. Succession may eventually occur because early colonizing species, such as *Ulva*, are more susceptible to the rigors of the physical environment and to attacks by herbivores, such as crabs (*Pachygrapsus crassipes*), than later successional species, such as *Gigartina*.

Tolerance Suggests That Early Colonists Neither Facilitate nor Inhibit Later Colonists

The huge differences between facilitation and inhibition as mechanisms of succession prompted Joseph Connell and Ralph Slatyer to view the two as extremes on a continuum. In 1977, the two researchers proposed a third mechanism of succession, which they termed **tolerance**. In this process, any species can start the succession, but the eventual climax community is reached in a somewhat orderly fashion. Early species neither facilitate nor inhibit subsequent colonists. Connell and Slatyer found the best evidence for the tolerance model in Frank Egler's earlier work on floral succession. In the 1950s, Egler showed that succession in plant communities is determined largely by species that already exist in the ground as buried seeds or old roots. Whichever species germinates first, or regenerates from roots, initiates the succession sequence.

The key distinction between the three models is in the manner in which succession proceeds. In the facilitation model, species replacement is facilitated by previous colonists; in the inhibition model, it is inhibited by the action of previous colonists; and in the tolerance model, species may be unaffected by previous colonists but they do not require them (**Figure 58.13**). Subsequent research has suggested that other factors may also influence succession, especially on islands. The study of succession on islands is often referred to as island biogeography, which is described next.

58.6 Island Biogeography

In the 1960s, two eminent ecologists, Robert MacArthur and E. O. Wilson, developed a comprehensive model to explain the process of succession on new islands, where a gradual buildup of species proceeds from a sterile beginning. Their findings, termed the **equilibrium model of island biogeography**, hold that the number of species on an island tends toward an equilibrium number that is determined by the balance between two factors: immigration rates and extinction rates. In this section, we explore island biogeography and how well the model's predictions are supported by data, particularly that provided by classic experiments in the Florida Keys.

The Island Biogeography Model Suggests That During Succession Losses from Extinction Are Balanced by Gains in Immigration

MacArthur and Wilson's model of island biogeography suggests that species repeatedly arrive on an island and either thrive or become extinct. Thus, the number of species tends towards an equilibrium number, Ŝ, which reflects a balance between the rate of immigration and the rate of extinction. The rate of immigration of new species is highest when no species are present on the island, so that each species that invades the island is a new species. As species accumulate, subsequent immigrants are less likely to represent new species. The rate of extinction is low at the time of first colonization, because few species are present and many have large populations. With the addition of new species, the populations of some species diminish, so the probability of extinction by chance alone increases. Species may continue to arrive and go extinct, but the number of species on the island remains approximately the same.

MacArthur and Wilson reasoned that both the immigration and extinction lines would be curved, for several reasons (**Figure 58.14a**). First, species arrive on islands at different rates. Some organisms, including plants with seed dispersal mechanisms and winged animals, are more mobile than others and will arrive quickly. Other organisms will arrive more slowly. This pattern causes the immigration curve to start off steep but get progressively shallower. On the other hand, extinctions rise at accelerating rates, because as later species arrive, competition increases and more species are likely to go extinct. As noted previously, earlier arriving species tend to be *r*-selected species, which are better dispersers, whereas later arriving species are generally *K*-selected species, which are better competitors. Later arriving species usually outcompete earlier arriving ones, causing an increase in extinctions.

The strength of the island biogeography model was that it generated several falsifiable predictions:

1. The number of species should increase with increasing island size. This is also known as the species-area effect (see Figure 58.4). Extinction rates would be greater on smaller islands because population sizes would be smaller and more susceptible to extinction (**Figure 58.14b**).
2. The number of species should decrease with increasing distance of the island from the mainland, or the **source pool**, the pool of species that is available to colonize the

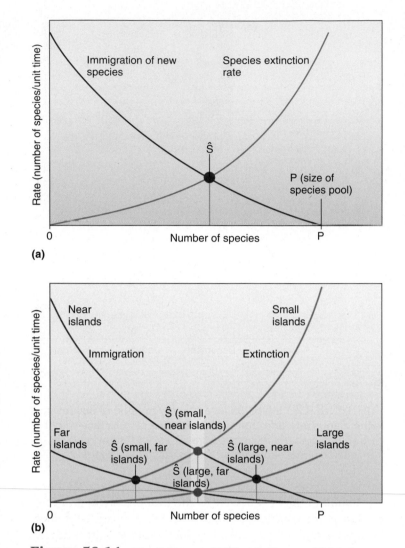

(a)

(b)

Figure 58.14 **MacArthur and Wilson's theory of island biogeography.** (a) The interaction of immigration rate and extinction rate produces an equilibrium number of species on an island, Ŝ. Ŝ varies from 0 species to *P* species, the total number of species available to colonize. (b) Ŝ varies according to the island's size and distance from the mainland. An increase in distance (near to far) lowers the immigration curve, while an increase in island area (small to large) lowers the extinction rate.

island. Immigration rates would be greater on islands near the source pool because species do not have as far to travel (Figure 58.14b).

3. The turnover of species should be considerable. The number of species on an island might remain the same, but the composition of the species should change continuously as new species colonize the island and others become extinct.

Let's examine these predictions one by one and see how well the data support them.

The Number of Species Increases with Increasing Island Size

The West Indies has traditionally been a key location for ecologists studying island biogeography. This is because the physical geography and the plant and animal life of the islands are well known. Furthermore, the Lesser Antilles, from Anguilla in the north to Grenada in the south, enjoy a similar climate and are surrounded by deep water (**Figure 58.15a**). In 1999, Robert Ricklefs and Irby Lovette summarized the available data on the richness of species of four groups of animals—birds, bats, reptiles and amphibians, and butterflies—over 19 islands that varied in area over two orders of magnitude (13–1,510 km²). In each case, a possible correlation occurred between area and species richness (**Figure 58.15b**). Note that these relationships are traditionally plotted on a double logarithmic scale, a so-called log-log plot, in which the horizontal axis is the logarithm to the base 10 of the area and the vertical axis is the logarithm to the base 10 of the number of species. A linear plot of the area versus the number of species would be difficult to produce, because of the wide range of area and richness of species

involved. Logarithmic scales condense this variation to manageable limits.

Apart from Ricklefs and Lovette's study, species-area relationships exist for birds of the East Indies, beetles on West Indian islands, ants in Melanesia, and land plants of the Galápagos, providing strong support for this prediction of the equilibrium model of island biogeography.

The Number of Species Decreases with Increasing Distance from the Source Pool

MacArthur and Wilson also provided evidence for the effect of the distance of an island from a source pool of colonists, usually the mainland. In studies of the numbers of lowland forest bird species in Polynesia, they found that the number of species decreased with the distance from New Guinea, the source pool in this case (**Figure 58.16**). They expressed the richness of bird species on the islands as a percentage of the number of bird species found on New Guinea. There was a significant decline in this percentage with increasing distance, with more distant islands containing lower numbers of species than nearer islands.

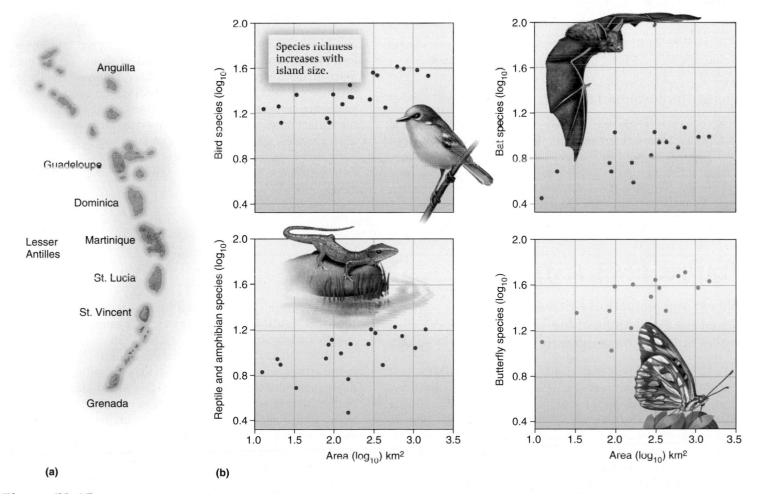

(a)   **(b)**

Figure 58.15 **Species richness increases with island size.** (a) The Lesser Antilles extend from Anguilla in the north to Grenada in the south. (b) The number of bird, bat, reptile and amphibian, and butterfly species increases with the area of an island.

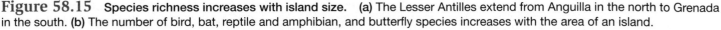

Biological inquiry: How large is the change in bird species richness across islands in the Lesser Antilles?

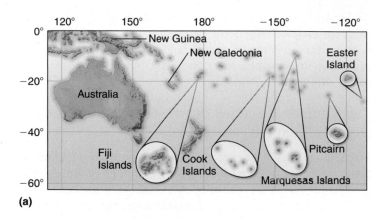

(a)

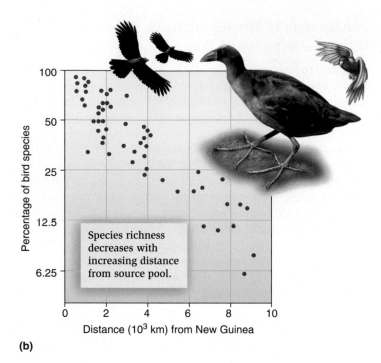

(b)

Figure 58.16 **Species richness decreases with distance from the source pool.** (a) Map of Australia, New Guinea, and these Polynesian Islands: New Caledonia, Fiji Islands, Cook Islands, Marquesas Islands, Pitcairn, and Easter Island. (b) The numbers of bird species on islands with increasing distance from the source pool, New Guinea, expressed as the percentage of bird species on New Guinea.

This research substantiated the prediction of species richness declining with increasing distance from the source pool.

Species Turnover on Islands Is Generally Low

Studies involving species turnover on islands are difficult to perform because detailed and complete species lists are needed over long periods of time, usually many years and often decades. The lists that do exist are often compiled in a casual way and are not usually suitable for comparison with more modern data. In 1980, British researcher Francis Gilbert reviewed 25 investigations carried out to demonstrate turnover and found a lack of this type of rigor in nearly all of them. Furthermore, most of the observed turnover in these studies, usually less than 1% per year, or less than one species per year, appeared to be due to immigrants that never became established, not due to the extinction of well-established species. More recent studies have revealed similar findings. The take-home message from most studies is that recorded rates of turnover are low, giving little conclusive support to this prediction of the equilibrium model of island biogeography.

FEATURE INVESTIGATION

Wilson and Simberloff's Experiments Tested the Predictions of the Equilibrium Model of Island Biogeography

E. O. Wilson and Daniel Simberloff conducted possibly the best test of the equilibrium model of island biogeography ever performed using islands in the Florida Keys. They surveyed four small red mangrove (*Rhizophora mangle*) islands, 11–25 m in diameter, for all terrestrial arthropods. They then enclosed each island with a plastic tent and had the islands fumigated with methyl bromide, a short-acting insecticide, to remove all arthropods on them. The tents were removed, and periodically thereafter Wilson and Simberloff surveyed the islands to examine recolonization rates. At each survey, they counted all the species present, noting any species not there at the previous census and the absence of others that were previously there but

had presumably gone extinct (results for four of the islands shown in **Figure 58.17**). In this way, they estimated turnover of species on islands.

After 250 days, all but one of the islands had a similar number of arthropod species as they had to begin with, even though population densities were still low. The data indicated that colonization rates were higher on islands nearer to the mainland than on far islands—as the island biogeography model predicts. However, the data, which consisted of lists of species on islands before and after extinctions, provided little support for the prediction of high turnover. Rates of turnover were low, only 1.5 extinctions per year, compared to the 15 to 40 species found on the islands within a year. Wilson and Simberloff concluded that turnover probably involves only a small subset of transient or unimportant species, with the more important species being permanent after colonization.

Figure 58.17   Wilson and Simberloff's experiments on the equilibrium model of biogeography.

HYPOTHESIS Island biogeography model predicts substantial turnover of species on islands.

STARTING LOCATION Mangrove islands in the Florida Keys

	Experimental level	Conceptual level

1 Take initial census of all terrestrial anthropods on 4 mangrove islands. Erect framework over each mangrove island.

Each mangrove island is isolated.

2 Cover framework with tent and fumigate with methyl bromide to defaunate island.

Methyl bromide is a low-persistent insecticide that at low levels will not kill plant life.

3 Remove tents and conduct censuses every month to monitor recolonization.

Mangrove islands are recolonized.

4 **THE DATA** Island E2 was closest to the mainland and supported the highest number of species. E3 and ST2 were at an intermediate distance from the mainland, and E1 was the most distant.

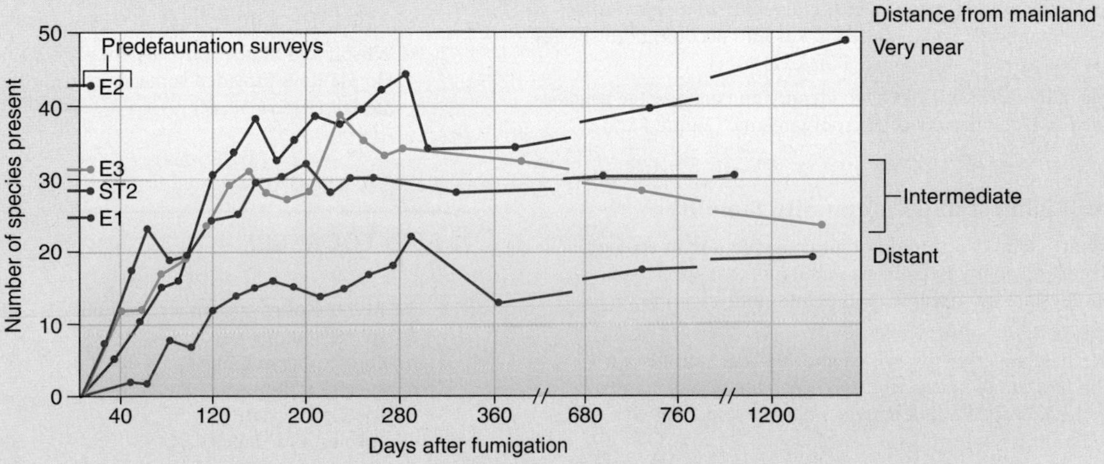

In summary, MacArthur and Wilson's equilibrium model of island biogeography has stimulated much research that confirms the strong effects of area and distance on species richness. However, species turnover appears to be low rather than considerable, which suggests that succession on most islands is a fairly orderly process. This means that colonization is not a random process and that the same species seem to always colonize first and other species gradually reappear in the same order. It is also important to note that the principles of island biogeography can be applied not only to a body of land surrounded by water, but also to fragments of habitat such as wildlife preserves and sanctuaries. For example, wildlife preserves are essentially islands in a sea of developed land, either agricultural fields or urban sprawl. Conservationists have therefore utilized island biogeography modeling in the concept of preserve design, a topic we will explore in Chapter 60.

In this chapter, we have seen how important species richness is to community function. It affects not only stability but, as we shall see, many other features of a community such as nutrient uptake, biomass, and productivity. In order to know how diversity affects these properties, we need to understand the basic processes of energy flow and chemical cycling within a community and its surrounding environment, and for that we turn to a discussion of ecosystems ecology.

CHAPTER SUMMARY

58.1 Differing Views of Communities

- Community ecology studies how groups of species interact and form functional communities. Ecologists have differing views on the nature of a community. In one view, communities are tightly organized groups of mutually dependent species, and in another, they are loose assemblies of species that happen to live in the same place at the same time. (Figure 58.1)

58.2 Patterns of Species Richness

- While many observations support the idea that communities are loose assemblages of species, sharp boundaries between groups of species do exist, especially related to physical differences that cause distinct communities to develop. (Figure 58.2)
- The number of species of most taxa varies according to geographic location, generally increasing from polar areas to tropical areas. (Figure 58.3)
- Varying hypotheses for the polar-equatorial gradient have been advanced, including the time hypothesis, the area hypothesis, and the productivity hypothesis. (Figures 58.4, 58.5)

58.3 Calculating Species Diversity

- The most widely used measure of the diversity of a community, called the Shannon diversity index, takes into account both species richness and species abundance. (Table 58.1)
- The field of metagenomics seeks to identify and analyze the genomes contained in a community of microorganisms. (Figure 58.6)

58.4 Species Richness and Community Stability

- Community stability is an important consideration in ecology. The diversity-stability hypothesis maintains that species-rich communities are more stable than communities with fewer species.
- Tilman's field experiments, which showed that year-to-year variation in plant biomass decreased with increasing species diversity, established a link between diversity and stability. (Figure 58.7)

- The intermediate-disturbance hypothesis proposes that higher species diversity is maintained in communities with moderate levels of disturbance. (Figure 58.8)

58.5 Succession: Community Change

- Succession describes the gradual and continuous change in community structure over time. Primary succession refers to succession on a newly exposed site not previously occupied by soil; secondary succession refers to succession on a site that has already supported life but has undergone a disturbance. (Figures 58.9, 58.10)
- Three mechanisms have been proposed for succession: in facilitation, each species facilitates or makes the environment more suitable for subsequent species. In inhibition, initial species inhibit later colonists. In tolerance, any species can start the succession, and species replacement is unaffected by previous colonists. (Figures 58.11, 58.12, 58.13)

58.6 Island Biogeography

- In the equilibrium model of island biogeography, the number of species on an island tends toward an equilibrium number determined by the balance between immigration rates and extinction rates. (Figure 58.14)
- The model predicts that the number of species increases with increasing island size, and that the number of species decreases with distance from the source pool. (Figures 58.15, 58.16)
- Wilson and Simberloff's experiments on mangrove islands in the Florida Keys provided some support for the island biogeography model. (Figure 58.17)

TEST YOURSELF

1. The number of species in a community is called
   a. species complexity.
   b. community complexity.
   c. species richness.
   d. species diversity.
   e. species abundance.

2. Which of the following statements best represents the productivity hypothesis regarding species richness?
   a. The larger the area, the greater the number of species that will be found there.
   b. Temperate regions have a lower species richness due to the lack of time available for migration after the last ice age.
   c. The number of species in a particular community is directly related to the amount of plant biomass available for consumers.
   d. As invertebrate productivity increases, species richness will increase.
   e. Species richness is not related to primary productivity.

3. The Shannon diversity index is a measure of
   a. the number of different species in a community.
   b. the abundance of a species in a community.
   c. the types of species found in a typical climate.
   d. the number of different species and their relative abundance in a community.
   e. the distribution of members of a species in a community.

4. Metagenomics is a field of study that
   a. determines the similarities of the genomes of all species in a community.
   b. focuses on the microbial genomes contained in a community.
   c. compares the genomes of similar species in different communities.
   d. none of the above.
   e. both a and b.

5. Extreme fluctuations in species abundance
   a. lead to more diverse communities.
   b. are usually seen in early stages of community development.
   c. may increase the likelihood of extinction.
   d. have very little effect on species richness.
   e. are characteristic of stable communities.

6. Which of the following statements best represents the relationship between species diversity and community disturbance?
   a. Species diversity and community stability have no relationship.
   b. Communities with high levels of disturbance are more diverse.
   c. Communities with low levels of disturbance are more diverse.
   d. Communities with intermediate levels of disturbance are more diverse.
   e. Communities with intermediate levels of disturbance are less diverse.

7. The process of primary succession occurs
   a. around a recently erupted volcano.
   b. on a newly plowed field.
   c. on a hillside that has suffered a mudslide.
   d. on a recently flooded riverbank.
   e. none of the above.

8. When the early colonizers exclude subsequent colonists from moving into a community, this is referred to as
   a. facilitation.
   b. competitive exclusion.
   c. secondary succession.
   d. inhibition.
   e. natural selection.

9. Which of the following statements is not true concerning island biogeography?
   a. Diversity on islands is directly related to the size of the island.
   b. Diversity on islands is inversely related to the distance of the island from the source for colonizing species.
   c. The number of species on an island represents the balance between immigration and extinction rates.
   d. The number of species on an island increases with increasing distance from the source pool.
   e. The number of species on an island increases with increasing island size.

10. The equilibrium model of island biogeography
    a. is a method of mapping islands in the Pacific Ocean.
    b. is a method of identifying particular species on particular islands.
    c. describes factors that influence succession on new islands.
    d. describes the pattern of species types that first colonize an island.
    e. predicts that the first colonizers of an island will become extinct.

CONCEPTUAL QUESTIONS

1. Define a community and community ecology.
2. Explain the productivity hypothesis
3. When is a community in equilibrium?

EXPERIMENTAL QUESTIONS

1. What was the purpose of Wilson and Simberloff's study?
2. Why did the researchers conduct a thorough species survey of arthropods before experimental removal of all the arthropod species?
3. What did the researchers conclude about the relationship between island size and species richness and about species turnover?

COLLABORATIVE QUESTIONS

1. Discuss the concept of ecological succession.
2. What are the three most important hypotheses used to predict species richness?

www.brookerbiology.com
This website includes answers to the Biological Inquiry questions found in the figure legends and all end-of-chapter questions.

59

ECOSYSTEMS ECOLOGY

CHAPTER OUTLINE

59.1   Food Webs and Energy Flow

59.2   Energy Production in Ecosystems

59.3   Biogeochemical Cycles

Mandara Lake Oasis, Libya.

The term **ecosystem** was coined in 1935 by the British plant ecologist A. G. Tansley to include not only the biotic community of organisms in an area but also the abiotic environment affecting that community. **Ecosystems ecology** is concerned with the movement of energy and materials through organisms and their communities. Just like the concept of a community, the ecosystem concept can be applied at any scale: A small pond inhabited by protozoa and insect larvae is an ecosystem, and an oasis with its plants, frogs, fish, and birds constitutes another. Most ecosystems cannot be regarded as having definite boundaries. Even in a clearly defined pond ecosystem, amphibians may be moving in and out (**Figure 59.1**). Neverthe-

less, studying ecosystems ecology allows us to use the common currency of energy and chemicals, or nutrients, to compare the functions between and within ecosystems.

In investigating the different processes of an ecosystem, at least three major constituents can be measured: energy flow, biomass production, and biogeochemical cycling. We begin the chapter by exploring **energy flow**, the movement of energy through the ecosystem. In examining energy flow, the main tasks are to document the complex networks of feeding relationships, called food webs, and to measure the efficiency of energy transfer between organisms in an ecosystem. We also consider the tendency of chemical elements to accumulate in organisms, a process called biomagnification.

Next, we focus on the measurement of **biomass**, a quantitative estimate of the total mass of living matter in a given area, usually measured in grams or kilograms per square meter. We will examine the amount of biomass produced through photosynthesis, termed primary production, and the amount of biomass produced by the organisms that are the consumers of primary production. Attaching too much importance to biomass alone, however, may lead to erroneous conclusions. For example, in the applied science of timber technology, a small biomass may indicate a small potential harvest or yield, but this is not necessarily true if the tree species in question has a high growth rate, because new biomass will be produced rapidly and a high rate of harvest could be sustained.

The functioning of an ecosystem can sometimes be most limited by the availability of a scarce chemical or mineral. In the last section, we examine the movement of limiting chemicals through ecosystems, called **biogeochemical cycles**. We explore the cycling of chemical elements such as nitrogen, carbon, sulfur, and phosphorus and the effect that human activities are having on these ecosystem-wide processes.

Figure 59.1   A small ecosystem.   Even in this pond ecosystem, frogs or other species such as birds may move in and out, importing or exporting nutrients and energy with them.

59.1 Food Webs and Energy Flow

Most organisms either make their own food using energy from sunlight or feed on other organisms. Simple feeding relationships between organisms can be characterized by an unbranched **food chain**, a linear depiction of energy flow, with each organism feeding on and deriving energy from the preceding organism. Each feeding level in the chain is called a **trophic level** (from the Greek *trophos*, feeder), and different species feed at different levels. In a food chain diagram, an arrow connects each trophic level with the one above it (**Figure 59.2**).

In this section, we will examine trophic relationships and the flow of energy in a food chain and a food web, a more complex model of interconnected food chains. We will then explore two of the most important features of food webs—chain length and the pyramid of numbers—and learn how the passage of nutrients through food webs can result in the concentration of harmful chemicals in the tissues of organisms at higher trophic levels.

The Main Trophic Levels Within Food Chains Consist of Primary Producers, Primary Consumers, and Secondary Consumers

Food chains typically consist of organisms that obtain energy in different ways. **Autotrophs** harvest light or chemical energy and store that energy in carbon compounds. Most autotrophs, including plants, algae, and photosynthetic prokaryotes, use sunlight for this process. These organisms, called **primary producers**, are at the base of the food chain. They produce the energy-rich tissue upon which nearly all other organisms depend. Note that not all primary producers utilize sunlight; some organisms, called chemoautotrophs, obtain their energy by oxidizing inorganic compounds such as sulfides (**Figure 59.3**).

Organisms in trophic levels above the primary producers are termed **heterotrophs**. These organisms receive their nutrition by eating other organisms. Organisms that obtain their food by eating primary producers are **primary consumers** and include animals, most protists, and even some plants. They are also called **herbivores**. Organisms that eat primary consumers are **secondary consumers**, also called **carnivores** (from the Latin, *carn*, flesh). Organisms that feed on secondary consumers are called **tertiary consumers**, and so on. Thus, energy enters a food chain through primary producers, via photosynthesis, and is passed up the food chain to primary, secondary, and tertiary consumers (see Figure 59.2).

Much energy from the first trophic level, the plants, goes unconsumed by herbivores. Instead, unconsumed plants die and decompose in place. This material, along with dead remains of animals and waste products, is called **detritus**. Consumers that get their energy from detritus, called **detritivores** or **decomposers**, break down dead organisms from all trophic levels (**Figure 59.4**). Detritivores probably carry out 80–90% of the consumption of plant matter, with different species working in concert to extract most of the energy. Detritivores may in turn support a community of predators that feed on them.

Usually, many herbivore species feed on the same plant species. For example, one may find many insect species and vertebrate grazers all feeding on one type of plant. Also, many species of herbivores eat several different plant species. Such branching of food chains also occurs at other trophic levels.

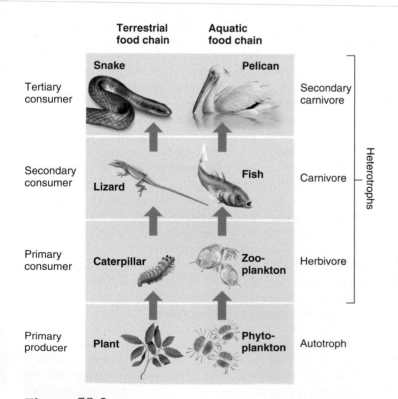

Figure 59.2 Food chains. Two examples of the flow of food energy up the trophic levels: a terrestrial food chain, and an aquatic food chain.

Figure 59.3 Chemoautotrophic organisms living around deep-sea hydrothermal vents. Some bacteria obtain their energy from the sulfur-rich emissions of hydrothermal vents on the ocean floor, also called black smokers. Deep-sea crabs and tube worms may also be found here.

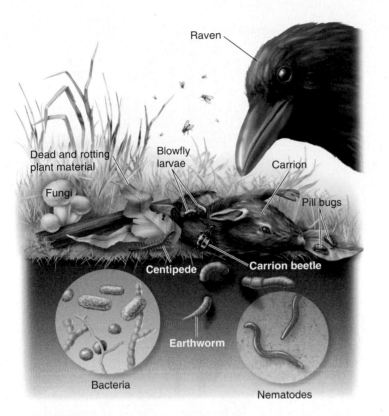

Figure 59.4 Decomposers (detritivores) feeding on dead plant and animal matter. Many dead plants and animals are eaten by a variety of organisms. Here, pill bugs and fungi feed on dead wood and rotting plant material, while earthworms, nematodes, and bacteria decompose organic matter in the soil. Dead animals, or carrion, may be fed upon by blowfly larvae, carrion beetles, and other scavengers. These may, in turn, support a variety of predators, including centipedes or larger predators such as this raven, which will also feed on the animal carcass.

For instance, on the African savanna, cheetahs, leopards, lions, and wild dogs all eat a variety of prey including giraffes, gazelles, baboons, zebras, and wildebeest. These, in turn, eat a variety of grasses and trees. It is more correct, then, to draw relationships between these plants and animals not as a simple chain but as a more elaborate interwoven **food web**, in which there are multiple links between species (**Figure 59.5**).

In Most Food Webs, Chain Lengths Are Short and a Pyramid of Numbers Exists

Now that we have introduced the structure of food webs, let's examine some of their characteristics in more detail. In food webs, the concept of chain length refers to the number of links between the trophic levels involved. For example, if a lion feeds on a zebra, and a zebra feeds on grass, the chain length would be two. For most food webs, chain lengths tend to be short, usually less than five. The main reason why they are short comes from the well-established laws of physics and chemistry that we discussed in Chapter 2. The first law of thermodynamics states that energy cannot be created or destroyed, only transformed. We can thus construct energy budgets for food webs that trace energy flow from green plants to tertiary consumers (and if needed beyond). The second law of thermodynamics states that energy conversions are not 100% efficient. In any transfer process, some useful energy, which can do work, is lost (**Figure 59.6**). This suggests that we can compare the efficiency of energy transfer through trophic levels in different types of food webs. The two main measures of the efficiency of consumers as energy transformers are production efficiency and trophic-level transfer efficiency.

Production Efficiency Production efficiency is defined as the percentage of energy assimilated by an organism that becomes incorporated into new biomass.

Trophic level

Secondary consumers

Primary consumers

Primary producers

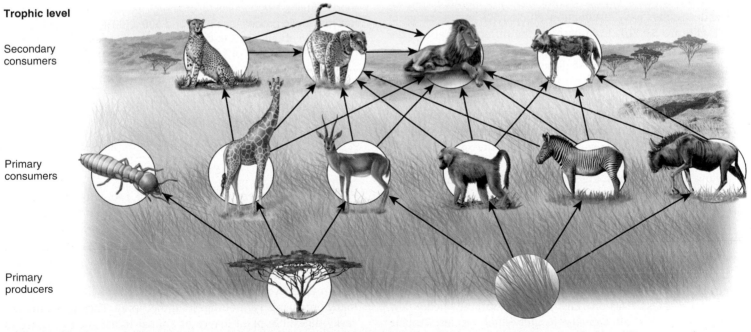

Figure 59.5 A simplified food web from an African savanna ecosystem. Each trophic level is occupied by different species. Generally, each species feeds on, or is fed upon by, more than one species.

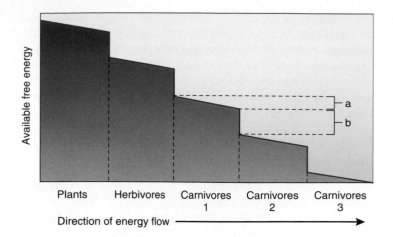

Figure 59.6 **Energy flow through an ecosystem.** **(a)** Energy lost as heat in a single trophic level. **(b)** Energy lost in the conversion between one trophic level and another.

$$\text{Production efficiency} = \frac{\text{Net productivity}}{\text{Assimilation}} \times 100$$

Here, net productivity is the energy, stored in biomass, that has accumulated over a given time span, and assimilation is the total amount of energy taken in by an organism over the same time span. Invertebrates generally have high production efficiencies that average about 10–40% (**Figure 59.7a**). Microorganisms also have relatively high production efficiencies. Vertebrates tend to have lower production efficiencies than invertebrates, because they devote more energy to sustaining their metabolism than to new biomass production. Even within vertebrates, much variation occurs. Fish, which are ectotherms, typically have production efficiencies of around 10%, and birds and mammals, which are endotherms, have production efficiencies in the range of 1–2% (**Figure 59.7b**). In large part, the difference reflects the energy cost of maintaining a constant body temperature.

One consequence of these differences is that sparsely vegetated deserts can support healthy populations of snakes and lizards while mammals might easily starve. The largest living lizard known, the Komodo dragon, eats the equivalent of its own weight every two months, whereas a cheetah consumes approximately four times its own weight in the same period. It is also interesting to note that production efficiencies are higher in young animals, which are rapidly accruing biomass, than in older animals, which are not. This is the main reason behind the practice of raising young chickens and calves for meat.

Trophic-Level Transfer Efficiency The second measure of efficiency of consumers as energy transformers is **trophic-level transfer efficiency**, which is the amount of energy at one trophic level that is acquired by the trophic level above and incorporated into biomass. This provides a way to examine energy flow between trophic levels, not just in individual species. Trophic level transfer efficiency is calculated as:

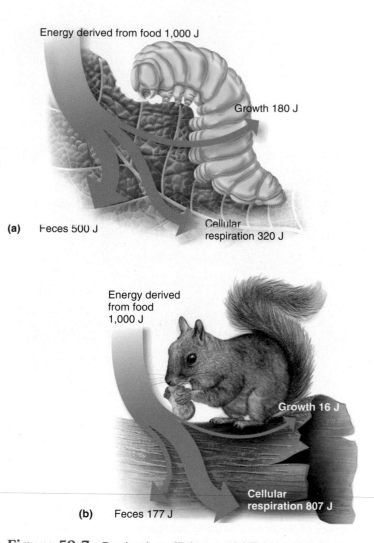

Figure 59.7 **Production efficiency.** **(a)** This caterpillar, an invertebrate, chews leaves to obtain its energy. If a mouthful of food contains 1,000 joules (J) of energy, about 320 J is used to fuel metabolic processes (32%) and 500 J (50%) is lost in feces. This leaves about 180 J to be converted into insect biomass, a production efficiency of 18%. **(b)** The production efficiency of this squirrel, a mammal, is much lower.

Biological inquiry: What is the production efficiency of the squirrel, using the numbers in the figure?

$$\text{Trophic-level transfer efficiency} = \frac{\text{Production at trophic level } n}{\text{Production at trophic level } n-1} \times 100$$

For example, if there were 14 g/m² of zooplankton in a lake (trophic level n) and 100 g/m² of phytoplankton production (trophic level $n - 1$), the trophic level efficiency would be 14%. Trophic-level transfer efficiency appears to average around 10%, though there is much variation. In some marine food chains, for example, it can exceed 30%.

Trophic-level transfer efficiency is low for two reasons. First, many organisms cannot digest all their prey. They take only the easily digestible plant leaves or animal tissue such as muscles and guts, leaving the hard wood or energy-rich bones behind.

Second, much of the energy assimilated by animals is used in maintenance, so most energy is lost from the system as heat. The 10% average transfer rate of energy from one trophic level to another also necessitates short food webs of no more than four or five levels. Relatively little energy is available for the higher levels.

The Pyramid of Numbers Trophic-level transfer efficiencies can be expressed in a graphical form called an Eltonian pyramid, named after the British ecologist Charles Elton. The best-known pyramid, and the one described by Elton in 1927, is the **pyramid of numbers**, in which the number of individuals decreases at each trophic level, with a huge number of individuals at the base and fewer individuals at the top. Elton's example was that of a small pond where the numbers of protozoa may run into the millions and those of *Daphnia*, their predators, number in the hundreds of thousands. Hundreds of beetle larvae may feed on *Daphnia*, and tens of fish feed on the beetles (**Figure 59.8a**). Many other examples of this type of pyramid are known. For example, in a grassland, there may be hundreds of individual plants per square meter, dozens of insects that feed on the plants, and a few spiders feeding on the insects.

However, one can also think of several exceptions to this pyramid. An oak tree, one single producer, supports hundreds of herbivorous beetles, caterpillars, and other primary consumers, which in turn may support thousands of predators and parasites (**Figure 59.8b**). This is called an inverted pyramid of numbers. The best way to reconcile this apparent exception is to weigh the organisms in each trophic level, creating a **pyramid of biomass**. The oak tree weighs 30,000 kg, all the herbivores on the tree total 5 kg, and the predators about 1 kg. Looking at the biomass at each trophic level rather than at numbers of organisms shows an upright pyramid (**Figure 59.8c**).

Inverted pyramids can still occur, albeit rarely, even when biomass is used as the measure. In some marine systems, the biomass of phytoplankton supports a higher biomass of zooplankton, which in turn is eaten by a higher biomass of carnivorous fish (**Figure 59.8d**). This is possible because the production rate of phytoplankton is much higher than that of zooplankton, and the small phytoplankton **standing crop** (the total biomass in an ecosystem at any one point in time) processes large amounts of energy. By expressing the pyramid in terms of production, it is no longer inverted. The **pyramid of production**, which shows rates of production rather than standing crop, is never inverted (**Figure 59.8e**). The laws of thermodynamics ensure that the highest amounts of free energy are found at the lowest trophic levels.

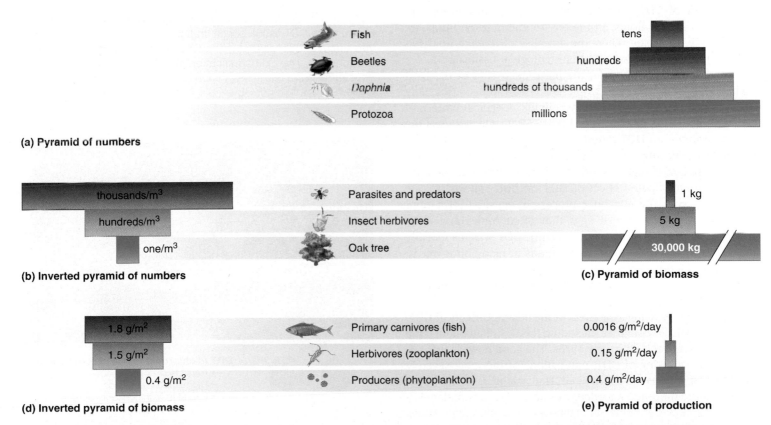

Figure 59.8 **Ecological pyramids in food webs.** **(a)** In this pyramid of numbers, the abundance of species in a typical British pond decreases with increasing trophic level. **(b)** An inverted pyramid of numbers, based on organisms living on an oak tree. **(c)** When amount of biological material is used instead of numbers of individuals, the pyramid is termed a pyramid of biomass. **(d)** An inverted pyramid of biomass in the English Channel. **(e)** A pyramid of production in the English Channel.

Biomagnification Can Occur in Higher Trophic Levels

Thus far, we have considered how available biomass and energy flow can influence the properties of food chains and webs. An issue that faces organisms is the tendency of certain chemical elements to accumulate or build up within food chains. This increase in the concentration of a substance in living organisms is called **biomagnification**, and the passage of DDT in food chains provides a startling example.

Dichlorodiphenyltrichloroethane (DDT) was first synthesized by chemists in 1874. In 1939, its insecticidal properties were recognized by Paul Müller, a Swiss scientist who won a Nobel Prize in 1948 for his discovery and subsequent research on the uses of the chemical. The first important application of DDT was in human health programs during and after World War II, particularly to control mosquito-borne malaria, and at that time its use in agriculture also began. The global production of DDT peaked in 1970, when 175 million kg of the insecticide were manufactured.

DDT has several chemical and physical properties that profoundly influence the nature of its ecological impact. First, DDT is persistent in the environment; it is not easily degraded to other, less toxic chemicals by microorganisms or by physical agents such as light and heat. The typical persistence in soil of DDT is about 10 years, which is two to three times longer than the persistence of most other insecticides. Another important characteristic of DDT is its low solubility in water and its high solubility in fats or lipids. In the environment, most lipids are present in living tissue. Therefore, because of its high lipid solubility, DDT tends to concentrate in biological tissues.

Because biomagnification occurs at each step of the food chain, organisms at higher tropic levels can amass especially large concentrations of DDT in their lipids. A typical pattern of biomagnification is illustrated in **Figure 59.9**, which shows the relative amounts of DDT found in a Lake Michigan food chain. The largest concentration of the insecticide was found in gulls, tertiary consumers that feed on fish that, in turn, eat small insects. An unanticipated effect of DDT on bird species was its interference with the metabolic process of eggshell formation. The result was thin-shelled eggs that often broke under the weight of incubating birds (**Figure 59.10**). DDT was responsible for a dramatic decrease in the populations of many birds due to failed reproduction. Relatively high levels of the chemical were also found to be present in some game fish, which became unfit for human consumption.

Because of growing awareness of the adverse effects of DDT, most industrialized countries, including the U.S., had banned the use of the chemical by the early 1970s. The good news is that following the outlawing of DDT, populations of the most severely affected bird species have recovered. Had scientists initially possessed a more thorough knowledge of how DDT accumulated in food chains, however, some of the damage to the bird populations might have been prevented. DDT is still used in some developing countries to control malaria, although there has been significant movement toward the use of alternative pest control technologies.

Figure 59.9   **Biomagnification in a Lake Michigan food chain.** The DDT tissue concentration in gulls, a tertiary consumer, was about 240 times that in the small insects sharing the same environment. The biomagnification of DDT in lipids causes its concentration to increase at each successive link in the food chain.

DDT (dichlorodiphenyltrichloroethane)
• Persists in environment
• High solubility in lipids
• Found in high concentrations at higher trophic levels

Figure 59.10   **Thinning of eggshells caused by DDT.** These ibis eggs are thin-shelled and have been crushed by the incubating adult.

59.2 Energy Production in Ecosystems

In this section, we will take a closer look at energy production. Because the bulk of the Earth's biosphere, 99.9% by mass, consists of primary producers, when we measure ecosystem energy production, we are primarily interested in plants, algae, or cyanobacteria. Since plants represent the first, or primary, trophic level, we measure plant production as **gross primary production (GPP)**. Gross primary production is equivalent to the carbon fixed during photosynthesis. **Net primary production (NPP)** is gross primary production minus the energy released in plant cellular respiration (R).

$$NPP = GPP - R$$

Net primary production is thus the amount of energy that is available to primary consumers. From now on, the term primary production refers to net primary production.

To measure primary production, calories can be used as a common currency, and organisms can be viewed as caloric equivalents. For example, 100 g of rye grass seeds (*Secale cereale*) has a calorific equivalent of about 380 calories, whereas 100 g of lead tree leaves (*Leucaena leucocephala*) has a calorific content of only 68 calories. Energy content is generally measured using dry biomass. Dry weight is used because the bulk of living matter in most species is water, and water content fluctuates widely, often according to wet or dry seasons. Of the dry weight, 95% is made up of carbon compounds, so that measuring energy flow in ecosystems is in many ways equivalent to examining the carbon cycle (see Section 59.3), and ecologists often measure NPP in terms of carbon fixed per square meter or per hectare.

Understanding what factors limit primary production is of vital importance if we are to examine ecosystems as energy transformers. Furthermore, by determining these factors, we can understand how primary production varies globally. We can also examine the effects of primary production on **secondary production**, the gain in the biomass of heterotrophs and decomposers.

Primary Production Is Influenced in Terrestrial Ecosystems by Water, Temperature, and Nutrient Availability

In terrestrial systems, water is a major determinant of primary production, and primary production shows an almost linear increase with annual precipitation, at least in arid regions. Likewise, temperature, which affects production primarily by slowing or accelerating plant metabolic rates, is also important. Ecologist Michael Rosenzweig noted that the evapotranspiration rate could predict the aboveground primary production with good accuracy in North America (**Figure 59.11**). Recall from Chapter 58 that the evapotranspiration rate is a measure of the amount of water entering the atmosphere from the ground through the process of evaporation from the soil and transpiration of plants, so it is a measure of both temperature and available water. For example, a desert will have a low evapotrans-

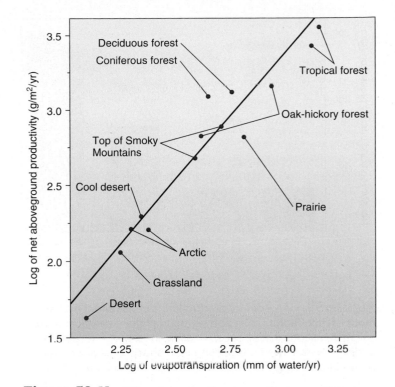

Figure 59.11 Primary production is positively correlated with the evapotranspiration rate. Warm, humid environments are ideal for plant growth. Dots represent different ecosystems.

piration rate because water availability is low despite high temperature. Rates of evapotranspiration are maximized when both temperature and moisture are at high levels, as in tropical rain forests.

A lack of **nutrients**, key elements in useable form, particularly nitrogen and phosphorus, can also limit primary production in terrestrial ecosystems, as agricultural practitioners know only too well. Fertilizers are commonly used to boost the production of annual crops. In 1984, Stewart Cargill and Rob Jefferies showed how a lack of both nitrogen and phosphorus was limiting to salt marsh sedges and grasses in subarctic conditions in Hudson Bay, Canada (**Figure 59.12**). Of the two nutrients, nitrogen was the most limiting; without it, the addition of phosphorus did not increase production. However, once nitrogen was added, phosphorus became the **limiting factor**, that is, the one in shortest supply for growth. Once nitrogen was added and was no longer limiting, the addition of phosphorus increased production. The addition of nitrogen and phosphorus together increased production the most. This result supports a principle known as **Liebig's law of the minimum**, named for Justus von Liebig, a 19th-century German chemist, which states that species biomass or abundance is limited by the scarcest factor. This factor can change, as the Hudson Bay experiment showed: When sufficient nitrogen is available, phosphorus becomes the limiting factor. Once phosphorus becomes abundant, then productivity will be limited by another nutrient.

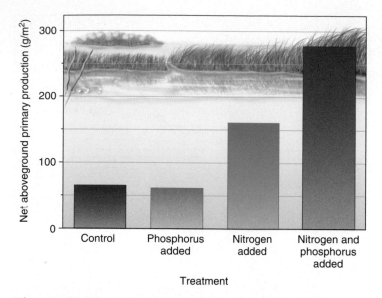

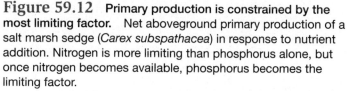

Figure 59.12 **Primary production is constrained by the most limiting factor.** Net aboveground primary production of a salt marsh sedge (*Carex subspathacea*) in response to nutrient addition. Nitrogen is more limiting than phosphorus alone, but once nitrogen becomes available, phosphorus becomes the limiting factor.

Table 59.1	Net Primary Production for Earth's Ecosystems

Ecosystem type	Mean net primary production g/m²/yr)
Terrestrial	
Tropical rain forest	2,500
Tropical deciduous forest	1,600
Temperate deciduous forest	1,550
Savanna	1,080
Prairie	750
Cultivated land	610
Taiga	380
Tundra	140
Hot desert	90
Aquatic	
Algal beds and coral reefs	2,500
Wetlands	2,000
Estuaries	1,500
Upwelling zones	500
Continental shelf	360
Lake and stream	250
Open ocean	125

Primary Production in Aquatic Ecosystems Is Limited Mainly by Light and Nutrient Availability

Of the factors limiting primary production in aquatic ecosystems, the most important are available light and available nutrients. Light is particularly likely to be in short supply because water readily absorbs light. At 1-m depth, more than half the solar radiation has been absorbed. By 20 m, only 5–10% of the radiation is left. The decrease in light is what limits the depth of water to which algae are restricted.

The most important nutrients affecting primary production in aquatic systems are nitrogen and phosphorus, because they occur in very low concentrations. While soil contains about 0.5% nitrogen, seawater contains only 0.00005% nitrogen. Enrichment of the aquatic environment by the addition of nitrogen and phosphorus can result in large, unchecked growths of algae called algal blooms. Such enrichment occurs naturally in areas of upwellings, where cold, deep, nutrient-rich water containing sediment from the ocean floor is brought to the surface by strong currents, resulting in very productive ecosystems and plentiful fish. Some of the largest areas of upwelling occur in the Antarctic and along the coasts of Peru and California.

Primary Production Is Greatest in Areas of Abundant Warmth and Moisture

Knowing which factors limit primary production helps ecologists understand why mean net primary production varies across the different ecosystems on Earth (**Table 59.1**). In general, primary production is highest in tropical rain forests and decreases progressively toward the poles (**Figure 59.13**). As we saw in Chapter 58, many ecologists suggest this is the primary cause for the polar-equatorial gradient of species richness. This primary productivity gradient occurs because temperatures decrease toward the poles, and, as we have just learned, temperatures affect primary production greatly. Wetlands also tend to be extremely productive, primarily because water is not limiting and nutrient levels are high. Productivity of the open ocean is very low, falling somewhere between the productivity of deserts and that of the Arctic tundra. Marine production is high on coastal shelves, particularly in upwelling zones. However, the greatest marine production occurs on algal beds and coral reefs, where temperatures are high and water levels are not so deep that light becomes limiting.

Secondary Production Is Generally Limited by Available Primary Production

What factors control secondary production, the productivity of herbivores, carnivores, and decomposers? This is a complex question, but it is generally thought to be limited largely by available primary production. A strong relationship exists between primary production in a variety of ecosystems and the biomass of herbivores (**Figure 59.14**). This means that more plant biomass, and thus more primary production, leads to increased herbivore biomass.

Figure 59.13 Annual net primary productivity on Earth.   Primary productivity generally increases from the poles to the equator.

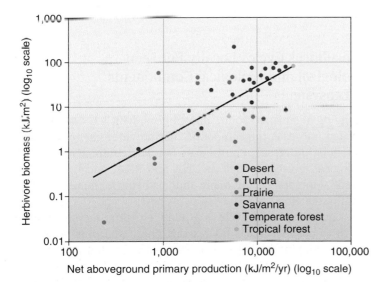

Figure 59.14 Herbivore biomass is positively correlated with net aboveground primary production.   These data are taken from a variety of case studies from different biomes. Herbivore biomass can be considered a surrogate for secondary production.

As we have noted before, trophic-level transfer efficiency averages about 10%. Thus, after one link in the food web, only 1/10 of the energy captured by plants is transferred to herbivores, and after two links in the food web, only 1/100 of the energy fixed by plants goes to carnivores. Thus, secondary production is much smaller than that of primary production. We can see this in the work of John Teal, whose 1962 work examined energy flow in a Georgia salt marsh (**Figure 59.15**). Salt marshes are among the most productive habitats on Earth in terms of amount of vegetation they produce. In salt marshes, most of the energy from the sun (incident sunlight) goes to two types of organisms: *Spartina* plants and marine algae. The *Spartina* plants are rooted in the ground, whereas the algae float on the water surface or live on the mud or on *Spartina* leaves at low tide. These photosynthetic organisms absorb about 6% of the incident sunlight. Most of the plant energy, 77.6%, is used in plant and algal cellular respiration. Of the energy that is accumulated in plant biomass, most dies in place and rots on the muddy ground, to be consumed by bacteria. Bacteria are the major decomposers in this system, followed distantly by nematodes and crabs, which feed on tiny food particles as they sift through the mud. Some of this dead material is also removed from the system (exported) by the tide. The herbivores take very little of the plant production, eating only a small proportion of the *Spartina* and none of the algae. Overall, if we view the species in ecosystems as transformers of energy, then plants and algae are by far the most important organisms on the planet, bacteria are next, and animals are a distant third.

59.3 Biogeochemical Cycles

A unit of energy passes through a food web only once. In contrast, elements such as nitrogen or carbon are recycled, moving from the physical environment to organisms and back to the

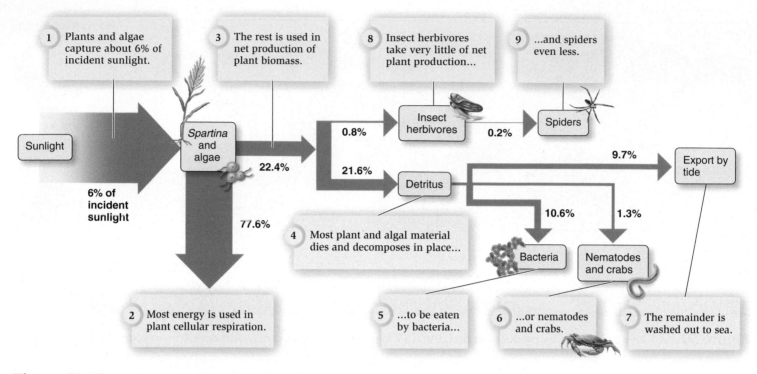

Figure 59.15 Energy-flow diagram for a Georgia salt marsh. Numbers reflect percentages of gross primary production that flows into different trophic levels or is used in plant respiration.

environment, where the cycle begins again. While energy dissipates as heat, chemical elements are available in limited amounts and are continually recycled. Because these cycles involve biological, geological, and chemical transport mechanisms, they are termed **biogeochemical cycles**. Biological mechanisms involve the absorption of chemicals by living organisms and their subsequent release back into the environment. Geological mechanisms include weathering and erosion of rocks, and elements transported by surface and subsurface drainage. Chemical transport mechanisms include dissolved matter in rain and snow, atmospheric gases, and dust blown by the wind.

In addition to the basic building blocks of hydrogen, oxygen, and carbon, which we discussed in Chapter 2, the elements required in the greatest amounts by living organisms are nitrogen, phosphorus, and sulfur. In this section, we take a detailed look at the cycles of these nutrients. These cycles can be divided into two broad types: local cycles, such as the phosphorus cycle, which involve elements with no mechanism for long-distance transfer; and global cycles, which involve an interchange between the atmosphere and the ecosystem. Global nutrient cycles, such as the carbon, nitrogen, and sulfur cycles, unite the Earth and its living organisms into one giant interconnected ecosystem called the **biosphere**. In our discussion, we will take a particular interest in the alteration of these cycles through human activities that increase nutrient inputs, such as the burning of fossil fuels.

Phosphorus Cycles Locally Between Geological and Biological Components of Ecosystems

All living organisms require phosphorus, which becomes incorporated into ATP, the compound that provides energy for most metabolic processes. Phosphorus is a key component of other biological molecules such as DNA and RNA, and it is also an essential mineral that in many animals helps maintain a strong, healthy skeleton.

The phosphorus cycle is a relatively simple cycle (**Figure 59.16**). Phosphorus has no gaseous phase and thus no atmospheric component; that is, it is not moved around by the wind or rain. As a result, phosphorus tends to cycle only locally. The Earth's crust is the main storehouse for this element. Weathering and erosion of rocks release phosphorus into the soil. Plants have the metabolic means to absorb dissolved ionized forms of phosphorus, the most important of which occurs as phosphate (HPO_4^{2-} or $H_2PO_4^-$). Plants can take up phosphate rapidly and efficiently. In fact, they can do this so quickly that they often reduce soil concentrations of phosphorus to extremely low levels, so that phosphorus becomes limiting (see Figure 59.12). Herbivores obtain their phosphorus only from eating plants, and carnivores obtain it by eating herbivores. When plants and animals excrete wastes or die, the phosphorus becomes available to decomposers, which release it back to the soil.

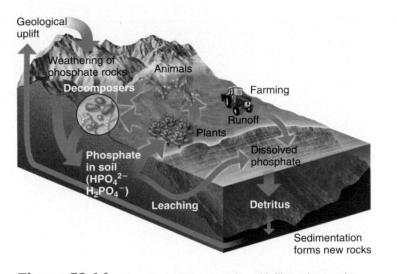

Figure 59.16 **The phosphorus cycle.** Unlike other major biogeochemical cycles, the phosphorus cycle does not have an atmospheric component and thus cycles only locally. The widths of the lines indicate the relative importance of each process.

Leaching and runoff eventually wash much phosphate into aquatic systems, where plants and algae utilize it. Phosphate that is not taken up into the food chain settles to the ocean floor or lake bottom, forming sedimentary rock. Phosphorus can remain locked in sedimentary rock for millions of years, becoming available again through the geological process of uplift.

As noted previously, phosphorus is a limiting element in most aquatic systems. The more phosphorus that is added, the more that aquatic productivity increases (**Figure 59.17a**). In a pivotal 1974 study, biologist David Schindler showed that an overabundance of phosphorus caused the rapid growth of algae and plants in an experimental lake in Canada (**Figure 59.17b**). The process by which elevated nutrient levels lead to an overgrowth of algae and the subsequent depletion of water oxygen levels is known as **eutrophication**. Cultural eutrophication refers to the enrichment of water with nutrients derived from human activities such as fertilizer use and sewage dumping.

Lake Erie became eutrophic in the 1960s due to the fertilizer runoff from farms rich in phosphorus and to the industrial and domestic pollutants released from the many cities along its shores. Fish species such as blue pike, white fish, and lake trout became severely depleted. The U.S. and Canada teamed together to reduce the levels of discharge by 80%, primarily through eliminating phosphorus in laundry detergents and maintaining strict controls on the phosphorus content of wastewater from sewage treatment plants. Fortunately, lake systems have great potential for recovery after phosphorous inputs are reduced, and Lake Erie has experienced a reduction in the occurrence of algal blooms, an improvement in water clarity, and an increase in fish stocks.

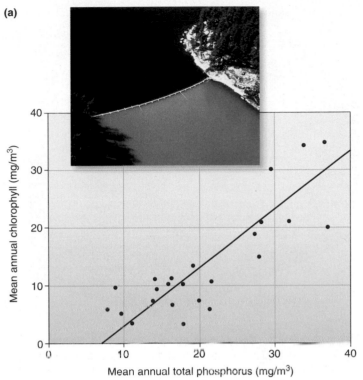

(a)

(b)

Figure 59.17 **Primary production increases with an increase in total phosphorus concentration.** (a) This aerial photograph shows the contrast in water quality of two basins of an experimental lake in Canada, separated by a plastic curtain. The lower basin received additions of carbon, nitrogen, and phosphorus, while the upper basin received only carbon and nitrogen. The bright green color is from a surface film of algae that resulted from the added phosphorus. (b) The increase in primary production is measured as an increase in chlorophyll concentration. A higher chlorophyll concentration in the water means more algae are present.

Carbon Cycles Among Biological, Geological, and Atmospheric Pools

The movement of carbon from the atmosphere into organisms and back again is known as the carbon cycle (**Figure 59.18**). Carbon dioxide is present in the atmosphere at a level of about 380 parts per million (ppm), or about 0.04%. Autotrophs, primarily plants and algae, acquire carbon dioxide from the atmosphere and incorporate it into the organic matter of their own biomass via photosynthesis. In the process, each year, plants and algae remove approximately one-seventh of the CO_2 from the atmosphere. At the same time, the decomposition of plants recycles a similar amount of carbon back into the atmosphere as CO_2. Herbivores can also return some carbon dioxide to the atmosphere, eating plants and breathing out CO_2, but the amount flowing through this part of the cycle is minimal.

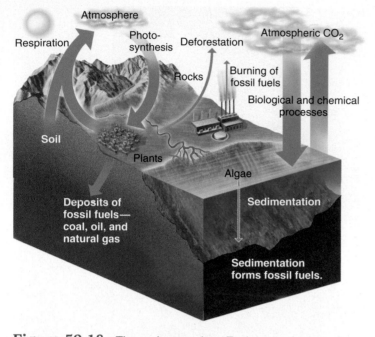

Figure 59.18   **The carbon cycle.**   Each year, plants and algae remove about one-seventh of the CO₂ in the atmosphere. Animal respiration is so small it is not represented. The width of the arrows indicates the relative importance of each process.

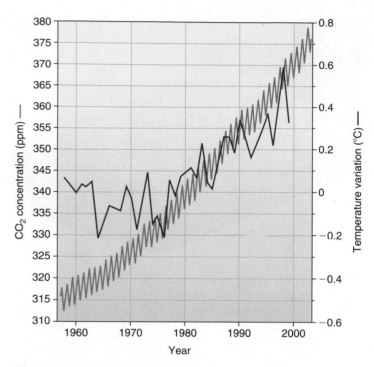

Figure 59.19   **The burning of fossil fuels has greatly increased atmospheric CO₂ levels and temperatures.**   From 1958 to 2004, atmospheric CO₂ shows an increase of nearly 20%. In addition, the graph shows a seasonal variation in CO₂. Temperatures are annual deviations from the 1961–1990 average. Measurements were recorded at Mauna Loa Observatory in Hawaii.

Biological inquiry: Why does the amount of CO₂ fluctuate seasonally in the graph?

Over time, much carbon is also incorporated into the shells of marine organisms, which eventually form huge limestone deposits on the ocean floor. Natural sources of CO₂ such as volcanoes, hot springs, and fires also release large amounts of CO₂. In addition, human activities, primarily the burning of **fossil fuels**—coal, petroleum, and natural gas—are increasingly causing large amounts of CO₂ to enter the atmosphere together with large volumes of particulate matter. Recently, such particulate matter has been shown to increase heritable mutations (see Genomes & Proteomes).

Direct measurements over the past five decades show a steady rise in atmospheric CO₂ (**Figure 59.19**), a pattern that shows no sign of slowing. Because of its increasing concentration in the atmosphere, CO₂ is the most troubling of the greenhouse gases, which are a primary source of global warming (see Chapter 54). Elevated atmospheric CO₂ has other dramatic environmental effects, boosting plant growth but lowering the densities of herbivores that feed on them (see Feature Investigation).

The amount of carbon dioxide in the atmosphere varies with the seasons in temperate environments. Concentrations of atmospheric carbon dioxide are lowest during the Northern Hemisphere's summer and highest during the winter, when photosynthesis is minimal. This is because there is more land in the Northern Hemisphere than in the Southern Hemisphere and therefore more vegetation. The vegetation has a maximum photosynthetic activity during the summer, reducing the global

amount of carbon dioxide. During the winter, photosynthesis is low and decomposition is relatively high, causing a global increase in the gas.

GENOMES & PROTEOMES

Pollution Can Cause Heritable Mutations

Carbon dioxide is one of many pollutants emitted from the burning of fossil fuels; others include sulfur dioxide and nitrous oxides. In addition to these atmospheric gases, the combustion of fossil fuel also results in the release of particulate matter, tiny solid and liquid particles, into the air. Larger particles are visible as smoke or soot. The small particles, when inhaled, can penetrate deeply into the lungs and become distributed throughout the body. Such exposure, whether from transportation sources, factories, power plants, or tobacco smoke, can lead to the formation of lung tumors and induce mutations in somatic cells.

A study by biologist James Quinn and colleagues in 2004 showed that polluted air can also induce genetic changes in mouse sperm. In the study, the researchers put two groups of mice into separate sheds that were downwind of steel mills in Hamilton, an urban-industrial center of Ontario, Canada (**Figure 59.20a**). One shed was fitted with high-efficiency particulate air (HEPA) filters, which removed particulate matter from the air, while the other was not. The researchers discovered changes in the size of noncoding DNA sequences in the offspring of the mice in the filterless shed. They concluded that the increased rate of mutations was paternally derived. Offspring of mice from the unfiltered shed inherited twice as many mutations as offspring of mice in the filtered shed. The study concluded that the pollutant particles, or some chemical compound associated with them, were responsible for the observed, heritable DNA changes.

What is the mechanism behind these genetic changes? Quinn and colleagues hypothesized that the inhaled pollutants are transported to the liver and metabolized to DNA-reactive compounds, and then transported to the testes and finally to the sperm stem cells where the mutations occurred (**Figure 59.20b**). The damaged DNA would then be transmitted to first-generation offspring.

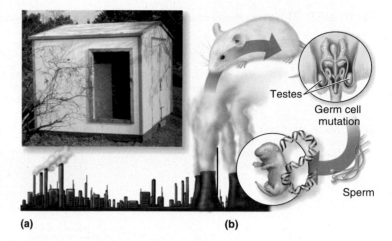

(a)　　　　　　　　　　　　　　　　**(b)**

Figure 59.20　**Mutations caused by air pollution.**　**(a)** In a study by Quinn and colleagues, mice were kept in sheds with and without HEPA air filters in Hamilton, Ontario. **(b)** The study concluded that mice kept in the filterless sheds inhaled airborne particles into their lungs, and compounds from the particles induced genetic mutations that were transmitted through the sperm cells into the next generation.

Feature Investigation

Stiling and Drake's Experiments with Elevated CO_2 Showed an Increase in Plant Growth but a Decrease in Herbivory

How will forests of the future respond to elevated CO_2? To begin to answer such a question, ecologists ideally would enclose large areas of forests with chambers, increase the CO_2 content within the chambers, and measure the responses. This has proven to be difficult for two reasons. First, it is hard to enclose large trees in chambers, and second, it is expensive to increase CO_2 levels over such a large area. However, in a discovery-based investigation, ecologists Peter Stiling and Bert Drake were able to increase CO_2 levels around small patches of forest at the Kennedy Space Center in Cape Canaveral, Florida. In much of Florida forest, trees are small, only 3–5 m when mature, because frequent lightning-initiated fires prevent the growth of larger trees. Stiling and Drake teamed up with NASA engineers to create 16 circular, open-topped chambers (**Figure 59.21**), and in 8 of these they increased atmospheric CO_2 to double their ambient levels, from 360 to 720 ppm, the latter of which is the atmospheric level predicted for the end of the 21st century. The experiments were initiated in 1996 and lasted until 2004. Plants produced more biomass in elevated CO_2, because carbon dioxide is limiting to plant growth, but the data revealed much more.

Because the chambers were open-topped, insect herbivores could come and go. Insect herbivores cause the largest amount of herbivory in North American forests, because vertebrate herbivores cannot access the high foliage. Censuses were conducted of all insect herbivores but focused on leaf miners, the most common type of herbivore at this site, which are small moths whose larvae are small enough to live between the surfaces of plant leaves and create blister-like "mines" on leaves.

Densities of all insects, including leaf miners, was lower in elevated CO_2 in every year studied. Part of the reason for the decline was that even though plants increased in mass, the existing soil nitrogen was diluted over a greater volume of plant material, so that the nitrogen level in leaves decreased. This increased insect mortality by two means. First, poorer leaf quality directly increased insect death because leaf nitrogen levels may have been too low to support the normal development of the leaf miners. Second, lower leaf quality increased the length of time insects had to feed to gain sufficient nitrogen. Increased feeding times in turn led to increased exposure to natural enemies such as parasitoids and predators like spiders and ants, and top-down mortality also increased (see the data of Figure 59.21). Thus, in a world of elevated CO_2, plant growth may increase but herbivore densities could decrease. If some herbivores, such as butterflies, are already rare in a habitat, then an elevation in CO_2 could threaten them with extinction.

Figure 59.21 The effects of elevated atmospheric CO_2 on insect herbivores.

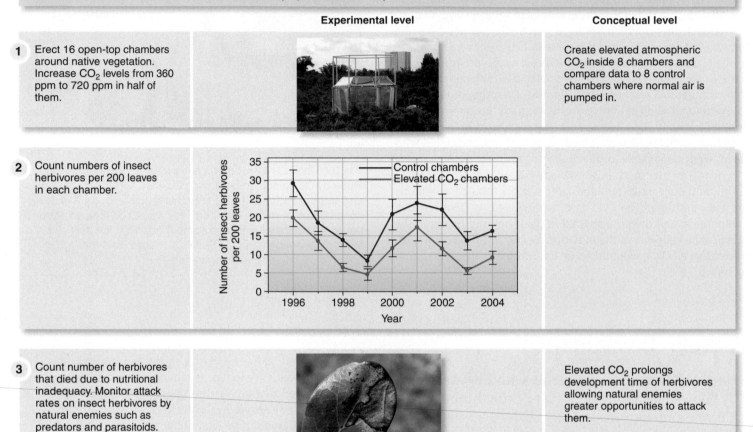

GOAL To determine the effects of elevated CO_2 on a forest ecosystem; effects on herbivores are highlighted here.

STUDY LOCATION Patches of forest at the Kennedy Space Center in Cape Canaveral, Florida.

	Experimental level	Conceptual level
1 Erect 16 open-top chambers around native vegetation. Increase CO_2 levels from 360 ppm to 720 ppm in half of them.		Create elevated atmospheric CO_2 inside 8 chambers and compare data to 8 control chambers where normal air is pumped in.
2 Count numbers of insect herbivores per 200 leaves in each chamber.		
3 Count number of herbivores that died due to nutritional inadequacy. Monitor attack rates on insect herbivores by natural enemies such as predators and parasitoids.		Elevated CO_2 prolongs development time of herbivores allowing natural enemies greater opportunities to attack them.

4 THE DATA

Source of mortality*	Elevated CO_2 (% mortality)	Control (% mortality)
Nutritional inadequacy	10.2	5.0
Predators	2.4	2.0
Parasitoids	10.0	3.2

* Data refer only to mortality of larvae within leaves and do not sum to 100%. Mortality of eggs on leaves, pupae in the soil, and flying adults is unknown.

The Nitrogen Cycle Is Strongly Influenced by Biological Processes That Transform Nitrogen into Usable Forms

Nitrogen is a limiting nutrient because it is an essential component of proteins, nucleic acids, and chlorophyll. Because 78% of the Earth's atmosphere consists of nitrogen gas (N_2), it may seem that nitrogen should not be in short supply for organisms. However, N_2 molecules must be broken apart before nitrogen atoms are available to combine with other elements. Because of its triple bond, nitrogen gas is very stable and only certain bacteria can break it apart into usable forms. This process, called nitrogen fixation, is a critical component of the five-part nitrogen cycle, which also includes nitrification, assimilation, ammonification, and denitrification (**Figure 59.22**):

1. Only a few species of bacteria can accomplish **nitrogen fixation**, that is, convert atmospheric nitrogen to forms

usable by other organisms. The bacteria that fix nitrogen are fulfilling their own metabolic needs, but in the process, they release excess ammonia (NH_3) or ammonium (NH_4^+), which can be used by some plants. The most important of these bacteria in the soil are called *Rhizobium*, which live in nodules on the roots of legumes, including peas, beans, lentils, and peanuts, and some woody plants.

2. In the process of **nitrification**, soil bacteria convert NH_3 or NH_4^+ to nitrate (NO_3^-), a form of nitrogen commonly used by plants. The bacteria *Nitrosomonas* and *Nitrococcus* first oxidize the forms of ammonia to nitrite (NO_2^-), after which the bacteria *Nitrobacter* converts NO_2^- to NO_3^-.

3. **Assimilation** is the process by which plants and animals incorporate the NH_3, NH_4^+, and NO_3^- formed through nitrogen fixation and nitrification. Plant roots take up these forms of nitrogen through their roots, and animals assimilate nitrogen by eating plant tissue.

4. Ammonia can also be formed in the soil through the decomposition of plants and animals and the release of animal waste. **Ammonification** is the conversion of organic nitrogen to NH_3 and NH_4^+. This process is carried out by bacteria and fungi. Most soils are slightly acidic and, because of an excess of H^+, the NH_3 rapidly gains an additional H^+ to form NH_4^+. Because many soils lack nitrifying bacteria, ammonification is the most common pathway for nitrogen to enter the soil.

5. **Denitrification** is the reduction of nitrate (NO_3^-) to gaseous nitrogen (N_2). Denitrifying bacteria, which are anaerobic and use NO_3^- in their metabolism instead of oxygen, perform the reverse of their nitrogen-fixing counterparts by delivering nitrogen to the atmosphere. This process only delivers a relatively small amount of nitrogen to the atmosphere.

In terms of the global nitrogen budget, industrial fixation of nitrogen for the production of fertilizer makes a significant contribution to the pool of nitrogen-containing material in the soils and waters of agricultural regions. Human alterations of the nitrogen cycle have approximately doubled the rate of nitrogen input to the cycle. One problem is that fertilizer runoff can cause eutrophication of rivers and lakes, and, as the resultant algae die, decomposition by bacteria depletes the oxygen level of the water, resulting in fish die-offs. Excess nitrates in surface or groundwater systems used for drinking water are also a health hazard, particularly for infants. In the body, nitrate is converted to nitrite, which then combines with hemoglobin to form methemoglobin, a type of hemoglobin that does not carry oxygen. In infants, the production of large amounts of nitrites can cause methemoglobinemia, a dangerous condition in which the level of oxygen carried through the body decreases.

Finally, burning fossil fuels releases not only carbon but also nitrogen in the form of nitrogen oxides, which can contribute to air pollution. Nitrogen oxides in turn can react with rain water to form nitric acid (HNO_3), which contributes to acid rain, decreasing the pH of lakes and streams and increasing fish mortality (see Chapter 54). While much of the acid rain problem can be traced to the sulfur cycle, nitrogen oxides are also partially to blame.

The dramatic effects of human activities on nutrient cycles in general and the nitrogen cycle in particular were illustrated by a famous long-term study by ecosystem ecologists Gene Likens, Herbert Bormann, and their colleagues at Hubbard Brook Experimental Forest in New Hampshire in the 1960s. Hubbard Brook is a 3,160-hectare reserve that consists of six catchments along a mountain ridge. A catchment is an area of land where all water eventually drains to a single outlet. In Hubbard Brook, each outlet is fitted with a permanent concrete dam that enables researchers to monitor the outflow of water and nutrients (**Figure 59.23a**). In this large-scale experiment, researchers felled all of the trees in one of the Hubbard Brook catchments (**Figure 59.23b**). The catchment was then sprayed with herbicides for three years to prevent regrowth of vegetation. An untreated catchment was used as a control.

Researchers monitored the concentrations of key nutrients in the streams exiting the two catchments for over three years. Their results revealed that the overall export of dissolved nutrients from the disturbed catchment rose to many times the normal rate (**Figure 59.23c**). The researchers determined that two phenomena were responsible. First, the enormous reduction in plants reduced water uptake by vegetation and led to 40% more precipitation passing through the groundwater to be discharged to the streams. This increased outflow caused greater rates of chemical leaching and rock and soil weathering.

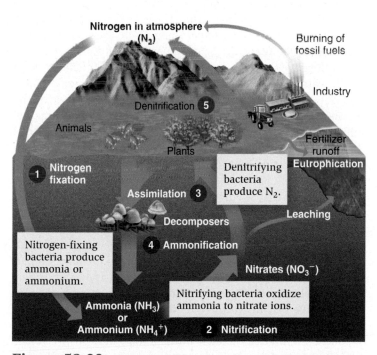

Figure 59.22 **The nitrogen cycle.** The five main parts of the nitrogen cycle are: (1) nitrogen fixation, (2) nitrification, (3) assimilation, (4) ammonification, and (5) denitrification. The recycling of nitrogen from dead plants and animals into the soil and then back into plants is of paramount importance. The width of the arrows indicates the relative importance of each process.

(a) Hubbard Brook dam and weir

(b) Hubbard Brook Experimental Forest, New Hampshire

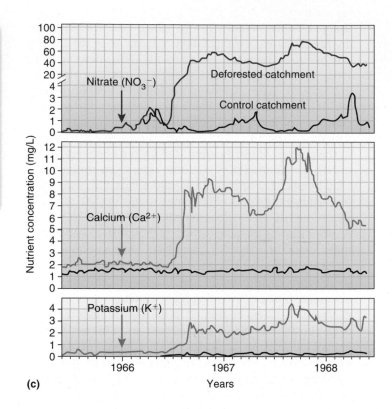

(c)

Figure 59.23 **The effects of deforestation on nutrient concentrations.** (a) Concrete dam and weir used to monitor nutrient flow from a Hubbard Brook catchment. (b) Deforested catchment at Hubbard Brook. (c) Nutrient concentrations in stream water from the experimentally deforested catchment and a control catchment at Hubbard Brook. The timing of deforestation is indicated by arrows.

Second, and more significantly, in the absence of nutrient up-take in spring, when the deciduous trees would have started production, the inorganic nutrients released by decomposer activity were simply leached in the drainage water. Similar processes operate in the majority of terrestrial ecosystems where deforestation is significant.

The Sulfur Cycle Is Heavily Influenced by Human Activities

Most naturally produced sulfur in the atmosphere comes from the gas hydrogen sulfide (H_2S), which is released from volcanic eruptions and during decomposition, especially in wetland environments, where sulfur is very common (**Figure 59.24**). The H_2S quickly oxidizes into sulfur dioxide (SO_2). Because SO_2 is soluble in water, it returns to Earth as weak sulfuric acid (H_2SO_4), or natural acid rain, making the pH of natural rainwater slightly acidic, about 5.6 (see also Chapter 54). The sulfate ions, SO_4^{2-}, thus enter the soil, where sulfate-reducing bacteria may release sulfur as H_2S, or the sulfate may be incorporated by plants into their tissue.

In the presence of iron, sulfur can precipitate as ferrous sulfide, FeS_2, and be incorporated in pyritic rocks. The weathering of rocks and the decomposition of organic matter therefore releases sulfur to solution, which runs through rivers to the sea. Because such rocks commonly overlay coal deposits, mining exposes them to the air and water, resulting in a discharge of sulfuric acid and other sulfur-containing compounds into aquatic

ecosystems. Mining has polluted hundreds of kilometers of streams and rivers in mid-Atlantic states such as West Virginia, Kentucky, and Pennsylvania in this way.

Interestingly, certain marine algae and a few salt marsh plants produce relatively large amounts of the sulfurous gas dimethyl sulfide (CH_3SCH_3). Small particles of dimethyl sulfide that diffuse to the atmosphere often form the nuclei around which water vapor can condense and form the water droplets making up clouds. Because of the sheer global extent of the oceans, changes in algal abundance and thus global dimethyl sulfide levels have the potential to alter cloud cover and thus climate. Because of its ability to cool the climate, some researchers are investigating how dimethyl sulfide production might offset global warming.

Human activity involving the combustion of fossil fuels has altered the sulfur cycle more than any of the other nutrient cycles. The burning of coal and oil to provide energy for heating or to fuel electric power stations produces huge amounts of sulfur dioxide (SO_2). This reacts with rain or snow to make human-produced acid rain (see Chapter 54). One of the main differences between human-produced acid rain and natural acid rain is its relative pH. In North America, for example, natural acid rain has a pH of about 5.6, while measurements of rain falling in southern Ontario, Canada, in the 1980s showed pH values in the range of 4.1–4.5. Huge areas of the industrial northeast U.S. and Europe were affected by acid rain in the 1950s through the 1980s, but a reduction in the use of high-sulfur coal and

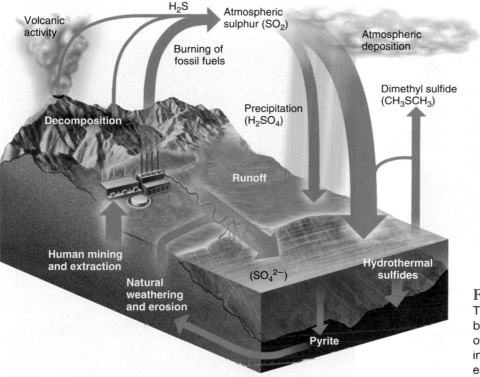

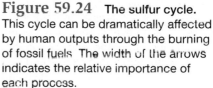

Figure 59.24 **The sulfur cycle.** This cycle can be dramatically affected by human outputs through the burning of fossil fuels. The width of the arrows indicates the relative importance of each process.

the use of scrubbers to prevent sulfur dioxide from passing through smokestacks reduced the problem in more recent times (**Figure 59.25**).

The Water Cycle Is Largely a Physical Process of Evaporation and Precipitation

The water cycle, or hydrological cycle, differs from the cycles of other nutrients in that very little of the water that cycles through ecosystems is chemically changed by any of the cycle's components (**Figure 59.26**). It is a physical process, fueled by the sun's energy, rather than a chemical one, because it consists of essentially two phenomena: evaporation and precipitation. Even so, the water cycle has important biological components. Over land, 90% of the water that reaches the atmosphere is moisture that has passed through plants and exited from the leaves via evapotranspiration. However, only about 2% of the total volume of Earth's water is found in the bodies of organisms or is held frozen or in the soil. The rest cycles between bodies of water, the atmosphere, and the land.

As we noted in Chapter 54, water is limiting to the abundance of many organisms, including humans. It takes 228 L of water to produce a pound of dry wheat, and 9,500 L of water to support the necessary vegetation to produce a pound of meat. Industry is also a heavy user of water, with goods such as oil, iron, and steel requiring up to 20,000 L of water per ton of product. Humans have therefore interrupted the hydrological cycle

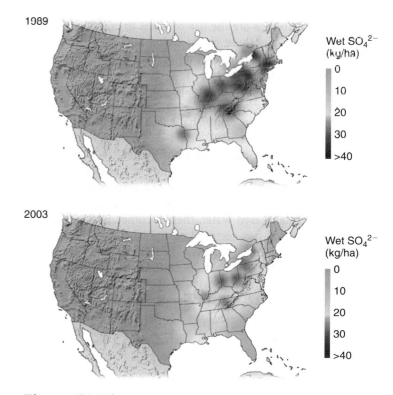

Figure 59.25 **The extent of acid rain in the U.S. has decreased.** Data show wet sulfate deposition due to acid rain in kilograms per hectare.

4. Biomagnification is the result of
   a. an increase in the population size of the primary consumers.
   b. increased levels of pollutants that are released into the environment.
   c. the buildup of chemicals in organisms at a rate greater than the chemicals can be broken down.
   d. genetic mutations that are passed down from one generation to the next.
   e. none of the above.

5. The amount of energy that is fixed during photosynthesis is
   a. net primary production.
   b. biomagnification.
   c. trophic-level transfer efficiency.
   d. gross primary production.
   e. production efficiency.

6. The chain lengths in food webs are limited by
   a. the amount of energy such as sunlight available in an ecosystem.
   b. the efficiency of energy transfers that occur between trophic levels.
   c. the efficiency by which primary consumers assimilate energy into biomass.
   d. all of the above.
   e. a and b only.

7. The evapotranspiration rate
   a. can be used as a predictor for primary production.
   b. is increased when temperature decreases.
   c. is not affected by temperature.
   d. is highest in deserts.
   e. can be predicted by measuring only the water content of the soil.

8. Eutrophication is
   a. caused by an overabundance of nitrogen, which leads to an increase in bacteria populations.
   b. caused by an overabundance of nutrients, which leads to an increase in algal populations.
   c. the normal breakdown of algal plants following a pollution event.
   d. normally seen in dry, hot regions of the world.
   e. none of the above.

9. Primary producers acquire the carbon necessary for photosynthesis from
   a. decomposing plant material.
   b. carbon monoxide released from the burning of fossil fuels.
   c. carbon dioxide in the atmosphere.
   d. carbon sources in the soil.
   e. both a and d.

10. Nitrogen fixation is the process
    a. that converts organic nitrogen to ammonia.
    b. by which plants and animals take up nitrates.
    c. by which bacteria convert nitrate to gaseous nitrogen.
    d. by which atmospheric nitrogen is converted to ammonia or ammonium ions.
    e. all of the above.

CONCEPTUAL QUESTIONS

1. Define autotrophs and heterotrophs.
2. Explain why chain lengths are short in food webs.
3. Describe the carbon cycle.

EXPERIMENTAL QUESTIONS

1. What was the hypothesis of the Stiling and Drake experiment?
2. What was the purpose of increasing the carbon dioxide levels in only half of the chambers in the experiment and not all of the chambers?
3. What were the results of the experiment?

COLLABORATIVE QUESTIONS

1. Outline the main trophic levels within a food chain.
2. Discuss several factors that influence primary productivity in terrestrial and aquatic ecosystems.

www.brookerbiology.com
This website includes answers to the Biological Inquiry questions found in the figure legends and all end-of-chapter questions.

60

CONSERVATION BIOLOGY AND BIODIVERSITY

CHAPTER OUTLINE

60.1 Why Conserve Biodiversity?

60.2 The Causes of Extinction and Loss of Biodiversity

60.3 Conservation Strategies

Giant Panda at Wolong Reserve, southwest China.

Biological diversity, or **biodiversity**, can be examined at three levels: genetic diversity, species diversity, and ecosystem diversity. Each level of biodiversity provides valuable benefits to humanity. Genetic diversity consists of the amount of genetic variation that occurs within and between populations. Maintaining genetic variation in the ancient relatives of crops may be vital to the continued success of crop-breeding programs. For example, in 1977 Rafael Guzman, a Mexican biologist, discovered a previously unknown ancient relative of corn, *Zea diploperennis*, that is resistant to many of the viral diseases that infect domestic corn, *Zea mays*. Genetic engineers believe that this relative has valuable genes that can improve current corn crops. Because corn is the third-largest crop on Earth, the discovery of *Z. diploperennis* may well turn out to be critical to the global food supply.

The second level of biodiversity concerns species diversity, an area on which much public attention is focused. You may be familiar with the U.S. Endangered Species Act (1973), which was designed to protect both endangered and threatened species. **Endangered species** are those species that are in danger of extinction throughout all or a significant portion of their range. **Threatened species** are those likely to become endangered in the future. Many species are currently threatened. According to the International Union for Conservation of Nature and Natural Resources (IUCN), more than 25% of the fish species that live on coral reefs are threatened with extinction, and 23% of all mammals, 12% of birds, and 31% of amphibians are threatened.

The last level of biodiversity is ecosystem diversity, the diversity of structure and function within an ecosystem. While conservation at the level of species diversity has focused attention on species-rich ecosystems such as tropical forests, some scientists have argued that other relatively species-poor ecosystems are highly threatened and similarly need to be conserved. In North America, many of the native prairies have been converted to agricultural use, especially in states such as Illinois and Iowa. In some counties, remnants of prairie often exist only inside cemetery plots, which have been spared from the plow (**Figure 60.1**).

Conservation biology uses principles and knowledge from molecular biology, genetics, and ecology to protect the biological diversity of life at each of these three levels. Because it draws from nearly all chapters of this textbook, a discussion of conservation biology is an apt way to conclude our study of biology. In this chapter, we begin by examining the question of why biodiversity should be conserved, and explore how much diversity is needed for ecosystems to function properly. We then survey the main threats to the world's biodiversity. For many species, there are multiple threats, ranging from habitat loss, hunting, and the effects of introduced species to climate change and pollution. Even if species are not exterminated, many may exist only at very small population sizes. We will see how small populations contribute to special problems such as inbreeding and genetic drift, emphasizing the importance of genetics in conservation biology.

Last, we consider what can be done and has been done to help conserve the world's endangered biota. This includes identifying global areas rich in species and establishing parks and refuges of the appropriate size, number, and connectivity. We also discuss conservation of particularly important types of species, and outline how ecologists have been active in restoring damaged habitats to their natural condition. We then examine how captive breeding programs have been useful in building up populations of rare species prior to their release back into the wild.

Figure 60.1 **Ecosystem biodiversity.** This small cemetery in Bureau County, Illinois, contains the remains of a natural prairie ecosystem. Most of the prairie has been plowed under for agriculture.

Some programs have also used modern genetic techniques such as cloning to help breed and perhaps eventually increase populations of endangered species.

60.1 Why Conserve Biodiversity?

Biologists Paul Ehrlich and E. O. Wilson have suggested that the loss of biodiversity should be of concern to everyone for at least three reasons. First, they proposed that we have an ethical responsibility to protect what are our only known living companions in the universe. Second, humanity has obtained enormous benefits from foods, medicines, and industrial products derived from plants, animals, and microorganisms, and we have the potential to gain many more. The third reason to preserve biodiversity focuses on preserving the array of essential services provided by ecosystems, such as clean air and water. In this section, we examine some of the primary reasons why preserving biodiversity matters and explore the link between biodiversity and ecosystem functioning.

The Preservation of Biological Diversity Can Be Justified Based on the Ecological and Economic Values of Diversity as Well as on Ethical Grounds

During the latter half of the 20th century, the reduction of the Earth's biological diversity emerged as a critical issue, one with implications for public policy. One major concern was that loss of plant and animal resources would impair future development of important products and processes in agriculture, medicine, and industry. For example, as previously noted, *Z. diploperennis*, the ancient relative of corn discovered in Mexico, is resistant to many corn viruses. Its genes are currently being used to develop virus-resistant types of corn. However, *Z. diploperennis* occurs naturally in only a few small areas of Mexico and could easily have been destroyed by development or cultivation of the land. If we allow such species to go extinct, we may unknowingly threaten the food supply on which much of the world depends.

The pharmaceutical industry is heavily dependent on information that is stored in plants. About 25% of the prescription drugs in the U.S. alone are derived from plants, and the 2006 market value of such drugs was estimated to be $19 billion. Many medicines come from plants found only in tropical rain forests, including quinine, a drug from the bark of the Cinchona tree (*Cinchona officinalis*) that is used for malaria, and vincristine, derived from rosy periwinkle (*Catharanthus roseus*), which is a treatment for leukemia and Hodgkin's disease. There are likely many plant chemicals of therapeutic importance in the thousands of rain forest plant species that have not been fully analyzed. The continued destruction of rain forests thus could mean the loss of billions of dollars in potential plant-derived pharmaceutical products.

On a smaller scale, individual species often thought worthless can actually be valuable for research purposes. The blood of the horseshoe crab (*Limulus polyphemus*) clots when exposed to toxins produced by some bacteria. Pharmaceutical industries use the blood enzyme responsible for this clotting to ensure that their products are free of bacterial contamination. Desert pupfishes, found in isolated ponds in the U.S. Southwest, tolerate salinity twice that of seawater and are valuable models for research on human kidney diseases. The technology does not exist to re-create individual species, let alone biomes such as rain forests. Once a species or an ecosystem is gone, it is lost forever.

Beyond this, humans benefit not only from individual species but also from the processes of natural ecosystems (**Table 60.1**). Forests soak up carbon dioxide, maintain soil fertility, and retain water, preventing floods; estuaries provide water filtration and protect rivers and coastal shores from excessive erosion. The loss of biodiversity could disrupt an ecosystem's ability to carry out such functions. Other ecosystem functions include the maintenance of populations of natural predators to regulate pest outbreaks and reservoirs of pollinators to pollinate crops and other plants.

A 1997 paper in the journal *Nature* by economist Robert Constanza and colleagues made an attempt to calculate the monetary value of ecosystems to various economies. They came to the conclusion that, at the time, the world's ecosystems were worth more than $33 trillion a year, nearly twice the gross national product of the world's economies combined ($19 trillion) (**Table 60.2**). If we were to include the value of these services in the cost of goods, most goods would cost a lot more than they currently do. While the open ocean has the greatest total global value of all ecosystems, perhaps a more meaningful statistic is the per hectare value of different ecosystems. This statistic reveals that shallow aquatic ecosystems, such as estuaries and swamps, are extremely valuable because of their role in nutrient cycling, water supply, and disturbance regulation. They also serve as nurseries for aquatic life. These habitats, once thought of as useless wastelands, are among the ecosystems most endangered by pollution and development.

Table 60.1	Examples of the World's Ecosystem Services
Service	**Example**
Atmospheric gas supply	Regulation of carbon dioxide, ozone, and oxygen levels
Climate regulation	Regulation of carbon dioxide, nitrogen dioxide, and methane levels
Water supply	Irrigation, water for industry
Pollination	Pollination of crops
Biological control	Pest population regulation
Wilderness and refuges	Habitat for wildlife
Food production	Crops, livestock
Raw materials	Fossil fuels, timber
Genetic resources	Medicines, genes for plant resistance
Recreation	Ecotourism
Cultural	Aesthetic and educational value
Disturbance regulation	Storm protection, flood control
Waste treatment	Sewage purification
Soil erosion control	Retention of topsoil, reduction of accumulation of sediments in lakes
Nutrient recycling	Nitrogen, phosphorus, carbon, and sulfur cycles

Table 60.2	Valuation of the World's Ecosystem Services		
Biome	**Total global value* ($ trillion)**	**Total value (per ha) ($)**	**Main ecosystem service**
Open ocean	8,381	252	Nutrient cycling
Coastal shelf	4,283	1,610	Nutrient cycling
Estuaries	4,100	22,832	Nutrient cycling
Tropical forest	3,813	2,007	Nutrient cycling/raw materials
Seagrass and algal beds	3,801	19,004	Nutrient cycling
Swamps	3,231	19,580	Water supply/disturbance regulation
Lakes and rivers	1,700	8,498	Water regulation
Tidal marsh	1,648	9,990	Waste treatment/ disturbance regulation
Grasslands	906	232	Waste treatment/food production
Temperate forest	894	302	Climate regulation/waste treatment/lumber
Coral reefs	375	6,075	Recreational/disturbance regulation
Cropland	128	92	Food production
Desert	0	0	
Ice and rock	0	0	
Tundra	0	0	
Urban	0	0	
Total	33,260		

*In 1997 values.

Arguments can be made against the loss of biodiversity on ethical grounds. As only one of many species, it has been argued that humans have no right to destroy other species and the environment around us. Philosopher Tom Regan suggests that animals should be treated with respect because they have a life of their own and therefore have value apart from anyone else's interests. Law professor Christopher Stone, in an influential article titled "Should Trees Have Standing?" has argued that entities such as nonhuman natural objects like trees or lakes should be given legal rights just as corporations are treated as persons for certain purposes. As E. O. Wilson proposed in a 1984 concept known as biophilia, humans have deep attachment with natural habitats and species because of our close association for over millions of years.

How Much Diversity Is Needed for Ecosystems to Function Properly?

Because biodiversity has an impact on the health of ecosystems, ecologists have explored the question of how much diversity is needed for ecosystems to function properly. Recall that in the 1950s, ecologist Charles Elton had proposed in the diversity-stability hypothesis that the more species present, the more stable the community (see Chapter 58). If we use stability as a surrogate for ecosystem function, Elton's hypothesis suggests a linear correlation between diversity and ecosystem function (**Figure 60.2a**).

In 1981, ecologists Paul and Anne Ehrlich proposed an alternative called the **rivet hypothesis** (**Figure 60.2b**). In this model, species are like the rivets on an airplane, with each species playing a small but critical role in keeping the plane (the ecosystem) airborne. The loss of a rivet weakens the plane and causes it to lose a little airworthiness. The loss of a few rivets could probably be tolerated, while the loss of more rivets would prove critical to the airplane's function.

A decade later, Australian ecologist Brian Walker proposed an extension of this idea, termed the **redundancy hypothesis** (or passenger hypothesis) (**Figure 60.2c**). According to this hypothesis, most species are more like passengers on a plane—they take up space but do not add to the airworthiness. The species are said to be redundant because they could simply be eliminated or replaced by others with no loss in function. Airworthiness is primarily affected by the activity of a few crucial species, in this case, the pilot or copilot, that are called keystone species (a concept we will discuss in detail later in this chapter). In this scenario, the loss of each species does not affect airworthiness unless the species is of critical importance.

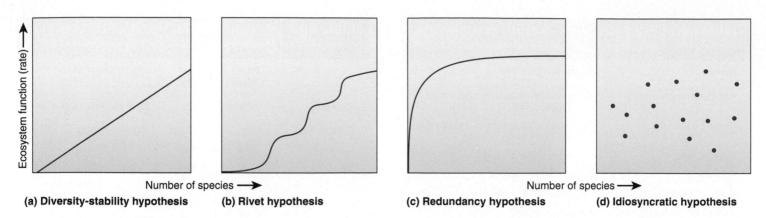

(a) Diversity-stability hypothesis   (b) Rivet hypothesis   (c) Redundancy hypothesis   (d) Idiosyncratic hypothesis

Figure 60.2   Four main models that describe the relationship between ecosystem function and biodiversity.   The relationship is strongest in **(a)** and weakest in **(d)**.

In another alternative, British ecologist John Lawton included the possibility that ecosystem function changes as the number of species increases or decreases, but that the direction of change is not predictable. He called this the idiosyncratic hypothesis (**Figure 60.2d**). Determining which model is most correct is very important, as our understanding of the effect of species loss on community function can greatly affect the way we manage our environment. As we will discuss next, experimental studies have provided data showing that reduced biodiversity does lead to reduced ecosystem functioning.

FEATURE INVESTIGATION

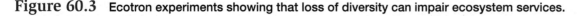

Ecotron Experiments Showed the Relationship Between Biodiversity and Ecosystem Function

In the early 1990s, Shahid Naeem and colleagues used a series of 14 environmental chambers in a facility termed the Ecotron, at Silwood Park, England, to determine how biodiversity affects ecosystem functioning. These chambers contained terrestrial communities that differed only in their level of biodiversity (**Figure 60.3**). The number of species in each chamber was manipulated to create high-, medium-, and low-diversity ecosystems, each with four trophic levels. The trophic levels consisted of primary producers (annual plants), primary consumers (insects, snails, and slugs), secondary consumers (parasitoids that fed on the herbivores), and decomposers (earthworms and

Figure 60.3   Ecotron experiments showing that loss of diversity can impair ecosystem services.

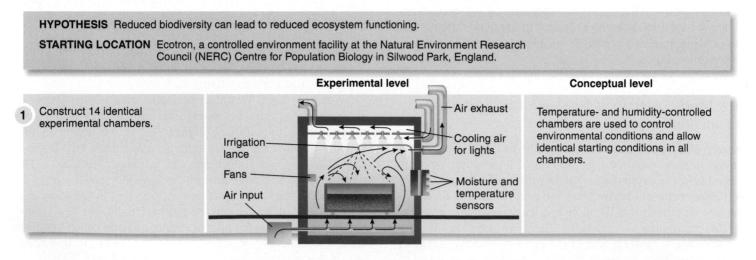

HYPOTHESIS   Reduced biodiversity can lead to reduced ecosystem functioning.

STARTING LOCATION   Ecotron, a controlled environment facility at the Natural Environment Research Council (NERC) Centre for Population Biology in Silwood Park, England.

Experimental level   Conceptual level

1   Construct 14 identical experimental chambers.

Air exhaust

Irrigation lance

Cooling air for lights

Fans

Air input

Moisture and temperature sensors

Temperature- and humidity-controlled chambers are used to control environmental conditions and allow identical starting conditions in all chambers.

2 Add different combinations of species to the 14 chambers. The species added were based on 3 types of model communities (food webs), each with 4 trophic levels but with varying degrees of species richness.

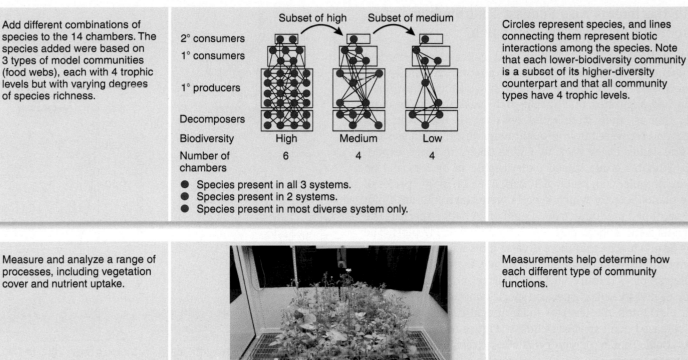

2° consumers
1° consumers
1° producers
Decomposers

Subset of high   Subset of medium

Biodiversity	High	Medium	Low
Number of chambers	6	4	4

● Species present in all 3 systems.
● Species present in 2 systems.
● Species present in most diverse system only.

Circles represent species, and lines connecting them represent biotic interactions among the species. Note that each lower-biodiversity community is a subset of its higher-diversity counterpart and that all community types have 4 trophic levels.

3 Measure and analyze a range of processes, including vegetation cover and nutrient uptake.

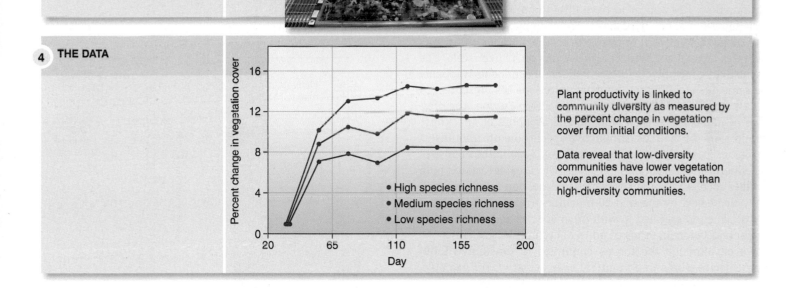

Measurements help determine how each different type of community functions.

4 THE DATA

● High species richness
● Medium species richness
● Low species richness

Plant productivity is linked to community diversity as measured by the percent change in vegetation cover from initial conditions.

Data reveal that low-diversity communities have lower vegetation cover and are less productive than high-diversity communities.

soil insects). The experiment ran for just over six months, and species were added only after the trophic level below them was established. For example, parasitoids were not added until herbivores were abundant.

Researchers monitored and analyzed a range of measures of ecosystem function, including community respiration, decomposition, nutrient retention rates, and community productivity. The result was that community productivity, expressed as percent change in vegetation cover (the amount of ground covered by leaves of plants), increased as species richness increased. This occurred because of a greater variety of plant growth forms that could utilize light at different levels of the plant canopy. A larger ground cover also meant a larger plant biomass and greater community productivity, and increased decomposition and nutrient uptake rates. For the first time, ecologists had provided an experimental demonstration that the loss of biodiversity can alter or impair the functioning of an ecosystem.

Field Experiments Also Suggest That Biodiversity Is Important for Ecosystem Function

In the mid-1990s, David Tilman and colleagues performed experiments in the field to determine how much biodiversity was necessary for proper ecosystem functioning. Previously, Tilman had suggested that species-rich grasslands were more stable; that is, they were more resistant to the ravages of drought and recovered from drought more quickly than species-poor grasslands (refer back to Figure 58.7). In these experiments, Tilman's group sowed plots, each 3 m by 3 m and on comparable soils, with seeds of 1, 2, 4, 6, 8, 12, or 24 species of prairie plants. Exactly which species were sown into each plot was determined randomly from a pool of 24 native species. The treatments were replicated 21 times, for a total of 147 plots. The results again showed that more diverse plots had increased productivity and used nutrients, such as nitrate, more efficiently than less diverse plots (**Figure 60.4a,b**). Furthermore, both the frequency of invasive plant species (species not originally planted in the plots) and the level of foliar fungal disease decreased with increased plant species richness (**Figure 60.4c,d**).

Although Tilman's experiments show a relationship between diversity and ecosystem function, they also suggest that a point may be reached at which function is maximized, beyond which additional species appear to have little to no impact. This supports the redundancy hypothesis. For example, uptake of nitrogen remains relatively unchanged as the number of species increases beyond six. We can also see this on a larger scale. The productivity of temperate forests in different continents is roughly the same despite different numbers of tree species present—729 in East Asia, 253 in North America, and 124 in Europe. The presence of more tree species may ensure a supply of "backups" should some of the most productive species die off from insect attack or disease. This can happen, as was seen in the demise of the American chestnut and elm trees. Diseases devastated both of these species, and their presence in American forests dramatically decreased by the mid-20th century (refer back to Figure 57.18). The forests filled in with other species and continued to function as before in terms of nutrient cycling and gas exchange. However, although the forests continued to function without these species, some important changes occurred. For example, the loss of chestnuts deprived bears and other animals of an important source of food and may have affected their reproductive capacity and hence the size of their populations.

60.2 The Causes of Extinction and Loss of Biodiversity

In light of research showing that the loss of species influences ecosystem function, the importance of understanding and preventing species loss takes on particular urgency. Throughout the history of life on Earth, **extinction**—the process by which species die out—has been a natural phenomenon. The average timespan of a typical animal or plant species in the fossil record is about 4 million years. To calculate the current extinction rate,

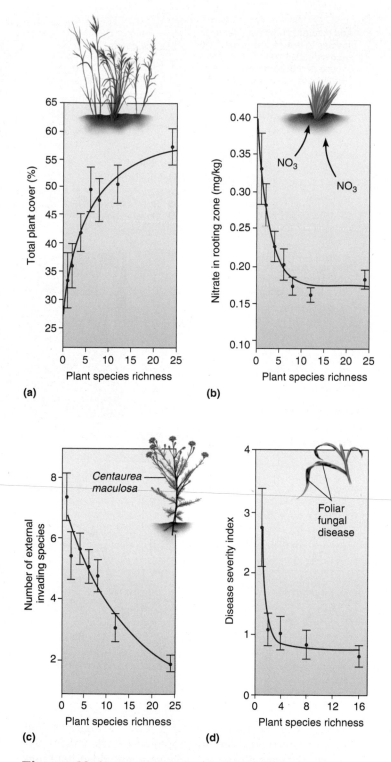

Figure 60.4   The functions of species richness.   The relationships between species richness and **(a)** percent plant cover, the area of ground covered by leaves, and **(b)** uptake of nitrogen in experimental plots in Tilman's biodiversity experiments. With increasing species richness, biomass increases and nitrogen is used up, leaving less nitrogen in the soil with more species of plants. **(c)** A decrease in susceptibility to invasion of temperate grasslands by exotic species with an increase in plant species richness. **(d)** A decrease in disease severity index for plant species in temperate grasslands with an increase in plant species richness.

we could take the total number of species estimated to be alive on Earth at present, around 10 million, and divide it by 4 million, giving an average extinction rate of 2.5 species each year. For the 5,500 species of living mammals, using the same average life span of around 4 million years, we would expect about one species to go extinct every 1,000 years; this is termed the background extinction rate. However, it can be argued that the fossil record is heavily biased toward successful, often geographically wide-ranging species, which undoubtedly have a longer than average persistence time. The fossil record is also biased toward vertebrates and marine mollusks, both of which fossilize well because of their hard body parts. If background extinction rates were 10 times higher than the rates perceived from the fossil record, then extinctions among the living mammals today would be expected to occur at a rate of about one every 100 years. For birds, the background extinction rate would be two species every 100 years.

It is indisputable, however, that the extinction rate for species in recent times has been far higher than this. In the past 100 years, around 20 species of mammals and over 40 species of birds went extinct (**Figure 60.5a**). The rates of species extinctions on islands in the recent past confirm the dramatic effects of human activity. The Polynesians, who colonized Hawaii in the 4th and 5th centuries, appear to have been responsible for the extinction of half of the 100 or so species of endemic land birds in the period between their arrival and that of the Europeans in the late 18th century. A similar impact was felt in New Zealand, which was colonized by European settlers some

500 years later than Hawaii. In New Zealand, an entire avian megafauna, consisting of huge land birds, was exterminated over the course of the century, probably through a combination of hunting and large-scale habitat destruction through burning. The term **biodiversity crisis** is often used to describe this elevated loss of species. Many scientists believe that the rate of loss is higher now than during most of geological history, and most suggest that the growth in the human population has led to the increase in the number of extinctions of other species (**Figure 60.5b**).

To understand the process of extinction in more modern times, it is essential for ecologists to examine the role of human activities and their environmental consequences. In this section, we examine why species have gone extinct in the past and look at the factors that are currently threatening species with extinction.

The Main Causes of Extinction Are Introduced Species, Direct Exploitation, and Habitat Destruction

While all causes of extinctions are not known, introduced species, direct exploitation, and habitat destruction have been major human-induced threats. Increasingly, climate change is viewed as a significant threat to species. As mentioned at the beginning of Chapter 54, human-induced climate change, or global warming, has been implicated in the dramatic decrease in the population sizes of frog species in Central and South America.

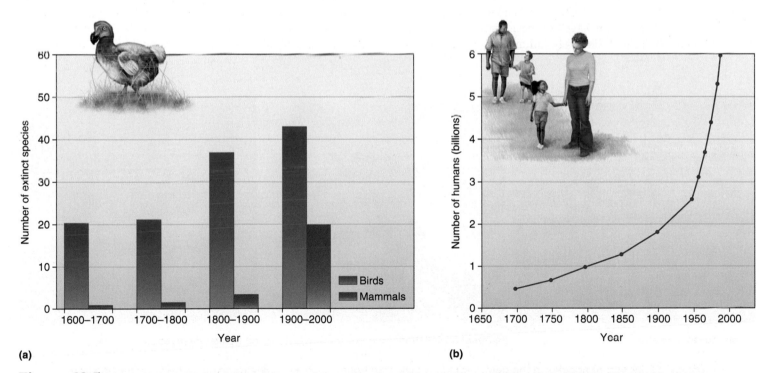

(a)   (b)

Figure 60.5 **Animal extinctions and human population growth.** (a) Increasing numbers of known extinctions in birds and mammals are concurrent with (b) exponential increases in the global human population. These figures suggest that as human numbers increase, more and more species will go extinct.

Biological inquiry: Why might the increasing human population result in an increase in the extinction rate of other species?

Indeed, a recent study of six biodiversity-rich regions employed computer models to simulate the movement of species' ranges in response to changing climate conditions. The models predicted that unless greenhouse gas emissions are cut drastically, climate change will cause 15–37% of the species in those regions to go extinct by the year 2050.

Introduced Species   **Introduced species**, often called exotic species, are those species moved by humans from a native location to another location. Most often the species are introduced for agricultural purposes or as sources of timber, meat, or wool, and these species need humans for their continued survival. Others, such as plants, insects, or marine organisms, are unintentionally transported via the movement of cargo by ships or planes. Regardless of their method of introduction, some introduced species become **invasive**, spreading and outcompeting native species for space and resources.

We can categorize the interactions between introduced and native species into competition, predation, and disease. Competition may eliminate local populations and cause huge reductions in the densities of native species, but it has not yet been clearly shown to exterminate entire species. On the other hand, there have been many recorded cases of extinction from predation. Introduced predators such as rats, cats, and mongooses have accounted for at least 43% of recorded extinctions of birds on islands. Lighthouse keepers' cats have annihilated populations of ground-nesting birds on small islands around the world. The brown tree snake, which was accidentally introduced onto the island of Guam, has decimated the country's native bird populations (refer back to Figure 57.12). Parasitism and disease carried by introduced organisms have also been important in causing extinctions. Avian malaria in Hawaii, spread by introduced mosquito species, is believed to have contributed to the demise of up to 50% of native Hawaiian birds (**Figure 60.6a**).

Direct Exploitation   Direct exploitation, particularly the hunting of animals, has been the cause of many extinctions in the past. Two remarkable North American bird species, the passenger pigeon and the Carolina parakeet, were hunted to extinction by the early 20th century. The passenger pigeon (*Ectopistes migratorius*) was once the most common bird in North America, probably accounting for over 40% of the entire bird population (**Figure 60.6b**). Flock sizes were estimated to be over 1 billion birds. It may seem improbable that the most common bird on the continent could be hunted to extinction for its meat, but that is just what happened. The flocking behavior of the birds made them relatively easy targets for hunters, who used special firearms to harvest the birds in quantity. In 1876 in Michigan alone, over 1.6 million birds were killed and sent to markets in the eastern U.S. The Carolina parakeet (*Conuropsis carolinensis*), the only species of parrot native to the eastern U.S., was similarly hunted to extinction by the early 1900s.

Many whale species were driven to the brink of extinction prior to a moratorium on commercial whaling issued in 1988 (refer back to Figure 57.13). Steller's sea cow (*Hydrodamalis gigas*), a 9-meter-long manatee-like mammal, was hunted to extinction in the Bering Straits only 27 years after its discovery in 1740. A poignant example of human excess in hunting was the dodo (*Raphus cucullatus*), a flightless bird native only to the island of Mauritius that had no known predators. A combination of over-exploitation and introduced species led to its extinction within 200 years of the arrival of humans. Sailors hunted it for its meat, and the rats, pigs, and monkeys brought to the island by humans destroyed the dodo's eggs and chicks in their ground nests.

(a) Introduced parasites   **(b) Direct exploitation**   **(c) Habitat destruction**

Figure 60.6   Extinction of species in the past.   **(a)** Many Hawaiian honeycreepers were exterminated by avian malaria from introduced mosquito species. This 'i'iwi (*Vestiaria coccinea*) is one of the few remaining honeycreeper species. **(b)** The passenger pigeon, which may have once been among the most abundant bird species on Earth, was hunted to extinction for its meat. **(c)** The ivory-billed woodpecker, the third-largest woodpecker in the world, was long thought to be extinct in the southeastern U.S. because of habitat destruction, but a possible sighting occurred in 2004. This nestling was photographed in Louisiana in 1938.

Habitat Destruction Habitat destruction through **deforestation**, the conversion of forested areas to nonforested land, has historically been a prime cause of the extinction of species. About one-third of the world's land surface is covered with forests, and much of this area is at risk of deforestation. While tropical forests are probably the most threatened forest type, with rates of deforestation in Africa, South America, and Asia varying between 0.6% and 0.9% per year, the destruction of forests is a global phenomenon.

Among North American terrestrial wildlife, about half of birds (272 species) and more than 10% of mammals (49 species) have an obligatory relationship with forest cover, meaning that they depend on trees for food and nesting sites. In terms of wildlife use, the oaks, whose acorns occur in the diet of at least 100 species of birds and mammals, are among the most valuable trees in North America. For many species of wildlife, the annual acorn crop is a major determinant of their abundance. Most woodpeckers nest in holes that they excavate in trees, and their food usually consists of insects collected on or in trees. The ivory-billed woodpecker (*Campephilus principalis*), the largest in North America and an inhabitant of wetlands and forests of the southeastern U.S., was widely assumed to have gone extinct in the 1950s due to destruction of its habitat by heavy logging (**Figure 60.6c**). Incredibly, in 2004, the woodpecker was sighted in the Big Woods area of eastern Arkansas, though this has not yet been confirmed.

Deforestation is not the only form of habitat destruction. The scouring of land to plant agricultural crops can create soil erosion, increased flooding, declining soil fertility, silting of the rivers, and desertification. While the average area of land under cultivation worldwide averages about 11%, with an additional 24% given over to rangeland, this amount varies tremendously between regions. For example, Europe uses 28% of its land for crops and pasturelands, with the result that many of its native species went extinct long ago. Wetlands also have been drained for agricultural purposes. Others have been filled in for urban or industrial development. In the U.S., as much as 90% of the freshwater marshes and 50% of the estuarine marshes have disappeared. Urbanization, the development of cities on previously natural or agricultural areas, is the most human-dominated and fastest-growing type of land use worldwide and devastates the land more severely than practically any other form of habitat degradation.

Small Populations Are Threatened by the Loss of Genetic Diversity

Knowing why species have gone extinct helps us to recognize the threats facing living species, which in turn may make it easier to protect endangered and threatened species in the future. As we have discussed, most of the factors currently threatening species—introduced species, direct exploitation, and habitat destruction—are derived from human activities. However, even if habitats are not destroyed, many become fragmented, leading to the development of small, isolated populations that can become less viable over time. The threats to species from small population size relate to a reduction of genetic diversity that can result from three factors: inbreeding, genetic drift, and limited mating, which reduces effective population size.

Inbreeding **Inbreeding**, which is mating among genetically related relatives, is more likely to take place in nature when population size becomes very small and there are a limited number of mates to choose from (see Chapter 24). In many species, survivorship of offspring declines as populations become more inbred. This phenomenon was shown as long ago as the 19th century. Litter size in inbred rats declined from 7.5 in 1887 to 3.2 in 1892. Furthermore, over the same time span, the level of nonproductive matings, those where no offspring were born, rose from 2% to 50%. Generally, the more inbred the population, the more severe these types of problems become.

One of the most striking examples of the effects of inbreeding in conservation biology involves the greater prairie chicken (*Tympanuchus cupido*). The male birds have a spectacular mating display that involves inflating the bright orange air sacs on their throat, stomping their feet, and spreading their tail feathers. The prairies of the Midwest were once home to millions of these birds, but as the prairies were converted to farmland, the range and population sizes of the bird shrank dramatically. The population of prairie chickens in Illinois decreased from 25,000 in 1933 to less than 50 in 1989. At that point, according to studies by Ronald Westemeier and colleagues, only 10 to 12 males existed. Because of the decreasing numbers of males, inbreeding in the population had increased. This was reflected in the steady reduction in the hatching success of eggs (**Figure 60.7**). The prairie chicken population had entered a downward spiral toward extinction from which it could not naturally recover, a phenomenon called an **extinction vortex**. In the early 1990s, conservation biologists began trapping prairie chickens in Kansas and Nebraska, where populations remained larger and more genetically diverse, and moved them to Illinois, bringing an infusion of new genetic material into the population. This translocation resulted in a rebounding of the egg-hatching success rate to 90% by 1993.

Genetic Drift In small populations, there is a good chance that some individuals will fail to mate successfully purely by chance, for example, because of the failure to find a mate. This is known as the **Allee effect**, after ecologist W. C. Allee, who first described it. If an individual that fails to mate possesses a rare gene, that genetic information will not be passed on to the next generation, resulting in a loss of genetic diversity from the population. As noted in Chapter 24, **genetic drift** refers to the random change in allele frequencies attributable to chance. Because the likelihood of an allele being represented in just one or a few individuals is higher in small populations than in large populations, small, isolated populations are particularly vulnerable to this type of reduction in genetic diversity. Such isolated populations will lose a percentage of their original diversity over time, approximately at the rate of $1/(2N)$ per generation, where N = population size. As described next, this has a greater effect in smaller versus larger populations.

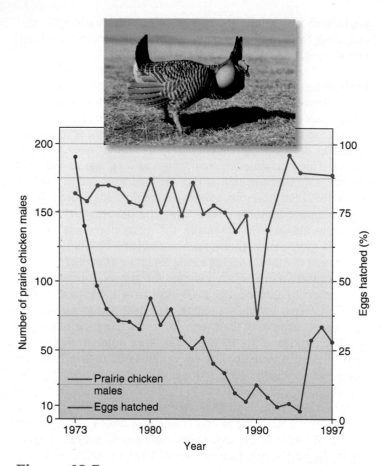

Figure 60.7 **Changes in the abundance and egg-hatching success rate of prairie chickens.** As the number of males decreased, inbreeding increased, resulting in a decrease in fertility, as indicated by a reduced egg-hatching rate. An influx of males in the early 1990s increased the egg-hatching success rate dramatically.

If $N = 500$, then $\frac{1}{2N} = 1/1{,}000 = 0.001$, or 0.1% genetic diversity lost per generation

If $N = 50$, then $\frac{1}{2N} = 1/100 = 0.01$, or 1% genetic diversity lost per generation

Due to genetic drift, a population of 500 will lose only 0.1% of its genetic diversity in a generation, while a population of 50 will lose 1%. Such losses become magnified over many generations. After 20 generations, the population of 500 will lose 2% of its original genetic variation, but the population of 50 will lose about 20%! For organisms that breed annually, this would mean a substantial loss in genetic variation over 20 years. Once again, this effect becomes more severe as the population size decreases.

As with inbreeding, the effects of genetic drift can be countered by immigration of individuals into a population. Even relatively low immigration rates of about one immigrant per generation (or one individual moved from one population to another) can be sufficient to counter genetic drift in a population of 100 individuals.

Limited Mating In many populations, the **effective population size**, the number of individuals that contribute genes to future populations, may be smaller than the number of individuals in the population, particularly in animals with a harem mating structure in which only a few dominant males breed. For example, dominant elephant seal bulls control harems of females, and a few males command all the matings (refer back to Figure 55.22). If a population consists of breeding males and breeding females, the effective population size is given by:

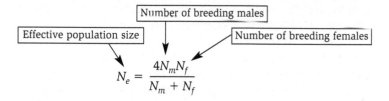

$$N_e = \frac{4N_mN_f}{N_m + N_f}$$

In a population of 500, a 50:50 sex ratio, and all individuals breeding, $N_e = (4 \times 250 \times 250)/(250 + 250) = 500$, or 100% of the actual population size. However, if 250 females breed with 10 males, $N_e = (4 \times 10 \times 250)/(10 + 250) = 38.5$, or 8% of the actual population size.

Knowledge of effective population size is vital to ensuring the success of conservation projects. One notable project in the U.S. involved planning the sizes of reserves designed to protect grizzly bear populations in the contiguous 48 states. The grizzly bear (*Ursus arctos*) has declined in numbers from an estimated 100,000 in 1800 to less than 1,000 at present. The range of the species is now less than 1% of its historical range and is restricted to six separate populations in four states (**Figure 60.8**). Research by biologist Fred Allendorf has indicated that the effective population size of grizzly populations is generally only about 25% of the actual population size because not all bears breed. Thus, even fairly large, isolated populations, such as the 200 bears in Yellowstone National Park, are vulnerable to the harmful effects of loss of genetic variation because the effective population size may be as small as 50 individuals. Allendorf and his colleagues proposed that an exchange of grizzly bears between populations or zoo collections would help tremendously in promoting genetic variation. Even an exchange of two bears per generation between populations would greatly reduce the loss of genetic variation.

60.3 Conservation Strategies

In their efforts to maintain the diversity of life on Earth, conservation biologists are currently active on many fronts and employ many strategies. We begin this section by discussing how conservation biologists identify the global habitats richest in species. We next explore the concept of nature reserves and consider questions such as how large conservation areas should be and how far apart they should be situated. These questions are

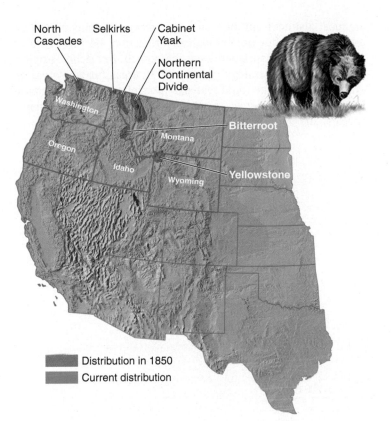

North Cascades · Selkirks · Cabinet Yaak · Northern Continental Divide

Washington · Oregon · Idaho · Montana · Wyoming · Bitterroot · Yellowstone

■ Distribution in 1850
■ Current distribution

Figure 60.8 Past and current ranges of the grizzly bear. The range of the grizzly bear is currently less than 1% of its historical range. The current range in the continental U.S. has contracted to just six populations in four states, as the population size has shrunk from 100,000 before the West was settled to about 1,000 today.

Biological inquiry: If only 500 male and 500 female grizzlies exist today, but only 25% of the males breed, what is the effective population size?

within the realm of **landscape ecology**, which studies the spatial arrangement of communities and ecosystems in a geographic area. Next we discuss how conservation efforts often focus on certain species that can have a disproportionate impact on their ecosystem. We will also examine the field of restoration ecology, focusing on how wildlife habitats can be established from degraded areas and how captive breeding programs have been used to re-establish populations of threatened species in the wild. We conclude by returning to the theme of genomes and proteomes to show how modern molecular techniques of cloning can be useful in the fight to save critically endangered species.

Habitat Conservation Focuses on Identifying Countries Rich in Species, Areas Rich in Endemics, or Representative Habitats

Conservation biologists often must make decisions regarding which species and habitats should be protected. Many conservation efforts have focused on saving habitats in so-called megadiversity countries, because they often have the greatest number of species, but more recent strategies have promoted preservation of certain key areas with the highest levels of unique species, or preservation of representative areas of all types of habitat, even relatively species-poor areas in temperate grasslands (prairies).

Megadiversity Countries One method of targeting areas for conservation is to identify those countries with the greatest numbers of species, the **megadiversity countries**. Using the number of plants, vertebrates, and selected groups of insects as criteria, biologist Russell Mittermeier and colleagues have determined that just 17 countries are home to nearly 70% of all known species. Brazil, Indonesia, and Colombia top the list, followed by Australia, Peru, Mexico, Madagascar, China, and nine other countries. The megadiversity country approach suggests that conservation efforts should be focused on the most biologically rich countries.

As we learned in Chapter 58, large areas generally have the greatest number of species, and proponents of the megadiversity country concept believe it would protect the greatest number of species. Large countries, such as Brazil, would garner most of the available international funds. Perhaps the greatest drawback of the megadiversity approach, however, is that although megadiversity areas may contain the most species, they do not necessarily contain the most unique species. The mammal species list for Peru is 344 and for Ecuador, it is 271; of these, however, 208 species are common to both.

Areas Rich in Endemic Species Another method of setting conservation priorities, one adopted by the organization Conservation International, takes into account the number of species that are **endemic**, or found only in a particular place or region and nowhere else. This approach suggests that conservationists focus their efforts on geographic **hot spots**. To qualify as a hot spot, a region must meet two criteria: It must contain at least 1,500 species of vascular plants as endemics, and have lost at least 70% of its original habitat. Vascular plants were chosen as the primary group of organisms to determine whether or not an area qualifies as a hot spot mainly because most other terrestrial organisms depend on them to some extent.

Conservationists Norman Myers, Russell Mittermeier, and colleagues identified 34 hot spots that together occupy a mere 2.3% of the Earth's surface but contain 150,000 endemic plant species, or 50% of the world's total (**Figure 60.9**). Of these areas, the Tropical Andes and Sundaland (the region including Malaysia, Indonesia, and surrounding islands) have the most endemic plant species (**Table 60.3**). This approach posits that protecting geographic hot spots will prevent the extinction of a larger number of endemic species than would protecting areas of a similar size elsewhere. The main argument against using hot spots as the criterion for targeting conservation efforts is that the areas richest in endemics—tropical rain forests—would receive the majority of attention and funding, perhaps at the expense of other areas.

Representative Habitats In a third approach to prioritizing areas for conservation, scientists have recently argued that we need to conserve representatives of all major habitats. Thus, while the Pampas region of South America, which is arguably the most threatened habitat on the continent because of conversion of its grasslands to agriculture, does not compare well in richness or endemics to the rain forests, it is a unique area that without preservation could disappear. By selecting habitats that are most distinct from those already preserved, many areas that are threatened but not biologically rich may be preserved in addition to the less immediately threatened, but richer, tropical forests. The best strategy of identifying areas for conservation efforts might be one that creates a "portfolio" of areas containing those with high species richness, large numbers of endemic species, and various habitat types.

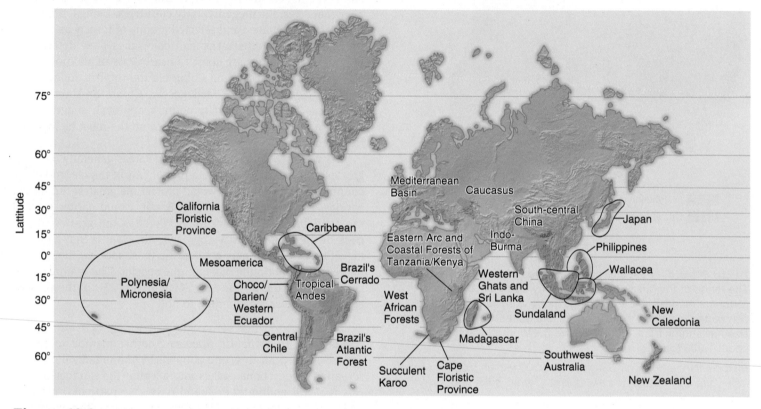

Figure 60.9 Location of major biodiversity hot spots around the world. Hot spots, shown in orange, have high numbers of endemic species.

Table 60.3	Numbers of Endemic Species Present in the Top 10 Hot Spots of the World, Ranked by the Numbers of Endemic Plants					
Rank	Hot spot	Plants	Birds	Mammals	Reptiles	Amphibians	Freshwater fish
1	Tropical Andes	15,000	584	75	275	664	131
2	Sundaland	15,000	146	173	244	172	350
3	Mediterranean Basin	11,700	32	25	77	27	63
4	Madagascar	11,600	183	144	367	226	97
5	Brazil's Atlantic Forest	8,000	148	71	94	286	133
6	Indo-Burma	7,000	73	73	204	139	553
7	Caribbean	6,550	167	41	468	164	65
8	Cape Floristic Province	6,210	6	4	22	16	14
9	Philippines	6,091	185	102	160	74	67
10	Brazil's Cerrado	4,400	16	14	33	26	200

The Theory and Practice of Reserve Design Incorporate Principles of Island Biogeography and Landscape Ecology

After having identified areas to preserve, conservationists must determine the size, arrangement, and management of the protected land. Among the questions conservationists ask is this: Is one large reserve preferable to an equivalent area composed of smaller reserves? Since reserves can be viewed as islands of habitat, this question theoretically can be answered with the help of the theory of island biogeography. Other questions concern whether parks should be close together or far apart, and whether they should be connected by strips of suitable habitat to allow the movement of plants and animals between them. These large-scale questions concerning the spatial arrangement of communities and ecosystems in a geographic area fall within the field of landscape ecology. Conservationists also need to consider the fact that park design is often contingent on economic factors. Let's examine some of the many issues that conservationists address in the creation and management of protected land.

The Role of Island Biogeography In our exploration of the equilibrium model of island biogeography (see Chapter 58), we noted that it could be applied not only to a body of land surrounded by water but also to isolated fragments of habitat. Seen this way, wildlife reserves and sanctuaries are in essence islands in a sea of human-altered land. One question for conservationists is how large a protected area should be (Figure 60.10a). According to island biogeography, the number of species should increase with increasing area (the species-area effect); thus, the larger the area, the greater the number of species would be protected. In addition, larger parks have other benefits. For example, they are beneficial for organisms that require large spaces, including migrating species and species with extensive territories, such as lions and tigers.

Another question is whether it is preferable to protect one single, large reserve or several smaller ones (Figure 60.10b). This is called the **SLOSS debate** (for single large or several small). Proponents of the single, large reserve claim that a larger reserve is better able to preserve more and larger populations than an equal area divided into small areas. According to island biogeography, a larger block of habitat should support more species than any of the smaller blocks.

However, many empirical studies suggest that multiple small sites of equivalent area will contain more species, because a series of small sites is more likely to contain a broader variety of habitats than one large site. Looking at a variety of sites, researchers Jim Quinn and Susan Harrison concluded that animal life was richer in collections of small parks than in fewer, larger parks. In their study, having more habitat types outweighed the effect of area on biodiversity. In addition, another benefit of a series of smaller parks is a reduction of extinction risk by a single event such as a wildfire or the spread of disease.

Landscape Ecology **Landscape ecology** is a subdiscipline of ecology that examines the spatial arrangement of elements in communities and ecosystems. In the design of nature reserves, one question that needs to be addressed is how close to situate reserves to each other, such as whether to have three or four small reserves close to each other or farther apart. A similar question is whether to have a linear or cluster arrangement of small reserves. Island biogeography suggests that if an area must be fragmented, the sites should be as close as possible to permit dispersal (Figure 60.10c,d). In practice, however, having small sites far apart may preserve more species than having them close together, since once again, distant sites are likely to incorporate slightly different habitats and thus species.

Landscape ecologists have also suggested that small reserves should be linked together by biotic corridors, or **movement corridors**, which are thin strips of land that may permit the movement of species between patches (Figure 60.10e).

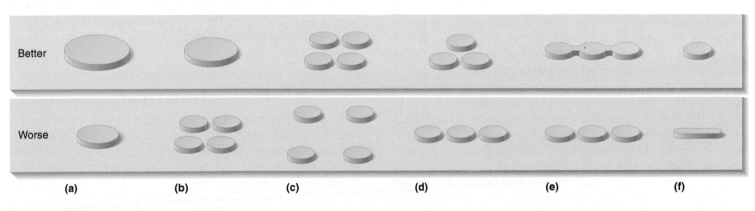

Figure 60.10 The theoretical design of nature reserves, based on the tenets of island biogeography. (a) A larger reserve will hold more species and have low extinction rates. (b) Given a certain area available, it should be fragmented into as few pieces as possible. (c) If an area must be fragmented, the pieces should be as close as possible to permit dispersal. (d) To enhance dispersal, a cluster of fragments is preferable to a linear arrangement. (e) Maintaining or creating corridors between fragments may also enhance dispersal. (f) Circular-shaped areas will minimize the amount of undesirable edge habitat.

Figure 60.11 Habitat corridors.

Biological inquiry: Why would these European hedgerows act as habitat corridors?

Such corridors may facilitate movements of organisms that are vulnerable to predation outside of their natural habitat or that have poor powers of dispersal between habitat patches. In this way, if a disaster befalls a population in one small reserve, immigrants from neighboring populations can more easily re-colonize it. This avoids the need for humans to physically move new plants or animals into an area.

Several types of habitat function as corridors, including hedgerows in Europe, which facilitate movement and dispersal of species between forest fragments (**Figure 60.11**). In China, corridors of habitat have been established to link small, adjacent populations of giant pandas. Riparian habitats, vegetated corridors bordering watercourses, are thought to help facilitate movement of species between lakes or rivers. In Florida, debate continues about whether to establish new populations of the Florida panther in refuges in both central and north Florida and whether to have these populations connect with the current population in south Florida via riparian movement corridors. In theory, this would lessen the threat of extinction of the Florida panther via inbreeding or a catastrophic event such as a hurricane. However, disadvantages are associated with corridors. First, corridors also can facilitate the spread of disease, invasive species, and fire between small reserves, and second, it is not yet clear if species, such as female Florida panthers, would actually use such corridors.

Finally, parks are often designed to minimize **edge effects**, the special physical conditions that exist at the boundaries or "edges" of ecosystems (see **Figure 60.10f**). Habitat edges, particularly those between natural habitats such as forests and developed land, are often different in physical characteristics from the habitat core. The center of a forest is shaded by trees and has less wind and light than the forest edge, which is unprotected. Many forest-adapted species thus shy away from forest edges and prefer forest centers. Circular-shaped parks are preferred over long, skinny parks because the amount of edge is minimized.

Economic Considerations in Conservation While the principles of island biogeography theory and landscape ecology are useful in illuminating conservation issues, in reality there is often little choice as to the size, shape, and location of nature reserves. Management practicalities, costs of acquisition and management, and politics often override ecological considerations, especially in developing countries, where costs for large reserves may be relatively high. Economic considerations often enter into the choice of which areas to preserve. Typically, many countries protect areas in those regions that are the least economically valuable rather than choosing areas to ensure a balanced representation of the country's biota. In the U.S., most national parks have been chosen for their scenic beauty, not because they preserve the richest habitat for wildlife.

When designing nature reserves, countries should also consider how to finance their management. Interestingly, the amount of money spent to protect nature reserves may better determine species extinction rates than reserve size. According to island biogeography theory, large areas minimize the risk of extinctions because they contain sizable populations. In Africa, several parks, such as Serengeti and Selous in Tanzania, Tsavo in Kenya, and Luangwa in Zambia, are large enough to fulfill this theoretical ideal. However, in the 1980s, populations of black rhinoceroses and elephants declined dramatically within these areas because of poaching, showing that a wide gap may exist between theory and reality. In reality, the rates of decline of rhinos and elephants, largely a result of poaching, have been related directly to conservation efforts and spending (**Figure 60.12**). The remaining black rhinos, lowland gorillas, and pygmy chimpanzees in Africa and the vicuna, a llama-like animal in South America, have all shown the greatest stabilization of numbers in areas that have been heavily patrolled and where resources have been concentrated.

The Single-Species Approach Focuses Conservation Efforts on Particular Types of Species

Some plants and animals are, for reasons we will explore in this section, of more interest to conservation biologists than others. The single-species approach to conservation focuses on saving particularly important species, including indicator, umbrella, flagship, and keystone species.

Indicator Species Some conservation biologists have suggested that certain organisms can be used as **indicator species**, or species whose status provides information on the overall health of an ecosystem. Corals are good indicators of marine processes such as siltation, the accumulation of sediments transported by water. Because siltation reduces the availability of light, the abundance of many marine organisms decreases in such situations, with corals among the first to display a decline in health. A proliferation of the dark variety of the peppered moth (*Biston betularia*) has been shown to be a good indicator of polluted air. Polar bears (*Ursus maritimus*) are thought to be an indicator species for global climate change (**Figure 60.13a**). Scientists believe that global warming is causing the ice in the Arctic to melt earlier in the spring than in the past. Because polar bears rely on the ice to hunt for seals, the earlier breakup of the ice is leaving the bears less time to feed and build the fat that enables them to sustain themselves and their young. Scientists have noted a decrease in the weight of polar bears and in the survival of their cubs.

Umbrella Species **Umbrella species** are species whose habitat requirements are so large that protecting them would protect many other species existing in the same habitat. The Northern

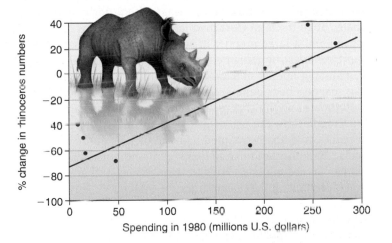

Figure 60.12   **The economics of conservation.**   A positive relationship is seen between change in black rhinoceros numbers between 1980 and 1984, and conservation spending in various African countries.

spotted owl (*Strix occidentalis*) of the Pacific Northwest is considered to be an important umbrella species (**Figure 60.13b**). A pair of birds needs at least 800 hectares of old-growth forest for survival and reproduction, so maintaining healthy owl populations is thought to help ensure survival of many other forest-dwelling species. In the southeast area of the U.S., the red-cockaded woodpecker (*Picoides borealis*) is often seen as the equivalent of the spotted owl, since it requires large tracks of old-growth long-leaf pine (*Pinus palustris*), including old diseased trees in which it can excavate its nests. Some species, such as *Zea diploperennis*, which we discussed at the beginning of the chapter, fall into this category. For example, to protect *Z. diploperennis* in its natural environment, the state of Jalisco in Mexico bought the land where it grows and established a nature reserve and research facility there.

Flagship Species In the past, conservation resources were often allocated to a **flagship species**, a single large or instantly recognizable species. Such species were typically chosen because they were attractive and thus more readily engendered support from the public for their conservation. The concept of the flagship species, which are usually charismatic vertebrates such as the American buffalo (*Bison bison*), is often used to raise awareness for conservation in general. The giant panda (*Ailuropoda melanoleuca*) is the World Wildlife Fund's emblem for endangered species, and the Florida panther (*Puma concolor*) has become a symbol of the state's conservation campaign (**Figure 60.13c**).

Keystone Species A different, perhaps more effective conservation strategy focuses on **keystone species**, species within a community that have a role out of proportion to their abundance. The beaver, a relatively small animal, can completely alter a community by building a dam and flooding an entire river valley (**Figure 60.14**). The resultant lake may become a home to fish species, wildfowl, and aquatic vegetation. A decline in the number of beavers could have serious ramifications for the remaining community members, promoting fish die-offs, waterfowl loss, and the death of vegetation adapted to waterlogged soil.

(a) Polar bear   **(b) Northern spotted owl**   **(c) Florida panther**

Figure 60.13   Indicator, umbrella, and flagship species.   (a) Polar bears have been called an indicator species of global climate change. (b) The Northern spotted owl is considered an umbrella species for the old-growth forest in the Pacific Northwest. (c) The Florida panther has become a flagship species for Florida.

Figure 60.14 Keystone species. The American beaver creates large dams across streams, and the resultant lakes provide habitats for a great diversity of species.

Tropical ecologist John Terborgh considers palm nuts and figs to be keystone species because they produce fruit during otherwise fruitless times of the year and are thus critical resources for tropical forest fruit-eating animals, including primates, rodents, and many birds. Together, these fruit-eaters account for as much as three-quarters of the tropical forest animal biomass. Without the fruit trees, wholesale extinction of these animals could occur. Note that a keystone species is not the same as a **dominant species**, one that has a large effect in a community because of its abundance or high biomass. For example, *Spartina* cordgrass is a dominant species in a salt marsh because of its large biomass (refer back to Figure 59.15), but it is not a keystone species.

In the southeastern U.S., gopher tortoises can be regarded as keystone species because the burrows they create provide homes for an array of other animals, including mice, opossums, frogs, snakes, and insects. Many of these creatures depend on the gopher tortoise burrows and would be unable to survive without them. Gopher tortoises and some other keystone species, including beavers, are also called **ecosystem engineers**,

because they create, modify, and maintain habitats. African elephants act as ecosystem engineers through their browsing activity, destroying small trees and shrubs and changing woodland habitats into grasslands.

Few studies have analyzed the community importance of keystone species, and no set criteria have been established for designating a keystone species. Nevertheless, such species do seem to affect species diversity. The conservation community is eager to identify keystone species, because by managing a keystone species, they may also ensure the survival of many other species in the ecosystem.

Restoration Ecology Attempts to Rehabilitate Degraded Ecosystems

Restoration ecology is the full or partial repair or replacement of biological habitats and/or their populations that have been damaged. It can focus on restoring or rehabilitating a habitat, or it can involve attempting to return species to the wild following captive breeding. Following opencast mining for coal or phosphate, huge tracts of disturbed land must be replenished with topsoil and a large number of species such as grasses, shrubs, and trees must be replanted. Aquatic habitats may be restored by reducing human impacts and replanting vegetation. In Florida, seagrass beds damaged by boat propellers are closed off to motorboats and the area replanted. Restoration can also involve **bioremediation**, the use of living organisms, usually microbes or plants, to detoxify polluted habitats such as dump sites or oil spills. Some bacteria can detoxify contaminants, while certain plants can accumulate toxins in their tissues and are then harvested, removing the poison from the system.

Habitat Restoration The three basic approaches to habitat restoration are complete restoration, rehabilitation, and ecosystem replacement. In complete restoration, conservationists attempt to put back exactly what was there prior to the disturbance. The University of Wisconsin pioneered the restoration of prairie habitats as early as 1935, converting agricultural land back to species-rich prairies (**Figure 60.15a**). The second approach aims to return

(a) Complete restoration   **(b) Rehabilitation**   **(c) Ecosystem replacement**

Figure 60.15 Habitat restoration. (a) As long ago as 1935, the University of Wisconsin was involved in returning agricultural land to native prairies. (b) In Florida, phosphate mines are so degraded that even after restoration, some exotic species such as cogongrass often invade. (c) These old open-pit mines in Middlesex, England, have been converted to valuable freshwater habitats, replacing the wooded area that was originally present.

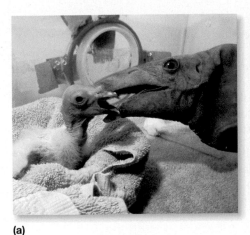

(a)   **(b)**

Figure 60.16   Captive breeding programs. The California condor, the largest bird in the U.S., with a wingspan of nearly 3 m, has been bred in captivity in California. **(a)** A researcher at the San Diego Wild Animal Park feeds a chick with a puppet, so that the birds will not become habituated to the presence of humans. **(b)** This captive-bred condor soars over the Grand Canyon. Note the tag on the underside of its wing.

the habitat to something similar to, but a little less than, full restoration, a goal called rehabilitation. In Florida, phosphate mining involves removing a layer of topsoil or "overburden," mining the phosphate-rich layers, returning the overburden, and replanting the area. Exotic species such as cogongrass (*Imperata cylindrica*), an invasive Southeast Asian species, often invade these disturbed areas, and the biodiversity of the restored habitat is usually not comparable to that of unmined areas (**Figure 60.15b**). The third approach makes no attempt to restore what was originally present but instead replaces the original ecosystem with a different one. The replacement could be an ecosystem that is simpler but more productive, as when deciduous forest is replaced after mining by simple grassland to be used for public recreation.

While any of these approaches can be employed in the habitat restoration process, a common viewpoint among environmentalists is that the endpoint of habitat restoration should be complete restoration. While in some cases, full restoration is appropriate, there are also many cases where restoration is so difficult or expensive as to be impractical. Ecosystem replacement is particularly sensible for land that has been significantly damaged by past activities. It would be nearly impossible to re-create the original landscape of an area that was mined for stone or gravel. In these situations, however, wetlands or lakes may be created in the open pits (**Figure 60.15c**).

Captive Breeding Captive breeding, the propagation of animals and plants outside their natural habitat to produce stock for subsequent release into the wild, has proved valuable in re-establishing breeding populations following extinction or near extinction. Peter Stiling has reintroduced greenhouse-propagated populations of a rare *Opuntia* cactus in the Florida Keys following the decimation of the original population by *Cactoblastis cactorum*, an invasive moth from the Caribbean. Zoos, aquariums, and botanical gardens often play a key role in captive breeding, propagating species that are highly threatened in the wild. They also play an important role in public education about the loss of biodiversity and the use of restoration programs.

Several classic programs illustrate the value of captive breeding and reintroduction. The peregrine falcon (*Falco peregrinus*) became extinct in nearly all of the eastern U.S. by the mid-1960s, a decline that was linked to the effects of DDT (refer back to Figure 59.9). In 1970, Tom Cade gathered falcons from other parts of the country to start a captive breeding program at Cornell University. Since then, the program has released thousands of birds into the wild, and in 1999 the peregrine falcon was removed from the Endangered Species List. A captive breeding program is also helping save the California condor (*Gymnogyps californicus*) from extinction. In the 1980s, there were only 25 known birds, some in captivity and some in the wild. Scientists made the decision to capture the remaining wild birds in order to protect and breed them (**Figure 60.16a**). By 2006, the captive population numbered 156 individuals, and 128 birds were living in the wild, 60 in California and 68 in Arizona (**Figure 60.16b**). A milestone was reached in 2002, when a pair of captive-reared California condors bred in the wild.

Because the number of individuals in any captive breeding program is initially small, care must be exercised to avoid inbreeding. Matings are usually carefully arranged to maximize resultant genetic variation in offspring. The use of genetic engineering to clone endangered species is a new area that may eventually help bolster populations of captive-bred species.

GENOMES & PROTEOMES

Can Cloning Save Endangered Species?

In 1997, Ian Wilmut and colleagues at Scotland's Roslin Institute announced to the world that they had cloned a now-famous sheep, Dolly, from mammary cells of an adult ewe (see Chapter 20). Since then, interest has arisen among conservation biologists about whether the same technology might be used to save species on the verge of extinction. Scientists were encouraged that in January 2001, an Iowa farm cow called Bessie gave birth to a cloned Asian gaur (*Bos gaurus*), an endangered species. The gaur, an oxlike animal native to the jungles of India and Burma, was cloned from a single skin cell taken from a dead animal. To clone the gaur, scientists removed the nucleus from a cow's egg and replaced it with a nucleus from the gaur's cell.

Figure 60.17 **Cloning may help save endangered species.** In 2004, this 8-month-old cloned Javan banteng made its public debut at the San Diego Zoo.

The treated egg was then placed into the cow's womb. Unfortunately, the gaur died from dysentery two days after birth, although scientists believe this was unrelated to the cloning procedure. In 2003, another type of endangered wild cattle, the Javan banteng (*Bos javanicus*), was successfully cloned (**Figure 60.17**).

Despite the promise of cloning, a number of issues remain unresolved:

1. Scientists would have to develop an intimate knowledge of different species' reproductive cycles. For sheep and cows this was routine, based on the vast experience in breeding these species, but eggs of different species, even if they could be harvested, often require different nutritive media in laboratory cultures.
2. Because it is desirable to leave natural mothers available for breeding, scientists will have to identify surrogate females of similar but more common species that can carry the fetus to term.
3. Some argue that cloning does not address the root causes of species loss, such as habitat fragmentation or poaching,

and that resources would be better spent elsewhere, for example, in preserving the endangered species' remaining habitat.

4. Cloning might not be able to increase the genetic variability of the population. However, if it were possible to use cells from deceased animals, for example, from their hair or feathers, these clones could theoretically reintroduce lost genes back into the population.

Many biologists believe that while cloning may have a role in conservation, it is only part of the solution and that we should address what made the species go extinct in the first place before attempting to restore it.

Conservation is clearly a matter of great importance, and a failure to value and protect our natural resources adequately could be a grave mistake. Some authors, most recently the ecologist and geographer Jared Diamond, have investigated why many great societies of the past—including Angkor Wat, Easter Island, and Norse Greenland—collapsed or vanished, leaving behind monumental ruins. Diamond has concluded that the collapse of these societies occurred partly because people inadvertently destroyed the ecological resources on which their societies depended. Modern nations such as Rwanda face similar issues. The country's population density is the highest in Africa, and it has a limited amount of land that can be used for growing crops. By the late 1980s, the need to feed a growing population led to the wholesale clearing of Rwanda's forests and wetlands, with the result that little additional land was available to farm. Increased population pressure, fueled by environmental scarcity, was likely a contributing factor in igniting the genocide of 1994.

As we've seen throughout this textbook, an understanding of biology is vital to understand and help solve many of society's problems. Within this large field, genomics and proteomics may have a huge potential for improving people's lives and society at large. These disciplines offer the opportunity to unlock new diagnoses and treatments for diseases, to improve nutrition and food production, and even to help us to restore biological diversity.

CHAPTER SUMMARY

60.1 Why Conserve Biodiversity?

- Conservation biology uses knowledge from molecular biology, genetics, and ecology to protect the biological diversity of life. (Figure 60.1)
- The preservation of biodiversity has been justified because of its economic value, because of the value of ecosystem services, and on ethical grounds. (Tables 60.1, 60.2)
- Four models exist that describe the relationship between biodiversity and ecosystem function; diversity-stability, rivet, redundancy, and idiosyncratic. (Figure 60.2)

- Experiments both in the lab and in the field have shown that increased biodiversity results in increased ecosystem function. (Figures 60.3 and 60.4)

60.2 The Causes of Extinction and Loss of Biodiversity

- Extinction—the process by which species die out—has been a natural phenomenon throughout the history of life on Earth. Extinction rates in recent times, however, have been much higher than in the past, a phenomenon called the diversity crisis. (Figure 60.5)

- The main causes of extinctions have been and continue to be introduced species, direct exploitation, and habitat destruction. (Figure 60.6)

- Reduced population size can lead to a reduction of genetic diversity through inbreeding, genetic drift, and limited mating, which reduces effective population size.

- Inbreeding, mating among genetically related relatives, can lead to a reduction in fertility. (Figure 60.7)

- Knowledge of a species' effective population size is vital to ensure the success of conservation projects. (Figure 60.8)

60.3 Conservation Strategies

- Habitat conservation strategies commonly target megadiversity countries, countries with the largest number of species; biodiversity hot spots, areas with the largest number of endemic species, those unique to the area; and representative habitats, areas that represent the major habitats. (Figure 60.9, Table 60.3)

- Conservation biologists employ many strategies in protecting biodiversity. Principles of the equilibrium model of island biogeography and landscape ecology are used in the theory and practice of park reserve design to determine, for example, whether the park should take the form of one single or several small reserves. (Figures 60.10, 60.11)

- Economic considerations also play an important role in reserve creation, and it has been shown that conservation spending is positively related to population size. (Figure 60.12)

- The single-species approach focuses conservation efforts on indicator species, umbrella species, flagship species, and keystone species. Indicator species are species whose status can provide information on the overall health of an ecosystem. Umbrella species are species whose habitat requirements are so large that preserving them would also preserve many other species. Flagship species are usually charismatic or instantly recognizable species; and keystone species have an effect out of proportion to their abundance. (Figures 60.13, 60.14)

- Restoration ecology seeks to repair or replace populations and their habitats. Three basic approaches to habitat restoration are complete restoration, rehabilitation, and ecosystem replacement. (Figure 60.15)

- Captive breeding is the propagation of animals outside their natural habitat. Several programs have illustrated the success of captive breeding and reintroduction to the wild, including those for California condors. (Figure 60.16)

- Cloning of endangered species has been accomplished on a very small scale and despite its limitations may have a role in conservation biology. (Figure 60.17)

TEST YOURSELF

1. Which of the following statements best describes an endangered species?
   a. a species that is likely to become extinct in a portion of its range
   b. a species that has disappeared in a particular community but is present in other natural environments
   c. a species that is extinct
   d. a species that is in danger of becoming extinct throughout all or a significant portion of its range
   e. both b and d

2. Biological diversity is important and should be preserved because
   a. food, medicines, and industrial products are all benefits of biodiversity.
   b. ecosystems provide valuable services to us in many ways.
   c. many species can be used as valuable research tools.
   d. we have an ethical responsibility to protect our environment.
   e. all of the above are correct.

3. The impact on an ecosystem of a particular species is dependent on how critical the species is to the ecosystem. This statement defines the _____ hypothesis.
   a. rivet
   b. redundancy
   c. idiosyncratic
   d. diversity-stability
   e. community ecology

4. The research conducted by Tillman and colleagues demonstrated that
   a. as diversity increases, productivity increases.
   b. as diversity decreases, productivity increases.
   c. areas with higher diversity demonstrate less efficient use of nutrients.
   d. species richness increases lead to an increase in invasive species.
   e. increased diversity results in increased susceptibility to disease.

5. The process by which species die out is termed
   a. endangerment.
   b. death.
   c. extant.
   d. extinction.
   e. conservation biology.

6. Which of the following is not a prime cause of extinction?
   a. predation by introduced species
   b. habitat destruction
   c. direct exploitation
   d. disease brought by invasive species
   e. bioremediation

7. The negative result of inbreeding is
   a. an increase in genetic diversity that leads to new genetic diseases.
   b. a decrease in genetic diversity that limits how a species can adapt.
   c. a decrease in number of mates.
   d. mating between related individuals.
   e. all of the above.

8. The number of individuals that contribute genes to future populations is called
   a. the effective population size.
   b. the adult population size.
   c. the breeding coefficient.
   d. the gene pool.
   e. the female population size.

9. Small strips of land that connect and allow organisms to move between small patches of natural habitat are called
   a. biological conduits.
   b. edge effects.
   c. movement corridors.
   d. migration pathways.
   e. landscape breaks.

10. Bioremediation is
    a. a process that restores a disturbed habitat to its original state.
    b. a process that uses microbes or plants to detoxify contaminated habitats.
    c. the legislation requiring rehabilitation of a disturbed habitat.
    d. a process of capturing all of the living individuals of species for breeding purposes.
    e. the process of removing tissue from a dead organism in the hopes of cloning it.

CONCEPTUAL QUESTIONS

1. What are the three levels at which biodiversity can be examined?
2. Define conservation biology.
3. Distinguish between an umbrella species, a flagship species, and a keystone species.

EXPERIMENTAL QUESTIONS

1. What was the goal of Shahid Naeem and colleagues in their experiment at Silwood Park, England?
2. What was the hypothesis tested by the researchers?
3. How did the researchers test for ecosystem functioning?

COLLABORATIVE QUESTIONS

1. Discuss several causes of species extinction.
2. Discuss several species-specific approaches to conservation biology.

www.brookerbiology.com

This website includes answers to the Biological Inquiry questions found in the figure legends and all end-of-chapter questions.

GLOSSARY

A

A band A wide, dark band produced by the orderly parallel arrangement of the thick filaments in the middle of each sarcomere.

abiotic The term used to describe interactions between organisms and their nonliving environment.

aboral Refers to the region opposite the mouth.

abortion Procedures or circumstances that cause the death of an embryo or fetus after implantation.

abscisic acid One of several plant hormones that help a plant to cope with environmental stress.

absorption The process in which digested nutrients are transported from the digestive cavity into an animal's circulatory system.

absorption spectrum A diagram that depicts the wavelengths of electromagnetic radiation that are absorbed by a pigment.

absorptive nutrition The process whereby an organism secretes enzymes into food substrates, breaking down complex organic molecules into small organic molecules that are absorbed as food.

absorptive state One of two alternating phases in the utilization of nutrients; occurs when ingested nutrients enter the blood from the gastrointestinal tract. The other phase is the postabsorptive state.

acclimatization The process of fine tuning an animal's adaptive mechanisms to a changing environment.

accommodation The process in which contraction and relaxation of the ciliary muscles adjust the lens according to the angle at which light enters the eye.

acid A molecule that releases hydrogen ions in solution.

acid hydrolase A hydrolytic enzyme found in lysosomes that functions at acidic pH and uses a molecule of water to break a covalent bond.

acid rain Precipitation with a pH of less than 5.6; acid rain results from the burning of fossil fuels, which releases sulfur dioxide and nitrogen oxide into the atmosphere, which in turn react with oxygen in the air to form sulfuric acid and nitric acid, and falls to the surface in rain or snow.

acidic A solution that has a pH below 7.

acoelomate An animal that lacks a body cavity entirely.

acquired antibiotic resistance The common phenomenon of a previously susceptible strain becoming resistant to a specific antibiotic.

acquired immune deficiency syndrome (AIDS) A disease caused by the human immunodeficiency virus (HIV) that leads to a defect in the immune system of infected individuals.

acrocentric A chromosome in which the centromere is near one end.

acromegaly A condition in which a person's GH levels remain elevated after growth has ceased, and the continued excess GH causes many bones, like those of the hands and feet, to thicken and enlarge.

acrosomal reaction An event in fertilization in which the binding of a sperm cell to proteins located in the egg cell plasma membrane triggers a series of events producing the fast block to polyspermy and the entry of the sperm cell's nucleus into the egg cell.

acrosome A special structure at the tip of a sperm's head that contains proteolytic enzymes that help break down the plasma membrane of the ovum at fertilization.

actin A cytoskeletal protein.

actin filament A thin type of protein filament composed of actin proteins that forms part of the cytoskeleton and supports the plasma membrane and plays a key role in cell strength, shape, and movement.

action potential The movement of an electrical impulse along the plasma membrane, which occurs in animal nerve axons and some plant cells.

action spectrum The rate of photosynthesis plotted as a function of different wavelengths of light.

activation energy An initial input of energy in a chemical reaction that allows the molecules to get close enough to cause a rearrangement of bonds.

activator A transcription factor that binds to DNA and increases the rate of transcription.

active immunity The acquired response to exposure to any type of antigen.

active site The location in an enzyme where the chemical reaction takes place.

active transport The transport of a solute across a membrane against its gradient—that is, from a region of low concentration to higher concentration. In the case of ions, active transport is against an electrochemical gradient.

adaptation The processes and structures by which organisms adjust to short-term or long-term changes in their environment.

adaptive radiation The process whereby a single ancestral species evolves into a wide array of descendant species that differ greatly in their habitat, form, or behavior.

adenine (A) A purine base found in DNA and RNA.

adenosine triphosphate (ATP) A nucleotide that is a common energy source of all cells.

adenylyl cyclase An enzyme in the plasma membrane that synthesizes cAMP from ATP.

adiabatic cooling The process in which increasing elevation leads to a decrease in air temperature.

adventitious root A root that is produced on the surfaces of stems (and sometimes leaves) of vascular plants; also, roots that develop at the bases of stem cuttings.

aerenchyma Spongy plant tissue with large air spaces.

aerobic Refers to a process that occurs in the presence of oxygen; a form of metabolism that requires oxygen.

aerobic respiration During this type of respiration, O_2 is consumed and CO_2 is released.

aerotolerant anaerobe A microorganism that does not use oxygen but is not poisoned by it either.

afferent arterioles Blood vessels that provide a pathway for blood into a tissue or organ. For example, afferent arterioles in the kidney that supplies each glomerulus with blood.

aflatoxins Fungal toxins that cause liver cancer and are a major health concern worldwide.

age structure The relative numbers of individuals of each defined age group in a population.

age-specific fertility rate The rate of offspring production for females of a certain age; used to help calculate how a population grows.

AIDS Acquired immune deficiency syndrome. AIDS reduces the body's immunity by killing helper T cells.

air sac A component of the avian respiratory system; air sacs—not lungs—expand when a bird inhales and shrink when it exhales, and they do not participate in gas exchange.

akinete A thick-walled cell used to survive unfavorable conditions in a dormant state.

aldosterone A steroid hormone made by the adrenal glands.

algae A term that applies to about 10 phyla of protists that include both photosynthetic and nonphotosynthetic species.

alimentary canal The single elongated tube of a digestive system with an opening at either end, through which food and eventually wastes pass from one end to the other.

alkaline A solution that has a pH above 7.

alkaloids A group of structurally related secondary metabolites that all contain nitrogen and usually have a cyclic, ringlike structure.

allantois An extraembryonic membrane in the amniotic egg that serves as a disposal sac for metabolic wastes.

allee effect The phenomenon that some individuals will fail to mate successfully purely by chance, for example, because of the failure to find a mate.

allele   A variant form of a gene.

allele frequency   The number of copies of an allele divided by the total number of alleles in a population.

allelochemical   A powerful plant chemical, often a root exudates, that kills other plant species.

allelopathy   The suppressed growth of one species due to the release of toxic chemicals by another species.

allodiploid   An alloploid that has only one set of chromosomes from two different species.

allometric growth   The pattern whereby different parts of the body grow at different rates with respect to each other.

allopatric   The term used to describe species occurring in different geographic areas.

allopatric speciation   A form of speciation that occurs when a population becomes geographically isolated from other populations and evolves into one or more new species.

alloploid   An organism that contains at least one set of chromosomes from two or more different species.

allopolyploid   An organism that contains two or more complete sets of chromosomes from two or more different species.

allosteric site   A site where a molecule can bind noncovalently and affect the function of the active site. The binding of a molecule to an allosteric site causes a conformational change in the enzyme that inhibits its catalytic function.

allotetraploid   A type of allopolyploid that contains two complete sets of chromosomes from two species for a total of four sets.

alternation of generations   The phenomenon that occurs in plants (and some protists) in which the life cycle alternates between multicellular diploid organisms, called sporophytes, and multicellular haploid organisms, called gametophytes.

alternative splicing   The splicing of pre-mRNA in more than one way to create two or more different polypeptides.

altruism   Behavior that appears to benefit others at a cost to oneself.

alveolus   A saclike structure of the lungs where gas exchange occurs.

Alzheimer's disease (AD)   The leading worldwide cause of dementia; characterized by a loss of memory and intellectual and emotional function.

amensalism   One-sided competition, where the interaction is detrimental to one species but not to the other.

Ames test   A test that helps ascertain whether or not an agent is a mutagen by using a strain of a bacterium, *Salmonella typhimurium*.

amino acid   The building block of proteins. Amino acids have a common structure in which a carbon atom, called the α-carbon, is linked to an amino group (NH_2) and a carboxyl group (COOH). The α-carbon also is linked to a hydrogen atom and a particular side chain.

amino terminus   *See* N-terminus.

aminoacyl site (A site)   One of the three sites for tRNA binding to the ribosome; the others are the peptidyl site (P site) and the exit site (E site). The A site is the site where incoming tRNA molecules bind to the mRNA (except for the initiator tRNA).

aminoacyl tRNA   *See* charged tRNA.

aminoacyl-tRNA synthetase   An enzyme that catalyzes the attachment of amino acids to tRNA molecules.

ammonia (NH_3)   One of the most highly toxic of the nitrogenous wastes because it disrupts pH, ion electrochemical gradients, and many chemical reactions that involve oxidations and reductions; typically produced in many aquatic species.

ammonification   The conversion of organic nitrogen to NH_3 and NH_4^+.

amnion   An innermost extraembryonic membrane in the amniotic egg; it protects the developing embryo in a fluid-filled sac called the amniotic cavity.

amniotes   A group of tetrapods with amniotic eggs that includes turtles, lizards, snakes, crocodiles, birds, and mammals.

amniotic egg   The structure that contains the developing embryo and the four separate extraembryonic membranes that it produces: the amnion, the yolk sac, the allantois, and the chorion.

amoeba   A protist cell that moves by pseudo-podia, which involves extending cytoplasm into filaments or lobes.

amoebocyte   A mobile cell within a sponge's mesohyl that absorbs food from choanocytes, digests it, and carries the nutrients to other cells.

amphibian   A tetrapod that has successfully invaded the land but must return to the water to reproduce.

amphipathic   In molecules, meaning they have a hydrophobic (water-fearing) region and a hydrophilic (water-loving) region.

ampulla   A muscular sac at the base of each tube foot of a echinoderm that stores water.

amygdala   An area of the brain known to be critical for understanding and remembering emotional situations.

amylase   A digestive enzyme involved in the digestion of starch.

anabolic reaction   A metabolic pathway that promotes the synthesis of larger molecules from smaller precursor molecules.

anabolism   The synthesis of cellular molecules and macromolecules, which usually requires an input of energy.

anaerobic   Refers to a process that occurs in the absence of oxygen; a form of metabolism that does not require oxygen.

anaerobic respiration   The breakdown of organic molecules in the absence of oxygen.

anagenesis   The pattern of speciation in which a single species is transformed into a different species over the course of many generations.

analogous structure   A trait that is the result of convergent evolution; structures have arisen independently, two or more times, because species have occupied similar types of environments on the Earth.

anaphase   The phase of mitosis during which the sister chromatids separate from each other and move to opposite poles; the poles themselves also move farther apart.

anatomy   The study of the morphology of living organisms, such as plants and animals.

anchoring junctions   A junction found in between animal cells that attaches cells to each other and to the ECM.

androgens   Steroid hormones produced by the male testes that affect most aspects of male reproduction.

anemia   A condition characterized by lower than normal levels of hemoglobin, which reduces the amount of oxygen that can be stored in the blood.

aneuploidy   An alteration in the number of particular chromosomes so that the total number of chromosomes is not an exact multiple of a set.

angina pectoris   Chest pain during exertion due to the tissues being deprived of oxygen.

angiosperm   A flowering plant; the term means "enclosed seed," which reflects the presence of seeds within fruits.

animal cap assay   A type of experiment used extensively to identify factors (proteins) secreted by embryonic cells that induce cells in the animal pole to differentiate into mesoderm.

animal pole   In triploblast organisms, the pole of the egg where less yolk and more cytoplasm are concentrated.

Animalia   One of the four traditional eukaryotic kingdoms of the domain Eukarya.

anion   An ion that has a net negative charge.

annual   A plant that dies after producing seed during its first year of life.

anther   A cluster of microsporangia in a flower that produces pollen and then opens to release it.

antheridia   Round or elongate gametangia that produce plant sperm.

antagonist   A muscle or group of muscles that produces oppositely directed movements at a joint.

antenna complex   *See* light-harvesting complex.

anterior   Refers to the end of an animal where the head is found.

anteroposterior axis   One of the three axes along which the adult body pattern is organized; the others are the dorsoventral axis and the right-left axis.

anthropoid   A member of a class of primates that includes the monkeys and the hominoids; species are larger-brained and diurnal.

antibiotic   A chemical, usually made by microorganisms, that inhibits the growth of certain microorganisms.

antibody   A protein secreted by plasma cells that is part of the immune response; antibodies travel all over the body to reach antigens identical to those that stimulated their production, and then they combine with these antigens and guide an attack that eliminates the antigens or the cells bearing them.

anticodon A three-nucleotide sequence in tRNA that is complementary to a codon in mRNA.

antidiuretic hormone (ADH) A hormone secreted by the posterior pituitary gland that acts on kidney cells to decrease urine production.

antigen Any foreign molecule that the host does not recognize as self and that triggers a specific immune response.

antigen-presenting cells (APCs) Cells bearing fragments of antigen, called antigenic determinants or epitopes, complexed with the cell's MHC proteins.

antiparallel An arrangement in DNA where one strand runs in the 5′ to 3′ direction while the other strand is oriented in the 3′ to 5′ direction.

antiporter A type of transporter that binds two or more ions or molecules and transports them in opposite directions.

apical-basal-patterning genes A category of genes that are important in early stages of plant development during which the apical and basal axes are formed.

apical-basal polarity An architectural feature of plants in which they display an upper, apical pole and a lower, basal pole; shoot apical meristem occurs at the apical pole, and the root apical meristem occurs at the basal pole.

apical constriction A cellular process during gastrulation that is crucial to development; a reduction in the diameter of the actin rings connected to the adherens junctions causes the cells to elongate toward their basal end.

apical region The region of a plant that projects upwards, usually from the soil, and produces the leaves and flowers.

apomixis A natural asexual reproductive process in which plant fruits and seeds are produced in the absence of fertilization.

apoplast The continuum of water-soaked cell walls and intercellular spaces in a plant.

apoplastic transport The movement of solutes through cell walls and the spaces between cells.

apoptosis Programmed cell death.

aposematic coloration Warning coloration that advertises an organism's unpalatable taste.

appendix A finger-like projection in the gastrointestinal tract of animals having no known essential function but that may at one time have been an important part of the body's defense mechanisms.

aquaporin A three-dimensional cell pore that allows water to diffuse through the membrane.

aqueous humor A thin liquid in the anterior cavity behind the cornea of the vertebrate eye.

aqueous solution A solution made with water.

aquifer An underground water supply.

arbuscular mycorrhizae Symbiotic associations between glomalean fungi and the roots of vascular plants.

Archaea One of the three domains of life; the other two are Bacteria and Eukarya.

archaea When not capitalized refers to a cell or species within the domain Archaea.

archegonia Flask-shaped plant gametangia that enclose an egg cell.

archenteron A cavity formed by the embryo during gastrulation that will become the organism's digestive tract.

area hypothesis The proposal that larger areas contain more species than smaller areas because they can support larger populations and a greater range of habitats.

arteriole A single-celled layer of endothelium surrounded by one or two layers of smooth muscle and connective tissue that delivers blood to the capillaries and distributes blood to regions of the body in proportion to metabolic demands.

artery A blood vessel that carries blood away from the heart.

artificial selection *See* selective breeding.

asci Fungal sporangia shaped like sacs that produce and release sexual ascospores.

ascocarp The type of fruiting body produced by ascomycete fungi.

ascomycetes A phylum of fungi that produce sexual spores in saclike asci located at the surfaces of fruiting bodies known as ascocarps.

ascospore The type of sexual spore produced by the ascomycete fungi.

aseptate The condition of not being partitioned into smaller cells, usually refers to fungal cells.

asexual reproduction A reproductive strategy that occurs when offspring are produced from a single parent, without the fusion of gametes from two parents. The offspring are therefore clones of the parent.

assimilation In the case of nitrogen, the process by which plants and animals incorporate the ammonia and NO_3^- formed through nitrogen fixation and nitrification.

associative learning A change in behavior due to an association between a stimulus and a response.

assort The process of distributing.

asthma A disease in which the smooth muscles around the bronchioles contract more than usual, increasing resistance to airflow.

AT/CC rule Refers to the phenomenon that an A in one DNA strand always hydrogen bonds with a T in the opposite strand, while a G in one strand bonds with a C.

atherosclerosis The condition in which large plaques may occlude (block) the lumen of an artery.

atmospheric (barometric) pressure The pressure exerted by the gases in air on the body surfaces of animals.

atom The smallest functional unit of matter that forms all chemical substances and cannot be further broken down into other substances by ordinary chemical or physical means.

atomic mass An atom's mass relative to the mass of other atoms. By convention, the most common form of carbon, which has six protons and six neutrons, is assigned an atomic mass of exactly 12.

atomic nucleus The center of an atom.

atomic number The number of protons in an atom.

ATP synthase An enzyme that utilizes the energy stored in an H^+ electrochemical gradient for the synthesis of ATP via chemiosmosis.

ATP-dependent chromatin remodeling enzyme An enzyme that catalyzes a loosening in the compaction of chromatin; this loosening facilitates the ability of RNA polymerase to recognize and transcribe a gene.

ATP-driven pump A common category of pump found in all living cells; this transporter has a binding site for ATP and hydrolyzes ATP to actively transport solutes against a gradient.

atrial natriuretic peptide (ANP) A peptide secreted from the atria of the heart whenever blood levels of sodium increase; ANP causes a natriuresis by decreasing sodium reabsorption in the kidney tubules.

atrioventricular (AV) node Specialized cardiac cells that sit near the junction of the atria and ventricles and conduct the electrical events from the atria to the ventricles.

atrioventricular (AV) valve A one-way valve into the ventricles of the vertebrate heart through which blood moves from the atria.

atrium In the heart, a single filling chamber to collect blood from the tissues.

atrophy A reduction in the size of a structure, such as a muscle.

atropine A potent toxin derived from the deadly nightshade plant.

audition The sense of hearing.

autoimmune disease In humans and many other vertebrates, the situation in which the body's normal state of immune tolerance breaks down, with the result that antibody-mediated or T-cell mediated attacks are directed against the body's own cells and tissues.

autonomic nervous system The division of the peripheral nervous system that regulates homeostasis and organ function; also called the visceral nervous system.

autophagosome Cellular material enclosed in a double membrane, produced by the process of autophagy.

autophagy Meaning "the eating of one's self." A process whereby cellular material, such as a worn-out organelle, becomes enclosed in a double membrane and degraded.

autosomes All of the chromosomes found in the cell nucleus of eukaryotes except for the sex chromosomes.

autotomy In echinoderms, the ability to intentionally detach a body part, such as a limb, that will later regenerate.

autotroph An organism that has metabolic pathways that directly harness energy from either inorganic molecules or light.

auxin One of several types of hormones considered to be the "master" plant hormones because they influence plant structure, development, and behavior in many ways, often working with other hormones.

auxin efflux carrier   One of several types of PIN proteins, which transport auxin out of plant cells.

auxin influx carrier   A plasma membrane protein that transports auxin into plant cells.

avirulence gene (*Avr* gene)   A gene that encodes a virulence-enhancing elicitor, which causes plant disease.

Avogadro's number   As first described by Italian physicist Amedeo Avogadro, 1 mole of any element contains the same number of particles—6.022×10^{23}.

axillary bud   A bud that occurs in the axil, the upper angle where a twig or leaf joins the stem.

axon   An extension of the plasma membrane that is involved in sending signals from a neuron to neighboring cells.

axon hillock   The part of the axon closest to the cell body.

axoneme   The internal structure of eukaryotic cilia and flagella consisting of microtubules, the motor protein dynein, and linking proteins.

B

B cell   A type of lymphocyte responsible for specific immunity.

bacilli   Rods; one of the five major shapes of prokaryotic cells.

backbone   The linear arrangement of phosphates and sugar molecules in a DNA or RNA strand.

Bacteria   One of the three domains of life; the other two are Archaea and Eukarya.

bacteria   When not capitalized, refers to a cell or species within the domain Bacteria.

bacterial artificial chromosome (BAC)   A cloning vector in bacteria that can contain large DNA inserts.

bacterial colony   A clone of genetically identical cells formed from a single cell.

bacteriophage   A virus that infects bacteria.

bacteroid   A modified bacterial cell of the type known as rhizobia present in mature root nodules of some plants.

balanced polymorphism   The phenomenon in which two or more alleles are kept in balance, and therefore are maintained in a population over the course of many generations.

balancing selection   A type of natural selection that maintains genetic diversity in a population.

balloon angioplasty   A common treatment to restore blood flow through a blood vessel. A thin tube with a tiny, inflatable balloon at its tip is threaded through the artery to the diseased area; inflating the balloon compresses the plaque against the arterial wall, widening the lumen.

baroreceptor   A pressure-sensitive region within the walls of certain arteries that contains the endings of nerve cells; these regions help to maintain blood pressure in the normal range for an animal.

Barr body   A highly condensed X chromosome.

basal body   A site at the base of flagella or cilia from which microtubules grow. Basal bodies are anchored on the cytosolic side of the plasma membrane.

basal metabolic rate (BMR)   The metabolic rate of an animal under resting conditions, in good health, and not under stress of any kind.

basal nuclei   Clusters of neuronal cell bodies that surround the thalamus on both sides and lie beneath the cerebral cortex; involved in planning and learning movements and also function via a complex circuitry to initiate or inhibit movements.

basal region   The region of a plant that produces the roots.

basal transcription   A low level of transcription resulting from the core promoter.

basal transcription apparatus   For a eukaryotic structural gene, refers to the complex of GTFs, RNA polymerase II, and a DNA sequence containing a TATA box.

base   1. A component of nucleotides that is a single or double ring of carbon and nitrogen atoms. 2. A molecule that when dissolved in water lowers the H^+ concentration.

base pairs   The structure in which two bases in opposite strands of DNA hydrogen bond with each other.

base substitution   A mutation that involves the substitution of a single base in the DNA for another base.

basic local alignment search tool (BLAST)   A computer program that can identify homologous genes that are found in a database.

basidiocarp   The type of fruiting body produced by basidiomycete fungi.

basidiomycetes   A group of fungi whose sexual spores are produced on the surfaces of club-shaped structures (basidia).

basidiospore   A sexual spore of the basidiomycete fungi.

basophil   A cell that secretes the anticlotting factor heparin at the site of an infection, which helps the circulation flush out the infected site; basophils also secrete histamine, which attracts infection-fighting cells and proteins.

Batesian mimicry   The mimicry of an unpalatable species (the model) by a palatable one (the mimic).

behavior   The observable response of organisms to external or internal stimuli.

behavioral ecology   A subdiscipline of organismal ecology that focuses on how the behavior of an individual organism contributes to its survival and reproductive success, which in turn eventually affects the population density of the species.

benign tumor   A precancerous condition.

bidirectional replication   In DNA replication, the two DNA strands unwind, and DNA replication proceeds outward from the origin in opposite directions.

biennial   A plant that does not reproduce during the first year of life but may reproduce within the following year.

bilateral   Refers to living organisms, such as animals, or parts of living organisms, such as leaves, that can be divided along a vertical plane at the midline to create two halves.

bile   A substance produced by the liver that contains bicarbonate ions, cholesterol, phospholipids, a number of organic wastes, and a group of substances collectively termed bile salts.

bile salts   A group of substances produced in the liver that solubilize dietary fat and increase its accessibility to digestive enzymes.

binary fission   The process of cell division in bacteria and archaea in which the cells divide into two cells.

binocular vision   A type of vision in animals with both eyes located at the front of the head; the overlapping images coming into both eyes are processed together in the brain to form one perception.

binomial   A two-part description used by biologists to provide each species with a unique scientific name. For example, the scientific name of the jaguar is *Panthera onca*. The first part is the genus and the second part is the specific epithet or species descriptor.

binomial nomenclature   The standard method for naming species. Each species has a genus name and species epithet.

biochemistry   The study of the chemistry of living organisms.

biodiversity   Biological diversity, including genetic diversity, species diversity, and ecosystem diversity.

biodiversity crisis   The idea that there is currently an elevated loss of species on Earth, far beyond the normal extinction rate of species.

biofilm   An aggregation of microorganisms that secrete adhesive mucilage, thereby gluing themselves to surfaces.

biogeochemical cycle   The continuous movement of nutrients such as phosphorus, carbon, and nitrogen from the physical environment to organisms and back.

biogeography   The study of the geographic distribution of extinct and modern species.

bioinformatics   A field of study that uses computers to study biological information.

biological evolution   The phenomenon that populations of organisms change over the course of many generations. As a result, some organisms become more successful at survival and reproduction.

biological species concept   A species is a group of individuals whose members have the potential to interbreed with one another in nature to produce viable, fertile offspring but cannot successfully interbreed with members of other species.

biology   The study of life.

bioluminescence   A phenomenon that results from chemical reactions that give off light rather than heat.

biomagnification   The increase in the concentration of a substance in living organisms with each increase in trophic level in a food web.

biomass A quantitative estimate of the total mass of living matter in a given area, usually measured in grams or kilograms per square meter.

biome A major type of habitat characterized by distinctive plant and animal life.

bioremediation The use of living organisms, usually microbes or plants, to detoxify polluted habitats such as dump sites or oil spills.

biosphere The regions on the surface of the Earth and in the atmosphere where living organisms exist.

biosynthetic reaction Also called an anabolic reaction; a chemical reaction to make larger molecules and macromolecules.

biotechnology The use of living organisms or the products of living organisms for human benefit.

biotic The term used to describe interactions among organisms.

biparental inheritance A pattern in which both the male and female gametes contribute particular genes to the offspring.

bipedal Having the ability to walk on two feet.

bipolar cells Cells in the eye that make synapses with the photoreceptors and relay responses to the ganglion cells.

bivalent Homologous pairs of sister chromatids associated with each other, lying side by side.

blastocyst The mammalian counterpart of a blastula.

blastoderm An early stage of embryonic development in animals that is composed of a mass of cells with an internal cavity.

blastomeres The two half-size daughter cells produced by each cell division during cleavage.

blastopore A small opening created when a band of tissue invaginates during gastrulation. It forms the primary opening of the archenteron to the outside.

blastula An animal embryo at the stage when it forms an outer epithelial layer and an inner cavity.

blending inheritance An early hypothesis of inheritance that stated that the seeds that dictate hereditary traits blend together from generation to generation, and the blended traits are then passed to the next generation.

blood A fluid connective tissue consisting of cells and (in mammals) cell fragments suspended in a solution of water containing dissolved nutrients, proteins, gases, and other molecules. Blood has four components: plasma, leukocytes, erythrocytes, and thrombocytes or platelets.

blood-doping An example of hormone misuse in which the number of red blood cells in the circulation is boosted to increase the oxygen-carrying capacity of the blood.

blood pressure The force exerted by blood on the walls of blood vessels; blood pressure is responsible for moving blood through the vessels.

blood vessels A system of hollow tubes within the body through which blood travels.

body mass index (BMI) A method of assessing body fat and health risk that involves calculating the ratio of weight compared to height; weight in kilograms is divided by the square of the height in meters.

bottleneck effect A form of genetic drift in which a population size is dramatically reduced and then rebounds. Genetic drift is common when the population size is small.

Bowman's capsule A sac at the beginning of the tubular component of a nephron in the mammalian kidney.

brain The structure in the head of animals that controls sensory and motor functions of the entire body.

brassinosteroid One of several plant hormones that help a plant to cope with environmental stress.

bronchiole A thin-walled, small tube that can dilate or constrict to prevent foreign particles from reaching delicate lung tissue.

bronchodilator A compound that binds to the muscles around bronchioles and causes them to relax and widen.

bronchus A tube branching from the trachea and leading to a lung.

brown fat A specialized tissue in small mammals such as hibernating bats, small rodents living in cold environments, and many newborn mammals including humans that can help to generate heat and maintain body temperature.

brush border The combination of villi and microvilli in the small intestine, which increase the surface area about 600-fold over that of a flat-surfaced tube having the same length and diameter.

bryophytes Liverworts, mosses, and hornworts, the modern non vascular land plants.

buccal pumping A form of breathing in which animals open their mouths to let in air and then close and raise the floor of the mouth, creating a positive pressure and pumping water or air across the gills or into the lungs; found in fish and amphibians.

bud A miniature plant shoot having a dormant shoot apical meristem.

budding A form of asexual reproduction in which a portion of the parent organism pinches off to form a complete new individual.

buffer A compound that acts to minimize pH fluctuations in the fluids of living organisms. Buffer systems can raise or lower pH as needed.

bulbourethral glands Paired accessory glands that secrete an alkaline mucus that protects sperm by neutralizing the acidity in the urethra.

bulk flow The mass movement of liquid in a plant caused by pressure, gravity, or both.

C

C_3 plant A plant that can only incorporate CO_2 into organic molecules via RuBP to make 3PG, a three-carbon molecule.

C_4 plant A plant that uses PEP carboxylase to initially fix CO_2 into a 4-carbon molecule and later uses rubisco to fix CO_2 into simple sugars.

cadherin A cell adhesion molecule found in animal cells that promotes cell-to-cell adhesion.

callose A carbohydrate that plays crucial roles in plant development, and plugging wounds in plant phloem.

calorie The amount of heat required to raise the temperature of 1 gram of water 1 degree Celsius.

Calvin cycle The cycle that includes carbon fixation, reduction and carbohydrate production, and regeneration of ribulose bisphosphate (RuBP). During this process, ATP is used as a source of energy and NADPH is used as a source of high-energy electrons so that CO_2 can be incorporated into carbohydrate.

calyx The sepals that form the outermost whorl of a flower.

CAM plants C_4 plants that take up carbon dioxide at night.

Cambrian explosion An event during the Cambrian period (543 to 490 mya) in which there was an abrupt increase (on a geological scale) in the diversity of animal species.

cancer A disease caused by gene mutations that lead to uncontrolled cell growth.

canopy The uppermost layer of tree foliage.

CAP site One of two regulatory sites near the *lac* promoter; this site is a DNA sequence recognized by the catabolite activator protein (CAP).

capillary A thin-walled vessel that is the site of gas and nutrient exchange between the blood and interstitial fluid.

capping A 7-methylguanosine covalently attached at the 5' end of mature mRNAs of eukaryotes.

capsid A protein coat enclosing a virus's genome.

capsule A very thick, gelatinous glycocalyx produced by certain strains of bacteria that invade animals' bodies that may help them avoid being destroyed by the animal's immune (defense) system.

carapace The hard protective covering of a crustacean.

carbohydrate An organic molecule with the general formula, $C(H_2O)$; a carbon-containing compound that is hydrated (that is, contains water).

carbon fixation In this process, inorganic CO_2 is incorporated into an organic molecule such as a carbohydrate.

carboxyl terminus *See* C-terminus.

carcinogen An agent that increases the likelihood of developing cancer, usually a mutagen.

carcinoma A cancer of epithelial cells.

cardiac cycle The events that produce a single heartbeat, which can be divided into two phases, diastole and systole.

cardiac muscle A type of muscle tissue found only in hearts in which physical and electrical connections between individual cells enable many of the cells to contract simultaneously.

cardiac output (CO) The amount of blood the heart pumps per unit time, usually expressed in units of L/min.

cardiovascular disease Diseases affecting the heart and blood vessels.

cardiovascular system A system containing three components: blood or hemolymph, blood vessels, and one or more hearts.

carnivore An animal that consumes animal flesh or fluids.

carotenoid A type of pigment found in chloroplasts that imparts a color that ranges from yellow to orange to red.

carpel A flower shoot organ that produces ovules that contain female gametophytes.

carrier *See* transporter.

carrying capacity (K) The upper boundary for a population size.

Casparian strips Suberin ribbons on the walls of endodermal cells of plant roots; prevent apoplastic transport of ions into vascular tissues.

caspase An enzyme that is activated during apoptosis.

catabolic reaction The breakdown of a molecule into smaller components, usually releasing energy.

catabolism A metabolic pathway that results in the breakdown of larger molecules into smaller molecules. Such reactions are often exergonic.

catabolite activator protein (CAP) An activator protein also known as the cAMP receptor protein (CRP). CAP is needed for activation of the *lac* operon.

catabolite repression In bacteria, a process whereby transcriptional regulation is influenced by the presence of glucose.

catabolized Broken down.

catalase An enzyme within peroxisomes that breaks down hydrogen peroxide to water and oxygen gas.

catalyst An agent that speeds up the rate of a chemical reaction without being consumed during the reaction.

cataract An accumulation of protein in the lens of the eye; causes blurring, poor night vision.

cation An ion that has a net positive charge.

cation exchange With regard to soil, the process in which hydrogen ions are able to replace mineral cations on the surfaces of humus or clay particles.

cDNA library A collection of recombinant vectors that have cDNA inserts.

cecum The first portion of the large intestine in humans and other similarly sized animals and mammals.

celiac sprue A common genetic disorder in humans that results in a loss of intestinal surface area due to an allergic sensitivity to the wheat protein gluten.

cell The simplest unit of a living organism.

cell adhesion The phenomenon in which cells adhere to each other. Cell adhesion provides one way to convey positional information between neighboring cells.

cell adhesion molecule (CAM) A membrane protein found in animal cells that promotes cell adhesion.

cell biology The study of individual cells and their interactions with each other.

cell body A part of a neuron that contains the cell nucleus and other organelles.

cell coat Also called the glycocalyx, the carbohydrate-rich zone on the surface of certain animal cells that shields the cell from mechanical and physical damage.

cell communication The process through which cells can detect and respond to signals in their extracellular environment. In multicellular organisms, cell communication is also needed to coordinate cellular activities within the whole organism.

cell cycle The series of phases eukaryotic cells progress through to divide.

cell differentiation Refers to the phenomenon in which cells become specialized into particular cell types.

cell doctrine *See* cell theory.

cell fate The ultimate morphological features that a cell or group of cells will adopt.

cell junctions Specialized structures that adhere cells to each other and to the ECM.

cell-mediated immunity A type of specific immunity in which cytotoxic T cells directly attack and destroy infected body cells, cancer cells, or transplanted cells.

cell nucleus The membrane-bounded area of a eukaryotic cell in which the genetic material is found.

cell plate In plant cells, a structure that forms a cell wall between the two daughter cells.

cell signaling A vital function of the plasma membrane that involves cells sensing changes in their environment and interacting with each other.

cell surface receptor A receptor found in the plasma membrane that enables a cell to respond to different kinds of signaling molecules.

cell theory A theory that states that all organisms are made of cells. Cells come from pre-existing cells by cell division.

cell wall A relatively rigid, porous structure that supports and protects the plasma membrane and cytoplasm of prokaryotic, plant, fungal, and certain protist cells.

cellular differentiation The process by which different cells within a developing organism acquire specialized forms and functions, due to the expression of cell-specific genes.

cellular respiration A process by which living cells obtain energy from organic molecules.

cellular response Adaptation at the cellular level that often times involves a cell responding to signals in its environment.

cellulose The main macromolecule of the primary cell wall of plants and green algae; a polymer made of repeating molecules of glucose attached end to end.

centimorgan (cM) *See* map unit (mu).

central cell In the female gametophyte of a flowering plant, a large cell that contains two nuclei; after double fertilization it forms the first cell of the nutritive endosperm tissue.

central dogma Refers to the steps of gene expression at the molecular level. DNA is transcribed into mRNA and mRNA is translated into a polypeptide.

central region The region of a plant apical meristem that produces stem tissue.

central vacuole An organelle that often occupies 80% or more of the cell volume of plant cells and stores a large amount of water, enzymes, and inorganic ions.

central zone The area of a plant apical meristem where undifferentiated stem cells are maintained.

centrioles A pair of structures within the centrosome of animal cells. Most plant cells and many protists lack centrioles.

centromere The region where the two sister chromatids are tightly associated; the centromere binds to the kinetochore.

centrosome A single structure often near the cell nucleus of eukaryotic cells that forms a nucleating site for the growth of microtubules.

cephalization The localization of sensory structures at the anterior end of the body of animals.

cephalothorax The fused head and thorax structure in species of the class Arachnida and Crustacea.

cerebellum The part of the hindbrain, along with the pons, responsible for monitoring and coordinating body movements.

cerebral cortex The surface layer of gray matter that covers the cerebrum of the brain.

cerebral ganglia In flatworms, a paired structure that receives input from photoreceptors in eyespots and sensory cells.

cerebrospinal fluid Fluid that surrounds the exterior of the brain and spinal cord and absorbs physical shocks to the brain resulting from sudden movements or blows to the head.

cerebrum A group of structures in the forebrain that are responsible for the higher functions of conscious thought, planning, and emotion in vertebrates.

cervix A fibrous structure at the end of the female vagina that forms the opening to the uterus.

channel A transmembrane protein that forms an open passageway for the direct diffusion of ions or molecules across a membrane.

character A visible characteristic, such as the appearance of seeds, pods, flowers, and stems.

character displacement The tendency for two species to diverge in morphology and thus resource use because of competition.

charophyceans The lineage of freshwater green algae that is most closely related to the land plants.

charged tRNA A tRNA with its attached amino acid; also called aminoacyl tRNA.

checkpoint One of three critical regulatory points found in the cell cycle of eukaryotic cells. At these checkpoints, a variety of proteins act as sensors to determine if a cell is in the proper condition to divide.

checkpoint protein A protein that senses if a cell is in the proper condition to divide and prevents a cell from progressing through the cell cycle if it is not.

chemical element Each specific type of atom—nitrogen, hydrogen, oxygen, and so on.

chemical energy The potential energy contained within covalent bonds in molecules.

chemical equilibrium In a chemical reaction, occurs when the rate of formation of products equals the rate of formation of reactants.

chemical mutagen A chemical that causes mutations.

chemical reaction A reaction that occurs when one or more substances are changed into other substances. This can happen when two or more elements or compounds combine with each other to form a new compound, when one compound breaks down into two or more molecules, or when electrons are added to or taken away from an atom.

chemical synapse A synapse in which a chemical called a neurotransmitter is released from the nerve terminal and acts as a signal from the presynaptic to the postsynaptic cell.

chemiosmosis A process for making ATP in which energy stored in an ion electrochemical gradient is used to make ATP from ADP and P_i.

chemoautotroph An organism able to use energy obtained by chemical modifications of inorganic compounds to synthesize organic compounds.

chemoorganotroph An organism that must obtain organic molecules both for energy and as a carbon source.

chemoreceptor Specialized cells located in the vertebrate aorta, carotid arteries, and brainstem that detect the circulating levels of hydrogen ions and the partial pressures of carbon dioxide and oxygen, and relay that information through nerves or interneurons to the respiratory centers.

chiasma The connection at a crossover site of two chromosomes.

chimeric gene A gene formed from the fusion of two gene fragments to each other.

chitin A tough, nitrogen-containing polysaccharide that forms the external skeleton of many insects and the cell walls of fungi.

chlorophyll A green pigment found in photosynthetic plants, algae, and bacteria.

chlorophyll *a* Type of chlorophyll pigment found in cyanobacteria, and photosynthetic algae and plants.

chlorophyll *b* Type of chlorophyll pigment found in green algae and plants.

chloroplast genome The chromosome found in chloroplasts.

chlorosis The yellowing of plant leaves caused by various types of mineral deficiency.

choanocyte A specialized cell of sponges that functions to trap and eat small particles.

chondrichthyans Members of the class Chondrichthyes, including sharks, skates, and rays.

chordate An organism with a spinal cord.

chorion An extraembryonic membrane in the amniotic egg that, along with the allantois, exchanges gases between the embryo and the surrounding air.

chorionic gonadotropin (CG) An LH-like hormone made by the placenta that maintains the corpus luteum.

chromatin Refers to the biochemical composition of chromosomes, which contain DNA and many types of proteins.

chromosome territory A distinct, nonoverlapping area where each chromosome is located within the cell nucleus of eukaryotic cells.

chromosome theory of inheritance An explanation of how the steps of meiosis account for the inheritance patterns observed by Mendel.

chromosome A unit of genetic material composed of DNA and associated proteins. Each cell has a characteristic number of chromosomes. Eukaryotes have chromosomes in their cell nuclei and in plastids and mitochondria.

chylomicrons Large fat droplets coated with amphipathic proteins that perform an emulsifying function similar to that of bile salts; they are formed in intestinal epithelial cells from absorbed fats in the diet.

chyme A solution of food particles in the stomach that contains water, salts, molecular fragments of proteins, nucleic acids, polysaccharides, droplets of fat, and various other small molecules.

chymotrypsin A protease involved in the breakdown of proteins in the small intestine.

chytrids Simple, early-diverging lineages of fungi; commonly found in aquatic habitats and moist soil, where they produce flagellate reproductive cells.

ciliate A protist that moves by means of cilia, which are tiny hairlike extensions on the outsides of cells.

cilium (plural, **cilia**) A cell appendage that functions like flagella to facilitate cell movement; cilia are shorter and more numerous on cells than are flagella.

circulatory system A system that transports necessary materials to all cells of an animal's body, and transports waste products away from cells. The three basic types of circulatory systems are gastrovascular cavities, open systems, and closed systems.

***cis*-acting element** See *cis*-effect.

***cis*-effect** A DNA segment that must be adjacent to the gene(s) that it regulates. The *lac* operator site is an example of a cis-acting element.

cisternae Flattened, fluid-filled tubules within the cell.

citric acid cycle A cycle that results in the breakdown of carbohydrates to carbon dioxide; also known as the Krebs cycle.

clade See monophyletic group.

cladistic approach An approach that reconstructs a phylogenetic tree by comparing primitive and shared derived characters.

cladogenesis A pattern of speciation in which a species is divided into two or more species.

cladogram A phylogenetic tree based on a cladistic approach.

clamp connection In basidiomycete fungi, a structure that helps distribute nuclei during cell division.

clasper An extension of the pelvic fin of a chondrichthyan, used by the male to transfer sperm to the female.

class A subdivision of a phylum.

classical conditioning A type of associative learning in which an involuntary response comes to be associated positively or negatively with a stimulus that did not originally elicit the response.

cleavage A succession of rapid cell divisions with no significant growth that produces a hollow sphere of cells called a blastula.

cleavage furrow In animal cells, an area that constricts like a drawstring to separate the cells.

climate The prevailing weather pattern in a given region.

climax community A distinct end point of succession.

clitoris Located at the anterior part of the labia minora, erectile tissue that becomes engorged with blood during sexual arousal and is very sensitive to sexual stimulation.

clonal deletion A process that explains why individuals normally lack active lymphocytes that respond to self components; T cells with receptors capable of binding self proteins are destroyed by apoptosis.

clonal inactivation A process that explains why individuals normally lack active lymphocytes that respond to self components; the process occurs outside the thymus and causes potentially self-reacting T cells to become nonresponsive.

clonal selection The process by which certain cells are selected to proliferate as clones.

cloning Methods that produce many copies of something. These can be copies of genes, copies of genetically identical cells, or copies of genetically identical organisms. The cloning of mammals can be achieved by fusing a somatic cell with an egg that has had its nucleus removed. Plants can be cloned simply by removing cells, and growing them in particular mixtures of hormones.

closed circulatory system A circulatory system in which blood flows throughout an animal entirely within a series of vessels and is kept separate from the interstitial fluid.

closed conformation Tightly packed chromatin that cannot be transcribed into RNA.

clumped The most common pattern of dispersion within a population, in which individuals are gathered in small groups.

cnidocil On the surface of a cnidocyte, a hairlike trigger that detects stimuli.

cnidocyte A characteristic feature of cnidarians; a stinging cell that functions in defense or the capture of prey.

coacervates Droplets that form spontaneously from the association of charged polymers such as proteins, carbohydrates, or nucleic acids.

coat protein A protein that surrounds a membrane vesicle and facilitates vesicle formation.

cocci Spheres; one of the five major shapes of prokaryotic cells.

cochlea A coiled structure containing the hair cells and other structures that generate the responses that travel via the auditory nerve to the brain.

coding strand The DNA strand opposite to the template (or noncoding strand).

codominance The phenomenon in which a single individual expresses both alleles.

codon A sequence of three nucleotide bases that specifies a particular amino acid or a stop codon; codons function during translation.

coefficient of relatedness The probability that any two individuals will share a copy of a particular gene is a quantity, r.

coelom A fluid-filled body cavity in an animal.

coelomate An animal with a true coelom.

coenzyme An organic molecule that participates in the chemical reaction but is left unchanged after the reaction is completed.

coevolution The process by which two or more species of organisms influence each other's evolutionary pathway.

cofactor Usually an inorganic ion that temporarily binds to the surface of an enzyme and promotes a chemical reaction.

cognitive learning The ability to solve problems with conscious thought and without direct environmental feedback.

cohesion-tension theory The explanation for long-distance water transport as the combined effect of the cohesive forces of water and evaporative tension.

cohort A group of organisms of the same age.

coleoptile A protective sheath that encloses the first bud of an epicotyl in a mature monocot embryo.

coleorhiza A protective envelope that encloses the monocot hypocotyl.

colinearity rule The phenomenon whereby the order of homeotic genes along the chromosome correlates with their expression along the anteroposterior axis of the body.

collagen A protein secreted from animal cells that forms large fibers in the extracellular matrix.

collecting duct A tubule in the human vertebrate kidney that collects urine from nephrons.

collenchyma cells Flexible cells that make up collenchyma tissue.

collenchyma tissue A tissue that provides support to plant organs.

colligative property Properties of water that depend on the amounts of dissolved substances. For example, the colligative properties of water cause certain solutes to function as antifreeze in certain organisms, and thereby lower the freezing point.

colloid A gel-like substance in the follicles of the thyroid gland.

colon A part of the large intestine consisting of three relatively straight segments—the ascending, transverse, and descending portions. The terminal portion of the descending colon is S-shaped, forming the sigmoid colon, which empties into the rectum.

colony hybridization A method that uses a labeled probe that recognizes a specific gene to identify that gene in a DNA library.

combinatorial control The phenomenon whereby a combination of many factors determines the expression of any given gene.

commensalism An interaction that benefits one species and leaves the other unaffected.

communication The use of specially designed visual, chemical, auditory or tactile signals to modify the behavior of others.

community An assemblage of many populations that live in the same place at the same time.

community ecology The study of how populations of species interact and form functional communities.

compartmentalization A characteristic of eukaryotic cells that is defined by many organelles that separate the cell into different regions. Cellular compartmentalization allows a cell to carry out specialized chemical reactions in different places.

competent The term used to describe bacterial strains that have the ability to take up DNA from the environment.

competition An interaction that affects two or more species negatively, as they compete over food or other resources.

competitive exclusion hypothesis The proposal that two species with the same resource requirements cannot occupy the same niche.

competitive inhibitor A molecule that binds to the active site of an enzyme and inhibits the ability of the substrate to bind.

complement The family of plasma proteins that provides a means for extracellular killing of microbes without prior phagocytosis.

complementary DNA (cDNA) DNA molecules that are made from mRNA as a starting material.

complementary In DNA, you can predict the sequence in one DNA strand if you know the sequence in the opposite strand according to the AT/GC rule.

complete flower A flower that possesses all four types of flower organs.

complete metamorphosis A dramatic change in body form in the majority of insects, from larva to a very different looking adult.

compound A molecule composed of two or more different elements.

compound eyes Image-forming eyes in arthropods and some annelids consisting of several hundred to several thousand light detectors called ommatidia.

computational molecular biology An area of study that uses computers to characterize the molecular components of living things.

concentration The amount of a solute dissolved in a unit volume of solution.

condensation reaction *See* dehydration reaction.

conditioned response The response that is created by a new, conditioned stimulus.

conditioned stimulus A new stimulus that is delivered at the same time as the old stimulus, and that over time, is sufficient to elicit the same response.

condom A sheathlike membrane worn over the penis that collects the ejaculate; in addition to their contraceptive function, condoms significantly reduce the risk of STDs such as HIV infection, syphilis, gonorrhea, chlamydia, and herpes.

conduction The process in which the body surface loses or gains heat through direct contact with cooler or warmer substances.

cone pigments The several types of visual pigments in cones.

cones 1. Photoreceptors found in the vertebrate eye; they are less sensitive to low levels of light but can detect color. Cones are used in daylight by most diurnal vertebrate species and by some insects. 2. The reproductive structures of conifer plants.

congestive heart failure The condition resulting from the failure of the heart to pump blood normally; this results in fluid build-up in the lungs (congestion).

conidia A type of asexual reproductive cell produced by many fungi.

Conifers A phylum of gymnosperm plants, Coniferophyta.

conjugation A type of genetic transfer between bacteria that involves a direct physical interaction between two bacterial cells.

connective tissue Clusters of cells that connect, anchor, and support the structures of an animal's body; derived from mesenchyme and include blood, adipose (fat-storing) tissue, bone, cartilage, loose connective tissue, and dense connective tissue.

connexon A channel that forms gap junctions consisting of six connexin proteins in one cell aligned with six connexin proteins in an adjacent cell.

conservation biology The study that uses principles and knowledge from molecular biology, genetics, and ecology to protect the biological diversity of life at all levels.

conservative mechanism In this incorrect model for DNA replication, both parental strands of DNA remain together following DNA replication. The original arrangement of parental strands is completely conserved, while the two newly made daughter strands are also together following replication.

constant regions In immunology, the amino acid sequences of the Fc domains, which are identical for all immunoglobulins of a given class.

constitutive gene An unregulated gene that has essentially constant levels of expression in all conditions over time.

contig A series of clones that contain overlapping pieces of chromosomal DNA.

continental drift The phenomenon whereby, over the course of billions of years, the major landmasses, known as the continents, have shifted their positions, changed their shapes, and in some cases have become separated from each other.

continuous trait A trait that shows continuous variation over a range of phenotypes.

contraception The use of birth control procedures to prevent fertilization or implantation of a fertilized egg.

contractile vacuole A small, membrane enclosed, water-filled space that eliminates excess liquid from the cells of certain protists.

contrast In microscopy, relative differences in the lightness, darkness, or color between adjacent regions in a sample. Contrast improves the ability to discern adjacent objects.

control sample The sample in an experiment that is treated just like an experimental sample except that it is not subjected to one particular variable. For example, the control and experiment samples may be treated identically except that the temperature may vary for the experimental sample.

convection The transfer of heat by the movement of air or water next to the body.

convergent evolution The process whereby two different species from different lineages show similar characteristics because they occupy similar environments.

convergent extension A cellular process during gastrulation that is crucial to development; two rows of cells merge to form a single elongated layer.

convergent trait See analogous structure.

copulation The process of sperm being deposited within the reproductive tract of the female.

coral reef A type of aquatic biome found in warm, marine environments.

core promoter For a eukaryotic structural gene, refers to the transcriptional start site and TATA box.

corepressor A small effector molecule that binds to a repressor protein to inhibit transcription.

Coriolis effect The effect of the Earth's rotation on the surface flow of wind.

cork cambium A secondary meristem in a plant that produces cork tissue.

cornea A thin, clear layer on the front of the vertebrate eye.

corolla The petals of a flower, which occur in the whorl to the inside of the calyx and the outside of the stamens.

corona The ciliated crown of members of the phylum Rotifera.

coronary artery An artery that carries oxygen and nutrients to the heart muscle.

coronary artery bypass A common treatment to restore blood flow through a blood vessel. A small piece of healthy blood vessel is removed from one part of the body and surgically grafted onto the coronary circulation in such a way that blood bypasses the diseased artery.

coronary artery disease A condition that occurs when plaques form in the coronary vessels.

corpus callosum The major tract that connects the two hemispheres of the cerebrum.

corpus luteum A structure that is responsible for secreting hormones that stimulate the development of the uterus needed for sustaining the embryo in the event of a pregnancy. If pregnancy does not occur, the corpus luteum degenerates.

correlation A meaningful relationship between two variables.

cortex The area of a plant stem or root beneath the epidermis that is largely composed of parenchyma tissue.

cortical reaction An event in fertilization in which calcium and IP_3 produce additional barriers to more than one sperm cell binding to and uniting with an egg, a process called the slow block to polyspermy.

cotranslational sorting The sorting process in which the synthesis of certain eukaryotic proteins begins in the cytosol and then halts temporarily until the ribosome has become bound to the ER membrane. After this occurs, translation resumes and the polypeptide is synthesized into the ER lumen or ER membrane.

cotransporter See symporter.

cotyledon An embryonic seed leaf.

countercurrent exchange mechanism An arrangement of water and blood flow in which water enters a fish's mouth and flows between the lamellae of the gills in the opposite direction to blood flowing through the lamellar capillaries.

countercurrent heat exchange A method of regulating heat loss to the environment; many animals conserve heat by returning it to the body's core and keeping the core much warmer than the extremities.

covalent bond A chemical bond that occurs when atoms share pairs of electrons.

CpG island A cluster of CpG sites. CpG refers to the nucleotides of C and G in DNA that are connected by a phosphodiester linkage.

cranial nerve A nerve in the peripheral nervous system that is directly connected to the brain; cranial nerves are located in the head and transmit incoming and outgoing information between the peripheral nervous system and the brain.

craniate A chordate that has a brain encased in a skull and possesses a neural crest.

cranium A protective bony or cartilaginous housing that encases the brain of a craniate.

crenation The process of cell shrinkage that occurs if animal cells are placed in a hypertonic medium—water exits the cells via osmosis and equalizes solute concentrations on both sides of the membrane.

critical period A limited time period of development in which many animals develop species-specific patterns of behavior.

crop A storage organ that is a dilation of the lower esophagus; found in most birds and many invertebrates, including insects and some worms.

cross-bridge cycle During muscle contraction, the sequence of events that occurs between the time when a cross-bridge binds to a thin filament and when it is set to repeat the process.

cross-bridge A region of myosin molecules that extend from the surface of the thick filaments toward the thin filaments in a skeletal muscle.

cross-fertilization The fusion of gametes formed by different individuals.

crossing over The exchange of genetic material between homologous chromosomes, which allows for increased variation in the genetic information each parent may pass to the offspring.

cross-pollination The process in which a stigma receives pollen from a different plant of the same species.

cryptic coloration The blending of an organism with the background color of its habitat; also known as camouflage.

cryptochrome A type of blue-light receptor in plants and protists.

CT scan Computerized tomography, which is a technique for examining the structure and activity level of the brain without anesthesia or surgery. An X-ray beam and a series of detectors rotate around the head, producing slices of images that are reconstructed into three-dimensional images based on differences in the density of brain tissue.

C-terminus The location of the last amino acid in a polypeptide.

cupula A gelatinous structure within the lateral line organ that helps an organism to detect changes in water movement.

cuticle A coating of wax and cutin that helps to reduce water loss from plant surfaces. Also, a nonliving covering that serves to both support and protect an animal.

cutin A polyester polymer produced at the surfaces of plants; helps to prevent attack by pathogens.

cycads A phylum of gymnosperm plants, Cycadophyta.

cyclic adenosine monophosphate (cAMP) A second messenger molecule.

cyclic AMP (cAMP) Cyclic adenosine monophosphate; a small effector molecule that is produced from ATP via an enzyme known as adenylyl cyclase.

cyclic electron flow See cyclic photophosphorylation.

cyclic photophosphorylation A pattern of electron flow in the thylakoid membrane that is cyclic and generates ATP alone.

cyclin A protein responsible for advancing a cell through the phases of the cell cycle by binding to a cyclin-dependent kinase.

cyclin-dependent kinase (cdk) A protein responsible for advancing a cell through the phases of the cell cycle. Its function is dependent on the binding of a cyclin.

cyst A one-to-few celled structure that often has a thick, protective wall and can remain dormant through periods of unfavorable climate or low food availability.

cytogenetics   The field of genetics that involves the microscopic examination of chromosomes.

cytokines   A family of proteins that function in both nonspecific and specific immune defenses by providing a chemical communication network that synchronizes the components of the immune response.

cytokinesis   The division of the cytoplasm to produce two distinct daughter cells.

cytokinin   A type of plant hormone; promotes cell division in addition to other effects.

cytoplasm   The region of the cell that is contained within the plasma membrane.

cytoplasmic inheritance   *See* extranuclear inheritance.

cytoplasmic streaming   A phenomenon in which the cytoplasm circulates throughout the cell to distribute resources efficiently in large cells, such as algal or plant cells.

cytosine (C)   A pyrimidine base found in DNA and RNA.

cytoskeleton   In eukaryotes, a network of three different types of protein filaments called microtubules, intermediate filaments, and actin filaments.

cytosol   The region of a eukaryotic cell that is inside the plasma membrane and outside the organelles.

cytotoxic T cell   A type of lymphocyte that travels to the location of its target, binds to the target by combining with an antigen on it, and directly kills the target via secreted chemicals.

D

dalton (Da)   One-twelfth the mass of a carbon atom, or about the mass of a proton or a hydrogen atom.

Darwinian fitness   The relative likelihood that a genotype will contribute to the gene pool of the next generation as compared with other genotypes.

database   A large number of computer data files that are collected, stored in a single location, and organized for rapid search and retrieval.

daughter strand   The newly made strand in DNA replication.

day-neutral plant   A plant that flowers regardless of the night length, as long as day length meets the minimal requirements for plant growth.

deafness   Hearing loss, usually caused by damage to the hair cells within the cochlea, although some cases result from functional problems in brain areas that process sound or in nerves that carry information to the brain from the hair cells.

decomposer   *See* detritivore.

defecation   The expulsion of feces that occurs through the final portion of the digestive canal, the anus; contractions of the rectum and relaxation of associated sphincter muscles expel the feces.

defensive mutualism   A mutually beneficial interaction often involving an animal defending a plant or a herbivore in return for food or shelter.

deficiency   The term used to describe when a segment of chromosomal material is missing.

deforestation   The conversion of forested areas by humans to nonforested land.

degenerate   In the genetic code, this means that more that one codon can specify the same amino acid.

dehydration   A reduction in the amount of water in the body.

dehydration reaction   A reaction that involves the removal of a water molecule, and the formation of a covalent bond between two separate molecules.

delayed implantation   A reproductive cycle in which a fertilized egg reaches the uterus but does not implant until later, when environmental conditions are more favorable for the newly produced young.

delayed ovulation   A reproductive cycle in which the ovarian cycle in females is halted before ovulation, and sperm are stored and nourished in the female's uterus over the winter. Upon arousal from hibernation in the spring, the female ovulates one or more eggs, which are fertilized by the stored sperm.

deletion   The term used to describe a missing region of a chromosome.

demographic transition   The shift in birth and death rates accompanying human societal development.

demography   The study of birth rates, death rates, age distributions, and the sizes of populations.

dendrite   A type of extension or projection that arises from the cell body; chemical and electrical messages from other neurons are received by the dendrites, and electrical signals move toward the cell body.

dendritic cell   A type of cell derived from bone marrow stem cells that plays an important role in nonspecific immunity; these cells are scattered throughout most tissues, where they perform various macrophage functions.

denitrification   The reduction of nitrate to gaseous nitrogen.

density   In the context of populations, the numbers of organisms in a given unit area.

density-dependent factor   A mortality factor whose influence varies with the density of the population.

density-independent factor   A mortality factor whose influence is not affected by changes in population size or density.

deoxynucleoside triphosphates   Individual nucleotides with three phosphate groups.

deoxyribonucleic acid (DNA)   One of two classes of nucleic acids; the other is ribonucleic acid (RNA). A DNA molecule consists of two strands of nucleotides coiled around each other to form a double helix, held together by hydrogen bonds according to the AT/GC rule.

deoxyribose   A five-carbon sugar found in DNA.

depolarization   The change in the membrane potential that occurs when the cell becomes less polarized, that is, less negative relative to the surrounding fluid.

dermal tissue   The covering on various parts of a plant.

descent with modification   Darwin's theory that existing life-forms on our planet are the product of the modification of pre-existing life-forms.

desertification   The overstocking of land with domestic animals that can greatly reduce grass coverage through overgrazing, turning the area more desert-like.

determinate cleavage   A characteristic of protostome development in which the fate of each embryonic cell is determined very early.

determined   The term used to describe a cell that is destined to differentiate into a particular cell type.

detritivore   A consumer that gets its energy from the remains and waste products of organisms.

detritus   Unconsumed plants that die and decompose, along with the dead remains of animals and animal waste products.

deuterostome   An animal exhibiting radial cleavage, indeterminate cleavage and where the blastopore becomes the anus; includes echinoderms and vertebrates.

development   In biology, a series of changes in the state of a cell, tissue, organ, or organism; the underlying process that gives rise to the structure and function of living organisms.

developmental genetics   A field of study aimed at understanding how gene expression controls the process of development.

diaphragm   A large muscle that subdivides the thoracic cavity from the abdomen in mammals.

diastole   The first phase of the cardiac cycle, in which the ventricles fill with blood coming from the atria through the open AV valves.

diazotroph   A bacterium that fixes nitrogen.

dideoxy chain-termination method   The most common method of DNA sequencing that utilizes dideoxynucleotides as a reagent.

dideoxy sequencing   *See* dideoxy chain-termination method.

differential gene regulation   The phenomenon in which the expression of genes is altered. Differential gene expression allows cells to adapt to environmental conditions, change during development, and differentiate into particular cell types.

differentiated   The term used to describe the actual alteration of a cell's morphology and physiology.

diffusion   For dissolved substances, occurs when a solute moves from a region of high concentration to a region of lower concentration.

digestion   The process of breaking down nutrients in food into smaller molecules that can be directly used by cells.

digestive system   In a vertebrate, this system consists of the alimentary canal plus several associated structures, not all of which are found in all vertebrates: the tongue, teeth, salivary glands, liver, gallbladder, and pancreas.

dihybrid   An offspring that is a hybrid with respect to two traits.

dikaryotic   The occurrence of two genetically distinct nuclei in the cells of fungal hyphae after mating has occurred.

dinosaur   A term meaning "terrible lizard" used to describe some of the extinct fossil reptiles.

dioecious   The term to describe plants that produce staminate and carpellate flowers on separate plants.

diploblastic   Having two distinct germ layers—ectoderm and endoderm but not mesoderm.

diploid cell   A cell that carries two sets of chromosomes.

diploid-dominant species   Species in which the diploid organism is the prevalent organism in the life cycle. Animals are an example.

direct calorimetry   A method of determining basal metabolic rate that involves quantifying the amount of heat generated by the animal.

direct repair   Refers to a DNA repair system in which an enzyme finds an incorrect structure in the DNA and directly converts it back to the correct structure.

directional selection   A pattern of natural selection that favors individuals at one extreme of a phenotypic distribution that have greater reproductive success in a particular environment.

directionality   In a DNA or RNA strand, refers to the orientation of the sugar molecules within that strand. Can be 5' to 3' or 3' to 5'.

disaccharide   A carbohydrate composed of two monosaccharides.

discontinuous trait   A trait with clearly defined phenotypic variants.

discovery science   *See* discovery-based science.

discovery-based science   The collection and analysis of data without the need for a preconceived hypothesis. Also called discovery science.

dispersion   A pattern of spacing in which individuals in a population are clustered together or spread out to varying degrees.

dispersive mechanism   In this incorrect model for DNA replication, segments of parental DNA and newly made DNA are interspersed in both strands following the replication process.

dispersive mutualism   A mutually beneficial interaction often involving plants and pollinators that disperse their pollen, and plants and fruit eaters that disperse the plant's seeds.

disruptive selection   A pattern of natural selection that favors the survival of two or more different genotypes that produce different phenotypes.

dissociation constant   An equilibrium constant between a ligand and a protein, such as a receptor or an enzyme.

distal convoluted tubule   A structure in the tubule of the nephron through which fluid flows into one of the many collecting ducts in the kidney.

diversity-stability hypothesis   The proposal that species-rich communities are more stable than those with fewer species.

DNA (deoxyribonucleic acid)   The genetic material that provides a blueprint for the organization, development, and function of living things.

DNA fingerprinting   A technology that identifies particular individuals using properties of their DNA.

DNA helicase   An enzyme that uses ATP and separates DNA strands.

DNA library   A collection of vectors each containing a particular fragment of chromosomal DNA or cDNA.

DNA ligase   An enzyme that catalyzes the formation of a covalent bond between nucleotides in adjacent DNA fragments to complete the replication process in the lagging strand.

DNA methylase   An enzyme that attaches methyl groups to bases in DNA.

DNA methylation   A process in which methyl groups are attached to bases in DNA. This usually inhibits gene transcription by preventing the binding of activator proteins or by promoting the compaction of chromatin.

DNA microarray   A technology used to monitor the expression of thousands of genes simultaneously.

DNA polymerase   An enzyme responsible for covalently linking nucleotides together to form DNA strands.

DNA primase   An enzyme that synthesizes a primer for DNA replication.

DNA repair systems   One of several systems to reverse DNA damage before a permanent mutation can occur.

DNA replication   The mechanism by which DNA can be copied.

DNA sequencing   A method to determine the base sequence of DNA.

DNA topoisomerase   An enzyme that affects the level of DNA supercoiling.

DNase   An enzyme that digests DNA.

domain   1. A defined region of a protein with a distinct structure and function. 2. One of the three major categories of life: Bacteria, Archaea, and Eukarya.

dominant   A term that describes the displayed trait in a heterozygote.

domestication   A process that involves artificial selection of plants or animals for traits desirable to humans.

dominant species   A species that has a large effect in a community because of its high abundance or high biomass.

dormancy   A phase of metabolic slowdown in a plant.

dorsal   Refers to the upper side of an animal.

dorsoventral axis   One of the three axes along which the adult body pattern is organized; the others are the anteroposterior axis and the right-left axis.

dosage compensation   The phenonomen that gene dosage is compensated between males and females. In mammals, the inactivation of one X chromosome in the female reduces the number of expressed copies (doses) of X-linked genes from two to one.

double bond   A bond that occurs when the atoms of a molecule share two pairs of electrons.

double fertilization   In angiosperms, the process in which two different fertilization events occur, producing both a zygote and a nutritive endosperm tissue.

double helix   Two strands of DNA hydrogen-bonded with each other. In a DNA double helix, two DNA strands are twisted together to form a structure that resembles a spiral staircase.

Down syndrome   A human disorder caused by the inheritance of three copies of chromosome 21.

duplication   The term used to describe when a section of a chromosome occurs two or more times.

dynamic instability   The oscillation of a single microtubule between growing and shortening phases; important in many cellular activities including the sorting of chromosomes during cell division.

E

ecdysis   The periodic shedding and re-formation of the exoskeleton.

ecdysone   A steroid hormone synthesized and secreted by the prothoracic glands of certain invertebrates such as arthropods; in response to this hormone, a larva molts and begins a new growth period until it must shed its skin again in response to another episode of ecdysone secretion.

Ecdysozoa   A clade of moulting animals that encompasses primarily the arthropods and nematodes.

echolocation   The phenomenon in which certain species generate high-frequency sound waves in order to determine the distance and location of an object.

ecological footprint   The amount of productive land needed to support each person on Earth.

ecological niche   The unique set of habitat resources that a species requires, as well as its influence on the environment and other species.

ecological species concept   A species concept that considers a species within its native environment. Each species occupies its own ecological niche.

ecology   The study of interactions among organisms and between organisms and their environments.

ecosystem   The biotic community of organisms in an area as well as the abiotic environment affecting that community.

ecosystem engineer   A species that creates, modifies, and maintains habitats.

ecosystems ecology   The study of the flow of energy and cycling of nutrients among organisms within a community and between organisms and the environment.

ectoderm   The outer most layer of cells formed during gastrulation that covers the surface of the embryo and differentiates into the epidermis and nervous system.

ectomycorrhizae Beneficial interactions between temperate forest trees and soil fungi whose hyphae coat tree-root surfaces and grow into the spaces between root cells.

ectoparasite A parasite that lives on the outside of the host's body.

ectotherm An animal whose body temperature changes with the environmental temperature.

edge effect A special physical condition that exists at the boundary or "edge" of an area.

effective population size The number of individuals that contribute genes to future populations, often smaller than the actual population size.

effector A molecule that directly influences cellular responses.

effector cell A component of the immune response; these cells carry out the attack response.

egg Also, egg cell. The female gamete.

ejaculation The movement of semen through the urethra by contraction of muscles at the base of the penis.

ejaculatory duct The structure within the male penis through which semen is released.

elastin A protein that makes up elastic fibers in the extracellular matrix of animals.

electrical synapse A synapse that directly passes electric current from the presynaptic to the postsynaptic cell via gap junctions.

electrocardiogram (ECG or EKG) A record of the electrical impulses generated during the cardiac cycle.

electrochemical gradient The combined effect of both an electrical and chemical gradient; determines the direction that an ion will move.

electrogenic pump A pump that generates an electrical gradient across a membrane.

electromagnetic receptor A sensory receptor that detects radiation within a wide range of the electromagnetic spectrum, including visible, ultraviolet, and infrared light, as well as electrical and magnetic fields in some animals.

electromagnetic spectrum All possible wavelengths of electromagnetic radiation, from relatively short wavelengths (gamma rays) to much longer wavelengths (radio waves).

electron A negatively charged particle found in orbitals around the nucleus. For atoms, the number of protons is equal to the number of electrons.

electron microscope A microscope that uses an electron beam for illumination.

electron transport chain A group of protein complexes and small organic molecules embedded in the inner mitochondrial membrane. The components accept and donate electrons to each other in a linear manner. The movement of electrons produces an H^+ electrochemical gradient.

electronegativity A measure of an atom's ability to attract electrons to its outer shell from another atom.

elicitor A compound produced by bacterial and fungal pathogens that promotes virulence.

elongation factor In translation, a protein that is needed for the growth of a polypeptide.

elongation stage The second step in transcription or translation where RNA strands or polypeptides are made, respectively.

embryo The early stages of development in a multicellular organism during which the organization of the organism is largely formed.

embryogenesis The process by which embryos develop from single-celled zygotes by mitotic divisions.

embryonic development The process by which a fertilized egg is transformed into an organism with distinct physiological systems and body parts.

embryonic germ cell (EG cell) At the early fetal stage of development, the cells that later give rise to sperm or eggs cells. These cells are pluripotent.

embryonic stem cell (ES cell) A cell in the early mammalian embryo that can differentiate into almost every cell type of the body.

embryophyte A synonym for the land plants.

emerging virus A newly arising virus.

empirical thought Thought that relies on observation to form an idea or hypothesis, rather than trying to understand life from a nonphysical or spiritual point of view.

emulsification A process during digestion that disrupts the large lipid droplets into many tiny droplets, thereby increasing their total surface area and exposure to lipase action.

enantiomer A type of stereoisomer that exists as a mirror image of another molecule.

endangered species Those species that are in danger of extinction throughout all or a significant portion of their range.

endemic The term to describe organisms that are naturally found only in a particular location.

endergonic Refers to a reaction that has a positive free energy change and does not proceed spontaneously.

endocrine disruptor A chemical found in water and soil exposed to pollution runoff that has a molecular structure that in some cases resembles estrogen sufficiently to bind to estrogen receptors.

endocrine gland A structure that contains epithelial cells that secrete hormone molecules into the bloodstream, where they circulate throughout the body.

endocrine system All the endocrine glands and other organs with hormone-secreting cells.

endocytic pathway A pathway to take substances into the cell; the reverse of the secretory pathway.

endocytosis A process in which the plasma membrane invaginates, or folds inward, to form a vesicle that brings substances into the cell.

endoderm The inner most layer of cells formed during gastrulation that lines the gut and gives rise to many internal organs.

endomembrane system A network of membranes that includes the nuclear envelope, which encloses the nucleus, and the endoplasmic reticulum, Golgi apparatus, lysosomes, vacuoles, and plasma membrane.

endomycorrhizae Partnerships between plants and fungi in which the fungal hyphae grow into the spaces between root cell walls and plasma membranes.

endoparasite A parasite that lives inside the host's body.

endophyte A mutualistic fungus that lives compatibly within the tissues of various types of plants.

endoplasmic reticulum (ER) A convoluted network of membranes in the cell's cytoplasm that forms flattened, fluid-filled tubules or cisternae.

endoskeleton An internal skeleton covered by soft tissue; composed of calcareous plates overlaid by a thin skin in echinoderms and composed of a bony skeleton overlaid by muscles in invertebrates.

endosperm A nutritive tissue that increases the efficiency with which food is stored and used in the seeds of flowering plants.

endospore A cell with a tough coat that is produced inside the cells of certain bacteria and then released when the enclosing cell dies and breaks down.

endosporic gametophyte A plant gametophyte that grows within the confines of microspore and megaspore walls.

endosymbiosis A symbiotic relationship in which the smaller species—the symbiont— lives inside the larger species.

endosymbiosis theory A theory that mitochondria and chloroplasts originated from bacteria that took up residence within a primordial eukaryotic cell.

endosymbiotic Describes a relationship in which one organism lives inside the other.

endothelium A single-celled inner layer of a blood vessel, which forms a smooth lining in contact with the blood.

endotherm An animal that generates its own internal heat.

endothermic A term to describe that ability of an organism to generate and retain body heat through its metabolism.

energy The ability to promote change.

energy expenditure The amount of energy an animal uses in a given period of time to power all of its metabolic requirements.

energy flow The movement of energy through an ecosystem.

energy intermediate A molecule such as ATP and NADH that is directly used to drive endergonic reactions in cells.

enhancement effect The phenomenon whereby maximal activation of the pigments in photosystems I and II is achieved when organisms are exposed to two wavelengths of light.

enhancer A response element in eukaryotes that increases the rate of transcription.

enterocoelous In deuterostomes, a pattern of development in which a layer of mesoderm cells forms outpockets that bud off from the developing gut to form the coelom.

enthalpy (H)   The total energy of a system.

entomology   The study of insects.

entropy   The degree of disorder of a system.

environmental science   The application of ecology to real-world problems.

enzyme   A protein responsible for speeding up a chemical reaction in a cell.

enzyme-linked receptor   A receptor found in all living species that typically has two important domains: an extracellular domain, which binds a signaling molecule, and an intracellular domain, which has a catalytic function.

enzyme-substrate complex   The binding between an enzyme and substrate.

eosinophil   A type of phagocyte found in large numbers in mucosal surfaces lining the gastrointestinal, respiratory, and urinary tracts, where they fight off parasitic invasions.

epicotyl   The portion of an embryonic plant stem with two tiny leaves in a first bud that is located above the point of attachment of the cotyledons.

epidermis   A layer of dermal tissue that helps protect a plant from damage.

epididymis   A coiled, tubular structure located on the surface of the testis; it is here that the sperm complete their differentiation by becoming motile and gaining the capacity to fertilize ova.

epigenetic inheritance   An inheritance pattern in which modification of a gene or chromosome during egg formation, sperm formation, or early stages of embryo growth alters gene expression in a way that is fixed during an individual's lifetime.

epinephrine   A hormone secreted by the adrenal glands; also known as adrenaline.

episome   A plasmid that can integrate into the bacterial chromosome.

epistasis   A gene interaction in which the alleles of one gene mask the expression of the alleles of another gene.

epithelial placodes   Regions of slightly thickened epithelial cells.

epithelial tissue   A sheet of densely packed cells that covers the body or individual organs or lines the walls of various cavities inside the body.

epitopes   Antigenic determinants; the peptide fragments of the antigen that are complexed to the MHC proteins and presented to the helper T cell.

equilibrium   1. In a chemical reaction, occurs when the rate of the forward reaction is balanced by the rate of the reverse reaction. 2. In a population, the situation in which the population size stays the same.

equilibrium model of island biogeography   A model to explain the process of succession on new islands that states that the number of species on an island tends toward an equilibrium number that is determined by the balance between immigration rates and extinction rates.

equilibrium potential   In membrane physiology, the membrane potential at which the flow of an ion is at equilibrium—no net movement in either direction.

ER lumen   A single compartment enclosed by the ER membrane.

ER signal sequence   A sorting signal in a polypeptide usually located near the amino terminus that is recognized by SRP (signal recognition particle) and directs the polypeptide to the ER membrane.

erythrocyte   *See* red blood cell.

erythropoietin (EPO)   A hormone made by the liver and kidneys in response to any situation where additional blood cells are required, such as when animals lose blood following an injury; when abused, as in blood-doping, the concentration of red blood cells reaches such high levels that the blood becomes much thicker than normal.

esophagus   The tubular structure that forms a pathway from the throat to the stomach.

essential amino acids   Those amino acids that are required in the diet of particular organisms. In humans, they include isoleucine, leucine, lysine, methionine, phenylalanine, threonine, tryptophan, and valine.

essential fatty acid   A polyunsaturated fatty acid such as linoleic acid that cannot be synthesized by animal cells and must therefore be consumed in the diet.

essential nutrient   A compound that cannot be synthesized from any ingested or stored precursor molecule and so must be obtained in the diet in its complete form.

estradiol   The major estrogen in many animals, including humans.

estrogens   Steroid hormones produced by the female ovaries that affect most aspects of reproduction.

ethology   Scientific studies of animal behavior.

ethylene   A plant hormone that is particularly important in coordinating plant developmental and stress responses.

euchromatin   The less condensed regions of a chromosome; areas that are capable of gene transcription.

eudicots   One of the two largest lineages of flowering plants; the embryo possesses two seed leaves.

Eukarya   One of the three domains of life; the other two are Bacteria and Archaea.

eukaryote   Cells or organisms from the domain Eukarya. The distinguishing feature is cell compartmentalization, which includes a cell nucleus. "Eukaryote" means "true nucleus," and includes protists, fungi, plants, and animals.

eukaryotic   Refers to organisms having cells with internal compartments that serve various functions; includes all members of the domain Eukarya.

Eumetazoa   A subgroup of animals having more than one type of tissue and, for the most part, different types of organs.

euphotic zone   A fairly narrow zone close to the surface of an aquatic environment, where light is sufficient to allow photosynthesis to exceed respiration.

euphyll   A leaf with branched veins.

euphyllophytes   The clade that includes pteridophytes and seed plants.

euploid   An organism that has a chromosome number that is a multiple of a chromosome set ($1n$, $2n$, $3n$, etc.).

eusociality   The phenomenon whereby sterile castes evolve in social insects in which the vast majority of females, known as workers, rarely reproduce themselves but instead help one reproductive female (the queen) to raise offspring.

Eustachian tube   A connection from the middle ear to the pharynx, which maintains the pressure in the middle ear at atmospheric pressure.

eustele   A ring of vascular tissue arranged around a central pith of nonvascular tissue; typical of progymnosperms, gymnosperms, and angiosperms.

eutherian   A placental mammal and member of the subclass Eutheria.

eutrophic   Waters that contain relatively high levels of nutrients such as phosphate or nitrogen and typically exhibit high levels of primary productivity and low levels of biodiversity.

eutrophication   The process by which elevated nutrient levels lead to an overgrowth of algae or aquatic plants and the subsequent depletion of water oxygen levels.

evaporation   The transformation of water from the liquid to the gaseous state.

evapotranspiration rate   The rate at which water moves into the atmosphere through the processes of evaporation from the soil and transpiration of plants.

evolutionarily conserved   The term used to describe DNA sequences that are very similar or identical between different species.

evolutionary developmental biology (evo-devo)   A field of biology that compares the development of different organisms in an attempt to understand ancestral relationships between organisms and the developmental mechanisms that bring about evolutionary change.

evolutionary species concept   A species is derived from a single lineage that is distinct from other lineages and has its own evolutionary tendencies and historical fate.

excitable cell   The term used to describe neurons and muscle cells, because they have the capacity to generate electrical signals.

excitation-contraction coupling   The sequence of events by which an action potential in the plasma membrane of a muscle fiber leads to cross-bridge activity.

excitatory postsynaptic potential (EPSP)   The response from an excitatory neurotransmitter depolarizing the postsynaptic membrane; the depolarization brings the membrane potential closer to the threshold potential that would trigger an action potential.

excurrent siphon   A tunicate structure used to expel water.

exercise   Any physical activity that increases an animal's metabolic rate.

exergonic   Refers to reactions that release free energy and occur spontaneously.

exit site (E site)   A site for the tRNA binding in the ribosome; the other two are the peptidyl site (P site) and the aminoacyl site (A site). This is the site where the uncharged tRNA exits.

exocytosis   A process in which material inside the cell is packaged into vesicles and excreted into the extracellular medium.

exon shuffling   A form of mutation in which exons are inserted into genes and thereby create proteins with additional functional domains.

exon   A portion of RNA that is found in the mature RNA molecule after splicing is finished.

exoskeleton   An external skeleton that surrounds and protects most of the body surface of animals such as insects.

exotic species   Species moved from a native location to another location, usually by humans.

expansin   A protein that occurs in the plant cell wall and fosters cell enlargement.

experimental sample   The sample in an experiment that is subjected to some type of variation that does not occur for the control sample.

exponential growth   J-shaped, rapid population growth that occurs when the per capita growth rate remains above zero.

extensor   A muscle that straightens a limb.

external fertilization   Fertilization that occurs in aquatic environments, when eggs and sperm are released into the water in close enough proximity for fertilization to occur.

extinction   The end of the existence of a species or group of species.

extinction vortex   A downward spiral toward extinction from which a species cannot naturally recover.

extracellular fluid   The fluid in an organism's body that is outside of the cells.

extracellular matrix (ECM)   A network of material that is secreted from cells and forms a complex meshwork outside of cells. The ECM provides strength, support, and organization.

extranuclear inheritance   In eukaryotes, the transmission of genes that are located outside the cell nucleus.

extremophile   An organism that occurs primarily in extreme habitats.

eye   The visual organ that detects light and sends signals to the brain.

eyecup   In planaria, a primitive eye that detects light and its direction.

F

F factor   A segment of DNA called a fertility factor that plays a role in bacterial conjugation.

F$_1$ generation   The first filial generation in a genetic cross.

F$_2$ generation   The second filial generation in a genetic cross.

facilitated diffusion   A method of passive transport that involves the aid of a transport protein.

facilitation   A mechanism for succession in which a species facilitates or makes the environment more suitable for subsequent species.

facultative aerobe   A microorganism that can use oxygen in aerobic respiration, obtain energy via anaerobic fermentation, or use inorganic chemical reactions to obtain energy.

facultative mutualism   An interaction that is beneficial but not essential to the survival and reproduction of either species.

family   A subdivision of an order.

fast block to polyspermy   A depolarization of the egg that blocks other sperm from binding to the egg membrane proteins.

fast fiber   A muscle fiber containing myosin with high ATPase activity.

fast-glycolytic fiber   A skeletal muscle fiber that has high myosin activity but cannot make as much ATP as oxidative fibers, because its source of ATP is glycolysis; best suited for rapid, intense actions.

fast-oxidative fiber   A skeletal muscle fiber that has high myosin activity and can make large amounts of ATP; used for long-term activities.

fat   A molecule formed by bonding glycerol to three fatty acids.

fate mapping   A technique in which a small population of cells within an embryo is specifically labeled with a harmless dye, and the fate of these labeled cells is followed to a later stage of embryonic development.

feedback inhibition   A form of regulation in which the product of a metabolic pathway inhibits an enzyme that acts early in the pathway, thus preventing the over accumulation of the product.

feedforward regulation   The process by which an animal's body begins preparing for a change in some variable before it even occurs.

female-enforced monogamy hypothesis   The suggestion that males are monogamous because females stop their male partners from being polygynous.

female gametophyte   In plants, a haploid generation that produces one or more eggs, but does not produce sperm cells.

fermentation   The breakdown of organic molecules to produce energy without any net oxidation of an organic molecule.

ferrell cell   The middle cell in the three-cell circulation of wind in each hemisphere.

fertilization   The union of two gametes, such as an egg cell with a sperm cell, to create a zygote.

fertilizer   A soil addition that enhances plant growth by providing essential elements.

fetus   The maturing embryo, after the eighth week of gestation in humans.

fibrous root system   The root system of monocots, which consists of multiple adventitious roots that grow from the stem base.

fibrin   A protein that forms a meshwork of threadlike fibers that wrap around and between platelets and blood cells, enlarging and thickening a blood clot.

fidelity   Refers to the high level of accuracy in DNA replication.

filament   The elongate portion of a flower's stamen; contains vascular tissue that delivers nutrients from parental sporophytes to anthers.

filtrate   In the process of filtration, the material that passes through the filter and enters the excretory organ for either further processing or excretion.

filtration   The passive removal of water and small solutes from the blood.

first law of thermodynamics   The first law states that energy cannot be created or destroyed; it is also called the law of conservation of energy.

5′ cap   The 7-methylguanosine cap structure found on mRNA in eukaryotes.

fixed action pattern (FAP)   An animal behavior that, once initiated, will continue until completed.

flagellate   A protist that uses one or more flagella to move in water or cause water motions useful in feeding.

flagellum (plural, **flagella**)   A relatively long cell appendage that facilitates cellular movement or the movement of extracellular fluids.

flagship species   A single large or instantly recognizable species.

flame cell   A cell that exists primarily to maintain osmotic balance between an organism's body and surrounding fluids; present in flatworms.

flatus   Intestinal gas, which is a mixture of nitrogen and carbon dioxide, with small amounts of hydrogen, methane, and hydrogen sulfide.

flavonoid   A type of phenolic secondary metabolite that provides plants with protection from UV damage or colors organs such as flower petals.

flexor   A muscle that bends a limb at a joint.

florigen   The hypothesized flowering hormone, now identified as the mRNA that produces FT (flowering time) protein in the shoot apex.

flower   A reproductive shoot; a stem branch that produces reproductive organs instead of leaves.

flowering plants   The angiosperms, which produce ovules within the protective ovaries of flowers; when ovules develop into seeds, angiosperm ovaries develop into fruits, which function in seed dispersal.

flow-through system   A form of ventilation in fish in which water moves unidirectionally such that the gills are constantly in contact with fresh, oxygenated water.

fluid-feeder   An animal that licks or sucks fluid from plants or animals and does not need teeth except to puncture an animal's skin.

fluidity   A quality of biomembranes that means that individual molecules remain in close association yet have the ability to move laterally or rotationally within the plane of the membrane. Membranes are semi-fluid.

fluid-mosaic model   The accepted model of the plasma membrane; its basic framework is the semi-fluid phospholipid bilayer with a mosaic of proteins. Carbohydrates may be attached to lipids or proteins.

follicle   A structure within the ovary where each ovum undergoes growth and development before it is ovulated.

follicle-stimulating hormone (FSH)   A gonadotropin that stimulates follicle development.

food chain   A linear depiction of energy flow between organisms, with each organism feeding on and deriving energy from the preceding organism.

food vacuole   *See* phagocytic vacuole.

food web   A complex model of interconnected food chains in which there are multiple links between species.

food-induced thermogenesis   A rise in metabolic rate for a few hours after eating that produces heat.

foot   In mollusks, a muscular structure usually used for movement.

forebrain   One of three major divisions of the vertebrate brain; the other two divisions are the midbrain and hindbrain.

fossil   Recognizable remains of past life on Earth.

fossil fuel   A fuel formed in the Earth from protist, plant or animal remains, such as coal, petroleum, and natural gas.

founder effect   A small group of individuals separates from a larger population and establishes a colony in a new location; genetic drift is common due to the small population size.

fovea   A small area on the retina directly behind the lens that is responsible for the sharpness with which we and other animals see in daylight.

frameshift mutation   A mutation that involves the addition or deletion of nucleotides that are not in multiples of three nucleotides.

free energy (G)   In living organisms, the usable energy, that is, the amount of available energy that can be used to do work.

free radical   A molecule containing an atom with a single, unpaired electron in its outer shell. A free radical is unstable and interacts with other molecules by "stealing" electrons from their atoms.

frequency   In regard to sound, the number of complete wavelengths that occur in 1 second, measured in number of waves per second, or Hertz (Hz).

frugivore   An herbivore that is adapted primarily to feed on fruits.

fruit   A structure that develops from flower organs, encloses seeds, and fosters seed dispersal in the environment.

fruiting bodies   The visible fungal reproductive structures; composed of densely packed hyphae that typically grow out of the substrate.

functional genomics   Genomic methods aimed at studying the expression of a genome.

functional group   A group of atoms with chemical features that are functionally important. Each functional group exhibits the same properties in all molecules in which it occurs.

fundamental niche   The optimal range in which a particular species best functions.

Fungi   One of the four traditional eukaryotic kingdoms of the domain Eukarya.

fungus-like protist   A heterotrophic organism that often resembles true fungi in having threadlike, filamentous bodies and absorbing nutrients from its environment.

G

G_0   A stage in which cells exit the cell cycle and postpone making the decision to divide.

G_1   The first gap phase of the cell cycle.

G_2   The second gap phase of the cell cycle.

G banding   A staining procedure for chromosomes that produces an alternating pattern of G bands that is unique for each type of chromosome.

G protein   An intracellular protein that binds guanosine triphosphate (GTP) and guanosine diphosphate (GDP) and participates in intracellular signaling pathways.

gallbladder   A small sac underneath the liver that is a storage site for bile that allows the release of large amounts of bile to be precisely timed to the consumption of fats.

gametangia   Specialized structures produced by many land plants in which developing gametes are protected by a jacket of tissue.

gamete   A cell that is involved with sexual reproduction, such as a sperm or egg cell.

gametic life cycle   In this type of life cycle, all cells except the gametes are diploid, and gametes are produced by meiosis.

gametogenesis   The formation of gametes.

gametophyte   In plants and many multicellular protists, the haploid stage that produces gametes by mitosis.

ganglia   Groups of neuronal cell bodies that perform basic functions of integrating inputs from sense organs and controlling motor outputs, usually in the peripheral nervous system.

ganglion cells   Cells that send their axons out of the eye into the optic nerve.

gap gene   A type of segmentation gene; when a mutation inactivates a gap gene, several adjacent segments are missing in the larva.

gap junction   A type of junction between animal cells that provides a passageway for intercellular transport.

gas exchange   The process of moving oxygen and carbon dioxide in opposite directions between cells and blood, and between blood and the environment.

gas vesicle   A cytoplasmic structure used by cyanobacteria and some other bacteria that live in aquatic habitats to adjust their buoyancy.

gastrovascular cavity   A body cavity with a single opening to the outside; it functions as both a digestive system and circulatory system.

gastrula   An embryo that is the result of gastrulation, which has three cellular layers called the ectoderm, endoderm, and mesoderm.

gastrulation   A process in which an area in the blastula invaginates and folds inward, creating different embryonic cell layers called germ layers.

gated channel   A channel that can open to allow the diffusion of solutes and close to prohibit diffusion.

gel electrophoresis   A technique used to separate macromolecules by using an electric field that causes them to pass through a gel matrix.

gene   A unit of heredity that contributes to the characteristics or traits of an organism. At the molecular level, a gene is composed of organized sequences of DNA.

gene amplification   An increase in the copy number of a gene.

gene cloning   The process of making multiple copies of a gene.

gene expression   Gene function both at the level of traits and at the molecular level.

gene family   A group of homologous genes within a single species.

gene flow   Occurs when individuals migrate between different populations and cause changes in the genetic composition of the resulting populations.

gene interaction   A situation in which a single trait is controlled by two or more genes.

gene knockout   An organism in which both copies of a functional gene have been replaced with nonfunctional copies. Experimentally, this can occur via gene replacement.

gene mutation   A relatively small change in DNA structure that alters a particular gene.

gene pool   All of the genes in a population.

gene regulation   The ability of cells to control their level of gene expression.

gene replacement   The phenomenon in which a cloned gene recombines with the normal gene on a chromosome and replaces it.

gene silencing   The ability of one gene to silence the effect of another via small RNA molecules called microRNAs.

gene therapy   The introduction of cloned genes into living cells in an attempt to cure disease.

general transcription factors (GTFs)   Five different proteins that play a role in initiating transcription at the core promoter of structural genes in eukaryotes.

generative cell   In seed plants, the male gametophyte cell that divides to produce sperm cells.

genetic code   A code that specifies the relationship between the sequence of nucleotides in the codons found in mRNA and the sequence of amino acids in a polypeptide.

genetic drift   The random change in a population's allele frequencies from one generation to the next that is attributable to chance. It occurs more quickly in small populations.

genetic engineering The direct manipulation of genes for practical purposes.

genetic linkage map A diagram that describes the linear arrangement of genes that are linked to each other along the same chromosome.

genetic mosaic An individual with somatic regions that are genetically different from each other.

genetic transfer The process by which genetic material is transferred from one bacterial cell to another.

genetically modified organisms (GMOs) *See* transgenic.

genome The complete genetic composition of a cell or a species.

genome database A specialized database that focuses on the genetic characteristics of a single species.

genomic imprinting A phenomenon in which a segment of DNA is imprinted, or marked, in a way that affects gene expression throughout the life of the individual who inherits that DNA.

genomic library A type of DNA library in which the inserts are derived from chromosomal DNA.

genomics Techniques that are used in the molecular analysis of the entire genome of a species.

genotype The genetic composition of an individual.

genotype frequency The number of individuals with a given genotype divided by the total number of individuals.

genus In taxonomy, a subdivision of a family.

geological timescale A time line of the Earth's history from its origin about 4.55 billion years ago to the present.

germ layer An embryonic cell layer such as ectoderm, mesoderm, or endoderm.

germ line Cells that give rise to gametes such as egg and sperm cells.

germ plasm Cytoplasmic determinants that help define and specify the primordial germ cells in the gastrula stage.

gestation *See* pregnancy.

giant axon A very large axon in certain species such as squid that facilitates high-speed nerve conduction and rapid responses to stimuli.

gibberellic acid A type of gibberellin.

gibberellin A plant hormone that stimulates both cell division and cell elongation.

gills Specialized filamentous organs in aquatic animals that aid in obtaining oxygen and eliminating carbon dioxide.

ginkgos A phylum of gymnosperms; Ginkgophyta.

gizzard In the stomach of a bird, the muscular structure with a rough inner lining capable of grinding food into smaller fragments.

glaucoma A condition in which drainage of aqueous humor in the eye becomes blocked and the pressure inside the eye increases as the fluid level rises. If untreated, damages cells in the retina and leads to irreversible loss of vision.

glia Cells that surround the neurons; a major class of cells in nervous systems that perform various functions.

global warming A gradual elevation of the Earth's surface temperature caused by the greenhouse effect.

glomerulus A cluster of interconnected, fenestrated capillaries in the renal corpuscle of the kidney; the site of filtration in the kidneys.

glucocorticoid A steroid hormone that regulates glucose balance and helps prepare the body for stress situations.

gluconeogenesis A mechanism for maintaining blood glucose levels; enzymes in the liver convert noncarbohydrate precursors into glucose, which are then secreted into the blood.

glucose sparing A metabolic adjustment that reserves the glucose produced by the liver for use by the nervous system.

glycocalyx 1. An outer viscous covering surrounding a bacterium. The glycocalyx, which is secreted by the bacterium, traps water and helps protect bacteria from drying out. 2. A carbohydrate covering that is found outside of animal cells.

glycogen A polysaccharide found in animal cells and sometimes called animal starch.

glycogenolysis A mechanism for maintaining blood glucose levels; stored glycogen can be broken back down into molecules of glucose by hydrolysis.

glycolipid A lipid that has carbohydrate attached to it.

glycolysis A metabolic pathway that breaks down glucose to pyruvate.

glycolytic fiber A skeletal muscle fiber that has few mitochondria but possesses both a high concentration of glycolytic enzymes and large stores of glycogen.

glycoprotein A protein that has carbohydrate attached to it.

glycosaminoglycan (GAG) The most abundant type of polysaccharide in the extracellular matrix (ECM) of animals, consisting of repeating disaccharide units that give a gel-like character to the ECM of animals.

glycosylation The attachment of carbohydrate to a protein or lipid, producing a glycoprotein or glycolipid.

glyoxysome A specialized organelle within plant seeds that contains enzymes needed to convert fats to sugars.

gnathostomes All vertebrate species that possess jaws.

gnetophytes A phylum of gymnosperms; Gnetophyta.

Golgi apparatus A stack of flattened, membrane-bounded compartments that performs three overlapping functions: secretion, processing, and protein sorting.

gonadotropins Hormones secreted by the anterior pituitary gland that are the same in both sexes; gonadotropins influence the ability of the testes and ovaries to produce the sex steroids.

gonads The testes in males and the ovaries in females, where the gametes are formed.

G-protein-coupled receptors (GPCRs) A common type of receptor found in the cells of eukaryotic species that interacts with G proteins to initiate a cellular response.

gradualism A concept that suggests that species evolve continuously over long spans of time.

grain The characteristic single-seeded fruit of cereal grasses such as rice, corn, barley, and wheat.

granum A structure composed of stacked tubules within the thylakoid membrane of chloroplasts.

gravitropism Plant growth in response to the force of gravity.

gray matter Brain tissue that consists of neuronal cell bodies, dendrites, and some unmyelinated axons.

grazer An herbivore that feeds almost constantly on grasses.

greenhouse effect The process in which short-wave solar radiation passes through the atmosphere to warm the Earth but is radiated back to space as long-wave infrared radiation. Much of this radiation is absorbed by atmospheric gases and reradiated back to Earth's surface, causing its temperature to rise.

groove In the DNA double helix, an indentation where the atoms of the bases make contact with the surrounding water.

gross primary production (GPP) The measure of biomass production by photosynthetic organisms; equivalent to the carbon fixed during photosynthesis.

ground meristem A type of primary plant tissue meristem that gives rise to ground tissue.

ground tissue Most of the body of a plant, which has a variety of functions, including photosynthesis, storage of carbohydrates, and support. Ground tissue can be subdivided into three types: parenchyma, collenchyma, and sclerenchyma.

group selection The premise that natural selection produces outcomes beneficial for the whole group or species rather than for individuals.

growth An increase in weight or size.

growth factors A group of proteins in animals that stimulate certain cells to grow and divide.

growth hormone (GH) A hormone produced in vertebrates by the anterior pituitary gland; GH acts on the liver to produce insulin-like growth factor-1 (IGF-1).

guanine (G) A purine base found in DNA and RNA.

guard cell A specialized plant cell that allows epidermal pores (stomata) to close when conditions are too dry, and open under moist conditions, allowing the entry of CO_2 needed for photosynthesis.

gustation The sense of taste.

gut The gastrointestinal (GI) tract.

gymnosperm A plant that produces seeds that are exposed rather than enclosed in fruits.

gynoecium The aggregate of carpels that forms the innermost whorl of flower organs.

H

H zone   A narrow, light region in the center of the A band of the sarcomere that corresponds to the space between the two sets of thin filaments in each sarcomere.

H⁺ electrochemical gradient   A transmembrane gradient for H⁺ composed of both a membrane potential and a concentration difference for H⁺ across the membrane.

habituation   The form of nonassociative learning in which an organism learns to ignore a repeated stimulus.

Hadley cell   The most prominent of the three cells in the three-cell circulation of wind in each hemisphere.

hair cell   A mechanoreceptor that is a specialized epithelial cell with deformable stereocilia.

half-life   1. In the case of organic molecules in a cell, refers to the time it takes for 50% of the molecules to be broken down. 2. In the case of radioisotopes, the time it takes for half the molecules to decay and emit radiation.

halophyte   A plant that can tolerate higher than normal salt concentrations in the cell sap, and thus can occupy coastal salt marshes or saline deserts.

Hamilton's rule   The proposal that an altruistic gene will be favored by natural selection when $r > C/B$ where r is the coefficient of relatedness of the donor (the altruist) to the recipient, B is the benefit received by the recipients of the altruism, and C is the cost incurred by the donor.

haplodiploidy   A genetic system in which females develop from fertilized eggs and are diploid but males develop from unfertilized eggs and are haploid.

haploid   Containing one set of chromosomes; or $1n$.

haploid-dominant species   Species in which the haploid organism is the prevalent organisms in the life cycle. Examples include fungi and some protists.

Hardy-Weinberg equation   An equation ($p^2 + 2pq + q^2 = 1$) that relates allele and genotype frequencies; the equation predicts an equilibrium if the population size is very large, mating is random, the populations do not migrate, no natural selection occurs, and no new mutations are formed.

heart attack   *See* myocardial infarction (MI).

heart   A muscular structure that pumps blood through blood vessels.

heat of fusion   The amount of heat energy that must be withdrawn or released from a substance to cause it to change from the liquid to the solid state.

heat of vaporization   The heat required to vaporize 1 mole of any substance at its boiling point under standard pressure.

heat shock protein   A protein that helps to protect other proteins from heat damage and refold them to their functional state.

heavy chain   A part of an immunoglobulin molecule.

helper T cell   A type of lymphocyte that assists in the activation and function of B cells and cytotoxic T cells.

hematocrit   The volume of blood that is composed of red blood cells, usually between 40 and 65% among vertebrates.

hemiparasite   A parasitic organism that generally photosynthesizes, but lacks a root system to draw water and thus depends on its hosts for that function.

hemispheres   The two halves of the cerebrum.

hemizygous   The term used to describe the single copy of an X-linked gene in a male.

hemocyanin   A copper-containing pigment that gives the blood or hemolymph a bluish tint.

hemodialysis   A medical procedure used to artificially perform the kidneys' excretory filtration and cleansing functions.

hemoglobin   An oxygen-binding protein found within the cytosol of red blood cells.

hemolymph   Blood and interstitial fluid combined in one fluid compartment in many invertebrates.

hemorrhage   A loss of blood from a ruptured blood vessel.

herbaceous plant   A plant that produces little or no wood and is composed mostly of primary vascular tissues.

herbivore   An animal that eats only plants.

herbivory   Refers to herbivores feeding on plants.

hermaphrodite   An individual that can produce both sperm and eggs.

hermaphroditism   A form of sexual reproduction in which individuals have both male and female reproductive systems.

heterochromatin   The highly compacted regions of chromosomes; in general, these regions are transcriptionally inactive because of their tight conformation.

heterocyst   A specialized cell of some cyanobacteria in which nitrogen fixation occurs.

heterodimer   The structure that results when two different proteins come together.

heterokaryon   In fungi, a mycelium having nuclei of two or more genetic types.

heterospory   In plants, the formation of two different types of spores, microspores and megaspores; microspores produce male gametophytes and megaspores produce female gametophytes.

heterotherm   An animal that has a body temperature that varies with the environment.

heterotroph   Organisms that cannot produce their own organic food and thus must obtain organic food from other organisms.

heterotrophic   Requiring organic food from the environment.

heterozygote advantage   A phenomenon in which a heterozygote has a higher Darwinian fitness compared to the corresponding homozygotes.

heterozygous   An individual with two different alleles of the same gene.

hibernation   The state of torpor in an animal over months.

high affinity   Refers to the binding of an ion or molecule to a protein very tightly. The substance will bind at a very low concentration.

highly repetitive sequence   A DNA sequence found tens of thousands or even millions of times throughout a genome.

hindbrain   One of three major divisions of the vertebrate brain; the other two divisions are the midbrain and forebrain.

hippocampus   The area of the brain whose main function appears to be establishing memories for spatial locations, facts, and sequences of events; composed of several layers of cells that are connected together in a circuit.

histone acetyltransferase   An enzyme that attaches acetyl groups to histone proteins.

histone code hypothesis   Refers to the pattern of histone modification recognized by particular proteins. The pattern of covalent modifications of amino terminus tails provides binding sites for proteins that subsequently affect the degree of chromatin compaction.

histones   A group of proteins involved in the formation of nucleosomes.

holoblastic cleavage   A complete type of cell cleavage in certain animals in which the entire zygote is bisected into two equal-sized blastomeres.

holoparasite   A parasitic organism that lacks chlorophyll and is totally dependent on the host plant for its water and nutrients.

homeobox   A 180-bp sequence within the coding sequence of homeotic genes.

homeodomain   A region of a protein that functions in binding to the DNA.

homeostasis   The process whereby living organisms regulate their cells and bodies to maintain relatively stable internal conditions.

homeotherm   An animal that maintains its body temperature within a narrow range.

homeotic   A term that describes changes in which one body part is replaced by another.

homeotic gene   A gene that controls the developmental fate of particular segments or regions of an animal's body.

hominoid   A gibbon, gorilla, orangutan, chimpanzee, or human.

hominin   Either an extinct or modern form of humans.

homodimer   The structure that results when two identical proteins come together.

homologous genes   Genes derived from the same ancestral gene that have accumulated random mutations that make their sequences slightly different.

homologous structures   Structures that are similar to each other because they are derived from the same ancestral structure.

homologue   A member of a pair of chromosomes in a diploid organism that are evolutionarily related.

homology   A fundamental similarity that occurs due to descent from a common ancestor.

homozygous   An individual with two identical copies of an allele.

horizontal gene transfer   The transfer of genes between different species.

hormone   A chemical messenger that is produced in a gland or other structure and acts on distant target cells in one or more parts of the body.

hornworts   A phylum of bryophytes; Anthocerophyta.

host   The prey organism in a parasitic association.

host cell   A cell that is infected by a virus, fungus, or a bacterium.

host plant resistance   The ability of plants to prevent herbivory.

host range   The number of species and cell types that a virus or bacterium can infect.

hot spot   A human-impacted geographic area with a large number of endemic species. To qualify as a hot spot, a region must contain at least 1,500 species of vascular plants as endemics and have lost at least 70% of its original habitat.

Hox **complex**   A group of adjacent homeotic genes in vertebrates that controls the formation of structures along the anteroposterior axis.

Hox **genes**   A class of genes involved in pattern formation in early embryos.

Human Genome Project   The largest genome project in history, which was a 13-year effort coordinated by the U.S. Department of Energy and the National Institutes of Health. The goals of the project were to identify all human genes, to sequence the entire human genome, to address the legal and ethical implications resulting from the project, and to develop programs to manage the information gathered from the project.

human immunodeficiency virus (HIV)   A retrovirus that is the causative agent of acquired immune deficiency syndrome (AIDS).

humoral immunity   A type of specific immunity in which plasma cells secrete antibodies that bind to antigens.

humus   A collective term for the organic constituents of soils.

hybrid zone   An area where two populations can interbreed.

hybridization   A situation in which two individuals with different characteristics are mated or crossed to each other; the offspring are referred to as hybrids.

hydrocarbon   Molecules with predominantly hydrogen-carbon bonds.

hydrogen bond   Electrostatic attraction between a hydrogen atom of a polar molecule and an electronegative atom of another polar molecule.

hydrolysis   The process in which reactions utilize water to break apart other molecules.

hydrophilic   "Water-loving"—generally, ions and molecules that contain polar covalent bonds will dissolve in water and are said to be hydrophilic.

hydrophobic   "Water-fearing"—molecules that are not attracted to water molecules. Such molecules are composed predominantly of carbon and hydrogen and are relatively insoluble in water. Because carbon-carbon and carbon-hydrogen bonds are nonpolar, the atoms in such compounds are electrically neutral.

hydrostatic skeleton   A fluid-filled body cavity surrounded by muscles that gives support and shape to the body of organisms.

hydroxide ion   An anion with the formula, OH^-.

hypermutation   A process that primarily involves numerous C to T point mutations that is crucial to enabling lymphocytes to produce a diverse array of immunoglobulins capable of recognizing many different antigens.

hyperpolarization   The change in the membrane potential that occurs when the cell becomes more polarized.

hypersensitive response (HR)   A plant's local defensive response to pathogen attack.

hyperthyroidism   A hyperactive thyroid gland.

hypertonic   When the solute concentration inside the cell is higher relative to the outside of the cell.

hypha   A microscopic, branched filament of the body of a fungus.

hypocotyl   The portion of an embryonic plant stem located below the point of attachment of the cotyledons.

hypothalamus   A gland located below the thalamus at the floor of the forebrain; it controls functions of the gastrointestinal and reproductive systems, and it regulates many basic behaviors such as eating and drinking.

hypothesis   In biology, a proposed explanation for a natural phenomenon based on previous observations or experimental studies.

hypothesis testing   Also known as the scientific method, a strategy for testing the validity of a hypothesis.

hypothyroidism   An underactive thyroid gland.

hypotonic   When the solute concentration outside the cell is lower relative to the inside of the cell.

I

I band   A light band that lies between the A bands of two adjacent sarcomeres.

immune system   The cells and organs within an animal's body that contribute to immune defenses.

immune tolerance   The process by which the body distinguishes between self and nonself components.

immunity   The ability of an animal to ward off internal threats, including the invasion of potentially harmful microorganisms such as bacteria, the presence of foreign molecules such as the products of microorganisms, and the presence of abnormal cells such as cancer cells.

immunization   *See* vaccination.

immunoglobulin   A Y-shaped protein with two heavy chains and two light chains that provide immunity to foreign substances; antibodies are a type of immunoglobulin.

immunological memory   The immune system's ability to produce a secondary immune response.

imperfect flower   A flower that lacks either stamens or carpels.

implantation   The first event of pregnancy, when the blastocyst embeds within the uterine endometrium.

imprinted   In genetics, a marked segment of DNA.

imprinting   1. The development of a species-specific pattern of behavior. A form of learning, with a large innate component, within a limited time period. 2. In genetics, the marking of DNA that occurs differently between males and females.

in vitro   Literally, "in glass." An approach to studying a process in living cells that involves isolating and purifying cellular components, outside the cell.

in vivo   Meaning, "in life." An approach to studying a process in living cells.

inactivation gate   A string of amino acids that juts out from a channel protein into the cytosol.

inborn error of metabolism   A genetic defect in the ability to metabolize certain compounds.

inbreeding   Mating among genetically related relatives.

inbreeding depression   The phenomenon whereby inbreeding produces homozygotes that are less fit, thereby decreasing the reproductive success of a population.

inclusive fitness   The term used to designate the total number of copies of genes passed on through one's relatives, as well as one's own reproductive output.

incomplete dominance   The phenomenon in which a heterozygote that carries two different alleles exhibits a phenotype that is intermediate between the corresponding homozygous individuals.

incomplete flower   A flower that lacks one or more of the four flower organ types.

incomplete metamorphosis   A gradual change in body form in some insects from different nymphal stages, called instars, into a similar looking adult.

incurrent siphon   A tunicate structure used to draw water through the mouth.

indeterminate cleavage   A characteristic of deuterostome development in which each cell produced by early cleavage retains the ability to develop into a complete embryo.

indeterminate growth   Growth in which plant shoot apical meristems continuously produce new stem tissues and leaves as long as conditions remain favorable.

indicator species   A species whose status provides information on the overall health of an ecosystem.

indirect calorimetry   A method of determining basal metabolic rate in which the rate at which an animal uses oxygen is measured.

individual selection   The proposal that adaptive traits generally are selected for because they benefit the survival and reproduction of the individual rather than the group.

individualistic model   A view of the nature of a community that considers it to be an assemblage of species coexisting primarily because of similarities in their physiological requirements and tolerances.

induced fit Occurs when a substrate(s) binds to an enzyme and the enzyme undergoes a conformational change that causes the substrate(s)s to bind more tightly to the enzyme.

induced mutation A mutation brought about by environmental agents that enter the cell and then alter the structure of DNA.

inducer In transcription, a small effector molecule that increases the rate of transcription.

inducible operon In this type of operon, the presence of a small effector molecule causes transcription to occur.

induction 1. In development, the process by which a cell or group of cells governs the developmental fate of neighboring cells. 2. In molecular genetics, refers to the process by which transcription has been turned on by the presence of a small effector molecule.

industrial nitrogen fixation The human activity of producing nitrogen fertilizers.

infertility The inability to produce viable offspring.

inflammation An innate local response to infection or injury characterized by local redness, swelling, heat, and pain.

inflorescence A cluster of flowers on a plant.

infundibular stalk The structure that is physically connected to a multilobed endocrine gland sitting directly below the hypothalamus, called the pituitary gland.

ingroup A monophyletic group in a cladogram of interest.

inheritance of acquired characteristics Jean-Baptiste Lamarck's hypothesis that species change over the course of many generations by adapting to new environments. He thought behavioral changes modified traits, and he hypothesized that such modified traits were inherited by offspring.

inhibition A mechanism for succession in which space is all-important, and order of colonization determines subsequent community structure.

inhibitory postsynaptic potential (IPSP) The response from an inhibitory neurotransmitter hyperpolarizing the postsynaptic membrane, which reduces the likelihood of an action potential.

initiation stage In transcription or translation, the first step that initiates the process.

initiator tRNA A specific tRNA that recognizes the start codon AUG in mRNA and binds to it.

innate The term used to describe behaviors that seem to be genetically programmed.

inner bark The thin layer of phloem that conducts most of the sugar transport in a woody stem.

inner ear One of the three main compartments of the mammalian ear; composed of the bony cochlea and the vestibular system, which plays a role in balance.

inorganic chemistry The study of the nature of atoms and molecules, with the exception of those that contain rings or chains of carbon.

instar A stage of growth in an insect with incomplete metamorphosis.

insulin-like growth factor-1 (IGF-1) A hormone that in mammals stimulates the elongation of bones, especially during puberty, when mammals become reproductively mature.

integral membrane protein A protein that cannot be released from the membrane unless it is dissolved with an organic solvent or detergent—in other words, you would have to disrupt the integrity of the membrane to remove it.

integrase An enzyme, sometimes encoded by viruses, that catalyzes the integration of the viral genome into a host-cell chromosome.

integrin A cell-surface receptor protein found in animal cells that connects cells and the extracellular matrix.

integument In plants, a modified leaf that encloses the megasporangium to form an ovule.

intercostal muscles Muscles that surround and connect the ribs in the chest.

interference competition Competition in which individuals interact directly with one another by physical force or intimidation.

interferon A protein that generally inhibits viral replication inside host cells.

intermediate-disturbance hypothesis The proposal that moderately disturbed communities are more diverse than undisturbed or highly disturbed communities.

intermediate filament A type of protein filament within the cytoskeleton that helps maintain cell shape and rigidity.

internal fertilization Fertilization that occurs in terrestrial animals, in which sperm are deposited within the reproductive tract of the female during the act called copulation.

interneuron A type of neuron that forms interconnections between other neurons.

internode The region of stem on a plant between adjacent nodes.

interphase The G_1, S, and G_2 phases of the cell cycle. It is phase of the cell cycle during which the chromosomes are decondensed and found in the nucleus.

intersexual selection Sexual selection between members of the opposite sex.

interspecies hybrid The offspring resulting from two species mating.

interspecific competition The term used to describe competition between individuals of different species.

interstitial Refers to the fluid between cells.

intertidal zone The area where the land meets the sea, which is alternately submerged and exposed by the daily cycle of tides.

intracellular fluid The fluid inside cells.

intranuclear spindle A spindle that forms within an intact nuclear envelope during nuclear division.

intrasexual selection Sexual selection between members of the same sex.

intraspecific competition The term used to describe competition between individuals of the same species.

intrinsic rate of increase The situation in which conditions are optimal for a population, and the per capita growth rate is at its maximum rate.

introduced species A species moved by humans from a native location to another location.

intron Intervening DNA sequences that are found in between the coding sequences of genes.

invasive The term used to describe introduced species that spread on their own, often outcompeting native species for space and resources.

invasive cell A cell that can invade healthy tissues.

inversion A change in the direction of the genetic material along a single chromosome.

invertebrate An animal that lacks vertebrae.

iodine-deficient goiter An overgrown gland that is incapable of making thyroid hormone.

ion At atom or molecule that gains or loses one or more electrons and acquires a net electric charge.

ion electrochemical gradient A dual gradient for an ion that is composed of both an electrical gradient and chemical gradient for that ion.

ionic bond The bond that occurs when a cation binds to an anion.

ionotropic receptor A ligand-gated ion channel that opens in response to binding of a neurotransmitter molecule.

iris The circle of pigmented smooth muscle that is responsible for eye color.

iron regulatory element (IRE) A response element within the ferritin mRNA to which the iron regulatory protein binds.

iron regulatory protein (IRP) An RNA-binding protein that regulates the translation of the mRNA that encodes ferritin.

islets of Langerhans Spherical clusters of endocrine cells that are scattered in great numbers throughout the endocrine pancreas.

isomers Two structures with an identical molecular formula but different structures and characteristics.

isotonic When the solute concentrations on both sides of the plasma membrane are equal.

isotope An element that exists in multiple forms that differ in the number of neutrons they contain.

iteroparity The pattern of repeated reproduction at intervals throughout the life cycle.

J

joint The juncture where two or more bones of a vertebrate endoskeleton come together.

K

K/T event An ancient cataclysm that involved at least one large meteorite or comet that crashed into the Earth near the present-day Yucatan Peninsula in Mexico about 65 million years ago.

karyogamy The process of nuclear fusion.

karyotype   A photographic representation of the chromosomes that reveals how many chromosomes are found within an actively dividing cell.

K_D   The dissociation constant between a ligand and its receptor.

ketones   Small compounds generated from carbohydrates, fatty acids, or amino acids. Ketones are made in the liver and released into the blood to provide an important energy source during prolonged fasting for many tissues, including the brain.

keystone species   A species within a community that has a role out of proportion to its abundance.

kilocalorie (kcal)   One thousand calories; the common unit of measurement when measuring biological activities.

kin selection   Selection for behavior that lowers an individual's own fitness but enhances the reproductive success of a relative.

kinesis   A movement in response to a stimulus, but one that is not directed toward or away from the source of the stimulus.

kinetic energy   Energy associated with movement.

kinetic skull   A defining characteristic of the class Lepidosauria, in which the joints between various parts of the skull are extremely mobile.

kinetochore   A group of proteins necessary for sorting each chromosome that binds to the centromere.

kingdom   A taxonomic group that contains one or more phyla.

knowledge   The awareness and understanding of information.

Koch's postulates   A series of steps used to determine whether a particular organism causes a specific disease.

K-selected species   A type of life history strategy, where species have a low rate of per capita population growth but good competitive ability.

L

labia majora   In the female genitalia, large outer folds that surround the external opening of the reproductive tract.

labia minora   In the female genitalia, smaller, inner folds near the external opening of the reproductive tract.

labor   A three-stage process that includes (1) dilation and thinning of the cervix to allow passage of the fetus out of the uterus; (2) the movement of the fetus through the cervix and the vagina and out into the world; and (3) the contraction of blood vessels within the placenta and umbilical cord, blocking further blood flow, making the newborn independent from the mother; the placenta detaches from the uterine wall and is delivered a few minutes after the birth of the baby.

lac operon   An operon in the genome of *E. coli* that contains the genes for the enzymes that allow it to metabolize lactose.

lac repressor   A repressor protein that regulates the lac operon.

lactation   In most mammals, a period after birth in which the young are nurtured by milk produced by the mother.

lacteal   A vessel in the center of each intestinal villus; lipids are absorbed by the lacteals, which eventually empty into the circulatory system.

lagging strand   A strand of DNA made as a series of small Okazaki fragments that are eventually connected to each other to form a continuous strand. The synthesis of these DNA fragments occurs in the direction away from the replication fork.

lamellae   Platelike structures in the internal gills of fish that branch from structures called filaments.

landscape ecology   The study of the influence of large-scale spatial patterns of land use or habitat type on ecological processes.

larva   A free-living organism that is morphologically very different from the embryo and adult.

larynx   The area beyond the throat where the vocal cords lie.

latent   The term used to describe when a prophage or provirus remains inactive for a long time.

lateral line system   A sensory system in fish and some toads that allows them to detect changes in their environment; hair cells detect changes in water currents brought about by waves, nearby moving objects, and low-frequency sounds traveling through the water.

law of independent assortment   The alleles of different genes assort independently of each other during gamete formation.

law of segregation   The phenomenon that the two copies of a gene segregate from each other during gamete formation and transmission from parent to offspring.

leaching   The dissolution and removal of inorganic ions as water percolates through materials such as soil.

leading strand   A DNA strand made in the same direction that the replication fork is moving. The strand is synthesized as one long continuous molecule.

leaf   A flattened plant organ that emerges from stems and functions in photosynthesis or other ways.

leaf abscission   The process by which a leaf drops after the formation of an abscission zone at the point where a leaf petiole connects with the stem. The abscission zone consists of an inner protective layer of cork cells whose tough walls help to prevent pathogen attack and dehydration, and an outer layer of cells having thin walls that break easily.

leaf primordia   Small bumps that occur at the sides of a shoot apical meristem and develop into young leaves.

leaflet   1. Half of a phospholipid bilayer. 2. A portion of a compound leaf.

learning   The ability of an animal to make modifications to a behavior based on previous experience.

legume   A member of the pea (bean) family. Also their distinctive fruits, dry pods that develop from one carpel and open down both sides when seeds are mature.

lek   A designated communal courting area.

lens   1. A structure of the eye that focuses light. 2. The glass components of a light microscope or electromagnetic parts of an electron microscope that allow the production of magnified images of microscopic structures.

lentic   Referring to a freshwater habitat characterized by standing water.

leptin   A molecule produced by adipose cells in proportion to fat mass; controls appetite and metabolic rate.

leukocyte   A white blood cell; involved in nonspecific immunity.

lichens   Mutualistic partnerships of particular fungi and certain photosynthetic green algae or cyanobacteria, and sometimes both to form a body distinctive from that of either partner alone.

Liebig's law of the minimum   Species biomass or abundance is limited by the scarcest factor.

life cycle   The sequence of events that characterize the steps of development of the individuals of a given species.

leukocyte   A white blood cell.

life table   A table that provides data on the number of individuals alive in particular age classes.

ligand   An ion or molecule that binds to a protein, such as an enzyme or a receptor.

ligand-gated channel   A channel controlled by the noncovalent binding of small molecules—called ligands—such as hormones or neurotransmitters.

ligand-gated ion channels   A type of cell surface receptor that binds a ligand and functions as an ion channel. Ligand binding either opens or closes a channel.

light chain   A part of an immunoglobulin molecule.

light-harvesting complex   A component of photosystem II and photosystem I composed of several dozen pigment molecules that are anchored to proteins. The role of these complexes is to absorb photons of light.

light microscope   A microscope that utilizes light for illumination.

light reactions   One of two stages in the process of photosynthesis. During the light reactions, photosystem II and photosystem I absorb light energy and produce O_2, ATP, and NADPH.

lignin   A tough polymer that adds strength and decay resistance to cell walls of tracheids, vessel elements, and other cells of plants.

lignophytes   Modern and fossil seed plants, and seedless ancestors that produced wood.

limbic system   The system primarily involved in the formation and expression of emotions, and also plays a role in learning, memory, and the perception of smells; includes certain areas of the telencephalon and parts of the diencephalon.

limiting factor   The factor that is most scarce in relation to need.

lineage   The genetic relationship between an individual or group of individuals and its ancestors. A series of ancestors in a population that shows a progression of changes.

linkage group   A group of genes that usually stay together during meiosis.

linkage   The phenomenon of two genes that are close together on the same chromosome tending to be transmitted as a unit.

lipase   The major digestive fat-digesting enzyme from the pancreas.

lipid   A molecule composed predominantly of hydrogen and carbon atoms. Lipids are nonpolar and therefore very insoluble in water. They include fats, phospholipids, and steroids.

lipid anchor   A way for proteins to associate with the plasma membrane; involves the covalent attachment of a lipid to an amino acid side chain within a protein.

lipid exchange protein   A protein that extracts a lipid from one membrane, diffuses through the cell, and inserts the lipid into another membrane.

lipopolysaccharides   Lipids having covalently bound carbohydrates. Major components of the thin, outer envelope that encloses the cell walls of Gram-negative bacteria.

liposome   A vesicle surrounded by a lipid bilayer.

liver   An organ in vertebrates that performs diverse metabolic functions and is the site of bile production.

liverworts   A phylum of bryophytes; formally called Hepatophyta.

lobe fins   The Actinistia (coelacanths) and Dipnoi (lungfish) and tetrapods; also called Sarcopterygii.

lobe-finned fish   Fish in which the fins are part of the body, and they are supported by skeletal extensions of the pectoral and pelvic areas that are moved by muscles residing in the fins.

locomotion   The movement of an animal from place to place.

locus   The physical location of a gene on a chromosome.

logistic equation $dN/dt = rN(K - N)/K$, where dN/dt is the rate of population change, r is the per capita rate of population growth, N is the population size, and $(K - N)/K$ represents the proportion of unused resources remaining.

logistic growth   The S-shaped pattern in which the growth of a population typically slows down as it approaches carrying capacity.

long-day plant   A plant that flowers in spring or early summer, when the night period is shorter (and thus the day length is longer) than a defined period.

long-term potentiation (LTP)   The long-lasting strengthening of the connection between neurons.

loop domain   A chromosomal segment that is folded into loops by the attachment to proteins; a method of compacting chromosomes.

loop of Henle   A sharp, hairpin-like loop in the tubule of the nephron of the kidney consisting of a descending limb coming from the proximal tubule and an ascending limb leading to the distal convoluted tubule.

lophophore   A horseshoe-shaped crown of tentacles used for feeding.

Lophotrochozoa   A clade that encompasses the annelids, mollusks, and several other phyla; they are distinguished by two morphological features—the lophophore, a crown of tentacles used for feeding, and the trochophore larva, a distinct larval stage.

lotic   Referring to a freshwater habitat characterized by running water.

lumen   The internal space of an organelle.

lung   A structure used to bring oxygen into the circulatory system and remove carbon dioxide.

lungfish   The Dipnoi; fish with primitive lungs which live in oxygen-poor freshwater swamps and ponds.

luteinizing hormone (LH)   A gonadotropin.

lycophyll   A relatively small leaf having a single unbranched vein; the type of leaf produced by lycophytes.

lycophytes   Members of a phylum of vascular land plants whose leaves are lycophylls; Lycopodiophyta.

lymphatic system   A system of vessels along with a group of organs and tissues where most leukocytes reside. The lymphatic vessels collect excess interstitial fluid and return it to the blood.

lymphocytes   A type of leukocyte that is responsible for specific immunity; the two types are B cells and T cells.

Lyon hypothesis   *See* X inactivation.

lysogenic cycle   The growth cycle of a bacteriophage consisting of integration, prophage replication, and excision.

lysogeny   Latency in bacteriophages.

lysosome   A small organelle found in animal cells that contains acid hydrolases that degrade macromolecules.

lytic cycle   The growth cycle of a bacteriophage in which the production and release of new viruses lyses the host cell.

M

M line   A narrow, dark band in the center of the H zone that corresponds to proteins that link together the central regions of adjacent thick filaments.

M phase   The sequential events of mitosis and cytokinesis.

macroevolution   Evolutionary changes that create new species and groups of species.

macromolecule   Many molecules bonded together to form a polymer. Carbohydrates, proteins, and nucleic acids (for example, DNA

and RNA) are important macromolecules found in living organisms.

macronutrient   An element required by plants in amounts of at least 1 g/kg of plant dry matter.

macroparasite   A parasite that lives in the host but releases infective juvenile stages outside the host's body.

macrophage   A type of phagocyte capable of engulfing viruses and bacteria; strategically located where it will encounter invaders.

macular degeneration   A condition in which photoreceptor cells in and around the fovea of the retina are lost; associated with loss of sharpness and color vision.

madreporite   A sieve-like plate on the surface of an echinoderm where water enters the water vascular system.

magnification   The ratio between the size of an image produced by a microscope and its actual size.

major groove   A groove that spirals around the DNA double helix. The major groove provides a location where a protein can bind to a particular sequence of bases and affect the expression of a gene.

major histocompatibility complex (MHC)   A gene family that encodes the plasma membrane self proteins that must be complexed with the antigen in order for T-cell recognition to occur.

malabsorption   Any interference with the secretion of bile or the action of bile salts in the intestine that decreases the absorption of fats, including fat-soluble vitamins.

male assistance hypothesis   A hypothesis to explain the existence of monogamy that maintains that males remain with females to help them rear their offspring.

male gametophyte   A haploid plant life cycle phase that produces sperm.

malignant tumor   A growth of cells that has progressed to the cancerous stage.

Malpighian tubules   Delicate projections from the digestive tract of insects and some other taxa that protrude into the hemolymph and function as an excretory organ.

mammal   A vertebrate that is a member of the class Mammalia that nourishes its young with milk secreted by mammary glands. Another distinguishing feature is hair.

mammary gland   A gland in female members of mammal species that secretes milk.

manganese cluster   A site where the oxidation of water occurs in photosystem II.

mantle   A fold of skin draped over the visceral mass of a mollusk that secretes a shell in those species that form shells.

mantle cavity   The chamber in a mollusk mantle that houses delicate gills.

many eyes hypothesis   The idea that increased group size decreases predators' success because of increased predator detection ability.

map distance   The distance between genes along chromosomes which is calculated as the number of recombinant offspring divided by the total number of offspring times 100.

mapping The process of determining the relative locations of genes or other DNA segments along a chromosome.

map unit (mu) A unit of distance equivalent to a 1% recombination frequency.

mark-recapture technique The capture and tagging of animals so they can be released and recaptured, allowing an estimate of population size.

marsupial A member of a group of seven mammalian orders and about 280 species found in the subclass Metatheria.

mass extinction When many species become extinct at the same time.

mass-specific BMR The amount of energy expended per gram of body mass.

mass spectrometry A method to determine the masses of molecules such as short peptide fragments within proteins. Tandem mass spectrometry can be used to determine the amino acid sequences of proteins.

mast cell A type of cell derived from bone marrow stem cells that plays an important role in nonspecific immunity; these cells are found throughout connective tissues, and secrete many locally acting molecules, notably histamine.

mastax The circular muscular pharynx in the mouth of rotifers.

masting The synchronous production of many progeny by all individuals in a population to satiate predators and thereby allow some progeny to survive.

mate-guarding hypothesis A hypothesis to explain the existence of monogamy that theorizes that males stay with a female to prevent her from being fertilized by other males.

maternal effect gene A gene that follows a maternal effect inheritance pattern.

maternal effect An inheritance pattern in which the genotype of the mother determines the phenotype of her offspring.

maternal inheritance A phenomenon in which offspring inherit particular genes only from the female parent (through the egg).

maturation promoting factor (MPF) The factor that causes oocytes to progress (or mature) from G_2 to M phase.

mature mRNA In eukaryotes, transcription produces a longer RNA, pre-mRNA, which undergoes certain processing events before it exits the nucleus; mature mRNA is the final functional product.

mean fitness of the population The average reproductive success of members of a population.

mechanosensitive channel A channel that is sensitive to changes in membrane tension.

mechanoreceptor A sensory receptor that transduces mechanical energy such as pressure, touch, stretch, movement, and sound.

mediator A large protein complex that plays a role in initiating transcription at the core promoter of structural genes in eukaryotes.

medulla oblongata The part of the hindbrain that coordinates many basic reflexes and bodily functions, such as breathing, that maintain the normal homeostatic processes of the animal.

megadiversity country Those countries with the greatest numbers of species; used in targeting areas for conservation.

megaspore In seed plants and some seedless plants, a large spore that produces a female gametophyte within the spore wall.

meiosis The process by which haploid cells are produced from a cell that was originally diploid.

meiosis I The first division of meiosis when the homologues are separated into different cells.

meiosis II The second division of meiosis in which sister chromatids are separated to different cells.

meiotic nondisjunction An occurrence during meiosis I or meiosis II that produces haploid cells that have too many or too few chromosomes.

Meissner's corpuscles Structures that sense touch and light pressure and that lie just beneath the skin surface.

membrane attack complex (MAC) A multiunit protein formed by the activation of complement proteins; the complex creates water channels in the microbial plasma membrane and causes the microbe to swell and burst.

membrane potential The difference between the electric charges inside and outside the cell; also called a potential difference (or voltage).

membrane transport The movement of ions or molecules across a cell membrane.

memory The retention of information over time.

memory cells A component of the immune response; these cells remain poised to recognize the antigen if it returns in the future.

Mendelian inheritance The inheritance patterns of genes that segregate and assort independently.

meninges A protective structure in the central nervous system consisting of three layers of sheathlike membranes.

meningitis A potentially life-threatening infectious disease in which the meninges become inflamed.

menopause The event during which a woman permanently stops having ovarian cycles.

menstrual cycle Also called the uterine cycle; the cyclical changes in the lining of the uterus that occur in parallel with the ovarian cycle.

menstruation A period of bleeding at the beginning of the menstrual cycle.

meristem In plants, an organized tissue that includes actively dividing cells and a reservoir of stem cells.

meroblastic cleavage An incomplete type of cell cleavage, in which only the region of the egg containing cytoplasm at the animal pole undergoes cell division.

merozygote A strain of bacteria containing an F′ factor.

mesoderm A layer of cells formed during gastrulation that develops between the ectoderm and endoderm; gives rise to skeleton, muscles and much of the circulatory system.

mesoglea A gelatinous substance that connects the two germ layers in the Radiata.

mesohyl A gelatinous, protein-rich matrix in between the choanocytes and the epithelial cells of a sponge.

mesophyll The internal tissue of a plant leaf whose cells carry out photosynthesis.

messenger A molecule that transmits messages from many types of activated sensors to effector molecules.

messenger RNA (mRNA) RNA that contains the information to specify a polypeptide with a particular amino acid sequence; its job is to carry information from the DNA to the ribosome.

metabolic cycle A biochemical cycle in which particular molecules enter while others leave; the process is cyclical because it involves a series of organic molecules that are regenerated with each turn of the cycle.

metabolic enzyme A protein that accelerates chemical reactions within the cell.

metabolic pathway In living cells, a series of chemical reactions; each step is catalyzed by a specific enzyme.

metabolic rate The total energy expenditure of an organism per unit of time.

metabolism The sum total of all chemical reactions that occur within an organism. Also, a specific set of chemical reactions occurring at the cellular level.

metabotropic receptor A G-protein-coupled receptor that is coupled to an intracellular signaling pathway that initiates changes in a postsynaptic cell.

metacentric A chromosome in which the centromere is near the middle.

metagenomics A field of study that seeks to identify and analyze the collective microbial genomes contained in a community of organisms, including, for microbial genomes, those that are not easily cultured in the laboratory.

metamerism The division of the body into identical subunits called segments.

metamorphosis The process in which a pupal or juvenile organism changes into a mature adult with very different characteristics.

metanephridia Excretory organs found in a variety of invertebrates; a type of tubular nephridium.

metanephridial system The filtration system used by annelids to cleanse the blood.

metaphase The phase of mitosis during which the chromosomes are aligned along the metaphase plate.

metaphase plate A plane halfway between the poles on which the sister chromatids align during metaphase.

metastatic cell A cancer cell that can migrate to other parts of the body.

Metazoa The collective term for animals.

methanogens Several groups of anaerobic archaea that convert CO_2, methyl groups, or acetate to methane and release it from their cells.

methanotroph An aerobic bacterium that consumes methane.

methyl-CpG-binding protein A protein that binds methylated sequences.

methyl-directed mismatch repair A DNA repair system that involves the participation of several proteins that detect the mismatch and specifically remove a segment from the newly made strand.

micelle The sphere formed by long amphipathic molecules when mixed with water. In animals, micelles aid in the absorption of poorly soluble products during digestion; they consist of bile salts, phospholipids, fatty acids, and monoglycerides clustered together.

microclimate Local variations of the climate within a given area.

microevolution The term used to describe changes in a population's gene pool from generation to generation.

microfilament *See* actin filament.

micronutrient An element required by plants in amounts at or less than 0.1 g/kg of plant dry mass.

microparasite A parasite that multiplies within its host, usually within the cells.

micropyle A small opening in the integument of a seed plant ovule through which pollen tubes grow.

microRNAs (miRNAs) Small RNA molecules, typically 22 nucleotides in length, that silence the expression of specific mRNAs, either by inhibiting translation or by promoting the degradation of mRNAs.

microscope A magnification tool that enables researchers to study the structure and function of cells.

microsphere A small water-filled vesicle surrounded by a macromolecular boundary.

microspore In seed plants and some seedless plants, a relatively small spore that produces a male gametophyte within the spore wall.

microtubule A type of hollow protein filament composed of tubulin proteins that is part of the cytoskeleton and is important for cell organization, shape, and movement.

microtubule-organizing center *See* centrosome.

microvilli Small projections in the surface membranes of epithelial cells in the small intestine.

midbrain One of three major divisions of the vertebrate brain; the other two divisions are the hindbrain and forebrain.

middle ear One of the three main compartments of the mammalian ear; contains three small bones called ossicles that connect the eardrum with the oval window.

middle lamella A layer composed primarily of carbohydrate that cements adjacent plant cell walls together.

migration Long-range seasonal movement among animals in order to feed or breed.

mimicry The resemblance of an organism (the mimic) to another organism (the model).

mineral An inorganic ion required by a living organism. Minerals are used to build skeletons, maintain the balance of salts and water in the body, provide a source of electric current across plasma membranes, and provide a mechanism for exocytosis and muscle contraction, among other functions.

mineralization The general process by which phosphorus, nitrogen, CO_2, and other minerals are released from organic compounds.

mineralocorticoid A steroid hormone such as aldosterone that regulates the balance of certain minerals in the body.

minor groove A smaller groove that spirals around the DNA double helix.

missense mutation A base substitution that changes a single amino acid in a polypeptide sequence.

mitochondrial genome The chromosome found in mitochondria.

mitochondrial matrix A compartment inside the inner membrane of a mitochondrion.

mitochondrion Literally, "thread granule." An organelle found in eukaryotic cells that supplies most of the cell's ATP.

mitosis In eukaryotes, the process in which nuclear division results in two nuclei; each daughter cell receives the same complement of chromosomes.

mitotic cell division A process whereby a eukaryotic cell divides to produce two new cells that are genetically identical to the original cell.

mitotic spindle *See* mitotic spindle apparatus.

mitotic spindle apparatus The structure responsible for organizing and sorting the chromosomes during mitosis.

mixotroph An organism that is able to use autotrophy as well as phagotrophy or osmotrophy to obtain organic nutrients.

moderately repetitive sequence A DNA sequence found a few hundred to several thousand times in a genome.

modern synthesis of evolution Within a given population of interbreeding organisms, natural variation exists that is caused by random changes in the genetic material. Such genetic changes may affect the phenotype of an individual in a positive, negative, or neutral way. If a genetic change promotes an individual's reproductive success, natural selection may increase the prevalence of that trait in future generations.

molar An adjective to describe the number of moles of a solute dissolved in 1 L of water.

molarity The number of moles of a solute dissolved in 1 L of water.

mole The amount of any substance that contains the same number of particles as there are atoms in exactly 12 g of carbon: 12 grams of carbon equals 1 mole, while 1 g of hydrogen equals 1 mole.

molecular biology A field of study spawned largely by genetic technology that looks at the structure and function of the molecules of life.

molecular clock A clock on which to measure evolutionary time.

molecular evolution The molecular changes in genetic material that underlie the phenotypic changes associated with evolution.

molecular formula A representation of a molecule that consists of the chemical symbols for all of the atoms present and subscripts that tell you how many of those atoms are present.

molecular homologies Similarities that indicate that living species evolved from a common ancestor or interrelated group of common ancestors.

molecular machine A machine that is measured in nanometers, which has moving parts and does useful work.

molecular mass The sum of the atomic masses of all the atoms in a molecule.

molecular pharming An avenue of research that involves the production of medically important proteins in agricultural crops or animals.

molecular recognition The process whereby surfaces on various protein subunits recognize each other in a very specific way, causing them to bind to each other and promote the assembly process.

molecular systematics A field of study that involves the analysis of genetic data, such as DNA sequences, to identify and study genetic homology and construct phylogenetic trees.

molecule Formed from two or more atoms that are connected by chemical bonds.

monocots One of the two largest lineages of flowering plants; the embryo produces a single seed leaf.

monocular vision A type of vision in animals with eyes on the sides of the head; the animal sees a wide area at one time, though depth perception is reduced.

monocyte A type of phagocyte that circulates in the blood for only a few days, after which it takes up permanent residence as a macrophage in different organs.

monoecious The term to describe plants that produce carpellate and staminate flowers on the same plant.

monogamy A mating system in which one male mates with one female, and most individuals have mates.

monohybrid The F_1 offspring, also called single-trait hybrids, of true-breeding parents that differ with regard to a single trait.

monomorphic gene A gene that exists predominantly as a single allele in a population.

monophagous The term used to define parasites that feed on one species or two or three closely related hosts.

monophyletic group A group of species, a taxon, consisting of the most recent common ancestor and all of its descendants.

monosaccharide A simple sugar.

monosomic An aneuploid organism that has one too few chromosomes.

monotreme One of three species in the mammalian order Monotremata, which are found in Australia and New Guinea: the duck-billed platypus and two species of echidna.

morphogen   A molecule that imparts positional information and promotes developmental changes at the cellular level.

morphogenetic field   A group of embryonic cells that ultimately produce a specific body structure.

morula   The mammal embryo in species having undergone holoblastic cleavage.

mosaic   An individual in which some cells throughout the body show genetic differences. For example, in female mammals, about half of the somatic cells will express one X-linked allele, while the rest of the somatic cells will express the other allele.

mosses   A phylum of bryophytes; Bryophyta.

motif   The structure that occurs when a domain or portion of a domain has a very similar structure in many different proteins.

motor neuron   A neuron that sends signals away from the central nervous system and elicits some type of response.

motor protein   A category of cellular proteins that uses ATP as a source of energy to promote movement; consists of three domains called the head, hinge, and tail.

movement corridor   Thin strips of habitat that may permit the movement of individuals between larger habitat patches.

mucigel   A gooey plant substance that lubricates roots, aiding in their passage through the soil; helps in water and mineral absorption, prevents root drying, and provides an environment hospitable to beneficial microbes.

Müllerian mimicry   A type of mimicry in which many noxious species converge to look the same, thus reinforcing the basic distasteful design.

multicellular   Consisting of more than one cell, with cells attached to each other; cells able to communicate with each other by chemical signaling, and some cells able to specialize.

multimeric protein   A protein with more than one polypeptide chain; also said to have a quarternary structure.

multiple alleles   Refers the occurrence of a gene that exists as three or more alleles in a population.

multiple sclerosis (MS)   A disease in which the patient's own body, for reasons that are unknown, attacks and destroys myelin as if it were a foreign substance. Eventually, these repeated attacks leave scarred (sclerotic) areas of tissue in the nervous system and impair the function of myelinated neurons that control movement, speech, memory, and emotion.

multipotent   A term used to describe a stem cell that can differentiate into several cell types, but far fewer than pluripotent cells.

muscle   A grouping of muscle tissue bound together by a succession of connective tissue layers.

muscle tissue   Clusters of cells that are specialized to contract, generating the mechanical forces that produce body movement, exert pressure on a fluid-filled cavity, or decrease the diameter of a tube.

mutagen   An agent known to cause mutation.

mutant allele   An allele that has been altered by mutation.

mutation   A heritable change in the genetic material.

mutualism   An interaction in which both species benefit.

mycelium   A fungal body composed of microscopic branched filaments known as hyphae.

mycorrhizae   Associations between the hyphae of certain fungi and the roots of most plants.

myelin sheath   In the nervous system, an insulating layer made up of specialized glial cells wrapped around the axons.

myocardial infarction (MI)   The death of cardiac muscle cells, which can occur if a region of the heart is deprived of blood for an extended time.

myofibrils   Cylindrical bundles within muscle fibers, each of which contains thick and thin filaments.

myogenic bHLH genes   A small group of genes that initiates muscle development in animals.

myogenic heart   A heart in which the signaling mechanism that initiates contraction resides within the cardiac muscle itself.

myoglobin   An oxygen-binding protein that increases the availability of oxygen in the muscle fiber by providing an intracellular reservoir of oxygen.

myosin   A motor protein in muscle.

N

nacre   The smooth, iridescent lining of the shells of oysters, mussels, abalone, and other mollusks.

NAD$^+$ Nicotinamide adenine dinucleotide; a dinucleotide that functions as an energy intermediate molecule. It combines with two electrons and H$^+$ to form NADH.

NADP$^+$ Nicotinamide adenine dinucleotide phosphate; a dinucleotide that functions as an energy intermediate molecule in chloroplasts. It combines with two electrons and H$^+$ to form NADPH.

natural killer (NK) cells   A type of leukocyte that is part of the body's nonspecific defenses because it recognizes general features on the surface of cancer cells or any virus-infected cells.

natural selection   The process that culls out those individuals that are less likely to survive and reproduce in a particular environment, while allowing other individuals with traits that confer greater reproductive success to increase in numbers.

nauplius   The first larval stage in a crustacean.

navigation   A mechanism of migration that involves the ability not only to follow a compass bearing but also to set or adjust it.

nectar   A sugar-rich substance produced by many flowers that serves as a food reward for pollinators; in some plants, nectar is produced by other plant parts as a reward for insects that protect the plant.

negative control   Transcriptional regulation by repressor proteins.

negative feedback loop   A system in which a change in the variable being regulated brings about responses that move the variable in the opposite direction.

negative frequency-dependent selection   A pattern of natural selection in which the fitness of a genotype decreases when its frequency becomes higher; the result is a balanced polymorphism.

negative pressure filling   The mechanism by which reptiles, birds, and mammals ventilate their lungs.

nekton   Free-swimming animals in the open ocean that can swim against the currents to locate food.

nematocyst   In a cnidarian, a powerful capsule with an inverted coiled and barbed thread that functions to immobilize small prey so they can be passed to the mouth and ingested.

neocortex   The layer of the brain that evolved most recently in mammals.

nephron   One of several million single-cell-thick tubules that are the functional units of the kidney.

Nernst equation   The formula that gives the equilibrium potential for an ion at any given concentration gradient.

nerve   A structure found in the peripheral nervous system that is composed of multiple neurons and makes contact with structures outside the central nervous system and transmits signals that enter or leave the CNS.

nerve cord   In more complex invertebrates, a structure that extends from the anterior end of the animal to the tail.

nerve impulse   A way that neurons communicate, involving changes in the amount of electric charge across a cell's plasma membrane.

nerve net   Interconnected neurons with no central control organ.

nervous system   Groups of cells that sense internal and environmental changes and transmit signals that enable an animal to respond in an appropriate way.

nervous tissue   Clusters of cells that initiate and conduct electrical signals from one part of an animal's body to another part.

net primary production (NPP)   Gross primary production minus the energy lost in plant cellular respiration.

net reproductive rate   The population growth rate per generation.

neural crest   A cell lineage that gives rise to all neurons and supporting cells of the peripheral nervous system in vertebrates; in addition, it gives rise to melanocytes and to cells that form facial cartilage and parts of the adrenal gland.

neural tube   In chordates, a structure formed from ectoderm located dorsal to the notochord; all neurons and their supporting cells in the central nervous system originate from neural precursor cells derived from the neural tube.

neurogenesis The production of new neurons by cell division.

neurogenic heart A heart that will not beat unless it receives regular electrical impulses from the nervous system.

neurohormone A hypothalamic releasing hormone made in and secreted by neurons whose cell bodies are in the hypothalamus.

neuromodulator Short chains of amino acids that can alter or modulate the response of the postsynaptic neuron to other neurotransmitters.

neuromuscular junction The junction between a motor neuron's axon and a muscle fiber.

neuron Another name for a nerve cell. A highly specialized cell that communicates with another cell of its kind and with other types of cells by electrical or chemical signals.

neuroscience The scientific study of nervous systems.

neurotransmitter A small signaling molecule that is synthesized and stored in nerve cells.

neurulation The embryological process responsible for initiating central nervous system formation.

neutral mutation A mutation that does not affect the function of the encoded protein.

neutral theory of evolution States that most genetic variation is due to the accumulation of neutral mutations that have attained high frequencies in a population via genetic drift.

neutral variation Variation that does not favor any particular genotype.

neutralism The phenomenon in which two species occur together but in fact do not interact in any measurable way.

neutron A neutral particle found in the center of the atom.

neutrophil A type of phagocyte and the most abundant type of leukocyte; neutrophils engulf bacteria by endocytosis.

niche The physical distribution and ecological role of an organism.

nitrification The conversion by soil bacteria of NH_3 or NH_4^+ to nitrate (NO_3^-), a form of nitrogen commonly used by plants.

nitrogen fixation A specialized metabolic process in which certain prokaryotes use the enzyme nitrogenase to convert inert atmospheric nitrogen gas into ammonia; also, the industrial process by which humans produce ammonia fertilizer from nitrogen gas.

nitrogenase An enzyme used in the biological process of fixing nitrogen.

nitrogen-limitation hypothesis The proposal that organisms select food based on its nitrogen content.

nitrogenous wastes Molecules that include nitrogen from amino groups; these wastes are toxic at high concentrations and must be eliminated from the body.

nociceptor A sensory receptor that responds to extreme heat, cold, and pressure, as well as to certain molecules such as acids; also known as a pain receptor.

nocturnal enuresis Bed-wetting.

Nod factor Nodulation factor; a substance produced by nitrogen-fixing bacteria in response to flavonoids secreted from the roots of potential host plants; the Nod factors bind to receptors in plant root membranes, starting a process that allows the bacteria to invade roots.

node The region of a plant stem from which one or more leaves, branches, or buds emerge.

nodes of Ranvier Exposed areas in the axons of myelinated neurons that contain many voltage-gated Na^+ channels.

nodule A small swelling on a plant root that contains nitrogen-fixing bacteria.

nodulin One of several plant proteins that foster root nodule development.

noncoding strand *See* template.

noncompetitive inhibitor A molecule that binds to an enzyme at a location that is outside the active site and inhibits the enzyme's function.

noncyclic electron flow The combined action of photosystem II and photosystem I in which electrons flow in a linear manner to produce NADPH.

non-Darwinian evolution Also "survival of the luckiest" to contrast it with Darwin's "survival of the fittest" theory; the idea that much of the modern variation in gene sequences is explained by neutral variation rather than adaptive variation.

nondisjunction An event in which the chromosomes do not sort properly during cell division.

nonparental type *See* recombinant.

nonpolar molecule A molecule composed predominantly of nonpolar bonds.

nonrandom mating The phenomenon that individuals choose their mates based on their genotypes or phenotypes.

nonrecombinant An offspring whose combination of traits has not changed from the parental generation.

nonsense codon *See* stop codon.

nonsense mutation A mutation that changes a normal codon into a stop codon; this causes translation to be terminated earlier than expected, producing a truncated polypeptide.

nonspecific (innate) immunity The body's defenses that are present at birth and act against foreign materials in much the same way regardless of the specific identity of the invading material; includes the body's external barriers, plus a set of cellular and chemical defenses that oppose substances that breach those barriers.

nonvascular plant A plant that does not produce lignified vascular tissue, such as a modern bryophyte or extinct pretracheophyte polysporangiophytes.

norepinephrine A neurotransmitter; also known as noradrenaline.

norm of reaction A description of how a trait may change depending on the environmental conditions.

notochord A single flexible rod that lies between the digestive tract and the nerve cord in a chordate.

N-terminus The location of the first amino acid in a polypeptide.

nuclear envelope A double-membrane structure that encloses the cell's nucleus.

nuclear genome The chromosomes found in the nucleus of the cell.

nuclear lamina A collection of filamentous proteins that line the inner nuclear membrane.

nuclear matrix A filamentous network of proteins that is found inside the nucleus and lines the inner nuclear membrane. The nuclear matrix serves to organize the chromosomes.

nuclear pore A passageway for the movement of molecules and macromolecules into and out of the nucleus; formed where the inner and outer nuclear membranes make contact with each other.

nucleic acid An organic molecule composed of nucleotides. The two types of nucleic acids are deoxyribonucleic acid (DNA) and ribonucleic acid (RNA).

nucleoid A region of a bacterial cell where the genetic material (DNA) is located.

nucleolus A prominent region in the nucleus of nondividing cells where ribosome assembly occurs.

nucleosome A structural unit of eukaryotic chromosomes composed of an octamer of histones (eight histone proteins) wrapped with DNA.

nucleotide An organic molecule having three components: a phosphate group, a five-carbon sugar (either ribose or deoxyribose), and a single or double ring of carbon and nitrogen atoms known as a base.

nucleotide excision repair (NER) A common type of DNA repair system that removes (excises) a region of the DNA where the damage occurs. This system can fix many different types of DNA damage, including UV-induced damage, chemically modified bases, missing bases, and various types of cross-links.

nucleus (plural, **nuclei**) 1. In cell biology, an organelle found in eukaryotic cells that contains most of the cell's genetic material. The primary function involves the protection, organization, and expression of the genetic material. 2. In chemistry, the region of an atom that contains protons and neutrons. 3. In neurobiology, a group of neuronal cell bodies in the brain that are devoted to a particular function.

nutrient Any substance taken up by a living organism that is needed for survival, growth, development, repair, or reproduction.

O

obese According to current National Institutes of Health guidelines, a person having a BMI of 30 or more.

obligatory mutualism An interaction in which neither species can live without the other.

ocelli Photosensitive organs.

octet rule The phenomenon that atoms are most stable when their outer shell is full. For many, but not all, types of atoms, their outer shell is full when they contain eight electrons, an octet.

Okazaki fragments Short segments of DNA synthesized in the lagging strand during DNA replication.

olfaction The sense of smell.

oligotrophic The term used to describe a young lake that starts off clear and with little plant life.

ommatidium An independent visual unit in the eye of insects that functions as a separate photoreceptor capable of forming an independent image.

omnivore An animal that has the ability to eat and survive on both plant and animal products.

oncogene A type of mutant gene derived from a protooncogene; an oncogene is overactive, thus contributing to uncontrolled cell growth and promoting cancer.

one gene–one polypeptide theory The concept that one structural gene codes for one polypeptide.

oogenesis Gametogenesis, which results in the production of egg cells.

oogonia In animals, diploid germ cells that give rise to the female gametes, the eggs.

open circulatory system A circulatory system in which hemolymph, which is not different than the interstitial fluid, flows throughout the body and is not confined to special vessels.

open complex Also called the transcription bubble; a small bubble-like structure between two DNA strands that occurs during transcription.

open conformation Loosely packed chromatin that can be transcribed into RNA.

operant conditioning A form of behavior modification; a type of associative learning in which an animal's behavior is reinforced by a consequence, either a reward or a punishment.

operator A DNA sequence in bacteria that is recognized by activator or repressor proteins that regulate the level of gene transcription.

operculum A protective flap that covers the gills of a bony fish.

operon An arrangement of two or more genes in bacteria that are under the transcriptional control of a single promoter.

opsin A protein that is a component of visual pigments.

opportunistic A term used to describe animals that have a strong preference for one type of food but can adjust their diet if the need arises.

opposable thumb A thumb that can be placed opposite the fingers of the same hand, which gives animals a precision grip that enables the manipulation of small objects.

optic nerve A structure of the eye that carries the electrical signals to the brain.

optimal foraging The concept that in a given circumstance, an animal seeks to obtain the most energy possible with the least expenditure of energy.

optimality theory The theory that predicts that an animal should behave in a way that maximizes the benefits of a behavior minus its costs.

oral Refers to the region of an animal where the mouth is located; refers to the top side of a radial animal.

orbital The region surrounding the nucleus of an atom where the probablility is high of finding a particular electron.

order In taxonomy, a subdivision of a class.

organ Two or more types of tissue combined to perform a common function. For example, the heart is composed of several types of tissues, including muscle, nervous, and connective tissue.

organ system Different organs that work together to perform an overall function in an organism.

organelle A subcellular structure or membrane-bounded compartment with its own unique structure and function.

organelle genome In eukaryotes, the genetic material found in mitochondria and plastids.

organic chemistry The study of carbon-containing molecules.

organic farming The production of crops without the use of commercial inorganic fertilizers, growth substances, and pesticides.

organic molecule A carbon-containing molecule, so named because they are found in living organisms.

organism A living thing that maintains an internal order that is separated from the environment.

organismal ecology The investigation of how adaptations and choices by individuals affect their reproduction and survival.

organismic model A view of the nature of a community that considers it to be equivalent to a superorganism; individuals, populations, and communities have a relationship to each other that resembles the associations found between cells, tissues, and organs.

organizing center A group of cells that ensures the proper organization of the meristem and preserves the correct number of actively dividing stem cells.

organogenesis The developmental stage during which cells and tissues form organs.

orientation A mechanism of migration in which animals have the ability to follow a compass bearing and travel in a straight line.

origin of replication A site within a chromosome that serves as a starting point for DNA replication.

ortholog A homologous gene in different species.

osculum A large opening at the top of a sponge.

osmoconformer An animal whose osmolarity conforms to that of its environment.

osmolarity The solute concentration of a solution of water, expressed as milliOsmoles/liter (mOsm/L).

osmoregulator An animal that maintains constant stable internal salt concentrations and osmolarities, even when living in water with very different osmolarities than its body fluids.

osmosis The movement of water across membranes to balance solute concentrations. Water diffuses from a solution that is hypotonic

(lower solute concentration) into a solution that is hypertonic (higher solute concentration).

osmotic adjustment The process by which a plant increases the solute concentration of its cytosol.

osmotic pressure The hydrostatic pressure required to stop the net flow of water across a membrane due to osmosis.

osmotroph An organism that relies on osmotrophy (uptake of small organic molecules via osmosis) as a form of nutrition.

ostia Small openings in the heart of an arthropod, through which hemolymph re-enters the heart.

ostracoderms An umbrella term for several classes of primitive, heavily armored fish, now extinct, that lacked a jaw.

otoliths Granules of calcium carbonate found in the gelatinous substance that embeds hair cells in the ear.

outer bark Protective layers of mostly dead cork cells that cover the outside of woody stems and roots.

outer ear One of the three main compartments of the mammalian ear; consists of the external ear, or pinna, and the auditory canal.

outer segment The highly convoluted plasma membranes found in the rods and cones of the eye.

outgroup A species or group of species that is most closely related to an ingroup.

ovaries In animals, the female gonads where eggs are formed.

overweight According to current National Institutes of Health guidelines, a person having a BMI of 25 or more.

oviduct A thin tube with undulating fimbriae (fingers) that extend out to the ovary; also called the fallopian tube.

oviparity Development of the embryo within an egg outside the mother.

ovoviparous The term used to describe an organism that retains eggs inside the body, where the young hatch.

ovoviviparity Development of the embryo involving aspects of both viviparity and oviparity; eggs covered with a thin shell are produced and hatch inside the mother's body, but the offspring receive no nourishment from the mother.

ovulation The release of the ovum from the ovary.

ovule In a plant, a megaspore-producing megasporangium and enclosing tissues known as integuments.

ovum *See* egg.

oxidation A process that involves the removal of electrons; occurs during the breakdown of small organic molecules.

oxidative fiber A skeletal muscle fiber that contains numerous mitochondria and has a high capacity for oxidative phosphorylation.

oxidative phosphorylation A process during which NADH and $FADH_2$ are oxidized to make more ATP via the phosphorylation of ADP.

oxytocin   A hormone secreted by the posterior pituitary gland that stimulates contractions of the smooth muscles in the uterus of a pregnant mammal, which facilitates the birth process; after birth, it is important in milk secretion.

P

P generation   The parental generation in a genetic cross.

P protein   Phloem protein; the proteinaceous material produced in sieve tube elements of plant phloem as a response to wounding.

pacemaker   *See* sinoatrial (SA) node.

Pacinian corpuscles   Structures located deep beneath the surface of the skin that respond to deep pressure or vibration.

paedomorphosis   The retention of juvenile traits in an adult organism.

pair-rule gene   A type of segmentation gene; a defect in this gene may cause alternating segments or parts of segments to be deleted.

paleontologist   A scientist who studies fossils.

palisade parenchyma   Photosynthetic ground tissue in a plant that consists of closely packed, elongate cells adapted to absorb sunlight efficiently.

pancreas   An elongated gland located behind the stomach that secretes digestive enzymes and a fluid rich in bicarbonate ions.

pangenesis   An idea proposed by the ancient Greek physician Hippocrates that suggested that "seeds" produced by all parts of the body are collected and transmitted to offspring at the time of conception, and that these seeds cause offspring to resemble their parents.

parabronchi   A series of parallel air tubes that make up the lungs and are the regions of gas exchange in birds.

paracrine   Refers to a type of cellular communication in which molecules are released into the interstitial fluid and act on nearby cells.

paralogs   Homologous genes within a single species.

paraphyletic taxon   A group that contains a common ancestor and some, but not all, of its descendants.

parapodia   Fleshy, footlike structures in the polychaetes that are pushed into the substrate to provide traction during movement.

parasite   A predatory organism that feeds off another organism but does not normally kill it outright.

parasitism   An association in which one organism feeds off another, but does not normally kill it outright.

parasympathetic division   The division of the autonomic nervous system that is involved in maintaining and restoring body functions.

parathyroid hormone (PTH)   A hormone that acts on bone to stimulate the activity of cells that dissolve the mineral part of bone.

Parazoa   A subgroup of animals that are not generally thought to possess specialized tissue types or organs, although they may have

several distinct types of cells; the one phylum in this group is the Porifera (sponges).

parenchyma cell   A type of plant cell that is thin-walled and alive at maturity.

parenchyma tissue   Plant tissue that is composed of parenchyma cells.

parental strand   The original strand in DNA replication.

parental type   *See* nonrecombinant.

parthenogenesis   An asexual process in which an offspring develops from an unfertilized egg.

partial pressure   The individual pressure of each gas in the air; the sum of these pressures is known as atmospheric pressure.

particulate inheritance   The idea that the determinants of hereditary traits are transmitted intact from one generation to the next.

parturition   The birth of an organism.

passive diffusion   Diffusion that occurs through a membrane without the aid of a transport protein.

passive immunity   A type of acquired immunity that confers protection against disease through the direct transfer of antibodies from one individual to another.

passive transport   The diffusion of a solute across a membrane in a process that is energetically favorable and does not require an input of energy.

paternal inheritance   A pattern in which only the male gamete contributes particular genes to the offspring.

pathogen   A microorganism that causes disease symptoms in its host.

pattern formation   The process that gives rise to a plant or animal with a particular body structure.

pedal glands   Glands in the foot of a rotifer that secrete a sticky substance that aids in attachment to the substrate.

pedicel   A narrow, waistlike point of attachment in a spider or insect body.

pedicellariae   The spines and jawlike pincers that cover the skeleton of an echinoderm and deter the settling of animals such as barnacles.

pedigree analysis   An examination of the inheritance of human traits in families.

pedipalps   In spiders, a pair of appendages that have various sensory, predatory, or reproductive functions.

peduncle   The tip of a flower stalk.

pelagic zone   The open ocean, where water depth averages 4,000 m and nutrient concentrations are typically low.

penis   A male external accessory sex organ found in many animals that is involved in copulation.

pentadactyl limb   A limb ending in five digits.

PEP carboxylase   An enzyme in C_4 plants that adds CO_2 to phosphoenolpyruvate (PEP) to produce the four-carbon compound oxaloacetate.

pepsin   An active enzyme in the stomach that begins the digestion of protein.

peptide bond   The covalent bond that links together amino acids in a polypeptide.

peptidoglycan   A polymer composed of protein and carbohydrate that is an important component of the cell walls of most bacteria.

peptidyl site (P site)   One of the three sites for tRNA binding to the ribosome; the others are the aminoacyl site (A site) and the exit site (E site) The polypeptide is usually in the P site.

peptidyl transfer reaction   As a peptide bond is formed, the polypeptide is removed from the tRNA in the P site and transferred to the amino acid at the A site.

per capita growth rate   The per capita birth rate minus the per capita death rate; the rate that determines how populations grow over any time period.

perception   An awareness of the sensations that are experienced.

perennial   A plant that lives for more than two years, often producing seeds each year after it reaches reproductive maturity.

perfect flower   A flower that has both stamens and carpels.

perianth   The term that refers to flower petals and sepals collectively.

pericarp   The wall of a plant's fruit.

pericycle   A cylinder of plant tissue having cell division (meristematic) capacity that encloses the root vascular tissue.

peripheral membrane protein   A protein that is noncovalently bound to regions of integral membrane proteins that project out from the membrane, or they are noncovalently bound to the polar head groups of phospholipids.

peripheral zone   The area of a plant that contains dividing cells that will eventually differentiate into plant structures.

peristalsis   Rhythmic, spontaneous waves of muscle contraction that begin near the mouth in the esophagus and move toward the stomach.

peritubular capillaries   Capillaries near the junction of the cortex and medulla in the nephron of the kidney.

permafrost   A layer of permanently frozen soil.

peroxisome   A relatively small organelle found in all eukaryotic cells that catalyzes detoxifying reactions.

petal   A flower organ that usually serves to attract insects or other animals for pollen transport.

phage   *See* bacteriophage.

phagocyte   A cell capable of phagocytosis; phagocytes provide nonspecific defense against pathogens that enter the body.

phagocytic vacuole   A vacuole that is formed by the process of phagocytosis.

phagocytosis   A form of endocytosis that involves the formation of a membrane vesicle called a phagosome, or phagocytic vacuole, that engulfs a large particle such as a bacterium.

phagotroph   An organism that specializes in phagotrophy (particle feeding) by means of phagocytosis as a form of nutrition.

phagotrophy   The use of phagocytosis as a feeding mechanism.

pharyngeal slit   A filter-feeding device in primitive chordates.

pharynx The area at the back of the throat where inhaled air from the mouth and nose converges.

phenolic Refers to compounds that contain a cyclic ring of carbon with three double bonds, known as a benzene ring, that is covalently linked to a single hydroxyl group.

phenotype The characteristics of an organism that are the result of the expression of its genes.

phenotypic plasticity The phenomenon in which individual members of the same plant species that experience different environmental conditions may display considerable variation in structure or behavior.

pheromone A powerful chemical attractant used to manipulate the behavior of others.

phloem A specialized conducting tissue at the center of a plant's stem.

phloem loading The process of conveying sugars to sieve-tube elements for long-distance transport.

phoresy A form of commensalism in which individuals of one species use individuals of a second species for transportation.

phosphodiester linkage Refers to a double linkage (two phosphoester bonds) that holds together adjacent nucleotides in DNA and RNA strands.

phosphodiesterase An enzyme that breaks down cAMP into AMP.

phospholipid A class of lipids that are similar in structure to triglycerides, but the third hydroxyl group of glycerol is linked to a phosphate group instead of a fatty acid. They are a key component of biological membranes.

phospholipid bilayer The basic framework of the cellular membrane, consisting of two layers of lipids.

photoautotroph An organism that uses the energy from light to make organic molecules from inorganic sources.

photoheterotroph An organism that is able to use light energy to generate ATP, but which must take in organic compounds from the environment.

photon A massless particle traveling in a wavelike pattern and moving at the speed of light.

photoperiodism A plant's ability to measure and respond to night length, and indirectly, day length, as a way of detecting seasonal change.

photoreceptor An electromagnetic receptor that responds to visible light energy.

photorespiration The metabolic process occurring in C_3 plants that results when the enzyme rubisco combines with oxygen instead of carbon dioxide and produces only one molecule of PGA instead of two PGA, thereby reducing photosynthetic efficiency.

photosynthesis The process whereby light energy is captured by plant, algal, or cyanobacterial cells and used to synthesize organic molecules from CO_2 and H_2O (or H_2S).

photosystem I (PSI) One of two distinct complexes of proteins and pigment molecules in the thylakoid membrane that absorbs light.

photosystem II (PSII) The complex of proteins and pigment molecules in the thylakoid membrane that generates oxygen from water.

phototropin The main blue-light sensor involved in phototropism.

phototropism The tendency of a plant to grow toward a light source.

phragmoplast A plant organelle involved in the construction of cell plate that produces an intervening cell wall between two dividing plant cells.

phyla The subdivisions of a kingdom.

phylogenetic species concept The members of a single species are identified by having a unique combination of characteristics.

phylogenetic tree A diagram that describes a phylogeny; such a tree is a hypothesis of the evolutionary relationships among various species, based on the information available to and gathered by systematists.

physical mutagen A physical agent, such as UV light, that causes mutations.

physiological ecology A subdiscipline of organismal ecology that investigates how organisms are physiologically adapted to their environment and how the environment impacts the distribution of species.

physiology The study of the functions of cells and body parts of living organisms.

phytochrome A red- and far-red-light receptor.

phytoplankton Microscopic algae and cyanobacteria that float or actively move through water.

phytoremediation The process of removing harmful metals from soils by growing hyperaccumulator plants on metal-contaminated soils, then harvesting and burning the plants to ashes for disposal and/or metal recovery.

pigment A molecule that can absorb light energy.

pili (singular, **pilus**) Threadlike surface appendages that allow prokaryotes to attach to each other during mating, or to move across surfaces.

piloting A mechanism of migration in which an animal moves from one familiar landmark to the next.

pinocytosis A form of endocytosis that involves the formation of membrane vesicles from the plasma membrane as a way for cells to internalize the extracellular fluid.

pistil A flower structure that may consist of a single carpel or multiple, fused carpels, and is differentiated into stigma, style, and ovary.

pit A small cavity in a plant cell wall where secondary wall materials such as lignin are absent.

pitch The tone of a sound wave.

pituitary dwarfism A condition in which a person's anterior pituitary gland fails to make adequate amounts of GH during childhood, so the concentrations of GH and IGF-1 in the blood will be lower than normal and growth is stunted.

pituitary giant A person who has a tumor of the GH-secreting cells of the anterior pituitary gland and thus produces excess GH during childhood and adulthood; the person can grow very tall.

pituitary gland A multilobed endocrine gland sitting directly below the hypothalamus.

placenta A structure through which humans and other eutherian mammals retain and nourish their young within the uterus via the transfer of nutrients and gases.

placental transfer tissue In plants, a nutritive tissue that aids in the transfer of nutrients from maternal parent to embryo.

plant A member of the kingdom Plantae.

plant tissue culture A laboratory process to produce thousands of identical plants having the same desirable characteristics.

Plantae A eukaryotic kingdom of the domain Eukarya; includes multicellular organisms having cellulose-rich cell walls and plastids, and which are adapted in many ways to terrestrial habitats (or if aquatic, derived from terrestrial ancestors).

plaque 1. A deposit of lipids, fibrous tissue, and smooth muscle cells that may develop in blood vessels. 2. A bacterial biofilm that may form on the surfaces of teeth.

plasma The fluid part of blood.

plasma cell A cell that synthesizes and secretes antibodies.

plasma membrane The biomembrane that separates the internal contents of a cell from its external environment.

plasmid A small circular piece of DNA found naturally in many strains of bacteria and also occasionally in eukaryotic cells.

plasmodesma (plural, **plasmodesmata**) A membrane-lined, ER-containing channel that connects the cytoplasm of adjacent plant cells.

plasmogamy The fusion of the cytoplasm between two gametes.

plasmolysis The shrinkage of algal or plant cytoplasm that occurs when water leaves the cell by osmosis, with the result that the plasma membrane no longer presses on the cell wall.

plastid A general name given to organelles found in plant and algal cells, which are bound by two membranes and contain DNA and large amounts of chlorophyll (chloroplasts), carotenoids (chromoplasts), or starch (amyloplasts).

platelets Cell fragments in the blood of mammals that play a crucial role in the formation of blood clots.

pleiotropy The phenomenon in which a mutation in a single gene can have multiple effects on an individual's phenotype.

pleural sac A double layer of sheathlike moist membranes that encases each lung.

pluripotent Refers to the ability of embryonic stem cells to differentiate into almost every cell type of the body.

point mutation A mutation that affects only a single base pair within the DNA or that involves the addition or deletion of a single base pair to a DNA sequence.

polar cell The highest latitude cell in the three-cell circulation of wind in each hemisphere.

polar covalent bond The bond that forms when two atoms with different electronegativities form a covalent bond and the shared electrons are closer to the atom of higher electronegativity rather than the atom of lower electronegativity. The distribution of electrons around the atoms creates a polarity, or difference in electric charge, across the molecule.

polar molecule A molecule containing significant numbers of polar bonds.

polar transport The process whereby auxin primarily flows downward in shoots and into roots.

polarized 1. In cell biology, refers to cells that have different sides, such as the apical and basal sides of epithelial cells. 2. In neuroscience, refers to the electrical gradient across the membrane. A neuron with a large electrical gradient across its plasma membrane is said to be highly polarized.

pole A structure of the spindle apparatus defined by each centrosome.

pole cells The primordial germ cells that are the first cells to form at the posterior end of the embryo in certain species such as *Drosophila*.

pollen Tiny male gametophytes enclosed by sporopollenin-containing microspore walls.

pollen coat A layer of material that covers the sporopollenin-rich pollen wall.

pollen grain An immature male gametophyte.

pollen tube A mature male gametophyte consisting of a germinated pollen grain with a long, thin pollen tube that carries haploid sperm cells.

pollen wall A tough, sporopollenin wall at the surface of a pollen grain.

pollination The process in which pollen grains are transported to an angiosperm flower or a gymnosperm cone by means of wind or animal pollinators.

pollination syndromes The pattern of coevolved traits between particular types of flowers and specific pollinators.

pollinator An animal that carries pollen between angiosperm flowers (or cones of gymnosperms).

poly A tail A string of adenine nucleotides at the 3′ end of most mature mRNAs in eukaryotes.

polyandry A form of mating in which one female mates with several males.

polycistronic mRNA An mRNA that contains the coding sequences for two or more structural genes.

polycythemia A condition of increased hemoglobin due to increased hematocrit.

polygenic A trait in which several or many genes contribute to the outcome of the trait.

polygyny A form of mating in which one male mates with more than one female in a single breeding season, but females mate only with one male.

polyketides A group of secondary metabolites that are produced by bacteria, fungi, plants, insects, dinoflagellates, mollusks, and sponges.

polymer A large molecule formed by linking together many smaller molecules called monomers.

polymerase chain reaction (PCR) A technique to make many copies of a gene in vitro; primers are used that flank the region of DNA to be amplified.

polymorphic gene A gene that commonly exists as two or more alleles in a population.

polymorphism The phenomenon that many traits or genes may display variation within a population.

polypeptide A linear sequence of amino acids; the term denotes structure.

polyphagous The term used to define parasites that feed on many host species.

polyphyletic taxon A group that consists of members of several evolutionary lines and does not include the most recent common ancestor of the included lineages.

polyploid An organism that has three or more sets of chromosomes.

polysaccharide Many monosaccharides linked together to form long polymers.

pons The part of the hindbrain, along with the cerebellum, responsible for monitoring and coordinating body movements.

population A group of individuals of the same species that can interbreed with one another.

population ecology The study of how populations grow and what promotes and limits growth.

population genetics The study of genes and genotypes in a population.

portal vein A vein that not only collects blood from capillaries—like all veins—but also forms another set of capillaries, as opposed to returning the blood directly to the heart like other veins.

positional information Molecules that are provided to a cell that allow it to determine its position relative to other cells.

positive control Transcriptional regulation by activator proteins.

positive feedback loop The acceleration of a process, leading to what is sometimes called an explosive system.

positive pressure filling The method by which amphibians ventilate their lungs—the animals gulp air and force it under pressure into the lungs, as if inflating a balloon.

postabsorptive state One of two alternating phases in the utilization of nutrients; occurs when the gastrointestinal tract is empty of nutrients and the body's own stores must supply energy. The other phase is the absorptive state.

posterior Refers to the rear (tail-end) of an animal.

postsynaptic cell The cell that receives the electrical or chemical signal sent from a neuron.

post-translational covalent modification A process of changing the structure of a protein, usually by covalently attaching functional groups.

post-translational sorting The uptake of proteins into the nucleus, mitochondria, chloroplasts, and peroxisomes that occurs after the protein is completely made (that is, completely translated).

postzygotic isolating mechanism A mechanism that prevents interbreeding by blocking the development of a viable and fertile individual after fertilization has taken place.

potential energy The energy that a substance possesses due to its structure or location.

power stroke The process in which an energized myosin cross-bridge binds to actin and triggers the movement of the bound cross-bridge.

prebiotic soup The medium formed by the slow accumulation of molecules in the early oceans over a long period of time prior to the existence of life.

predation An interaction where the action of the predator results in the death of the prey.

predator An animal that kills live prey.

pregnancy The time during which a developing embryo and fetus grows within the uterus of the mother. The period of pregnancy is also known as gestation.

preinitiation complex The structure of the completed assembly of GTFs and RNA polymerase II at the TATA box prior to transcription of eukaryotic structural genes.

pre-mRNA The RNA transcript prior to any processing.

presynaptic cell The cell that sends the electrical or chemical signal from a neuron to another cell.

prezygotic isolating mechanism A mechanism that stops interbreeding by preventing the formation of a zygote.

primary active transport A type of transport that involves pumps that directly use energy and generate a solute gradient.

primary cell wall In plants, a cell wall that is synthesized first between the two newly made daughter cells.

primary consumer An organism that obtains its food by eating primary producers; also called an herbivore.

primary electron acceptor The molecule to which a high-energy electron from an excited pigment molecule such as P680* is transferred.

primary endosymbiosis The process by which a eukaryotic host cell acquires prokaryotic endosymbionts or plastids or mitochondria derived from a prokaryotic endosymbiont.

primary immune response The response to an initial exposure to an antigen.

primary meristem A meristematic tissue that increases plant length and produces new organs.

primary metabolism The synthesis and breakdown of molecules and macromolecules that are found in all forms of life and are essential for cell structure and function.

primary oocytes   In animals, the first stage of producing female gametes by meiosis.

primary plastid   A plastid that arose from a prokaryote as the result of endosymbiosis.

primary producer   An autotroph, which typically harvests light energy from the sun; located at the base of the food chain.

primary spermatocytes   In animals, the first stage of producing sperm by meiosis.

primary structure   The linear sequence of amino acids of a polypeptide. One of four levels of protein structure.

primary succession   Succession on newly exposed sites that were not previously occupied by soil and vegetation.

primary tissue   Plant tissue generated as the result of primary growth at apical meristems.

primary vascular tissue   Plant tissue composed of primary xylem and phloem.

primer   A short segment of RNA, typically 10 to 12 nucleotides in length, that is needed to begin DNA replication.

principle of parsimony   The preferred hypothesis is the one that is the simplest.

principle of species individuality   A view of the nature of a community; each species is distributed according to its physiological needs and population dynamics; most communities intergrade continuously and competition does not create distinct vegetational zones.

probability   The chance that an event will have a particular outcome.

proboscis   The coiled tongue of a butterfly or moth, which can be uncoiled, enabling it to drink nectar from flowers.

procambium   A type of primary plant tissue meristem that produces vascular tissue.

producer   An organism that synthesizes the organic compounds used by other organisms for food.

product   During a chemical reaction, the reactants are converted to products.

product rule   The probability that two or more independent events will occur is equal to the product of their individual probabilities.

production efficiency   The percentage of energy assimilated by an organism that becomes incorporated into new biomass.

productivity hypothesis   The proposal that greater production by plants results in greater overall species richness.

progesterone   A hormone secreted by the female ovaries that plays a key role in pregnancy.

progymnosperms   An extinct group of plants having wood but not seeds, that evolved before the gymnosperms.

prokaryote   One of the two categories into which all forms of life can be placed, based on cell structure; the other is eukaryote. Prokaryotic cells lack a nucleus having an envelope with pores; includes bacteria and archaea.

prometaphase   A phase of mitosis during which the mitotic spindle is completely formed.

promoter   The site in the DNA where transcription begins.

proofreading   The ability of DNA polymerase to identify a mismatched nucleotide and remove it from the daughter strand.

prophage   Refers to the DNA of a phage that has become integrated into the bacterial chromosome.

prophase   A phase of mitosis during which the chromosomes condense and the nuclear membrane fragments.

proplastid   Unspecialized structures that form plastids.

prosimian   A member of a class of primates that includes the smaller species such as bush babies, lemurs, pottos, and tarsiers.

prostate gland   A structure that secretes a thin fluid that protects sperm once they are deposited within the female reproductive tract.

prosthetic group   Small molecules that are permanently attached to the surface of an enzyme and aid in catalysis.

protease   An enzyme that cuts proteins into smaller polypeptides.

proteasome   A molecular machine that is the primary pathway for protein degradation in archaea and eukaryotic cells.

protein   A functional unit composed of one or more polypeptides. Each polypeptide is composed of a linear sequence of amino acids.

protein kinase   An enzyme that transfers phosphate groups from ATP to a protein.

protein phosphatase   An enzyme responsible for removing phosphate groups from proteins.

protein-protein interactions   Specific interactions between proteins that can carry out cellular processes that occur as a series of steps or build larger structures that provide organization to the cell.

proteoglycan   A glycosaminoglycan in the extracellular matrix linked to a core protein.

proteolysis   A processing event within the cell in which enzymes called proteases cut proteins into smaller polypeptides.

proteome   All of the types and relative amounts of proteins that are made in a particular cell at a particular time and under specific conditions. The proteome largely determines a cell's structure and function.

proteomics   Techniques used to identify and study groups of proteins.

protist   A eukaryotic organism that is not a member of the animal, plant, or fungal kingdoms; lives in moist habitats, and is typically microscopic in size.

Protista   In traditional classification systems, a eukaryotic kingdom of the domain Eukarya.

protobiont   The term used to describe the first nonliving structures that evolved into living cells.

protoderm   A type of primary plant tissue meristem that generates the outermost dermal tissue.

proton   A positively charged particle found in the nucleus of the atom. The number of protons in an atom is called the atomic number and defines each type of element.

protonephridia   The simplest filtration mechanism for cleansing the blood; used in flatworms.

proton-motive force   See H$^+$ electrochemical gradient.

proto-oncogene   A normal gene that, if mutated, can become an oncogene.

protostome   An animal that exhibits spiral determinate cleavage, and where the blastopore becomes the mouth; includes mollusks, annelid worms, and arthropods.

provirus   Refers to viral DNA that has become incorporated into a eukaryotic chromosome.

protozoa   A term commonly used to describe diverse heterotrophic protists.

proventriculus   The glandular portion of the stomach of a bird.

proximal convoluted tubule   The segment of the tubule of the nephron in the kidney that drains Bowman's capsule.

proximate cause   A specific genetic and physiological mechanism of behavior.

pseudocoelom   A coelom that is not completely lined by tissue derived from mesoderm.

pseudocoelomate   An animal with a pseudocoelom.

pteridophytes   A phylum of of vascular plants having euphylls, but not seeds; Pteridophyta.

pulmocutaneous circulation   The routing of blood from the heart through different vessels to the gas exchange organs (lungs and skin) of frogs and some other amphibians.

pulmonary circulation   The pumping of blood from the right side of the heart to the lungs to release carbon dioxide and pick up oxygen from the atmosphere.

pulse-chase experiment   A procedure in which researchers administer a pulse of radioactively labeled materials to cells so that they make radioactive products. This is followed by the addition of nonlabeled materials called a chase.

pump   A transporter that directly couples its conformational changes to an energy source, such as ATP hydrolysis.

punctuated equilibrium   A concept that suggests that the tempo of evolution is more sporadic than gradual. Species rapidly evolve into new species followed by long periods of equilibrium with little evolutionary change.

Punnett square   A common method for predicting the outcome of simple genetic crosses.

pupa   The organism resulting after the larval stages in insects that gives rise to an adult organism via metamorphosis.

pupil   A small opening in the eye of a vertebrate that transmits different patterns of light emitted from images in the animal's field of view.

purine   The bases adenine (A) and guanine (G), with double (fused) rings of nitrogen and carbon atoms.

pyramid of biomass   A measure of efficiency in which the organisms at each trophic level are weighed.

pyramid of numbers   An expression of trophic-level transfer efficiency, in which the number of individuals decreases at each trophic level, with a huge number of individuals at the base and fewer individuals at the top.

pyramid of production   A measure of efficiency in which rates of production are shown rather than standing crop; the laws of thermodynamics ensure that the highest amounts of free energy are found at the lowest trophic levels.

pyrimidine   The bases cytosine (C), thymine (T), and uracil (U) with a single ring.

Q

quantitative trait   *See* continuous trait.

quaternary structure   The association of two or more polypeptides to form a protein. One of four levels of protein structure.

quorum sensing   A mechanism by which prokaryotic cells are able communicate when they reach a critical population size.

R

radial cleavage   A mechanism of development in which the cleavage planes are either parallel or perpendicular to the vertical axis of the egg.

radial loop domain   A loop of chromatin, often 25,000 to 200,000 base pairs in size, that is anchored to the nuclear matrix.

radial pattern   A characteristic of the body pattern of plants.

radial symmetry   1. In plants, an architectural feature of plants in which plant embryos display a cylindrical shape, which is retained in the stems and roots of seedlings and mature plants. In addition, new leaves or flower parts are produced in circular whorls, or spirals, around shoot tips. 2. In animals, an architectural feature in which the body can be divided into symmetrical halves by many different longitudinal planes along a central axis.

Radiata   Radially symmetric animals, which means they can be divided equally by a longitudinal plane passing through the central axis; includes cnidarians and ctenophores.

radiation   The emission of electromagnetic waves by the surfaces of objects.

radicle   An embryonic root, which extends from the plant hypocotyl.

radioisotope   An isotope found in nature that is inherently unstable and does not exist for long periods of time. Such isotopes emit energy called radiation in the form of subatomic particles.

radioisotope dating   A common way to estimate the age of a fossil by analyzing the elemental isotopes within the accompanying rock.

radula   A unique, protrusible, tonguelike organ in a mollusk that has many teeth and is used to eat plants, scrape food particles off rocks, or bore into shells of other species and tear flesh.

rain shadow   An area where precipitation is noticeably less, such as on the side of the mountain sheltered from air movement.

ram ventilation   A mechanism used by fish to ventilate their gills; fish swim or face upstream with their mouths open, allowing water to enter into their buccal cavity and from there across their gills.

random   A pattern of dispersion within a population, in which individuals do not appear to be specially positioned relative to anyone else.

random genetic drift   A change in allele frequencies due to random sampling error.

random sampling error   The deviation between the observed and expected outcomes.

rate-limiting step   The slowest step in a pathway.

ray-finned fish   The Actinopterygii, which includes all bony fish except the coelacanths and lungfish.

reabsorption   In the production of urine, the process in which useful solutes in the filtrate are recaptured and transported back into the body fluids of an animal.

reactant   A substance that participates in a chemical reaction and becomes changed by that reaction.

reaction mechanism   In the case of the Na^+/K^+-ATPase, a molecular roadmap of the steps that direct the pumping of ions across the plasma membrane.

reading frame   Refers to the idea that codons are read from the start codon in groups of three bases each.

receptacle   The enlarged region at the tip of a flower peduncle to which flower parts are attached.

receptor   A cellular protein that recognizes a signaling molecule.

receptor-mediated endocytosis   A common form of endocytosis in which a receptor is specific for a given cargo.

receptor potential   The membrane potential in a sensory receptor cell.

recessive   A term that describes a trait that is masked by the presence of a dominant trait in a heterozygote.

reciprocal translocation   The process in which two different types of chromosomes exchange pieces, thereby producing two abnormal chromosomes carrying translocations.

recombinant   An offspring that has a different combination of traits from the parental generation.

recombinant DNA   Any DNA molecule that has been manipulated so that it contains DNA from two or more sources.

recombinant DNA technology   The use of laboratory techniques to isolate and manipulate fragments of DNA.

recombinant vector   A vector containing a piece of chromosomal DNA.

recombination frequency   The frequency of crossing over between two genes.

red blood cell   A cell that serves the critical function of transporting oxygen throughout the body; also known as an erythrocyte.

redox reaction   A type of reaction in which the electron that is removed during the oxidation of an atom or molecule must be transferred to another atom or molecule, which becomes reduced; short for a reduction-oxidation reaction.

reduction   A process that involves the addition of electrons to an atom or molecule.

reductionism   An approach that involves reducing complex systems to simpler components as a way to understand how the system works. In biology, reductionists study the parts of a cell or organism as individual units.

redundancy hypothesis   An extension of the rivet hypothesis. In this model, also called the passenger hypothesis, most species are more like passengers on a plane—they take up space but do not add to the airworthiness. The species are said to be redundant because they could simply be eliminated or replaced by others with no loss in function.

reflex arc   A simple circuit that allows an organism to respond rapidly to inputs from sensory neurons and consists of only a few neurons.

refractory   The term used to describe a cell that is unresponsive to another stimulus.

regeneration   A form of asexual reproduction in which a complete organism forms from small fragments of its body.

regulatory gene   A gene whose function is to regulate the expression of other genes.

regulatory sequence   In the regulation of transcription, a sequence that functions as a site for genetic regulatory proteins. Regulatory sequences control whether a gene is turned on or off.

regulatory transcription factor   A protein that binds to DNA in the vicinity of a promoter and affects the rate of transcription of one or more nearby genes.

relative abundance   The frequency of occurrence of species in a community.

relative water content (RWC)   The property often used to gauge the water content of a plant organ or entire plant; RWC integrates the water potential of all cells within an organ or plant and is thus a measure of relative turgidity.

release factor   A protein that recognizes the three stop codons in the termination stage of translation and promotes the termination of translation.

renal corpuscle   A filtering component in the nephron of the kidney.

repetitive sequence   Short DNA sequences that are present in many copies in a genome.

replica plating   A technique in which a replica of bacterial colonies is transferred to a new petri plate.

replication   1. The performing of experiments several or many times. 2. The copying of DNA strands.

replication fork   The area where two DNA strands have separated and new strands are being synthesized.

repressible operon   In this type of operon, a small effector molecule inhibits transcription.

repressor A transcription factor that binds to DNA and inhibits transcription.

reproduction The process by which organisms produce offspring.

reproductive isolating mechanisms The mechanisms that prevent interbreeding between different species.

reproductive isolation Mechanisms that prevent one species from successfully interbreeding with other species.

resistance (R) The tendency of blood vessels to slow down the flow of blood through their lumens.

resistance gene (R gene) A plant gene that has evolved as part of a defense system in response to pathogen attack.

resolution In microscopy, the ability to observe two adjacent objects as distinct from one another; a measure of the clarity of an image.

resonance energy transfer The process by which the energy (not the electron itself) can be transferred to adjacent pigment molecules.

resource competition Competition in which organisms compete indirectly through via the consumption of a limited resource, with each obtaining as much as it can.

resource partitioning The differentiation of niches, both in space and time, that enables similar species to coexist in a community.

respiratory centers Several regions of the brainstem in vertebrates that initiate expansion of the lungs.

respiratory chain *See* electron transport chain.

respiratory distress syndrome of the newborn The situation in which a human baby is born prematurely, before sufficient surfactant is produced, and many alveoli quickly collapse.

respiratory pigment A large protein that contains one or more metal atoms that bind to oxygen.

respiratory system All components of the body that contribute to the exchange of gas between the external environment and the blood; in mammals, includes the nose, mouth, airways, lungs, and muscles and connective tissues that encase these structures within the thoracic (chest) cavity.

response elements DNA sequences that are recognized by regulatory transcription factors and regulate the expression of genes.

response regulator In bacteria, a protein that interacts with a sensor kinase and regulates the expression of many genes.

resting potential The difference in charges across the plasma membrane in an unstimulated cell.

restoration ecology The full or partial repair or replacement of biological habitats and/or their populations that have been damaged.

restriction enzyme An enzyme that recognizes particular DNA sequences and cleaves the DNA backbone at two sites.

restriction point A point in the cell cycle in which a cell has become committed to divide.

restriction sites The base sequences recognized by restriction enzymes.

retina A sheetlike layer of photoreceptors at the back of the eye.

retinal A derivative of vitamin A that is capable of absorbing light energy.

retroelement A type of transposable element that moves via an RNA intermediate.

retrotransposon *See* retroelement.

retrovirus An RNA virus that utilizes reverse transcription to produce viral DNA that can be integrated into the host cell genome.

reverse transcriptase A viral enzyme that catalyzes the synthesis of viral DNA starting with viral RNA as a template.

rhizobia The collective term for proteobacteria involved in the nitrogen-fixation symbioses with plants that are important in nature and to agriculture.

rhizomorphs Fungal mycelia that have the shape of roots.

rhodopsin The visual pigment in rods.

ribonucleic acid (RNA) One of two classes of nucleic acids; the other is deoxyribonucleic acid (DNA). RNA consists of a single strand of nucleotides.

ribose A five-carbon sugar found in RNA.

ribosomal initiation factor In the initiation stage of translation, a protein that facilitates the interactions between mRNA, the first tRNA, and the ribosomal subunits.

ribosomal RNA (rRNA) An RNA that forms part of ribosomes, which provide the site where translation occurs.

ribosome A structure composed of proteins and rRNA that provides the site where protein synthesis occurs.

ribozyme A biological catalyst that is an RNA molecule.

right-left axis In bilateral animals, one of the three axes along which the adult body pattern is organized; the others are the dorsoventral axis and the anteroposterior axis.

ring canal A structure in the water vascular system of echinoderms.

rivet hypothesis An alternative to the diversity-stability hypothesis. In this model, species are like the rivets on an airplane, with each species playing a small but critical role in keeping the plane (the ecosystem) airborne. The loss of a rivet weakens the plane and causes it to lose a little airworthiness. The loss of a few rivets could probably be tolerated, while the loss of more rivets would prove critical to the airplane's function.

RNA *See* ribonucleic acid.

RNA-induced silencing complex (RISC) The complex that mediates RNA interference via microRNAs.

RNA interference (RNAi) Refers to a type of mRNA silencing; miRNA interferes with the proper expression of an mRNA.

RNA polymerase The enzyme that synthesizes strands of RNA during gene transcription.

RNA processing A step in gene expression between transcription and translation; the RNA transcript, termed pre-mRNA, is modified in ways that make it a functionally active mRNA.

RNA world A hypothetical period on primitive Earth when both the information needed for life and the enzymatic activity of living cells were contained solely in RNA molecules.

RNase An enzyme that digests RNA.

rods Photoreceptors found in the vertebrate eye; they are very sensitive to low-intensity light but do not readily discriminate different colors. Rods are utilized mostly at night, and they send signals to the brain that generate a black-and-white visual image.

"Roid" rage Extreme aggressive behavior brought about by androgen administration.

root A plant organ that provides anchorage in the soil and also fosters efficient uptake of water and minerals.

root apical meristem (RAM) The region where new root tissues of plants are produced.

root cap A protective covering on the root tip that is produced by the root apical meristem of a plant.

root cortex A region of ground parenchyma located between the epidermis and vascular tissue of mature plant roots.

root hair A long, thin root epidermal cell that functions to absorb water and minerals, usually from soil.

root meristem The collection of cells at the root tip that generate all of the tissues of a plant root.

root pressure A process whereby plants are able to refill embolized vessels.

root system The collection of roots and root branches produced by root apical meristems.

rough endoplasmic reticulum (rough ER) The part of the ER that is studded with ribosomes; this region plays a key role in the initial synthesis and sorting of proteins that are destined for the ER, Golgi apparatus, lysosomes, vacuoles, plasma membrane, or outside of the cell.

r-selected species A type of life history strategy, where species have a high rate of per capita population growth but poor competitive ability.

rubisco The enzyme that that catalyzes the first step in the Calvin cycle in which CO_2 is incorporated into an organic molecule.

S

S The synthesis phase of the cell cycle.

saltatory conduction A type of conduction in which sodium ions move into the cell and the charge moves rapidly through the cytosol to the next node, where the action potential continues.

sarcoma A tumor of connective tissue such as bone or cartilage.

sarcomere One compete unit of the repeating pattern of thick and thin filaments within a myofibril.

sarcoplasmic reticulum A cellular organelle that provides a muscle fiber's source of the cytosolic calcium involved in muscle contraction.

satiety A feeling of fullness.

satiety signal   A response to eating that removes the sensation of hunger and sets the time period before hunger returns again.

saturated fatty acid   A fatty acid in which all the carbons are linked by single covalent bonds.

scaffold   An area in metaphase chromosomes formed from proteins that holds the radial loops in place.

scanning electron microscopy (SEM)   A type of microscopy that utilizes an electron beam to produce an image of the three-dimensional surface of biological samples.

scavenger   An animal that eats the remains of dead animals.

schizocoelous   In protostomes, a pattern of development in which a solid mass of mesoderm cells splits to form the cavity that becomes the coelom.

Schwann cells   The glial cells that form myelin on axons that travel outside the brain and spinal cord.

science   In biology, the observation, identification, experimental investigation, and theoretical explanation of natural phenomena.

scientific method   A series of steps to test the validity of a hypothesis. The experimentation often involves a comparison between control and experimental samples.

sclera   The white of the vertebrate eye, a strong outer sheet that in the front is continuous with a thin, clear layer known as the cornea.

sclereid   Star- or stone-shaped plant cells having tough, lignified cell walls.

sclerenchyma tissue   Rigid plant tissue composed of tough-walled fibers and sclereids.

scurvy   A potentially fatal vitamin C deficiency in humans characterized by weakness, bleeding gums, and tooth loss.

seaweed   A multicellular protist that occurs in marine habitats and may display a relatively large and complex body.

second law of thermodynamics   The second law states that the transfer of energy or the transformation of energy from one form to another increases the entropy, or degree of disorder of a system.

second messengers   Small molecules or ions that relay signals inside the cell.

secondary active transport   A type of transport that involves the utilization of a pre-existing gradient to drive the active transport of another solute.

secondary cell wall   A plant cell wall that is synthesized and deposited between the plasma membrane and the primary cell wall after a plant cell matures and has stopped increasing in size.

secondary consumer   An organism that eats primary consumers; also called a carnivore.

secondary endosymbiosis   A process that occurs when a eukaryotic host cell acquires a eukaryotic endosymbiont having a primary plastid.

secondary immune response   An immediate and heightened production of additional specific antibodies against a particular antigen that elicited a primary immune response.

secondary metabolism   Involves the synthesis of chemicals that are not essential for cell structure and growth and are usually not required for cell survival, but are advantageous to the organism.

secondary metabolite   Molecules that are produced by secondary metabolism.

secondary oocyte   In animals, the large egg cell that is the result of meiosis I.

secondary phloem   The inner bark of a woody plant.

secondary plastid   A plastid that has originated by the endosymbiotic incorporation into a host eukaryote of a eukaryotic cell having a primary plastid.

secondary production   The measure of production of heterotrophs and decomposers.

secondary spermatocytes   The haploid cells produced in the primary spermatocyte by meiosis I.

secondary structure   The bending or twisting of proteins into helices or β sheets. One of four levels of protein structure.

secondary succession   Succession on a site that has already supported life but that has undergone a disturbance.

secretion   In the production of urine, the process in which some solutes are actively transported from the interstitial fluid surrounding the epithelial cells of the tubules of an excretory organ, into the tubule lumens.

secretory pathway   A pathway for the movement of larger substances, such as proteins and carbohydrates, out of the cell.

secretory vesicle   A membrane vesicle carrying different types of materials that later fuses with the cell's plasma membrane to release the contents extracellularly.

seed   A reproductive structure produced by flowering plants and other seed plants, usually as the result of sexual reproduction.

seed coat   A hard and tough covering that develops from the ovule's integument and protects the plant's embryo.

seed plant   The informal name for gymnosperms and angiosperms.

segment   1. The portion of the eye cell that contains the cell nucleus and other cytoplasmic organelles. 2. A body part in animals that is repeated many times in a row along an anteroposterior axis.

segmentation gene   A gene that controls the segmentation pattern of an animal embryo.

segment-polarity gene   A type of segmentation gene; a mutation in this gene causes portions of segments to be missing either an anterior or a posterior region and for adjacent regions to become mirror images of each other.

segregate   To separate, as in chromosomes during mitosis.

selectable marker   A gene whose presence can allow organisms (such as bacteria) to grow under a certain set of conditions. For example, an antibiotic resistance gene is a selectable marker that allows bacteria to grow in the presence of the antibiotic.

selective breeding   Programs and procedures designed to modify traits in domesticated species.

selectively permeable   The phenomenon that membranes allow the passage of certain ions or molecules but not others.

self-fertilization   Fertilization that involves the union of a female gamete and male gamete from the same individual.

self-incompatibility (SI)   Rejection of pollen that is genetically too similar to the pistil of a plant.

selfish DNA hypothesis   The hypothesis that transposable elements exist because they have the characteristics that allow them to insert themselves into the host cell DNA but do not provide any advantage.

self-pollination   The process in which pollen from the anthers of a flower is transferred to the stigma of the same flower, or between flowers of the same plant.

self-splicing   The phenomenon that RNA itself can catalyze the removal of its own intron(s).

semelparity   A reproductive pattern in which organisms produce all of their offspring in a single reproductive event.

semen   A mixture containing fluid and sperm that is released during ejaculation.

semicircular canals   Structures of the ear that can detect circular motions of the head.

semiconservative mechanism   In this model for DNA replication, the double-stranded DNA is half conserved following the replication process such that the new double-stranded DNA contains one parental strand and one daughter strand.

semifluid   A quality of motion within biomembranes; considered two-dimensional, which means that movement occurs within the plane of the membrane.

semilunar valves   A one-way valve into the systemic and pulmonary arteries through which blood is pumped.

seminal vesicles   Paired accessory glands that secrete fructose, the main nutrient for sperm, into the urethra to mix with the sperm.

seminiferous tubule   A tightly packed structure in the testis, where spermatogenesis takes place.

sense   A system that consists of sensory cells that respond to a specific type of chemical or physical stimulus and send signals to the central nervous system, where the signals are received and interpreted.

senescent   Cells that have doubled many times and have reached a point where they have lost the capacity to divide any further.

sensor kinase   An enzyme linked receptor that recognizes a signal found in the environment and also has the ability to hydrolyze ATP and phosphorylate itself.

sensory neuron   A neuron that detects or senses information from the outside world, such as light, sound, touch, and heat; sensory neurons also detect internal body conditions such as blood pressure and body temperature.

sensory receptor   A specialized cell whose function it is to receive sensory inputs.

sensory transduction   The process by which incoming stimuli are converted into neural signals.

sepal   A flower organ that often functions to protect the unopened flower bud.

septum (plural, **septa**)   A cross wall; examples include the cross walls that divide the hyphae of most fungi into many small cells, and the structure that separates the old and new chambers of a nautilus.

sere   Each phase of succession in a community; also called a seral stage.

setae   Chitinous bristles in the integument of many invertebrates.

sex chromosomes   A distinctive pair of chromosomes that are different in males and females.

sex linked   Refers to genes that are found on one sex chromosome but not on the other.

sex pili   Hair-like structures made by bacterial F$^+$ cells that bind specifically to F$^-$ cells.

sex-influenced inheritance   The phenomenon in which an allele is dominant in one sex but recessive in the other.

sexual reproduction   A process that requires a fertilization event in which two gametes unite to create a cell called a zygote.

sexual selection   A type of natural selection that is directed at certain traits of sexually reproducing species that make it more likely for individuals to find or choose a mate and/or engage in successful mating.

Shannon diversity index (H$_S$)   A means of measuring the diversity of a community; H$_S$ = −Σp$_i$ ln p$_i$.

shared derived character   A trait that is shared by a group of organisms but not by a distant common ancestor.

shared primitive character   A trait shared with a distant ancestor.

shattering   The process by which ears of wild grain crops break apart and disperse seeds.

shell   A tough, protective covering that is impermeable to water and prevents the embryo from drying out.

shivering thermogenesis   Rapid muscle contractions in an animal, without any locomotion, in order to raise body temperature.

shoot apical meristem (SAM)   The region of rapidly dividing plant cells at plant shoot apices.

shoot meristem   The tissue that produces all aerial parts of the plant, which include the stem as well as lateral structures such as leaves and flowers.

shoot system   The collection of plant organs produced by shoot apical meristems.

short-day plant   A plant that flowers only when the night length is longer than a defined period. Such night lengths occur in late summer, fall, or winter, when days are short.

short tandem repeat sequences (STRs)   Sequences found in multiple sites in the genome of humans and other species, and they vary in length among different individuals.

shotgun DNA sequencing   A strategy for sequencing an entire genome by randomly sequencing many different DNA fragments.

sickle-cell anemia   A disease due to a genetic mutation in a hemoglobin gene in which sickle-shaped red blood cells are less able to move smoothly through capillaries and can block blood flow, resulting in severe pain and cell death of the surrounding tissue.

sieve elements   Thin-walled living cells that are arranged end to end in a plant to form pipelines.

sieve plate   The perforated end wall of a mature sieve-tube element.

sieve plate pore   One of many perforations in a plant's sieve plate.

sieve-tube elements   A component of the phloem tissues of flowering plants; these structures are arranged end to end to form transport pipes.

sigma factor   A protein that plays a key role in bacterial promoter recognition and recruits RNA polymerase to the promoter.

sign stimulus   A factor that initiates a fixed-action pattern of behavior.

signal recognition particle (SRP)   A protein/RNA complex that recognizes the ER signal sequence and pauses translation; then, SRP binds to a receptor in the ER membrane, which docks the ribosome over a translocation channel.

signal transduction pathway   A group of proteins that convert an initial signal to a different signal inside the cell.

signal   An agent that can influence the properties of cells.

silencer   A response element that prevents transcription of a given gene when its expression is not needed.

silent mutation   A gene mutation that does not alter the amino acid sequence of the polypeptide, even though the nucleotide sequence has changed.

simple Mendelian inheritance   The inheritance pattern of traits affected by a single gene that is found in two variants, one of which is completely dominant over the other.

simple translocation   A single piece of chromosome that is attached to another chromosome.

single-factor cross   A cross in which an experimenter follows the variants of only one trait.

single-strand binding protein   A protein that binds to both of the single strands of parental DNA and prevents them from re-forming a double helix.

sinoatrial (SA) node   A collection of modified cardiac cells that spontaneously and rhythmically generates action potentials that spread across the entire atria; also known as the pacemaker of the heart.

sister chromatids   The two duplicated chromatids that are still joined to each other after DNA replication.

skeletal muscle   A type of muscle tissue that is attached to bones in vertebrates and to the exoskeleton of invertebrates.

skeleton   A structure or structures that serve one or more functions related to support, protection, and locomotion.

sliding filament mechanism   The way in which a muscle fiber shortens during muscle contraction.

SLOSS debate   In conservation biology, the debate over whether it is preferable to protect one single, large reserve or several smaller ones (SLOSS stands for single large or several small).

slow block to polyspermy   Events that produce barriers to more sperm penetrating an already fertilized egg.

slow fiber   A skeletal muscle fiber with a low rate of myosin ATP hydrolysis.

slow-oxidative fiber   A skeletal muscle fiber that has a low rate of myosin ATP hydrolysis but has the ability to make large amounts of ATP; used for prolonged, regular activity.

small effector molecule   With regard to transcription, refers to a molecule that exerts its effects by binding to a regulatory transcription factor and causing a conformational change in the protein.

small intestine   A tube that leads from the stomach to the large intestine where nearly all digestion of food and absorption of food nutrients and water occur.

smooth endoplasmic reticulum (smooth ER)   The part of the ER that is not studded with ribosomes. This region is continuous with the rough ER and functions in diverse metabolic processes such as detoxification, carbohydrate metabolism, accumulation of calcium ions, and synthesis and modification of lipids.

smooth muscle   A type of muscle tissue that surrounds hollow tubes and cavities inside the body's organs, such that their contraction can propel the contents of those organs.

soil horizon   Layers of soil, ranging from topsoil to bedrock.

solute   A substance dissolved in a liquid.

solute potential (S)   The osmotic potential; an element in the water potential equation.

solution   A liquid that contains one or more dissolved solutes.

solvent   The liquid in which a solute is dissolved.

soma   See cell body.

somatic cell   The type of cell that constitutes all cells of the body excluding the germ-line cells. Examples include skin cells and muscle cells.

somatic embryogenesis   The production of plant embryos from body (somatic) cells.

somatic nervous system   The division of the peripheral nervous system that senses the external environmental conditions and controls skeletal muscles.

somites   Blocklike structures of mesoderm that are formed during neurulation.

soredia Small clumps of hyphae surrounding a few algal cells that can disperse in wind currents; an asexual reproductive structure produced by lichens.

sorting signal A short amino acid sequence in a protein's structure that directs the protein to its correct location.

source pool The pool of species on the mainland that is available to colonize an island.

Southern blotting A method in which a labeled probe, which is a strand of DNA from a specific gene, is used to identify that gene in a mixture of many chromosomal DNA fragments.

speciation The formation of new species.

species A group of related organisms that share a distinctive form in nature.

species-area effect The relationship between the amount of available area and the number of species present.

species concepts Different approaches for distinguishing species, including the phylogenetic, biological, evolutionary, and ecological species concepts.

species interactions A part of the study of population ecology which focuses on interactions such as predation, competition, parasitism, mutualism and commensalism.

species richness The numbers of species in a community.

specific (acquired) immunity An immunity defense that develops only after the body is exposed to foreign substances.

Spemann's organizer An extremely important morphogenetic field in the early gastrula; the organizer secretes morphogens responsible for inducing the formation of a new embryonic axis.

sperm Refers to a "male" gamete that is generally smaller than the female gamete (egg); the male gamete is often motile or in many species of plants transported to the egg via a pollen tube.

sperm storage A method of synchronizing the production of offspring with favorable environmental conditions; female animals store and nourish sperm in their reproductive tract for long periods of time.

spermatids In animals, the haploid cells produced by the secondary spermatocytes by meiosis II; these cells eventually differentiate into sperm cells.

spermatogenesis The formation of sperm.

spermatogonia In animals, diploid germ cells that give rise to the male gametes, the spermatozoa.

spermatophytes All of the living and fossil seed plant phyla.

spinal cord In vertebrates and simpler chordates, the structure that connects the brain to all areas of the body and constitutes the central nervous system.

spinal nerve A nerve that connects the peripheral nervous system and the spinal cord.

spinneret A spider's abdominal silk gland; also found in the mouths of caterpillars.

spiracle A pore to the trachea that is found in the bodies of insects.

spiral cleavage A mechanism of development in which the planes of cell cleavage are oblique to the axis of the embryo.

spirilli Rigid, spiral-shaped prokaryotic cells.

spirochaetes Flexible, spiral-shaped prokaryotic cells.

spliceosome A complex of several subunits known as snRNPs that removes introns from eukaryotic pre-mRNA.

splicing The process whereby introns are removed from RNA and the remaining exons are connected to each other.

spongin A tough protein that lends skeletal support to a sponge.

spongocoel A central cavity in the body of a sponge.

spongy parenchyma Photosynthetic tissue of the plant leaf mesophyll that contains cells separated by abundant air spaces.

spontaneous mutation A mutation resulting from abnormalities in biological processes.

sporangia Structures that produce and disperse the spores of plants, fungi, or protists.

spore Single-celled reproductive structure that is dispersed into the environment and is able to grow into new fungal hyphae, plant gametophytes, or protists if they find suitable habitats.

sporic life cycle *See* alternation of generations.

sporophyte The diploid generation of plants or multicellular protists that have a sporic life cycle; this generation produces haploid spores by the process of meiosis.

sporopollenin The tough material that composes much of the walls of plant spores and helps to prevent cellular damage during transport in air.

stabilizing selection A pattern of natural selection that favors the survival of individuals with intermediate phenotypes.

stamen A flower structure that makes the male gametophyte, pollen.

standard metabolic rate (SMR) A method for measuring the BMR of ectotherms at a standard temperature for each species—one that approximates the average temperature that a species normally encounters.

standing crop The total biomass in an ecosystem at any one point in time.

starch A polysaccharide produced by the cells of plants and some algal protists.

start codon A three-base sequence—usually AUG—that specifies the first amino acid in a polypeptide.

statocyst An organ of equilibrium found in many animal species.

statoliths 1. Tiny granules of sand or other dense objects that aid in equilibrium in many vertebrates. 2. In plants, a starch-heavy plastid that allows both roots and shoots to detect gravity.

stem A plant organ that produces buds, leaves, branches and reproductive structures.

stem cell A cell that divides and supplies the cells that construct the bodies of all animals and plants.

stereoisomers Isomers with identical bonding relationships, but the spatial positioning of the atoms differs in the two isomers.

sternum The breastbone.

steroid A lipid molecule with a chemical structure containing four interconnected rings of carbon atoms.

steroid receptor A transcription factor that recognizes a steroid hormone and usually functions as a transcriptional activator.

steroid A lipid containing four interconnected rings of carbon atoms; function as hormones in animals and plants.

sticky ends Single-stranded ends of DNA fragments that will hydrogen-bond to each other due to their complementary sequences.

stigma 1. A topmost portion of the pistil, which receives and recognizes pollen of the appropriate species or genotype. 2. In many protists, a red-colored cellular structure.

stomach A saclike organ that most likely evolved as a means of storing food. It partially digests some of the macromolecules in food and regulates the rate at which the contents empty into the small intestine.

stomata Surface pores on plant surfaces that can be closed to retain water or open to allow the entry of CO_2 needed for photosynthesis and the exit of oxygen and water vapor.

stop codon One of three three-base sequences—UAA, UAG, and UGA—that signals the end of translation.

strain Within a given species, a lineage that has genetic differences compared to another strain.

strand A structure of DNA (or RNA) formed by the covalent linkage of nucleotides in a linear manner.

stretch receptor A type of mechanoreceptor found widely in organs and muscle tissues that can be distended.

striated muscle Skeletal and cardiac muscle with a series of light and dark bands perpendicular to the muscle's long axis.

stroke The condition that occurs when blood flow to part of the brain is disrupted.

stroke volume (SV) The amount of blood ejected with each beat, or stroke, of the heart.

stroma The fluid-filled region of the chloroplast between the thylakoid membrane and the inner membrane.

stromatolite A layered calcium carbonate structure in an aquatic environment produced by cyanobacteria.

strong acid An acid that completely ionizes in solution.

structural gene Refers to most genes, which produce an mRNA molecule that contains the information to specify a polypeptide with a particular amino acid sequence.

structural genomics Genomic methods aimed at the direct analysis of the DNA itself.

structural isomers Isomers that contain the same atoms but in different bonding relationships.

style The elongate pistil structure through which the pollen tube of a flower grows.

stylet A sharp, piercing organ in the mouth of nematodes and some insects.

submetacentric A chromosome in which the centromere is off center.

subsidence zones Areas of high pressure that are the sites of the world's tropical deserts, because the subsiding air is relatively dry, having released all of its moisture over the equator.

substrate-level phosphorylation A method of synthesizing ATP that occurs when an enzyme directly transfers a phosphate from one molecule to a different molecule.

substrates The reactant molecules and/or ions that bind to an enzyme at the active site and participate in a chemical reaction.

succession The gradual and continuous change in species composition and community structure over time.

sugar sink The plant tissues or organs in which more sugar is consumed than is produced by photosynthesis.

sugar source The plant tissues or organs that produce more sugar than they consume in respiration.

sum rule The probability that one of two or more mutually exclusive outcomes will occur is the sum of the probabilities of the possible outcomes.

supercoiling A method of compacting chromosomes; the phenomenon of forming additional coils around the long, thin DNA molecule.

supergroup A proposed way to organize eukaryotes into monophyletic groups.

surface area/volume (SA/V) ratio The ratio between a structure's surface area and the volume in which the structure is contained.

surfactant A mixture of proteins and amphipathic lipids produced in certain alveolar cells that acts to reduce surface tension in lungs.

survivorship curve A graphical plot of the numbers of surviving individuals at each age.

suspension feeder An aquatic animal that sifts water, filtering out the organic matter and expelling the rest.

suspensor A short chain of cells at the base of an early angiosperm embryo that provides anchorage and nutrients.

swim bladder A gas-filled, balloon-like structure that helps a fish to remain buoyant in the water even when it is completely stationary.

swimmeret An abdominal appendage in a crustacean that provides movement.

symbiosis An intimate association between two or more organisms.

symbiotic Describes a relationship in which two or more different species live in direct contact with each other.

sympathetic division The division of the autonomic system that is responsible for rapidly activating systems that provide immediate energy to the body in response to danger or stress.

sympatric The term used to describe species occurring in the same geographic area.

sympatric speciation A form of speciation that occurs when members of a species that initially occupy the same habitat within the same range diverge into two or more different species.

symplast All of a plant's protoplasts (the cell contents without the cell walls) and plasmodesmata.

symplastic transport The movement of a substance from the cytosol of one cell to the cytosol of an adjacent cell via membrane-lined channels called plasmodesmata.

symplesiomorphy *See* shared primitive character.

symporter A type of transporter that binds two or more ions or molecules and transports them in the same direction.

synapomorphy *See* shared derived character.

synapse A junction where a nerve terminal meets a target neuron, muscle cell, or gland and can communicate with other cells.

synapsis The process of forming a bivalent.

synaptic cleft The extracellular space between two neurons.

synaptic integration The summation of EPSPs generated at one time, which can bring the membrane potential to the threshold potential at the axon hillock for action potential firing.

synaptic plasticity A change in synapses that occurs as a result of learning.

synaptic signaling A specialized form of paracrine signaling that occurs in the nervous system of animals.

synergids In the female gametophyte of a flowering plant, the two cells adjacent to the egg cell that help to import nutrients from maternal sporophyte tissues.

systematics The study of biological diversity and evolutionary relationships among organisms, both extinct and modern.

systemic acquired resistance (SAR) A whole-plant defensive response to pathogenic microorganisms.

systemic circulation The pumping of blood from the left side of the heart to the body to drop off oxygen and nutrients and pick up carbon dioxide and wastes. The blood then returns to the right side of the heart.

systemic hypertension An arterial blood pressure above normal, which in humans ranges from systolic/diastolic pressures of about 90/60 to 120/80 mmHg; often called hypertension or high blood pressure.

systems biology A field of study in which researchers study living organisms in terms of their underlying network structure—groups of structural and functional connections—rather than their individual molecular components.

systole The second phase of the cardiac cycle, in which the ventricles contract and eject the blood through the open semilunar valves.

T

T cell A type of lymphocyte that directly kills infected, mutated, or transplanted cells.

tagmata The fusion of body segments into functional units.

tapetum lucidum A reflective layer of tissue located beneath the photoreceptors at the back of the eye of certain animals.

taproot system The root system of eudicots, which has one main root with many branch roots.

***Taq* polymerase** A heat-stable form of DNA polymerase; one of several reagents required for synthesis of DNA via polymerase chain reaction (PCR).

TATA box One of three features found in most promoters; the others are the transcriptional start site and response elements.

taxis A directed type of response either toward or away from a stimulus.

taxon A group of species that are evolutionarily related to each other. In taxonomy, each species is placed into several taxons that form a hierarchy from large (domain) to small (genus).

taxonomy The field of biology that is concerned with the theory, practice, and rules of classifying living and extinct organisms and viruses.

telocentric A chromosome in which the centromere is at the end.

telomerase An enzyme that catalyzes the replication of the telomere.

telomere A region at the ends of eukaryotic chromosomes where a specialized form of DNA replication occurs.

telophase The phase of mitosis during which the chromosomes decondense and the nuclear membrane re-forms.

temperate phage A bacteriophage that may spend some of its time in the lysogenic cycle.

template The DNA strand that is used as a template for RNA synthesis or DNA replication.

termination codon *See* stop codon.

termination stage The final stage of transcription or translation in which the process ends.

terminator A sequence that specifies the end of transcription.

territory A fixed area in which an individual or group excludes other members of its own species, and sometimes other species, by aggressive behavior or territory marking.

tertiary consumer An organism that feeds on secondary consumers.

tertiary endosymbiosis The acquisition by eukaryotic protist host cells of plastids from cells that possess secondary plastids.

tertiary plastid A plastid acquired by the incorporation into a host cell of an endosymbiont having a secondary plastid.

tertiary structure The three-dimensional shape of a single polypeptide. One of four levels of protein structure.

testcross A cross in which an individual with a dominant phenotype is mated with a homozygous recessive individual.

testes The male gonads of certain animals, where sperm are produced.

testosterone The primary androgen in vertebrates.

tetrad *See* bivalent.

tetraploid An organism or cell that has four sets of chromosomes.

tetrapod A vertebrate animal having four legs or leglike appendages.

thalamus In vertebrates, a gland that plays a major role in relaying sensory information to appropriate parts of the cerebrum and, in turn, sending outputs from the cerebrum to other parts of the brain. It receives input from all sensory systems and is involved in the perception of pain and the degree of mental arousal in the cortex.

theory In biology, a broad explanation of some aspect of the natural world that is substantiated by a large body of evidence. Biological theories incorporate observations, hypothesis testing, and the laws of other disciplines such as chemistry and physics. A theory makes valid predictions.

thermodynamics The study of energy interconversions.

thermoreceptor A sensory receptor that responds to cold and heat.

theropods A group of bipedal saurischian dinosaurs.

thick filament A skeletal muscle structure composed almost entirely of the motor protein myosin.

thigmotropism Touch responses in plants.

thin filament A skeletal muscle structure that contains the cytoskeletal protein actin, as well as two other proteins—troponin and tropomyosin—that play important roles in regulating contraction.

30-nm fiber Nucleosome units organized into a more compact structure that is 30 nm in diameter.

thoracic breathing Breathing in which coordinated contractions of muscles expand the rib cage, creating a negative pressure to suck air in and then forcing it out later; found in amniotes.

threatened species Those species that are likely to become endangered in the future.

threshold concentration The concentration above which a morphogen will exert its effects but below which it is ineffective.

thrifty genes Genes that boosted our ancestors' ability to store fat from each feast in order to sustain them through the next famine.

thrombocytes Intact cells in the blood of vertebrates other than mammals that play a crucial role in the formation of blood clots.

thylakoid A flattened, plate-like membranous region found in cyanobacterial cells and the chloroplasts of photosynthetic protists and plants; the location of the light reactions of photosynthesis.

thylakoid lumen The fluid-filled compartment within the thylakoid.

thylakoid membrane A membrane within the chloroplast that forms many flattened, fluid-filled tubules that enclose a single, convoluted compartment. It is the site where the light-dependent reactions of photosynthesis occurs.

thymine (T) A pyrimidine base found in DNA.

thymine dimer One type of pyrimidine dimer; a site where two adjacent thymine bases become covalently cross-linked to each other.

thyroglobulin A protein found in the colloid of the thyroid gland.

thyroxine A thyroid hormone that contains iodine and helps regulate metabolic rate.

Ti plasmid Tumor-inducing plasmid found in *Agrobacterium tumefaciens;* it is used as a cloning vector to transfer genes into plant cells.

tidal ventilation A type of breathing in which the lungs are inflated with air, and then the chest muscles and diaphragm relax and recoil back to their original positions as an animal exhales. During exhalation, air leaves via the same route that it entered during inhalation, and no new oxygen is delivered to the airways at that time.

tidal volume The volume of air that is normally breathed in and out at rest.

tight junction A type of junction that forms a tight seal between adjacent epithelial cells and thereby prevents molecules from leaking between cells; also called an occluding junction.

tissue The association of many cells of the same type, for example, muscle tissue.

tolerance A mechanism for succession in which any species can start the succession, but the eventual climax community is reached in a somewhat orderly fashion; early species neither facilitate nor inhibit subsequent colonists.

tonoplast The membrane of the central vacuole in a plant or algal cell.

torpor The strategy in the smallest endotherms of lowering internal body temperature to just a few degrees above that of the environment in order to conserve energy.

torus The nonporous, flexible central region of a conifer pit that functions like a valve.

total fertility rate The average number of live births a female has during her lifetime.

total peripheral resistance (TPR) The sum of all the resistance in all the arterioles.

totipotent The ability of a fertilized egg to produce all of the cell types in the adult organism; also the ability of unspecialized plant cells to regenerate an adult plant.

trachea 1. A sturdy tube arising from the spiracles of an insect's body involved in respiration. 2. The name of the tube leading to the lungs of air-breathing vertebrates.

tracheal system In insects, a series of finely branched air tubes called tracheae that lead into the body from pores called spiracles.

tracheid A type of dead, lignified plant cell that conducts water, along with dissolved minerals and certain organic compounds.

tracheophytes A term used to describe the vascular plants.

tract A parallel bundle of myelinated axons.

traffic signal *See* sorting signal.

trait *See* character.

transcription The use of a gene sequence to make a copy of RNA: transcription occurs in three stages called initiation, elongation, and termination.

transcription factor A protein that influences the ability of RNA polymerase to transcribe genes.

transcriptional start site The site in a promoter where transcription begins.

transduction A type of genetic transfer between bacteria in which a virus infects a bacterial cell and then transfers some of that cell's DNA to another bacterium.

trans-effect In both prokaryotes and eukaryotes, a form of genetic regulation that can occur even though two DNA segments are not physically adjacent. The action of the *lac* repressor on the *lac* operon is a trans-effect.

transfer RNA (tRNA) An RNA that carries amino acids and is used to translate mRNA into polypeptides.

transformation A type of genetic transfer between bacteria in which a segment of DNA from the environment is taken up by a competent cell and incorporated into the bacterial chromosome.

transgenic The term used to describe an organism that carries genes that were introduced using molecular techniques such as gene cloning.

transition state In a chemical reaction, a state in which the original bonds have stretched to their limit; once this state is reached, the reaction can proceed to the formation of products.

transitional form An organism that provides a link between earlier and later forms in evolution.

translation The process of synthesizing a specific polypeptide on a ribosome; a nucleotide sequence in mRNA is used to make an amino acid sequence of a polypeptide. The process of translation occurs in three stages: initiation, elongation, and termination.

translocation A phenomenon that occurs when one segment of a chromosome becomes attached to a different chromosome.

transmembrane gradient The phenomenon that the concentration of a solute is higher on one side of a membrane than on the other.

transmembrane protein A protein that has one or more regions that are physically embedded in the hydrophobic region of the cell's phospholipid bilayer.

transmembrane segment A region of a membrane protein that is a stretch of hydrophobic amino acids that spans or traverses the membrane from one leaflet to the other.

transmission electron microscopy (TEM) A type of microscopy in which a beam of electrons is differentially transmitted through a biological sample to form an image on a photographic plate or screen.

transpiration The evaporative loss of water from plant surfaces into sun-heated air.

transport protein Proteins embedded within the phospholipid bilayer that allow plasma membranes to be selectively permeable by providing a passageway for the movement of some but not all substances across the membrane.

transporter   A membrane protein that binds a solute and undergoes a conformational change to allow the movement of the solute across a membrane.

transposable element (TE)   A segment of DNA that can move from one site to another.

transposase   An enzyme that facilitates transposition.

transposition   The process in which a short segment of DNA moves within a cell from its original site to a new site in the genome.

transposon   A transposable element that moves via DNA that is removed from one site and inserted into a new site.

transverse tubules (T-tubules)   Invaginations of the plasma membrane that open to the extracellular fluid and conduct action potentials from the outer surface of a muscle fiber to the myofibrils.

trichome   A projection from the epidermal tissue of a plant that offers protection from excessive light, ultraviolet radiation, extreme air temperature, or attack by herbivores.

trichromatic color vision   The ability to distinguish blue, green, and red colors.

triiodothyronine   A thyroid hormone that contains iodine and helps regulate metabolic rate.

triplet   A group of three bases that function as a codon.

triploblastic   Having three distinct germ layers—endoderm, ectoderm, and mesoderm.

triploid   An organism or cell that has three sets of chromosomes.

trisomic   An aneuploid organism that has one too many chromosomes.

trochophore larva   A distinct larval stage of many invertebrate phyla.

trophic level   Each feeding level in a food chain.

trophic-level transfer efficiency   The amount of energy at one trophic level that is acquired by the trophic level above and incorporated into biomass.

tropomyosin   A protein that plays an important role in regulating muscle contraction.

troponin   A protein that plays an important role in regulating muscle contraction.

***trp* operon**   An operon of *E. coli* that encodes enzymes required to make the amino acid tryptophan, a building block of cellular proteins.

true-breeding line   A strain that continues to exhibit the same trait after several generations of self-fertilization or inbreeding.

trypsin   A protease involved in the breakdown of proteins in the small intestine.

t-snare   A protein in a target membrane that recognizes a v-snare in a membrane vesicle.

tube cell   In a plant, the cell that stores proteins and lipids that may be used during later stages of male gametophyte development, after pollen has germinated. This cell forms the pollen tube that after pollination grows to reach the female gametophyte.

tube feet   Echinoderm structures that function in movement, gas exchange, and feeding.

tumor   An overgrowth of cells that serves no useful purpose.

tumor-suppressor gene   A gene that when normal (that is, not mutant) encodes a protein that prevents cancer; however, when a mutation eliminates its function, cancer may occur.

tunic   A nonliving structure that encloses a tunicate, made of a protein and a cellulose-like material called tunicin.

turgid   The term used to describe a plant cell whose cytosol is so full of water that the plasma membrane presses right up against the cell wall; as a result, turgid cells are firm or swollen.

turgor pressure   *See* osmotic pressure.

two-component regulatory system   A signaling system found in bacteria and plants composed of an enzyme-linked receptor called a sensor kinase and a response regulator, which is usually a protein that regulates the expression of many genes.

two-dimensional (2D) gel electrophoresis   A technique to separate many different proteins that utilizes an isoelectric-focusing tube gel in the first dimension, and an SDS-slab gel in the second dimension.

two-factor cross   Crosses in which the inheritance of two different traits are followed.

type 1 diabetes mellitus (T1DM)   A disease in which the pancreas does not produce sufficient insulin; as a result, extracellular glucose cannot cross plasma membranes, and so glucose accumulates to very high concentrations in the blood.

type 2 diabetes mellitus (T2DM)   A disease in which the pancreas functions normally, but the cells of the body lose much of their ability to respond to insulin.

U

ubiquitin   A small protein in eukaryotic cells that directs unwanted proteins to a proteasome by its covalent attachment.

ultimate cause   The reason a particular behavior evolved, in terms of its effect on reproductive success.

umbrella species   A species whose habitat requirements are so large that protecting them would protect many other species existing in the same habitat.

unconditioned response   A response that is already elicited by an unconditioned stimulus.

unconditioned stimulus   A factor that originally elicits a response.

uniform   A pattern of dispersion within a population, in which individuals maintain a certain minimum distance between themselves to produce an evenly spaced distribution.

uniporter   A type of transporter that binds a single molecule or ion and transports it across the membrane.

unipotent   A term used to describe a stem cell found in the adult that can only produce daughter cells that differentiate into one cell type.

unsaturated   The quality of a lipid when a double bond is formed.

unsaturated fatty acid   A fatty acid that contains one or more C $=$ C double bonds.

upwelling   In the ocean, a phenomenon that carries mineral nutrients from the bottom waters to the surface.

uracil (U)   A pyrimidine base found in RNA.

urea   A nitrogenous waste commonly produced in many terrestrial species.

uremia   A condition characterized by the presence of nitrogenous wastes, such as urea, in the blood; typically results from kidney disease.

ureter   A structure in the kidney through which urine flows from the kidney into the urinary bladder.

urethra   The structure through which urine is eliminated from the body.

uric acid   A nitrogenous waste commonly produced in many terrestrial species.

urinary bladder   The structure that collects urine before it is eliminated.

urinary system   The structures which collectively act to filter blood or hemolymph and excrete wastes while recapturing useful compounds. In humans, it includes the two kidneys, two ureters, the urinary bladder, and the urethra.

uterine cycle   The cyclical changes that occur in the uterus in parallel with the ovarian cycle in a female mammal.

uterus   A large, conical-shaped structure that consists of an inner lining of glandular and secretory cells called the endometrium, and a thick muscular layer called the myometrium; the uterus is specialized for carrying the developing fetus.

V

vaccination   The injection into the body of small quantities of living or dead microbes, small quantities of toxins, or harmless antigenic molecules derived from a microorganism or its toxin, resulting in a primary immune response including the production of memory cells. Subsequent natural exposure to the immunizing antigen results in a response that can prevent or reduce the severity of disease.

vacuole   Specialized compartments found in eukaryotic cells that function in storage, the regulation of cell volume, and degradation.

vagina   A tubular, smooth muscle structure in the female into which sperm are deposited.

vaginal diaphragm   A barrier method of preventing fertilization in which the device is placed in the upper part of the vagina just prior to intercourse and blocks movement of sperm to the cervix.

valence electron   An electron in the outer shell that is available to combine with other atoms. Such electrons allow atoms to form chemical bonds with each other.

variable region   A domain within an immunoglobulin that serves as the antigen-binding site.

vas deferens   A muscular tube through which sperm leave the epididymis.

vasa recta capillaries   Capillaries in the medulla in the nephron of the kidney.

vascular bundle   Primary plant vascular tissues that occur in a group.

vascular cambium   A meristematic tissue of plants that produces both wood and inner bark.

vascular plant   A plant that can transport water, sugar, and salts throughout the plant body via xylem and phloem tissues.

vascular tissue   Plant tissue that makes up the vascular system, which conducts materials within the plant body and also provides support.

vasectomy   A surgical procedure in men that severs the vas deferens, thereby preventing the release of sperm at ejaculation.

vasoconstriction   A decrease in blood vessel radius; an important mechanism for directing blood flow away from specific regions of the body.

vasodilation   An increase in blood vessel radius; an important mechanism for directing blood flow to specific regions of the body.

vasotocin   A peptide that is responsible for regulating salt and water balance in the blood of nonmammalian vertebrates.

vector   A type of DNA that acts as a carrier of a DNA segment that is to be cloned.

vegetal pole   In triploblast organisms, the pole of the egg where the yolk is most concentrated.

vegetative growth   The production of new tissues by the shoot apical meristem and root apical meristem during seedling development and growth of mature plants.

vein   In animals, a blood vessel that returns blood to the heart.

veliger   A free-swimming larva that has a rudimentary foot, shell, and mantle.

ventral   Refers to the lower side of an animal.

ventricle   In the heart, a chamber to pump blood out of the heart.

venule   In animals, a small, thin-walled extension of a capillary that empties into larger vessels called veins that return blood to the heart for another trip around the circulation.

vernalization   The induction of flowering by cold treatment.

vertebrae   A bony or cartilaginous column of interlocking structures that provides support and also protects the nerve cord, which lies within its tubelike structure.

vertebrate   An organism with a backbone.

vertical evolution   A process that involves genetic changes in a series of ancestors, which form a lineage.

vesicle   A small membrane-enclosed sac within a cell.

vessel   In a plant, a pipeline-like file of dead, water-conducting vessel elements.

vessel element   A type of plant cell that conducts water, along with dissolved minerals and certain organic compounds.

vestibular system   The organ of balance in vertebrates, located in the inner ear next to the cochlea.

vestigial structure   An anatomical feature that has no apparent function but resembles a structure of a presumed ancestor.

vibrios   Comma-shaped prokaryotic cells.

villi   Finger-like projections extending from the luminal surface into the lumen of the small intestine; these are specializations that aid in digestion and absorption.

viral envelope   A structure enclosing the capsid that consists of a membrane derived from the plasma membrane of the host cell and embedded with virally encoded spike glycoproteins.

viral genome   The genetic material in a virus.

viral reproductive cycle   When a virus infects a cell, the series of steps that results in the production of new viruses.

viral vector   A type of vector used in cloning experiments that is derived from a virus.

viroid   An RNA particle that infects plant cells.

virulent phage   A phage that follows only the lytic cycle.

virus   A small infectious particle that consists of nucleic acid enclosed in a protein coat.

visceral mass   In mollusks, a structure that rests atop the foot and contains the internal organs.

vitamin   An important organic nutrient that serves as a coenzyme for metabolic and biosynthetic reactions.

vitamin D   A molecule that regulates calcium levels in the blood through an action on intestinal transport of calcium ions.

vitreous humor   A thick liquid in the large posterior cavity of the vertebrate eye, which helps maintain the shape of the eye.

viviparity   The process in which the embryo develops within the mother.

viviparous   The term used to describe an organism whose eggs develop within the uterus, receiving nourishment from the mother via a placenta.

volt   A unit of measurement of the difference in charge (the electrical force) such as between the interior and exterior of the cell.

voltage-gated channel   A channel that opens and closes in response to changes in the amount of electric charge across the membrane.

v-snare   A protein incorporated into the vesicle membrane during vesicle formation that is recognized by a t-snare in a target membrane.

W

water potential   The potential energy of water.

water vascular system   A network of canals powered by hydraulic power, that is, by water pressure generated by the contraction of muscles that enables the extension and contraction of the tube feet, allowing echinoderms to move slowly.

wavelength   The distance from the peak of one sound or light wave to the next.

waxy cuticle   A protective, water-proof layer of wax and polyester present on most surfaces of vascular plant sporophytes.

weak acid   An acid that only partially ionizes in solution.

weathering   The physical and chemical breakdown of rock.

white blood cell   A cell that develops from the inner parts (the marrow) of certain bones in vertebrates; all white blood cells (known as leukocytes) perform vital functions that defend the body against infection and disease.

white matter   Brain tissue that consists of myelinated axons that are bundled together in large numbers to form tracts.

whorls   In a flower, four concentric rings of sepals, petals, stamens, and carpels.

wild-type allele   One or more prevalent alleles in a population.

wood   A secondary plant tissue composed of numerous pipelike arrays of dead, empty, water-conducting cells whose walls are strengthened by an exceptionally tough secondary metabolite known as lignin.

X

X inactivation   The phenomenon in which one X chromosome in the somatic cells of female mammals is inactivated, meaning that its genes are not expressed.

X inactivation center (Xic)   A short region on the X chromosome known to play a critical role in X inactivation.

xenoestrogen   A synthetic compound that exerts estrogen-like actions or, in some cases, inhibits the actions of the body's own estrogen; the consequences are dramatic on fertility and development of embryos and fetuses.

X-linked gene   A gene found on the X chromosome but not on the Y.

X-linked inheritance   The pattern displayed by pairs of dominant and recessive alleles located on X chromosomes.

X-ray crystallography   A technique in which researchers must purify a molecule such as a protein or protein complex and expose it to conditions that cause the proteins molecules to associate with each other in an ordered array. In other words, the proteins form a crystal. When a crystal is exposed to X-rays, the resulting pattern can be analyzed mathematically to determine the three-dimensional structure of the crystal's components.

xylem   A plant vascular tissue that conducts water, minerals, and organic compounds.

xylem loading   The process by which root xylem parenchyma cells transport ions and water across their membranes into the xylem apoplast, which includes the vessel elements and tracheids.

Y

yeast   A fungus that can occur as a single cell and that reproduces by budding.

yolk sac   An extraembryonic membrane in the amniotic egg that encloses a stockpile of nutrients, in the form of yolk, for the developing embryo.

Z

Z line   A network of proteins that anchor thin filaments in a sarcomere.

Z scheme   According to this scheme, an electron absorbs light energy twice, and it loses some of that energy as it flows along the electron transport chain in the thylakoid membrane. The energy diagram of this process occurs in a zigzag pattern.

zero population growth   The situation in which no changes in population size occur.

zoecium   A nonliving case that houses a bryozoan.

zone of elongation   The area above the root apical meristem of a plant where cells extend by water uptake, thereby dramatically increasing root length.

zone of maturation   The area above the zone of elongation in a plant where root cell differentiation and tissue specialization occur.

zooplankton   Aquatic organisms including minute animals consisting of some worms, copepods, tiny jellyfish, and the small larvae of invertebrates and fish that graze on the phytoplankton.

zygomycete   A type of fungus that produces distinctive, large zygospores as the result of sexual reproduction.

zygospore   A dark-pigmented, thick-walled spore that matures within the zygosporangium of zygomycete fungi.

zygote   A diploid cell formed by the fusion of two gametes that divides and develops into an embryo, and eventually into an adult organism.

zygotic life cycle   In this type of life cycle, haploid cells transform into gametes.

zooplankton   Open-ocean organisms including minute animals consisting of some worms, copepods, tiny jellyfish, and the small larvae of invertebrates and fish that graze on the phytoplankton.

PHOTO CREDITS

Contents: UNIT I: © Kevin Schafer/Peter Arnold, Inc.; UNIT II: © Visuals Unlimited; **UNIT III:** © Schultz, S. C., Shields, G. C., and T. A. Steitz. (1991). "Crystal structure of a CAP-DNA complex: the DNA is bent by 90 degrees," *Science* 253:1001, August 30, 1991. Image by Daniel Gage, University of Connecticut.; **UNIT IV:** George Bernard Photo Researchers, Inc; **UNIT V:** © Steve Bloom/ stevebloom.com; **UNIT VI:** © Gerald & Buff Corsi/Visuals Unlimited; **UNIT VII:** Kim Taylor/ naturepl.com; **UNIT VIII:** © Erwin & Peggy Bauer/ Animals Animals Enterprises

Chapter 1 Opener: © Art Wolfe/Getty Images; **1.1a:** Photo W. Wüster, Courtesy Instituto Butantan; **1.1b:** © Charlie Heidecker/Visuals Unlimited; **1.1c:** © blickwinkel/Alamy; **1.1d:** © SciMAT/Photo Researchers, Inc.; **1.1e:** © Russ Bishop; **1.1f:** © Robert Winslow/Earth Sciences; **1.2a:** © Stephen J. Krasemann/Photo Researchers, Inc.; **1.2b:** © Tom Uhlman/Visuals Unlimited; **1.2c:** © Cathlyn Melloan/Stone/Getty Images; **1.2d:** © TAPANI RASANEN/WWI/Peter Arnold, Inc.; **1.2e:** © Patti Murray/Animals Animals; **1.2f:** © Paul Hanna/Reuters/CORBIS; **1.2g:** © Tom Brakefield/CORBIS; **1.4:** © Michael L. Smith/Photo Researchers Inc.; **1.9a:** © Dr. David M. Phillips/ Visuals Unlimited; **1.9b:** © B. Boonyaratanakornkit & D.S. Clark, G. Vrdoljak/EM Lab, U of C Berkeley/Visuals Unlimited; **1.9c(Bottom Left):** © Dr. Dennis Kunkel/Visuals Unlimited; **1.9c(Bottom Right):** © Carl Schmidt-Luchs/Photo Researchers, Inc; **1.9c(Middle Left):** © Fritz Polking/Visuals Unlimited; **1.9c(Middle Right):** © Kent Foster/Photo Researchers, Inc; **1.12b:** © Fabio Colombini/Animals Animals; **1.13a:** © Richard T. Nowitz/CORBIS; **1.13b:** © Olivier DIGOIT; **1.13c:** © Ton Koene/Visuals Unlimited; **1.13d:** © AP Images; **1.13e:** © Andrew Brookes/CORBIS; **1.13e(Inset):** Alfred Pasieka/Photo Researchers, Inc.; **1.14a:** © Robert Isear/Photo Researchers, Inc; **1.14b:** © Science VU/Visuals Unlimited; **1.17:** © AP/WideWorld Photo

Chapter 2 Opener: © Kevin Schafer/Peter Arnold, Inc.; **2.6:** © SIU/VU/Mediscan.; **2.12b:** © Charles D. Winters/Photo Researchers, Inc.; **2.17(Top):** © Jeremy Burgess/Photo Researchers, Inc.; **2.19a:** © Norbert Wu/Peter Arnold, Inc.; **2.19b:** © Jerome Wexler/Visuals Unlimited.; **2.20b:** © Aaron Haupt/Photo Researchers, Inc.; **2.20d:** © Rainer Drexel/Bilderberg/Peter Arnold, Inc.; **2.20e:** © Ethel Davies/Imagestate.; **2.20f:** © Anthony Bannister; Gallo Images/CORBIS.; **2.20g:** © OSF.

Chapter 3 Opener: de Vos, A. M., Ultsch, M. and A. A. Steitz (1992). "Human growth hormone and extracellular domain of its receptor: crystal structure of the complex. Science 255:306, January 17, 1992. Image by Daniel Gage, University of Connecticut.; **3.1a:** © SPL/Photo Researchers, Inc.; **3.1b:** © Alejandro Andreatta/Phototake.; **3.4a:** © Andrey Zvoznikov/ardrea.com.; **3.4b:** © B. Murton/Southampton Oceanography Centre/Photo Researchers, Inc.; **3.7(Right):** © Brian Hagiwara/ FoodPix/Jupiterimages.; **3.11a(Left), 3.11a(Middle):** © Tom Pantages.; **3.11b:** © Felicia Martinez/ PhotoEdit, Inc.; **3.13c,b:** © Adam Jones/Photo Researchers, Inc.

Chapter 4 Opener: © Biophoto Associates/ Photo Researchers, Inc.; **Table4.1(Left):** © Thomas Deerinck/Getty Images; **Table4.1(Middle), Table 4.1(Right):** © Dr. Gopal Murti/Visuals Unlimited.; **4.2a,c,d,e:** Courtesy Dr. Paul Letourneau, Neuroscience, University of Minnesota.; **4.2b:** © Ed Reschke/Peter Arnold, Inc.; **4.2f:** © Aumas et al., J Neurosci. 2001 Dec 15; 21(24):RC187. © 2001 Society for Neuroscience.; **4.3a:** © L. Zamboni, D.W. Fawcett/Visuals Unlimited; **4.3b:** © David M. Phillips/Photo Researchers, Inc.; **4.4b:** © Dr. Dennis Kunkel/Getty Images.; **4.6b(Left):** © Ed Reschke/ Peter Arnold, Inc.; **4.6b(Right):** © Eye of Science/ Photo Researchers, Inc.; **4.11a:** © Visuals Unlimited; **4.11b:** Courtesy of Robert A. Bloodgood, PhD, University of Virginia School of Medicine.; **4.11c:** © Dr. Dennis Kunkel/Visuals Unlimited/Getty Images.; **4.12a:** © Biophoto Associates/Photo Researchers, Inc.; **4.14(4):** Courtesy Dr. James Spudich. Figure 2c. Sheetz and Spudich (1983) Nature, 303, p. 31-35.; **4.16(Middle):** © Dr. Richard Kessel & Dr. Gene Shih/Visuals Unlimited.; **4.16(Top):** © Dr. Don W. Fawcett/ Visuals Unlimited.; **4.17:** Cremer, T. & Cremer, C. Chromosome territories, nuclear architecture and gene regulation in mamalian cells. Nature Reviews/Genetics Vol. 2, no. 4, f. 2b&d, p. 295; **4.18(Right):** © Richard Rodewald/Biological Photo Service.; **4.21a:** © E.H. Newcomb & S.E. Frederick/ Biological Photo Service.; **4.21b:** Courtesy Dr. Peter Luykx, Biology, University of Miami; **4.21c:** © Dr. David Patterson/Photo Researchers, Inc.; **4.24:** © Dr. Don W. Fawcett/Visuals Unlimited.; **4.25:** © Dr. Jeremy Burgess/Photo Researchers, Inc.; **4.26a:** © David Sieren/Visuals Unlimited; **4.26b:** Mitch Hrdlicka/Getty Images; **4.26c:** © Barry L. Runk/ Grant Heilman Photography, Inc.; **4.27:** © Dr. Donald Fawcett & Richard Wood/Visuals Unlimited.

Chapter 5 Opener: © Tom Pantages, **5.7:** © "Biochemistry: The Molecular Basis Of Life" 3rd Ed. by McKee and McKee. McGraw Hill; **5.8a:** © Warren Rosenberg/Biological Photo Service; **5.8b:** © Dr. Don W. Fawcett/Visuals Unlimited; **5.13a,b:** © James Strawser/Grant Heilman Photography, Inc.; **5.14a,b:** © Carolina Biological Supply/Visuals Unlimited; **5.15(4):** Courtesy Dr. Peter Agre. From Preston GM, Carroll TP, Guggino WP Agre P: Appearance of water channels in Xenopus oocytes expressing red cell CHIP28 protein. Science 256:385, 1992.; **5.16:** © Tom Pantages

Chapter 6 Opener: © Visuals Unlimited; **6.3a:** © Kiseleva and Donald Fawcett/Visuals Unlimited.; **6.3b(Middle):** © Eye of Science/Photo Researchers, Inc.; **6.3b(Right):** © Don W. Fawcett/Photo Researchers, Inc.; **6.4a:** © Dr. Stanley Flegler/ Visuals Unlimited; **6.11(6):** Photo from Caro and Palade (1964) Journal of Cell Biology 29, p. 479 (fig 3) The Rockefeller University Press; **6.16b:** © T. Kanaseki and Donald Fawcett/Visuals Unlimited

Chapter 7 Opener: © Claude Cortier/Photo Researchers, Inc.; **7.1a:** © Bob Daemmrich/The Image Works; **7.20(5):** Reprinted by permission from Macmillan Publishers Ltd: [Nature] (386, Noll et al, 299-303), copyright (1997).; **7.21:** © Tom Pantages; **7.22a:** © Bill Aron/Photo Edit, Inc.; **7.22b:** © Jeff Greenberg/The Image Works; **7.23a:** © Chris Hellier/SPL/Photo Researchers, Inc.; **7.23b:** © Michael P. Gadomski/Photo Researchers, Inc; **7.24:** © Naturfoto Honal/CORBIS; **7.25:** © James Randklev/Photographer's Choice/Getty Images; **7.26:** © SciMAT/Photo Researchers, Inc.

Chapter 8 Opener: © Digital Vision/ PunchStock.; **8.1(Bottom):** © Visuals Unlimited; **8.1(Middle):** © J. Michael Eichelberger/Visuals Unlimited; **8.1(Top):** © Norman Owen Tomalin/ Bruce Coleman, Inc.; **8.7b:** © Reproduction of Fig 1A from Ferreira, K.N., Iverson, T.M., Maghlaoui, K., Barber, J. and Iwata, S. (2004) Architecture of the photosynthetic oxygen evolving center. Science 303, 1831-1838 with permission.; **8.14(6):** © Reprinted from Chemistry 1942-1962, 1964, M. Calvin, Nobel Lectures, Elsevier Publishing Company, Copyright (1964).; **8.15a:** © David Noton Photography/Alamy; **8.15b:** © David Sieren/Visuals Unlimited; **8.17(Left):** © Walter H. Hodge/Peter Arnold, Inc.; **8.17(Right):** © John Foxx/Alamy

Chapter 9 Opener: © Steve Gschmeissner/ Photo Researchers, Inc.; **9.13(Left), (Right):** Courtesy of Brian J. Bacskai, from Bacskai et al., Science 260:222-226, 1993, reprinted with permission.; **9.19a:** Don W. Fawcett/Photo Researchers, Inc.

Chapter 10 Opener: © Stephen Alvarez/Getty Images; **10.6:** © Dr. Dennis Kunkel/Visuals Unlimited; **10.11:** © Dr. Daniel Friend; **10.12(Right):** © Daniel Goodenough; **10.14:** Lee W. Wilcox; **10.15(Top):** © E.H. Newcomb & W.P. Wergin/ Biological Photo Service; **10.20a:** © Carolina

Chapter 25 Opener: Frederic B. Siskind; **25.1(Left):** © Mark Smith/Photo Researchers, Inc; **25.1a(Right):** © Pascal Goetgheluck/ardea.com; **25.1b(Left):** © Gary Meszaros/Visuals Unlimited; **25.1b(Right):** R. Andrew Odum/Peter Arnold, Inc.; **25.3(Left):** C. Allan Morgan/Peter Arnold, Inc.; **25.3(Right):** © Bryan E. Reynolds; **25.4a:** Ron Austing/Photo Researchers, Inc; **25.4b:** Rod Planck/Photo Researchers, Inc; **25.5(Bottom):** Stephen L. Saks/Photo Researchers, Inc; **25.5(Top Left):** © Mark Boulton/Photo Researchers, Inc; **25.5(Top Right):** © Carolina Biological Society/Visuals Unlimited; **25.7(Left):** © Hal Beral/V&W/imagequestmarine.com; **25.7(Right):** © Guillen Photography/Alamy; **25.8(Bottom Left, Middle Right):** © Jim Denny; **25.8(Bottom Right, Middle Left, TopRight, Top Left):** Jack Jeffrey Photography; **25.8(Left):** FLPA/Alamy; **25.14a(Both):** Courtesy Ed Laufer; **25.14b,c(All):** Courtesy of Dr. J.M. Hurle (Originally published in Development. 1999 Dec; 126(23):5515-22.); **25.16b:** © Gary Nafis; **25.17a,b:** © Prof. Walter J. Gehring, University of Basel

Chapter 26 Opener: © Peter Weimann/Animals Animals - Earth Scenes; **26.3a(Bottom Left):** Courtesy T. A. Fernandes, A. K. Chaurasia and S. K. Apte; **26.3a(BottomRight):** © Friedrich Widdell/Visuals Unlimited; **26.3a(Center):** Jany Sauvanet/Photo Researchers, Inc; **26.3a(Top Left):** SciMAT/Photo Researchers, Inc; **26.3a(Top Right):** © Dennis Kunkel/Phototake; **26.3b(Left):** © Joe Sohm/The Image Works; **26.3b(Right):** D.Montpetit/Agriculture Canada; **26.4a:** © Runk/Schoenberger/Grant Heilman Photography, Inc.; **26.4b:** Dr. Stanley Flegler/Visuals Unlimited/Getty Images; **26.4c:** Photo by Andrew DeVogelaere, courtesy of John Pearse.; **26.5a:** © Gerald & Buff Corsi/Visuals Unlimited; **26.5b:** © Runk/Schoenberger/Grant Heilman Photography, Inc.; **26.5c:** SciMAT/Photo Researchers, Inc; **26.6a:** © Ken Wagner/Phototake; **26.6b:** © Biophoto Associates/Photo Researchers, Inc; **26.6c:** © Robert & Jean Pollock/Visuals Unlimited; **26.6d:** Patrick Frischknecht/Peter Arnold, Inc.; **26.7a:** © Wilfried Bay-Nouailhat/Association Mer et Littoral; **26.7b:** Darwin Dale/Photo Researchers, Inc; **26.7c:** Ted Kinsman/Photo Researchers, Inc; **26.7d:** Thomas & Pat Leeson/Photo Researchers, Inc; **26.7e:** © Kevin Schafer/CORBIS; **26.7f:** Gregory K. Scott/Photo Researchers, Inc.

Chapter 27 Opener: Dr. Jeremy Burgess/SPL/Photo Researchers, Inc; **27.2:** © Michael S. Lewis/CORBIS; **27.3:** © Runk/Schoenberger/Grant Heilman Photography, Inc.; **27.4a:** Linda E. Graham, OK; **27.4b:** © Wim van Egmond/Visuals Unlimited; **27.4c:** © Michael Abbey/Visuals Unlimited; **27.4d:** Lee W. Wilcox; **27.5:** © H. Stuart Pankratz/Biological Photo Service; **27.6:** Dr. Richard P. Blakemore, University of New Hampshire; **27.7a:** SciMAT/Photo Researchers, Inc, **27.7b,c,d:** Dennis Kunkel Microscopy, Inc.; **27.8a,b:** Courtesy of Chern Hsiung Lai, DND, PhD/Professor of University of Pennsylvania, USA and Kaohsiung Maedical University, Taiwan; **27.10a,b:** Lee W.

Wilcox; **27.11:** Fig. 1 from "A Fluorescent Gram Stain for Flow Cytometry and Epifluorescence Microscopy" Appl Environ Microbiol, July 1998, p. 2681–2685, Vol. 64, No.7. American Society for Microbiology.; **27.13a:** Dennis Kunkel/Phototake; **27.13b:** © Scientifica/Visuals Unlimited; **27.14:** © Dr. T. J. Beveridge/Visuals Unlimited; **27.15a:** © Scientifica/RML/Visuals Unlimited; **27.15b:** © Cytographics/Visuals Unlimited ; **27.15c:** Lee W. Wilcox; **27.16a:** Lee W. Wilcox; **27.16b:** © DR. Kari Lounatmaa/Photo Researchers, Inc; **27.17:** © Roger Ressmeyer/CORBIS; **Table27.1(Bottom):** © Prof. Dr. Hans Reichenbach/GBF; **Table27.1(Middle):** © Dr. David M. Phillips/Visuals Unlimited; **Table27.1 (Top):** Michael J. Daly, Uniformed Services University of the Health Sciences, Bethesda, MD, U.S.A.

Chapter 28 Opener: Photographs reproduced with permission of the copyright holder Dr. H. Canter-Lund; **28.2:** Photographs reproduced with permission of the copyright holder Dr. H. Canter-Lund; **28.3a:** Andrew Syred/Photo Researchers, Inc; **28.3b:** © Microfield Scientific LTD/Photo Researchers, Inc; **28.4a:** UTEX–The Culture Collection of Algae at The University of Texas at Austin; **28.4b:** Roland Birke/Phototake; **28.4c,d:** Linda E. Graham, OK; **28.4e:** © Claudia Lipke; **28.5:** © Dr. Stanley Flegler/Visuals Unlimited; **28.7a:** David Phillips/Mediscan; **28.7b:** © Jerome Paulin/Visuals Unlimited; **28.8a:** Eye of Science/Photo Researchers, Inc; **28.8b:** Ross Waller, University of Melbourne, Australia; **28.9a:** © Wim van Egmond/Visuals Unlimited; **28.10a:** Joe Scott, Department of Biology, College of William and Mary, Williamsburg, VA 23187; **28.10b:** © Dr. Brandon D. Cole/CORBIS; **28.10c:** © Walter H. Hodge/Peter Arnold, Inc.; **28.11a,b:** Lee W. Wilcox; **28.11c:** © Claude Taylor, III; **28.14a:** © Claude Nuridsany & Marie Perennou/Science Photo Library; **28.14b:** Manfred Kage/Peter Arnold, Inc.; **28.14c:** © Bob Krist/CORBIS; **28.15:** Stephen Fairclough, King Lab, University of California at Berkeley.; **28.16:** Lee W. Wilcox; **28.18:** Linda E. Graham, OK; **28.22a(Right):** © Dr. Dennis Kunkel/Visuals Unlimited

Chapter 29 Opener: © Peter Arnold, Inc./Alamy; **29.2a:** © Garry T. Cole/Biological Photo Service; **29.4:** © Matt Meadows/Peter Arnold, Inc.; **29.5a:** Fig 16, Kaminskyj SGW and Heath IB 1996 Studies on Saprolegnia ferax suggest the general importance if the cytoplasm in determining hyphal morphology. Mycologia 88: 20-37. The New York Botanical Garden, Bronx, NY; Allen Press, Lawrence Kansas.; **29.6a:** Micrograph courtesy of Timothy M Bourett, DuPont Crop Genetics, Wilmington, DE USA; **29.6b:** © Charles Mims; **29.7a:** © Agriculture and Agri-Food Canada, Southern Crop Protection and Food Research Centre, London ON.; **29.7b:** CDC; **29.8a:** Biophoto Associates/Photo Researchers, Inc; **29.8b:** © Dr. Dennis Kunkel/Visuals Unlimited; **29.10:** Fritz Polking/Peter Arnold, Inc.; **29.9a:** © Felix Labhardt/Getty Images; **29.9b:** Bob Gibbons/ardea.com; **29.11:** rob casey/Alamy; **29.12:** Hans

Pfletschinger/Peter Arnold, Inc.; **29.13:** Dennis Kunkel/Phototake; **29.14:** N. Allin & G.L. Barron, University of Guelph/BPS; **29.15a:** Nigel Cattlin/Photo Researchers, Inc; **29.15b:** Herve Conge/ISM/Phototake; **29.16a:** © Felix Labhardt/Getty Images; **29.16b:** Courtesy of Larry Peterson and hugues Massicotte; **29.17a:** Mark Brundrett; **29.18a:** © Joe McDonald/CORBIS; **29.18b:** Lee W. Wilcox; **29.18c:** Ed Reschke/Peter Arnold, Inc.; **29.18d:** Lee W. Wilcox; **29.19:** Linda E. Graham, OK; **29.21:** Thomas Kuster, USDA, FS, Forest Products Laboratory; **29.22:** Photographs reproduced with permission of the copyright holder Dr. H. Canter-Lund; **29.23a(2):** © Peres/Custom Medical Stock Photo; **29.23b(4):** © William E. Schadel/Biological Photo Service; **29.24:** Yolande Dalpé , Agriculture and Agri-Food Canada; **29.25b(3):** © Peter Arnold, Inc./Alamy; **29.26:** Nigel Cattlin; **29.27:** © David Scharf/Peter Arnold, Inc.; **29.28a:** Darlyne A. Murawski/Peter Arnold, Inc.; **29.28b:** Jupiterimages; **29.29(5):** Dr Jeremy Burgess / Photo Researchers, Inc; **29.29(8):** Biophoto Associates/Photo Researchers, Inc.

Chapter 30 Opener: © Craig Tuttle/CORBIS; **30.2a,d:** Lee W. Wilcox; **30.2b,c:** Linda E. Graham, OK; **30.3a:** Dr. Jeremy Burgess/Science Photo Library; **30.3b,c:** Lee W. Wilcox; **30.4:** G.R. "Dick" Roberto © Natural Sciences Image Library; **30.5:** Lee W. Wilcox; **30.7(1):** © L. West/Photo Researchers, Inc; **30.7(3):** Linda E. Graham, OK; **30.7(4):** © Claudia Lipke; **30.9:** Ed Reschke/Peter Arnold, Inc.; **30.10a:** Lee W. Wilcox; **30.10b:** Walter H. Hodge/Peter Arnold, Inc.; **30.10c:** © Patrick Johns/CORBIS; **30.10d:** © RICH REID/Animals Animals - Earth Scenes; **30.10e(1):** © Barrett&MacKay; **30.10e(2):** © Carolina Biological Supply Company/Phototake; **30.10e(3Left, Bottom Left):** Linda E. Graham; **30.10e(4,5 Top, 5Right):** Lee W. Wilcox; **30.10e(8):** © Dr. Richard Kessel & Dr. Gene Shih/Visuals Unlimited; **30.11:** Linda E. Graham, OK; **30.12:** © Grant Heilman/Grant Heilman Photography, Inc.; **30.13:** © Brand X Pictures/PunchStock; **30.14:** Botanical Society of America, St. Louis, MO., www.botany.org, Photo by Steven R. Manchester, University of Florida; **30.17(Left):** Lee W. Wilcox; **30.17(Middle):** © Charles McRae/Visuals Unlimited; **30.17(Right):** © David R. Frazier/The Image Works; **30.19:** Courtesy of Anna Carafa.; **30.22c:** Lee W. Wilcox

Chapter 31 Opener: © Gallo Images/CORBIS; **31.3a:** © Walter H. Hodge/Peter Arnold, Inc.; **31.3b:** © Ed Reschke/Peter Arnold, Inc.; **31.4a,b:** Lee W. Wilcox; **31.5a:** Karlene V. Schwartz; **31.5b:** © Wolfgang Kaehler/CORBIS; **31.5c:** © B. Runk/S. Schoenberger/Grant Heilman Photography; **31.6a:** © Fred Bruemmer/Peter Arnold, Inc.; **31.6b:** © Bryan Pickering; Eye Ubiquitous/CORBIS; **31.8a:** Zach Holmes Photography; **31.8b:** Duncan McEwan/naturepl.com; **31.8c:** Ed Reschke/Peter Arnold, Inc.; **31.10a:** © Unknown photographer/Grant Heilman Photography; **31.10b:** © Ken Wagner/Phototake; **31.10c:** Lee W. Wilcox; **31.11a:** Robert & Linda Mitchell; **31.11b:** © Walter H. Hodge/Peter Arnold, Inc.; **31.11c:** © Michael &

Q